Heath

TEACHER'S EDITION

General Mathematics

David W. Lowry
Earl G. Ockenga
Walter E. Rucker

D.C. Heath and Company
Lexington, Massachusetts/Toronto, Ontario

ART CREDITS

Leo Abbett, 7, 28, 37; Walter Fournier, 83, 93, 101, 121, 128, 191, 233, 240, 241, 273, 295, 313, 325, 387, 388, 401, 438, 439, 444, 445; George Ulrich, 97, 106, 171, 180, 195, 217, 221, 239, 249, 250, 272, 275, 286, 293, 299, 342, 423

From the Guinness Book of World Records © 1985 by Sterling Publishing Co., Inc., 15, 83, 103, 242, 243, 297

PHOTO CREDITS   Cover photos: Paul Johnson

Chapter 1: 1: Focus on Sports. 2: t © by Universal Pictures, a Division of Universal City Studios, Inc. Courtesy of MCA Publishing Rights, a Division of MCA Inc. b © 1983 Lucasfilm Ltd. all Rights Reserved. 4: Stuart Cohen/© D.C. Heath. 6: Stuart Cohen/© D.C. Heath. 8: Stuart Cohen/© D.C. Heath. 10: Stuart Cohen/© D.C. Heath. 11: The Granger Collection. 14: Focus on Sports. 15: Stuart Cohen. 16: t Stuart Cohen; bl, bm, br Stuart Cohen/D.C. Heath. 18: t Sports Illustrated; b Wide World Photos. 20: Stuart Cohen/© D.C. Heath. 22: Stuart Cohen/© D.C. Heath. 26: Stuart Cohen/© D.C. Heath.

Chapter 2: 31: Deborah Edelstein (CONTACT/Press Images). 32: John Des Jardins (Design Photographers International, Inc.). 33: The Granger Collection. 34: Stuart Cohen/© D.C. Heath. 36: Focus on Sports. 38: The Granger Collection. 40: Stuart Cohen/© D.C. Heath. 42, 43: Reprinted from GAMES MAGAZINE (810 7th Avenue, N.Y. 10019) © 1981. 44: Paul Johnson/© D.C. Heath. 45: Paul Johnson/© D.C. Heath. 46: Paul Johnson/© D.C. Heath. 50: t Paul Johnson/© D.C. Heath. b Stuart Cohen/© D.C. Heath.

Chapter 3: 55: Paul Johnson. 57: Stuart Cohen/© D.C. Heath. 58: Richard Haynes/© D.C. Heath. 60: Richard Haynes/© D.C. Heath. 61: Richard Haynes/© D.C. Heath. 62: Gregg Mancuso (Globe Photos). 64: Ken O'Donoghue/© D.C. Heath. 65: Ken O'Donoghue/© D.C. Heath. 66: Stuart Cohen/© D.C. Heath. 69: NASA. 70: 1 Stuart Cohen/© D.C. Heath; r Paul Johnson/© D.C. Heath. 71: Stuart Cohen/© D.C. Heath. 72: Stuart Cohen/© D.C. Heath. 74: Richard Haynes/© D.C. Heath. 76: Stuart Cohen/© D.C. Heath.

Chapter 4: 81: © Susan Lapides 1987. 82: © Susan Lapides 1987. 83: Richard Haynes/© D.C. Heath. 84: Richard Haynes/© D.C. Heath. 86: Stuart Cohen/© D.C. Heath. 88: Stuart Cohen/© D.C. Heath. 90: Stuart Cohen/© D.C. Heath. 92: Stuart Cohen/© D.C. Heath. 94: Stuart Cohen/© D.C. Heath. 96: Paul Johnson/© D.C. Heath. 98: Stuart Cohen/© D.C. Heath. 100: Richard Haynes/© D.C. Heath. 102: Stuart Cohen/© D.C. Heath. 103: Paul Johnson/© D.C. Heath. 104: Stuart Cohen (Stock Boston). 105: Stuart Cohen/© D.C. Heath. 107: Ken O'Donoghue/© D.C. Heath. 108: t Paul Johnson/© D.C. Heath; b Hilary Wallace/© D.C. Heath.

Chapter 5: 113: Stuart Cohen. 114: 1 Hilary Wallace/© D.C. Heath; r Robert Nese (Globe Photos). 116: 1 Paramount Pictures; ml Yoram Kahana (Shooting Star); mr Steve Schapiro (Sygma). r Yoram Kahana (Shooting Star). 118: Hilary Wallace/© D.C. Heath. 120: Stuart Cohen/© D.C. Heath. 122: Hilary Wallace/© D.C. Heath. 124: Stuart Cohen/© D.C. Heath. 125: Hilary Wallace/© D.C. Heath. 126: N. Cutler (Globe Photos); ml Mike Keza (Globe Photos); mr The Bettmann Archive, Inc.; r Ross Marino (Sygma). 127: Richard Haynes/© D.C. Heath. 129: Hilary Wallace/© D.C. Heath. 130: Richard Haynes/© D.C. Heath. 132: Paul Johnson/© D.C. Heath.

Chapter 6: 137: Paul Johnson. 140: Richard Haynes/© D.C. Heath. 141: Richard Haynes/© D.C. Heath. 143: Richard Haynes/© D.C. Heath. 146: Stuart Cohen/© D.C. Heath. 147: The Bettmann Archive, Inc. 149: © 1982 Alex Dunca (Taurus Photos). 151: Paul Johnson/© D.C. Heath. 152: Unknown. 154: Stuart Cohen/© D.C. Heath. 156: Paul Johnson/© D.C. Heath. 157: Stuart Cohen/© D.C. Heath. 159: Globe Photos. 160: t Paul Johnson/© D.C. Heath.. b Stuart Cohen/© D.C. Heath. 161: Paul Buddle. 162: Paul Johnson/© D.C. Heath. 163: Paul Johnson/© D.C. Heath. 166: t Ken O'Donoghue/© D.C. Heath; b Paul Johnson/© D.C. Heath. 168: Paul Johnson/© D.C. Heath. 169: Paul Johnson/© D.C. Heath. 170: Richard Haynes/© D.C. Heath. 172: Stuart Cohen/© D.C. Heath.

Chapter 7: 177: Hans Namuth (Photo Researchers). 178: Stuart Cohen/© D.C. Heath. 179: Stuart Cohen/© D.C. Heath. 182: Michael Rusnock. 184: Hilary Wallace/© D.C. Heath. 186: Paul Johnson/© D.C. Heath. 187: Paul Johnson/© D.C. Heath. 188: t P. Grant (Outdoor Adventure Images); bl © Susan Lapides 1987; br Paul Buddle. 190: Hilary Wallace/© D.C. Heath. 192: Peter Menzel (Stock Boston) 194: Richard Haynes/© D.C. Heath.

Chapter 8: 201: Russ Kinne (Photo Researchers). 202: Paul Johnson/© D.C. Heath. 205: Stuart Cohen/© D.C. Heath. 206: Paul Johnson/© D.C. Heath. 208: Paul Johnson/© D.C. Heath.209: t,m Paul Johnson/© D.C. Heath. b Bill Gillette (Stock Boston). 210: Hilary Wallace/© D.C. Heath. 212: Focus on Sports. 214: Stuart Cohen/© D.C. Heath. 216: Richard Haynes/© D.C. Heath. 218t: Paul Johnson/© D.C. Heath.

Chapter 9: 223: © Susan Lapides 1987. 224: Paul Johnson/© D.C. Heath. 225: Paul Johnson/© D.C. Heath. 226: Mickey Palmer (Focus on Sports). 228: Hilary Wallace/© D.C. Heath. 229: © Susan Palides 1987. 230: Paul Johnson/© D.C. Heath. 232: t Thomas E. Evans (Photo Researchers) b Leeanne Schmidt (Design Photographers International). 234: Stuart Cohen/© D.C. Heath. 238: Richard Haynes/© D.C. Heath. 240: Paul Johnson/© D.C. Heath. 242: Stuart Cohen/© D.C. Heath. 243: 1 Stuart Cohen/© D.C. Heath; r Paul Johnson/© D.C. Heath. 244: Paul Johnson/© D.C. Heath. 245: Stuart Cohen/© D.C. Heath. 246: t Hilary Wallace/© D.C. Heath; m Rick Rizzoto/© D.C. Heath; b Stuart Cohen/© D.C. Heath. 247: Paul Johnson/© D.C. Heath. 248: Richard Haynes/© D.C. Heath. 251: Lynn M. Stone (Animals Animals). 252: Paul Johnson/© D.C. Heath.

Note: Photo Credits are continued on page 518.

Copyright © 1989 by D.C. Heath and Company

All rights reserved. Certain portions of this publication copyright © 1985, by D.C. Heath and Company. No part of this publication may be reproduced or transmitted in any form or by any means, electronic or mechanical, including photocopy, recording, or any information storage or retrieval system, without permission in writing from the publisher.

Published simultaneously in Canada

Printed in the United States of America

International Standard Book Code Number: 0-669-16414-3

3 4 5 6 7 8 9 0

CONTENTS

Overview of
    Heath GENERAL MATHEMATICS   T6
Philosophy   T13
Program Goals   T14
Pacing Chart   T23
Learning Strands   T24

1 Adding Whole Numbers and Decimals

Reading standard numerals   2
Writing standard numerals in words   4
Adding whole numbers   6
- Problem Solving   8
- Cumulative Skill Practice   9

Rounding whole numbers   10
Estimating sums   12
Reading decimals   14
- Problem Solving   16
- Cumulative Skill Practice   17

Rounding decimals   18
Adding decimals   20
More on adding decimals   22
Problem Solving—too much/too little information   24
- Problem Solving   26
- Cumulative Skill Practice   27

Chapter Review   28
Chapter Test   29
Cumulative Test   30

2 Subtracting Whole Numbers and Decimals

Comparing whole numbers   32
Subtracting 2- and 3-digit numbers   34
Estimating differences   36
Subtracting larger numbers   38
- Problem Solving   40
- Cumulative Skill Practice   41

Comparing decimals   42
Subtracting decimals   44
More on subtracting decimals   46
Problem Solving—using logical reasoning   48
- Problem Solving   50
- Cumulative Skill Practice   51

Chapter Review   52
Chapter Test   53
Cumulative Test   54

3 Multiplying Whole Numbers and Decimals

Mental math—multiplying by multiples of 10, 100 or 1000   56
Multiplying by a 1-digit number   58
Estimating products   60
Multiplying by a 2-digit number   62
Multiplying larger numbers   64
- Problem Solving   66
- Cumulative Skill Practice   67

Multiplying decimals   68
Simplifying expressions   70
Mental math—multiplying decimals by 10, 100 or 1000   72
Problem Solving—using simpler problems to solve multistep problems   74
- Problem Solving   76
- Cumulative Skill Practice   77

Chapter Review   78
Chapter Test   79
Cumulative Test   80

4 Dividing Whole Numbers and Decimals

Dividing by a 1-digit number   82
Estimating quotients   84
More on dividing by a 1-digit number   86
Dividing by a 2-digit number   88
- Problem Solving   90
- Cumulative Skill Practice   91

Dividing by a 3-digit number   92
Order of operations   94
Dividing a decimal by a whole number   96
- Problem Solving   98
- Cumulative Skill Practice   99

Mental Math—dividing by 10, 100, or 1000   100
Dividing a decimal by a decimal   102
More on dividing a decimal by a decimal   104
Problem Solving—interpreting remainders   106
- Problem Solving   108
- Cumulative Skill Practice   109

Chapter Review   110
Chapter Test   111
Cumulative Test   112

First Quarter Test   112A

5 Graphs and Statistics

Organizing data   114
Using bar graphs   116
Using line graphs   118
Using circle graphs and picture graphs   120
- Problem Solving   122
- Cumulative Skill Practice   123

Analyzing data—finding the mean   124
Analyzing data—finding the median, mode, and range   126
Reading and constructing graphs   128
Problem Solving—choosing sensible answers   130
- Problem Solving   132
- Cumulative Skill Practice   133

Chapter Review   134
Chapter Test   135
Cumulative Test   136

6 Number Theory, Fractions, and Decimals

Exponents   138
Divisibility   140
Prime and composite numbers   142
Greatest common factor   144
Equivalent fractions   146
Writing fractions in lowest terms   148
Least common multiple   150
Comparing fractions   152
- Problem Solving   154
- Cumulative Skill Practice   155

Writing whole numbers and mixed numbers as fractions   156
Writing fractions as whole numbers or mixed numbers   158
Writing fractions and mixed numbers in simplest form   160
Writing quotients as mixed numbers   162
- Problem Solving   164
- Cumulative Skill Practice   165

Writing fractions and mixed numbers as decimals   166
Writing decimals as fractions or mixed numbers   168
Problem Solving—guess and check   170
- Problem Solving   172
- Cumulative Skill Practice   173

Chapter Review   174
Chapter Test   175
Cumulative Test   176

7 Adding and Subtracting Fractions and Mixed Numbers

Adding fractions with common denominators *178*
Adding fractions with different denominators *180*
Adding mixed numbers *182*
• Problem Solving *184*
• Cumulative Skill Practice *185*
Subtracting fractions with common denominators *186*
Subtracting fractions with different denominators *188*
Subtracting mixed numbers without regrouping *190*
Subtracting mixed numbers with regrouping *192*
Problem Solving—making a table *194*
• Problem Solving *196*
• Cumulative Skill Practice *197*

Chapter Review *198*
Chapter Test *199*
Cumulative Test *200*

8 Multiplying and Dividing Fractions and Mixed Numbers

Multiplying fractions *202*
A fraction of a whole number *204*
Multiplying mixed numbers *206*
More on a fraction of a whole number *208*
• Problem Solving *210*
• Cumulative Skill Practice *211*
Dividing fractions *212*
Dividing mixed numbers *214*
Problem Solving—making a list *216*
• Problem Solving *218*
• Cumulative Skill Practice *219*

Chapter Review *220*
Chapter Test *221*
Cumulative Test *222*

Second Quarter Test 222A

9 Measurement

Using a metric ruler *224*
Metric units of length *226*
Changing units in the metric system *228*
Liquid volume—metric system *230*
Weight—metric system *232*
• Problem Solving *234*
• Cumulative Skill Practice *235*
Temperature—degrees Celsius and degrees Fahrenheit *236*
Elapsed time *238*
Length—customary units *240*
Changing units of length—customary *242*
Liquid volume—customary units *244*
Weight—customary units *246*
Computing with customary units *248*
Problem Solving—working backward *250*
• Problem Solving *252*
• Cumulative Skill Practice *253*

Chapter Review *254*
Chapter Test *255*
Cumulative Test *256*

10 Ratio and Proportion

Ratios *258*
Proportions *260*
Solving proportions *262*
Rates *264*
• Problem Solving *266*
• Cumulative Skill Practice *267*
Scale drawings *268*
Similar figures *270*
Indirect measurement *272*
Problem Solving—applications *274*
• Problem Solving *276*
• Cumulative Skill Practice *277*

Chapter Review *278*
Chapter Test *279*
Cumulative Test *280*

11 Percent

Changing a percent to a fraction *282*
Changing a fraction to a percent *284*
Percents and decimals *286*
• Problem Solving *288*
• Cumulative Skill Practice *289*
Finding a percent of a number *290*
Estimating a percent of a number *292*
More on finding a percent of a number *294*
Finding the number when a percent is known *296*
More on percent *298*
Problem Solving—using estimation *300*
• Problem Solving *302*
• Cumulative Skill Practice *303*

Chapter Review *304*
Chapter Test *305*
Cumulative Test *306*

12 Consumer Mathematics

Earning money and payroll deductions *308*
Buying on sale *310*
Comparison buying *312*
Bargain buying *314*
• Problem Solving *316*
• Cumulative Skill Practice *317*
Checking accounts *318*
Savings account *320*
Borrowing money *322*
Paying bills *324*
Problem Solving—using a sales tax table *326*
• Problem Solving *328*
• Cumulative Skill Practice *329*

Chapter Review *330*
Chapter Test *331*
Cumulative Test *332*

Third Quarter Test 332A

13 Geometry—Perimeter and Area

Lines and angles *334*
Perpendicular and parallel lines *336*
Geometric constructions *338*
Polygons *340*
Perimeter *342*
Circumference *344*
● Problem Solving *346*
● Cumulative Skill Practice *347*
Area—squares and rectangles *348*
Area—parallelograms *350*
Area—triangles *352*
Area—circles *354*
Problem Solving—making a drawing *356*
● Problem Solving *358*
● Cumulative Skill Practice *359*

Chapter Review *360*
Chapter Test *361*
Cumulative Test *362*

14 Surface Area and Volume

Space figures *364*
More on space figures *366*
Surface area—rectangular prisms and cubes *368*
● Problem Solving *370*
● Cumulative Skill Practice *371*
Volume—rectangular prisms and cubes *372*
Volume—cylinders *374*
Volume—pyramids and cones *376*
Problem Solving—finding and using patterns *378*
● Problem Solving *380*
● Cumulative Skill Practice *381*

Chapter Review *382*
Chapter Test *383*
Cumulative Test *384*

15 Probability

A basic counting principle *386*
Permutations *388*
Probability *390*
Sample spaces *392*
● Problem Solving *394*
● Cumulative Skill Practice *395*
Probability—more than 1 event *396*
Odds *398*
Applying probability—expectation *400*
Problem Solving—using income tax tables *402*
● Problem Solving *404*
● Cumulative Skill Practice *405*

Chapter Review *406*
Chapter Test *407*
Cumulative Test *408*

16 Integers

Comparing integers and absolute value *410*
Adding integers *412*
Subtracting integers *414*
● Problem Solving *416*
● Cumulative Skill Practice *417*
Multiplying integers *418*
Dividing integers *420*
Properties of addition and multiplication *422*
Graphing ordered pairs *424*
Problem Solving—applications *426*
● Problem Solving *428*
● Cumulative Skill Practice *429*

Chapter Review *430*
Chapter Test *431*
Cumulative Test *432*

17 Algebra

Writing expressions *434*
Evaluating expressions *436*
Solving addition equations *438*
Solving subtraction equations *440*
● Problem Solving *442*
● Cumulative Skill Practice *443*
Solving multiplication equations *444*
Solving division equations *446*
Solving two-step equations *448*
Square roots *450*
The Pythagorean rule *452*
Problem Solving—using formulas *454*
● Problem Solving *456*
● Cumulative Skill Practice *457*

Chapter Review *458*
Chapter Test *459*
Cumulative Test *460*

Fourth Quarter Test *460A*
Final Test *460C*

 Resources

Skill Test *462*
Extra Practice *471*
Glossary *503*
Symbols and Formulas *511*
Index *512*
Time/Customary Measures/Metric Measures *Back Cover*

HEATH
General Mathematics

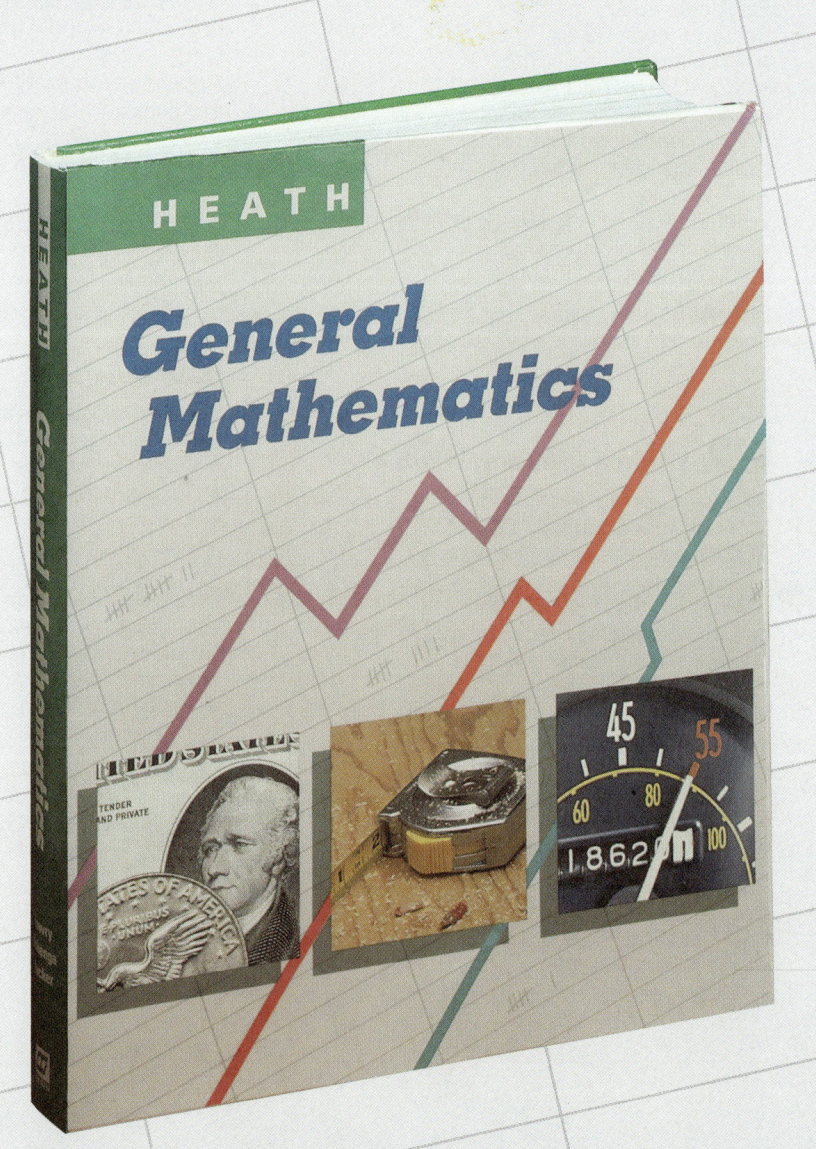

Actively involves students . . .

When students are interested in the material being presented they are more actively involved and more motivated to learn.

Heath GENERAL MATHEMATICS uses lesson settings and situations that are familiar and motivational to students. These settings help students to see a connection between the mathematics taught in the classroom and the mathematics used in everyday life.

Integrates problem solving . . .

Problem solving is more than solving word problems.

Heath GENERAL MATHEMATICS integrates problem solving into all lessons. By using higher order thinking skills, specific strategies, and collecting and analyzing data, students go beyond computation and focus on real applications of mathematical skills.

Teaches skills in context . . .

Skills taught in isolation seem irrelevant and uninteresting.

Computational skills in **Heath GENERAL MATHEMATICS** are taught in a situational context. Therefore, students become interested and motivated to learn the underlying concepts and to apply those skills.

Provides a variety of support materials . . .

Every class is different.

Teachers are provided with alternative methods of presenting the lessons. **Heath GENERAL MATHEMATICS** provides a variety of support materials for students and teachers in order to reinforce and extend the text lessons.

Lesson Format

1 *The lesson theme*
- provides motivational settings that hold student attention.
- gets students involved with data from daily experiences.
- includes such things as newspaper articles, charts, tables, advertisements, menus, price lists, and photographs.

2 *The first few questions*
- help students get started.
- help connect motivational settings with skills to be learned.

3 *The* Here's how *box provides*
- step-by-step examples worked out the same way teachers do them on the chalkboard.
- an easy reference for students who need reteaching later in the year.

4 *The follow-up questions*
- extend the *Here's how* instruction.
- provide additional examples when appropriate.

5 *The Exercises*
- include scrambled answers that help students get started.
- provide sufficient practice for every ability level.
- relate problem-solving exercises to the lesson setting.

6 *The lesson challenge*
- provides practice of nonstandard problem-solving skills.
- relates back to the lesson theme.
- appears at the end of every lesson.

T8

Problem Solving

Problem-Solving Lessons

- teach specific problem-solving strategies that can be used to help students solve problems in their everyday lives.
- use the same format as all the other two-page lessons.
- provides a 3-step framework in the *Here's how* to help students systematically approach problems.

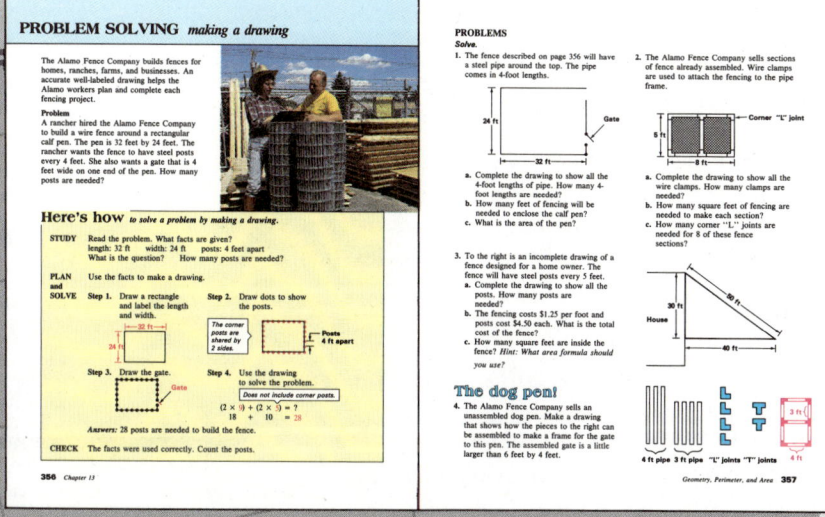

Lesson challenges

- use higher order thinking skills, specific strategies, and collecting and analyzing data activities to practice skills previously taught.
- integrate problem solving into every lesson.

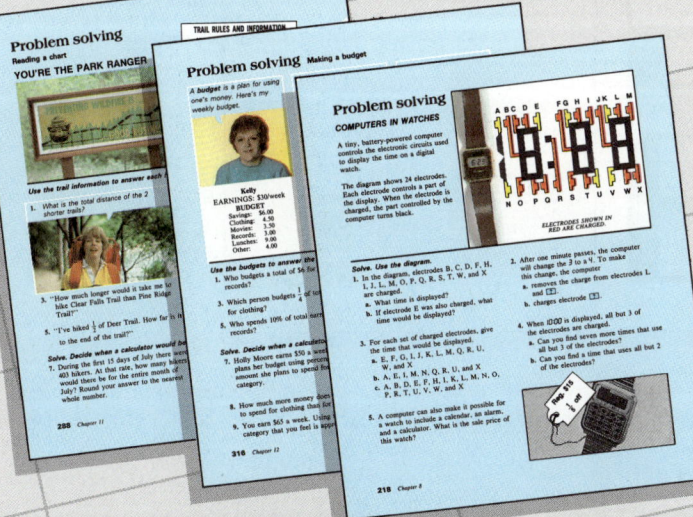

Problem-Solving Practice

- reviews previously taught skills in the context of everyday situations.
- emphasizes mathematics on the job, consumer mathematics, and computer applications.
- occurs after every three or four lessons.

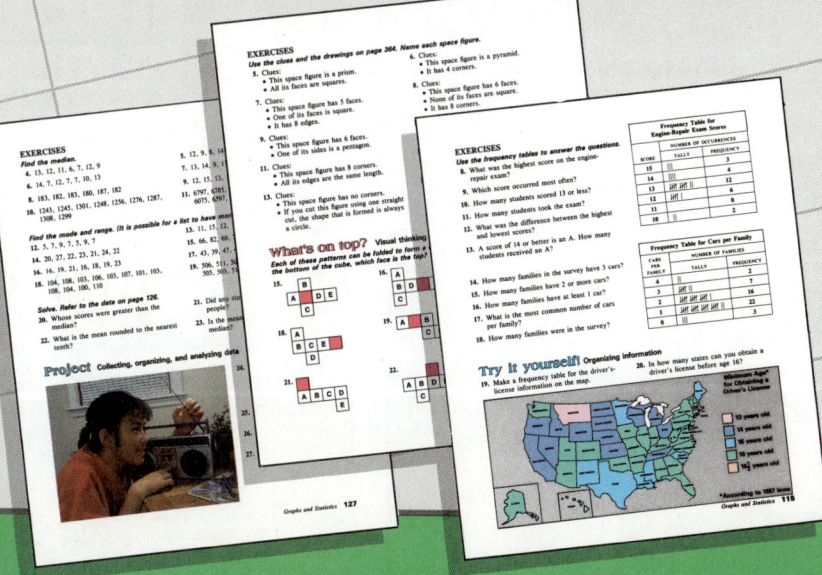

T9

Review and Evaluation

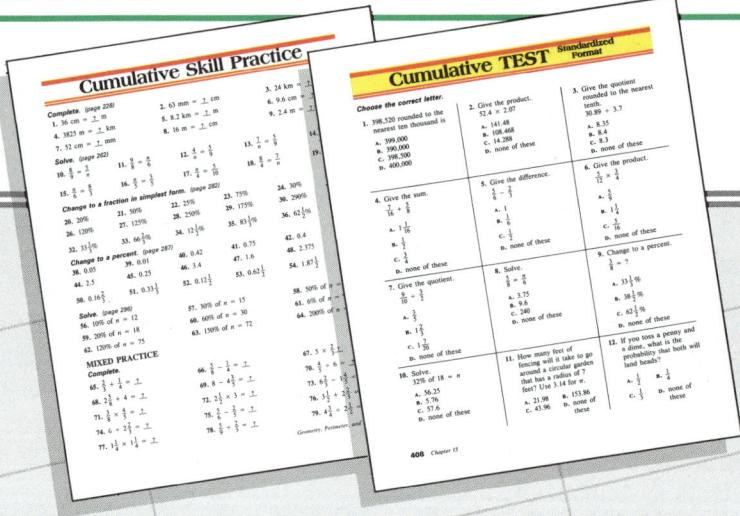

SKILL MAINTENANCE

Cumulative Skill Practice
- maintains skills previously taught.
- offers mixed practice.
- occurs after every three or four lessons.

Cumulative Test
- tests skills that were reviewed and practiced on the *Cumulative Skills Practice* pages in the chapter.
- is in standardized test format.
- occurs at the end of every chapter.

REVIEW AND TESTING

Chapter Review
- reviews the objectives of the chapter.
- uses a unique format to encourage students to think through all the steps of a computation process.

Chapter Test
- tests the objectives of the chapter.

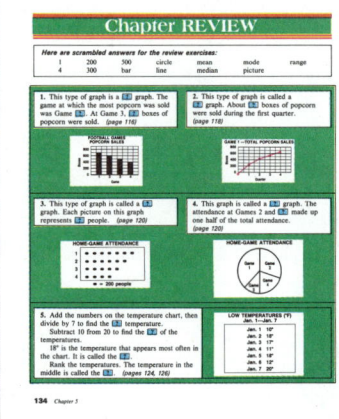
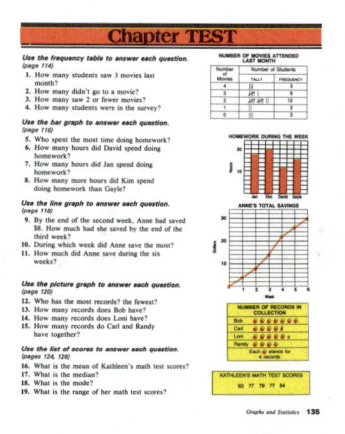

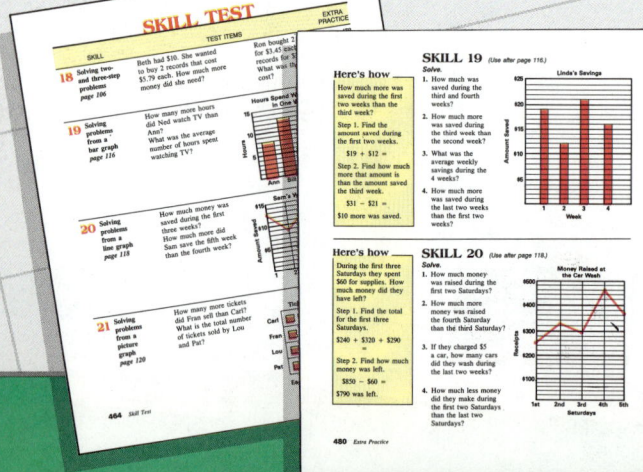

END-OF-BOOK RESOURCES

Skill Test
- tests 64 major skills taught in the text.
- is referenced to *Extra Practice*.

Extra Practice
- includes 64 practice sets correlated to the *Skill Test*.
- provides examples to reteach and review the skill.

Teaching Support

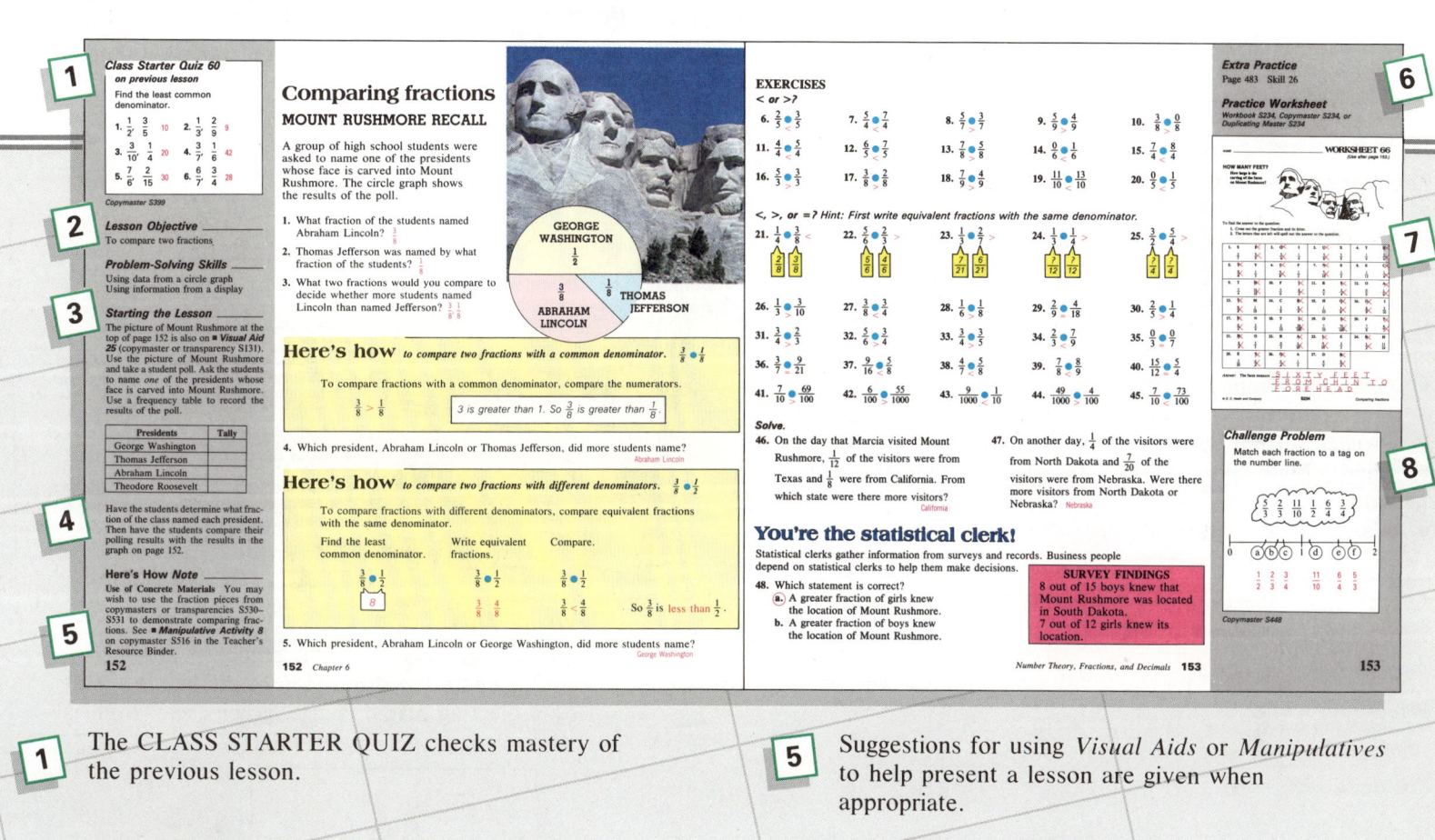

1 The CLASS STARTER QUIZ checks mastery of the previous lesson.

2 The LESSON OBJECTIVES and PROBLEM-SOLVING SKILLS are listed to help you plan your lesson.

3 STARTING THE LESSON provides suggestions on how to proceed. When appropriate the commentary offers alternate ways to approach the lesson.

4 HERE'S HOW and EXERCISE NOTES are helpful suggestions to the teacher.

5 Suggestions for using *Visual Aids* or *Manipulatives* to help present a lesson are given when appropriate.

6 Additional practice in the student text is referenced.

7 The reduced facsimile of the WORKSHEET that accompanies the lesson helps you decide whether to assign it to any of your students. Answers are overprinted for your convenience.

8 Either a CHALLENGE PROBLEM, a PROJECT, or a GROUP PROJECT is provided for use at the end of each lesson.

T11

Supplementary Resources

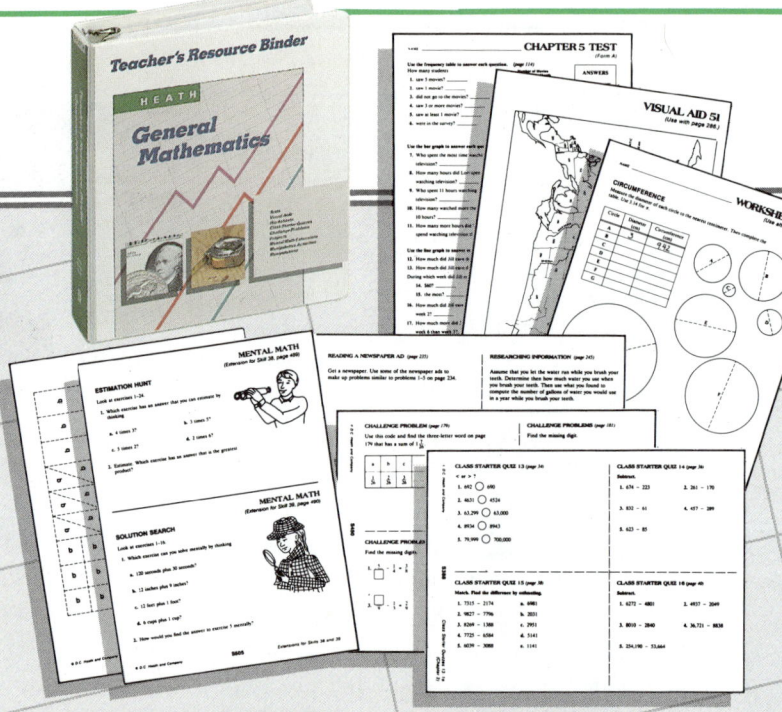

Teacher's Resource Binder (blackline masters) provides:
- Tests - chapter tests (2 forms), quarterly, and a final test
- Visual Aids - maps, charts, graphs, etc., from student text
- Worksheets - one for every lesson, uses self-checking, decoder, or problem-solving formats
- Class Starter Quizzes - from extended margin of the Teacher's Edition
- Challenge Problems - from extended margin of the Teacher's Edition
- Projects - from extended margin of the Teacher's Edition
- Mental Math Extensions - 1 for every computational *Extra Practice* set
- Manipulatives - alternate instructional materials

	Blackline Masters	Duplicating Masters	Workbook	Transparencies
Tests	●	●		
Visual Aids	●			●
Worksheets	●	●	●	
Manipulatives	●			●

Visual Aids

The Visual Aids direct the students' attention and facilitate the discussion of the lesson theme by projecting maps, charts, graphs, etc. from the student text.

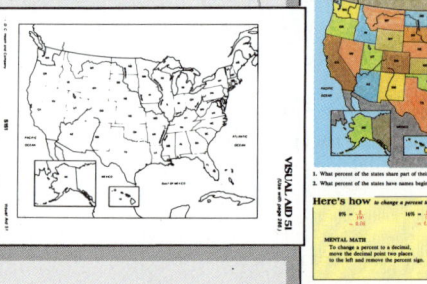

Manipulatives

The manipulative package can be used to supplement lesson concepts in the text, or to explore various topics through specified activities.

- Powers-of-ten tiles
- Fraction pieces
- Area tiles
- 10-by-10 grid
- Tangram pieces
- Tangram shapes
- Area grid
- Area-formula pieces
- Positive and negative charges

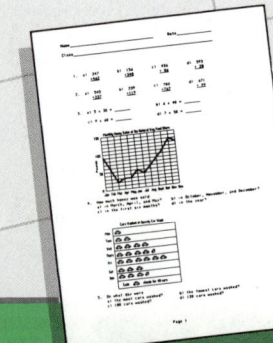

Computer Gernerated Practice and Tests

- generates worksheets of specific skills.
- customizes individual objective tests, specific chapter tests, or cumulative tests.
- utilizes a complete range of graphic capabilities.

Philosophy

How did the authors develop the program design?
The authors used recent research on teaching general mathematics, a national survey of general mathematics teachers conducted by D.C. Heath, and a wide variety of other resources including their combined 70 years of teaching experience.

What insights were gained from Heath's national survey?
Teachers are developing exciting new approaches to teaching general mathematics because they feel that existing instructional materials do little more than practice computational skills.

How does *Heath General Mathematics* respond to recent research studies in general mathematics?
Researchers from the Institute for Research on Teaching at Michigan State University suggest teachers should use three basic strategies with their students.

(1) **Increase the quality and quantity of mathematics communication.**
   Heath General Mathematics provides students and teachers with an opportunity to engage in a dialogue about mathematics. The lesson settings are familiar to students, and most of the settings provide numerical data that teachers can use to create additional problems which will directly relate to the class. The teacher can encourage students to explain their thinking and reasoning. When appropriate, manipulative materials are provided so students and teachers have the opportunity to take a fresh look at concepts, which in turn stimulates additional discussion and interaction.

(2) **Use the social organization of the classroom to facilitate instruction.**
   Heath General Mathematics provides easy-to-use activities throughout. Teachers who use small group instruction and other interactive teaching strategies will find these especially useful. In addition, most lessons simulate real-life applications which help students develop an understanding and appreciation of the role mathematics plays in their daily life. Also, all lessons involve various problem-solving techniques that students will use outside the classroom. Students use higher order thinking skills, learn specific strategies, and collect and analyze data to solve everyday problems.

(3) **Teach new topics or modify existing ones and assign interesting and challenging problems.**
   Heath General Mathematics uses a fresh approach to teaching traditional content and includes new content for the curriculum of the 1990's. Estimation is used throughout, graphing is introduced early and practiced often, probability and statistics topics are included, and basic algebra topics are covered in sufficient depth to prepare students for further study in mathematics.

How does the program promote effective teaching?
Research suggests that lessons are more effective when they are constructed around a core of activities. *Heath General Mathematics* provides supplementary materials that will help teachers develop a wide variety of learning activities for accommodating a broad range of interests, aptitudes, and learning styles. Each teaching/learning aid is pictured or referenced in the teacher's edition at its point of use.

Research also suggests that teachers are more effective when they gather continuous information on students and give continuous feedback to students. *Heath General Mathematics* provides materials to help the teacher accomplish this in an efficient way. In addition, the program provides special help in dealing with short attention spans and reading difficulties.

It should also be pointed out that the author team wrote *every* component of the program. Thus every supplementary learning activity correlates properly and is written at the appropriate level of difficulty.

Program Goals

This table presents an overview of the major goals achieved by students when using **Heath General Mathematics**.

The *Lesson Objectives* column lists the mathematics skill(s) and explicit problem-solving strategies taught in lessons of the text. In addition, this column references skills-practice where relevant.

The *Problem-Solving Skills* column lists any higher-order thinking skills or other problem-solving skills utilized within lessons.

Chapter 1 Adding Whole Numbers and Decimals

Pages	Lesson Objectives	Problem-Solving Skills
2-3	To read standard numerals less than 1 billion	Reading a chart Using logical reasoning
4-5	To write standard numerals (through the billions) in words	Reading a chart
6-7	To add whole numbers	Reading a chart Reading road signs
8-9	To practice skills previously taught	Reading a map
10-11	To round whole numbers	Following instructions Number sense
12-13	To estimate sums	Finding information in an ad Using a guess-and-check strategy
14-15	To read decimals	Interpreting information Number sense
16-17	To practice skills previously taught	Reading a sign Choosing the appropriate information
18-19	To round decimals	Reading a chart
20-21	To add decimals (addends having the same number of decimal places)	Reading a chart Using a guess-and-check strategy
22-23	To add decimals (addends not having the same number of decimal places)	Selecting information from a newspaper article Reading a chart
24-25	To solve problems with too much or too little information	Reading a map Following directions
26-27	To practice skills previously taught	Using computer-displayed information

Chapter 2 Subtracting Whole Numbers and Decimals

Pages	Lesson Objectives	Problem-Solving Skills
32-33	To compare whole numbers	Reading a chart Using logical reasoning
34-35	To subtract 2- and 3-digit numbers	Reading a chart Choosing the correct operation
36-37	To estimate differences	Reading a chart Using logical reasoning
38-39	To subtract 3- and 4-digit numbers To subtract across zeros	Using a guess-and-check strategy Choosing the correct operation
40-41	To practice skills previously taught	Reading a newspaper ad
42-43	To compare decimals	Reading a chart Using visual thinking
44-45	To subtract decimals (with the same number of decimal places)	Reading a chart Choosing the correct operation Using logical reasoning
46-47	To subtract decimals (with a different number of decimal places)	Finding information in a display Using a guess-and-check strategy
48-49	To solve problems using logical reasoning	Making up a problem
50-51	To practice skills previously taught	Using computer-displayed information

Chapter 3 Multiplying Whole Numbers and Decimals

Pages	Lesson Objectives	Problem-Solving Skills
56-57	To multiply mentally by multiplying multiples of 10, 100, or 1000	Reading a chart
58-59	To multiply by a 1-digit number	Reading a chart Using logical reasoning
60-61	To estimate products	Reading a graph Using estimation
62-63	To multiply by a 2-digit number	Reading a newspaper ad Using a guess-and-check strategy
64-65	To multiply by a 3-digit number	Finding information in a display Reading a chart
66-67	To practice skills previously taught	Finding information in an ad Recognizing when information is missing Choosing the correct operation
68-69	To multiply decimals	Finding information in a table Solving a multistep problem Using logical reasoning
70-71	To simplify expressions	Choosing appropriate information Choosing the correct operation Solving a multistep problem
72-73	To multiply decimals mentally by 10, 100, or 1000	Finding information in a table Checking answers
74-75	To solve multistep problems using simpler numbers	Number sense
76-77	To practice skills previously taught	Using information on a menu and a cash register display Choosing the correct operation Using a guess-and-check strategy

Program Goals **T15**

Chapter 4 Dividing Whole Numbers and Decimals

Pages	Lesson Objectives	Problem-Solving Skills
82-83	To divide by a 1-digit number (with a 3-digit dividend)	Finding information in a newspaper article Choosing correct operations Solving a multistep problem Number sense
84-85	To estimate quotients	Reading a mileage chart Using estimation
86-87	To divide by a 1 digit number (with a 4 digit dividend and zeros in the quotient)	Reading an ad Choosing correct operations
88-89	To divide by a 2-digit number To check division	Reading a table Using a guess-and-check strategy
90-91	To practice skills previously taught	Reading a mileage chart Choosing correct operations
92-93	To divide by a 3-digit number To round quotients to the nearest whole number	Choosing appropriate information Choosing correct operations
94-95	To apply rules of order of operations to simplify expressions To write mathematical expressions	Finding information in a store coupon
96-97	To divide a decimal by a whole number	Reading a chart Choosing correct operations Collecting, organizing, and analyzing data
98-99	To practice skills previously taught	Choosing correct operations Solving a two-step problem
100-101	To divide mentally by 10, 100, or 1000	Choosing appropriate information Choosing correct operations Following directions
102-103	To divide a decimal by a decimal	Finding information in an ad Choosing correct operations
104-105	To divide a decimal by a decimal To write a remainder as a decimal	Reading a chart Selecting data from a drawing Choosing correct operations Making a drawing
106-107	To interpret remainders in the solution to a problem	Using visual thinking
108-109	To practice skills previously taught	Finding information on a computer printout

Chapter 5 Graphs and Statistics

Pages	Lesson Objectives	Problem-Solving Skills
114-115	To read data from frequency table To make a frequency table from a set of data	Reading and interpreting a table Organizing and tabulating information
116-117	To read data from bar graphs	Reading and interpreting bar graphs
118-119	To read data from line graphs	Reading and interpreting line graphs
120-121	To read data from circle graphs and picture graphs	Reading and interpreting circle graphs and picture graphs
122-123	To practice skills previously taught	Reading a graph Choosing the correct operation
124-125	To compute the mean of a set of data	Reading a chart Choosing the correct operation Analyzing data
126-127	To compute the median, mode, and range of a set of data	Reading and interpreting a chart Collecting, organizing, and analyzing data
128-129	To interpret bar, line, circle, and picture graphs To construct bar, line, circle, and picture graphs	Finding information from graphs Making a graph
130-131	To choose the most sensible answer when selecting an answer to a problem	Making up a problem
132-133	To practice skills previously taught	Reading and interpreting graphs

Chapter 6 — Number Theory, Fractions, and Decimals

Pages	Lesson Objectives	Problem-Solving Skills
138-139	To write numbers with exponents as standard numerals To use exponents to write products	Finding a pattern
140-141	To use rules to test for divisibility	Estimating quantity
142-143	To write the prime factorization of a composite number	Making a list
144-145	To find the greatest common factor of a pair of numbers	Using a guess-and-check strategy
146-147	To find equivalent fractions for a given fraction	Finding information in a display Using logical reasoning
148-149	To write fractions in lowest terms	Using logical reasoning
150-151	To identify the LCM To identify the least common denominator of a pair of fractions	Drawing a diagram Using logical reasoning
152-153	To compare two fractions	Using data from circle graphs Using display information
154-155	To practice skills previously taught	Finding information in a ad Solving a multistep problem
156-157	To write whole numbers and mixed numbers as fractions	Using logical reasoning
158-159	To write fractions as whole numbers or mixed numbers	Finding information in an ad Following instructions
160-161	To write fractions and mixed numbers in simplest form	Using logical reasoning
162-163	To write quotients as mixed numbers	Finding information in a display Choosing correct operations Using logical reasoning
164-165	To practice skills previously taught	Reading a map Choosing correct operations Solving a multistep problem
166-167	To write fractions and mixed numbers as decimals	Finding information in a recipe Choosing correct operations Finding a pattern
168-169	To write decimals as fractions or mixed numbers	Using a guess-and-check strategy
170-171	To use a guess-and-check strategy	Using logical reasoning
172-173	To practice skills previously taught	Reading a cash register receipt Choosing correct operations

Chapter 7 — Adding and Subtracting Fractions and Mixed Numbers

Pages	Lesson Objectives	Problem-Solving Skills
178-179	To add fractions with common denominators	Choosing information from a display Making a list
180-181	To add fractions with different denominators	Reading a map
182-183	To add mixed numbers	Selecting information from a chart Following instructions and checking answers
184-185	To practice skills previously taught	Making a drawing Choosing the correct operation
186-187	To subtract fractions with common denominators	Finding information in a display Choosing the correct operation Making a drawing
188-189	To subtract fractions with different denominators	Reading circle graphs Checking answers
190-191	To subtract mixed numbers without regrouping	Selecting information from a display Using logical reasoning
192-193	To subtract mixed numbers with regrouping	Selecting information from a chart Reading a map and a map scale
194-195	To solve problems by making a table	Using visual thinking
196-197	To practice skills previously taught	Using information on a display Choosing the correct operation

Program Goals T17

Chapter 8   Multiplying and Dividing Fractions and Mixed Numbers

Pages	Lesson Objectives	Problem-Solving Skills
202-203	To multiply a fraction by a fraction	Finding information in a display Solving a multistep problem
204-205	To find a fraction of a whole number	Selecting information from a newspaper ad Choosing the correct operation Using logical reasoning
206-207	To multiply mixed numbers	Finding information in a recipe Choosing the correct operation Following directions and checking answers
208-209	To relate fractions to customary units of measure	Analyzing a sequence of events
210-211	To practice skills previously taught	Using a table Solving a multistep problem
212-213	To identify reciprocals To divide by a fraction To divide a fraction by a whole number	Using a picture to solve a problem Reading a chart
214-215	To divide mixed numbers	Selecting information from a display Following directions and checking answers
216-217	To solve problems by making a list	Finding a pattern
218-219	To practice skills previously taught	Using a drawing Following instructions

Chapter 9   Measurement

Pages	Lesson Objectives	Problem-Solving Skills
224-225	To measure lengths with centimeters and millimeters	Finding information in a display Reading a metric ruler Choosing correct operations
226-227	To become familiar with metric units of length	Reading a map Collecting, organizing, and analyzing data
228-229	To make conversions between metric units of length	Utilizing metric relationships Using logical reasoning
230-231	To become familiar with and make conversions between metric units of liquid volume	Selecting information in an ad
232-233	To become familiar with and make conversions between metric units of weight (mass)	Utilizing metric relationships
234-235	To practice skills previously taught	Choosing correct operations Solving a multistep problem
236-237	To read the scale on a Fahrenheit or a Celsius thermometer To find the difference between two temperatures	Reading a scale Reading a graph
238-239	To find elapsed time	Reading a time card Collecting, organizing, and analyzing data
240-241	To measure and draw lengths in inches	Reading an inch ruler Following directions
242-243	To make conversions between customary units of length	Utilizing relationships among customary units of length Using logical reasoning
244-245	To make conversions between customary units of liquid volume	Selecting information from a recipe Choosing correct operations Using logical reasoning
246-247	To make conversions between customary units of weight	Selecting information from a display Solving a multistep problem Using logical reasoning
248-249	To compute with customary units	Reading a table
250-251	To solve problems by working backward	Making up a problem
252-253	To practice skills previously taught	Finding data in a computer display Choosing correct operations

Program Goals

Chapter 10 Ratio and Proportion

Pages	Lesson Objectives	Problem-Solving Skills
258-259	To give the ratio of two quantities To change ratios to higher or lower terms	Using information in a display
260-261	To compare the cross products to tell whether the ratios are equal or not equal	Finding information in a table Using logical reasoning
262-263	To solve proportions	Using information in a display Checking answers
264-265	To solve rate problems using proportions	Using a proportion to solve a problem Reading scales
266-267	To practice skills previously taught	Finding information in a display Choosing the correct operation Using a proportion to solve a problem
268-269	To solve problems that involve scale drawings by using proportions	Reading a map Using a ruler and a map scale Choosing the correct operation Solving a multistep problem
270-271	To use proportions to find missing measurements of sides of similar figures	Finding information in a drawing Using a proportion to solve a problem Using scale drawings
272-273	To use similar triangles to solve indirect measurement problems	Selecting information from a drawing Using a drawing and proportion to solve a problem
274-275	To solve problems using facts from more than one source	Acting it out
276-277	To practice skills previously taught	Reading a computer display Using a proportion to solve a problem

Chapter 11 Percent

Pages	Lesson Objectives	Problem-Solving Skills
282-283	To change a percent to a fraction	Finding information in a circle graph Interpreting directions
284-285	To change a fraction to a percent	Reading a chart Solving a simpler problem
286-287	To change a percent to a decimal To change a decimal to a percent	Reading a map
288-289	To practice skills previously taught	Selecting information from a sign Choosing the correct information Solving a multistep problem
290-291	To find a percent of a number by multiplying by a fraction or decimal equivalent to a percent	Selecting information from a sale ad Using a guess-and-check strategy
292-293	To estimate a percent of a number	Estimating quantity
294-295	To find a percent of a number by solving a proportion	Reading a chart Making a list
296-297	To use a proportion to find a number when a percent of the number is known	Choosing the correct operation Following directions
298-299	To solve percent problems by using proportions	Reading a sale ad Setting up a proportion to solve any kind of percent problem Interpreting directions
300-301	To solve problems using estimation	Using estimation
302-303	To practice skills previously taught	Finding information in a computer display Choosing the correct operation

Program Goals T19

Chapter 12 Consumer Mathematics

Pages	Lesson Objectives	Problem-Solving Skills
308-309	To compute earnings To learn about some kinds of payroll deductions	Reading a help-wanted ad Reading a statement of withholdings and deductions
310-311	To compute discounts and sales prices	Reading a sale ad Checking information in an ad
312-313	To compute unit prices To determine which of two sizes is more expensive in terms of unit price	Selecting information from an ad Working backward
314-315	To solve problems by relating discounts and coupons to real situations	Selecting information from sales ads and coupons Choosing the correct operation
316-317	To practice skills previously taught	Selecting information from a chart
318-319	To understand checking accounts and checks To balance a checking account	Selecting information from checks and a check register
320-321	To understand savings accounts and passbooks To understand compound interest	Selecting information from a savings-account passbook Reading a graph Following directions
322-323	To compute simple interest	Selecting information from a display Reading an auto loan application
324-325	To understand monthly bills and payments	Finding information in a bill Choosing the correct operation Reading a telephone rate table
326-327	To solve problems using sales tax tables	Solving a multistep problem
328-329	To practice skills previously taught	Reading a computer display Choosing the correct operation Using a guess-and-check strategy

Chapter 13 Geometry—Perimeter and Area

Pages	Lesson Objectives	Problem-Solving Skills
334-335	To name and draw lines and angles To use a protractor to measure angles To classify angles To use a protractor to draw angles	Estimating measurements
336-337	To identify perpendicular and parallel lines and segments	Finding information in a drawing Using logical reasoning
338-339	To use a straightedge and compass to copy segments To use a straightedge and compass to bisect segments and angles	Following directions
340-341	To classify polygons according to their sides and angles	Reading a chart Using logical reasoning Making a drawing
342-343	To compute the perimeter of a polygon, given the length of its sides	Finding information in a drawing Using a formula
344-345	To compute the circumference of a circle, given its diameter or radius	Finding information in a drawing Using a formula Checking answers
346-347	To practice skills previously taught	Finding information in an ad Using a drawing
348-349	To compute the areas of squares and other rectangles	Finding information in a drawing Using a formula Checking correct operations
350-351	To compute the area of a parallelogram	Finding information in a drawing Using a formula Using visual thinking
352-353	To compute the area of a triangle	Finding information in a drawing Using a formula
354-355	To compute the area of a circle	Using a formula Using a drawing Using a guess-and-check strategy
356-357	To solve problems by making a drawing	Making a drawing
358-359	To practice skills previously taught	Selecting information from a computer drawing

Chapter 14 — Surface Area and Volume

Pages	Lesson Objectives	Problem-Solving Skills
364-365	To classify space figures by their faces, corners, and edges	Finding information in a drawing Using logical reasoning Making geometric visualizations
366-367	To visualize how triangular, square, and rectangular faces are used to build three-dimensional models	Selecting information from a display Making geometric visualizations
368-369	To compute the surface area of rectangular prisms and cubes	Finding information in a drawing Making geometric visualizations
370-371	To practice skills previously taught	Finding information in an ad Using a drawing to solve a problem Solving a multistep problem
372-373	To compute the volume of rectangular prisms and cubes	Finding information in a drawing Using a formula Choosing the correct operation Making geometric visualizations
374-375	To compute the volume of a cylinder	Finding information in a drawing Using a formula Choosing the correct formula
376-377	To compute the volume of a pyramid or cone	Finding information in a drawing Using a formula Using a guess-and-check strategy
378-379	To solve problems using a pattern	Using visual thinking
380-381	To practice skills previously taught	Making geometric visualizations

Chapter 15 — Probability

Pages	Lesson Objectives	Problem-Solving Skills
386-387	To find the total number of outcomes by using a tree diagram To use a basic counting principle to determine the number of outcomes of a compound event	Drawing a tree diagram
388-389	To compute the number of permutations (possible arrangements of things in a definite order)	Selecting information from a display
390-391	To determine the probabilities of outcomes of simple events	Selecting information from a display Conducting an experiment and collecting data
392-393	To compute the probability of an outcome	Selecting information from a display Using a tree diagram
394-395	To practice skills previously taught	Reading an ad Choosing the correct operation Solving a multistep problem
396-397	To compute the probability of a compound outcome	Using a tree diagram Selecting information from a display
398-399	To compute the odds of an outcome	Reading a chart
400-401	To compute the expectation of an outcome	Selecting information from a display Using a guess-and-check strategy
402-403	To solve problems using income tax tables	Using a table
404-405	To practice skills previously taught	Reading a blueprint Choosing the correct operation Solving a multistep problem

Program Goals

Chapter 16 Integers

Pages	Lesson Objectives	Problem-Solving Skills
410-411	To compare integers To find the absolute values of integers	Reading a map
412-413	To add integers	Finding information in a display Applying an addition model Working backward
414-415	To subtract integers	Finding information in a display Applying a subtraction model Working backward
416-417	To practice skills previously taught	Finding information in a price list Choosing the correct operation Solving a multistep problem
418-419	To multiply integers	Finding information in a display Applying a multiplication model Using a guess-and-check strategy
420-421	To divide integers	Discovering numerical relationships Working backward
422-423	To recognize and use the basic properties of addition and multiplication	Collecting, organizing, and analyzing data
424-425	To give the coordinates of a point on a graph To draw the graph of an ordered pair of integers	Reading a graph Interpreting directions
426-427	To solve problems using scoring rules	Finding information in a display
428-429	To practice skills previously taught	Identifying differences in two photographs

Chapter 17 Algebra

Pages	Lesson Objectives	Problem-Solving Skills
434-435	To write algebraic expressions for word phrases	Interpreting and checking information Using logical reasoning
436-437	To substitute numbers for variables and then evaluate the resulting expression	Finding information in a display Comparing prices Choosing the correct operation Using a guess-and-check strategy
438-439	To solve addition equations	Using equations to solve problems
440-441	To solve subtraction equations	Choosing the correct equation
442-443	To practice skills previously taught	Choosing the correct equation Writing equations Solving a multistep problem
444-445	To solve multiplication equations	Choosing the correct equation
446-447	To solve division equations	Choosing the correct equation
448-449	To solve two-step equations	Choosing the correct equation
450-451	To find the square root of a perfect square To use the divide-and-average method To find a decimal approximation of a square root	Using a guess-and-check strategy
452-453	To use the Pythagorean rule to find the lengths of the sides of a right triangle	Finding information in a drawing
454-455	To solve problems using formulas	Reading a map
456-457	To practice skills previously taught	Following computer commands Making geometric visualizations

T22 *Program Goals*

Pacing Chart

Planning for Class Needs

Heath General Mathematics can be used by students who have varying levels of ability. The *Pacing Chart* at the right suggests how the text can be used for a comprehensive course in general mathematics based on 160 school days.

Heath General Mathematics can also be used as a text to prepare students for competency tests that many schools require for graduation. In this case, be sure to include the lessons appropriate for your competency tests and use the *Cumulative Skill Practices* and *Cumulative Tests* to reinforce major skills generally found on competency tests.

Chapter	Suggested number of days to spend on each chapter — Comprehensive Course
1 Adding Whole Numbers and Decimals	7
2 Subtracting Whole Numbers and Decimals	7
3 Multiplying Whole Numbers and Decimals	7
4 Dividing Whole Numbers and Decimals	7
5 Graphs and Statistics	10
6 Number Theory, Fractions, and Decimals	14
7 Adding and Subtracting Fractions and Mixed Numbers	7
8 Multiplying and Dividing Fractions and Decimals	7
9 Measurement	8
10 Ratio and Proportion	10
11 Percent	8
12 Consumer Mathematics	12
13 Geometry—Perimeter, and Area	13
14 Surface Area and Volume	10
15 Probability	10
16 Integers	10
17 Algebra	13

Learning Strands

Skill	Teach page	Practice page	Review page	Test page	Apply page item	Maintain page item	Retest page item	Reteach/Extra Practice page skill	Supplement*
1 Adding Whole Numbers	6	6-7	28	29	**7** 52-59 **8** 1-12 **24** all **25** 1-4, 9, 10 **50** 8 **98** 1, 5 **115** 10, 11, 13, 15, 16, 18 **125** 28, 29 **164** 1 **227** 17, 19, 20 **343** 15-17 **346** all	**9** 24-35, 40-43 **17** 1-16, 52-55 **41** 21-26 **51** 54 **67** 1-6, 46, **50** 77 **46** 91 17-21 **99** 43 **109** 54	**30** 4-5 **54** 3 **80** 1 **112** 2 **462** sk 1	**471** 1	Worksheets, S171, S174 Tests, S1, S3 Visual Aids, S114, S122, S134
2 Rounding Whole Numbers	10	11	28	29	**11** 60-61 **12** 6, 7 **13** 40-51 **33** 25 **37** 30-33 **85** 41, 42 **93** 45, 46, 48	**17** 17-34 **41** 27-38 **51** 52 **67** 7-16 **123** 10-17	**30** 6 **54** 4 **80** 2 **136** 2 **408** 1 **462** sk 2	**471** 2	Worksheets, S173, S180 Tests, S1, S3 Visual Aids, S115
3 Rounding Decimals	18	19	28	29	**19** 78, 79 **61** 42-44 **69** 51 **75** 1-8 **105** 38 **269** 19-21 **291** 45, 46	**27** 1-30, 56, 57 **41** 39-50 **51** 52, 53 **67** 25-34 **109** 43-53 **112** 11 **123** 24-31 **133** 28-36, 46, 47 **165** 25-36 **173** 53-73 **185** 37-51 **197** 1-9 **211** 13-24 **219** 1-14 **235** 13-22 **253** 47-49 **267** 13-22, 33-46 **289** 13-28 **329** 40-45 **371** 10-32 **381** 32-40 **395** 28-39 **417** 19-32 **429** 44-49 **443** 13-24 **457** 31-36	**30** 8, 9 **54** 5 **80** 4 **136** 4, 10 **176** 7, 11 **200** 5 **222** 2, 6 **256** 2 **280** 2, 4 **306** 2 **384** 2, 3 **408** 3 **432** 3 **460** 2 **462** sk 3	**472** 3	Worksheets, S177 Tests, S2, S4, S13, S14, S15, S16 Visual Aids, S116
4 Adding Decimals	20, 22	20-23	28	29	**21** 54-59 **25** 5, 8 **75** 3-5, 8, 9 **79** 52, 55 **83** 75 **111** 51, 54 **171** 1-11 **288** 8 **343** 6, 8, 9, 14, 18	**27** 31-54, 58-63 **41** 54-59 **51** 1-9, 54, 56, 58, 60 **67** 35-43, 47, 49, 52 **77** 48, 55 **91** 22-46, 52, 53 **99** 44, 48 **109** 59 **123** 32-37, 46, 48 **133** 48, 50, 56 **155** 16-27 **173** 75, 80 **185** 1-9 **211** 60, 67 **277** 43, 48 **317** 39, 46 **405** 52, 59	**30** 10-12 **54** 6, 11 **80** 5 **112** 3 **136** 5 **176** 2 **200** 1 **462** sk 4	**472** 4	Worksheets, S178, S179, S199 Tests, S2, S4, S10, S12 Visual Aids, S136, S150
5 Comparing Whole Numbers	32	33	52	53	**33** 23, 24 **40** 1, 6 **122** 5 **127** 23, 25 **184** 10	**51** 10-21	**54** 7 **462** sk 5	**473** 5	Worksheets, S190 Tests, S5, S7 Visual Aids, S118
6 Subtracting Whole Numbers	34, 38	35, 38, 39	52	53	**39** 39-44 **40** 7-9, 11, 12 **50** 6, 9 **59** 53, 54 **90** 5-7 **187** 50 **227** 18 **237** 20-22	**51** 22-27, 55, 61 **67** 48, 53 **77** 1-6, 49, 57 **91** 48, 50 **99** 1-10	**54** 8 **80** 6 **112** 4 **462** sk 6	**473** 6	Worksheets, S184, S185 Tests, S5, S7 Visual Aids, S118, S120, S121, S122

* Worksheets are available as copymasters in the TRB, as duplicating masters, and in a workbook.
Tests are available as copymasters in the TRB and as duplicating masters.
Visual aids are available as copymasters in the TRB and as transparencies.

Skill	Teach page	Practice page	Review page	Test page	Apply page item	Maintain page item	Retest page item	Reteach/ Extra Practice page skill	Supplement*
7 Comparing Decimals	42	43	52	53	**43** 42-47 **97** 45 **105** 38	**51** 28-39 **77** 7-15 **99** 11-22 **155** 28-39 **219** 64-66	**54** 9 **80** 7 **112** 5 **176** 3 **462** sk 7	**474** 7	Worksheets, S187 Tests, S6, S8 Visual Aids, S150
8 Subtracting Decimals	44, 46	44-47	52	53	**45** 54, 55 **47** 48-51 **65** 52-54 **66** 6, 7, 9 **71** 33, 34 **87** 39, 40 **105** 39-41 **111** 54 **154** 5 **164** 4 **211** 8, 11, 12 **234** 1, 2 **288** 4, 8 **315** 9 **416** 6	**51** 40-51, 57, 59, 62 **67** 51, 54 **77** 16-24 51, 53 **91** 51, 55 **99** 23-34 **109** 56 **123** 38-43, 50, 52 **133** 49, 53, 54 **155** 40-51, 61, 62 **173** 77, 79 **185** 10-18 **211** 61, 65, 66, 69, 70 **219** 62, 63 **277** 45 **317** 41, 43, 51, 52 **405** 54, 56, 64, 65	**54** 10 **80** 8 **112** 6 **136** 6 **176** 4 **200** 2 **462** sk 8	**474** 8	Worksheets, S189, S199 Tests, S6, S8, S10 Visual Aids, S120, S121, S134
9 Multiplying by Multiples of 10, 100, and 1000	56	57	78	79	**57** 45-56 **59** 51	**133** 10-12 **165** 13-15 **173** 74 **277** 42 **317** 40 **405** 53	**463** sk 9	**475** 9	Worksheets, S192, S200 Tests, S9, S11
10 Multiplying Whole Numbers	58, 62, 64	59, 63, 65	78	79	**16** 1-7 **57** 45-56 **59** 51-53 **89** 43-50 **196** 1 **225** 24, 27 **234** 33-36 **275** 5-8 **455** 6	**77** 25-30, 52 **99** 35-40 **109** 55, 57 **185** 53	**80** 9 **112** 7 **463** sk 10	**475** 10	Worksheets, S193, S194, S195, S196 Tests, S9, S11 Visual Aids, S121, S150
11 Multiplying Decimals	68	69	78	79	**69** 45-50 **71** 32-34 **73** 56-63 **75** 1-5, 7 **79** 53-55 **83** 77, 78 **103** 60 **111** 52, 54 **163** 56, 57 **164** 5 **196** 2, 4 **231** 44, 45 **234** 7, 9 **239** 30, 31 **245** 29-31 **275** 3, 4 **315** 10 **416** 6 **455** 8	**77** 31-39, 47, 54, 56 **91** 47, 49, 54 **109** 1-12 **123** 44, 49 **133** 1-9 **155** 54, 56, 58 **165** 1-12 **173** 76 **185** 19-27 **211** 1-12 **219** 69 **235** 1-12 **267** 1-12 **277** 44 **289** 1-12 **317** 42 **371** 1-9 **395** 16-27 **405** 55 **443** 1-12	**80** 10 **112** 8 **136** 7 **176** 5 **200** 3 **222** 1 **256** 1 **280** 1 **306** 1 **384** 1 **408** 2 **460** 1 **463** sk 11	**476** 11	Worksheets, S198 Tests, S9, S11 Visual Aids, S121, S134, S141
12 Simplifying Expressions	70	71	78	79 S15	**71** 31-34 **75** 3-8 **79** 55	**77** 40-45, 58-60 **91** 56, 57, 59-61 **99** 49-51 **109** 34-36 **155** 60-62 **173** 86-88 **211** 69-74 **253** 50-55 **277** 54-56 **317** 51-53 **405** 64-66	**80** 11 **463** sk 12	**476** 12	Worksheets, S199 Tests, S10, S12, S13 S15,
13 Multiplying Decimals by 10, 100, or 1000	72	72, 73	78	79	**90** 11 **73** 56-63 **229** 35, 37	**133** 13-18 **165** 16-24 **173** 81, 84 **211** 62, 64 **277** 49, 52 **317** 45, 48 **405** 58, 61 **417** 1-9	**136** 8 **176** 6 **432** 1 **463** sk 13	**477** 13	Worksheets, S200 Tests, S10, S12

Learning Strands T25

Skill	Teach page	Practice page	Review page	Test page	Apply page item	Maintain page item	Retest page item	Reteach/ Extra Practice page skill	Supplement*
14 Dividing Whole Numbers	82, 86, 88, 92	82, 83, 87, 89, 93	110	111	85 41, 42 90 8, 9, 12 93 45, 46, 48 107 1-8 125 28-30 154 4, 8 163 59 225 25 455 5, 7, 9, 11, 12	109 13-27, 32 123 53	112 9 463 sk 14	477 14	Worksheets, S203, S204, S205, S206, S208, S239 Tests, S13, S15, S22, S24 Visual Aids, S122, S124
15 Applying rules of order of operations	94	95	110	111	95 39-42 98 1-8	109 28-42 123 53-58 185 52-57	112 10 463 sk 15	478 15	Worksheets, S209 Tests, S15
16 Dividing Decimals by 10, 100, or 1000	100	100, 101	110	111	100 61, 62 229 34, 36	133 19-27 173 83, 85 185 28-36 211 63, 68 277 51, 53 317 49, 50 405 62, 63 417 10-18	136 9 200 4 432 2 463 sk 16	478 16	Worksheets, S212 Tests, S14, S16 Visual Aids, S123
17 Dividing Decimals	102, 104	103, 105	110	111	97 44, 45 103 58, 59 105 38 107 5-8 111 53 163 58 187 51 196 2 210 9, 10, 12 231 43 234 3 247 37 313 14-21 455 10	109 43-51 123 45, 47, 51 133 28-36 155 55, 57, 59 165 25-36 173 78, 82 185 37-48 211 13-24 219 68 235 13-22 267 13-22 277 46, 47, 50 289 13-22 317 44, 47 371 10-18 395 28-39 405 57, 60 443 13-24	112 11 136 10 176 7 200 5 222 2 256 2 280 2 306 2 384 2 408 3 460 2 463 sk 17	479 17	Worksheets, S213 Tests, S13, S14, S15, S16 Visual Aids, S141, S150
18 Solving Two- and Three-Step Problems	74, 250, 274, 356, 378	75, 251, 275, 357, 379	330	53, 79, 111, 331	16 2-9 71 31-34 83 78 98 all 154 2-9 164 3-10 210 8-13 234 all 247 34-37 269 11-21 288 all 327 all 370 3-6 394 3-10 404 8-11 416 3-10 442 4, 5, 9-13 449 40-43	381 38-40 457 37-39	30 12 54 11, 12 112 12 136 12 200 12 222 12 256 12 280 12 332 11, 12 362 11 384 10 460 9 464 sk 18	479 18	Worksheets, S201, S277 Tests, S6, S8, S10, S12, S14, S16, S45, S47 Visual Aids, S147
19 Solving Problems from a Bar Graph	116	117	134	135	117 8-17 128 1-7	119 14 128 1-4 132 1-6	176 12 256 5 464 sk 19	480 19	Worksheets, S218 Tests, S17, S19 Visual Aids, S126, S130
20 Solving Problems from a Line Graph	118	119	134	135	119 7-13 122 1-10 129 8-10	132 3-6	200 11 464 sk 20	480 20	Worksheets, S219 Tests, S17, S19 Visual Aids, S127, S128, S129, S130

Skill	Teach page	Practice page	Review page	Test page	Apply page item	Maintain page item	Retest page item	Reteach/ Extra Practice page skill	Supplement*
21 Solving Problems from a Picture Graph	120	121	134	135	121 16-19	129 8-11	222 11 464 sk 21	481 21	Worksheets, S220, S224 Tests, S18, S20 Visual Aids, S130
22 Solving Problems from a Chart	124, 126, 128	125, 127, 129	134	135	2 all 4 all 6 1-4 19 78, 79 20 1-4 23 60-65 33 23-24 34 1-4 36 1-5 42 1-5 44 1-5 45 52-57 48 all 49 all 57 45-52 65 48-52 68 1-4 72 1-3 88 1-3 90 all 104 1-5 114 1-7 115 8-18 182 1-3 192 1, 2 210 all 213 56-58 249 27-38 260 1-6 275 1-8 284 1-4 288 all 294 all 295 31-38 316 7-9 398 all 402 all 403 all 416 all	See Apply.	465 sk 22	481 22	Worksheets, S176, S216, S253, S262 Tests, S17, S18, S19, S20 Visual Aids, S125, S140, S142
23 Finding Equivalent Fractions	146	147	174	175	147 43, 44 153 46-48 181 42	Maintained throughout fraction lessons.	465 sk 23	482 23	Worksheets, S231 Tests, S21, S23
24 Writing Fractions in Lowest Terms	148	149	174	175	234 4, 8	173 11-31	176 10 465 sk 24	482 24	Worksheets, S232 Tests, S21, S22, S23, S24
25 Finding the Least Common Denominator	150	151	174	175	181 37-40, 42, 43 183 34, 35 184 1-4 189 28-33	165 53 197 52 211 35-44 267 61 289 51 371 53	222 4 465 sk 25	483 25	Worksheets, S233 Tests, S21, S23 Visual Aids, S131, S137
26 Comparing Fractions	152	153	174	175	153 46-48 181 40 184 6	173 1-10 211 45-59 267 23-32	176 9 222 5 280 3 465 sk 26	483 26	Worksheets, S234 Tests, S21, S23 Visual Aids, S131

Learning Strands T27

Skill	Teach page	Practice page	Review page	Test page	Apply page item	Maintain page item	Retest page item	Reteach/ Extra Practice page skill	Supplement*
27 Writing Whole and Mixed Numbers as Fractions	156	157	174	175	189 33 203 68	See Multiplying Mixed Numbers (skill 38).	465 sk 27	484 27	Worksheets, S236 Tests, S21, S23 Visual Aids, S133
28 Writing Fractions as Whole and Mixed Numbers	158	159	174	175	163 59 181 38, 43 183 35	173 12, 14, 17, 18, 25, 27, 28	465 sk 28	484 28	Worksheets, S237 Tests, S22, S24 Visual Aids, S133
29 Writing Fractions and Mixed Numbers in Simplest Form	160	161	174	175	163 59 181 38, 43 203 67 234 4, 8	173 11-31	466 sk 29	485 29	Worksheets, S238 Tests, S22, S23, S24, S25
30 Writing Fractions and Mixed Numbers as Decimals	166	167	174	175	167 67	173 32-37 197 10-23 219 1-14 235 23-36 267 33-46 303 58 347 1-14 371 19-32	176 11 200 7 256 3 362 1 384 3 466 sk 30	485 30	Worksheets, S241 Tests, S22, S24
31 Writing Decimals as Fractions or Mixed Numbers	168	169	174	175	See Writing Percents as Fractions or Mixed Numbers (skill 46).	197 24-27 235 37-48 303 59	200 8 466 sk 31	486 31	Worksheets, S242 Tests, S22, S24
32 Adding Fractions	178, 180	179, 181	198	199	179 34-39 181 37-40 199 45 203 68	219 15-26 235 55, 57 267 47-56 289 29-38 329 46, 53 347 15-24 359 65 381 41 395 40-49 429 50 443 25-36	222 7 280 5 306 4 362 2 408 4 460 3 466 sk 32	486 32	Worksheets, S245, S246 Tests, S25, S27 Visual Aids, S137
33 Adding Mixed Numbers	182	183	198	199	183 34, 35 184 1-4 215 47 221 54-56 288 2, 6	197 38-42 235 49-54 317 1-6 329 48, 55 359 68, 76 381 44, 52 417 33-38 429 53, 61 457 50, 54, 59, 63	200 9 256 6 332 1 432 4 466 sk 33	487 33	Worksheets, S247 Tests, S25, S27

T28 *Learning Strands*

Skill	Teach page	Practice page	Review page	Test page	Apply page item	Maintain page item	Retest page item	Reteach/Extra Practice page skill	Supplement*
34 Subtracting Fractions	186, 188	187, 189	198	199	189 28, 29, 32, 33	219 27-38 235 58, 60 277 1-10 289 39-48 329 50 60 347 25-34 359 66, 75 381 42, 51 395 50-59 429 51, 60 443 37-48	222 8 280 6 306 5 362 3 408 5 460 4 466 sk 34	487 34	Worksheets, S249, S250 Tests, S26, S28
35 Subtracting Mixed Numbers	190, 192	191, 193	198	199	191 36-38 199 46 200 12 203 68 207 37 234 4 288 3	197 43-47 253 1-6 317 7-12 329 56, 58 359 69, 73 371 33-38 381 45, 49 429 54, 58 457 51, 55, 57, 61	200 10 256 7 332 2 384 4 466 sk 35	488 35	Worksheets, S251, S252 Tests, S26, S28
36 Multiplying Fractions	202	203	220	221	203 67	235 56, 61, 66 253 7-16 303 1-12 329 54 347 35-44 359 71 381 43, 47 405 1-10	256 8 306 6 362 4 408 6 467 sk 36	488 36	Worksheets, S255, S260 Tests, S29, S31 Visual Aids, S141
37 Finding a Fraction of a Number	204	205	220	221	205 47-50 210 13	277 11-18	280 7 467 sk 37	489 37	Worksheets, S256, S258 Tests, S29, S31
38 Multiplying Mixed Numbers	206	207	220	221	207 36, 38, 40, 41 210 4, 6, 8, 12	219 39-48 317 13-22 329 47, 52 359 72, 77 381 48, 53 417 39-46 429 57, 62	332 3 432 5 467 sk 38	489 38	Worksheets, S257 Tests, S29, S31
39 Finding a Fraction of a Unit of Measure	208	209	220	221	205 47-50 234 5 288 5	See Finding a Percent of a Number (skills 50 and 51).	467 sk 39	490 39	Worksheets, S258 Tests, S30, S32
40 Dividing Fractions	212	213	220	221	213 54, 55 245 31	235 65, 69 253 17-26 303 13-24 329 51, 57 347 45-54 359 78 381 54 405 11-20 429 63 457 1-12	256 9 306 7 362 5 408 7 460 6 467 sk 40	490 40	Worksheets, S260 Tests, S30, S32
41 Dividing Mixed Numbers	214	215	220	221	215 46 221 57 222 12	219 49-58 317 23-32 329 49, 59 359 70, 74, 79 371 39-48 381 46, 50, 55 429 59, 64 457 53, 58, 62	222 10 280 8 332 4 467 sk 41	491 41	Worksheets, S261 Tests, S30, S32

Learning Strands T29

Skill	Teach page	Practice page	Review page	Test page	Apply page item	Maintain page item	Retest page item	Reteach/ Extra Practice page skill	Supplement*
42 Solving Measurement Problems in the Metric System	226, 228, 230, 232	227, 229, 231, 233	254	255	225 24-27 227 9-18 229 34-37 231 42-45 233 43	253 27-38 277 27-32 303 25-30 317 33-38 359 1-9 381 1-9 429 1-9	256 10 280 9 306 8 332 5 362 6 384 6 432 6 467 sk 42	491 42	Worksheets, S265, S266, S267, S268 Tests, S33, S35 Visual Aids, S143, S144, S145, S148
43 Solving Measurement Problems in the Customary System	240, 242, 244, 246, 248	240, 241, 243, 245, 247, 248, 249	254	255	209 14-37 237 20, 21, 23 241 25 243 35, 36 245 29-31 247 34-37 249 25, 26	253 39-46 277 33-36 303 31-34 329 1-8	256 11 280 10 306 9 332 6 467 sk 43	492 43	Worksheets, S270, S272, S273, S274, S275, S276 Tests, S34, S36 Visual Aids, S146
44 Solving Proportions	262	263	278	279	271 1-10 273 1-8 279 36-43	277 37-41 303 35-44 359 10-19 405 21-30 429 10-19	280 11 306 10 362 7 408 8 432 7 468 sk 44	492 44	Worksheets, S279, S280, S281 Tests, S38, S40 Visual Aids, S148
45 Solving Ratio and Proportion Problems	264, 268	265, 269	278	279	259 58-60 261 47-52 263 23 265 1-6	See Apply.	280 12 332 11 468 sk 45	493 45	Worksheets, S282, S284, S286 Tests, S38, S40 Visual Aids, S148, S149
46 Writing Percents as Fractions or Mixed Numbers	282	283	304	305	283 67-70 301 1, 2, 5, 6	303 45-54 329 9-20 359 20-37 429 20-31	306 11 332 7 362 8 432 8 468 sk 46	493 46	Worksheets, S289 Tests, S41, S43
47 Writing Fractions as Percents	284	285	304	305	285 49-52 287 51, 52 297 29, 30 299 31	329 21-27 405 31-42	332 8 408 9 468 sk 47	494 47	Worksheets, S290, S291 Tests, S41, S42, S43, S44 Visual Aids, S151
48 Writing Percents as Decimals	286	287	304	305	257 41-46 259 31-38 261 31-34 273 11-14 283 7-12 285 13, 14	381 10-19 291 41-46 295 31-38 297 31-34 311 11-14 321 7-12 323 13, 14	384 7 468 sk 48	494 48	Worksheets, S291 Tests, S41, S43

T30 *Learning Strands*

Skill	Teach page	Practice page	Review page	Test page	Apply page item	Maintain page item	Retest page item	Reteach/ Extra Practice page skill	Supplement*
49 Writing Decimals as Percents	287	287	304	305	See Writing Fractions as Percents (skill 47).	359 38-55	362 9 468 sk 49	495 49	Worksheets, S291 Tests, S41, S43
50 Finding a Percent of a Number	290	291	304	305	297 31, 32 299 32, 34 301 3, 4, 7, 8 323 13, 14 327 5-10 416 5, 8	329 28-33 347 55, 57, 59, 60, 62, 66, 69 381 20-22 395 60, 62, 64, 65, 67, 71, 74 417 47, 49, 51, 52, 54, 58, 61 429 32-37 457 13-21	332 9 408 10 432 9 468 sk 50	495 50	Worksheets, S293, S294, S295, S297 Tests, S42, S44
51 Finding a Percent of a Number by Proportion	294	295	304	305	291 41-44 295 31-38 311 11-15 315 5, 12 316 7-9 394 3-6	347 64, 67 381 20-25 395 69, 72 417 56, 59 429 37 457 19-21	384 8 460 7 468 sk 51	496 51	Worksheets, S296 Tests, S42, S44
52 Finding the Number When a Percent is Known	296	297	304	305	297 33, 34, 299 33	329 34-45 347 56, 58, 61, 63, 65, 68 359 56-64 381 26-37 395 61, 63, 66, 68, 70, 73 417 48, 50, 53, 55, 57, 60 429 38-49 457 22-30	332 10 362 10 384 9 432 10 460 8 468 sk 52	496 52	Worksheets, S296 Tests, S42, S44
53 Solving Percent Problems	298	299	304	305	285 49-52 287 51, 52	381 38-40 457 37-39	306 12 332 12 362 11 384 10 460 9 469 sk 53	497 53	Worksheets, S301 Tests, S42, S44
54 Solving Percent Problems Using Circle Graphs	300	301	304	305	301 3, 4, 7, 8, 13-17	See Apply.	469 sk 54	497 54	Worksheets, S298 Tests, S42, S44
55 Solving Personal Finance Problems	310, 312, 314, 318, 320, 322, 326	311, 313, 315, 319, 321, 323, 327	330	331	316 7-9 321 7-12 323 13, 14 327 5-14 403 1-11	381 38-40 457 37-39	322 12 362 11 384 10 460 9 469 sk 55	498 55	Worksheets, S301, S302, S303, S305, S306, S307, S308, S309, S341 Tests, S45, S46, S47, S48 Visual Aids, S152, S153, S154, S155, S156, S157, S158

Learning Strands **T31**

Skill	Teach page	Practice page	Review page	Test page	Apply page item	Maintain page item	Retest page item	Reteach/Extra Practice page skill	Supplement*
56 Finding an Area	348, 350, 352, 354	349-351, 353-355	360	361	349 7-15, 19 350 3-23 353 5-19 354 5-7 355 8-17 369 5-14 370 5, 6	368-369 all 372-377 all	362 12 384 11 469 sk 56	498 56	Worksheets, S318, S319, S320, S321, S325, S326, S340 Tests, S50, S52, S53, S54, S55, S56
57 Finding a Volume	372, 374, 376	373, 375, 377	382	383	373 7-15 375 5-13 377 5-22	See Apply.	384 12 460 11 470 sk 57	499 57	Worksheets, S328, S329 Tests, S53, S54, S55, S56
58 Solving Probability Problems	396	396, 397	406	407	390 5-16 391 19-48 392 6-10 393 14, 15 396 6-9 397 10-34	398-399 all 400-401 all	408 12 470 sk 58	499 58	Worksheets, S333, S334, S335, S336, S338, S339 Tests, S57, S58, S59, S60
59 Comparing Integers	410	411	430	431	411 31-35	427 7	470 sk 59	500 59	Worksheets, S343 Tests, S61, S63 Visual Aids, S164
60 Adding Integers	412	413	430	431	413 61-66 416 2, 3 427 all	443 61, 65, 69, 73 457 40-49	460 10 470 sk 60	500 60	Worksheets, S344 Tests, S61, S63
61 Subtracting Integers	414	415	430	431	416 4-6	443 62, 66, 70, 74	432 12 470 sk 61	501 61	Worksheets, S345 Tests, S61, S63
62 Multiplying Integers	418	419	430	431	425 60	443 63, 67, 71, 75	470 sk 62	501 62	Worksheets, S347 Tests, S62, S64
63 Dividing Integers	420	421	430	431	421 103-105	443 64, 68, 72	470 sk 63	502 63	Worksheets, S348 Tests, S62, S64
64 Solving Equations	438, 440, 444, 446, 448	439, 441, 445, 447, 449	458	459	441 50-54 442 all 447 51-53 449 40-43 455 1-13	449 4-38 453 7-14	460 12 470 sk 64	502 64	Worksheets, S355, S356, S358, S359, S360 Tests, S66, S68 Visual Aids, S166, S167

T32 Learning Strands

Adding Whole Numbers and Decimals

Chapter 1
Adding Whole Numbers and Decimals

Resources

- **Class Starter Quizzes 1-12** *(Copymasters S385-S387)*
- **Visual Aids 1-7** *(Copymasters or Transparencies S111-S117)*
- **Manipulatives**
  Manipulative Activity 1 *(Copymaster S515)*
  Powers-of-ten tiles *(Copymaster or Transparency S529)*
- **Worksheets 1-13** *(Copymasters, Duplicating Masters, or Workbook pages S169-S181)*
- **Challenge Problems** for pages 9, 19, 21, 25, 27 *(Copymasters S439-S440)*
- **Projects** for pages 3, 5, 7, 11, 23 *(Copymasters S473-S474)*
- **Mental Math Extensions** for Skills 1-4 *(Copymasters S489-S490)*
- **Tests** *(Copymasters or Duplicating Masters S1-S4)*

Lesson Objective
To read standard numerals less than 1 billion

Problem-Solving Skills
Reading a chart
Using logical reasoning

Starting the Lesson
Take a Survey Write these movie titles on the chalkboard:

  E.T.
  Star Wars
  Star Trek, The Motion Picture

Before the students open their books, take a survey. Ask them to guess which of these movies cost the most to produce. (*Star Trek*) Record the guesses on the chalkboard. Then say: "Open your book to page 2. Use the chart at the top of the page to check your guess."

Go over the *Here's how*. The chart on page 2 is also on ■ **Visual Aid 1** (copymaster or transparency S111). Use the chart to discuss questions 1–7.

Exercise Note
Problem Solving Exercises 8–15 require logical reasoning. Some students may benefit from the following: "Look at exercise 8. List all the movies that fit the first clue. (*Star Wars, Close Encounters*) Cross out the movie that does not fit the second clue." (*Star Wars, Close Encounters*)

Reading standard numerals

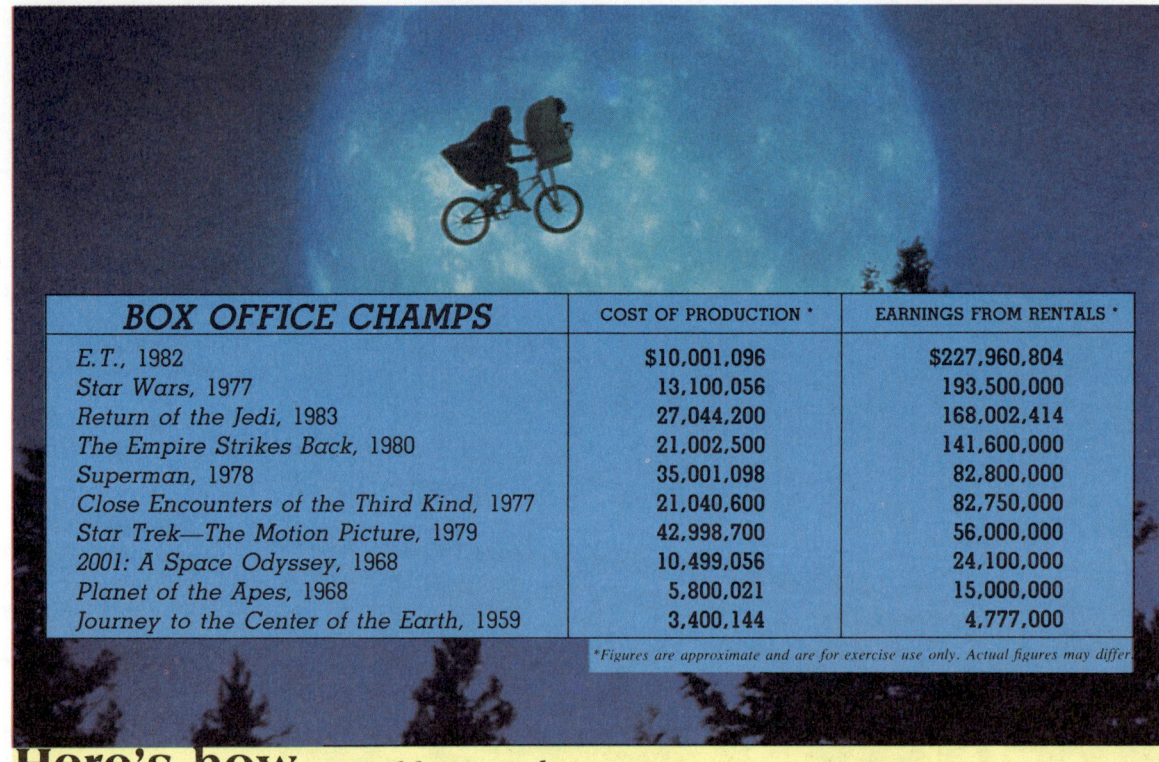

BOX OFFICE CHAMPS	COST OF PRODUCTION*	EARNINGS FROM RENTALS*
E.T., 1982	$10,001,096	$227,960,804
Star Wars, 1977	13,100,056	193,500,000
Return of the Jedi, 1983	27,044,200	168,002,414
The Empire Strikes Back, 1980	21,002,500	141,600,000
Superman, 1978	35,001,098	82,800,000
Close Encounters of the Third Kind, 1977	21,040,600	82,750,000
Star Trek—The Motion Picture, 1979	42,998,700	56,000,000
2001: A Space Odyssey, 1968	10,499,056	24,100,000
Planet of the Apes, 1968	5,800,021	15,000,000
Journey to the Center of the Earth, 1959	3,400,144	4,777,000

*Figures are approximate and are for exercise use only. Actual figures may differ.

Here's how to read large numbers.

13,100,056   193,500,000
[million] [thousand]   [million] [thousand]

Star Wars cost 13 million, 100 thousand, 56 dollars to produce. The film earned 193 million, 500 thousand dollars.

1. Which film cost 10 million, 1 thousand, 96 dollars to produce?
   E.T.
2. Which film earned 24 million, 100 thousand dollars?
   2001: A Space Odyssey
3. *Star Wars* was released in 1977. Which film was released in 1983?
   Return of the Jedi
4. Which films earned more than 100 million dollars?
   The Empire Strikes Back, Return of the Jedi, Star Wars, E.T.
5. Which films cost less than 10 million dollars to produce?
   Planet of the Apes, Journey to the Center of the Earth
6. Which film earned about 141 million dollars?
   The Empire Strikes Back
7. Which film earned about 120 million dollars more than it cost?
   The Empire Strikes Back

2 Chapter 1

EXERCISES

Study the clues. Use the chart on page 2 to name the movie.

8. Clues:
   - This movie was released in 1977.
   - It earned more than 100 million dollars.

   Star Wars

9. Clues:
   - This movie earned less than 90 million dollars.
   - It was released in 1977.

   Close Encounters of the Third Kind

10. Clues:
    - This movie cost less than 10 million dollars to produce.
    - It earned more than 10 million dollars.

    Planet of the Apes

11. Clues:
    - This movie earned more than 100 million dollars.
    - It cost more than 25 million dollars to produce.

    Return of the Jedi

12. Clues:
    - This movie earned less than 70 million dollars.
    - It was released before 1979.
    - It cost more than 10 million dollars to produce.

    2001: A Space Odyssey

13. Clues:
    - This movie was released after 1977.
    - It cost more than 30 million dollars to produce.
    - It earned less than 60 million dollars.

    Star Trek

14. Clues:
    - This movie cost less than 15 million dollars to produce.
    - It was released in 1968.
    - It earned less than 20 million dollars.

    Planet of the Apes

15. Clues:
    - This movie earned less than 80 million dollars.
    - It cost more than 20 million dollars to produce.
    - It was released after 1978.

    Star Trek

Show time — Reading an ad

Use the ad to answer the questions.

LITTLETOWN CINEMA
- CINEMA 1: E.T. THE EXTRA-TERRESTRIAL (PG) — 2:15, 4:30, 7:15, 9:30
- CINEMA 2: SUPERMAN THE MOVIE (PG) — 1:45, 4:15, 7:00, 9:30

16. What movie is showing at Cinema 2? *Superman*
17. What time is the first showing of *E.T.*? *2:15*
18. Sonya went to see *E.T.* She arrived at the theater at 3:10. How long did she wait for the next movie to begin? *1 hour 20 minutes*
19. You live 15 minutes from Cinema 2. What time should you leave home to get to the theater 5 minutes before the start of the last showing? *9:10*

Adding Whole Numbers and Decimals **3**

Practice Worksheet
Workbook S169, Copymaster S169, or Duplicating Master S169

WORKSHEET 1 (Use after page 3.)

NAME _____

BIG NUMBER HUNT
Circle the hidden number.

1. Find the largest number that is less than 1 million.
   9 8 5 8 4 9 8 1 9 7 (9 8 5 9 7 2) 9 8 4 9 8 2 1 7 6

2. Find the largest number that is less than 20 million.
   3 1 9 2 1 9 3 (1 9 4 2 1 9 1 4) 0 6 1 9 3 8 1 9 7 5

3. Find the largest number that is less than 50 million.
   4 4 6 4 8 3 (4 8 9 4 7 4 8 8) 6 4 8 4 7 3 2 1 0 6 2

4. Find the largest number that is less than 100 million.
   9 8 9 7 6 7 8 9 4 0 3 (9 8 9 8 9 2 9 8) 9 9 8 7 7 2 6

5. Find the largest number that is less than 30 thousand.
   2 6 3 1 2 6 8 4 2 6 8 5 0 9 2 1 0 (2 6 8 5 6) 6 2 8

6. Find the largest number that is less than 600 thousand.
   5 7 6 3 5 7 5 2 8 (5 7 6 4 1 0) 5 7 6 3 2 5 7 6 1 7

7. Find the smallest number that is greater than 4 million.
   4 3 4 0 6 7 1 4 0 5 1 2 5 0 4 0 9 (4 0 4 5 1 3 4 2)

8. Find the smallest number that is greater than 60 million.
   6 0 8 6 2 6 3 4 6 0 9 2 (6 0 7 5 8 6 0 7) 6 7 0 1 2

9. Find the smallest number that is greater than 75 thousand.
   8 7 6 4 6 7 6 8 2 (7 6 3 8 2) 7 6 4 2 1 2 7 6 8 1 0

10. Find the smallest number that is greater than 10 million.
    1 0 6 5 1 0 7 2 1 0 6 3 1 0 6 2 (1 0 6 1 4 7 1 0) 4

11. Find the smallest number that is greater than 800 million.
    7 8 1 0 5 8 1 1 2 7 8 1 0 4 3 (8 0 1 0 4 4 8 1 0) 4

© D. C. Heath and Company S169 Reading standard numerals

Project

Reading a newspaper ad

Get a newspaper. Use the newspaper movie ads to make up problems similar to exercises 16–19 on page 3.

Copymaster S473

Class Starter Quiz 1
on previous lesson

Match.
1. 4 million, 942 thousand c
2. Less than 4 million b
3. More than 5 million a
4. About 4 million d

a. 5,978,932
b. 3,052,193
c. 4,942,000
d. 4,000,942

Copymaster S385

Lesson Objective
To write standard numerals (through the billions) in words

Problem-Solving Skill
Reading a chart

Starting the Lesson
Estimation Before the students open their books, have them guess how long it would take to slice 2,400,000,000 slices of cheese, slicing one slice per second, 24 hours per day. Have the students write their guesses. Record the high and low guesses. Then say: "Open your book to page 4. Read the first paragraph. What answer does the book give?" (More than 75 years)

The chart on page 4 is also on ■ **Visual Aid 2** (copymaster or transparency S112). Use the chart to discuss exercises 1–8 and the *Here's how*.

Exercise Note
Exercises 30 and 34 require the students to write 4-digit numbers. Remind the students that it is not necessary to use a comma when writing a 4-digit number.

Answers for page 5.
9. 47 thousand, 258
10. 16 thousand, 234
11. 776 thousand, 39
12. 14 thousand, 732
13. 520 thousand, 66
14. 177 thousand, 406
15. 6 million, 835 thousand, 270
16. 93 million, 427 thousand, 600
17. 74 million, 50
18. 75 million
19. 60 million, 600 thousand, 600
20. 275 million, 675 thousand, 834

4

Writing standard numerals in words

A famous fast-food chain served 2,400,000,000 slices of cheese last year. If you sliced one piece of cheese per second, 24 hours per day, it would take you more than 75 years to slice this much cheese.

Jack's fast food FACTS

FOOD SERVED LAST YEAR

	BILLIONS	MILLIONS	THOUSANDS	
Slices of cheese	2	400	000	000
Pounds of fish		46	205	500
Pounds of potatoes		542	840	000
Eggs		378	210	400

1. How many pounds of potatoes did the fast-food chain serve last year? 542,840,000
2. How many pounds of fish were served? 46,205,500
3. How many eggs were served? 378,210,400

Here's how *to write the standard numeral 46,205,500 in words.*

Short word-name: 46 million, 205 thousand, 500
Long word-name: forty-six million, two hundred five thousand, five hundred

4. Write the short word-name for the number of eggs the fast-food chain served.
   378 million, 210 thousand, 400
5. Write the long word-name for the number of eggs served.
   three hundred seventy-eight million, two hundred ten thousand, four hundred
6. Write the short word-name for the number of slices of cheese served.
   2 billion, 400 million
7. Write the long word-name for the number of slices of cheese served.
   two billion, four hundred million
8. Write the long word-name for the number of pounds of potatoes served.
   five hundred forty-two million, eight hundred forty thousand

4 Chapter 1

EXERCISES

Write the short word-name. *Hint: Study the Here's how.*

9. 47,258
10. 16,234
11. 776,039
12. 14,732
13. 520,066
14. 177,406
15. 6,835,270
16. 93,427,600
17. 74,000,050
18. 75,000,000
19. 60,600,600
20. 275,675,834
21. 14,360,220,000
22. 842,000,000,000
23. 5,000,600,000

Write the standard numeral.

24. 225 thousand, 16 225,016
25. 14 million 14,000,000
26. 14 thousand, 616 14,616
27. 543 billion 543,000,000,000
28. 8 million, 800 thousand 8,800,000
29. 999 thousand, 50 999,050
30. six thousand two hundred four 6204
31. fifty-nine thousand, eight hundred 59,800
32. four million, three hundred eleven thousand, one hundred thirty-seven 4,311,137
33. twenty-one million, sixty-three thousand, three hundred 21,063,300

Write each short word-name as a standard numeral.

34. A fast-food chain serves **860 million** ounces of orange juice yearly. This is enough juice to fill **1 thousand 200** home-size swimming pools.
860,000,000; 1200

35. The same fast-food chain has served a total of **40 billion** hamburgers. This is enough hamburgers to make a stack **473 thousand, 500** miles high.
40,000,000,000; 473,500

The check is in the mail! Writing checks

This check was mailed to the winner of Jack's Pot-of-Gold Sweepstakes. The amount of the check is written as a standard numeral and in words.

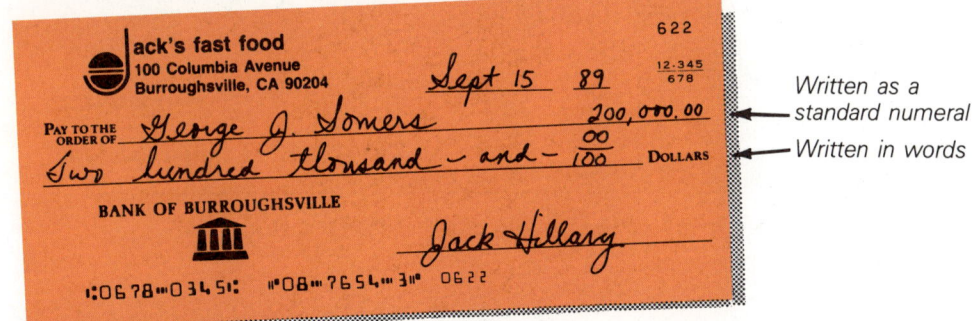

Written as a standard numeral
Written in words

Write the amount of each check in words.

36. $350
37. $1200
38. $14,000
39. $5710
40. $26,010
41. $12,900
42. $9999
43. $48,600
44. $125,800
45. $132,002

Adding Whole Numbers and Decimals **5**

Practice Worksheet
Workbook S170, Copymaster S170, or Duplicating Master S170

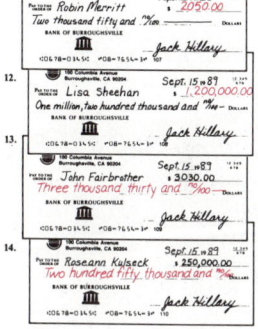

Project

Using library resources

Use a book of world records, such as the *Guinness Book of World Records*, to find the amount of the largest check ever written. Then write the amount in words.

Answers will vary. Records may change from year to year.

Copymaster S473

More answers for page 5.

21. 14 billion, 360 million, 220 thousand
22. 842 billion
23. 5 billion, 600 thousand
36. three hundred fifty
37. one thousand two hundred
38. fourteen thousand
39. five thousand seven hundred ten
40. twenty-six thousand, ten
41. twelve thousand, nine hundred
42. nine thousand nine hundred ninety-nine
43. forty-eight thousand, six hundred
44. one hundred twenty-five thousand, eight hundred
45. one hundred thirty-two thousand, two

Class Starter Quiz 2
on previous lesson

Write the short word-name.

1. 74,260
   74 thousand, 260
2. 126,800
   126 thousand, 800
3. 6,900,000
   6 million, 900 thousand
4. 15,925,800
   15 million, 925 thousand, 800
5. 8,400,500
   8 million, 400 thousand, 500

Copymaster S385

Lesson Objective
To add whole numbers

Problem-Solving Skills
Reading a chart
Using road signs

Starting the Lesson
Visual Thinking Sketch these shapes on the chalkboard:

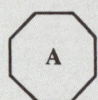 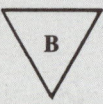 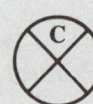

Ask the students to identify the road sign suggested by each shape. Find out how many students correctly identified each sign.

Here's How Note
Use lined notebook paper to assist the students in aligning the digits. You may wish to use ■ *Visual Aid 3* (copymaster or transparency S113) to demonstrate this.

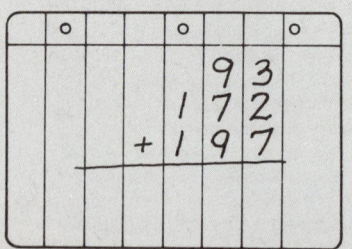

Adding whole numbers

Do you know these road signs? Two hundred people in each of three age groups were surveyed. This chart shows the number of people who identified each sign.

Age	Sign A	Sign B	Sign C
Under 10	93	16	65
10 to 15	172	73	115
16 and over	197	157	183

1. How many of the people 10 to 15 years old knew that Sign B was a yield sign? *73*
2. How many of the people 16 and over knew that Sign C was a railroad-crossing sign? *183*
3. Which three numbers would you add to find out how many of the people surveyed knew that Sign A was a stop sign? *93, 172, and 197*

SIGNS OF THE TIMES

Here's how *to add whole numbers.*   93 + 172 + 197 = ?

Line up the digits vertically.	Add ones. Regroup.	Add tens. Regroup.	Add hundreds.
93 172 +197	¹93 172 +197 ――― 2	²¹93 172 +197 ――― 62	¹²93 172 +197 ――― 462

Look for the sums of 10.

The answer is called the **sum**.

4. Look at the *Here's how*. How many of the people could identify a stop sign? *462*

EXERCISES
Add. Here are scrambled answers for the next row of exercises: 775   621   1183   504   709   522

5. 247
   +462
   709

6. 156
   +348
   504

7. 436
   + 86
   522

8. 593
   + 28
   621

9. 297
   +886
   1183

10. 623
   +152
   775

Chapter 1

11. 6375 + 298 = 6673
12. 4378 + 774 = 5152
13. 7409 + 4907 = 12,316
14. 62,237 + 8,073 = 70,310
15. 92,876 + 38,846 = 131,722

16. 7234 + 186 + 2145 = 9565
17. 483 + 2964 + 192 = 3639
18. 9263 + 4063 + 812 = 14,138
19. 12,610 + 8,715 + 24,025 = 45,350
20. 37,096 + 492 + 15,405 = 52,993

21. $2.78 + 3.18 + 6.92 = $12.88
22. $6.99 + 2.08 + 9.36 = $18.43
23. $71.24 + 9.76 + 43.08 = $124.08
24. $28.09 + 75.34 + 38.68 = $142.11
25. $34.76 + 52.64 + 93.28 = $180.68

26. 1231 + 875   2106
27. 176 + 2874   3050
28. 2538 + 57   2595
29. 945 + 3465   4410

30. 236 + 61 + 9   306
31. 81 + 914 + 39   1034
32. 92 + 7 + 163   262
33. 721 + 86 + 375   1182

34. 68 + 219 + 6   293
35. 301 + 98 + 24   423
36. 68 + 9 + 952   1029
37. 385 + 99 + 37   521

MENTAL MATH Add. Write answers only.

38. 40 + 80   120

> 4 tens + 8 tens equals 12 tens, or 120.

39. 70 + 40   110
40. 60 + 30   90
41. 40 + 90   130
42. 600 + 300   900
43. 200 + 700   900
44. 800 + 400   1200
45. 60 + 10 + 20   90
46. 30 + 10 + 40   80
47. 10 + 20 + 50   80
48. 80 + 20 + 30   130
49. 70 + 20 + 20   110
50. 20 + 90 + 10   120
51. 60 + 60 + 60   180

Solve. Use the survey information on page 6.

52. How many of the people surveyed identified the railroad-crossing sign? 363
53. How many people identified the yield sign? 246
54. How many people 10 or over identified the stop sign? 369
55. How many people under 16 identified the railroad-crossing sign? 180

On the road again — Reading road signs

56. How many miles is it from Abilene to Odessa? 146
57. How far is it from Big Spring to Pecos? 133 miles
58. When you are at Big Spring, how far are you from Odessa? 61 miles
59. Which city is 218 miles from Pecos? Abilene

ABILENE 110 miles
BIG SPRING 25 miles
ODESSA 36 miles
PECOS 108 miles

Adding Whole Numbers and Decimals

Extra Practice
Page 471 Skill 1

Practice Worksheet
Workbook S171, Copymaster S171, or Duplicating Master S171

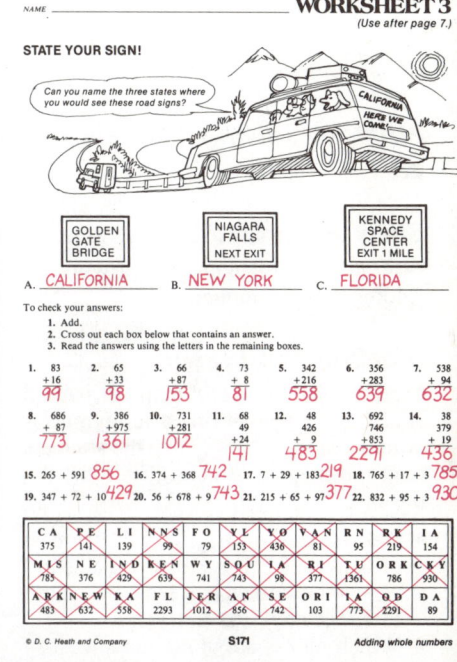

Project

Researching information

Make a color sketch of each of these road signs:
- No Passing Zone
- Do Not Enter
- Hospital
- No U Turn

Copymaster S473

Class Starter Quiz 3
on previous lesson

Add.

1.	6808 + 24,578	31,386
2.	57,819 + 4468	62,287
3.	675 + 32,910	33,585
4.	567 + 3492 + 13,672	17,731
5.	78 + 5902 + 35,896	41,876

Copymaster S385

Problem-Solving Skill
Reading a map

Skills Reviewed
Writing standard numerals in words
Adding whole numbers

Starting the Lesson
Problem Solving The map on page 8 is also on ■ *Visual Aid 4* (copymaster or transparency S114). Help the students read the map by asking questions like these:

- Which cities on Highway 62 are east of Burr Oak? (Elk Horn and Shelby)
- How many miles is it from Fairmont to Ridgeway? (63 miles)
- Which city is about 60 miles west of Burr Oak? (Garber)

Cumulative Skill Practice Write these six answers on the chalkboard:

92 thousand, 57
86 million, 100 thousand
51,967
930
43,903
1619

Challenge the students to an answer hunt by saying: "Look at exercises 1–35 on page 9. Find the six exercises that have these answers. You have five minutes to find as many of the exercises as you can." (Exercises 6, 12, 20, 25, 29, and 34) Do not expect any one student to find all six answers.

Problem solving Reading a map

**Use the map to answer the CB users' questions.
Decide when a calculator would be useful.**

1. "I know it's 14 miles from Red Oak to Gray. How far is it from Gray to Fairmont?" 80 miles

2. "How many miles is it from Conway to Garber?" 85

3. "Which city is about 180 miles east of Brooks?" Elk Horn

4. "I'm driving east on Highway 34. I'm 90 miles east of Red Oak. Which city will I drive through next?" Fairmont

5. "I'm just crossing Willow River, traveling west on Highway 34. How far am I from Red Oak?" 94 miles

6. "I'm at Red Oak on my way to Shelby. I plan to drive about 100 miles before lunch. At which city on Highway 34 should I stop and eat?" Fairmont

7. "How far is it from Ridgeway to Bristow if I take Highways 14 and 62? Is that the shortest route?" 155 miles, No

8. "I'm 150 miles west of Elk Horn on Highway 62. I'm headed for Conway. Is the next town Bristow or Brooks?" Bristow

9. "I want to take the shortest route from Elk Horn to Fairmont. What highways should I take?" 62 and 77

10. "I'm now traveling west on Highway 34. I just passed a sign that says Ridgeway is 15 miles ahead. How many miles am I from Goodell?" 129

11. "I'm 16 miles south of Bristow, going north on Highway 25. How far am I from Burr Oak?" 129 miles

12. "I'm at Grant's Truckstop, 40 miles west of Burr Oak. How far am I from Shelby?" 122 miles

Cumulative Skill Practice

Write the short word-name. *(page 4)*
Hint: Think about where the commas go.

1. 35198
2. 4354
3. 12326
4. 36402
5. 443031
6. 92057
7. 180043
8. 200001
9. 1426005
10. 63128
11. 42000261
12. 86100000

Write the standard numeral. *(page 4)*

13. 347 thousand, 172   347,172
14. 18 million   18,000,000
15. 62 billion   62,000,000,000
16. 19 million, 418 thousand   19,418,000
17. 219 million, 76 thousand   219,076,000
18. 7 million, 3 thousand, 4   7,003,004
19. six hundred thirty-seven thousand, two hundred sixteen   637,216
20. fifty-one thousand, nine hundred sixty-seven   51,967
21. eighteen million, one hundred sixteen thousand   18,116,000
22. four hundred twenty-three thousand, fourteen   423,014
23. three billion, one hundred ten   3,000,000,110

Add. *(page 6)*

24. 83 + 92 = 175
25. 698 + 232 = 930
26. 7861 + 6573 = 14,434
27. 8250 + 4948 = 13,198
28. 53,816 + 29,754 = 83,570
29. 24,319 + 19,584 = 43,903

30. 267 + 384 + 37 + 115 = 803
31. 628 + 35 + 275 + 56 = 994
32. 57 + 629 + 38 + 115 = 839
33. 342 + 708 + 56 + 86 = 1192
34. 628 + 52 + 395 + 544 = 1619
35. 291 + 35 + 426 + 184 = 936

MIXED PRACTICE
Complete.

36. The short word-name for 53270 is ?   53 thousand 270
37. The short word-name for 13680120 is ?   13 million 680 thousand 120
38. The standard numeral for six million, two hundred eighty-one thousand is ?   6,281,000
39. The standard numeral for five billion, four hundred eight thousand is ?   5,000,408,000
40. 538 + 653 = ?   1191
41. 5820 + 17,283 = ?   23,103
42. 35 + 78 + 153 = ?   266
43. 592 + 86 + 179 + 57 = ?   914

Adding Whole Numbers and Decimals **9**

Problem-Solving Worksheet
Workbook S172, Copymaster S172, or Duplicating Master S172

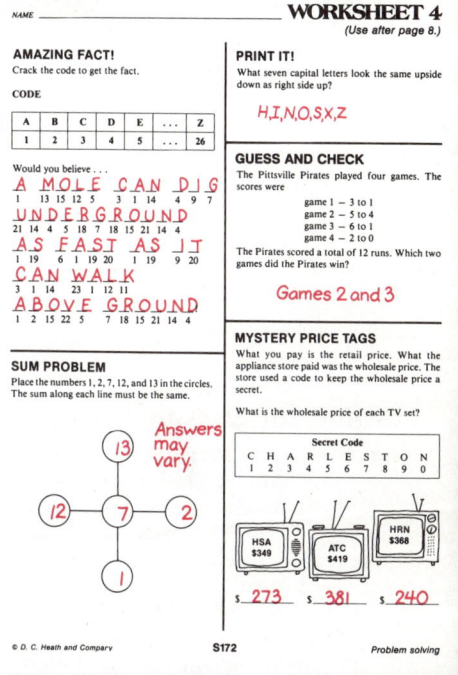

Challenge Problem

Look on page 9. Find the largest number (not an answer) less than 700. Find the smallest number greater than 700. Add them. Did you get a sum of 1406?
698, 708, yes

Copymaster S439

Answers for page 9.
1. 35 thousand, 198
2. 4 thousand, 354
3. 12 thousand, 326
4. 36 thousand, 402
5. 443 thousand, 31
6. 92 thousand, 57
7. 180 thousand, 43
8. 200 thousand, 1
9. 1 million, 426 thousand, 5
10. 63 thousand, 128
11. 42 million, 261
12. 86 million, 100 thousand

Class Starter Quiz 4
on previous lesson

Solve. Use the map on page 8.
1. How many miles is it from Bristow to Burr Oak? **113**
2. What city is about 140 miles east of Gray? **Ridgeway**

Copymaster S385

Lesson Objective
To round whole numbers

Problem-Solving Skills
Following instructions
Number sense

Starting the Lesson
Estimation Have the students guess the price of the car pictured at the top of page 10. (This car is also shown on ■ **Visual Aid 5,** copymaster or transparency S115.) Tell the students that this car is "loaded" with special features such as a 5-liter engine, automatic transmission, tilt steering wheel, removable hatch roof, AM/FM radio with front and rear speakers, custom air-conditioning, dual exhaust system, hand-rubbed lacquer paint, and a security system. Have the students write their guesses. Then list the guesses on the chalkboard and determine whose guess was closest to the exact price, $19,574.

Here's How *Note*
You may wish to present these additional examples before assigning the exercises.

Round to the nearest ten.
 3 (0) 197 (200) 6998 (7000)
Round to the nearest hundred.
 43(0) 4985 (5000) 12,999 (13,000)

10

Rounding whole numbers

1. Carlos guessed $19,500. Joan guessed $19,600. The exact price of the sports car is $19,574. Whose guess was nearer the exact price? **Joan's**
2. Is $19,570 or $19,580 nearer the exact price? **$19,570**

Rounded numbers are often used in place of exact numbers.

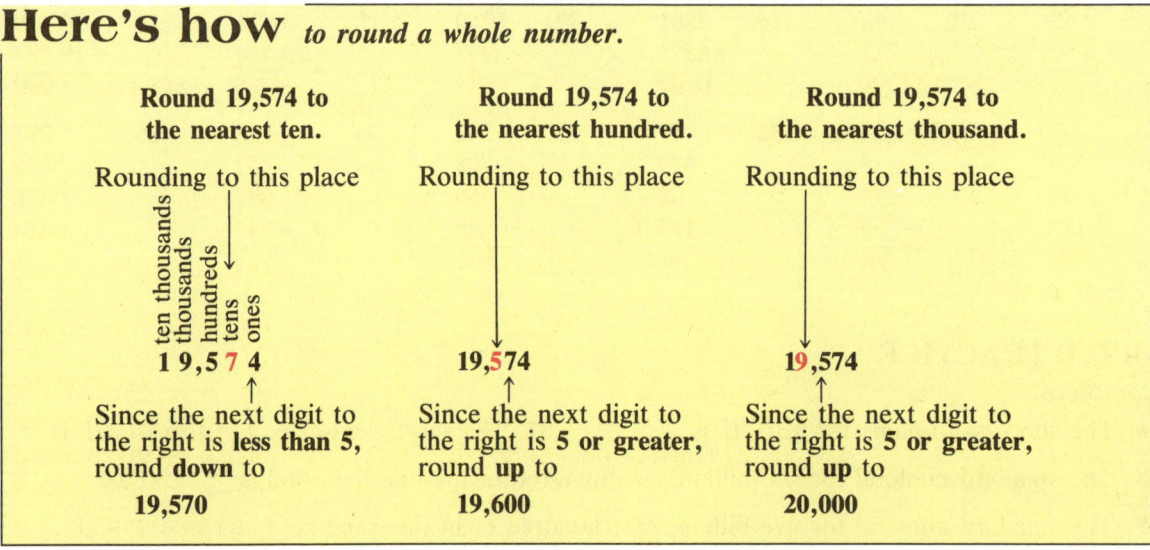

3. Look at the *Here's how*. In 19,574, which digit is in the ten thousands place? Is the next digit to the right 5 or greater? **Yes**
4. Round 19,574 to the nearest ten thousand. **20,000**

10   Chapter 1

EXERCISES

Round to the nearest ten.

5. 73 70
6. 65 70
7. 6 10
8. 497 500
9. 521 520
10. 3 0
11. 2653 2650
12. 4708 4710
13. 6222 6220
14. 6803 6800

Round to the nearest hundred.

15. 378 400
16. 450 500
17. 99 100
18. 3692 3700
19. 4987 5000
20. 2509 2500
21. 5621 5600
22. 7770 7800
23. 9050 9100
24. 3021 3000
25. 16,405 16,400
26. 25,980 26,000
27. 41,912 41,900
28. 53,950 54,000
29. 6092 6100

Round to the nearest thousand.

30. 5732 6000
31. 8026 8000
32. 741 1000
33. 8500 9000
34. 203 0
35. 26,332 26,000
36. 41,582 42,000
37. 64,398 64,000
38. 50,225 50,000
39. 15,432 15,000
40. 236,479 236,000
41. 183,500 184,000
42. 379,199 379,000
43. 829,602 830,000
44. 699,999 700,000

Round to the nearest ten thousand.

45. 37,168 40,000
46. 42,600 40,000
47. 63,911 60,000
48. 9830 10,000
49. 17,302 20,000
50. 58,502 60,000
51. 47,300 50,000
52. 62,499 60,000
53. 92,888 90,000
54. 65,898 70,000
55. 239,100 240,000
56. 468,492 470,000
57. 623,619 620,000
58. 745,000 750,000
59. 99,999 100,000

Round.

60. During the first day of the grand opening at East Meadow, **13,721** shoppers entered the Guess the Sports-Car Price contest. Round the number to the nearest hundred. 13,700

61. A total of **173,517** shoppers entered the sports-car contest. Round the number to the nearest thousand. 174,000

Changing times

62. In 1921, you could buy this Model T for about $400. In an antique-car auction in 1982, this car sold for $[?].

    To find [?], write a *6* in the tens place, a *9* in the hundreds place, a *4* in the ones place, and a *5* in the thousands place. 5964

63. If the buyer at the auction used hundred-dollar bills to pay for the car, how many bills did he use? (*Hint:* the buyer got back some bills in change.) 60

Adding Whole Numbers and Decimals **11**

Extra Practice
Page 471 Skill 2

Practice Worksheet
Workbook S173, Copymaster S173, or Duplicating Master S173

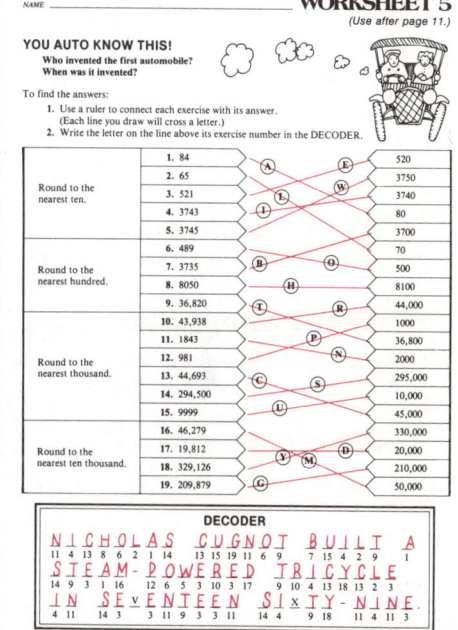

Project

Using library resources

Use a book of world records to find the greatest price ever paid for a used car. Then write the price in words.

Answers will vary. Records may change from year to year.

Copymaster S473

Class Starter Quiz 5
on previous lesson

Round to the nearest ten.
1. 38   40
2. 624   620
3. 1625   1630

Round to the nearest thousand.
4. 8561   9000
5. 24,858   25,000
6. 345,487   345,000

Copymaster S386

Lesson Objective
To estimate sums

Problem-Solving Skills
Finding information in an ad
Using a guess-and-check strategy

Starting the Lesson
What are the facts? Have the students study the sale ad on page 12 for 30 seconds and then close their books. Challenge the students to answer these questions from memory:
- What four items are on sale? (Camera, TV, radio, and calculator)
- What is the most expensive item? (Camera)
- Which item costs about $20? (Calculator)
- Which item costs about $100? (TV or Radio)
- Are the items in a summer, fall, or winter sale catalog? (Fall)

Exercise Note
Problem Solving Exercises 40–51 involve rounding and mental arithmetic. Some students may need to use a guess-and-check approach.

Estimating sums

1. Which item costs about $300? *Camera*
2. Which item costs about $90? *Radio*
3. Which two prices would you round to estimate the total cost of the camera and the personal radio? *$329 and $87*

Here's how to estimate sums.

Round to the nearest ten dollars.

Ray
$329 → 330
$87 → + 90
   $420

Round to the nearest hundred dollars.

Cindy
$329 → 300
$87 → +100
   $400

4. What is the actual total cost of the camera and the personal radio? *$416*
5. Look at the *Here's how*. Whose estimate was closer to the actual total cost, Ray's or Cindy's? *Ray's*
6. Use Ray's method. Which two items cost about $110? *Radio and calculator*
7. Use Cindy's method. Which two items cost about $200? *Radio and TV*

EXERCISES

Which estimate would Cindy give? Hint: Study the Here's how.

8. $325 + $479 a. $600 (b.) $800 c. $1000
9. $281 + $94 + $319 (a.) $700 b. $1000 c. $1300
10. $631 + $477 + $819 a. $1300 b. $1600 (c.) $1900

Which estimate would Ray give?

11. $789 + $42 a. $810 (b.) $830 c. $850
12. $37 + $86 + $129 a. $230 (b.) $260 c. $290
13. $29 + $43 + $68 (a.) $140 b. $170 c. $200

Estimate each sum by rounding to the nearest hundred dollars.

14. $428 + 583 → $400 + $600 = $1000
15. $685 + 519 = $1200
16. $867 + 109 = $1000
17. $929 + 409 = $1300
18. $789 + 239 = $1000
19. $615 + $309 = $900
20. $599 + $112 = $700
21. $614 + $928 = $1500
22. $797 + $196 = $1000
23. $385 + $89 = $500
24. $996 + $217 = $1200
25. $1209 + $619 = $1800
26. $2999 + $399 = $3400

Estimate each sum by rounding to the nearest ten dollars.

27. $574 + 319 → $570 + $320 = $890
28. $281 + 113 = $390
29. $79 + 18 = $100
30. $149 + 53 = $200
31. $418 + 73 = $490
32. $81 + $78 = $160
33. $29 + $69 = $100
34. $329 + $19 = $350
35. $428 + $48 = $480
36. $27 + $16 + $39 = $90
37. $78 + $23 + $49 = $150
38. $96 + $32 + $19 = $150
39. $448 + $53 + $69 = $570

Jewelry juggle — Estimating costs

Use your estimation skills. Which two items cost about

40. $500?
41. $1100?
42. $300?
43. $110?
44. $700?
45. $900?

Which three items cost about

46. $700?
47. $500?
48. $300?
49. $1200?
50. $800?
51. $1300?

WALKER'S GRAND OPENING SALE

14K Gold Chain ....
Quartz Watch $209
Opal Ring $415
Diamond Pendant $88
Silver Charm $679
.... $19

East Meadow Mall

Adding Whole Numbers and Decimals 13

Practice Worksheet

Workbook S174, Copymaster S174, or Duplicating Master S174

WORKSHEET 6 (Use after page 13.)

NAME _____

SKILL DRILL

Estimate each sum by rounding to the nearest hundred dollars.

1. $429 + 79 → $400 + $100 = $500
2. $273 + 129 → $300 + $100 = $400
3. $809 + 333 → $800 + $300 = $1100
4. $726 + 133 → $700 + $100 = $800
5. $816 + 89 + 359 → $800 + $100 + $400 = $1300
6. $903 + 619 + 306 → $900 + $600 + $300 = $1800
7. $393 + 416 + 909 → $400 + $400 + $900 = $1700
8. $891 + 23 + 89 → $900 + $0 + $100 = $1000

Estimate each sum by rounding to the nearest ten dollars.

9. $319 + 89 → $320 + $90 = $410
10. $463 + 219 → $460 + $220 = $680
11. $708 + 434 → $710 + $430 = $1140
12. $516 + 142 → $520 + $140 = $660
13. $629 + 78 + 513 → $630 + $80 + $510 = $1220
14. $103 + 809 + 718 → $100 + $810 + $720 = $1630
15. $113 + 8 + 21 → $110 + $10 + $20 = $140
16. $4 + 17 + 103 → $0 + $20 + $100 = $120

17. $415 + $389 $810
18. $381 + $411 + $95 $890
19. $741 + $583 + $915 $2240
20. $32 + $585 + $472 $1090
21. $828 + $93 + $513 $1430
22. $19 + $609 + $759 $1390

Check yourself. Here are the scrambled answers:
$120 $140 $400 $410 $500 $660 $680 $800 $810
$890 $1000 $1090 $1100 $1140 $1220 $1300 $1390 $1430
$1630 $1700 $1800 $2240

© D. C. Heath and Company   S174   Estimating sums

Group Project

Using a catalog

Have a contest. The winner is the person who finds 3 items in a catalog that come closest to costing a total of $100.

Answers for page 13.
40. watch, ring
41. watch, pendant
42. chain, ring
43. ring, charm
44. pendant, charm
45. chain, pendant
46. chain, watch, ring
47. watch, ring, charm
48. chain, ring, charm
49. watch, ring, pendant
50. ring, pendant, charm
51. chain, watch, pendant

Class Starter Quiz 6
on previous lesson

Match. Find the sums by estimating.

1. 399 + 512   b   a. 451
2. 156 + 295   a   b. 911
3. 911 + 390   c   c. 1301
4. 623 + 114 + 384   f   d. 2081
5. 693 + 575 + 813   d   e. 710
6. 125 + 289 + 296   e   f. 1121

Copymaster S386

Lesson Objective
To read decimals

Problem-Solving Skills
Interpreting information
Number sense

Starting the Lesson
Number Sense Write this statement on the chalkboard:

A downhill skier, going 30 miles per hour, can travel 50 feet in 1136 seconds.

Have the students decide where to place the decimal point in the number 1136 so that the statement makes sense. (1.136 seconds)

Here's How Note
Use ■ **Visual Aid 6** (copymaster or transparency S116) or draw a place-value chart on the chalkboard to reinforce the idea of decimal place-value.

Answers for page 15.
25. 3 and 5 tenths
26. 12 and 35 hundredths
27. 125 thousandths
28. 17 and 3 thousandths
29. 9 and 2 hundredths
30. 25 thousandths
31. 14 and 9 tenths
32. 3 and 75 ten-thousandths
33. 634 ten-thousandths
34. 253 and 61 hundredths
35. 72 and 6 thousandths
36. 594 ten-thousandths
37. 631 and 74 hundredths
38. 3 and 1005 ten-thousandths

Reading decimals

The decimal shows the time (in seconds) that it took the skier to complete her first downhill run.

SPLIT SECONDS!

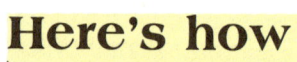

5 7 . 6 8

Tens Ones Tenths Hundredths

1. In what place is the digit 6? *tenths*
2. In what place is the last digit? *hundredths*

Here's how *to read decimals.*

Her first run took **57 and 68 hundredths** seconds. Notice that the decimal point is read as "and" and the place of the last digit is read last.

Here are some more examples of how to read decimals:

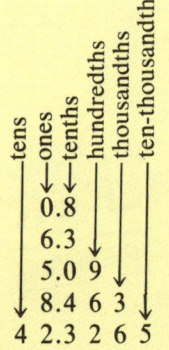

STANDARD NUMERAL	SHORT WORD-NAME
0.8	8 tenths
6.3	6 and 3 tenths
5.09	5 and 9 hundredths
8.463	8 and 463 thousandths
42.3265	42 and 3265 ten-thousandths

3. Look at the *Here's how*. To read 57.68, you say "57 and 68 [?]." *hundredths*
4. To read 8.463, you say "8 and 463 [?]." *thousandths*

EXERCISES

In what place is the last digit?

5. 16.3 tenths
6. 0.357 thousandths
7. 6.25 hundredths
8. 0.4216 ten-thousandths
9. 2.069 thousandths
10. 16.38 hundredths
11. 26.9 tenths
12. 0.0371 ten-thousandths
13. 19.6421 ten-thousandths
14. 58.4 tenths
15. 13.005 thousandths
16. 24.57 hundredths
17. 8.0007 ten-thousandths
18. 220.68 hundredths
19. 126.9 tenths
20. 8.594 thousandths
21. 1206.74 hundredths
22. 1.7241 ten-thousandths
23. 0.003 thousandths
24. 468.2 tenths

Write the short word-name.

25. 3.5 3 and 5 tenths
26. 12.35
27. 0.125
28. 17.003
29. 9.02
30. 0.025
31. 14.9
32. 3.0075
33. 0.0634
34. 253.61
35. 72.006
36. 0.0594
37. 631.74
38. 3.1005
39. 0.875
40. 3968.4
41. 860.2
42. 9002.11
43. 63.0004
44. 0.4005
45. 4216.9
46. 421.69
47. 42.169
48. 4.2169
49. 0.0062

Write the standard numeral.

50. 2 thousandths 0.002
51. 25 and 4 tenths 25.4
52. 12 and 3 hundredths 12.03
53. 2 thousandths 0.002
54. 34 and 32 hundredths 34.32
55. 164 and 58 hundredths 164.58
56. 452 thousandths 0.452
57. 27 and 148 thousandths 27.148
58. 9 and 75 thousandths 9.075
59. 8 and 6 thousandths 8.006
60. 4275 ten-thousandths 0.4275
61. 20 and 840 ten-thousandths 20.0840

Speed records

Here are my fastest speeds.

Bill Gomez — miles per hour
Bicycling 23.2
Skateboarding 13.8
Rollerskating 18.1

Copy the numeral and place a decimal point so that the statement makes sense. Hint: Use Bill's chart.

62. The world speed record for a skateboard is **718** miles per hour. 71.8
63. The world speed record for a bicycle is **152284** miles per hour. 152.284
64. Top speed for roller skates is **2578** miles per hour.

Records may change from year to year.

Adding Whole Numbers and Decimals **15**

Practice Worksheet
Workbook S175, Copymaster S175, or Duplicating Master S175

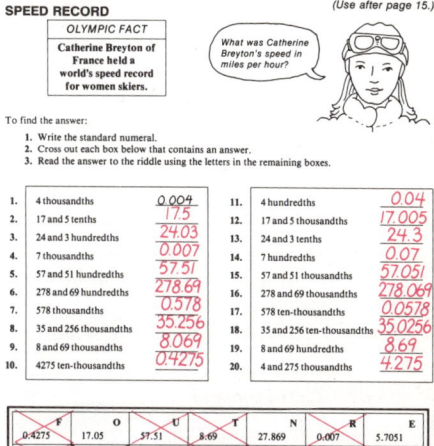

WORKSHEET 7 (Use after page 15.)
SPEED RECORD

OLYMPIC FACT
Catherine Breyton of France held a world's speed record for women skiers.

What was Catherine Breyton's speed in miles per hour?

To find the answer:
1. Write the standard numeral.
2. Cross out each box below that contains an answer.
3. Read the answer to the riddle using the letters in the remaining boxes.

1. 4 thousandths 0.004
2. 17 and 5 tenths 17.5
3. 24 and 3 hundredths 24.03
4. 7 thousandths 0.007
5. 57 and 51 hundredths 57.51
6. 278 and 69 hundredths 278.69
7. 578 thousandths 0.578
8. 35 and 256 thousandths 35.256
9. 8 and 69 thousandths 8.069
10. 4275 ten-thousandths 0.4275
11. 4 hundredths 0.04
12. 17 and 5 tenths 17.005
13. 24 and 3 tenths 24.3
14. 7 hundredths 0.07
15. 57 and 51 thousandths 57.051
16. 278 and 69 thousandths 278.069
17. 578 ten-thousandths 0.0578
18. 35 and 256 ten-thousandths 35.0256
19. 8 and 69 hundredths 8.69
20. 4 and 275 thousandths 4.275

Answer: ONE HUNDRED THREE miles per hour

© D. C. Heath and Company   S175   Reading decimals

Group Project
Using a newspaper
Have the students choose a page from a newspaper and circle every decimal they find. The student who finds the most decimals wins.

More answers for page 15.
39. 875 thousandths
40. 3968 and 4 tenths
41. 860 and 2 tenths
42. 9002 and 11 hundredths
43. 63 and 4 ten-thousandths
44. 4005 ten-thousandths
45. 4216 and 9 tenths
46. 421 and 69 hundredths
47. 42 and 169 thousandths
48. 4 and 2169 ten-thousandths
49. 62 ten-thousandths

Class Starter Quiz 7
on previous lesson

Write the standard numeral.

1. 4 thousandths — 0.004
2. 15 and 2 tenths — 15.2
3. 34 and 8 hundredths — 34.08
4. 6 and 35 thousandths — 6.035
5. 40 and 125 thousandths — 40.125

Copymaster S386

Problem-Solving Skills

Reading a sign
Solving problems involving more than one step
Choosing the appropriate information

Skills Reviewed

Adding whole numbers
Rounding whole numbers
Reading decimals
Writing standard numerals

Starting the Lesson

Problem Solving Have the students use the ticket price information to answer questions like these:

- How much would you charge a 20-year-old to ride the slide? ($2.25)
- Can a teenager ride the slide for $1? (No)
- Is $5 enough money to buy 2 adult tickets and 1 child's ticket? (No)

Cumulative Skill Practice Challenge the students to an estimation hunt by saying, "Pick the exercise that has the largest sum in the first row of exercises." (Exercise 5) Then have the students pick the exercise with the largest sum in the second, third, and fourth rows of exercises. (Exercises 9, 13, and 16)

16

Problem solving

Reading a sign

You have a summer job at the Lincoln Woods Mountain Slide. You sell tickets and work in the gift shop.

RIDE THE SLIDE!

TICKET PRICES
Adult (18 and over) ...$2.25
Student (12 to 18).....$1.50
Child (under 12).......$1.00

Use the chart to answer these customers' questions. Decide when a calculator would be useful.

1. *How much will 2 adult tickets cost?* $4.50

2. *How much for 2 adult tickets and 1 child's ticket?* $5.50

3. *5 students and 1 adult, please. How much do I owe you?* $9.75

4. "I have $5.00. Can I buy tickets for 2 adults and 1 child?" *Hint: Use your answer to problem 2.* No

5. "Is $10.00 enough money to buy 1 adult ticket and 5 student tickets?" Yes

6. "My sister has $4.75 and I have $3.50."
   a. "How much do we have altogether?" $8.25
   b. "Do we have enough money to buy 5 student tickets?" Yes

7. a. "What is the total cost of 1 student ticket and 1 child's ticket?" $2.50
   b. "My father decided not to ride. Can we trade his adult ticket for 1 student ticket and 1 child's ticket?" No

Solve.

8. Your ticket sales for today were 350 children's, 249 students', and 123 adults'.
   a. How many tickets did you sell in all? 722
   b. Yesterday you sold a total of 738 tickets. Did you sell at least that many tickets today? No

9. Bumper-sticker sales for today were 24 *LINCOLN WOODS*, 72 *ZIPPER*, and 63 *MOUNTAIN SLIDE*.
   a. Which bumper sticker had the highest number of sales? Zipper
   b. What were the combined sales of the *ZIPPER* and the *MOUNTAIN SLIDE*? 135 bumper stickers

16 Chapter 1

Cumulative Skill Practice

Add. *(page 6)*

1. 7406 + 1629 = 9035
2. 2917 + 2579 = 5496
3. 824 + 5081 = 5905
4. 1056 + 3780 = 4836
5. 3643 + 8561 = 12,204
6. 53,246 + 2,107 = 55,353
7. 38,529 + 4,266 = 42,795
8. 17,329 + 54,600 = 71,929
9. 51,083 + 74,291 = 125,374
10. 38,294 + 27,461 = 65,755
11. 3982 + 427 + 965 5374
12. 428 + 3461 + 2009 5898
13. 4721 + 3066 + 5814 13,601
14. 52,140 + 89 + 2417 54,646
15. 17 + 2573 + 43,880 46,470
16. 889 + 14 + 72,354 73,257

Round to the nearest ten. *(page 10)*

17. 62 60
18. 85 90
19. 151 150
20. 438 440
21. 395 400
22. 982 980
23. 216 220
24. 302 300
25. 461 460
26. 285 290
27. 373 370
28. 795 800
29. 5675 5680
30. 3502 3500
31. 6296 6300
32. 8743 8740
33. 9608 9610
34. 6304 6300

Write the standard numeral. *(page 14)*

35. 9 tenths 0.9
36. 6 hundredths 0.06
37. 4 thousandths 0.004
38. 15 hundredths 0.15
39. 36 thousandths 0.036
40. 147 thousandths 0.147
41. 6 and 3 tenths 6.3
42. 22 and 81 hundredths 22.81
43. 40 and 5 hundredths 40.05
44. 36 and 235 ten-thousandths 36.0235
45. 28 and 16 thousandths 28.016
46. 9 and 1374 ten-thousandths 9.1374
47. 45 and 4653 ten-thousandths 45.4653
48. 38 and 491 ten-thousandths 38.0491

MIXED PRACTICE
Complete.

49. The standard numeral for 57 thousand, 128 is __?__ 57,128
50. The standard numeral for 33 million is __?__ 33,000,000
51. The standard numeral for 9 million, 7 thousand, 42 is __?__ 9,007,042
52. 267 + 85 + 121 + 37 = __?__ 510
53. 29 + 3 + 8174 + 326 = __?__ 8532
54. 9175 + 812 + 6 + 8 = __?__ 10,001
55. 441 + 27 + 5 + 1831 = __?__ 2304

Adding Whole Numbers and Decimals **17**

Problem-Solving Worksheet
Workbook S176, Copymaster S176, or Duplicating Master S176

WORKSHEET 8 (Use after page 16.)

NAME _____

SCRAMBLED MATH
Unscramble the letters to get the answer.
ETN plus HIEGT plus REHTE equals **21**
TEN   EIGHT   THREE

OFF AND ON
You switched off a light. Then you switched it 15 more times. Then you switched it 13 more times. Was the light on or off when you stopped?
Off

LOGICAL REASONING
Each of Jim, Brian, Hal, and Tony scored a different number of points. Jim scored more points than Brian. Hal scored fewer points than Tony. Only Brian's number of points was between Hal's and Jim's.
Who scored the most points? **Tony**

LOOSE CHANGE
There is 55¢ in the purse. What 4 coins are in the purse?
**quarter
dime
dime
dime**

BOWLING SCORES

Name	FIRST GAME	SECOND GAME	THIRD GAME
Rick	118	101	158
Carrie	159	112	137
Jody	97	121	89
Eric	167	116	127

Which bowler had the lowest total score for all three games? **Jody**

MAKE A LIST
How many times would you use the digit 9 to number the pages of a 100-page book? *Hint: The answer is more than 18.* **20**

Which two bowlers had about the same total score for all three games? **Carrie and Eric**

© D. C. Heath and Company   S176   Problem solving

Group Project
Researching information
Have the students guess which four states in the United States have mountain peaks higher than 14,000 feet. Have an almanac or atlas available for the students to check their guesses.

Alaska, California, Colorado, and Washington

Class Starter Quiz 8
on previous lesson

Solve. Use the ticket prices on page 16.

1. How much will 1 adult and 3 student tickets cost? $6.75
2. Is $7.50 enough money for 1 child's and 3 adult tickets? No

Copymaster S386

Lesson Objective

To round decimals

Problem-Solving Skill

Reading a chart

Starting the Lesson

What are the facts? Have the students look at the picture on page 18, read the paragraph at the top of the page, and then close their books. Challenge the students to answer these questions from memory:

- Who holds the land speed record for women? (Kitty O'Neil)
- Did she average more or less than 500 miles per hour? (More)
- Did her car have more or less than 50,000 horsepower? (Less)
- How many wheels were on her rocket-powered car? (3)

Here's How Note

Use ■ **Visual Aid 6** (copymaster or transparency S116) or draw a place-value chart on the chalkboard to reinforce the idea of rounding decimals.

18

Rounding decimals

Kitty O'Neil holds the land speed record for women. Driving her 48,000-horsepower rocket-powered car, she averaged 512.715 miles per hour. It took her over 5 miles just to stop!

1. What was O'Neil's average speed?
   512.715 miles per hour

Here's how *to round a decimal.*

Rounding to this place
512.715

*Since the next digit to the right is **5 or greater**, round **up** to 513.*

Rounded to the nearest whole number, 512.715 is **513**.

Rounding to this place
512.715

*Since the next digit to the right is **less than 5**, round **down** to 512.7.*

Rounded to the nearest tenth, 512.715 is **512.7**.

Rounding to this place
512.715

*Since the next digit to the right is **5 or greater**, round **up** to 512.72.*

Rounded to the nearest hundredth, 512.715 is **512.72**.

2. Was O'Neil's speed closer to 512 miles per hour or 513 miles per hour? 513

18 Chapter 1

EXERCISES

Round to the nearest whole number.

3. 15.7 *16*
4. 47.2 *47*
5. 0.25 *0*
6. 35.34 *35*
7. 0.95 *1*
8. 52.19 *52*
9. 28.928 *29*
10. 69.523 *70*
11. 421.073 *421*
12. 99.786 *100*
13. 215.07 *215*
14. 429.562 *430*
15. 0.895 *1*
16. 76.0125 *76*
17. 25.21 *25*
18. 97.612 *98*
19. 10.72 *11*
20. 172.26 *172*
21. 51.347 *51*
22. 0.86 *1*

Round to the nearest tenth.

23. 1.62 *1.6*
24. 28.108 *28.1*
25. 7.28 *7.3*
26. 0.552 *0.6*
27. 7.342 *7.3*
28. 2.460 *2.5*
29. 69.169 *69.2*
30. 31.03 *31.0*
31. 9.381 *9.4*
32. 8.1106 *8.1*
33. 705.49 *705.5*
34. 913.91 *913.9*
35. 0.056 *0.1*
36. 43.012 *43.0*
37. 53.299 *53.3*
38. 4.295 *4.3*
39. 0.123 *0.1*
40. 72.575 *72.6*
41. 2.107 *2.1*
42. 41.0086 *41.0*

Round to the nearest hundredth.

43. 18.216 *18.22*
44. 38.107 *38.11*
45. 2.543 *2.54*
46. 61.625 *61.63*
47. 84.612 *84.61*
48. 83.581 *83.58*
49. 4.2029 *4.20*
50. 0.0254 *0.03*
51. 8.3815 *8.38*
52. 0.013 *0.01*
53. 6.3522 *6.35*
54. 0.005 *0.01*
55. 16.949 *16.95*
56. 234.789 *234.79*
57. 13.023 *13.02*
58. 2.1685 *2.17*
59. 12.825 *12.83*
60. 5.201 *5.20*
61. 398.166 *398.17*
62. 4.117 *4.12*

Round to the nearest dollar.

63. $7.77 *$8*
64. $14.48 *$14*
65. $234.61 *$235*
66. $67.52 *$68*
67. $24.87 *$25*
68. $35.92 *$36*
69. $35.50 *$36*
70. $129.79 *$130*
71. $99.89 *$100*
72. $3.02 *$3*
73. $179.79 *$180*
74. $42.49 *$42*
75. $2.55 *$3*
76. $34.09 *$34*
77. $48.91 *$49*

You're a reporter — Using a chart

Use the chart. Complete the story.

78. On August __?__, __?__, __?__ in the __?__ set a new one-mile speed record of about 410 miles per hour.
(date) (year) (driver's name) (car's name)
5, 63 Breedlove, Spirit of America

79. On October __?__, __?__, __?__ in the __?__ set a new record of nearly 540 miles per hour.
(date) (year) (driver's name) (car's name)
27, 64 Arfons, Green Monster

One-Mile Speed Records			
DATE	DRIVER	CAR	MPH
9/3/35	Campbell	Bluebird Special	301.13
9/16/38	Eyston	Thunderbolt 1	357.5
9/16/47	Cobb	Railton-Mobil	394.2
8/5/63	Breedlove	Spirit of America	407.45
10/27/64	Arfons	Green Monster	536.71
11/15/65	Breedlove	Spirit of America	600.601
10/23/70	Gabelich	Blue Flame	622.407

Extra Practice
Page 472 Skill 3

Practice Worksheet
Workbook S177, Copymaster S177, or Duplicating Master S177

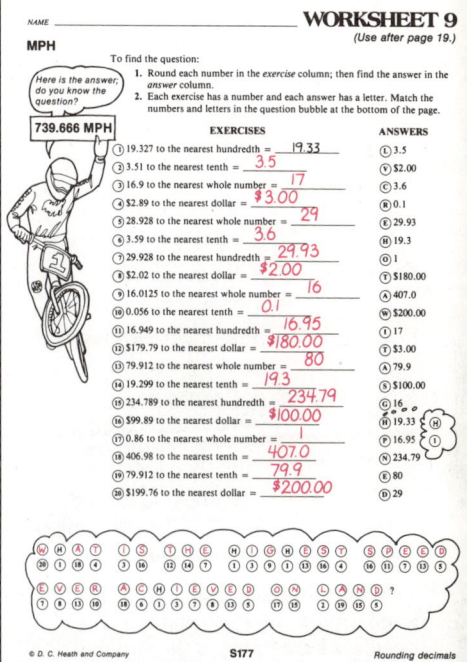

Challenge Problem

Find the number on page 19 that fits the clues.

Clues:
- It is less than 400.
- Rounded to the nearest hundred, it is 200.
- Rounded to the nearest ten, it is 220. *215.07 (exercise 13)*

Copymaster S439

Adding Whole Numbers and Decimals **19**

Class Starter Quiz 9
on previous lesson

Round to the nearest tenth.
1. 9.28   9.3   2. 0.681   0.7
3. 37.612   37.6   4. 0.75   0.8

Round to the nearest hundredth.
5. 28.316   28.32   6. 4.173   4.17
7. 0.125   0.13   8. 23.196   23.20

Copymaster S387

Lesson Objective
To add decimals (addends having the same number of decimal places)

Problem-Solving Skills
Reading a chart
Using a guess-and-check strategy

Starting the Lesson
Number Sense On the chalkboard, write these statements:

In 1964, the Beatles recorded, "A Hard Day's Night." The 247-minute song sold over 1000000 copies.

Tell the students one of the numbers is missing a decimal point. Have them decide where to place the decimal point so that the statement makes sense. (2.47)

Here's How Note
Use of Concrete Materials You may wish to use the powers-of-ten tiles from copymaster or transparency S529 to demonstrate adding decimals. See ■ **Manipulative Activity 1** on copymaster S515 in the Teacher's Resource Binder.

Exercise Note
Problem Solving Encourage the students to use a guess-and-check strategy for exercises 60–62. The thinking for exercise 60 might be as follows:

34.6 ⊕ 20 ⊜ 54.6 → too small
42.6 ⊕ 30 ⊜ 72.6 → too big
43.6 ⊕ 20 ⊜ 63.6 → perfect

20

Adding decimals

You are a disc jockey! You have a request to play some songs from the soundtracks of old Beatles movies. The list below shows some of the songs you plan to play.

SONG	PLAYING TIME (IN MINUTES)
A Hard Day's Night	2.47
I Am the Walrus	4.57
And I Love Her	2.45
Help!	2.28
Yellow Submarine	2.62
Let It Be	4.02
Ticket to Ride	3.10
The Long and Winding Road	3.60

1. How many minutes will it take to play *A Hard Day's Night*? 2.47
2. How many minutes will it take to play *I Am the Walrus*? 4.57
3. Would you add or subtract to find the number of minutes needed to play both songs? Add

Here's how to add decimals.   2.47 + 4.57 = ?

Line up the decimal points.	Add hundredths and regroup.	Add tenths and regroup.	Add ones.
2.47 +4.57	2.47 +4.57 ．4	¹ 2.47 +4.57 .04	¹ ¹ 2.47 +4.57 7.04

4. Study the *Here's how*. How long will it take you to play both songs? 7.04 minutes

EXERCISES
Add. Here are scrambled answers for the next row of exercises:
19.6   9.1   13.4   8.7   10.7   10.4

5. 6.3   6. 8.5   7. 9.8   8. 8.6   9. 6.5   10. 7.2
   +2.4   +0.6   +9.8   +4.8   +3.9   +3.5
    8.7   9.1   19.6   13.4   10.4   10.7

20   Chapter 1

11. 5.26 + 3.42 = 8.68
12. 6.74 + 3.19 = 9.93
13. 8.65 + 4.93 = 13.58
14. 5.99 + 0.86 = 6.85
15. 2.48 + 0.06 = 2.54
16. 7.91 + 3.14 = 11.05
17. 52.83 + 1.95 = 54.78
18. 73.47 + 8.61 = 82.08
19. 5.09 + 34.84 = 39.93
20. 641.1 + 74.9 = 716.0
21. 63.84 + 9.66 = 73.50
22. 2.43 + 1.11 = 3.54
23. 5.6 + 3.9 + 8.4 = 17.9
24. 9.4 + 5.9 + 4.3 = 19.6
25. 42.6 + 55.7 + 62.8 = 161.1
26. 8.96 + 3.74 + 5.09 = 17.79
27. 81.6 + 5.9 + 17.4 = 104.9
28. 87.2 + 1.3 + 15.2 = 103.7

29. 5.8 + 2.9 8.7
30. 9.4 + 3.7 13.1
31. 12.0 + 7.5 19.5
32. 19.8 + 6.5 26.3
33. 4.32 + 1.65 5.97
34. 0.83 + 9.07 9.90
35. 1.16 + 3.28 + 5.36 9.80
36. 4.07 + 0.35 + 1.68 6.10
37. 2.06 + 3.18 + 6.95 12.19
38. 16.3 + 0.8 + 5.7 22.8
39. 3.99 + 0.87 + 5.77 10.63
40. 2.74 + 3.95 + 6.05 12.74

MENTAL MATH Add. Write answers only.

41. 0.4 + 0.7
42. 0.6 + 0.6 1.2
43. 0.4 + 0.5 0.9
44. 0.4 + 0.6 1

4 tenths + 7 tenths equals 11 tenths, or 1.1

45. 0.9 + 0.2 1.1
46. 0.3 + 0.3 + 0.5 1.1
47. 0.5 + 0.2 + 0.4 1.1
48. 0.4 + 0.6 + 0.3 1.3
49. 0.5 + 0.5 + 0.7 1.7
50. 0.2 + 0.7 + 0.5 1.4
51. 0.2 + 0.8 + 0.5 1.5
52. 0.6 + 0.5 + 0.2 1.3
53. 0.9 + 0.1 + 0.8 1.8

Solve. Refer to the list on page 20.

54. How many minutes will it take you to play *A Hard Day's Night* and *Help!*? 4.75
55. How many minutes will it take you to play the two shortest songs on the list? 4.73
56. You play *And I Love Her*, read a 0.75-minute commercial, then play *Yellow Submarine*. How much program time do you use? 5.82 minutes
57. You have 10 minutes left in your show. Do you have time to play *Let It Be*, *Ticket to Ride*, and *The Long and Winding Road*? No
58. How many minutes will be needed to play the two longest songs on the list? 8.59
59. Which four songs can you play in less than 10 minutes? *A Hard Day's Night, And I Love Her, Help!, Yellow Submarine*

Key it in!
Guess and check
Find a way to push each marked key once to get the answer. Answers will vary.

Sample answers:

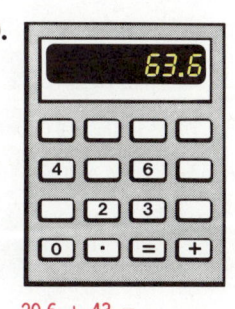
60. 20.6 + 43 =

61. 359 + 1.8 =

62. 13.4 + 25 =

Adding Whole Numbers and Decimals **21**

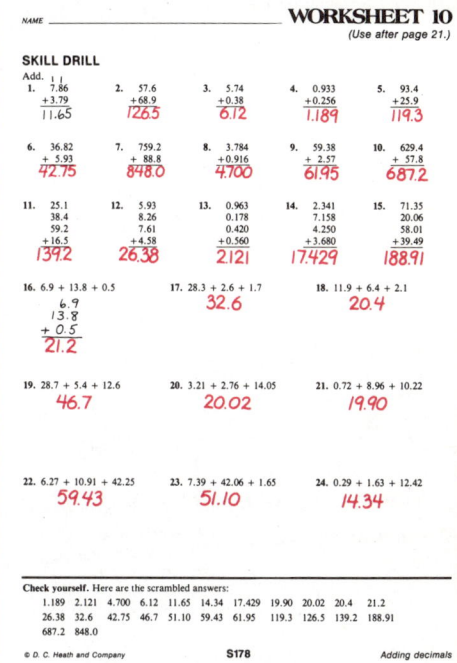

Challenge Problems

Find the missing digits.

1. [2].7 + 1.[4] = 4.1
2. [6]3.7 + 14.5 = 78.[2]
3. 2[4].1 + 3.[6] + 57.2 = [8]4.9

Copymaster S439

Class Starter Quiz 10
on previous lesson

Add.

1. 4.26 + 3.42	7.68
2. 6.91 + 2.14	9.05
3. 5.6 + 1.2 + 3.4	10.2
4. 61.2 + 4.9 + 14.3	80.4
5. 13.5 + 6.8 + 13.4 + 6.2	39.9

Copymaster S387

Lesson Objective
To add decimals (addends not having the same number of decimal places)

Problem-Solving Skills
Selecting information from a newspaper article
Reading a chart

Starting the Lesson
What are the facts? Have the students read the newspaper article and then close their books. Challenge the students to answer these questions from memory:

- What were the names of the two horses sold? (Western Dancer and Miss Smoothy)
- Who bought Western Dancer? (Alice Logan)
- Did Miss Smoothy cost more or less than $1,000,000? (More)
- In what state is Unicorn Acres Farms located? (Kentucky)

Here's How *Note*
Use ■ **Visual Aid 3** (copymaster or transparency S113) or lined notebook paper to assist the students in aligning the digits.

0.07 + 0.6 + 3 = ?

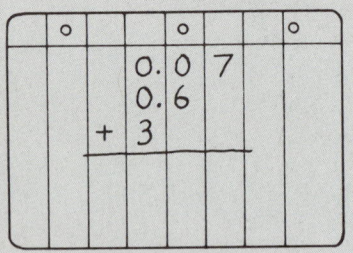

More on adding decimals

MULTIMILLION–DOLLAR DEAL

Rolling Hills, KY Unicorn Acres, the largest local horse farm, made two near-record sales at yesterday's thoroughbred auction. Western Dancer was sold for $2.835 million, and Miss Smoothy went for $1.39 million.

Western Dancer and his new owner, Alice Logan of Saratoga, NY.

1. Read the newspaper report. What was Western Dancer's sale price? $2.835 million
2. What was Miss Smoothy's sale price? $1.39 million
3. Would you add or subtract to find the amount of the total sale in millions of dollars? Add

Here's how to add decimals.   2.835 + 1.39 = ?

Line up the decimal points.   Add.

2.835
+ 1.39

2.835
+ 1.39
4.225

To estimate the sum, I first round each number to the nearest whole number and then add.

2.835 + 1.39 = ?
 ↓   ↓
 3   +   1   = 4

4. Look at the *Here's how*. Would $4 million be a good estimate? Yes

EXERCISES

5. Three of the calculator answers are wrong. Find them by estimating.

a. 58.07 + 9.784   *67.854*
b. 29.799 + 21.042   *50.841*
c. 8.0654 + 2.8152   *10.8806*
(d.) 34.968 + 12.141   *69.109*
e. 63.597 + 8.295   *71.892*
f. 6.9537 + 4.8806   *11.8343*
g. 43.952 + 14.231   *58.183*
(h.) 7.693 + 8.444   *161.35*
(i.) 53.809 + 20.298   *94.109*
j. 5.3975 + 0.7055   *6.103*

First estimate the sum and then add.

6. 18.33 + 9.40 = 27.73	7. 19.783 + 15.95 = 35.733	8. 26.3 + 24.0 = 50.3	9. 17 + 8.56 = 25.56	10. 16.33 + 38.994 = 55.324
11. $4.32 + 3.18 = $7.50	12. $7 + 4.35 = $11.35	13. $12.52 + 9 = $21.52	14. $18.06 + 7.29 = $25.35	15. $23.56 + 17 = $40.56
16. $32.21 + 19 = $51.21	17. $25 + 3.98 = $28.98	18. $52.06 + 8.35 = $60.41	19. $46.53 + 9.00 = $55.53	20. $8.57 + 2 = $10.57
21. 3.329 + 6.437 = 9.766	22. 9.509 + 7.388 = 16.897	23. 7.4206 + 0.7835 = 8.2041	24. 9.684 + 6.70 = 16.384	25. 16.942 + 9.77 = 26.712
26. 5.96 + 8.842 = 14.802	27. 16.543 + 8.92 = 25.463	28. 8.04 + 2.973 = 11.013	29. 16.295 + 12.03 = 28.325	30. 6.4928 + 9.653 = 16.1458
31. 5.6 + 3.84 + 2.9 = 12.34	32. 5.72 + 3.6 + 2.89 = 12.21	33. 8 + 2.74 + 3.6 = 14.34	34. 8.07 + 4 + 3.99 = 16.06	35. 7.4 + 3.75 + 16 = 27.15

36. 3.18 + 4 = 7.18
37. 5 + 2.63 = 7.63
38. 1.8 + 9.38 = 11.18
39. 7.04 + 2.9 = 9.94
40. 8.62 + 5 = 13.62
41. 7 + 4.6 = 11.6
42. 4.16 + 0.379 = 4.539
43. 27.6 + 9.28 = 36.88
44. 7.88 + 0.594 = 8.474
45. 3.21 + 0.853 = 4.063
46. 0.174 + 6.76 = 6.934
47. 3.0683 + 1.925 = 4.9933
48. 5.8 + 2.42 + 6.3 = 14.52
49. 19.4 + 31.6 + 8.74 = 59.74
50. 52 + 3.6 + 1.8 = 57.4
51. 2.5 + 17 + 0.8 = 20.3
52. 30 + 2.9 + 0.541 = 33.441
53. 2.6 + 4 + 0.75 = 7.35
54. 0.06 + 0.4 + 2 = 2.46
55. 0.8 + 0.36 + 0.04 = 1.20
56. 12 + 1.2 + 0.372 = 13.572
57. 5.36 + 0.1 + 0.4 = 5.86
58. 9 + 3.2 + 0.15 = 12.35
59. 6 + 7.2 + 9.184 = 22.384

And the winner is . . .

60. Which horse won $1.98 million? **Kelso**
61. Which horse won $1.46 million? **Buckpasser**
62. Which horse won $1.75 million? **Round Table**
63. Which horse won about $2.4 million? **Spectacular Bid**
64. Which horse won $1.18 million more than Seattle Slew? **Spectacular Bid**
65. Which horse won over $6 million? **John Henry**

Reading a chart

Thoroughbred Racing—Money Winners	
HORSE	TOTAL WINNINGS
John Henry	$6,590,000
Spectacular Bid	2,390,000
Kelso	1,980,000
Forego	1,940,000
Round Table	1,750,000
Buckpasser	1,460,000
Seattle Slew	1,210,000

Adding Whole Numbers and Decimals **23**

Extra Practice
Page 472 Skill 4

Practice Worksheet
Workbook S179, Copymaster S179, or Duplicating Master S179

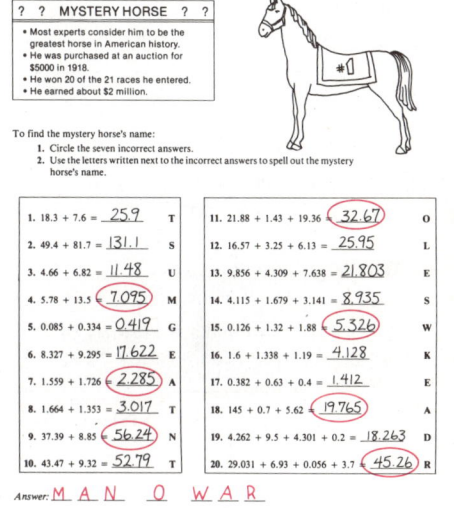

Project

Using library resources

Use a book of world records to find the name and dollar value of the most valuable horse. Then write the dollar value in words.

Answers will vary. Records may change from year to year.

Copymaster S474

Class Starter Quiz 11
on previous lesson

Two of these sums are incorrect. Find and correct the two wrong answers.

1. 17 + 9.56 = 26.56
2. 49.72 + 25 = 52.22   74.72
3. 3.597 + 0.31 = 3.907
4. 58.1 + 2.42 = 82.3   60.52
5. 3.068 + 1.92 = 4.988

Copymaster S387

Lesson Objective
To solve problems with too much or too little information

Starting the Lesson
Problem-Solving Cover Up Use the chalkboard or mask ■ *Visual Aid 7* (copymaster or transparency S117).

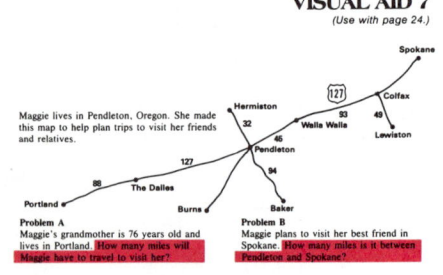

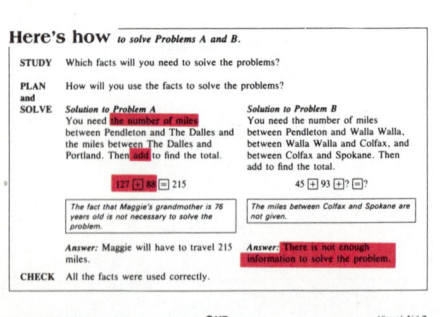

Have the students, working in small groups, study Problem A on page 24 and the problem-solving steps for several minutes. Then have them close their books, look at the visual aid, and tell what has been covered up.

24

PROBLEM SOLVING *too much/too little information*

Some problems give more information than needed to solve the problem. Other problems need additional information.

Maggie lives in Pendleton, Oregon. She made this map to help plan trips to visit her friends and relatives.

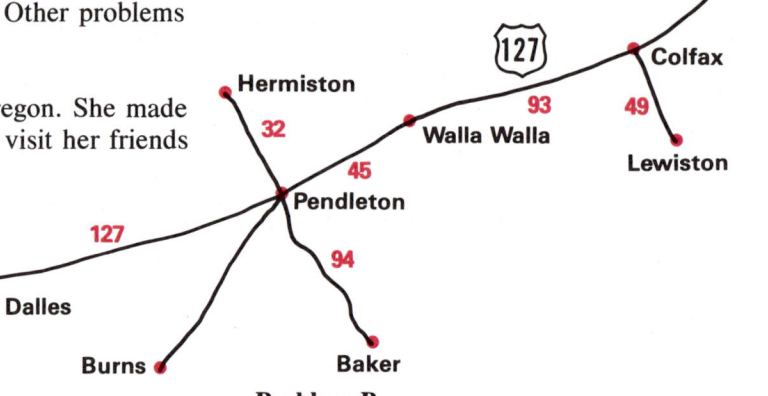

Problem A
Maggie's grandmother is 76 years old and lives in Portland. How many miles will Maggie have to travel to visit her?

Problem B
Maggie plans to visit her best friend in Spokane. How many miles is it between Pendleton and Spokane?

Here's how *to solve Problems A and B.*

STUDY Which facts will you need to solve the problems?

PLAN and SOLVE How will you use the facts to solve the problems?

Solution to Problem A
You need the number of miles between Pendleton and The Dalles and the miles between The Dalles and Portland. Then add to find the total.

127 ☐+☐ 88 ☐=☐ 215

The fact that Maggie's grandmother is 76 years old is not necessary to solve the problem.

Answer: Maggie will have to travel 215 miles.

Solution to Problem B
You need the number of miles between Pendleton and Walla Walla, between Walla Walla and Colfax, and between Colfax and Spokane. Then add to find the total.

45 ☐+☐ 93 ☐+☐ ? ☐=☐ ?

The miles between Colfax and Spokane are not given.

Answer: There is not enough information to solve the problem.

CHECK All the facts were used correctly.

24 Chapter 1

PROBLEMS

Use the map on page 24 to decide whether too little or too much information is given to solve the problem.

1. You are planning a trip from Walla Walla to Burns. What is the total number of miles between these two cities?   too little information

2. You traveled from Baker to Hermiston. You stopped in Pendleton for lunch. How many miles is the total trip?   too much information

3. It is 77 miles from Walla Walla to Hermiston. Is Pendleton closer to Walla Walla or Hermiston?   too much information

4. You are traveling from Burns to Pendleton and have gone 17 miles. How many miles are left to travel?   too little information

Solve if possible. If there is not enough information to solve the problem, tell what other facts you would need. If too much information is given, tell which facts were not needed.

5. Maggie listened to her Lionel Richie tape on her drive to Portland, 215 miles away. Her favorite songs play 2.47, 3.57, and 2.62 minutes. How long would it take to play all three songs?   8.66 minutes. Portland being 215 miles away could be eliminated.

6. It took 7.65 gallons of gasoline to fill Maggie's car before she left on her trip. She filled the tank again before she started home. How many gallons did Maggie use on the trip?   Have to know how much gasoline she added to fill the tank again.

7. Maggie spent $8.64 for gas, $1.75 for oil, and $1.19 for windshield washer fluid. She also bought new windshield wipers. What was her total expense on this trip?   Have to know the price of the windshield wipers.

8. Maggie kept track of how far she traveled each hour. She traveled 43.7, 48.3, 51.5, and 42.6 miles during the first four hours of her trip to Portland. How far did she travel during those four hours?   186.1 miles

On the road again

These directions are part of a travel game. Which city would these travelers be closest to?

9.
   - Start in Baker.
   - Travel 100 miles northwest.
   - Travel 6 miles southeast.
   - Turn left and travel 120 miles.
   - Stop.   The Dalles

10.
   - Start in Lewiston.
   - Go to Colfax and turn left.
   - Travel at the speed limit for two hours.
   - Stop.

Adding Whole Numbers and Decimals **25**

Practice Worksheet
Workbook S180, Copymaster S180, or Duplicating Master S180

NAME _____   **WORKSHEET 12**
(Use after page 25.)

OREGON

Solve, if possible. If there is not enough information to solve the problem, tell what information is needed. Underline any facts not used to solve the problem.

1. The population of Eugene, Oregon is 106,000. The population of Portland, Oregon is 273,003 more than that. What is the population of Portland? __379,003__

2. Medford, Oregon has a population of 41,975. If you round that number to the nearest thousand, you get 42,000. What number do you get if you round Eugene's population of 106,100 to the nearest thousand? __106,000__

3. The population of Oregon in 1980 was 2,633,105, making it the 30th most populous state in the USA. In 1985 the population was estimated to be about 2,687,000. What was the 1980 population rounded to the nearest thousand? __2,633,000__

4. Oregon has a land area of 96,184 square miles, and ranks as the 10th largest state. Alaska, the largest state, is 474,649 square miles larger than Oregon. What is the land area of Alaska? __570,833__

5. Oregon was organized as a territory in 1848. It entered the USA as a state 11 years later. What year did it enter the USA? __1859__

6. Washington state's coastline is 157 miles long. Oregon's coastline is much longer than Washington's, and California's coastline is 840 miles long. How long is Oregon's coastline? __Need the number of miles longer than Washington State's coastline that Oregon's coastline is.__

7. The Three Sisters are a series of three mountain peaks in Oregon. Each of the peaks is higher than 10,000 feet. The shortest peak has an elevation of 10,047 feet. The tallest peak is 311 feet above that. What is the elevation of the tallest Three Sisters peak? __10,358 feet__

8. Mt. McLoughlin, in southern Oregon, has an elevation of 9495 feet. Mt. Hood, with the highest point in the state, is 1740 feet higher than Mt. McLoughlin. How high is Mt. Hood? __11,235 feet__

9. The weather in Portland, Oregon is clear about 75 days each year. It is partly cloudy more days than it is clear. About how many partly cloudy days does Portland have each year? __Need the number of days it is more partly cloudy than it is clear.__

10. The University of Oregon in Eugene has an enrollment of 12,635 students. The enrollment at Oregon State University is 761 more than that. What is the enrollment at Oregon State University? __13,396 students__

© D.C. Heath and Company   S180   *Problem solving—too much/too little information*

Challenge Problem

Write a question that fits the answer.

Kathy borrowed $26. Then she earned $18. Then she borrowed $14 more.

__How much money did she borrow?__

Answer: $40

Copymaster S439

Class Starter Quiz 12
on previous lesson

Use the map on page 24. Solve, if possible. If there is not enough information, tell what information is missing.

1. Todd is traveling from Walla Walla through Pendleton to Baker. How far is he from Baker?
   139 miles
2. Phaedra leaves Burns and drives to Portland. How many miles will she have to travel?
   The mileage between Burns and Pendleton is missing

Copymaster S387

Problem-Solving Skill
Using computer-displayed information

Skills Reviewed
Rounding decimals
Adding decimals
Reading decimals

Starting the Lesson
Problem Solving Have the students look at the display screens to decide whether each of these statements is true or false.

- To make a payment, you push the white key in the bottom row. (True)
- To make a deposit, you push the white key in the top row. (False)
- If you want to make a withdrawal, you must type in an amount that is a multiple of $10.00. (True)

Cumulative Skill Practice Write these four answers on the chalkboard:

18.3   4.53   83.4   0.88

Challenge the students to an answer hunt by saying: "Look at exercises 1–54 on page 27. Find the four exercises that have these answers. You have four minutes to find as many of the exercises as you can." (Exercises 8, 26, 39, and 44)

Problem solving
COMPUTERS IN BANKING

Many banks have computers to operate machines that can be used by customers 24 hours a day. Customers can use these "24-hour tellers" to deposit and withdraw money, make payments, and check account balances. The customer follows directions that appear on the screen. The computer makes the transaction and prints a record of it.

When Justine inserts her card and enters her personal identification number, this message appears on the screen.

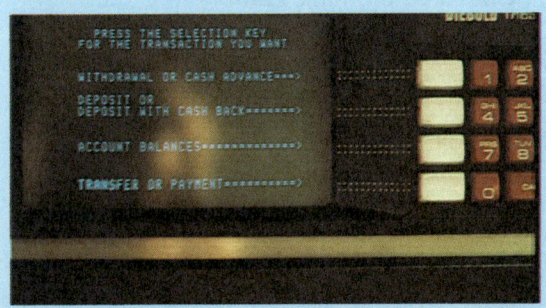

When she wants to make a withdrawal, this message appears.

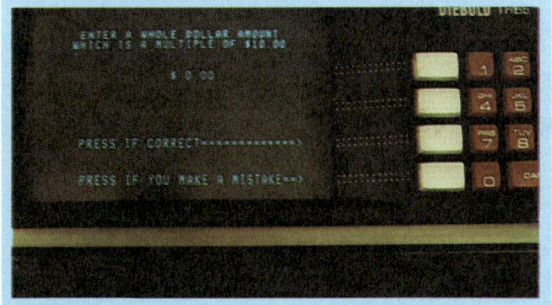

Solve. Decide when a calculator would be useful.

1. If Justine needs $27, she has to withdraw $30. How much would she withdraw if she needs
   a. $18.57? $20   b. $69.50? $70   c. $31.95? $40
   d. $185.99? $190 e. $ 5.92? $10   f. $299.30? $300

2. Justine requests her account balances. They are:
   Savings $695.43
   Checking $381.35
   What is her total balance? $1076.78

3. In October, Justine made these deposits to her checking account:
   October 5 $78.00
   October 14 $89.00
   October 20 $65.00
   October 31 $83.00
   a. What was the amount of her largest deposit in October? $89.00
   b. Did Justine deposit more than $350 in October? No

4. Justine made these withdrawals from her savings account in October:
   October 17 $20.00
   October 23 $60.00
   October 29 $90.00
   a. What was the amount of her smallest withdrawal? $20.00
   b. Did she withdraw less than $200 in October? Yes

Cumulative Skill Practice

Round to the nearest tenth. *(page 18)*

1. 0.42 0.4
2. 8.55 8.6
3. 6.69 6.7
4. 13.42 13.4
5. 50.98 51.0
6. 2.436 2.4
7. 8.761 8.8
8. 18.250 18.3
9. 27.342 27.3
10. 71.062 71.1
11. 3.8214 3.8
12. 6.3500 6.4
13. 9.6175 9.6
14. 14.0924 14.1
15. 26.9341 26.9

Round to the nearest hundredth. *(page 18)*

16. 2.726 2.73
17. 4.634 4.63
18. 4.205 4.21
19. 2.371 2.37
20. 8.082 8.08
21. 1.0314 1.03
22. 7.2936 7.29
23. 5.1171 5.12
24. 9.0853 9.09
25. 6.6152 6.62
26. 4.5347 4.53
27. 6.0821 6.08
28. 3.6349 3.63
29. 8.5981 8.60
30. 2.7436 2.74

Add. *(page 20)*

31. 53.6 + 29.4 = 83.0
32. 7.38 + 2.97 = 10.35
33. 5.33 + 0.46 = 5.79
34. 9.74 + 2.38 = 12.12
35. 5.75 + 5.75 = 11.50
36. 6.03 + 6.03 = 12.06
37. 5.6 + 2.9 + 3.4 = 11.9
38. 6.82 + 3.74 + 2.96 = 13.52
39. 15.4 + 38.3 + 29.7 = 83.4
40. 8.72 + 6.91 + 4.75 = 20.38
41. 5.93 + 8.41 + 1.09 = 15.43
42. 5.91 + 2.33 + 1.09 = 9.33

Give the sum. *(page 22)*

43. 0.05 + 0.8 + 6 6.85
44. 0.1 + 0.73 + 0.05 0.88
45. 15 + 1.5 + 0.15 16.65
46. 5 + 2.3 + 6.4 13.7
47. 5.9 + 2.7 + 5.4 14.0
48. 18 + 6.7 + 3.4 28.1
49. 9.23 + 6.04 + 5.8 21.07
50. 16 + 3.74 + 19 38.74
51. 22.8 + 3.5 + 31 57.3
52. 0.03 + 2 + 2.1 4.13
53. 6.3841 + 2.9871 + 4.0035 13.3747
54. 6.281 + 3.091 + 15.26 24.632

MIXED PRACTICE

Complete.

55. The standard numeral for nine billion, thirty-seven million is __?__ 9,037,000,000
56. 2.0531 rounded to the nearest tenth is __?__ 2.1
57. 6.5409 rounded to the nearest hundredth is __?__ 6.54
58. 2.1 + 3.46 = __?__ 5.56
59. 5.78 + 2.92 = __?__ 8.7
60. 8.14 + 35.6 = __?__ 43.74
61. 7.5 + 18 + 0.6 = __?__ 26.1
62. 50 + 4.7 + 0.184 = __?__ 54.884
63. 7.7 + 6 + 0.92 = __?__ 14.62

Adding Whole Numbers and Decimals **27**

Problem-Solving Worksheet
Workbook S181, Copymaster S181, or Duplicating Master S181

WORKSHEET 13 (Use after page 26.)

HAVE YOU HEARD THIS ONE?
Crack the code to get the answer.

CODE				
A	B	C	...	Z
1+1	2+2	3+3	4+4	26+26

Riddle: What do you get when you cross a parrot with a tiger?

Answer:
I DON'T KNOW
18 8 30 28 40   22 28 30 46
BUT WHEN IT
4 42 40 46 16 10 28 18 40
TALKS, YOU
40 2 24 22 38   50 30 42
HAD BETTER
16 2 8   4 10 40 40 36
LISTEN!
24 18 38 40 10 28

IT ALL ADDS UP
What is the smallest whole number you can add to 637 and get all 5's in the answer?
4918

MISSING DIGITS
Fill in the missing digits.

```
    5 2 7
    2 7 6
  + 4 4 5
  -------
  1 2 4 8
```

MAKE A LIST
You have these coins:

What different amounts could you pay?

1 c	5 c	6 c
10 c	11 c	15 c
16 c	25 c	26 c
30 c	31 c	35 c
36 c	40 c	41 c

HAPPY BIRTHDAY
Jeff's birthday was the day before yesterday. The day after tomorrow is Saturday. On what day was Jeff's birthday?
Tuesday

LOGICAL REASONING
Each of Sally, Susan, Shirley, and Sandra is a different age. Sally is younger than Susan. Shirley is younger than Susan. Only Susan's age is between Sally's and Sandra's.
Who is the oldest? Sandra
Who is the youngest? Shirley

© D. C. Heath and Company   S181   Problem solving

Challenge Problems

Look at exercises 1–15 on page 27. Estimate which two decimals in exercises 1–15 have a sum of about

1. 45.2   18.250 + 26.9341
2. 85.2   14.0924 + 71.062
3. 60.6   9.6175 + 50.98
4. 6.3   2.436 + 3.8214
5. 10.0   0.42 + 9.6175

If you have a calculator, use it to check your answers.

Copymaster S440

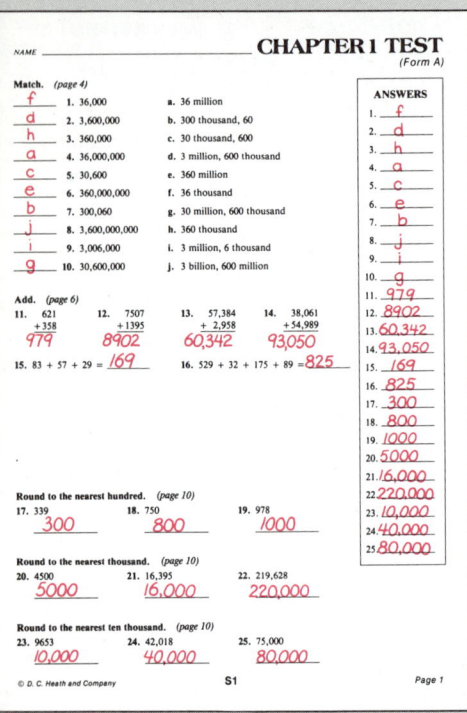

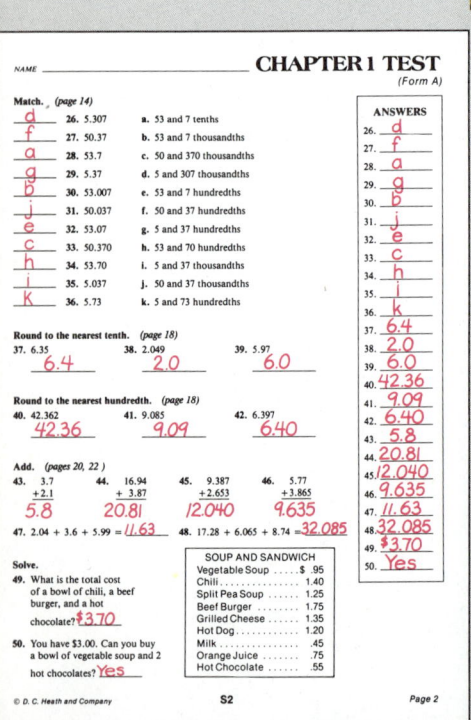

Copymasters S1 and S2
or Duplicating Masters S1 and S2

Chapter REVIEW

Here are scrambled answers for the review exercises:

| 4 | 13 | billion | greater | hundredths | sum | thousand |
| 8 | 80 | decimal | hundreds | million | tenths | thousandths |

1. million, thousand **2.** billion **3.** sum, hundreds

1. The short word-name for this number is 2 [?], 74 [?]. *(page 4)*

2,074,000

2. The long word-name for this number is two [?], six million, fifty-four thousand. *(page 4)*

2,006,054,000

3. The answer to an addition exercise is called the [?]. To complete this addition exercise, add the digits in the [?] place. *(page 6)*

$$\begin{array}{r}\overset{1}{3}28\\+154\\\hline 82\end{array}$$

4. greater, 4 **5.** 80 **6.** thousandths

4. To round this number to the nearest hundred, first look at the digit in the tens place. Since it is [?] than 5, round the hundreds digit up to [?]. *(page 10)*

16,382

5. To estimate this sum, round each amount to the nearest ten dollars. The estimated sum is [?] dollars. *(page 12)*

$54
+32

6. The short word-name for this number is 24 and 86 [?]. *(page 14)*

24.086

7. tenths, 8 **8.** hundredths **9.** 13, decimal

7. To round this number to the nearest whole number, you would first look at the digit in the [?] place. Since it is less than 5, you would round the number to [?]. *(page 18)*

8.247

8. To do this addition exercise, first add the digits in the [?] place. *(page 20)*

5.78
+2.95

9. Round each number to the nearest whole number and estimate the sum. The estimate is [?]. To find the sum, line up the [?] points and add. *(page 22)*

2.94 + 10.2 = ?

28 *Chapter 1*

25. 5 and 9 tenths **26.** 38 and 6 hundredths **27.** 9 and 274 thousandths
28. 63 thousandths **29.** 7 and 865 ten-thousandths **30.** 6 and 803 thousandths

Chapter TEST

Write the short word-name. (page 4)

1. 25,340 — 25 thousand, 340
2. 836,000 — 836 thousand
3. 19,046,000 — 19 million, 46 thousand
4. 25,000,260 — 25 million, 260
5. 6,330,000,000 — 6 billion, 330 million

Write the standard numeral. (page 4)

6. 9 thousand 420 9420
7. 63 million, 75 thousand, 436 63,075,436
8. five hundred nine thousand 509,000
9. sixteen million, eighty-three thousand 16,083,000

Add. (page 6)

10. 52 + 36 = 88
11. 349 + 142 = 491
12. 396 + 158 = 554
13. 6381 + 2974 = 9355
14. 26,352 + 8,968 = 35,320
15. 79,368 + 14,973 = 94,341
16. 74 + 39 + 98 = 211
17. 25 + 82 + 9 = 116
18. 236 + 95 + 381 + 74 = 786

Round to the nearest hundred. (page 10)

19. 634 600
20. 850 900
21. 961 1000
22. 13,439 13,400
23. 42,956 43,000
24. 908 900

Write the short word-name. (page 14)

25. 5.9
26. 38.06
27. 9.274
28. 0.063
29. 7.0865
30. 6.803

Write the standard numeral. (page 14)

31. 26 and 43 hundredths 26.43
32. 8 and 61 thousandths 8.061
33. 60 and 52 ten-thousandths 60.0052

Round to the nearest tenth. (page 18)

34. 6.38 6.4
35. 15.43 15.4
36. 9.75 9.8
37. 8.064 8.1
38. 6.98 7.0
39. 8.87 8.9

Add. (pages 20, 22)

40. 8.4 + 2.3 = 10.7
41. 15.9 + 8.7 = 24.6
42. 6.09 + 2.954 = 9.044
43. 15.936 + 8.742 = 24.678
44. 3.750 + 8.6 = 12.350
45. 6.36 + 1.2 = 7.56
46. 8.3 + 2.9 = 11.2
47. 2.7 + 3.45 = 6.15
48. 6.01 + 8.213 = 14.223
49. 3.06 + 2.784 + 5.39 = 11.234

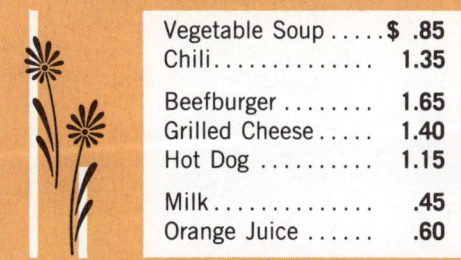

BARBARA'S KITCHEN

Vegetable Soup	$ .85
Chili	1.35
Beefburger	1.65
Grilled Cheese	1.40
Hot Dog	1.15
Milk	.45
Orange Juice	.60

Solve.

50. How much for a beefburger and milk? $2.10
51. What is the total cost of a bowl of chili, a grilled cheese sandwich, and orange juice? $3.35
52. You have $3.00. Can you buy 2 hot dogs and an orange juice? Yes

Adding Whole Numbers and Decimals **29**

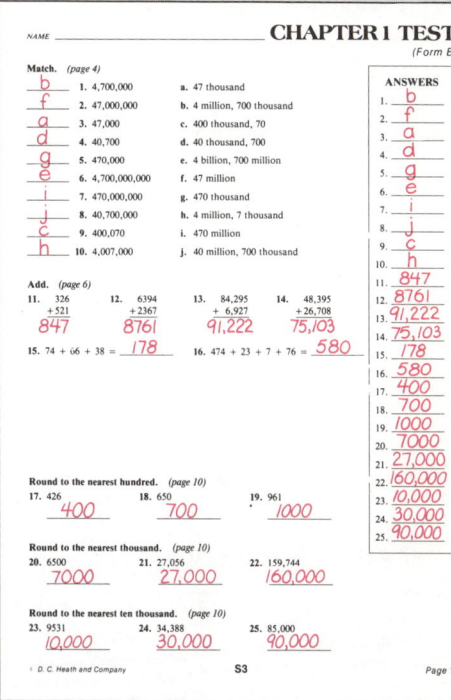

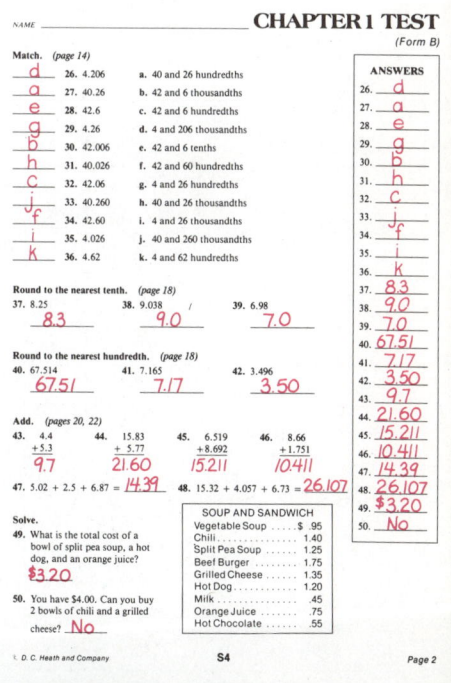

*Copymasters S3 and S4
or Duplicating Masters S3 and S4*

Cumulative Test
(Chapter 1)

Use Copymaster S109 to provide the student with an answer sheet in standardized test format.

Answers for Cumulative Test, Chapter 1

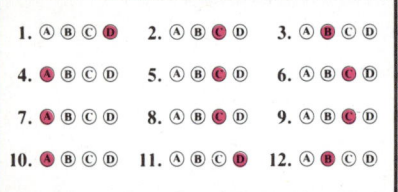

The table below correlates test items with student text pages.

Test Item	Page(s) Taught	Skill Practice
1	4	p. 9, exercises 1–12
2	4	p. 9, exercises 13–23
3	4	p. 9, exercises 13–23
4	6	p. 9, exercises 24–35
5	6	p. 17, exercises 1–16
6	10	p. 17, exercises 17–34
7	14	p. 17, exercises 35–48
8	18	p. 27, exercises 1–15
9	18	p. 27, exercises 16–30
10	20	p. 27, exercises 31–42
11	22	p. 27, exercises 43–54
12	20	

Cumulative TEST — Standardized Format

Choose the correct letter.

1. The short word-name for 5,060,000 is
- A. 5 million, 600 thousand
- B. 5 million, 6 thousand
- C. 5 billion, 60 million
- **D. none of these**

2. The standard numeral for five billion, twenty-five million is
- A. 5,025,000
- B. 5,000,025,000
- **C. 5,025,000,000**
- D. none of these

3. The standard numeral for thirty-two million is
- A. 32,000
- **B. 32,000,000**
- C. 32,000,000,000
- D. none of these

4. Add. 563
         291
    + 87
- **A. 941**
- B. 931
- C. 741
- D. none of these

5. Give the sum.

327 + 84 + 219
- A. 1386
- B. 610
- **C. 630**
- D. none of these

6. 2895 rounded to the nearest ten is
- A. 3000
- B. 2890
- **C. 2900**
- D. none of these

7. The standard numeral for 15 and 34 thousandths is
- **A. 15.034**
- B. 15.34
- C. 15.0034
- D. none of these

8. 36.952 rounded to the nearest tenth is
- A. 36.9
- B. 36.95
- **C. 37.0**
- D. none of these

9. 54.0349 rounded to the nearest hundredth is
- A. 54.035
- B. 54.04
- **C. 54.03**
- D. none of these

10. Add. 3.82
         2.09
    +6.57
- **A. 12.48**
- B. 11.38
- C. 11.48
- D. none of these

11. Give the sum.

36.3 + 8.09 + 5.96
- A. 17.68
- B. 39.25
- C. 49.35
- **D. none of these**

12. What is the total price of 2 adult tickets and 1 child's ticket?

- A. $6.25
- **B. $8.00**
- C. $4.75
- D. none of these

Chapter 1

Subtracting Whole Numbers and Decimals

Resources

- **Class Starter Quizzes 13-21** *(Copymasters S388-S390)*
- **Visual Aids 3, 8-10** *(Copymasters or Transparencies S113, S118-S120)*
- **Manipulatives**
  Manipulative Activities 2 and 3 *(Copymaster S515)*
  Powers-of-ten tiles *(Copymaster or Transparency S529)*
- **Worksheets 14-23** *(Copymasters, Duplicating Masters, or Workbook pages S182-S191)*
- **Challenge Problems** for pages 33, 37, 39, 43, 47, 49 *(Copymasters S440-S441)*
- **Projects** for pages 35, 41, 45, 51 *(Copymasters S474-S475)*
- **Mental Math Extensions** for Skills 5-8 *(Copymasters S491-S492)*
- **Tests** *(Copymasters or Duplicating Masters S5-S8)*

Lesson Objective
To compare whole numbers

Problem-Solving Skills
Reading a chart
Using logical reasoning

Starting the Lesson
What are the facts? Have the students study the two paragraphs on page 32 for 30 seconds and then close their books. Challenge the students to answer these questions:

- What were the names of the two balloons? (*High Rise, Explorer*)
- In what event did the balloons compete? (Cross-country event)
- How far did *High Rise* float? (8206 meters)
- How far did *Explorer* float? (8098 meters)
- Which balloon floated the greater distance? (*High Rise*)

Have the students open their books and look at the *Here's how* to check their answers to the last question.

Here's How *Note*
Remind the students that the smaller end of the symbols < and > always "points to" the smaller number.

Comparing whole numbers

Hot-air balloonists compete in several events. In the cross-country event, the balloon that floats farthest during a specified time wins the event.

High Rise floated 8206 meters, and *Explorer* floated 8098 meters. To determine the winner, the balloonists compared the two numbers.

Here's how to compare whole numbers.   8206 ● 8098

To compare two whole numbers, start at the left and compare the digits that are in the same place.

Step 1.

8206 ● 8098

same

Step 2.

8206 ● 8098

2 is greater than 0.

Step 3.

So, 8206 > 8098.
Read > as "is greater than."

1. Look at the *Here's how*. Which balloon won (floated the greater distance)?   High Rise

Study these examples.

Example A.

2 is less than 5.

So, 3628 < 3659.
Read < as "is less than."

Example B.

ten thousands place

No digit here!

53,402 > 6348

2. Look at Example B. Are the digits 5 and 6 in the same place?   No

EXERCISES

Less than (<) or greater than (>)?

3. 783 < 784
4. 593 > 590
5. 856 < 1200
6. 1342 > 819
7. 3621 > 3514
8. 5834 > 5741
9. 9834 < 9843
10. 6519 > 6514
11. 68,352 < 68,411
12. 39,436 > 39,400
13. 88,361 < 89,000
14. 29,361 < 30,362
15. 86,000 > 85,999
16. 74,399 > 74,000
17. 634,298 > 624,298
18. 714,362 < 714,459
19. 597,821 < 609,375
20. 560,000 > 559,000
21. 900,000 > 879,694
22. 89,999 < 900,000

Solve.

23. The results of the spot-landing event are shown in the table. The winner is the balloon that lands closest to a certain spot.
    a. Which balloon came in first (landed closest to the spot)? *Free Spirit*
    b. Which balloon came in last? *Easy Floater*
    c. List the balloons in order of finish, from first to last. *Free Spirit, High Flier, America, Up and Away, Big Apple, Easy Floater*

SPOT-LANDING EVENT

NAME OF BALLOON	METERS LANDED FROM SPOT
America	834
Big Apple	929
Easy Floater	938
Free Spirit	763
High Flier	771
Up and Away	840

24. Remember that the winner of the cross-country event is the balloon that floats the farthest.
    a. Which balloon came in first? *Up and Away*
    b. Which balloon came in last? *Big Apple*
    c. List the balloons in order of finish, from first to last. *Up and Away, America, Easy Floater, Free Spirit, High Flier, Big Apple*

CROSS-COUNTRY EVENT

NAME OF BALLOON	METERS TRAVELED
America	12,642
Big Apple	10,837
Easy Floater	11,134
Free Spirit	11,099
High Flier	10,909
Up and Away	12,800

Test pilots wanted
Logical reasoning

25. The first "test pilots" on a hot-air balloon were a duck, a rooster, and a sheep. The historic flight took place in the year ⬚. Study these clues to find the year. *1783*

    Clues:
    • If you round the year to the nearest ten, you get 1780.
    • If you add the digits of the year, you get 19.

Subtracting Whole Numbers and Decimals **33**

Extra Practice
Page 473 Skill 5

Practice Worksheet
Workbook S182, Copymaster S182, or Duplicating Master S182

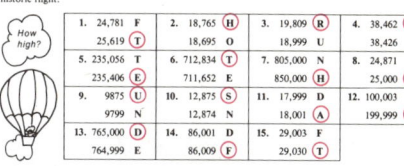

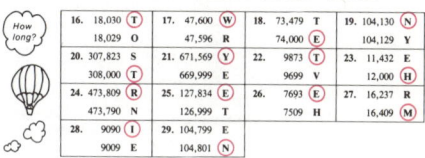

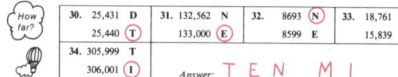

Challenge Problems

1. How many different 4-digit numbers can you build using these four digits?

   [8] [3] [6] [4]   *24*

2. How many of the numbers are greater than 5000? *12*

Copymaster S440

Class Starter Quiz 13
on previous lesson

< or >?

1. 692 ● 690   >
2. 4631 ● 4524   >
3. 63,299 ● 63,000   >
4. 8934 ● 8943   <
5. 79,999 ● 700,000   <

Copymaster S388

Lesson Objective
To subtract 2- and 3-digit numbers

Problem-Solving Skills
Reading a chart
Choosing the correct operation

Starting the Lesson
Take a Survey Write these listening times on the chalkboard:

Less than 1 hour
1 hour
2 hours
3 hours
4 hours
More than 4 hours

Before the students open their books, take a class survey. Ask the students to guess how long most people listen to the radio each day. Record their guesses on the chalkboard. Have the students compare their results with the data in the chart on page 34.

Exercise Note
Problem Solving Many students may assume that problems 41–44 will all require subtraction. Remind the students to read each problem carefully before solving it.

Subtracting 2- and 3-digit numbers

WHAT WOULD YOU SAY?

Twelve hundred people were asked how long they listened to the radio each day. Here are the results of the survey:

LISTENING TIME	NUMBER OF PEOPLE
Less than 1 hour	240
1 hour	456
2 hours	252
3 hours	96
4 hours	84
More than 4 hours	72

1. How many people listened 2 hours a day? 252
2. How many listened 3 hours? 96
3. Would you add or subtract to find how many more people listened 2 hours than listened 3 hours? Subtract

Here's how to subtract whole numbers.   252 − 96 = ?

Line up the digits that are in the same place.	Regroup. Subtract ones.	Regroup. Subtract tens.	Subtract hundreds.
252 − 96	2⁴5̸2 − 96 —— 6	¹2̸⁴5̸2 − 96 —— 56	¹2̸⁴5̸2 − 96 —— 156

The answer is called the **difference**.

4. Look at the *Here's how*. How many more people listened 2 hours than listened 3 hours? 156

EXERCISES

Subtract. Here are scrambled answers for the next row of exercises: 26 13 15 21 27 54

5. 56 −35 21	6. 68 −42 26	7. 42 −15 27	8. 73 −58 15	9. 90 −77 13	10. 81 −27 54
11. 256 − 28 228	12. 341 − 50 291	13. 722 − 65 657	14. 429 − 84 345	15. 536 − 98 438	16. 828 − 19 809
17. 429 −116 313	18. 638 −229 409	19. 514 −152 362	20. 923 −347 576	21. 752 −294 458	22. 541 −329 212

23. 93 − 21 72   24. 53 − 17 36   25. 598 − 375 223   26. 547 − 229 318

27. 243 − 30 213   28. 351 − 26 325   29. 633 − 59 574   30. 835 − 76 759

MENTAL MATH Subtract. Write answers only.

31. 120 − 80   32. 110 − 20 90   33. 70 − 50 20   34. 90 − 30 60

12 tens minus 8 tens equals 4 tens, or 40

35. 150 − 60 90   36. 120 − 60 60   37. 80 − 40 40

38. 800 − 400 400   39. 700 − 100 600   40. 900 − 500 400

Solve. Use the survey information on page 34.

41. How many more people listened 1 hour than listened 2 hours? 204

42. How many more people listened 1 hour than listened less than 1 hour? 216

43. How many people listened 1 hour or less than 1 hour? 696

44. How many people listened 3 hours or more? 252

You're a program director — Reading a schedule

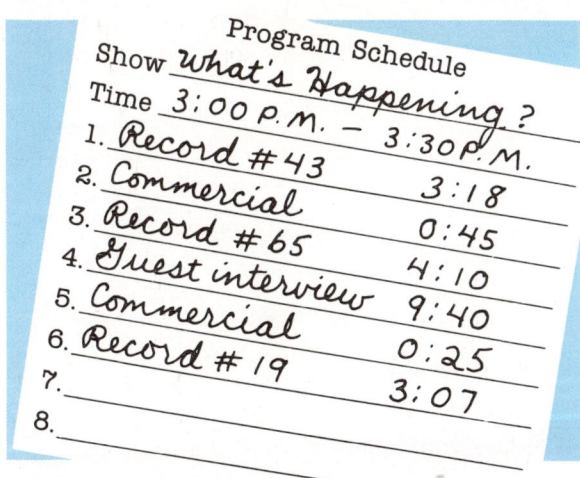

45. What is the name of the show? *What's Happening?*

46. How many minutes long is the show? 30

47. The schedule shows that the first record will take 3 minutes and 18 seconds to play. Study the schedule. How much more time must you fill to complete the show? *Hint: 1 minute = 60 seconds.* 8 minutes 35 seconds

Practice Worksheet
Workbook S183, Copymaster S183, or Duplicating Master S183

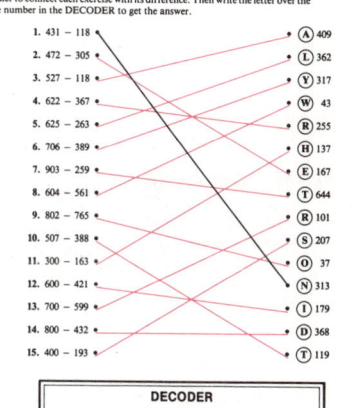

Project

Using library resources

Use a book of world records to find the number of hours of the longest continuous radio broadcast.

Answers will vary. Records may change from year to year.

Copymaster S474

Class Starter Quiz 14
on previous lesson

Subtract.

1. 674 − 223		451
2. 261 − 170		91
3. 832 − 61		771
4. 457 − 289		168
5. 623 − 85		538

Copymaster S388

Lesson Objective
To estimate differences

Problem-Solving Skills
Reading a chart
Using logical reasoning

Starting the Lesson
Take a Survey Before students open their books, ask them what sport each of these Hall of Fame players played.

Hank Aaron (Baseball)
Jim Brown (Football)
Bill Russell (Basketball)
Frank Gifford (Football)

Then say: "Look at the chart on page 36. Which Hall of Fame football player gained more yards rushing, Jim Brown or Frank Gifford?" (Jim Brown) Then use the *Here's how* to introduce estimating differences.

Estimating differences

Total Yards Gained		
NFL Hall of Fame Player	Yards Rushing	Yards Pass-Receiving
Jim Brown	12,312	2,499
Frank Gifford	3,069	5,434
Lenny Moore	5,174	6,039
Gale Sayers	4,957	1,309
Jim Taylor	8,597	225

1. Which Hall of Fame player gained about 3000 yards rushing? **Frank Gifford**
2. Which two numbers would you round to estimate how many more yards Lenny Moore gained rushing than Frank Gifford? **5174, 3069**

Here's how *to estimate differences*.

Round to the nearest thousand yards.

Randy

5174 →   5000
3069 → − 3000
         2000

Round to the nearest hundred yards.

Kathy

5174 →   5200
3069 → − 3100
         2100

3. Look at the *Here's how*.
   a. To estimate the difference, Randy rounded each number to the nearest ⬜. **thousand**
   b. Kathy rounded each number to the nearest ⬜. **hundred**
   c. Whose estimate was closer to the exact answer, Randy or Kathy? **Kathy**
4. Use Randy's method. How many more yards did Jim Brown gain rushing than Frank Gifford? **9000**
5. Use Kathy's method. Which player had about 4100 less yards pass-receiving than Frank Gifford? **Gale Sayers**

EXERCISES

Which estimate would Randy give?
Hint: Study the Here's how.

6. 6217 − 2356　　　a. 2000　　　(b.) 4000　　　c. 6000
7. 5807 − 1095　　　a. 1000　　　b. 3000　　　(c.) 5000
8. 13,298 − 2385　　a. 9000　　　(b.) 11,000　　c. 13,000

Which estimate would Kathy give?

9. 3821 − 2115　　　(a.) 1700　　　b. 1900　　　c. 2100
10. 8089 − 5093　　　a. 2800　　　(b.) 3000　　　c. 3200
11. 1735 − 615　　　a. 700　　　b. 900　　　(c.) 1100

Estimate each difference by rounding to the nearest thousand.

12.　6321　　　13.　7852　　　14.　4213　　　15.　8875　　　16.　7942
　　−1852　　　　　−1705　　　　　−2851　　　　　−6021　　　　　−4199
　　 4000 (6000/−2000)　6000　　　　 1000　　　　 3000　　　　 4000

17. 7102 − 5216　　18. 5816 − 3744　　19. 9026 − 7815　　20. 12,182 − 8216
　　　2000　　　　　　　2000　　　　　　　1000　　　　　　　　4000

Estimate each difference by rounding to the nearest hundred.

21.　7528　　　22.　4706　　　23.　8917　　　24.　1438　　　25.　2872
　　−1285　　　　　−1517　　　　　−2685　　　　　− 345　　　　　− 689
　　 6200 (7500/−1300)　3200　　　　 6200　　　　 1100　　　　 2200

26. 821 − 635　　27. 892 − 131　　28. 1106 − 812　　29. 1785 − 692
　　　200　　　　　　　800　　　　　　　300　　　　　　　　1100

Estimate. Use the chart on page 36.

30. Which player gained about 7000 less yards rushing than Jim Brown? Gale Sayers

31. Which player had about 3000 less yards pass-receiving than Lenny Moore? Jim Brown

32. Who had about 10,000 more yards rushing than pass-receiving? Jim Brown

33. Who had a total rushing and pass-receiving yardage of about 11,000 yards? Lenny Moore

Who wore this helmet?

Logical reasoning

34. Study the clues. Use the chart on page 36 to name the player.

    Clues:
    - This player gained more than 5000 yards pass-receiving.
    - This player gained less than 5000 yards rushing. Frank Gifford

Subtracting Whole Numbers and Decimals

Practice Worksheet
Workbook S184, Copymaster S184, or Duplicating Master S184

WORKSHEET 16
(Use after page 37.)

NAME _____

SKILL DRILL
Estimate each difference by rounding to the nearest thousand.

1. 3269 → 3000　2. 7369 → 7000　3. 7785 → 8000　4. 5068 → 5000
　− 985 → −1000　−2483 → −2000　−5012 → −5000　− 896 → −1000
　　　　　2000　　　　　5000　　　　　3000　　　　　4000

5. 8176 → 8000　6. 8768 → 9000　7. 12,265 → 12,000　8. 15,897 → 16,000
　−7268 → −7000　−2809 → −3000　− 4,995 → −5,000　−7,953 → −8,000
　　　　　1000　　　　　6000　　　　　　7,000　　　　　　8,000

Estimate each difference by rounding to the nearest hundred.

9. 1288 → 1300　10. 2647 → 2600　11. 638 → 600　12. 3775 → 3800
　− 413 → − 400　− 878 → − 900　−241 → −200　− 628 → − 600
　　　　　900　　　　　1700　　　　　400　　　　　3200

13. 926 → 900　14. 3267 → 3300　15. 3206 → 3200　16. 2785 → 2800
　− 633 → −600　−2183 → −2200　−1812 → −1800　− 692 → − 700
　　　　300　　　　　1100　　　　　1400　　　　　2100

17. 415 − 326　 100　　18. 828 − 93　 700　　19. 3897 − 1427　 2500

20. 6708 − 1537　 5200　　21. 936 − 85　 800　　22. 5872 − 1698　 4200

23. 7182 − 995　 6200　　24. 682 − 88　 600　　25. 8942 − 4199　 4700

Check yourself. Here are the scrambled answers:

100　300　400　600　700　800　900　1000　1100　1400
1700　2000　2100　2500　3000　3200　4000　4200　4700　5000
5200　6000　6200　7000　8000

© D.C. Heath and Company　　S184　　Estimating differences

Challenge Problem

Copy and complete this Magic Square so that the sums of the numbers along each row, column, and diagonal are the same.

23	28	21
22	24	26
27	20	25

Copymaster S440

Class Starter Quiz 15
on previous lesson

Match. Find the difference by estimating.

1. 7315 − 2174   d   a. 6981
2. 9827 − 7796   b   b. 2031
3. 8269 − 1388   a   c. 2951
4. 7725 − 6584   e   d. 5141
5. 6039 − 3088   c   e. 1141

Copymaster S388

Lesson Objectives
To subtract 3-, 4-, 5-, and 6-digit numbers
To subtract across zeros

Problem-Solving Skills
Choosing the correct operation
Using a guess-and-check strategy

Starting the Lesson
Write these inventions and dates on the chalkboard:

Phonograph	1877
Radio	1894
Telephone	1875

Before the students open their books, ask them to try matching each item with its date of invention. Then have them check their guesses with the information at the top of the page.

Here's How *Note*
Calculator You may wish to show students a calculator shortcut for checking the example in the *Here's how*.

Step 1. Subtract.
1901 ⊖ 1875 ⊜ ⟨26⟩

Step 2. To check, add 1875.
⊕ 1875 ⊜ ⟨1901⟩
↑ The last display should always be the starting number.

38

Subtracting larger numbers

Here are some things that you probably use each day. Notice that each was invented before the beginning of the 20th century (the year 1901).

1875 Telephone

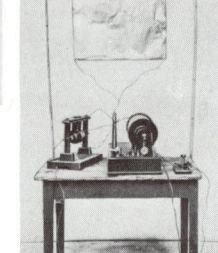
1894 Radio

1877 Phonograph

INVENTIONS

1. Which was invented first? second? third? Telephone, phonograph, radio
2. Was the telephone invented before or after the beginning of the 20th century? Before
3. Would you add or subtract to find how many years before the beginning of the 20th century the telephone was invented? Subtract

Here's how to subtract whole numbers.   1901 − 1875 = ?

First round each number to the nearest ten and then estimate the difference.

1901 − 1875 = ?
↓ ↓
1900 − 1880 = 20

Line up the digits that are in the same place.	No tens! Regroup 1 hundred for 10 tens.	Regroup 1 ten for 10 ones.	Subtract.	Check by adding.
1901 −1875	1⁸9̸01 −1875	1⁸9̸⁹0̸1 −1875	1⁸9̸⁹0̸1 −1875 26	26 +1875 1901

4. Look at the *Here's how*. Was 20 years a reasonable estimate? Yes
5. How many years before the beginning of the 20th century was the telephone invented? 26

EXERCISES

6. Three of the calculator answers are wrong. Find them by estimating.

a. 281 − 259   22
b. 329 − 197   132
c. 578 − 383   295 (circled)
d. 600 − 511   89
e. 902 − 750   252 (circled)
f. 800 − 311   489
g. 1249 − 993   256
h. 1680 − 1275   305 (circled)
i. 2000 − 1309   691

First estimate the difference. Then subtract.

7. 802 −238 564	8. 305 −157 148	9. 901 −396 505	10. 704 −429 275	11. 806 −638 168	12. 861 −249 612
13. 1883 − 351 1532	14. 1980 − 635 1345	15. 1704 −1216 488	16. 4700 −1453 3247	17. 1603 − 496 1107	18. 3302 − 861 2441
19. $35.00 − 13.94 $21.06	20. $29.00 − 16.82 $12.18	21. $23.02 − 13.44 $9.58	22. $42.00 − 29.37 $12.63	23. $96.00 − 47.01 $48.99	24. $98.02 − 12.99 $85.03

Subtract. Then check by addition.

25. 19,354 − 16,258 = 3096
26. 36,093 − 24,720 = 11,373
27. 68,391 − 26,493 = 41,898
28. 94,003 − 62,875 = 31,128
29. 36,902 − 12,118 = 24,784

30. 2761 − 1325 = 1436
31. 8834 − 4609 = 4225
32. 9000 − 1492 = 7508
33. 6280 − 750 = 5530
34. 7266 − 914 = 6352
35. 8000 − 625 = 7375
36. 59,061 − 12,652 = 46,409
37. 903,000 − 26,200 = 876,800
38. 608,010 − 52,325 = 555,685

Solve. Use the dates on page 38.

39. How many years after the invention of the telephone was the radio invented? 19

40. How many years before the beginning of the 20th century was the phonograph invented? 24

41. The telescope was invented 288 years before the radio. What year was that? 1606

42. The pendulum clock was invented 221 years before the telephone. What year was that? 1654

43. The airplane was invented 26 years after the phonograph. What year was that? 1903

44. The thermometer was invented 7 years before the beginning of the 17th century. What year was that? 1594

Key it in!

Guess and check.

Find a way to push each marked key once to get the answer.

45. 87 − 51 =  36

46. 93 − 51 =  42

47. 54 − 36 = or 63 − 45 = 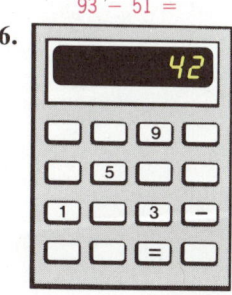 18

Subtracting Whole Numbers and Decimals **39**

Extra Practice
Page 473 Skill 6

Practice Worksheet
Workbook S185, Copymaster S185, or Duplicating Master S185

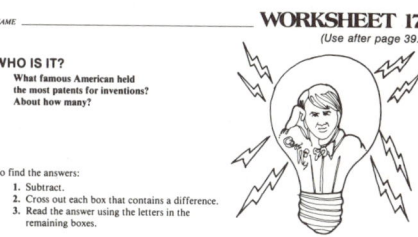

Challenge Problems

Find the missing digits.

1. 7 [8]
   − [2] 5
      5 3

2. 3 [6] 4
   − 2 5 [3]
      [1] 1 1

3. [8] 6 4
   − 6 3 [5]
      2 [2] 9

Copymaster S440

39

Class Starter Quiz 16
on previous lesson

Subtract.
1. 6272 − 4801   1471
2. 4937 − 2049   2888
3. 8010 − 2840   5170
4. 36,721 − 8838   27,883
5. 254,190 − 53,664   200,526

Copymaster S388

Problem-Solving Skill
Reading a newspaper ad

Skills Reviewed
Reading and writing standard numerals
Adding whole numbers and decimals
Rounding whole numbers and decimals

Starting the Lesson
Problem Solving The ad at the top of page 40 is also on ■ *Visual Aid 8* (copymaster or transparency S118). Use the ad and ask questions like these:

- What is the price of the '81 Cutlass? ($5895)
- Is $4500 enough money to buy the '79 Malibu? (No)
- How many miles has the '79 Mustang been driven? (25,000)

Cumulative Skill Practice Write these five answers on the chalkboard:

63 thousand, 201
17,063,090
12,823
700
7.4

Challenge the students to an answer hunt by saying, "Look at exercises 1–50 on page 41. Find the five exercises that have these answers. You have five minutes to find as many of the exercises as you can." (Exercises 4, 20, 23, 29, and 47)

Answers for page 41.
1. 46 thousand, 220
2. 9 thousand, 337
3. 11 thousand, 835
4. 63 thousand, 201
5. 186 thousand, 743

40

Problem solving
Buying a car

You have circled the ads for seven used cars you'd like to look at.

Solve. Decide when a calculator would be useful.

1. Which car costs the most?
   81 Cutlass Supreme
2. Which car costs the least?
   1978 Chevette
3. Which car is the newest model?
   81 Cutlass Supreme
4. Which car is the oldest?
   77 Olds Cutlass
5. Which car has been driven the most miles? *79 Chevette*
6. Which car has been driven the fewest miles? *1980 Chevy Monza*
7. How many more miles has the 1977 Cutlass been driven than the 1981 Cutlass? *29,689*
8. What is the difference in price between the two Cutlass cars? *$2100*
9. What is the difference in price between the cheapest and the most expensive 1979 model? *$1195*
10. You have $1835 in one savings account and $979 in another. Do you have enough money to pay cash for the cheapest car? *Yes*
11. You have $2814. How much will you need to borrow to buy the 1979 Mustang? *$1381*
12. Village Auto Sales guarantees its cars for 7500 miles after the purchase. How many miles will the Cutlass have when the guarantee expires? *24,811*

40 Chapter 2

Cumulative Skill Practice

Write the short word-name. *(page 4)*

1. 46220
2. 9337
3. 11835
4. 63201
5. 186743
6. 237105
7. 864319
8. 145623
9. 6723804
10. 5657129
11. 29000673
12. 1000235

Write the standard numeral. *(page 4)*

13. 246 thousand, 659
   246,659
14. 34 million
   34,000,000
15. 57 billion
   57,000,000,000
16. 28 million, 213 thousand
   28,213,000
17. 5 million, 18 thousand
   5,018,000
18. 4 million, 24 thousand, 73
   4,024,073
19. sixty-two thousand, two hundred fifty
   62,250
20. seventeen million, sixty-three thousand, ninety
   17,063,090

Add. *(page 6)*

21. 328 + 296 + 54 = 678
22. 927 + 58 + 629 = 1614
23. 3942 + 567 + 8314 = 12,823
24. 2715 + 6130 + 852 = 9697
25. 593 + 2577 + 4255 = 7425
26. 623 + 1586 + 29 = 2238

Round to the nearest hundred. *(page 10)*

27. 567 — 600
28. 824 — 800
29. 650 — 700
30. 471 — 500
31. 35 — 0
32. 98 — 100
33. 3607 — 3600
34. 6532 — 6500
35. 8880 — 8900
36. 9050 — 9100
37. 3982 — 4000
38. 1924 — 1900

Round to the nearest tenth. *(page 18)*

39. 3.53 — 3.5
40. 24.305 — 24.3
41. 9.67 — 9.7
42. 0.884 — 0.9
43. 1.750 — 1.8
44. 3.215 — 3.2
45. 52.04 — 52.0
46. 8.96 — 9.0
47. 7.390 — 7.4
48. 0.064 — 0.1
49. 27.95 — 28.0
50. 9.013 — 9.0

MIXED PRACTICE

Complete.

51. The standard numeral for five hundred forty-seven thousand, three hundred ninety-five is ___?___ 547,395
52. The standard numeral for thirty-one million, two hundred eighty-six is ___?___ 31,000,286
53. The standard numeral for sixteen billion, forty-nine million, seven thousand is ___?___ 16,049,007,000
54. 8.45 + 6.9 = ___?___ 15.35
55. 3.08 + 17 = ___?___ 20.08
56. 6.83 + 0.9 = ___?___ 7.73
57. 0.06 + 3 + 2.75 = ___?___ 5.81
58. 5.083 + 16 + 9.4 = ___?___ 30.483
59. 0.37 + 2.063 + 7 = ___?___ 9.433

Subtracting Whole Numbers and Decimals **41**

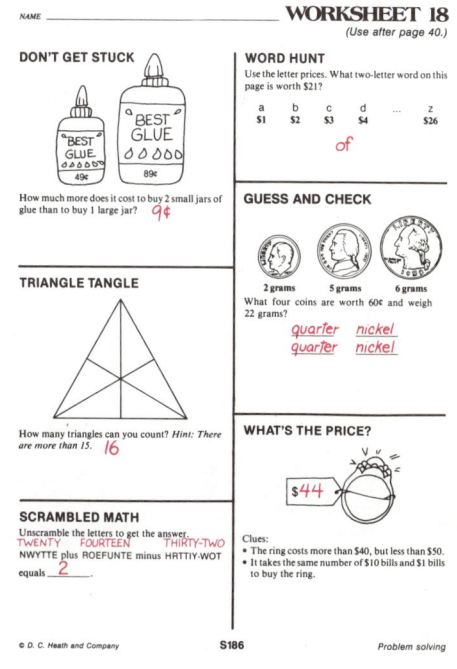

More answers for page 41.
6. 237 thousand, 105
7. 864 thousand, 319
8. 145 thousand, 623
9. 6 million, 723 thousand, 804
10. 5 million, 657 thousand, 129
11. 29 million, 673
12. 1 million, 235

Class Starter Quiz 17
on previous lesson

Solve. Use the used-car ad on page 40.

1. You have $1680 in one savings account and $750 in another. Do you have enough money to pay cash for the 1979 Chevette? No
2. You have $3450. How much will you need to borrow to buy the 1980 Chevy Monza? $1570

Copymaster S389

Lesson Objective
To compare decimals

Problem-Solving Skills
Reading a chart
Visual thinking

Starting the Lesson
Ask the students if they can identify the four common objects in the closeup photos at the top of the page. Have them write their answers. Challenge them to identify all four photos in less than 60 seconds. (Rubber bands, strawberry, crayon tip, broom)

Here's How Note
Use of Concrete Materials You may wish to use the powers-of-ten tiles from copymaster or transparency S529 to demonstrate comparing decimals. See
■ **Manipulative Activity 2** on copymaster S515 in the Teacher's Resource Binder.

Exercise Note
Some students may find it helpful to annex zeros when comparing some of the numbers in exercises 11–41.

42

Comparing decimals

Can you recognize the four common objects in these close-up photos?

Stretchers — Rubber bands

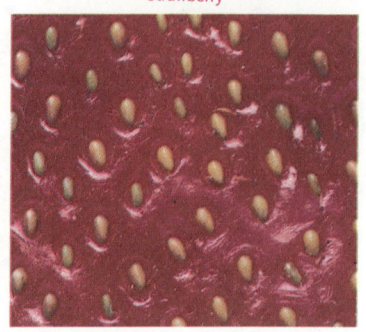
...fields forever — Strawberry

CLOSE-UP CONTEST

Draw, Pardner! — Crayon

Solve this for a clean sweep — Broom

CONTEST RESULTS
(Time needed to identify all four photos)

NAME	SECONDS
Peter	58.29
Tony	54.65
Stan	55.32
Janice	57.68
Rick	53.48
Gene	52.34
Linda	53.67
Donna	51.75

1. Who recognized all four close-up photos in 53.67 seconds? Linda
2. Who had a time of 53.48 seconds? Rick
3. Which two numbers would you compare to decide whether Rick or Linda had the better (shorter) time? 53.48 and 53.67

Here's how to compare decimals.   53.48 ● 53.67

Start at the left and compare digits that are in the same place.

Step 1. 5**3**.48 ● 5**3**.67   **Step 2.** 53.**4**8 ● 53.**6**7   **Step 3.** 53.48 < 53.67
    same   4 is less than 6.

4. Look at the *Here's how*. Who had the better time, Linda or Rick? Rick
5. Check each example. Have the decimals been compared correctly?

    a. 5**2**.34 ● 5**1**.68   2 is greater than 1.   b. 54.6**0** ● 54.6**5**   It helps to fill in a zero.
        52.34 > 51.68   Yes   54.6 < 54.65   Yes

42   Chapter 2

EXERCISES

Less than (<) or greater than (>)?

6. 0.4 < 0.5
7. 0.07 > 0.06
8. 0.009 > 0.008
9. 14.3 > 14.1
10. 6.75 > 6.57
11. 0.27 > 0.2
12. 0.005 < 0.03
13. 0.1 > 0.02
14. 8.23 < 8.32
15. 31.69 < 31.7
16. 2.1 > 1.98
17. 5.352 < 53.52
18. 0.725 < 1.1
19. 1.07 > 1.007
20. 0.815 < 0.82
21. 33.86 < 33.87
22. 0.34 < 0.43
23. 6.215 < 62.16
24. 23.78 > 21.88
25. 1.1 > 1.08
26. 782.1 < 783
27. 18.02 > 18.003
28. 53.06 < 53.2
29. 0.333 > 0.332
30. 3.504 < 3.54
31. 52.8 > 8.29
32. 0.021 < 0.12
33. 6.72 < 6.75
34. 6.153 < 6.2
35. 2.61 > 2.58
36. 0.04 > 0.006
37. 7.017 > 7.005
38. 3.53 < 3.55
39. 13.7 > 13.69
40. 38.06 < 38.7
41. 0.914 < 0.92

Solve. Use the chart on page 42.

42. Which girl recognized the four close-up photos in less than 52 seconds? **Donna**

43. Which boys took more than 55 seconds? **Peter, Stan**

44. Who had the shorter time, Tony or Gene? **Gene**

45. Who had the longer time, Rick or Janice? **Janice**

46. Who had the better combined time, Tony and Linda or Janice and Gene? **Tony and Linda**

47. a. List the times in order from best (shortest) to worst (longest).
    b. Who came in first? **Donna**
    c. Who came in last? **Peter**

a. 51.75, 52.34, 53.48, 53.67, 54.65, 55.32, 57.68, 58.29

More close-ups *Visual thinking*
Use the word clues to name these close-ups.

48.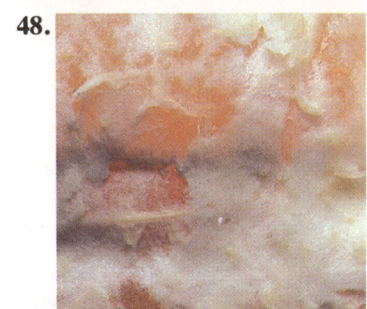
Lost appeal
Orange peel

49.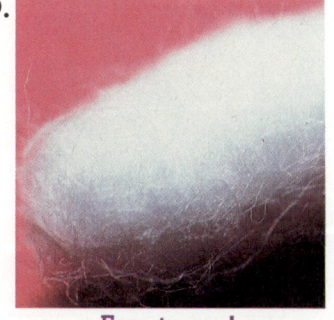
Ears to you!
Cotton swab

50.
Rubber soul
Sole of running shoe

Subtracting Whole Numbers and Decimals

Extra Practice
Page 474   Skill 7

Practice Worksheet
Workbook S187, Copymaster S187, or Duplicating Master S187

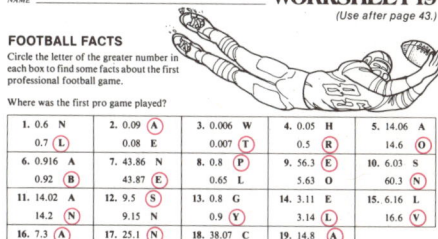

Challenge Problem

Write the numbers 1, 2, 3, 4, 5, 6, 7, and 8 in the circles so that no two consecutive numbers are connected by a line. (Consecutive numbers have a difference of 1.)

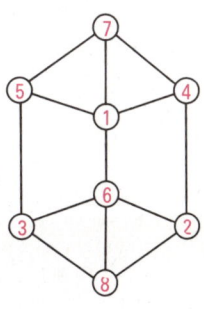

Answers may vary.

Copymaster S441

Class Starter Quiz 18
on previous lesson

< or >?
1. 0.6 ● 0.7   <
2. 0.35 ● 0.3   >
3. 1.05 ● 1.005 >
4. 2.302 ● 2.32 <
5. 36.08 ● 36.6 <

Copymaster S389

Lesson Objective
To subtract decimals (with the same number of decimal places)

Problem-Solving Skills
Reading a chart
Choosing the correct operation
Using logical reasoning

Starting the Lesson
Write these dates and market values on the chalkboard:

Date	Market Value
1900 nickel	$2.90
1920 nickel	$1.75
1940 nickel	$1.25
1950 nickel	$0.20

Before the students open their books, have them try matching each of the nickels to its market value. Tell them to check their guesses with the chart at the top of page 44.

Here's How Note
Use of Concrete Materials You may wish to use the powers-of-ten tiles from copymaster or transparency S529 to demonstrate subtracting decimals. See ■ **Manipulative Activity 3** on copymaster S515 in the Teacher's Resource Binder.

44

Subtracting decimals

1890   1900   1910

1920   1930   1940   1950

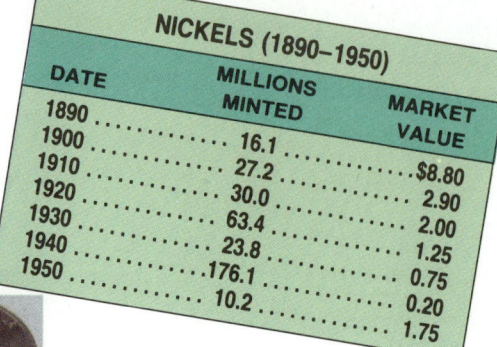

NICKELS (1890–1950)

DATE	MILLIONS MINTED	MARKET VALUE
1890	16.1	$8.80
1900	27.2	2.90
1910	30.0	2.00
1920	63.4	1.25
1930	23.8	0.75
1940	176.1	0.20
1950	10.2	1.75

1. A coin dealer will pay you $8.80 for one of these nickels. Which nickel is it? **1890**

2. Which nickel has a market value of $1.25? **1920**

3. How many million nickels were minted in 1920? in 1930? **63.4, 23.8**

4. Would you add or subtract to find how many more millions of nickels were minted in 1920 than in 1930? **Subtract**

Here's how to subtract decimals. $63.4 - 23.8 = ?$

Line up the decimal points.	Regroup and subtract.	Regroup and subtract.
63.4 −23.8	² 6̸3.4 −23.8 .6	5 ¹² 6̸3̸.4 −23.8 39.6

5. Look at the *Here's how*. How many more millions of nickels were minted in 1920 than in 1930? **39.6**

EXERCISES
Subtract.
Here are scrambled answers for the next row of exercises: 6.54 7.15 26.4 21.7 34.7 37.1

| 6. 49.6
−12.5
37.1 | 7. 7.42
−0.27
7.15 | 8. 42.6
−7.9
34.7 | 9. 9.18
−2.64
6.54 | 10. 75.3
−48.9
26.4 | 11. 43.5
−21.8
21.7 |

44 Chapter 2

12. 0.31 − 0.18 = 0.13
13. 80.2 − 38.9 = 41.3
14. 6.19 − 2.74 = 3.45
15. 5.91 − 0.26 = 5.65
16. 72.9 − 8.7 = 64.2
17. 8.42 − 3.54 = 4.88
18. 0.74 − 0.68 = 0.06
19. 646.5 − 89.3 = 557.2
20. 6.75 − 2.41 = 4.34
21. 87.3 − 1.6 = 85.7
22. 93.71 − 8.41 = 85.30
23. 104.5 − 25.6 = 78.9
24. $52.40 − 12.75 = $39.65
25. $7.29 − 5.36 = $1.93
26. $7.38 − .88 = $6.50
27. $110.60 − 85.20 = $25.40
28. $47.35 − 39.88 = $7.47
29. $87.65 − 4.82 = $82.83

30. 70.2 − 6.1 64.1
31. 16.3 − 5.7 10.6
32. 80.06 − 5.14 74.92
33. 55.06 − 2.14 52.92
34. 59.4 − 23.7 35.7
35. 23.16 − 4.10 19.06
36. 75.4 − 44.6 30.8
37. 35.2 − 8.4 26.8
38. 60.49 − 33.71 26.78
39. 63.3 − 9.6 53.7
40. 70.02 − 4.16 65.86
41. 63.14 − 7.21 55.93

MENTAL MATH Subtract. Write answers only.

42. 1.2 − 0.5
   12 tenths minus 5 tenths equals 7 tenths or 0.7
43. 1.3 − 0.9 0.4
44. 1.5 − 0.2 1.3
45. 1.1 − 0.5 0.6
46. 0.8 − 0.6 0.2
47. 0.9 − 0.7 0.2
48. 1.4 − 0.8 0.6
49. 2.6 − 0.4 2.2
50. 2.7 − 0.5 2.2
51. 3.8 − 0.6 3.2

Solve. Use the chart on page 44.

52. How many million nickels were minted in
   a. 1890? 16.1
   b. 1900? 27.2
   c. 1950? 10.2

53. The number of nickels minted in 1890 written as a whole number is 16,100,000. Write the number of nickels minted in the other years as whole numbers.

54. How many more nickels were minted in 1940 than in 1930? Give the answer as a whole number. 152,300,000

55. How much more is the market value of a 1900 nickel than the market value of a 1920 nickel? $1.65

56. What is the total market value of a 1910, a 1940, and a 1950 nickel? $3.95

57. Why do you think a 1950 nickel has a greater market value than a 1930 nickel?
   Fewer were minted.

Nickel mysteries

Study the clue. Use the pictures of the coins and the chart on page 44.

58. What date is on each nickel?
   a.    b.

   Clue:
   • The total market value of both nickels is $4.15.
   a. 1920   b. 1900

59. What date is on each nickel?
   a.    b.

   Clue:
   • 40.2 million of these nickels were minted altogether.
   a. 1950   b. 1910

Subtracting Whole Numbers and Decimals 45

Practice Worksheet
Workbook S188, Copymaster S188, or Duplicating Master S188

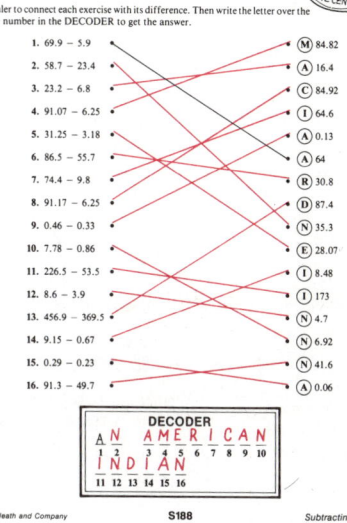

Project

Using library resources

Use a book of world records to find the highest price paid for a coin collection.

Answers will vary. Records may change from year to year.

Copymaster S474

Answer for page 45.
53.
Year	Count
1900	27,200,000
1910	30,000,000
1920	63,400,000
1930	23,800,000
1940	176,100,000
1950	10,200,000

Class Starter Quiz 19
on previous lesson

Subtract.

1. 0.64 − 0.38   0.26
2. 15.85 − 3.41   12.44
3. 83.61 − 7.31   76.3
4. 6.1 − 3.7   2.4
5. 60.02 − 4.16   55.86

Copymaster S389

Lesson Objective
To subtract decimals (with a different number of decimal places)

Problem-Solving Skills
Finding information in a display
Using a guess-and-check strategy

Starting the Lesson
Estimation If a digital watch that shows tenths or hundredths of seconds is available, have volunteers follow the contest rules and estimate 30 seconds. Record their times on the chalkboard to determine who came closest to guessing how long 30 seconds is.

Here's How *Note*
Use ■ **Visual Aid 3** (lined notebook paper on copymaster or transparency S113) to assist the students in aligning the digits.

30 − 24.29 = ?

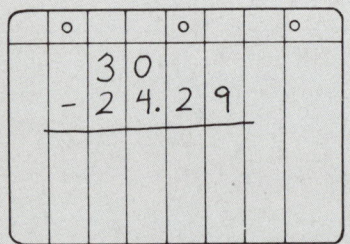

More on subtracting decimals

HOW LONG IS 30 SECONDS?

Cal and Barb had a contest to see who could come closer to guessing how long 30 seconds is. Their stopwatches show the results.

CONTEST RULES
- Use a stopwatch.
- Push the Start button.
- Don't look at the display.
- When you think 30 seconds have passed, push the Stop button.
- The person whose guess is closer to 30 seconds wins.

1. Read the stopwatches. Whose guess was off by 5.67 seconds? *Cal's*
2. What time is shown on Barb's stopwatch? *24.29 sec*
3. Which two numbers would you use to compute how far off Barb's guess was? *30 and 24.29*

Cal's guess

Barb's guess

Here's how *to subtract decimals.*   30 − 24.29 = ?

Line up the decimal points. Write in the 0's.

```
  30.00
− 24.29
```

Regroup.

```
   2 9 9
  30.00
− 24.29
```

Subtract.

```
   2 9 9
  30.00
− 24.29
   5.71
```

To estimate the difference, first round each number to the nearest whole number and then subtract.

30 − 24.29 = ?
↓   ↓
30 − 24 = 6

4. Look at the *Here's how*. Whose guess was off by 5.71 seconds? Who won the contest, Cal or Barb? *Barb's, Cal*

EXERCISES

5. Four of the calculator answers are wrong. Find them by estimating.

a. 37.42 − 4.25   33.17
(c.) 5.01 − 1.99   9.02
(e.) 43.8 − 2.12   4.16
g. 53.8 − 19.926   33.874
(i.) 29.79 − 21.046   18.744

(b.) 68.12 − 29.31   28.81
d. 8.2 − 6.95   1.25
f. 7.21 − 3.194   4.016
h. 4.78 − 0.8   3.98
j. 118.9 − 19.02   99.88

Subtract. Here are scrambled answers for the next row of exercises: 1.055 9.42 5.21 3.84 0.69 13.02

6. 8.63 − 3.42 = 5.21
7. 9.76 − 0.34 = 9.42
8. 24.1 − 11.08 = 13.02
9. 6.3 − 5.245 = 1.055
10. 3.29 − 2.6 = 0.69
11. 6.14 − 2.3 = 3.84

12. 3.74 − 2.5 = 1.24
13. 7.3 − 0.74 = 6.56
14. 7.86 − 1.59 = 6.27
15. 6.83 − 2.7 = 4.13
16. 8.7 − 6.25 = 2.45
17. 9.61 − 2.9 = 6.71

18. 9.6 − 2.64 = 6.96
19. 28.31 − 24.7 = 3.61
20. 17.0 − 8.95 = 8.05
21. 3.781 − 0.97 = 2.811
22. 7.52 − 4.083 = 3.437
23. 8.312 − 2.77 = 5.542

24. $4 − 2.98 = $1.02
25. $5 − 1.47 = $3.53
26. $10 − 5.78 = $4.22
27. $24 − 16.43 = $7.57
28. $16 − 11.29 = $4.71
29. $6 − 3.11 = $2.89

30. 6 − 2.7 = 3.3 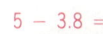 6.0 − 2.7 = 3.3
31. 8 − 4.2 = 3.8
32. 12.94 − 8.53 = 4.41
33. 16.4 − 5 = 11.4
34. 15.1 − 12.8 = 2.3
35. 23.4 − 2.89 = 20.51
36. 9.5 − 6 = 3.5
37. 14.5 − 12.8 = 1.7
38. 7 − 6.52 = 0.48
39. 9.72 − 0.865 = 8.855
40. 7.4 − 5.125 = 2.275
41. 123.7 − 101.4 = 22.3
42. 12.935 − 4.6 = 8.335
43. 9.323 − 1.747 = 7.576
44. 5.2 − 2.456 = 2.744
45. 25 − 8.2 = 16.8
46. 100 − 44.75 = 55.25
47. 75.25 − 16 = 59.25

Solve.

48. What is the difference in price between a $37 digital watch and a $29.85 alarm watch? **$7.15**

49. You have $17.50. How much more money do you need in order to buy a $29.95 video-game watch? **$12.45**

50. A customer paid for a $32.99 calculator watch with a $50 bill. How much change should she get? **$17.01**

51. You are the clerk. How much change should you give a customer who paid for a $27.97 stopwatch and a $24.95 alarm watch with a $100 bill? **$47.08**

Key it in!

Find a way to push each marked key once to get the answer.

52. 5 − 3.8 = 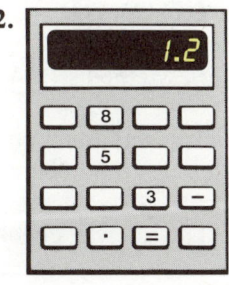 1.2

53. 3.5 − 2 = 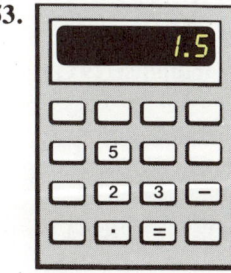 1.5

54. 9 − 6.7 =  2.3

Subtracting Whole Numbers and Decimals **47**

Extra Practice
Page 474 Skill 8

Practice Worksheet
Workbook S189, Copymaster S189, or Duplicating Master S189

Challenge Problem

Look at page 47.
- Find the largest number.
- Find the smallest number.
- Subtract your two numbers.
- Did you get a difference of 123.36?

123.7, 0.34, Yes

Copymaster S441

Class Starter Quiz 20
on previous lesson

Subtract.
1. 6.43 − 1.21   5.22
2. 17.4 − 6   11.4
3. 25.1 − 12.7   12.4
4. 8.27 − 1.125   7.145
5. 100 − 48.25   51.75

Copymaster S389

Lesson Objective
To solve problems using logical reasoning

Problem-Solving Skill
Making up a problem

Starting the Lesson
The Southwest Conference Schools facts are also on ■ **Visual Aid 9** (copymaster or transparency S119). You may want to use the visual aid to illustrate how to use the clues and facts to solve the problems on pages 48 and 49.

PROBLEM SOLVING *using logical reasoning*

Southwest Conference Schools

School	Nickname	City/State	Founded	Enrollment
Univ. of Arkansas	Razorbacks	Fayetteville, AR	1871	13,887
Baylor Univ.	Bears	Waco, TX	1845	11,481
Univ. of Texas	Longhorns	Austin, TX	1883	47,838
Rice Univ.	Owls	Houston, TX	1891	3,984
Southern Methodist Univ.	Mustangs	Dallas, TX	1911	9,048
Texas A&M Univ.	Aggies	College Station, TX	1876	35,675
Texas Christian Univ.	Horned Frogs	Fort Worth, TX	1873	6,925
Texas Technical Univ.	Red Raiders	Lubbock, TX	1923	23,589
Univ. of Houston	Cougars	Houston, TX	1927	31,213

Problem
Use the clues to name this Southwest Conference School.
  Clue 1: It was founded after 1900.
  Clue 2: There are more than 10,000 students enrolled.
  Clue 3: There are less than 30,000 students enrolled.

Here's how *to solve a problem using logical reasoning.*

STUDY Study the clues and the facts above.
What facts will you use to solve the problem?

PLAN and SOLVE

Use Clue 1.	Use Clue 2.	Use Clue 3.
List schools founded after 1900.	Cross off schools that do not fit the second clue.	Cross off schools that do not fit the third clue.
Texas Tech 1923	Texas Tech 23,589	Texas Tech 23,589
Houston 1927	Houston 31,213	~~Houston 31,213~~
Southern Methodist 1911	~~Southern Methodist 9048~~	

Answer: Texas Tech is the school that was founded after 1900 and has more than 10,000 students but less than 30,000 students.

CHECK All the facts were used correctly. The answer checks!

PROBLEMS

Clue 1 was used to make a list. Use the list and the other two clues to name the school.

1. *Clue 1:* It was founded before 1890.

   Univ. of Arkansas (1871)
   Baylor Univ. (1845)
   Univ. of Texas (1883)
   Texas A&M Univ. (1876)
   Texas Christian Univ. (1873)

   Clue 2: It has more than 30,000 students.
   Clue 3: It has less than 40,000 students.   **Texas A&M Univ.**

2. *Clue 1:* It has the word *Texas* in its name.

   Univ. of Texas
   Texas A&M Univ.
   Texas Christian Univ.
   Texas Technical Univ.

   Clue 2: It was founded after 1880.
   Clue 3: It has more than 30,000 students.   **Univ. of Texas**

Use the clues to name the school.

3. *Clue 1:* It has fewer students than Texas Technical Univ.
   Clue 2: Its population number has an 8 in the tens place.
   Clue 3: Its nickname is not a four-legged animal.   **Rice Univ.**

4. *Clue 1:* Its population number has a 1 in the thousands place.
   Clue 2: It is in the state of Texas.
   Clue 3: It has more students than the Univ. of Arkansas.   **Univ. of Houston**

5. *Clue 1:* It has fewer students than Baylor Univ.
   Clue 2: Its population number has an 8 in it.
   Clue 3: Its nickname is a four-legged animal.   **Southern Methodist Univ.**

6. *Clue 1:* Its population number has at least one 9 in it.
   Clue 2: It was founded before 1900.
   Clue 3: It has more students than Rice Univ.   **Texas Christian Univ.**

7. *Clue 1:* It does not have the word *Texas* in its name.
   Clue 2: It has at least one 8 and at least one 1 in its population number.
   Clue 3: It was founded after 1850.   **Univ. of Arkansas**

8. *Clue 1:* If you round its population to the nearest ten thousand, you get 10,000.
   Clue 2: It is in the state of Texas.
   Clue 3: It was founded before 1870.   **Baylor Univ.**

Give me a clue!

Making up a problem.

Use the facts on page 48 to make up a logical reasoning problem that fits the answer. Clue 1 is already given.   Answers will vary.

9. *Clue 1:* This school was founded before 1900.   *Clue 2:* ?   *Clue 3:* ?   *Answer:* Texas A&M Univ.

Subtracting Whole Numbers and Decimals **49**

Class Starter Quiz 21

on previous lesson

Use the clues and the facts on page 48 to name the school.

Clue 1: It was founded after 1900.
Clue 2: It has a higher enrollment than Baylor University.
Clue 3: It is not in Houston.

Texas Technical University

Copymaster S390

Problem-Solving Skills

Using computer-display information
Choosing the correct operation

Skills Reviewed

Adding whole numbers and decimals
Comparing whole numbers and decimals
Subtracting whole numbers and decimals
Rounding decimals

Starting the Lesson

Problem Solving The computer display at the top of page 50 is also on ■ **Visual Aid 10** (copymaster or transparency S120). Have the students use the computer display to answer questions like these:

- What time does Flight 65 arrive from Atlanta? (9:35)
- Which flight is arriving at Gate 2? (Flight 303)
- Which flight is scheduled to depart 55 minutes after it arrives? (Flight 110)

Cumulative Skill Practice Challenge the students to an estimation hunt by saying, "Pick the largest sum in the first row of exercises." (Exercise 3) Then have the students pick the largest sum in the second and third rows of exercises. (Exercises 4 and 9)

Problem solving

COMPUTERS AT THE AIRPORT

You're an airline agent. You use computers to write tickets, make reservations, assign seats, check luggage, and provide arrival and departure information.

Use the computer screen to answer these customers' questions.

1. What time does Flight 615 leave for Dallas? 10:10 From which gate? 15

2. Which flight is arriving at 9:35? Which city is it coming from? 65, Atlanta

3. My sister is arriving at 9:40 from Philadelphia. At which gate should I meet her? 20

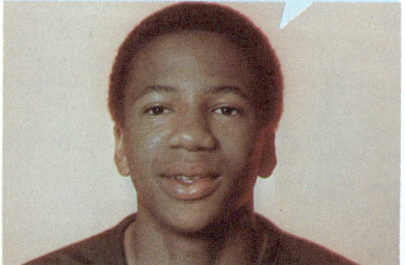

4. "My friend is arriving on Flight 65 and departing on Flight 45. Will he have more than a half hour between flights?" Yes

5. "My travel agent said you start assigning seats 45 minutes before each flight leaves. Will you be assigning seats for Flight 110 at 9:00?" No

Solve. Decide when a calculator would be useful.

6. A first-class ticket from Atlanta to Chicago is $201. Coach fare is $32 less. What is the price of a coach ticket? $169

7. Airfare from Los Angeles to Chicago is $372. How much would 2 tickets cost? $744

8. A 727 jet liner carries 135 passengers. Your computer shows 123 coach reservations and 18 first-class reservations. Has the flight been overbooked? Yes ("Overbooked" means there are more reservations than seats.)

9. A DC 9 has 139 passenger seats. The flight attendant counted 27 empty seats. Your computer shows you have assigned seats to 129 passengers. Are all the passengers on board? No

Cumulative Skill Practice

Give the sum. (page 22)
1. 7.6 + 0.82 + 5.3   13.72
2. 3.74 + 2.9 + 65.9   72.54
3. 39.82 + 52 + 96.5   188.32
4. 84.8 + 7.463 + 73.29   165.553
5. 7.564 + 7.3 + 68.83   83.694
6. 70.2 + 58.61 + 3.56   132.37
7. 52.14 + 0.89 + 24   77.03
8. 1.7 + 25.73 + 43.83   71.26
9. 88.9 + 14 + 72.35   175.25

Less than (<) or greater than (>)? (page 32)
10. 593 < 594
11. 786 > 768
12. 599 < 600
13. 895 > 885
14. 960 < 1000
15. 1501 > 999
16. 3431 > 3318
17. 2815 < 3815
18. 29,361 < 29,400
19. 40,000 > 38,652
20. 55,399 > 54,000
21. 4106 < 4214

Subtract. (page 38)
22. 2634 − 256 = 2378
23. 7972 − 388 = 7584
24. 3174 − 570 = 2604
25. 2063 − 1421 = 642
26. 3105 − 1638 = 1467
27. 4986 − 2897 = 2089

Less than (<) or greater than (>)? (page 42)
28. 0.008 > 0.007
29. 3.57 < 3.75
30. 0.005 < 0.03
31. 2.01 < 2.1
32. 3.1 > 2.97
33. 0.615 < 0.62
34. 4.215 < 42.15
35. 3.82 > 3.62
36. 1.1 > 1.07
37. 621.7 < 622
38. 0.031 < 0.13
39. 73.9 > 7.39

Give the difference. (pages 44, 46)
40. 53.4 − 29.6   23.8
41. 30.6 − 15.8   14.8
42. 9.00 − 5.74   3.26
43. 25.41 − 3.02   22.39
44. 8.6 − 3.5   5.1
45. 10.18 − 9.54   0.64
46. 20.34 − 15.95   4.39
47. 41.08 − 23.6   17.48
48. 26 − 13.5   12.5
49. 20 − 6.34   13.66
50. 9.8 − 6.99   2.81
51. 14.32 − 4.08   10.24

MIXED PRACTICE
Complete.
52. 5635 rounded to the nearest hundred is _?_   5600
53. 92.084 rounded to the nearest tenth is _?_   92.1
54. 36 + 184 = _?_   220
55. 450 − 136 = _?_   314
56. 8.3 + 9.42 = _?_   17.72
57. 15.92 − 8.7 = _?_   7.22
58. 5 + 6.93 + 8.4 = _?_   20.33
59. 23 − 7.48 = _?_   15.52
60. 11 + 5.67 + 0.94 = _?_   17.61
61. 9000 − 5231 = _?_   3769
62. 42 − 14.7 = _?_   27.3

Subtracting Whole Numbers and Decimals

Problem-Solving Worksheet
Workbook S191, Copymaster S191, or Duplicating Master S191

WORKSHEET 23 (Use after page 50.)

LOGICAL REASONING

- Brian is older than Susan.
- Todd is 8 years younger than Brian.
- Susan is 6 years older than Todd.
- Susan is 15 years old. How old is Brian?   17

MISSING DIGITS
Fill in the missing digits.
```
  4 2.7 6 5
−   1 7.2 9 8
    2 5.4 6 7
```

SOUP'S ON!

Al's Diner

HAMBURGER	$.99
HOT DOG	$.69
SOUP	$.79
MILK	$.40
JUICE	$.50

A hamburger and milk cost about $.10 less than soup and a _hot dog_.

WHAT'S THE ANSWER?
Use the code to answer the riddle.

CODE					
A	B	C	D	...	Z
26	25	24	23	...	1

Riddle: What's the difference between an excited skunk and a calm skunk?
Answer:
A N   E I G H T Y -
26 13   22 18 20 19 7 2
D O L L A R
23 12 15 15 26 9
L A U N D R Y
15 26 6 13 23 9 2
B I L L.
25 18 15 15

FIND THE PATTERN
This pyramid is 4 rows high. How many people are needed to build a human pyramid 8 rows high?   36

© D. C. Heath and Company   S191   Problem solving

Project
Using library resources
Use an almanac to find which airport in the United States has the most takeoffs and landings.

Chicago O'Hare (Records may change from year to year.)

Copymaster S475

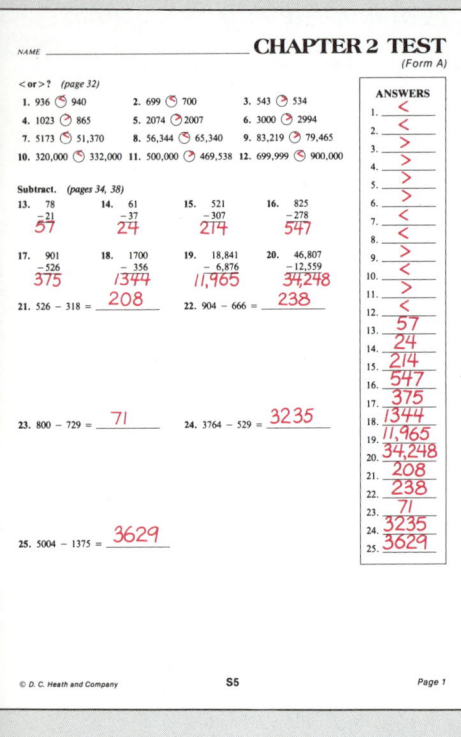
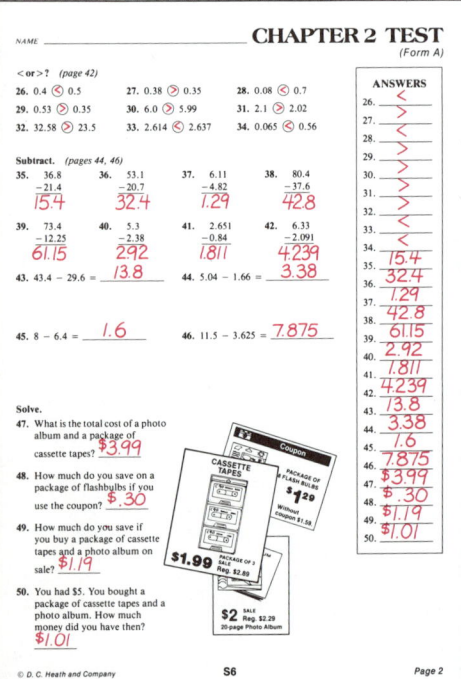

Chapter REVIEW

Here are scrambled answers for the review exercises:

2	add	digits	same
10	decimal	less	zero
900	difference	regroup	

1. same, less 2. regroup, difference 3. 900

1. To compare two whole numbers, start at the left and compare digits that are in the ⬚ place. The symbol < is read as "is ⬚ than." *(page 32)*

$$264 \bullet 486$$
$$264 < 486$$

2. To do this subtraction exercise, first ⬚ 1 ten for 10 ones and then subtract ones. The answer to a subtraction exercise is called the ⬚. *(page 34)*

$$\begin{array}{r} 324 \\ -\ 75 \end{array}$$

3. To estimate this difference, you can round each number to the nearest hundred. The estimate would be ⬚. *(page 38)*

$$1602 - 695$$

4. 10 5. add 6. digits

4. The next step in this subtraction exercise is to regroup 1 hundred for ⬚ tens. *(page 38)*

$$\begin{array}{r} 4 \\ \cancel{5}\cancel{10}4\ 6 \\ -\ 3\ 9\ 4 \\ \hline 2 \end{array}$$

5. To check this difference, ⬚ 1719 and 895. *(page 38)*

$$\begin{array}{r} 2614 \\ -\ 895 \\ \hline 1719 \end{array}$$

6. To compare these two numbers, start at the left and compare ⬚ that are in the same place. *(page 42)*

$$48.67 \bullet 48.52$$

7. decimal 8. zero 9. 2

7. To do this subtraction exercise, first line up the ⬚ points. *(page 44)*

$$6.29 - 0.58$$

8. To do this subtraction exercise, first write a ⬚ after 8.3 and then subtract. *(page 46)*

$$\begin{array}{r} 8.3 \\ -2.57 \end{array}$$

9. To estimate this difference, you can round each number to the nearest whole number. The estimate would be ⬚. *(page 46)*

$$4.18 - 2.09$$

52 Chapter 2

Chapter TEST

Less than (<) or greater than (>)? *(page 32)*

1. 843 > 840
2. 799 < 800
3. 653 > 635
4. 900 < 3482
5. 2836 < 2851
6. 5396 > 684
7. 6999 > 700
8. 248 < 1419
9. 39,426 > 39,420
10. 172,299 > 71,109
11. 48,746 < 58,764
12. 68,342 < 68,411

Subtract. *(pages 34, 38)*

13. 87 − 23 = 64
14. 76 − 39 = 37
15. 421 − 156 = 265
16. 604 − 275 = 329
17. 703 − 214 = 489
18. 800 − 158 = 642
19. 9008 − 3721 = 5287
20. 1503 − 829 = 674
21. 16,800 − 3,562 = 13,238
22. 35,216 − 21,628 = 13,588
23. 61,100 − 13,251 = 47,849
24. 25,112 − 11,004 = 14,108
25. 724 − 98 = 626
26. 803 − 229 = 574
27. 3921 − 766 = 3155
28. 32,407 − 8,959 = 23,448

Less than (<) or greater than (>)? *(page 42)*

29. 0.8 < 0.9
30. 0.05 < 0.4
31. 5.1 > 4.99
32. 6.3 > 2.77
33. 1.1 > 1.06
34. 18.3 > 18.26
35. 0.034 < 0.43
36. 0.011 < 0.22

Subtract. *(pages 44, 46)*

37. 37.3 − 15.1 = 22.2
38. 8.03 − 2.94 = 5.09
39. 15.23 − 4.7 = 10.53
40. 7.51 − 3.467 = 4.043
41. 8.12 − 3.345 = 4.775
42. 6.31 − 2.7 = 3.61
43. 16.4 − 8.7 = 7.7
44. 50.25 − 12.89 = 37.36
45. 7.51 − 3.861 = 3.649
46. 21 − 13.66 = 7.34

Solve.

47. What is the total cost of a baseball shirt and a package of notebook paper? $3.78

48. How much do you save on a package of batteries if you use the coupon? $.50

49. You had $5. You bought two packages of batteries using a coupon. How much money did you have then? $2.42

Subtracting Whole Numbers and Decimals 53

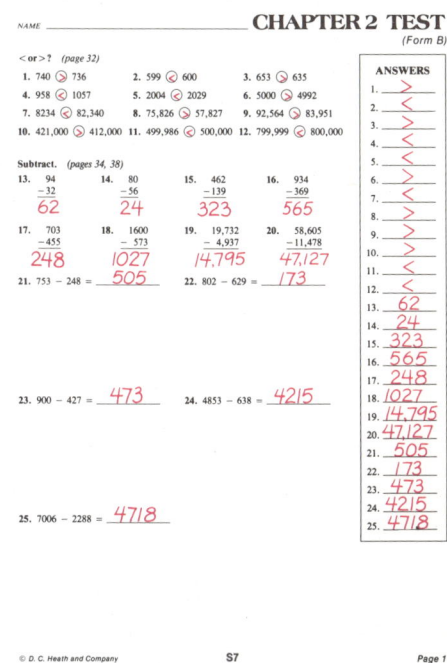

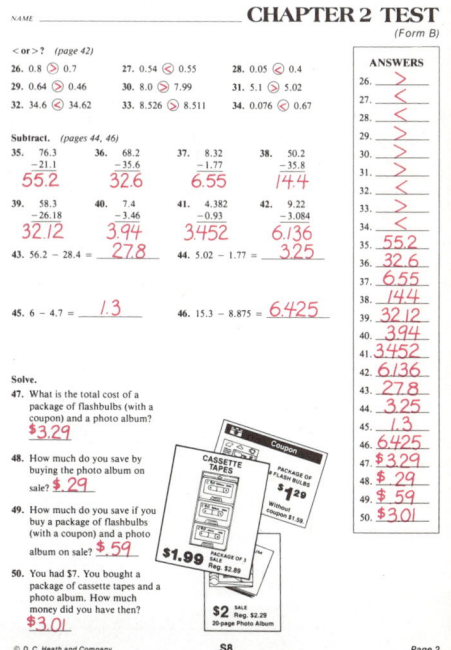

Copymasters S7 and S8 or Duplicating Masters S7 and S8

Cumulative Test
(Chapters 1–2)

Use Copymaster S109 to provide the students with an answer sheet in standardized test format.

Answers for Cumulative Test, Chapters 1–2

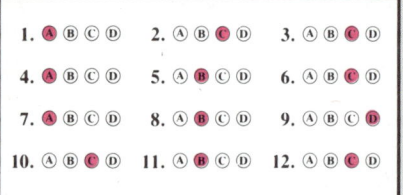

The table below correlates test items with student text pages.

Test Item	Page(s) Taught	Skill Practice
1	4	p. 41, exercises 1–12
2	4	p. 41, exercises 13–20
3	6	p. 41, exercises 21–26
4	10	p. 41, exercises 27–38
5	18	p. 41, exercises 39–50
6	22	p. 51, exercises 1–9
7	32	p. 51, exercises 10–21
8	38	p. 51, exercises 22–27
9	42	p. 51, exercises 28–39
10	44, 46	p. 51, exercises 40–51
11	20	
12	46	

54

Cumulative TEST — Standardized Format

Choose the correct letter.

1. The short word-name for 6,034,000 is
- **A.** 6 million, 34 thousand
- **B.** 6 billion, 34 million
- **C.** 6 billion, 34 thousand
- **D.** none of these

2. The standard numeral for two billion, fifty thousand is
- **A.** 2,050,000
- **B.** 2,050,000,000
- **C.** 2,000,050,000
- **D.** none of these

3. Add. 4435
  796
  +2074
- **A.** 6195
- **B.** 7205
- **C.** 7305
- **D.** none of these

4. 6984 rounded to the nearest hundred is
- **A.** 7000
- **B.** 6980
- **C.** 6990
- **D.** none of these

5. 36.0572 rounded to the nearest tenth is
- **A.** 36.06
- **B.** 36.1
- **C.** 36.05
- **D.** none of these

6. Give the sum.
42.5 + 3.18 + 6.8
- **A.** 41.48
- **B.** 14.23
- **C.** 52.48
- **D.** none of these

7. Which number is less than 43,057?
- **A.** 42,978
- **B.** 43,507
- **C.** 43,100
- **D.** none of these

8. Subtract. 1304
  − 295
- **A.** 1019
- **B.** 1009
- **C.** 1191
- **D.** none of these

9. Which number is greater than 0.54?
- **A.** 0.45
- **B.** 0.5
- **C.** 0.055
- **D.** none of these

10. Give the difference.
42.8 − 3.56
- **A.** 0.72
- **B.** 39.36
- **C.** 39.24
- **D.** none of these

11. Record album: $5.79
Single record: $1.35
What will 1 album and 2 single records cost?
- **A.** $12.93
- **B.** $8.49
- **C.** $7.14
- **D.** none of these

12. Record Album: $5.79
Single record: $1.35
You had $10. You bought 1 album and 1 single. How much money did you have left?
- **A.** $4.21
- **B.** $8.65
- **C.** $2.86
- **D.** none of these

54   Chapter 2

Multiplying Whole Numbers and Decimals

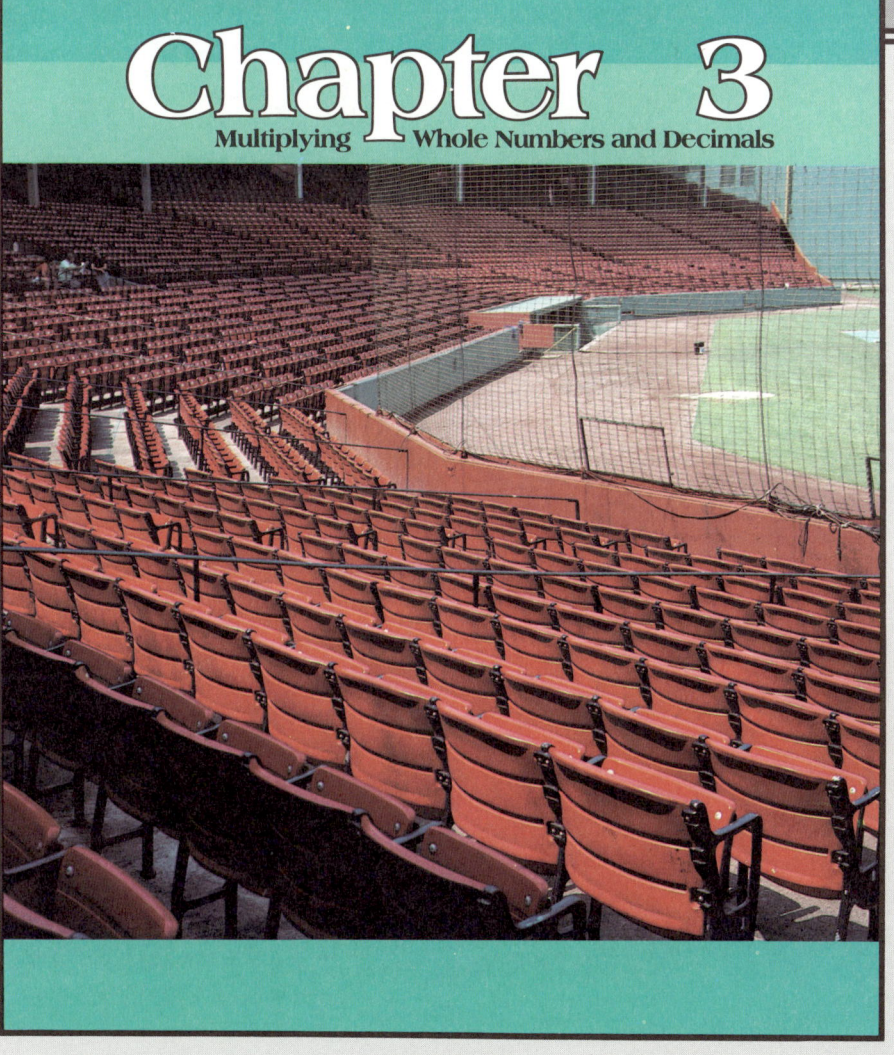

Chapter 3
Multiplying Whole Numbers and Decimals

Resources

- **Class Starter Quizzes 22-31** *(Copymasters S390-S392)*
- **Visual Aids 3, 11, and 12** *(Copymasters or Transparencies S113, S120, S121)*
- **Manipulatives**
  Manipulative Activity 4 *(Copymaster S515)*
  Powers-of-ten tiles *(Copymaster or Transparency S529)*
- **Worksheets 24-34** *(Copymasters, Duplicating Masters, or Workbook pages S192-S202)*
- **Challenge Problems** for pages 59, 61, 65, 67, 71, 75 *(Copymasters S441-S443)*
- **Projects** for pages 57, 63, 69, 73, 77 *(Copymasters S475-S476)*
- **Mental Math Extensions** for Skills 9-13 *(Copymasters S493-S495)*
- **Tests** *(Copymasters or Duplicating Masters S9-S11)*

Lesson Objective
To multiply mentally by multiples of 10, 100, or 1000

Problem-Solving Skill
Reading a chart

Starting the Lesson
Have the students number from 1 to 12 on scratch paper. Tell them you will read a multiplication fact and they are to write the product. Explain that you will read each fact only twice.

1. 5 × 8 (40) 2. 4 × 9 (36)
3. 7 × 7 (49) 4. 8 × 9 (72)
5. 6 × 8 (48) 6. 5 × 9 (45)
7. 6 × 9 (54) 8. 8 × 8 (64)
9. 9 × 9 (81) 10. 9 × 7 (63)
11. 7 × 8 (56) 12. 8 × 5 (40)

Have the students correct their own papers as you give the answers.
Before students open their books, ask them which is worth more money, a bill with Hamilton's portrait on it, a bill with Cleveland's portrait, or a bill with Franklin's portrait. Then have them open their books and find the answer at the top of page 56.

Exercise Note
Mental Math Use several of exercises 1–44 as oral exercises before making a written assignment. Have the students verbalize the method they used to mentally compute the products.

56

Mental math—multiplying by multiples of 10, 100, or 1000

FRANKLIN, HAMILTON, or CLEVELAND?

Which is worth more money, 6 Hamiltons, 4 Franklins, or 3 Clevelands? *Hint: Hamilton is on the $10 bill, Franklin is on the $100 bill, and Cleveland is on the $1000 bill.* 3 Clevelands

To answer the question, you will need to multiply by 10, 100, and 1000.

Here's how to multiply by multiples of 10, 100, or 1000.

6 × 10 = ?   4 × 100 = ?   3 × 1000 = ?

To multiply a whole number by 10, multiply by 1 and write 1 zero.

To multiply a whole number by 100, multiply by 1 and write 2 zeros.

To multiply a whole number by 1000, multiply by 1 and write 3 zeros.

6 × 10 = 60   4 × 100 = 400   3 × 1000 = 3000

Six Hamiltons are worth $60.   Four Franklins are worth $400.   Three Clevelands are worth $3000.

Study these examples.

a. 20   Use basic facts.
   ×30   Multiply 2 × 3.
   ―――   Write 2 zeros.
   600

b. 500   Multiply 7 × 5.
   × 70   Write 3 zeros.
   ―――
   35,000

56 Chapter 3

EXERCISES

MENTAL MATH—Multiply. Write answers only.
Here are scrambled answers for the next row of exercises: 800 12,000 120 6000

1. 12 × 10   120
2. 20 × 40   800
3. 60 × 100   6000
4. 60 × 200   12,000
5. 83 × 10   830
6. 9 × 1000   9000
7. 44 × 10   440
8. 60 × 40   2400
9. 81 × 10   810
10. 30 × 300   9000
11. 50 × 30   1500
12. 60 × 400   24,000
13. 3 × 70   210
14. 9 × 200   1800
15. 70 × 60   4200
16. 4 × 200   800
17. 8 × 3000   24,000
18. 50 × 70   3500
19. 23 × 100   2300
20. 8 × 2000   16,000
21. 20 × 4   80
22. 800 × 3   2400
23. 70 × 40   2800
24. 500 × 3   1500
25. 900 × 20   18,000
26. 10 × 30   300
27. 100 × 40   4000
28. 8000 × 2   16,000
29. 70 × 70   4900
30. 5000 × 50   250,000
31. 80 × 100   8000
32. 70 × 100   7000
33. 70 × 1000   70,000
34. 60 × 1000   60,000
35. 300 × 100   30,000
36. 5000 × 90   450,000
37. 400 × 80   32,000
38. 1000 × 600   600,000
39. 900 × 100   90,000
40. 600 × 800   480,000
41. 80 × 1000   80,000
42. 600 × 10   6,000
43. 700 × 100   70,000
44. 300 × 10   3000

Use the money facts. What is the value of

45. 14 Hamiltons? $140
46. 40 Jeffersons? $80
47. 23 Franklins? $2300
48. 60 Lincolns? $300
49. 7 Grants? $350
50. 9 Jacksons? $180
51. 10 Washingtons and 5 Jacksons? $110
52. 20 Jeffersons and 3 Grants? $190

MONEY FACTS

BILL	PORTRAIT
$1	Washington
$2	Jefferson
$5	Lincoln
$10	Hamilton
$20	Jackson
$50	Grant
$100	Franklin

You're the bank teller!

Answer these questions.

53. Are 9 Hamiltons the same amount of money as 90 Washingtons?   Yes

54. "Are 8 Grants the same amount of money as 80 Lincolns?"   Yes

55. "Are 7 Jacksons and 30 Jeffersons equal to 3 Franklins?"   No

56. "Would you give me 50 Hamiltons, 20 Jacksons, and 60 Grants for 15 Franklins? Why or why not?"   No. $3900 is more than $1500.

Multiplying Whole Numbers and Decimals **57**

Extra Practice
Page 475   Skill 9

Practice Worksheet
Workbook S192, Copymaster S192, or Duplicating Master S192

Project

Researching information
Find how many times the digit "1" or the word "one" appears on the front and back of a dollar bill.

16 or more

Copymaster S475

Class Starter Quiz 22
on previous lesson

Multiply.

1. 17 × 10   2. 50 × 30
     170   1500
3. 8 × 400   4. 1000 × 70
     3200   70,000
5. 500 × 70   6. 60 × 300
     35,000   18,000
7. 1000 × 500   8. 64 × 100
     500,000   6400
9. 8 × 800   10. 40 × 400
     6400   16,000

Copymaster S390

Lesson Objective
To multiply by a 1-digit number

Problem-Solving Skills
Reading a chart
Using logical reasoning

Starting the Lesson
What are the facts? Allow the students 60 seconds to study the information in the chart on page 58. Then have the students close their books and answer these questions from memory:

- Who burns more calories per hour, a 100-pound person or a 150-pound person? (150-pound person)
- For which activity do you burn the most calories per hour? (swimming)
- Do you burn more calories per hour jogging or dancing? (dancing)

Use exercises 1–4 and the example in the *Here's how* to introduce multiplying by a 1-digit number.

Multiplying by a 1-digit number

Activity	Calories Burned per Hour	
	100-pound person	150-pound person
Jogging	560	840
Swimming	1168	1752
Dancing	608	912
Sitting quietly	64	96

1. Read the chart. How many calories does a 100-pound person burn in 1 hour of dancing? **608 calories**

2. What two numbers would you multiply to find how many calories a 100-pound person burns in 3 hours of dancing? **608 and 3**

Here's how to multiply by a 1-digit number.   608 × 3 = ?

Multiply ones and regroup.	Multiply tens and add.	Multiply hundreds and regroup.
2 608 × 3 ――― 4	2 608 × 3 ――― 24	2 608 × 3 ――― 1824

The numbers that are multiplied are called **factors**.

The answer is called the **product**.

3. Look at the *Here's how*. How many calories does a 100-pound person burn in 3 hours of dancing? **1824 calories**

4. Check each example. Is the answer correct?

a.   6
870
× 9
――
7830 yes

b.   75
697
× 8
――
5576 yes

c.   154
2176
× 7
――
15,232 yes

d.   11
1023
× 5
――
5115 yes

EXERCISES

Multiply.
Here are scrambled answers for the next row of exercises:
256   352   477   486   435   76

5. 19 × 4 = 76
6. 81 × 6 = 486
7. 53 × 9 = 477
8. 87 × 5 = 435
9. 32 × 8 = 256
10. 44 × 8 = 352

11. 46 × 7 = 322
12. 71 × 3 = 213
13. 68 × 7 = 476
14. 90 × 6 = 540
15. 89 × 4 = 356
16. 29 × 5 = 145

17. 612 × 9 = 5508
18. 797 × 6 = 4782
19. 921 × 5 = 4605
20. 584 × 3 = 1752
21. 604 × 8 = 4832
22. 201 × 8 = 1608

23. 504 × 4 = 2016
24. 685 × 7 = 4795
25. 732 × 8 = 5856
26. 493 × 5 = 2465
27. 893 × 2 = 1786
28. 691 × 3 = 2073

29. $12.02 × 8 = $96.16
30. $21.93 × 3 = $65.79
31. $16.88 × 6 = $101.28
32. $43.07 × 9 = $387.63
33. $58.87 × 4 = $235.48
34. $23.81 × 5 = $119.05

35. 63 × 7 441
36. 88 × 4 352
37. 47 × 9 423
38. 31 × 6 186
39. 252 × 9 2268
40. 319 × 5 1595
41. 571 × 6 3426
42. 487 × 8 3896
43. 1023 × 5 5115
44. 4216 × 4 16,864
45. 3924 × 6 23,544
46. 4879 × 3 14,637
47. 2106 × 7 14,742
48. 8135 × 9 73,215
49. 2147 × 2 4294
50. 6627 × 8 53,016

Solve. Use the table on page 58.

51. How many calories does a 150-pound person burn in 4 hours of jogging? 3360

52. How many calories does a 100-pound person burn in 5 hours of swimming? 5840

53. How many more calories per hour does a 100-pound person burn swimming than dancing? 560

54. How many more calories does a 150-pound person burn in 3 hours of jogging than a 100-pound person? 840

Keep on dancing

Logical reasoning

55. The record for nonstop dancing is [?] 120 hours 30 minutes. To find how many hours, use the clues.

Clues:
- The digit in the hundreds place is 1.
- The ones digit is less than the tens digit.
- The sum of the digits is 3.

Multiplying Whole Numbers and Decimals **59**

Practice Worksheet

Workbook S193, Copymaster S193, or Duplicating Master S193

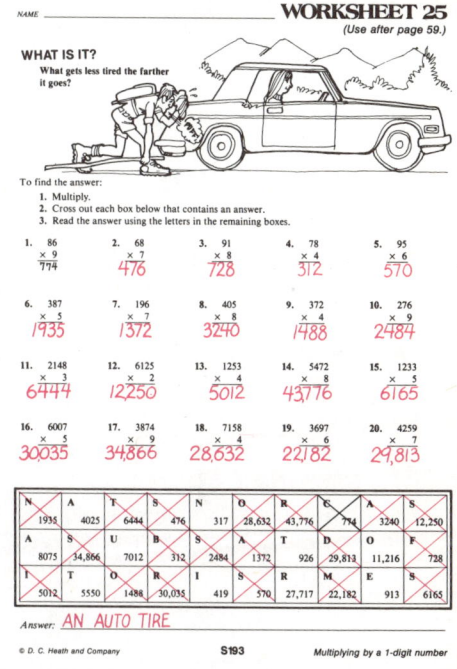

Challenge Problem

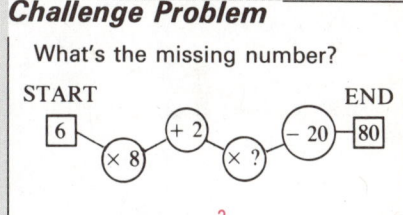

Copymaster S441

Class Starter Quiz 23
on previous lesson

Multiply.
1. 53 × 8   424
2. 37 × 9   333
3. 354 × 7   2478
4. 2105 × 6   12,630
5. 6739 × 5   33,695

Copymaster S390

Lesson Objective

To estimate products

Problem-Solving Skills

Reading a graph
Using estimation

Starting the Lesson

Provide a 30-second warmup of rounding numbers. Have the students round each of these numbers to the nearest ten.

33 (30)   27 (30)
84 (80)   61 (60)
97 (100)   92 (90)

Then use exercises 1–5 and the example in the *Here's how* to introduce estimating products.

Estimating products

1. Look at the teen-spending graph. Which age group spends about $40 per week on themselves? 18–19

2. Which number would you round to estimate how much money an average 16-year-old spends in 4 weeks? 28

Here's how *to estimate products.*

To estimate the product $28 × 4 = ?
of $28 × 4, round $28 to
the nearest ten dollars.   30 × 4 = $120

I spend about $120 a month.

3. Look at the *Here's how*. To estimate the product, $28 was rounded to the nearest [?] dollars. Then $30 was multiplied by [?].
   10   4

4. Estimate. About how much money does an average 18-year-old spend each month (4 weeks)? 40 × 4 = $160

5. Does an average 15-year-old spend more or less than $40 each month? more

Chapter 3

EXERCISES

Use estimation to find the correct answer.

6. 39 × 5
   a. 95
   b. 195
   c. 295

7. 83 × 7
   a. 581
   b. 681
   c. 781

8. 47 × 6
   a. 82
   b. 182
   c. 282

9. 33 × 8
   a. 164
   b. 264
   c. 364

10. 94 × 4
    a. 176
    b. 276
    c. 376

11. 317 × 6
    a. 902
    b. 1902
    c. 2902

12. 492 × 5
    a. 2460
    b. 3460
    c. 4460

13. 197 × 8
    a. 576
    b. 1576
    c. 2576

14. 812 × 7
    a. 3684
    b. 4684
    c. 5684

15. 796 × 4
    a. 2184
    b. 3184
    c. 4184

Estimate each product by rounding the 2-digit number to the nearest ten.

16. 78 × 5 [80 × 5] 400
17. 34 × 9 270
18. 19 × 7 140
19. 66 × 8 560
20. 93 × 6 540
21. 39 × 5 200
22. 72 × 4 280
23. 88 × 6 540
24. 88 × 3 270
25. 19 × 9 180
26. 74 × 7 490
27. 63 × 8 480
28. 98 × 5 500

Estimate each product by rounding the 3-digit number to the nearest hundred.

29. 319 × 7 [300 × 7] 2100
30. 726 × 3 2100
31. 495 × 5 2500
32. 198 × 6 1200
33. 423 × 8 3200
34. 687 × 4 2800
35. 718 × 3 2100
36. 117 × 9 900
37. 488 × 8 4000
38. 616 × 6 3600
39. 389 × 5 2000
40. 726 × 4 2800
41. 997 × 7 7000

Earning Money
Using estimation

I can earn $29 a week.

I can earn $123 a month.

ESTIMATE True or false?

42. Jim can earn more than $200 in 6 weeks. false
43. Kari can earn more than $600 in 6 months? true
44. In one month (4 weeks), Jim can earn more money than Kari. false

Multiplying Whole Numbers and Decimals

Practice Worksheet
Workbook S194, Copymaster S194, or Duplicating Master S194

WORKSHEET 26
(Use after page 61.)

NAME _____

SKILL DRILL
Estimate each product by rounding the two-digit number to the nearest ten.

1. 92 → 90 × 6 → × 6 = 540
2. 56 → 60 × 8 → × 8 = 480
3. 28 → 30 × 6 → × 6 = 180
4. 52 → 50 × 7 → × 7 = 350
5. 78 → 80 × 7 → × 7 = 560
6. 98 → 100 × 8 → × 8 = 800
7. 83 → 80 × 9 → × 9 = 720
8. 89 → 90 × 7 → × 7 = 630

Estimate each product by rounding the three-digit number to the nearest hundred.

9. 486 → 500 × 7 → × 7 = 3500
10. 612 → 600 × 8 → × 8 = 4800
11. 217 → 200 × 5 → × 5 = 1000
12. 623 → 600 × 9 → × 9 = 5400
13. 489 → 500 × 9 → × 9 = 4500
14. 812 → 800 × 8 → × 8 = 6400
15. 987 → 1000 × 5 → × 5 = 5000
16. 773 → 800 × 3 → × 3 = 2400

17. 4 × 891 3600
18. 6 × 165 1200
19. 3 × 899 2700
20. 7 × 695 4900
21. 9 × 213 1800
22. 5 × 489 2500
23. 7 × 576 4200
24. 6 × 128 600
25. 9 × 873 8100

Check yourself. Here are the scrambled answers:
180 350 480 540 560 600 630 720 800 1000
1200 1800 2400 2500 2700 3500 3600 4200 4500 4800
4900 5000 5400 6400 8100

© D.C. Heath and Company   S194   *Estimating products*

Challenge Problem

Use a calculator. Try this number trick:

Begin with your age in years. Multiply it by 4. Add 6. Multiply by 5. Subtract 10. Divide by 20. Subtract your age in years. What is the answer? 1

Now try this number trick with other people's ages. Is the answer always the same? Yes

Copymaster S442

Class Starter Quiz 24
on previous lesson

Match. Find the products by estimating.

1. 6 × 59   d   a. 424
2. 8 × 53   a   b. 2844
3. 4 × 76   f   c. 1564
4. 9 × 316   b   d. 354
5. 2 × 782   c   e. 594
6. 3 × 198   e   f. 304

Copymaster S390

Lesson Objective
To multiply by a 2-digit number

Problem-Solving Skills
Reading a newspaper ad
Using a guess-and-check strategy

Starting the Lesson
What are the facts? Have the students study the want ad and the photo at the top of page 62 for 30 seconds. Then have them close their books and challenge them to answer these questions from memory:

- To be a rodeo clown, how old must you be? (18)
- How much per hour does the job pay? ($12)
- If you took the job, would you work more or less than 30 hours per week? (Less)

Exercise Note
Problem Solving Encourage the students to use a guess-and-check-strategy for exercises 43–45. The thinking for exercise 43 might be as follows:

27 [×] 9 [=]   [243] → too big
29 [×] 7 [=]   [203] → perfect

Multiplying by a 2-digit number

WANTED:
RODEO CLOWN
Must be 18 or older
Earn $12/hour
Work 26 hours/week

1. Which operation would you use to find how many dollars a rodeo clown is paid per week? *Multiplication*

Here's how *to multiply by a 2-digit number.*   26 × 12 = ?

Line up the digits. Multiply by 2.	Multiply by 10.	Add.
26 × 12 ――― *52*	26 × 12 ――― *52* *260*	26 × 12 ――― *52* *260* ――― *312*

First round each number to the nearest ten and then estimate the product.

26 × 12 = ?
↓   ↓
*3*0 × *1*0 = *3*00

312 is near the estimate. So, the product seems reasonable.

2. Look at the *Here's how*. How much money is a rodeo clown paid per week? *$312*
Would you take the job for that much money?

3. Three of these calculator answers are wrong. Find them by estimating.

(**a.**) 24 × 30   *72*   **b.** 41 × 59   *2419*
c. 20 × 431   *862*   **d.** 51 × 603   *30753*
e. 39 × 114   *4446*   (**f.**) 40 × 882   *3528*

62 Chapter 3

EXERCISES

Multiply. Here are scrambled answers for the next row of exercises: 2576 2655 1980 1242 2176 1610

4. 68 ×32 2176	5. 59 ×45 2655	6. 46 ×56 2576	7. 70 ×23 1610	8. 33 ×60 1980	9. 27 ×46 1242	
10. 631 × 20 12,620	11. 168 × 52 8736	12. 246 × 19 4674	13. 604 × 31 18,724	14. 120 × 20 2400	15. 403 × 15 6045	
16. 146 × 32 4672	17. 487 × 51 24,837	18. 268 × 25 6700	19. 452 × 46 20,792	20. 283 × 49 13,867	21. 765 × 43 32,895	
22. 864 × 65 56,160	23. 398 × 56 22,288	24. 577 × 48 27,696	25. 903 × 52 46,956	26. 685 × 63 43,155	27. 492 × 53 26,076	
28. 2413 × 41 98,933	29. 2804 × 34 95,336	30. 6533 × 95 620,635	31. 1897 × 57 108,129	32. 4294 × 83 356,402	33. 2319 × 24 55,656	
34. $1.56 × 33 $51.48	35. $4.08 × 17 $69.36	36. $18.06 × 22 $397.32	37. $71.24 × 95 $6767.80	38. $38.65 × 79 $3053.35	39. $42.08 × 44 $1851.52	

Solve. Use the classified-ad information to tell which job each person is thinking about.

40. "If I took that job, I'd earn $48 a day." *Photographer*

41. "I would earn $440 a week doing that job." *Water-tower painter*

42. "If I took that job, I would earn $75 in three days." *Dishwasher*

DISHWASHER
Mon.–Sat.
5 hours a day $5/hour

WATER-TOWER PAINTER
40-hour week $11/hour

PHOTOGRAPHER
Mon.–Wed.
6-hour day $8/hour
Must furnish own camera.

Key it in!

Guess and check.

Find a way to push each marked key once to get the answer.

43. 29 × 7 = 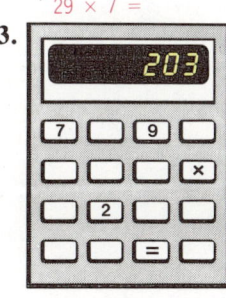 203

44. 82 × 6 = 492

45. 98 × 5 =  490

Multiplying Whole Numbers and Decimals **63**

Extra Practice
Page 475   Skill 10

Practice Worksheet
Workbook S195, Copymaster S195, or Duplicating Master S195

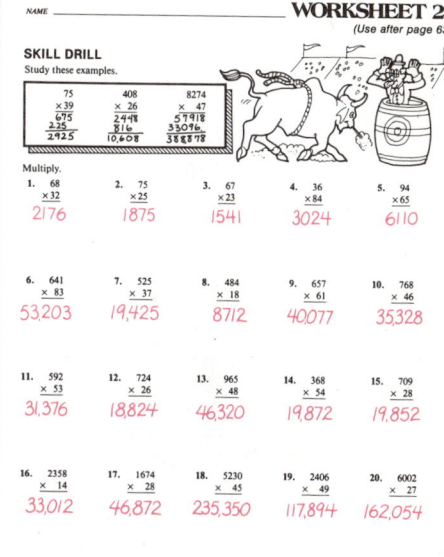

Project
Using library resources

Use a book of world records to find who holds the record for the longest industrial career in one job.

Answers will vary. Records may change from year to year.

Copymaster S475

Class Starter Quiz 25
on previous lesson

Multiply.
1. 58 × 42   2436
2. 90 × 37   3330
3. 147 × 51   7497
4. 6422 × 83 533,026
5. 3219 × 65 209,235

Copymaster S391

Lesson Objective
To multiply by a 3-digit number

Problem-Solving Skills
Finding information in a display
Reading a chart

Starting the Lesson
What are the facts? Allow the students 30 seconds to study the information at the top of page 64. Then have the students close their books and answer these questions from memory:
- How many buttons are in a box? (125)
- Are there more or less than 200 boxes of buttons in stock? (More)
- How many buttons are being ordered by Concert Tours, Inc.? (32,000)

Here's How Note
Use ■ **Visual Aid 3** (lined notebook paper, copymaster or transparency S113) to assist the students in aligning the partial products.

253 × 125 = ?

		2	5	3
	×	1	2	5
	1	2	6	5
	5	0	6	
2	5	3		
3	1	6	2	5

64

Multiplying larger numbers

You're a Shipping Clerk!

You have 253 boxes of assorted buttons. There are 125 buttons in each box.

1. How many buttons are in a box? 125
2. How many boxes do you have in stock? 253
3. What two numbers would you multiply to find out whether you have enough buttons to fill the order? 125 and 253

Here's how to multiply 3-digit numbers. 253 × 125 = ?

Start each product directly below the digit you are multiplying by.

```
    253
  × 125
   1265
    506
   253
  31,625
```

You don't have to write the 0's.

First round to the nearest hundred and then estimate the product.

253 × 125 = ?
↓   ↓
300 × 100 = 30,000

4. Look at the *Here's how*. Do you have enough buttons to fill the order? No
   Was 30,000 buttons a reasonable estimate? Yes

5. Check each example. Is the answer correct?

a. 372   b. 301   c. 807
  × 132   × 506   × 490
   744   1806   72630
  1116   1505   3228
   372   152,306   395,430 Yes
  49,104 Yes   Yes

64   Chapter 3

EXERCISES.

Multiply. Here are scrambled answers for the next row of exercises: 122,990 81,315 81,432 52,204 146,784 124,982

6. 348 ×234 **81,432**	7. 695 ×117 **81,315**	8. 834 ×176 **146,784**	9. 502 ×245 **122,990**	10. 506 ×247 **124,982**	11. 421 ×124 **52,204**
12. 523 ×402 **210,246**	13. 938 ×136 **127,568**	14. 734 ×274 **201,116**	15. 608 ×403 **245,024**	16. 909 ×123 **111,807**	17. 707 ×246 **173,922**
18. 1839 × 256 **470,784**	19. 1265 × 329 **416,185**	20. 2576 × 395 **1,017,520**	21. 8152 × 406 **3,309,712**	22. 6805 × 203 **1,381,415**	23. 7304 × 155 **1,132,120**
24. 1652 × 330 **545,160**	25. 3728 × 250 **932,000**	26. 9162 × 407 **3,728,934**	27. 8610 × 720 **6,199,200**	28. 1785 × 909 **1,622,565**	29. 8042 × 105 **844,410**
30. $65.23 × 200 **$130.46**	31. $13.75 × 400 **$5500**	32. $17.06 × 600 **$10,236**	33. $86.10 × 500 **$43,050**	34. $75.69 × 300 **$22,707**	35. $42.30 × 200 **$8460**

36. 546 × 324 **176,904**
37. 603 × 425 **256,275**
38. 582 × 515 **299,730**
39. 497 × 305 **151,585**
40. 1252 × 213 **266,676**
41. 8260 × 206 **1,701,560**
42. 1175 × 225 **264,375**
43. 1063 × 215 **228,545**
44. 8606 × 400 **3,442,400**
45. 7625 × 900 **6,862,500**
46. 5403 × 700 **3,782,100**
47. 9302 × 200 **1,860,400**

Button Prices	
NUMBER	COST
1	$ .60
2	$1.20
3	$1.80
6	$3.60
9	$5.40

Solve. Use the price list.

48. How many buttons can you buy for $1.20? **2**

49. What is the cost of 6 buttons? **$3.60**

50. How many buttons can you buy for $5.40? **9**

51. What is the cost of 10 buttons? **$6.00**

52. You gave the clerk a $10 bill for 9 buttons. How much change should you receive? **$4.60**

Make the change

53. Carlo made a $1.43 purchase. He gave the clerk $2. What 5 coins did he get in change? **2 pennies, 1 nickel, and 2 quarters**

54. Sandy made an $8.79 purchase. She gave the clerk $10.04. What coin and bill did Sandy receive as change?
A quarter and a dollar bill

Multiplying Whole Numbers and Decimals **65**

Practice Worksheet
Workbook S196, Copymaster S196, or Duplicating Master S196

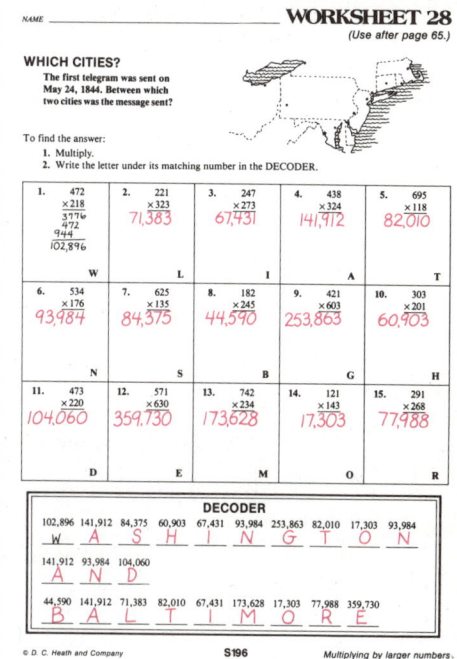

Challenge Problem

Where would four darts land in order to score 51?

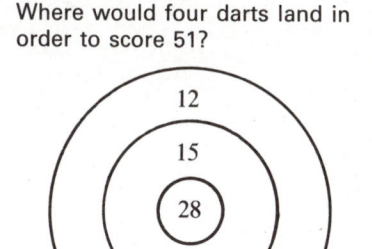

15 + 12 + 12 + 12

Copymaster S442

Class Starter Quiz 26
on previous lesson

Multiply.
1. 437 × 234   102,258
2. 402 × 245   98,490
3. 717 × 301   215,817
4. 4805 × 203   975,415
5. 2563 × 400   1,025,200

Copymaster S391

Problem-Solving Skills
Finding information in an ad
Recognizing when information is missing
Choosing the correct operation

Skills Reviewed
Adding whole numbers and decimals
Rounding whole numbers and decimals
Writing standard numerals (whole numbers and decimals)

Starting the Lesson
Cumulative Skill Practice Write these five answers on the chalkboard:

2924   8100   8.32   9.38   35.21

Challenge students to an answer hunt by saying, "Look at exercises 1–43 on page 67. Find the five exercises that have these answers. You have five minutes to find as many of the exercises as you can." (Exercises 5, 14, 20, 32, and 40)

Problem Solving The sale ad at the top of page 66 is also on ■ **Visual Aid 11** (copymaster or transparency S120). Use the ad to ask questions like these:

- How much does a roll of 110 film cost? ($1.49)
- How many exposures are on a roll of 126 film? (20)
- Which size enlargement costs $.99? (5 × 7)
- How many color prints can you get for $.55? (3)

Problem solving Reading an ad

You are a sales clerk at the Quick Print Photo Shop. The shop is running a sale on film and photo finishing.

Big Holiday Sale!

FILM SHOP Color Film			PHOTO FINISHING Color Prints	
SIZE	EXPOSURES	PRICE		
135	36	$3.00	Any negative	25¢
135	20	$2.15	Each additional print	15¢
126	20	$2.05	5 × 7 enlargement	99¢
110	12	$1.49	8 × 10 enlargement	$2.59

What questions would you have to ask each customer before you could complete the sale?

1. I'd like a print of this negative for each person in my family.

2. I'd like a roll of size 135 film for my camera.

3. My mother wants an enlargement of this picture of Niagara Falls. How much will it cost?

4. "I have $15. Can I buy 6 rolls of film for my camera?" What size film do you need? How many exposures would you like?

5. "I'd like to have a 5 × 7 enlargement made from each of these negatives. How much will they cost?" How many negatives do you have?

Solve. Use the sale prices. Decide when a calculator would be useful.

6. Garret had $7.50. He bought a roll of 20-exposure size 135 film. How much money did he have left? $5.35

7. Jenny bought a roll of film. She got $7.95 change from a $10 bill. What size film did she buy? 126

8. Mrs. Fisher bought 2 rolls of film. She spent $5.15. How many pictures will she be able to take? 56

9. How much more does an 8 × 10 enlargement cost than a 5 × 7 enlargement? $1.60

10. How many pictures will a customer be able to take if she buys 3 rolls of 36-exposure size 135 film? 108

11. Mr. Wilson wants 5 negatives made into color prints. How much will they cost? $1.25

66   Chapter 3

Cumulative Skill Practice

Add. *(page 6)*

| 1. | 462
329
83
+114
988 | 2. | 632
95
377
+216
1320 | 3. | 536
291
432
+ 51
1310 | 4. | 329
416
238
+627
1610 | 5. | 513
753
916
+742
2924 | 6. | 457
365
21
+914
1757 |

Round to the nearest hundred. *(page 10)*

7. 478 500   8. 609 600   9. 746 700   10. 250 300   11. 963 1000

12. 2748 2700   13. 3290 3300   14. 8062 8100   15. 5555 5600   16. 6350 6400

Write the standard numeral. *(page 14)*

17. 7 hundredths 0.07   18. 15 thousandths 0.015

19. 3 and 9 tenths 3.9   20. 8 and 32 hundredths 8.32

21. 15 and 4 hundredths 15.04 22. 7 and 26 thousandths 7.026

23. 9 and 3 thousandths 9.003 24. 24 and 396 ten-thousandths 24.0396

Round to the nearest hundredth. *(page 18)*

25. 0.296 0.30   26. 0.275 0.28   27. 3.6081 3.61   28. 4.002 4.00   29. 2.7450 2.75

30. 3.4382 3.44   31. 8.297 8.30   32. 9.3841 9.38   33. 7.5004 7.50   34. 6.8296 6.83

Give the sum. *(page 22)*

35. 3.28 + 0.56 3.84   36. 4.381 + 2.743 7.124   37. 15.829 + 6.542 22.371

38. 9.1 + 3.06 12.16   39. 17.2 + 23.4 40.6   40. 26.71 + 8.5 35.21

41. 6 + 3.7 + 2.94 12.64   42. 0.18 + 1.6 + 3 4.78   43. 52.5 + 6.21 + 7 65.71

MIXED PRACTICE
Complete.

44. The standard numeral for 23 million is ___?___ 23,000,000

45. The standard numeral for 4 million, 72 thousand is ___?___ 4,072,000

46. 38 + 193 + 7 = ___?___ 238   47. 5.62 + 9.54 = ___?___ 15.16   48. 4938 − 175 = ___?___ 4763

49. 2.6 + 0.9 + 1.46 = ___?___ 4.96   50. 46 + 9 + 278 = ___?___ 333   51. 19.6 − 12.7 = ___?___ 6.9

52. 5 + 3.082 + 6.77 = ___?___ 14.852 53. 7000 − 1396 = ___?___ 5604   54. 71 − 23.9 = ___?___ 47.1

Multiplying Whole Numbers and Decimals **67**

Problem-Solving Worksheet
Workbook S197, Copymaster S197, or Duplicating Master S197

WORKSHEET 29 (Use after page 66.)

GUESS AND CHECK
One of these clocks is 5 minutes fast, another is 10 minutes slow, and the other is 15 minutes slow. What is the correct time?
3:40

COIN PROBLEM
Todd has $1.25 in dimes and quarters. He has 8 coins in all. How many dimes does he have?
5

DOGGONE IT!
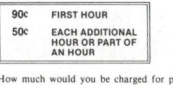
Lucky, Buddy, Dandy, and Flip are dogs. Flip is older than Lucky. Buddy is younger than Dandy, and Dandy is younger than Flip. What is the name of the oldest dog?
Flip

PARK YOUR CAR
90¢ FIRST HOUR
50¢ EACH ADDITIONAL HOUR OR PART OF AN HOUR
How much would you be charged for parking from 8:00 AM to 5:00 PM?
$4.90

ON THE ROAD
SIBLEY 10 miles
ROCK RAPIDS 20 miles
LARCHWOOD 30 miles
You are halfway between Rock Rapids and Larchwood. How many miles are you from Sibley?
35

WORK BACKWARD
If you multiply my age by 3 and add 20, you get 200. How old am I?
60

© D. C. Heath and Company S197 Problem solving

Challenge Problem

A customer traded some dollar bills for the same number of dimes as quarters. How many dollars did the customer have? *Hint: The customer had about $20.*

21 (60 dimes and 60 quarters)

Copymaster S442

Answers for page 66.
1. How many people are there in your family?
2. Would you like 36 or 20 exposures?
3. What size enlargement would you like?

Class Starter Quiz 27
on previous lesson

Solve. Use the sale ad on page 66.

1. A customer bought a roll of film. He got $2.95 change from a $5 bill. What size film did he buy? **126**
2. How many pictures can you take if you buy 5 rolls of 110 film? **60**

Copymaster S391

Lesson Objective
To multiply decimals

Problem-Solving Skills
Finding information in a table
Solving a multistep problem
Using logical reasoning

Starting the Lesson
On the chalkboard write this statement:

If you can jump one meter high on Earth, you could jump 263 meters on the planet Mars.

Have the students guess where to place the decimal point in the number so that the statement is true. Then have the students use the chart at the top of page 68 to check their guesses. (2.63 meters) Ask the students if the force of gravity on Mars is more or less than the force of gravity on Earth. (Less)

Here's How Note
Use of Concrete Materials You may wish to use the powers-of-ten tiles from copymaster or transparency S529 to demonstrate multiplying a decimal by a whole number. See ▪ **Manipulative Activity 4** on copymaster S515 in the Teacher's Resource Binder.

Multiplying decimals

DID YOU KNOW . . . The height that you can jump depends on the force of gravity. Since the force of gravity varies from planet to planet, you would jump different heights on different planets.

PLANET	HIGH-JUMP FACTOR
Earth	1.00
Mercury	3.57
Venus	1.16
Mars	2.63
Jupiter	0.38
Saturn	0.83
Uranus	1.09
Neptune	0.91
Pluto	1.43

1. What is the high-jump factor for the planet Mercury? **3.57**
2. Suppose that you high-jumped 1.63 m (meters) on the planet Earth. You could multiply 1.63 by the high-jump factor of Venus to find how high the jump would have been on Venus. What two numbers would you multiply to find how many meters you would have jumped on Venus? **1.63 and 1.16**

Here's how to multiply decimals. $1.63 \times 1.16 = ?$

First round each decimal to the nearest whole number and estimate the product.

$$1.63 \times 1.16 = ?$$
$$\downarrow \quad \downarrow$$
$$2 \times 1 = 2$$

Multiply as whole numbers.	Count the digits to the right of the decimal points.	Count off the same number of digits in the product.
1.63 ×1.16 978 163 163 18908	1.63 ×1.16 **4** 978 163 163 18908	1.63 ×1.16 978 163 163 1.8908

---

3. Look at the *Here's how*. How high would you have jumped on Venus? **1.8908 meters** Was 2 meters a reasonable estimate? **Yes**

4. Check each example. Is the answer correct?

   a. 4.6
       × 2
       ─────
       9.2 **Yes**

   b. 3.46
       ×0.02
       ──────
       0.0692 **Yes**
       You have to write a zero here to place the decimal point.

   c. 65
       ×0.3
       ─────
       19.5 **Yes**

68 Chapter 3

EXERCISES

Multiply. Here are scrambled answers for the next row of exercises: 0.0231 11.97 2.914 2.635 26.22 4.8

5. 3.8 × 6.9 = 26.22	6. 5.7 × 2.1 = 11.97	7. 2.4 × 2 = 4.8	8. 0.31 × 9.4 = 2.914	9. 2.31 × 0.01 = 0.0231	10. 4.25 × 0.62 = 2.635
11. 1.8 × 56 = 100.8	12. 7.4 × 10 = 74	13. 0.05 × 0.7 = 0.035	14. 0.85 × 16 = 13.6	15. 0.44 × 6.6 = 2.904	16. 321 × 8.4 = 2696.4
17. 3.28 × 3.5 = 11.48	18. 7.46 × 6.1 = 45.506	19. 5.26 × 4.2 = 22.092	20. 0.06 × 0.44 = 0.0264	21. 8.52 × 0.65 = 5.538	22. 4.16 × 0.06 = 0.2496
23. $7.52 × 89 = $669.28	24. $5.47 × 33 = $180.51	25. $2.83 × 17 = $48.11	26. $7.53 × 62 = $466.86	27. $3.91 × 53 = $207.23	28. $8.74 × 21 = $183.54

29. 5.7 × 0.42 2.394
30. 3.5 × 0.57 1.995
31. 0.29 × 0.05 0.0145
32. 2.4 × 0.6 1.44
33. 39 × 6.3 245.7
34. 0.51 × 8.2 4.182
35. 65 × 0.39 25.35
36. 77 × 1.4 107.8
37. 368 × 2.7 993.6
38. 4.06 × 10 40.6
39. 0.08 × 0.4 0.032
40. 0.25 × 0.3 0.075
41. 5.83 × 0.95 5.5385
42. 0.01 × 0.5 0.005
43. 79.5 × 0.36 28.62
44. 4.62 × 9.5 43.89

Solve. Use the table on page 68.

45. If you high-jumped 1.63 meters on Earth, how high would you have jumped on Pluto? 2.3309 meters

46. If you high-jumped 1.56 meters on Earth, how high would you have jumped on Mercury? Round the answer to the nearest hundredth of a meter. 5.57 meters

47. On which planet would you jump the lowest? the highest? Jupiter, Mercury

48. On which planets would you jump higher than on the planet Earth? On which planets would you jump lower?

49. Suppose that you could jump 1.54 meters on Earth. How much higher could you jump on Uranus than on Neptune? 0.2772 meter

50. Solve problem 49 in another way. 1.54 × (1.09 − 0.91)

Moon meet Logical reasoning

51. Study the clues to determine the high-jump factor for the moon. 5.88

   Clues:
   • There are two places to the right of the decimal point.
   • If you round it to the nearest tenth, you get 5.9.
   • It has two digits that are the same.
   • The sum of the digits is odd.

Multiplying Whole Numbers and Decimals **69**

Extra Practice
Page 476 Skill 11

Practice Worksheet
Workbook S198, Copymaster S198, or Duplicating Master S198

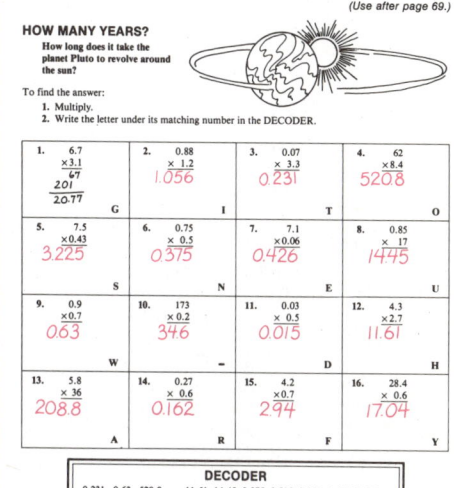

WORKSHEET 30 (Use after page 69.)

HOW MANY YEARS? How long does it take the planet Pluto to revolve around the sun?

To find the answer:
1. Multiply.
2. Write the letter under its matching number in the DECODER.

1. 6.7 ×3.1 = 20.77 **G**	2. 0.88 ×1.2 = 1.056 **I**	3. 0.07 ×3.3 = 0.231 **T**	4. 62 ×8.4 = 520.8 **O**
5. 7.5 ×0.43 = 3.225 **S**	6. 0.75 ×0.5 = 0.375 **N**	7. 7.1 ×0.06 = 0.426 **E**	8. 0.85 ×17 = 14.45 **O**
9. 0.9 ×0.7 = 0.63 **W**	10. 173 ×0.2 = 34.6 **−**	11. 0.03 ×0.5 = 0.015 **D**	12. 4.3 ×2.7 = 11.61 **H**
13. 5.8 ×36 = 208.8 **A**	14. 0.27 ×0.6 = 0.162 **R**	15. 4.2 ×0.7 = 2.94 **F**	16. 28.4 ×0.6 = 17.04 **Y**

DECODER

0.231 0.63 520.8   11.61 14.45 0.375   0.015 0.162 0.426 0.015
 T   W   O   H   U   N   D   R   E   D

2.94 520.8 0.162 0.231 17.04 34.6 0.426 1.056 20.77 11.61 0.231
 F   O   R   T   Y   −   E   I   G   H   T

17.04 0.426 208.8 0.162 3.225
 Y   E   A   R   S

© D. C. Heath and Company   S198   *Multiplying decimals*

Project

Using library resources

Use an almanac to find which planet, with an orbit speed of over 100,000 miles per hour, is the fastest planet.

Mercury

Copymaster S475

Answer for page 69.
48. higher—Mercury, Venus, Mars, Uranus, Pluto
    lower—Jupiter, Saturn, Neptune

69

Class Starter Quiz 28
on previous lesson

Multiply.

1. 2.4 × 3.1   7.44
2. 9.3 × 20   186
3. 0.06 × 0.8   0.048
4. 2.5 × 0.66   1.65
5. 89 × 0.4   35.6

Copymaster S391

Lesson Objective
To simplify expressions

Problem-Solving Skills
Choosing the appropriate information
Choosing the correct operation
Solving a multistep problem

Starting the Lesson
Write these expressions on the chalkboard:

Customer A's purchase:
$24.79 + (2 × $29.79)

Customer B's purchase:
2 × ($29.79 + $24.79)

Have the students describe, by using the clues on the chalkboard and the information at the top of page 70, what Customers A and B purchased. (Customer A bought an Office Mouse and 2 Graduate Owls; Customer B bought 2 Graduate Owls and 2 Office Mice.)

70

Simplifying expressions

You are a mail-order clerk for the Wee Forest Folk Gift Shop. Your job is to compute the total cost of the gifts you pack and ship.

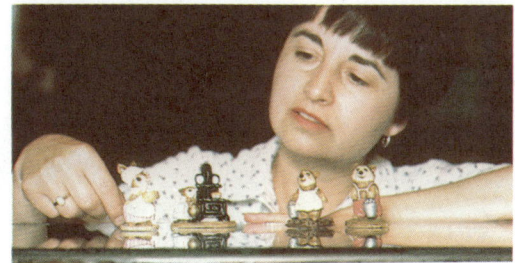

Travelin' Bear
$39.89

$29.79
Graduate Owl

Wee Forest Folk Gifts

Office Mouse
$24.79

1. What is the cost of an Office Mouse? $24.79
2. What is the cost of a Graduate Owl? $29.79
3. To find the total cost of 3 Graduate Owls and an Office Mouse, you would first multiply 3 times [?] and then add $24.79. $29.79

Here's how to simplify the expression (3 × $29.79) + $24.79.

If you estimate before you calculate, you will know whether your answer makes sense.

Remember to work inside the grouping symbols first.

Estimate this way:
(3 × 30) + 25 = ?
90 + 25 = 115

(3 × $29.79) + $24.79 = ?

First multiply.
$29.79
×   3
$89.37

Then add.
+ 24.79
$114.16

4. Look at the *Here's how*. What is the total cost of 3 Graduate Owls and an Office Mouse? Was the estimate near the answer? $114.16, Yes

5. Check each example. Which answer cannot be correct? b
   Hint: Estimate by rounding each decimal to the nearest whole number.

   80 − (25 + 30)   (30 + 40) × 4

   a. 79.78 − (24.79 + 29.79) = 25.20   b. (29.79 + 39.89) × 4 = 73.68

70 Chapter 3

EXERCISES

6. Two of these calculator answers are wrong. Find them by estimating.
Hint: Remember to work inside the grouping symbols first.

(82 + 60) − 10

a. (82.3 + 59.7) − 9.8   `132.2`
c. 29.7 + (19.8 × 3)   `198.5`
e. (59.2 − 10.9) + 21.8   `70.1`
g. (3.1 + 5.8) × 2.1   `18.69`

9 − (6 + 2)

b. 9.3 − (6.2 + 1.9)   `1.2`
d. (24.8 + 35.3) + 40.1   `100.2`
f. 20 − (5.8 × 2.1)   `51.06`
h. 3.1 + (5.8 × 2.1)   `15.28`

(c and f are circled)

Simplify. Here are scrambled answers for the next row of exercises: 2.2   12.4   74.4

7. 5.2 + (2.4 × 3) 12.4
8. 10 − (4.6 + 3.2) 2.2
9. 9.3 × (4.8 + 3.2) 74.4
10. (15 × 0.6) + 3.25 12.25
11. (0.3 × 4) + 1.45 2.65
12. 6 − (3.5 + 2.5) 0
13. (4.8 − 3.5) × 5 6.5
14. (2.8 + 4.2) + 5.9 12.9
15. 2.4 + (2.65 × 4) 13
16. 14.6 + (9.2 − 3.1) 20.7
17. 35 − (3.6 × 8) 6.2
18. (3.4 × 6) − 8.2 12.2
19. (9.15 + 6.45) − 12 3.6
20. (9.1 × 2) + 4.6 22.8
21. 8 − (1.5 × 3) 3.5
22. 1.45 + (2.15 × 3) 7.9
23. (7.7 + 5.2) − 6 6.9
24. 15 − (2.5 × 4) 5
25. (7.2 + 8.4) − 5.9 9.7
26. 2.8 + (6 × 5.4) 35.2
27. (6 × 2.5) − 7 8
28. 8.5 − (4.4 + 1.8) 2.3
29. (7.5 × 4) − 12 18
30. 22.25 − (5.25 × 3) 6.5

You decide!

Use the prices on page 70. Decide whether expression A, B, C, or D would be used to solve the problem. Then solve the problem.

Expression A: $39.89 + (3 × $24.79)
Expression B: $100 − (3 × $29.79)
Expression C: (3 × $39.89) − $100
Expression D: $24.79 + (3 × $39.89)

31. What's the total cost for an Office Mouse and 3 Travelin' Bears? D, $144.46

32. What is the total cost for a Travelin' Bear and 3 Office Mice? A, $114.26

33. A customer paid for 3 Graduate Owls with a $100 bill. How much change should the customer receive? B, $10.63

34. How much more money would a customer need in order to buy 3 Travelin' Bears if the customer had a $100 gift certificate? C, $19.67

Multiplying Whole Numbers and Decimals **71**

Extra Practice
Page 476   Skill 12

Practice Worksheet
Workbook S199, Copymaster S199, or Duplicating Master S199

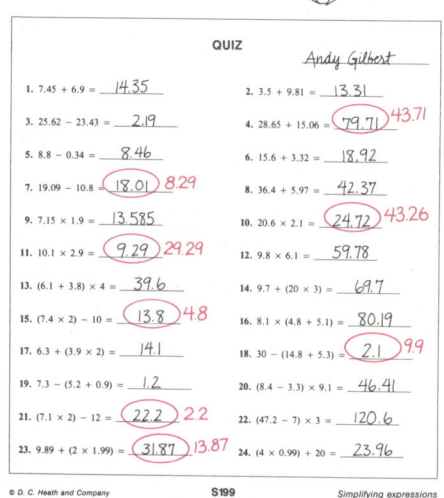

Challenge Problems

Put in parentheses to make the equation true.

1. 3 × (6 + 4) = 30
2. (6 + 2) × 3 − 4 = 20
3. (5 − 4) × 6 + 4 = 10

Copymaster S442

Class Starter Quiz 29
on previous lesson

Simplify.

1. 4.3 + (2.5 + 1.5)   8.3
2. 10 − (3.7 + 2.3)   4
3. (6.1 × 3) + 5.2   23.5
4. 38.5 − (6.5 × 3)   19
5. (6.4 × 5) + 3.5   35.5

Copymaster S392

Lesson Objective
To multiply decimals mentally by 10, 100, or 1000

Problem-Solving Skills
Finding information in a table
Checking answers

Starting the Lesson
Take a Survey Before the students open their books, take a class survey. Ask the students to guess how many pounds of paper products the average person throws away each day. Record the high and low guesses on the chalkboard. Then say, "Open your book to page 72. Look at the top of the page. What answer does the book give?" (2.48)

Exercise Note
Mental Math Use several of exercises 4–55 as oral exercises before making a written assignment. Have the students verbalize the method they use to mentally compute the products.

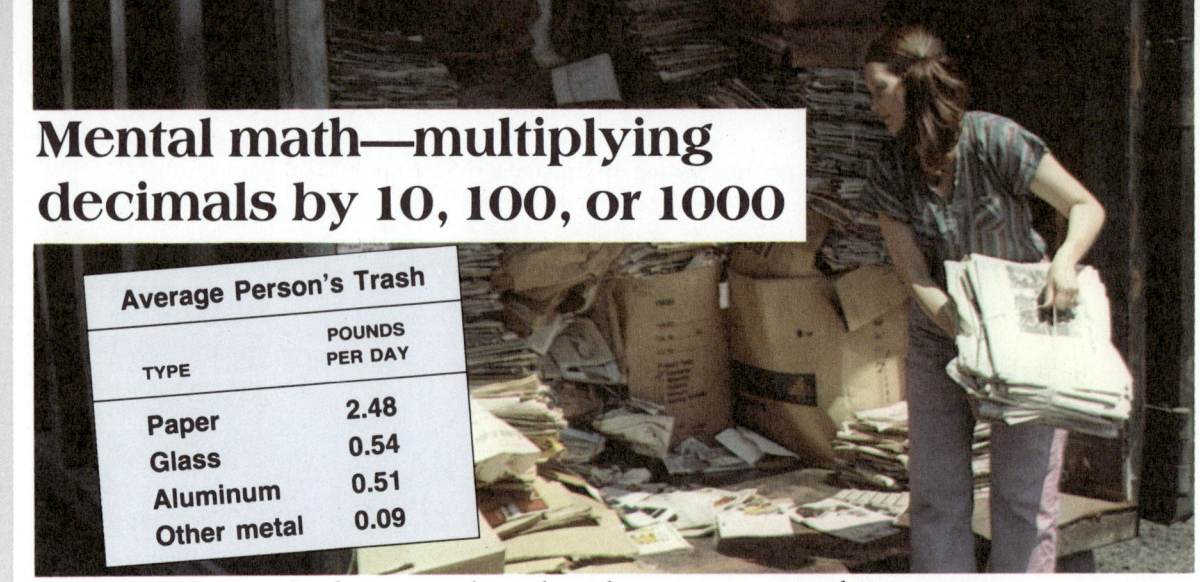

Mental math—multiplying decimals by 10, 100, or 1000

Average Person's Trash

TYPE	POUNDS PER DAY
Paper	2.48
Glass	0.54
Aluminum	0.51
Other metal	0.09

1. How many pounds of paper products does the average person throw away each day? 2.48

2. What two numbers would you multiply to find how many pounds of paper products the average person throws away in 10 days? 100 days? 1000 days?
2.48 and 10; 2.48 and 100; 2.48 and 1000

Here's how to multiply by 10, 100, or 1000.

When you multiply a number by 10, 100, or 1000, the product is greater than the number.

2.48 × 10 = ?	2.48 × 100 = ?	2.48 × 1000 = ?
Multiplying by 10 moves the decimal point 1 place to the right.	Multiplying by 100 moves the decimal point 2 places.	Multiplying by 1000 moves the decimal point 3 places.

$$\begin{array}{r} 2.48 \\ \times\ 10 \\ \hline 24.80 \end{array}$$

2.48 × 10 = 24.8   2.48 × 100 = 248.   2.48 × 1000 = 2480.

3. Look at the *Here's how*. How many pounds of paper products does the average person throw away in 10 days? 100 days? 1000 days?

EXERCISES

MENTAL MATH—Multiply. Write answers only. Here are scrambled answers for the next two rows of exercises: 53,600   680   42   420   536   5360   4.2   6800

4. 4.2 × 10   42
5. 42 × 10   420
6. 0.42 × 10   4.2
7. 5.36 × 100   536
8. 536 × 100   53,600
9. 5.36 × 1000   5360
10. 6.8 × 1000   6800
11. 0.68 × 1000   680
12. 68 × 1000   68,000
13. 0.396 × 1000   396
14. 3.96 × 10   39.6
15. 3.96 × 100   396

16. 3.96 × 1000 3960
17. 15.3 × 10 153
18. 15.3 × 100 1530
19. 15.3 × 1000 15,300
20. 4900 × 10 49,000
21. 4900 × 100 490,000
22. 4900 × 1000 4,900,000
23. 490 × 100 49,000
24. 38 × 10 380
25. 38 × 100 3800
26. 38 × 1000 38,000
27. 380 × 100 38,000
28. 0.67 × 1000 670
29. 0.67 × 100 67
30. 0.67 × 10 6.7
31. 6.7 × 1000 6700
32. 0.08 × 1000 80
33. 0.08 × 100 8
34. 0.08 × 10 0.8
35. 0.8 × 10 8
36. 142 × 10 1420
37. 142 × 1000 142,000
38. 142 × 100 14,200
39. 1.42 × 100 142
40. 346 × 100 34,600
41. 346 × 10 3460
42. 346 × 1000 346,000
43. 34.6 × 1000 34,600
44. 7.46 × 1000 7460
45. 7.46 × 10 74.6
46. 7.46 × 100 746
47. 746 × 100 74,600
48. 0.006 × 100 0.6
49. 0.006 × 1000 6
50. 0.006 × 10 0.06
51. 0.06 × 10 0.6
52. 6.088 × 1000 6088
53. 6.088 × 100 608.8
54. 6.088 × 10 60.88
55. 60.88 × 10 608.8

Solve. Use the table on page 72. Assume that your trash is the average amount.

56. How many pounds of glass would you throw away in 10 days? 5.4
57. How many pounds of aluminum would you throw away in 100 days? 51
58. What is the total number of pounds of trash that you throw away each day? 3.62
59. How many pounds of non-paper trash would you throw away in 100 days? 114
60. Suppose you save glass instead of throwing it away. How many pounds would you save for recycling in 100 days? 54
61. How many pounds of aluminum would you save for the recycling center in 100 days? 51
62. The recycling center pays $.02 a pound for paper products. How much would you earn if you sold 100 days' worth of paper products? $4.96
63. The price for recycled aluminum is $.15 per pound. How much would you earn if you recycled 100 days' worth of aluminum? $7.65

Check the products Checking answers

64. Find and correct the two wrong answers.

a. Three and sixth tenths times ten times four equals

144

b. Four and twelve hundredths times one hundred times two equals

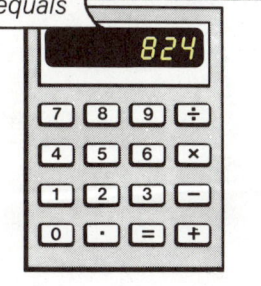

c. Six and one tenth times ten times two tenths equals

12.2

Multiplying Whole Numbers and Decimals **73**

Extra Practice
Page 477   Skill 13

Practice Worksheet
Workbook S200, Copymaster S200, or Duplicating Master S200

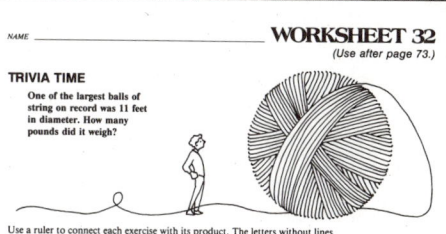

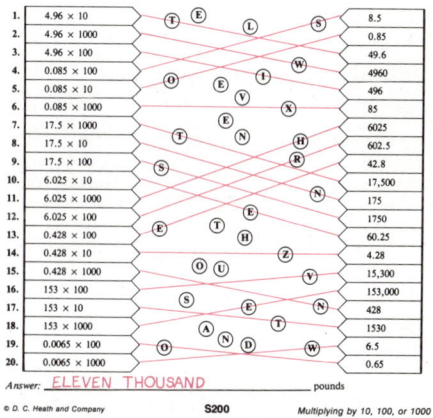

Project
Researching information
Collect all the newspapers brought to your home in a week and weigh them. Then estimate how many pounds of newspapers are brought to your home in a year.

Copymaster S476

PROBLEM SOLVING using simpler problems to solve multistep problems

Some problems look harder than they are. It may help you solve problems if you substitute simple numbers for the actual numbers.

Problem
Pfiefer's Auto Supply purchases a box of 4 spark plugs for $8.84. Pfiefer's sells the plugs for $11.56 a box. What is the total markup (difference between the selling price and the cost) on 29 boxes?

$11.56 (box of 4)

Here's how *to write a simpler problem.*

STUDY Find the facts that are needed. Then substitute simpler numbers for the actual numbers.

cost: $8.84 a box ($9)

selling price: $11.56 a box ($12)

boxes sold: 29 (30)

PLAN and SOLVE First solve the problem with simpler numbers. Then solve the problem using the actual numbers.

Using simpler numbers

$12 − $9 = $3 *Estimated markup on each box of spark plugs.*

$3 × 30 = $90 *Estimated total markup.*

Using actual numbers

11.56 − 8.84 = 2.72

3.08 × 29 = 89.32

Answer: The markup on 29 boxes of spark plugs is $89.32.

CHECK $89.32 is near the estimated total markup. So the answer seems reasonable.

PROBLEMS

First solve the simpler problem. Then solve the problem using the actual numbers.

1. Pfiefer's purchases a box of 6 spark plugs for $12.96. It sells 3 boxes of plugs for $50.28. What is the total markup on 3 boxes of spark plugs?

   > *Using simpler numbers*
   > $13 × 3 = $39
   > $50 − $39 = ?

   $11, $11.40

2. Motor oil that regularly sells for $1.88 a quart, is marked down $.29 a quart. What is the total sale price for 12 quarts of oil?

   > *Using simpler numbers*
   > $1.90 − $.30 = $1.60
   > $1.60 × 10 = ?

   $16, $19.08

Solve. Simplify the problem if you need to.

3. Alicia buys 4 shock absorbers for $9.88 each and 1 muffler for $28.98. How much money does she spend in all? $68.50

4. Angelo buys 6 cans of motor oil for $1.88 a can, 4 spark plugs for $2.79 each, and an oil filter for $7.99. What is his total cost? $30.43

5. Arturo buys 2 windshield wiper blades for $3.79 each and a wiper arm for $6.98. He gives the cashier a coupon for $.75 off the price of the wiper arm. How much does he spend in all? $13.81

6. Pfiefer's Auto Supply purchased 18 cases of oil filters. Each case contained 24 filters. After selling 178 of the oil filters, how many oil filters are left to be sold? 254

7. Linda buys 4 all-weather radial tires that regularly sell for $49.88 each at the sale price of $39.95 each. What is the total discount (amount of money Linda saves by buying the tires at the sale price)? $39.72

8. All car speakers that regularly sell for less than $50 each are marked down $9.99. Speakers that sell for $50 or more are marked down $19.99. What is the total sale price for 2 speakers that regularly sell for $39.95 and $79.95? $89.92

Mileage matchup Number sense

9. When Al left Pfiefer's Auto Supply at 8:00 A.M., the odometer on the delivery van read:

   | 3 | 7 | 6 | 8 |

   Al drove 68 miles to San Antonio to make a delivery and returned to Pfiefer's at 11:30 A.M. Which reading now showed on the odometer?

   a. | 3 | 8 | 7 | 3 |
   b. | 4 | 6 | 2 | 1 |
   (c.) | 3 | 9 | 6 | 1 |

Multiplying Whole Numbers and Decimals **75**

Practice Worksheet
Workbook S201, Copymaster S201, or Duplicating Master S201

Challenge Problem

Write a question that fits the answer.

> After the bell, 300 students came to the auditorium. The auditorium has 24 rows of seats with 16 seats in each row.
>
> How many students will not get seats?
>
> *Answer:* 12 students

Copymaster S443

Class Starter Quiz 31
on previous lesson

First solve a simpler problem. Then solve the problem using actual numbers.

Kurt buys 2 windshield-wiper blades for $3.79 each and an air filter for $7.99. What is his total cost? $15; $15.57

Copymaster S392

Problem-Solving Skills

Using information on a menu and a cash-register display
Choosing the correct operation
Using a guess-and-check strategy

Skills Reviewed

Adding, subtracting, and multiplying whole numbers and decimals
Comparing decimals
Simplifying expressions

Starting the Lesson

Problem Solving Have the students use the menu and cash-register display to answer questions like these:

- How much does a chef salad cost? ($2.85)
- What would the total be if you pushed the milk, tuna, and total keys? ($2.65)
- Which 2 items on the menu cost a total of $3.80? (Ham and tossed green salad)

Cumulative Skill Practice Challenge the students to an estimation hunt by saying, "Pick the largest difference in the first row of exercises." (Exercise 6) Then have students pick the smallest difference in the first row of exercises. (Exercise 1)

Answers for page 76.
3. a. $4.30 b. $6.15 c. $3.60
4. a. 1 dime, 3 $1 bills
   b. 2 dimes, 1 $1 bill, 1 $5 bill
   c. 2 dimes, 1 quarter, 1 $1 bill, 1 $5 bill
5. a. roast beef, chef salad, coffee or tea
   b. tuna plate, fruit salad, coffee or tea; or tuna plate, green salad, milk
   c. roast beef, green salad, juice; or ham, green salad, coffee

Problem solving

COMPUTERIZED CASH REGISTERS

Many cafeterias use computerized cash registers. The computer is programmed to calculate and display the total cost when the cashier keys in an order. When the cashier keys in the amount rendered (the amount the customer uses to pay), the computer will calculate the amount of change.

Use the information on the menu and the cash register to answer these questions. Decide when a calculator would be useful.

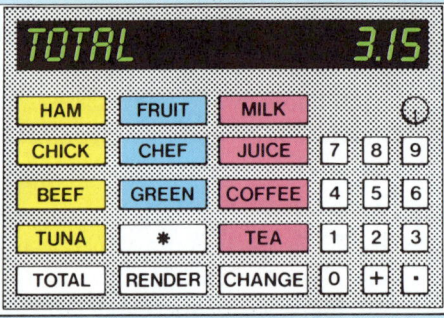

1. My last customer ordered an entree and a salad. What keys did I push? Chick, Fruit

2. He paid with a $5 bill. I gave him 4 coins and a bill for change. What were the coins? 3 quarters and 1 dime

3. What total does the computer display when these keys are pushed?

4. What coins and bills would you use to make change for these purchases? (Assume each customer paid with a $10 bill. Use the fewest coins possible.)

   a. TOTAL 6.90
   b. TOTAL 3.80
   c. TOTAL 3.55

5. Each of these customers bought an entree, a salad, and a drink. What did each customer order?

   a. TOTAL 5.90
   b. TOTAL 3.20
   c. TOTAL 4.10

6. Each of these customers bought two items and paid with a $5 bill. What did each customer buy?

   a. CHANGE 2.45
   b. CHANGE .95
   c. CHANGE .25

Cumulative Skill Practice

Subtract. *(page 38)*

1. 3897 − 142 = 3755
2. 6351 − 248 = 6103
3. 7293 − 1526 = 5767
4. 8374 − 2987 = 5387
5. 5328 − 1479 = 3849
6. 9742 − 2836 = 6906

Less than (<) or greater than (>)? *(page 42)*

7. 0.6 > 0.5
8. 0.03 < 0.04
9. 12.8 > 12.7
10. 0.1 > 0.03
11. 3.05 > 3.005
12. 2.6 > 2.599
13. 4.206 < 4.26
14. 1 > 0.99
15. 0.034 < 0.34

Give the difference. *(page 46)*

16. 74.3 − 38.5 35.8
17. 26.3 − 14.9 11.4
18. 6.00 − 3.81 2.19
19. 9.4 − 3.8 5.6
20. 20.15 − 6.83 13.32
21. 48.06 − 3.57 44.49
22. 29 − 15.8 13.2
23. 30 − 7.42 22.58
24. 6.3 − 5.88 0.42

Multiply. *(page 64)*

25. 223 × 115 = 25,654
26. 402 × 321 = 129,042
27. 336 × 152 = 51,072
28. 538 × 226 = 121,588
29. 629 × 427 = 268,583
30. 745 × 281 = 209,345

Give the product. *(page 68)*

31. 3.2 × 0.8 2.56
32. 4.6 × 0.31 1.426
33. 0.55 × 0.4 0.22
34. 28 × 1.4 39.2
35. 0.56 × 8.3 4.648
36. 4.21 × 1.3 5.473
37. 6.54 × 18 117.72
38. 26.8 × 7.3 195.64
39. 46.3 × 0.84 38.892

Simplify. *(page 70)*

40. 3.5 + (2.1 × 4) 11.9
41. 26 − (8.3 + 2.6) 15.1
42. (8.3 − 6) + 2.6 4.9
43. (9.4 × 0.5) − 1.4 3.3
44. (7.2 − 3.81) × 5 16.95
45. 4.6 + (4.2 × 3) 17.2

MIXED PRACTICE

Complete.

46. 38,256 + 5,992 = ? 44,248
47. 1.2 × 1.2 = ? 1.44
48. 5.63 + 2.41 + 0.93 = ? 8.97
49. 8000 − 3556 = ? 4444
50. 529 + 7 + 88 = ? 624
51. 38.2 − 17.6 = ? 20.6
52. 256 × 140 = ? 35,840
53. 43 − 21.68 = ? 21.32
54. 355 × 2.4 = ? 852
55. 7.29 + 8 + 5.4 = ? 20.69
56. 0.71 × 7.6 = ? 5.396
57. 4738 − 2179 = ? 2559
58. (5.4 + 3.86) × 2 = ? 18.52
59. 5.4 + (3.86 × 2) = ? 13.12
60. (7.5 − 4.1) × 0.7 = ? 2.38

Multiplying Whole Numbers and Decimals **77**

Problem-Solving Worksheet

Workbook S202, Copymaster S202, or Duplicating Master S202

NAME _____ **WORKSHEET 34**
(Use after page 76.)

SIGN TIME

Four of these road signs have a total of 15 edges. How many of the signs are SLOW signs? 3

LOGICAL REASONING

What is the price of the radio?
Clues:
- The radio costs more than $80, but less than $100.
- You can buy it with the same number of $10 bills, $5 bills, and $1 bills.

DON'T STRIKE OUT!

Use the code to get the answer.

CODE					
A	B	C	D	...	Z
60−1	59−2	58−3	57−4		35−26

Trivia question: Who was the only major league pitcher to strike out 10 batters in a row?
Answer:

T O M   S E A V E R
21 31 35   23 51 59 17 51 25
O F   T H E   N E W
31 49   21 45 51   33 51 15
Y O R K   M E T S
11 31 25 39   35 51 21 23

GUESS AND CHECK

Three of these coins are Rita's and the others are Terry's. If Rita gave one of her coins to Terry, both would have the same amount of money. How much money does Rita have? 40¢

NUMBER LETTERS

What number is the same as the number of letters in its name? Four

CALENDAR MATH

If December 5th is on a Monday, what day of the week is December 24th? Saturday

© D. C. Heath and Company   S202   Problem solving

Project

Researching information

Find how a clerk gives change. Use the clerk's method to tell the coins given as change for a dollar, when each of these amounts is charged: $.57, $.19, $.36, $.34, $.83

$.57: 3 pennies, 1 dime, 1 nickel, 1 quarter
$.19: 1 penny, 1 nickel, 3 quarters
$.36: 4 pennies, 1 dime, 2 quarters
$.34: 1 penny, 1 nickel, 1 dime, 2 quarters
$.83: 2 pennies, 1 nickel, 1 dime

Note: Answers will vary if half dollars are used.

Copymaster S476

More answers for page 76.
6. a. tuna and coffee or tea
   b. ham and juice
   c. chicken and tuna

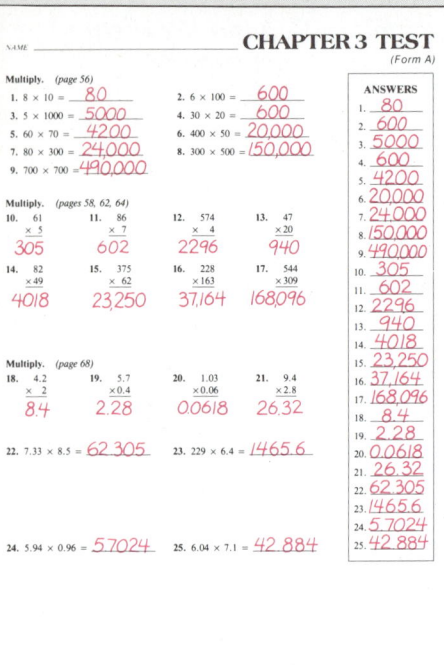
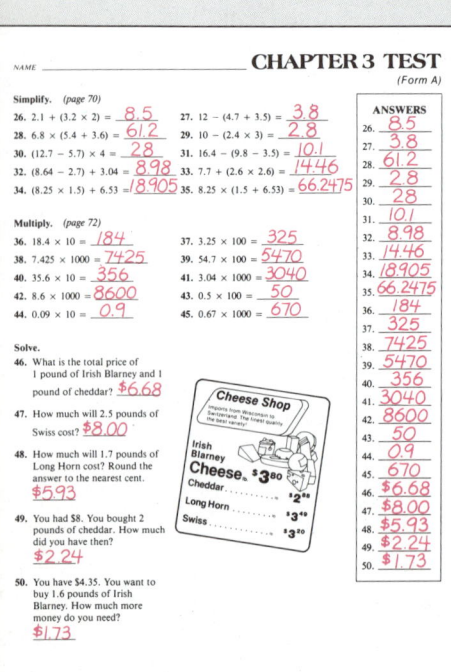

Chapter REVIEW

Here are scrambled answers for the review exercises:

3	5	12	add	right	zeros
4	7	80	product	whole	

1. 4   2. 5. product   3. 7

1. To do this multiplication exercise, you can multiply 3 × 5 and write ? zeros. *(page 56)*

300 × 500

2. To estimate this product, you could multiply 400 by ?. The answer to a multiplication exercise is called the ?. *(page 58)*

391 × 5

3. To complete this multiplication exercise, you would multiply 3 by 8 and then add ?. *(page 58)*

```
   7
  391
 ×  8
 ────
   28
```

4. 80   5. zeros   6. 12

4. To find this product, you would multiply 97 by 6 and 97 by ? and then add. *(page 62)*

```
  97
 ×86
```

5. To do this multiplication exercise, you don't have to write the 3 ?. *(page 64)*

```
    342
  × 127
  ─────
   2394
   6840
  34200
  ─────
  43,434
```

6. To estimate this product, you could round each decimal to the nearest whole number. The estimate would be ?. *(page 68)*

1.95 × 6.1

7. whole, 3   8. add   9. right

7. To find this product, you would multiply as if the factors were ? numbers and count off ? digits to put the decimal point in the product. *(page 68)*

1.95 × 6.1

8. To simplify this expression, you would first ? and then subtract. *(page 70)*

12.4 − (4.6 + 3.05)

9. Multiplying a decimal by 1000 moves the decimal point 3 places to the ?. *(page 72)*

6.2 × 1000 = 6200

78   Chapter 3

Chapter TEST

Multiply. (pages 56, 58)

1. 6 × 10   60
2. 9 × 100   900
3. 8 × 1000   8000
4. 30 × 30   900
5. 23 × 100   2300
6. 60 × 40   2400
7. 18 × 100   1800
8. 30 × 500   15,000
9. 700 × 800   560,000
10. 5000 × 90   450,000
11. 32 × 3   96
12. 76 × 8   608
13. 256 × 5   1280
14. 3906 × 6   23,436
15. 487 × 8   3896

Multiply. (pages 62, 64)

16. 24 × 12 = 288
17. 78 × 27 = 2106
18. 139 × 35 = 4865
19. 420 × 64 = 26,880
20. 268 × 25 = 6700
21. 685 × 63 = 43,155
22. 326 × 153 = 49,878
23. 439 × 540 = 237,060
24. 3561 × 209 = 744,249
25. 4298 × 753 = 3,236,394
26. 6533 × 95 = 620,635
27. 1806 × 22 = 39,732

Multiply. (page 68)

28. 3.4 × 1.5 = 5.1
29. 6.3 × 1.4 = 8.82
30. 0.84 × 0.6 = 0.504
31. 53.7 × 4.6 = 247.02
32. 0.06 × 0.44 = 0.0264
33. 3.28 × 3.5 = 11.48
34. 0.39 × 2.8   1.092
35. 0.12 × 6   0.72
36. 9.44 × 1.05   9.912
37. 82.1 × 0.7   57.47

Simplify. (page 70)

38. 2.41 + (0.04 × 0.2)   2.418
39. (31.5 − 4.6) + 8.9   35.8
40. (5.7 + 3.2) × 4.6   40.94
41. 10.42 − (8.63 − 2.48)   4.27
42. (60 − 5.4) × 5.6   305.76
43. 4 − (1.63 × 1.5)   1.555

Multiply. (page 72)

44. 27 × 10   270
45. 3.49 × 10   34.9
46. 5.4 × 100   540
47. 0.67 × 1000   670
48. 0.08 × 100   8
49. 15.6 × 1000   15,600
50. 0.04 × 1000   40
51. 2.0312 × 100   203.12

Solve.

52. What is the total price of 1 pound of macaroni salad and 1 pound of Bar-B-Q beef ribs? $3.97
53. How much will 1.5 pounds of honey-cured ham cost? $6.72
54. How much will 2.25 pounds of cheddar cheese cost? Round the answer to the nearest cent. $6.50
55. What is the total price of 1.5 pounds of beef ribs and 1.75 pounds of ham? $12.31

DAVID'S DELI

Fresh and Creamy
Macaroni Salad .... pound 99¢

Fresh Sliced
Honey-cured Ham .... pound $4.48

Bar-B-Q Beef Ribs .... pound $2.98

Fresh Sliced
Cheddar Cheese .... pound $2.89

Multiplying Whole Numbers and Decimals   79

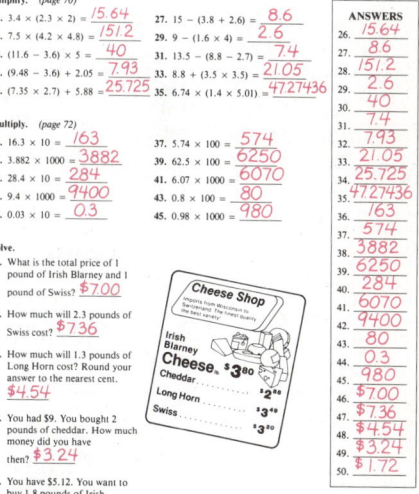

Copymasters S11 and S12
or Duplicating Masters S11 and S12

Cumulative Test
(Chapters 1–3)

Use Copymaster S109 to provide the students with an answer sheet in standardized test format.

Answers for Cumulative Test, Chapters 1–3

1. **A** B C D	2. A B **C** D	3. A B **C** D
4. A **B** C D	5. **A** B C D	6. A **B** C D
7. A **B** C D	8. A B **C** D	9. **A** B C D
10. A B C **D**	11. A **B** C D	12. **A** B C D

The table below correlates test items with student text pages.

Test Item	Page Taught	Skill Practice
1	6	p. 67, exercises 1–6
2	10	p. 67, exercises 7–16
3	14	p. 67, exercises 17–24
4	18	p. 67, exercises 25–34
5	22	p. 67, exercises 35–43
6	38	p. 77, exercises 1–6
7	42	p. 77, exercises 7–15
8	46	p. 77, exercises 16–24
9	64	p. 77, exercises 25–30
10	68	p. 77, exercises 31–39
11	70	p. 77, exercises 40–45
12	46	

80

Cumulative TEST — Standardized Format

Choose the correct letter.

1. Add.
$$\begin{array}{r} 326 \\ 435 \\ 284 \\ +175 \\ \hline \end{array}$$
 A. 1220
 B. 1000
 C. 1200
 D. none of these

2. 4950 rounded to the nearest hundred is
 A. 4950
 B. 4900
 C. 5000
 D. none of these

3. The standard numeral for 3 and 48 thousandths is
 A. 3.48
 B. 3.0048
 C. 3.048
 D. none of these

4. 8.954 rounded to the nearest hundredth is
 A. 9.0
 B. 8.95
 C. 8.75
 D. none of these

5. Give the sum.
 $31.4 + 3.89 + 4$
 A. 39.29
 B. 38.29
 C. 70.7
 D. none of these

6. Subtract.
$$\begin{array}{r} 4600 \\ -235 \\ \hline \end{array}$$
 A. 4375
 B. 4365
 C. 4435
 D. none of these

7. Which number is less than 0.89?
 A. 0.98
 B. 0.089
 C. 1
 D. none of these

8. Give the difference.
 $35.2 - 1.84$
 A. 16.8
 B. 33.44
 C. 33.36
 D. none of these

9. Multiply.
$$\begin{array}{r} 374 \\ \times 106 \\ \hline \end{array}$$
 A. 39,644
 B. 5984
 C. 5564
 D. none of these

10. Give the product.
 6.25×1.8
 A. 112.50
 B. 1125.0
 C. 5.625
 D. none of these

11. Simplify.
 $26.3 - (4.6 \times 3.8)$
 A. 35.12
 B. 8.82
 C. 82.46
 D. none of these

12. You're on a 10-mile hike. You hike 2.5 miles during the first hour. How many miles do you have left to hike?
 A. 7.5
 B. 6.5
 C. 4.5
 D. none of these

80 *Chapter 3*

Dividing Whole Numbers and Decimals

Resources

- **Class Starter Quizzes 32-44** *(Copymasters S392-S395)*
- **Visual Aids 3, 13-17** *(Copymasters or Transparencies S113, S122-S124)*
- **Manipulatives**
  Manipulative Activity 5 *(Copymaster S516)*
  Powers-of-ten tiles *(Copymaster or Transparency S529)*
- **Worksheets 35-48** *(Copymasters, Duplicating Masters, or Workbook pages S203-S216)*
- **Challenge Problems** for pages 83, 85, 89, 93, 95, 97, 99, 101, 103, 107 *(Copymasters S443-S445)*
- **Projects** for pages 87, 91, 105, 109 *(Copymasters S476-S477)*
- **Mental Math Extensions** for Skills 14-17 *(Copymasters S495-S497)*
- **Tests** *(Copymasters or Duplicating Masters S13-S16, S69-S72)*

Lesson Objective
To divide by a 1-digit number (with a 3-digit dividend)

Problem-Solving Skills
Finding information in a newspaper article
Choosing the correct operation
Solving a multistep problem
Number sense

Starting the Lesson
Have the students number from 1 to 12 on scratch paper. Tell them you will read a division fact and they are to write the quotient. Explain that you will read each fact only twice.

1. 42 ÷ 6 (7)   2. 54 ÷ 9 (6)
3. 45 ÷ 9 (5)   4. 72 ÷ 8 (9)
5. 56 ÷ 7 (8)   6. 49 ÷ 7 (7)
7. 81 ÷ 9 (9)   8. 63 ÷ 7 (9)
9. 48 ÷ 6 (8)   10. 40 ÷ 5 (8)
11. 36 ÷ 6 (6) 12. 24 ÷ 3 (8)

Have the students check their own papers as you give the answers.

What Are the Facts?
Allow the students 60 seconds to study the article on the top of page 82. Then have the students close their books and decide whether these statements are true or false:

- One hundred four cyclists raised $1456. (True)
- The cyclists were raising money for the Cancer Fund Drive. (False)
- Eight members of the Touring Club rode a total of 384 miles. (True)

82

Dividing by a 1-digit number

BIKE-FOR-BUCKS SUCCESSFUL!
One hundred four local cyclists raised a total of $1456 for this year's Heart Fund Drive. According to John Weaver, this year's organizer, 53 cyclists each rode more than 20 miles and the 8 members of the Touring Club rode a total of 384 miles.

1. How many members of the Touring Club rode a total of 384 miles? **8**
2. What two numbers would you use to compute the average number of miles ridden by the Touring Club members? **384 and 8**
3. Would you multiply or divide to compute the average? **Divide**

Here's how to divide by a 1-digit number.   384 ÷ 8 = ?

Not enough hundreds.
Think 38 tens.

$$8\overline{)384}$$

Divide tens.
Subtract.

$$\begin{array}{r}4\\8\overline{)384}\\-32\\\hline 6\end{array}$$

Think 64 ones.

$$\begin{array}{r}4\\8\overline{)384}\\-32\\\hline 64\end{array}$$

Divide ones.
Subtract.

$$\begin{array}{r}48\\8\overline{)384}\\-32\\\hline 64\\-64\\\hline 0\end{array}$$

The answer is called the **quotient.**

4. Look at the *Here's how*. What was the average distance ridden by the Touring Club members? **48 miles**

EXERCISES
Divide. Here are scrambled answers for the next row of exercises: 22 21 24 14 13 29

5. $2\overline{)48}$ **24**   6. $3\overline{)39}$ **13**   7. $4\overline{)56}$ **14**   8. $3\overline{)63}$ **21**   9. $4\overline{)88}$ **22**   10. $2\overline{)58}$ **29**

11. $2\overline{)38}$ **19**   12. $2\overline{)86}$ **43**   13. $5\overline{)75}$ **15**   14. $6\overline{)96}$ **16**   15. $2\overline{)66}$ **33**   16. $3\overline{)72}$ **24**

Chapter 4

17. 8)96 — 12
18. 4)92 — 23
19. 7)91 — 13
20. 2)96 — 48
21. 4)60 — 15
22. 5)65 — 13

23. 3)81 — 27
24. 5)85 — 17
25. 4)72 — 18
26. 2)84 — 42
27. 6)72 — 12
28. 4)96 — 24

29. 7)84 — 12
30. 9)99 — 11
31. 5)70 — 14
32. 3)75 — 25
33. 5)90 — 18
34. 7)77 — 11

35. 2)426 — 213
36. 4)484 — 121
37. 3)693 — 231
38. 5)555 — 111
39. 4)848 — 212
40. 6)456 — 76

41. 3)537 — 179
42. 6)738 — 123
43. 3)432 — 144
44. 7)994 — 142
45. 3)735 — 245
46. 8)432 — 54

47. 84 ÷ 6   14
48. 60 ÷ 5   12
49. 84 ÷ 3   28
50. 78 ÷ 6   13
51. 896 ÷ 8   112

52. 267 ÷ 3   89
53. 216 ÷ 9   24
54. 725 ÷ 5   145
55. 896 ÷ 7   128
56. 844 ÷ 4   211

57. 732 ÷ 6   122
58. 592 ÷ 8   74
59. 684 ÷ 9   76
60. 679 ÷ 7   97
61. 745 ÷ 5   149

MENTAL MATH Divide. Write answers only.

62. 180 ÷ 3

> 18 tens divided by 3 equals 6 tens, or 60.

63. 120 ÷ 6   20
64. 120 ÷ 4   30
65. 150 ÷ 5   30
66. 350 ÷ 7   50
67. 240 ÷ 4   60
68. 320 ÷ 8   40
69. 420 ÷ 6   70
70. 240 ÷ 3   80
71. 630 ÷ 9   70
72. 560 ÷ 8   70
73. 420 ÷ 7   60
74. 450 ÷ 5   90

Solve.

75. One cyclist had 4 pledges. They were $.60, $.85, $1.25, and $.55 per mile. What was her total pledge per mile? $3.25

76. The best time for a 68-mile course was 4 hours. How many miles per hour did the cyclist average? 17

77. One cyclist had a total pledge of $3.17 per mile. He rode 53 miles. How much money did he raise? $168.01

78. One Touring Club member had a total pledge of $3.35 per mile and another had a total pledge of $2.92 per mile. They both rode 56 miles. What was the difference in the amounts they raised? $24.08

An awesome average!

79. In 1973, Dr. Allan Abbot set a speed record for a bicycle. He rode behind a special windshield mounted on a car. Over a 1-mile course he averaged [?] miles per hour.

   To find [?], write the number that has
   8 ones   3 tens
   7 hundredths 6 tenths   138.674
   4 thousandths 1 hundred

Dividing Whole Numbers and Decimals **83**

Practice Worksheet
Workbook S203, Copymaster S203, or Duplicating Master S203

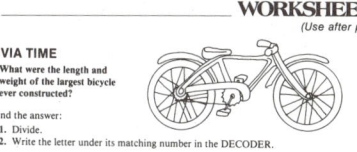

WORKSHEET 35 (Use after page 83.)

NAME _____

TRIVIA TIME What were the length and weight of the largest bicycle ever constructed?

To find the answer:
1. Divide.
2. Write the letter under its matching number in the DECODER.

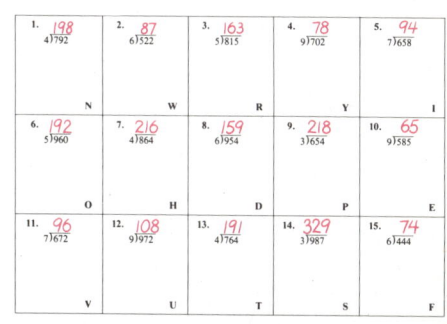

1. 4)792 — 198	2. 6)522 — 87	3. 5)815 — 163	4. 9)702 — 78	5. 7)658 — 94
N	W	R	Y	I
6. 5)960 — 192	7. 4)864 — 216	8. 6)954 — 159	9. 3)654 — 218	10. 9)585 — 65
O	H	D	P	E
11. 7)672 — 96	12. 9)972 — 108	13. 4)764 — 191	14. 3)987 — 329	15. 6)444 — 74
V	U	T	S	F

DECODER

The largest bicycle was
191 87 65 198 191 78 191 216 163 65 65
T W E N T Y - T H R E E

74 65 65 191   191 216 163 65
F E E T long and weighed T H R E E

216 108 198 159 163 65 74 94 96 65   218 192 108 198 159 329
H U N D R E D   F I V E   P O U N D S

© D. C. Heath and Company   S203   Dividing by a 1-digit number

Challenge Problems

Divide.

707 ÷ 7   101   9009 ÷ 9   1001
808 ÷ 8   101   424 ÷ 4   106
303 ÷ 3   101   848 ÷ 8   106
6006 ÷ 6   1001   7042 ÷ 7   1006
4008 ÷ 4   1002

• Do you notice a pattern? Yes
• Can you work these mentally?

5)505 — 101   3)309 — 103   7)714 — 102
6)6012 — 1002 2)2004 — 1002 5)5025 — 1005
9)9027 — 1003 8)816 — 102   4)432 — 108

Copymaster S443

Class Starter Quiz 32
on previous lesson
Divide.
1. 56 ÷ 4   14
2. 66 ÷ 3   22
3. 75 ÷ 5   15
4. 68 ÷ 2   34
5. 896 ÷ 7   128
6. 745 ÷ 5   149
7. 592 ÷ 8   74
8. 648 ÷ 9   72

Copymaster S392

Lesson Objective
To estimate quotients

Problem-Solving Skills
Reading a mileage chart
Using estimation

Starting the Lesson
What are the facts? Allow students 30 seconds to study the travel information at the top of page 84. Then have the students close their books and answer these questions from memory.

- Did the driver drive more or less than 200 miles? (Less)
- Did the car use more or less than 8 gallons of gasoline? (Less)
- Did the trip take more or less than 3 hours? (More)

Use exercises 1–5 and the example in the *Here's how* to introduce estimating quotients.

Estimating quotients

1. Read the road sign. How many miles did Maria drive? 192 miles
2. How many gallons of gasoline did the trip take? 6 gallons
3. To estimate how many miles per gallon the car averaged, would you multiply or divide? divide

I drove from El Paso to here in 4 hours. I made the trip on 6 gallons of gasoline.

Here's how *to estimate quotients.*

To estimate 192 ÷ 6, find the first digit in the quotient.

$$6\overline{)192}^{3}$$

Then write 0's in the remaining places.

$$6\overline{)192}^{30}$$

The quotient is more than 30.

4. Look at the *Here's how*. About how many miles per gallon did the car average? 30 miles per gallon
5. Estimate. Was the average driving speed for the trip more or less than 50 miles per hour? less

EXERCISES

Estimate. Choose the most reasonable answer.

6. 6)324
   a. 5
   b. **50** ✓
   c. 500

7. 3)426
   a. 1
   b. 10
   c. **100** ✓

8. 9)747
   a. 8
   b. **80** ✓
   c. 800

9. 4)812
   a. 2
   b. 20
   c. **200** ✓

10. 7)805
    a. 1
    b. 10
    c. **100** ✓

11. 3)945
    a. 3
    b. 30
    c. **300** ✓

12. 4)380
    a. 9
    b. **90** ✓
    c. 900

13. 2)976
    a. 4
    b. 40
    c. **400** ✓

14. 6)870
    a. 1
    b. 10
    c. **100** ✓

15. 8)504
    a. 6
    b. **60** ✓
    c. 600

Estimate by finding the first digit in the quotient. Then write 0's in the remaining places.

16. 4)292 → 70 Think: 4)29
17. 5)756 → 100 Think: 5)7
18. 7)224 → 30 Think: 7)22
19. 3)654 → 200 Think: 3)6
20. 6)258 → 40 Think: 6)25

21. 8)920 → 100
22. 4)144 → 30
23. 6)882 → 100
24. 3)972 → 300
25. 4)908 → 200

26. 7)217 → 30
27. 9)522 → 50
28. 3)645 → 200
29. 4)352 → 80
30. 2)956 → 400

31. 9)972 → 100
32. 3)135 → 40
33. 4)224 → 50
34. 7)805 → 100
35. 3)834 → 200

36. 4)4896 → 1000
37. 3)1374 → 400
38. 5)5620 → 1000
39. 7)1036 → 100
40. 6)7344 → 1000

You're the driver Using estimation

Estimate. True or false.

41. When you drive from Atlanta to Birmingham in 3 hours, you are averaging less than 50 miles per hour? **true**

42. When you drive from Cincinnati to Pittsburgh on 9 gallons of gasoline, your car is averaging less than 30 miles per gallon of gasoline. **false**

MILEAGE CHART	Birmingham, AL	New Orleans, LA	Pittsburgh, PA
Atlanta, GA	147	518	741
Cincinnati, OH	499	848	288
Nashville, TN	219	549	582

Dividing Whole Numbers and Decimals **85**

Practice Worksheet
Workbook S204, Copymaster S204, or Duplicating Master S204

WORKSHEET 36 (Use after page 85.)

NAME _____

CHECKING ANSWERS
Seven of Brenda's answers are wrong. Find and correct the seven wrong answers.

QUIZ — Brenda Anderson

1. 315 ÷ 7 = 45
2. 496 ÷ 8 = 62
3. 960 ÷ 3 = ~~32~~ **320**
4. 876 ÷ 6 = 146
5. 224 ÷ 7 = 32
6. 912 ÷ 4 = ~~22~~ **228**
7. 138 ÷ 3 = 46
8. 228 ÷ 4 = 57
9. 963 ÷ 9 = ~~17~~ **107**
10. 812 ÷ 7 = 116
11. 356 ÷ 4 = 89
12. 958 ÷ 2 = 479
13. 928 ÷ 8 = 116
14. 728 ÷ 7 = ~~14~~ **104**
15. 148 ÷ 4 = 37
16. 975 ÷ 3 = 325
17. 1632 ÷ 8 = ~~24~~ **204**
18. 1575 ÷ 5 = 315
19. 4275 ÷ 9 = 475
20. 1043 ÷ 7 = ~~14~~ **149**
21. 6324 ÷ 6 = ~~105~~ **1054**
22. 1800 ÷ 8 = 225

© D.C. Heath and Company   S204   Dividing by a 1-digit number

Challenge Problems

Use the digits. Fill in the boxes to get the quotient.

1. digits: 6, 7, 8, 1

   [8][6][1] ÷ [7] = 123

2. digits: 2, 8, 5, 6

   [8][5][2] ÷ [6] = 142

Copymaster S443

Class Starter Quiz 33
on previous lesson

Match. Find the quotients by estimating.

1. 408 ÷ 6   c
2. 224 ÷ 7   d
3. 459 ÷ 9   a
4. 1236 ÷ 3   b

a. 51
b. 412
c. 68
d. 32

Copymaster S393

Lesson Objectives

To divide by a 1-digit number (with a 3- or 4-digit dividend and zeros in the quotient)
To estimate quotients

Problem-Solving Skills

Reading an ad
Choosing the correct operation
Solving a simpler problem

Starting the Lesson

What are the facts? Have the students study the ad at the top of page 86 for 30 seconds. Then have them close their books and answer these questions from memory:

- What is the ad selling? (Home computer)
- How much down payment is needed? (No down payment)
- Do you have more or less than 4 months to pay? (Less)

Here's How Note

Use ■ *Visual Aid 3* (lined notebook paper, copymaster or transparency S113) to assist the students in aligning the digits.

Exercise Note

Problem Solving In exercise 41, the radio is off after an odd number of button pushes and on after an even number of pushes. If it starts on and then is switched 92 times (37 + 55), it will be on again.

86

More on dividing by a 1-digit number

"... BUY A HOME COMPUTER. PAY NO MONEY DOWN. PAY $615 IN THREE EQUAL MONTHLY PAYMENTS..."

1. Read the radio commercial. What is the price of the home computer? $615

2. How many monthly payments would you have to make? 3

3. To compute how many dollars you would pay each month, you would divide 615 by what number? 3

Here's how to divide by a 1-digit number.   615 ÷ 3 = ?

To estimate the quotient, find the first digit in the quotient.

```
   2
3)615
```

Then write 0's in the remaining places.

```
  200
3)615
```

Divide hundreds.
```
   2
3)615
  -6
```

Not enough tens.
Think 15 ones.
```
   20
3)615
  -6
   15
```
Don't forget the zero!

Divide ones.
```
   205
3)615
  -6
   15
  -15
    0
```

4. Look at the *Here's how*. Is $200 a good estimate for the monthly payment? Yes

5. How much is the monthly payment? $205

6. Find the missing quotient. Use an estimate to check your answer.

a.
```
   907
4)3628
 -36
    28
   -28
     0
```

b.
```
   420
6)2520
 -24
    12
   -12
     0
```

86 Chapter 4

EXERCISES

7. Three of the calculator answers are wrong. Find them by estimating.

a. $3\overline{)6243}$ **2081** b. $4\overline{)832}$ **208** (c.) $5\overline{)1020}$ **24**

(d.) $6\overline{)6456}$ **176** e. $9\overline{)873}$ **97** (f.) $4\overline{)1624}$ **46**

g. $7\overline{)2961}$ **423** h. $8\overline{)5608}$ **701** i. $6\overline{)3792}$ **632**

Divide. Here are scrambled answers for the next row of exercises: 31 2077 412 405 302

8. $5\overline{)2025}$ **405**
9. $8\overline{)248}$ **31**
10. $7\overline{)2884}$ **412**
11. $6\overline{)1812}$ **302**
12. $4\overline{)8308}$ **2077**

13. $3\overline{)597}$ **199**
14. $7\overline{)217}$ **31**
15. $9\overline{)450}$ **50**
16. $5\overline{)355}$ **71**
17. $6\overline{)2934}$ **489**

18. $2\overline{)4132}$ **2066**
19. $6\overline{)3672}$ **612**
20. $9\overline{)6390}$ **710**
21. $7\overline{)5621}$ **803**
22. $3\overline{)525}$ **175**

23. $4\overline{)732}$ **183**
24. $6\overline{)240}$ **40**
25. $5\overline{)2965}$ **593**
26. $7\overline{)4263}$ **609**
27. $9\overline{)207}$ **23**

28. $782 \div 2$ **391**
29. $592 \div 4$ **148**
30. $480 \div 3$ **160**
31. $1248 \div 6$ **208**
32. $830 \div 5$ **166**
33. $945 \div 9$ **105**
34. $1640 \div 2$ **820**
35. $405 \div 5$ **81**
36. $450 \div 9$ **50**
37. $1896 \div 4$ **474**

Solve. Use the information in the ad.

38. What was the regular price of the radio? **$79.99**

39. How much would you save if you bought the radio on sale? **$11.99**

40. After you made the first monthly payment, how much would you have left to pay on the radio? **$51**

RADIO SALE
Was $79⁹⁹
Now $68⁰⁰
No money down! 4 monthly payments

On or off? Solving a simpler problem

41. You turned a radio on. Then a friend pushed the off-and-on button 37 times. Then you pushed the off-and-on button 55 more times. Was the radio on or off when you stopped? *Hint:* look at the chart to do a simpler but similar problem. Start with the radio on. **on**

Number of pushes	Then radio is
1	off
2	on
3	off
4	on
5	off

Dividing Whole Numbers and Decimals **87**

Practice Worksheet
Workbook S205, Copymaster S205, or Duplicating Master S205

WORKSHEET 37
(Use after page 87.)

SKILL DRILL
Study these examples.

$$\begin{array}{r}109\\5\overline{)545}\\-5\\\hline 45\\-45\\\hline 0\end{array}\quad\begin{array}{r}320\\7\overline{)2240}\\-21\\\hline 14\\-14\\\hline 0\end{array}\quad\begin{array}{r}49\\9\overline{)441}\\-36\\\hline 81\\-81\\\hline 0\end{array}$$

Divide.
1. $6\overline{)2412}$ **402**
2. $8\overline{)2960}$ **370**
3. $5\overline{)3675}$ **735**
4. $9\overline{)5580}$ **620**
5. $7\overline{)3549}$ **507**

6. $4\overline{)1236}$ **309**
7. $3\overline{)6201}$ **2067**
8. $5\overline{)3900}$ **780**
9. $8\overline{)5320}$ **665**
10. $6\overline{)2454}$ **409**

11. $2\overline{)4616}$ **2308**
12. $9\overline{)5040}$ **560**
13. $7\overline{)4221}$ **603**
14. $5\overline{)5015}$ **1003**
15. $4\overline{)3636}$ **909**

16. $8\overline{)8064}$ **1008**
17. $6\overline{)6270}$ **1045**
18. $5\overline{)4175}$ **835**
19. $9\overline{)3150}$ **350**
20. $7\overline{)1050}$ **150**

Check yourself. Here are the scrambled answers:
150 309 350 370 402 409 507 560 603 620 665 735
780 835 909 1003 1008 1045 2067 2308

© D. C. Heath and Company **S205** Dividing by a 1-digit number

Project

Using library resources

Use a book of world records to find the highest price ever paid for a single page of advertising.

Answers will vary. Records may change from year to year.

Copymaster S476

Class Starter Quiz 34
on previous lesson

Divide.
1. 3045 ÷ 5   609
2. 4221 ÷ 7   603
3. 1830 ÷ 3   610
4. 344 ÷ 8   43
5. 4926 ÷ 6   821

Copymaster S393

Lesson Objectives
To divide by a 2-digit number
To check division

Problem-Solving Skills
Reading a table
Using a guess-and-check strategy

Starting the Lesson
Tell the students that after today's lesson they will be able to use mathematics to predict their adult height—but before beginning the lesson, you want them to guess what their adult height will be in inches.

Here's How Note
After discussing exercises 1–4, ask for volunteers to compute their adult-height predictions on the chalkboard.

Calculator You may want to show students how to find the remainder when you divide 5700 by 82 with a calculator.

Step 1. Divide.

5700 ÷ 82 = 69.512195

Step 2. Subtract the whole-number part of the quotient.

− 69 = 0.512195

Step 3. Multiply the divisor.

× 82 = 41.99999

The remainder is 42 (rounded to the nearest whole number).

Dividing by a 2-digit number

You can predict your adult height by first multiplying your present height (in inches) by 100 and then dividing the answer by the growth factor found in the table.

1. On his 12th birthday, John was 57 inches tall. To predict his adult height, you would first multiply 57 by what number? 100

2. Which number in the table is the growth factor for John? 82

HOW TALL WILL YOU BE?

Age (in years)	Growth Factor	
	Girl	Boy
12	91	82
13	95	86
14	98	90
15	99	94
16	100	97

3. To predict John's adult height, you would divide what number by 82? 5700

Here's how to divide by a 2-digit number.

$$5700 \div 82 = ?$$

Step 1. Think about dividing 57 by 8. So, try 7.

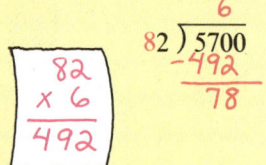

574 is too large.

Step 2. Try 6.

Step 3. Think about dividing 78 by 8. Try 9.

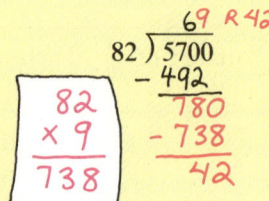

Notice that the division did not come out even. The **quotient** is 69 and the **remainder** is 42.

Here's how to check the division.

```
    69    ← quotient
  × 82
   138
   552
  5658
  +  42  ← remainder
  5700
```

4. Look at the *Here's how*. John's adult height would be between 69 and [?] inches. 70

EXERCISES

Divide and check.

5. 43)1333 → 31
6. 51)1337 → 26 R11
7. 34)1620 → 47 R22
8. 60)1690 → 28 R10
9. 63)1638 → 26
10. 81)2703 → 33 R30
11. 38)2812 → 74
12. 41)2500 → 60 R40
13. 54)2916 → 54
14. 73)2835 → 38 R61
15. 66)3823 → 57 R61
16. 92)3700 → 40 R20
17. 83)3206 → 38 R52
18. 71)3408 → 48
19. 46)3642 → 79 R8

Divide. Here are scrambled answers for the next row of exercises: 24 R37 36 25 44 R25 26

20. 63)1575 → 25
21. 56)1381 → 24 R37
22. 92)4073 → 44 R25
23. 58)1508 → 26
24. 78)2808 → 36
25. 12)5400 → 450
26. 39)4263 → 109 R12
27. 53)2204 → 41 R31
28. 61)1323 → 21 R42
29. 57)4640 → 81 R23
30. 58)74,693 → 1287 R47
31. 94)63,941 → 680 R21
32. 78)15,825 → 202 R69
33. 42)75,061 → 1787 R7
34. 44)23,561 → 535 R21

35. 1596 ÷ 38 42
36. 2974 ÷ 24 123 R22
37. 2624 ÷ 82 32
38. 3806 ÷ 17 223 R15
39. 50,000 ÷ 72 694 R32
40. 89,216 ÷ 82 1088
41. 53,819 ÷ 25 2152 R19
42. 74,281 ÷ 31 2396 R5

Solve. Use the table on page 88.

43. A 13-year-old girl is 60 inches tall. Her adult height (in inches) should be between which two whole numbers? **63 and 64**

44. A 15-year-old boy is 64 inches tall. His adult height (in inches) should be between which two whole numbers? **68 and 69**

45. a. What is your height to the nearest inch? *Answers will vary.*
    b. What is your growth factor? *Answers will vary.*

46. Your adult height (in inches) should be between what two whole numbers? *Answers will vary.*

47. At what age do girls usually attain their adult height? **16**

48. Do boys usually attain their adult height earlier or later than girls? **Later**

49. Who should be the taller adult, a 12-year-old girl who is 56 inches tall or a 13-year-old girl who is 58 inches tall? **The 12-year-old**

50. Who should be the taller adult, a 15-year-old girl who is 66 inches tall or a 15-year-old boy who is 65 inches tall? **The boy**

Key it in!

Guess and check.

Find a way to push each marked key once to get the answer.

51. 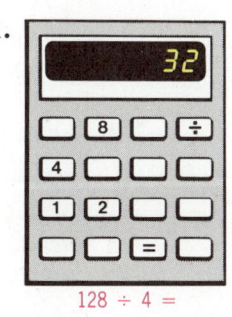 128 ÷ 4 =
52. 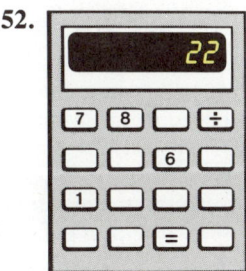 176 ÷ 8 =
53. 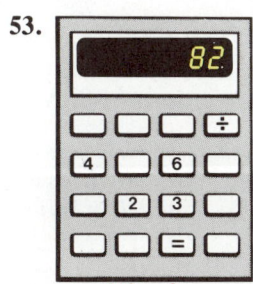 246 ÷ 3 =

Extra Practice
Page 477 Skill 14

Practice Worksheet
Workbook S206, Copymaster S206, or Duplicating Master S206

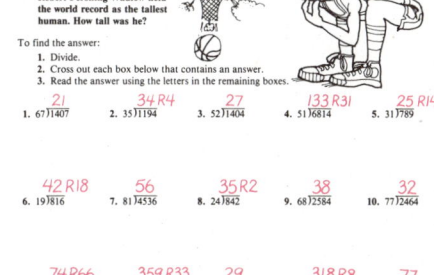

Challenge Problem

Use the chart on page 88.

If an adult is 6 feet (72 inches) tall, about how tall was he at age 14?

65 inches, or 5 feet 5 inches

Copymaster S443

Problem solving Reading a chart

Answer these truckers' questions. Decide when a calculator would be useful.

1. *I know it's 802 miles from New York to Chicago. How far is it from New York to Denver?* 1771 miles

2. *How far is it from New York to Los Angeles?* 2786 miles

3. "I first drive from Chicago to Denver and then from Denver to Los Angeles. What is the total mileage?" 2055

4. "What is the total mileage of a Houston–Denver–Los Angeles run?" 2078

5. "How much farther is Cleveland from Baltimore than from Chicago?" 8 miles

6. "I have driven the first 195 miles of a Houston–Cleveland run. How much farther do I have to go?" 1078 miles

7. "I am 380 miles from Chicago on a Baltimore–Chicago run. How many miles have I driven?" 288

8. "I leave Cleveland and plan to drive 55 miles per hour. Will I be in Baltimore in 7 hours?" Yes

9. "I want to drive from Chicago to Baltimore in less than 10 hours. Can I do it by traveling at 50 miles per hour?" No

10. "I leave Chicago at 8:00 A.M. and plan to drive 50 miles per hour on a run to Cleveland. Will I reach Cleveland by 3:00 P.M.?" Yes

11. "My truck holds 100 gallons of diesel fuel. If it averages 4.5 miles per gallon, can I drive from Denver to Cleveland without stopping for fuel?" No

12. "My truck gets about 5 miles per gallon of diesel fuel. How many gallons of fuel will I need to get from Chicago to Cleveland?" 67

MILEAGE CHART	Baltimore	Chicago	Cleveland	Denver	Houston	Los Angeles	New York
Baltimore		668	343	1621	1412	2636	196
Chicago	668		335	996	1067	2054	802
Cleveland	343	335		1321	1273	2367	473
Denver	1621	996	1321		1019	1059	1771
Houston	1412	1067	1273	1019		1538	1608
Los Angeles	2636	2054	2367	1059	1538		2786
New York	196	802	473	1771	1608	2786	

Cumulative Skill Practice

Write the short word-name. *(page 4)*

1. 26338
2. 9057
3. 36401
4. 23104
5. 419853
6. 64083
7. 250036
8. 100003
9. 9961240
10. 344082129
11. 31650000
12. 261418
13. 68000125
14. 5643088
15. 83000000
16. 10000000

Add. *(page 6)*

17. 21,345 + 18,466 = 39,811
18. 85,841 + 23,509 = 109,350
19. 11,295 + 27,832 = 39,127
20. 38,451 + 56,009 = 94,460
21. 93,816 + 26,957 = 120,773

22. $9.25 + .63 + .25 = $10.13
23. $7.63 + 1.25 + .18 = $9.06
24. $5.94 + 3.29 + 1.36 = $10.59
25. $8.32 + 4.73 + 5.19 = $18.24
26. $5.99 + 5.99 + 6.14 = $18.12

27. $21.32 + 18.56 + 9.27 = $49.15
28. $16.53 + 6.74 + 29.07 = $52.34
29. $12.34 + 12.34 + 12.34 = $37.02
30. $47.35 + 62.00 + 73.65 = $183.00
31. $56.00 + 37.52 + 83.36 = $176.88

Give the sum. *(page 22)*

32. 8.3 + 2.6   10.9
33. 5.74 + 3.96   9.7
34. 0.82 + 1.74   2.56
35. 3.521 + 2.806   6.327
36. 4.333 + 9.074   13.407
37. 12.02 + 9.08   21.1
38. 6 + 3.4   9.4
39. 8.2 + 9   17.2
40. 15 + 7.68   22.68
41. 2.6 + 3 + 8.04   13.64
42. 5 + 6.71 + 3.0   14.71
43. 0.83 + 2.7 + 6   9.53
44. 0.92 + 2.7 + 8.6   12.22
45. 9.34 + 21.6 + 9.8   40.74
46. 15.21 + 16 + 21.5   52.71

MIXED PRACTICE
Complete.

47. 28 × 4.1 = ?   114.8
48. 4738 − 1295 = ?   3443
49. 5.72 × 1.4 = ?   8.008
50. 503 − 287 = ?   216
51. 8.2 − 2.7 = ?   5.5
52. 18.45 + 27.88 = ?   46.33
53. 6.094 + 5 + 7.28 = ?   18.374
54. 2.9 × 3.7 = ?   10.73
55. 45.3 − 18.72 = ?   26.58
56. 5.2 + (3.6 × 3) = ?   16
57. 17 − (5.6 + 3.8) = ?   7.6
58. 9.4 − 5 + 8.7 = ?   13.1
59. (6.4 × 0.5) − 1.8 = ?   1.4
60. (12.6 − 4.96) × 6 = ?   45.84
61. 5.4 + (0.9 × 8) = ?   12.6

Dividing Whole Numbers and Decimals

Problem-Solving Worksheet
Workbook S207, Copymaster S207, or Duplicating Master S207

WORKSHEET 39 *(Use after page 91.)*

CORNY PROBLEM
How much would you have to pay for all the cans of corn? **$5**

BE A PRINTER
You can print B, C, I, J, M, O, and U without lifting your pencil or retracing a line. Which other capital letters can be written this way?
D G L N P Q R S V W Z

BE A WINNER!
If the digits on your coupon have a sum of 4, you are a winner! All coupons have 3 digits. How many winning coupons are there? **15**

SCAVENGER HUNT
Use the letter prices. Find the 2-letter word on this page that is worth $34. **SO**

GUESS AND CHECK
Which two numbers would you switch so that the sum along each side would be 21? **6 and 9**

TOOTHPICK PUZZLE
Can you move 4 toothpicks to make 3 squares? Make a drawing to show your answer.

Project
Using library resources

It is 2054 miles by truck from Chicago to Los Angeles. Use an almanac to find how many miles it is by airplane from Chicago to Los Angeles. How many miles more is it by truck?

About 300 miles

Copymaster S476

More answers for page 91.
5. 419 thousand, 853
6. 64 thousand, 83
7. 250 thousand, 36
8. 100 thousand, 3
9. 9 million, 961 thousand, 240
10. 344 million, 82 thousand, 129
11. 31 million, 650 thousand
12. 261 thousand, 418
13. 68 million, 125
14. 5 million, 643 thousand, 88
15. 83 million
16. 10 million

Dividing by a 3-digit number

Suppose that an eccentric millionaire asked you to enter this contest.

CARRY-THE-CASH CONTEST
1. Before touching any of the bags, choose a bag of money.
2. If you can carry the bag of money for 1 mile without resting, the money is yours.

Note: Each bag contains only $1 bills.

WHICH WOULD YOU CHOOSE?

1. How many $1 bills are in the largest bag? 100,000
2. The weight of 454 $1 bills is one pound. Suppose that you choose the $100,000 bag. To find the number of pounds that you would have to carry you would divide 100,000 by [?]. 454

Here's how to divide by a 3-digit number. 100,000 ÷ 454 = ?

Step 1. Think about dividing 10 by 4. Try 2.

Step 2. Think about dividing 9 by 4. Try 2.

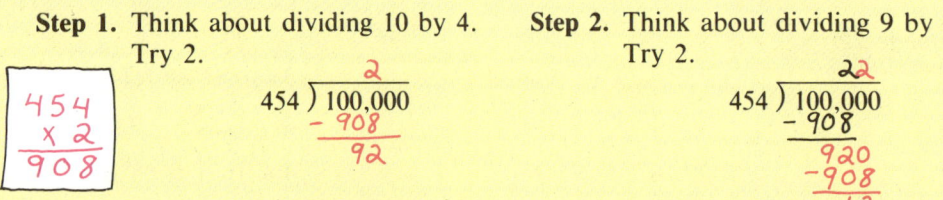

Step 3. Don't forget the zero!

Step 4. Think about dividing 12 by 4. Try 3.

Too large! So use 2.

You can write a 0 in the tenths place and carry out the division to that place.

3. Look at the *Here's how*. How many pounds would the $100,000 weigh? Round the answer to the nearest whole pound. 220
4. Could you have carried the $100,000 for a mile? No

EXERCISES

Divide. Use the multiplication facts.

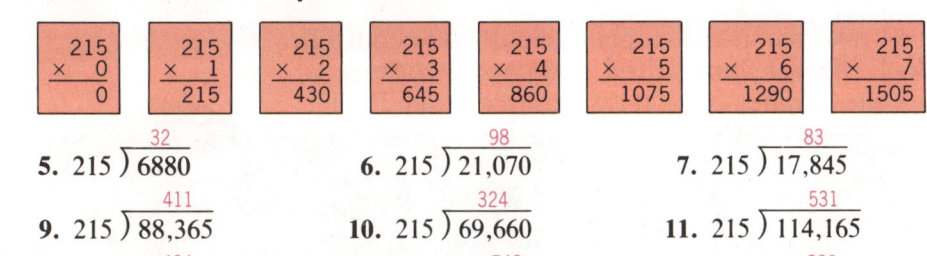

215 × 0 = 0	215 × 1 = 215	215 × 2 = 430
215 × 3 = 645	215 × 4 = 860	215 × 5 = 1075
215 × 6 = 1290	215 × 7 = 1505	215 × 8 = 1720
215 × 9 = 1935		

5. 215)6880 → 32
6. 215)21,070 → 98
7. 215)17,845 → 83
8. 215)13,115 → 61
9. 215)88,365 → 411
10. 215)69,660 → 324
11. 215)114,165 → 531
12. 215)175,010 → 814
13. 215)91,160 → 424
14. 215)159,745 → 743
15. 215)189,630 → 882
16. 215)444,620 → 2068

First carry out the division to the tenths place. Then round the quotient to the nearest whole number.
Here are scrambled answers for the next row of exercises: 139 82 43 101

17. 7)299 → 42.7 43
18. 9)741 → 82.3 82
19. 6)604 → 100.6 101
20. 4)555 → 138.7 139
21. 3)811 → 270.3 270
22. 8)507 → 63.3 63
23. 2)731 → 365.5 366
24. 9)847 → 94.1 94
25. 8)3421 → 427.6 428
26. 6)6539 → 1089.8 1090
27. 7)9603 → 1371.8 1372
28. 5)5702 → 1140.4 1140
29. 12)593 → 49.4 49
30. 15)974 → 64.9 65
31. 18)800 → 44.4 44
32. 16)777 → 48.5 49
33. 23)5012 → 217.9 218
34. 35)2906 → 83.0 83
35. 40)3582 → 89.5 90
36. 56)4711 → 84.1 84
37. 123)53,061 → 431.3 431
38. 130)29,438 → 226.4 226
39. 146)42,111 → 288.4 288
40. 193)70,629 → 365.9 366
41. 625)139,528 → 223.2 223
42. 834)374,200 → 448.6 449
43. 935)561,348 → 600.3 600
44. 753)829,005 → 1100.9 1101

Solve. Use the information on page 92.

45. How many pounds would the $60,000 bag weigh? Round the answer to the nearest whole number. **132**

46. How many pounds would the $25,000 bag weigh? Round the answer to the nearest whole number. **55**

47. Suppose that you could carry 48 pounds for 1 mile. How many $1 bills would that be? **21,792**

Million-dollar weigh-in

48. How much would 1 million dollars in $100 bills weigh? Round the answer to the nearest whole number. **22 pounds**

49. Could you carry 1 million dollars in $100 bills? **Answers may vary.**

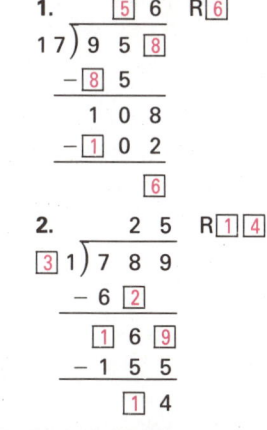

Class Starter Quiz 37
on previous lesson

Divide. Then round the quotient to the nearest whole number.

1. 234 ÷ 7   33
2. 5635 ÷ 4   1409
3. 600 ÷ 17   35
4. 2962 ÷ 140 21

Copymaster S394

Lesson Objectives
To apply the rules for order of operations in simplifying expressions
To write mathematical expressions

Problem-Solving Skill
Finding information in a store coupon

Starting the Lesson
Write this expression on the chalkboard:

$$36 + 2 \times 24$$

Have the students determine the value of the expression. List the various student answers on the chalkboard. Next to each answer, write the number of students who got that answer. Have the students open their books. Discuss exercises 1–5 and the *Here's how*. Go back to the list on the chalkboard to determine how many students got the correct answer.

Exercise Note
Calculator You may wish to show students how to use the calculator memory keys ⎡M+⎤ and ⎡MR⎤.

Example: $17 \times 4 + 360 \div 8$
$ 68 + 45 = 113$

Using a calculator:

17 ⎡×⎤ 4 ⎡=⎤ ⎡M+⎤ (68 is "saved" in memory)

360 ⎡÷⎤ 8 ⎡=⎤ ⎡M+⎤ (45 is added to 68 in memory)

⎡MR⎤ (113 is recalled to display)

Be sure to clear the memory for the next problem.

Order of operations

David, Carla, and Jane take pictures for their school's yearbook. In checking the film supply, David found a roll of 36 and 2 rolls of 24. Here is how he wrote the number of pictures that they could take:

NOTES
$36 + 2 \times 24$

1. When Carla saw the note, she claimed that they could take 912 pictures. How did she get 912? *She added 36 and 2, and then multiplied by 24.*

2. When Jane saw the note, she claimed that they could take 84 pictures. How did she get 84? *She multiplied 2 and 24, and then added 36.*

3. They got different answers because they did the operations in different orders. Who added first and then multiplied? *Carla*

4. Which operation did Jane do first? *Multiplication*

Here's how *to simplify expressions having more than one kind of operation.*

So that an expression has only one value, we use these rules for the **order of operations:**

Rule 1. First do the operations within the grouping symbols.

Rule 2. Next, work from left to right doing any multiplication and division.

Rule 3. Last, work from left to right doing any addition and subtraction.

5. Study the rules. Who was right, Carla or Jane? *Jane*

Chapter 4

EXERCISES

Simplify each expression.
Hint: The order of operations is shown by the numbered arrows.

6. $\overset{1}{8} \div \overset{2}{4} \times 2$ 4
7. $\overset{1}{10} - \overset{2}{6} + 3$ 7
8. $\overset{2}{7} + \overset{1}{3} \times \overset{3}{8} - 4$ 27
9. $\overset{1}{4} \times 2 + \overset{3}{20} \div \overset{2}{2}$ 18
10. $\overset{3}{5} + \overset{1}{(6 + 2)} \div \overset{2}{4}$ 7
11. $\overset{1}{(7 + 6)} \times \overset{2}{2} - \overset{3}{10}$ 16

12. $20 \div 5 - 1$ 3
13. $5 \times 6 - 3$ 27
14. $18 \times 3 \div 3$ 18
15. $50 - 10 - 4$ 36
16. $20 + 16 \div 4$ 24
17. $16 + 8 \div 4$ 18
18. $36 \div 9 \times 2$ 8
19. $43 + 17 - 20$ 40
20. $23 - 9 + 6$ 20
21. $(96 + 24) \div 6$ 20
22. $72 \times (8 - 2)$ 432
23. $16 \times (15 - 5)$ 160
24. $24 + 8 \div 4 + 4$ 30
25. $(24 + 8) \div 4 + 4$ 12
26. $24 + 8 \div (4 + 4)$ 25
27. $18 + 12 \times 6 - 1$ 89
28. $(18 + 12) \times 6 - 1$ 179
29. $18 + 12 \times (6 - 1)$ 78
30. $15 \times 3 + 5 \times 3$ 60
31. $(15 + 5) \times 3$ 60
32. $15 \times (3 + 5) \times 3$ 360
33. $24 - (14 + 4) \div 2$ 15
34. $24 - 14 + 4 \div 2$ 12
35. $(24 - 14) + 4 \div 2$ 12
36. $36 \div (2 + 4) \times 4 - 2$ 22
37. $36 \div 2 + 4 \times 4 - 2$ 32
38. $36 \div 2 + 4 \times (4 - 2)$ 26

Write an expression for the number of pictures that can be taken with

39. 3 rolls of 24. 3×24
40. 5 rolls of 36 and 12 on another roll. $5 \times 36 + 12$
41. 4 rolls of 24 with 7 pictures already taken on one of the rolls. $4 \times 24 - 7$
42. 2 rolls of 24 and 3 rolls of 36. $2 \times 24 + 3 \times 36$

Clip 'n' save

43. Two of these statements are false. Which two are they?
    a. With this coupon you can buy 6 rolls of film and get 2 rolls free.
    b. This coupon is good on the third Monday in March.
    c. The regular price for 4 rolls of film is $15.96.
    d. With this coupon you get 4 rolls of film for less than $10.

STORE COUPON
BUY 3 rolls of film and get 1 roll FREE.
Reg. **$3.99** a roll
COLOR FILM FOR PRINTS
Heathcofilm Inc.   36 exp.
Limit: 1 free roll per coupon
Coupon expires March 22.

Dividing Whole Numbers and Decimals   **95**

Extra Practice
Page 478   Skill 15

Practice Worksheet
Workbook S209, Copymaster S209, or Duplicating Master S209

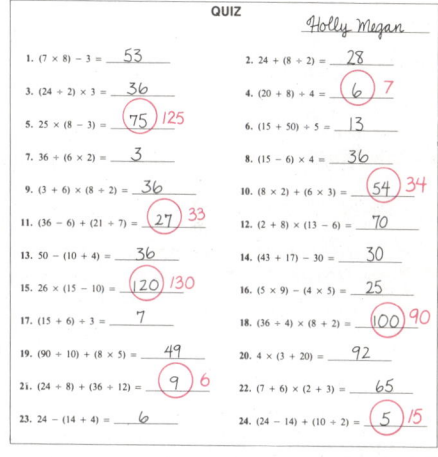

Challenge Problem

Look for the pattern.
Find the missing numbers.
$(1 \times 8) + 1 = 9$
$(12 \times 8) + 2 = 98$
$(123 \times 8) + 3 = 987$
⋮
$(? \times ?) + ? = 9,876,543$

1,234,567; 8; 7

Copymaster S444

95

Class Starter Quiz 38
on previous lesson

Simplify each expression.

1. $6 + 3 \times 5$ — 21
2. $30 \div 6 - 1$ — 4
3. $4 + (5 + 3) \div 2$ — 8
4. $16 + 8 \div 2$ — 20
5. $10 + 6 \times (3 + 2)$ — 40

Copymaster S394

Lesson Objective
To divide a decimal by a whole number

Problem-Solving Skills
Selecting data from a chart
Choosing the correct operation
Collecting, organizing, and analyzing data

Starting the Lesson
Group Activity You may wish to have groups of students conduct the project described on page 97. You will need a stopwatch. Some students may have stopwatches on their wristwatches. If not, you might try to borrow one from the physical education department. On the chalkboard, record the total reaction time (in seconds) and the number of students in each group. Discuss exercises 1–5 and the *Here's how*. Then have the students look at the information on the chalkboard and compute the average reaction time for each of their groups.

Here's How Note
Use of Concrete Materials You may wish to use the powers-of-ten tiles from copymaster or transparency S529 to demonstrate dividing a decimal by a whole number. See ■ **Manipulative Activity 5** on copymaster S516 in the Teacher's Resource Binder.

Dividing a decimal by a whole number

HOW FAST CAN YOU REACT?

Here is how some students answered the question. First they formed a circle by grasping hands. Then with their eyes closed, they "passed a hand squeeze" around the circle as quickly as possible. Their teacher timed how long it took for the squeeze to go around the circle.

1. How many students were in Experiment 1? 24
2. How long did it take the hand squeeze to go around the circle (the Total Reaction Time)? 10.32 seconds
3. For Experiment 1, what would you do to compute the average reaction time per student? Divide 10.32 by 24.

Here's how to divide a decimal by a whole number. $10.32 \div 24 = ?$

Place the decimal point in the quotient.

$$24 \overline{) 10 \overset{.}{\uparrow} 32}$$

Divide as you would whole numbers.

$$\begin{array}{r} 0.43 \\ 24 \overline{) 10.32} \\ -9\ 6 \\ \hline 72 \\ -72 \\ \hline 0 \end{array}$$

4. Look at the *Here's How*. What is the average reaction time for Experiment 1? 0.43 second
5. Study these examples. Round each quotient to the nearest hundredth.

a.
$$\begin{array}{r} 0.023 \\ 7 \overline{) 0.161} \\ -14 \\ \hline 21 \\ -21 \\ \hline 0 \end{array}$$

You have to write a zero here.

0.02

b.
$$\begin{array}{r} 0.541 \\ 6 \overline{) 3.250} \\ -30 \\ \hline 25 \\ -24 \\ \hline 10 \\ -6 \\ \hline 4 \end{array}$$

Sometimes the division does not come out even. You can write a zero here and carry out the division to the next place.

0.54

EXERCISES

Divide. Here are scrambled answers for the next row of exercises: 6.5   0.94   1.76   0.346

6. 8)14.08 → 1.76
7. 12)11.28 → 0.94
8. 6)2.076 → 0.346
9. 25)162.5 → 6.5
10. 34)0.714 → 0.021
11. 41)108.65 → 2.65
12. 7)2.492 → 0.356
13. 53)331.25 → 6.25
14. 78)$82.68 → $1.06
15. 81)$286.74 → $3.54
16. 66)$162.36 → $2.46
17. 35)$368.55 → $10.53
18. 90)367.20 → 4.08
19. 9)165.33 → 18.37
20. 87)3036.3 → 34.9
21. 8)821.6 → 102.7

First carry out the division to the thousandths place. Then round the quotient to the nearest hundredth.

22. 7)4.100
```
  0.585 ≈ 0.59
7)4.100
  -35
   60
  -56
   40
  -35
    5
```
23. 12)8 → 0.67
24. 18)42.16 → 2.34
25. 64)38.45 → 0.60
26. 11)6 → 0.55
27. 62)74.08 → 1.19
28. 23)4.278 → 0.19
29. 56.92 ÷ 25 2.28
30. 7.34 ÷ 14 0.52
31. 3.64 ÷ 90 0.04
32. 0.891 ÷ 8 0.11
33. 7.53 ÷ 38 0.20
34. 2.96 ÷ 26 0.11
35. 4.82 ÷ 68 0.07
36. 3.11 ÷ 42 0.07
37. 89.1 ÷ 94 0.95
38. 3.8 ÷ 9 0.42
39. 4.64 ÷ 12 0.39
40. 8.42 ÷ 15 0.56
41. 9.25 ÷ 13 0.71
42. 8.42 ÷ 15 0.56

Solve. Use the experiments on page 96.

43. What was the average reaction time for Experiment 2? Round the answer to the nearest hundredth of a second. 0.46 s
44. What was the average reaction time for Experiment 3? Round the answer to the nearest hundredth of a second. 0.40 s
45. In which experiment was the reaction time the fastest? Experiment 3
46. What was the difference between the fastest and slowest average reaction times? Work with times rounded to the nearest hundredth of a second. 0.06 s

Project Collecting, organizing, and analyzing data

Answers will vary for 47 and 48.

47. Find how long it takes for a hand squeeze to go about a circle of your classmates.
48. Compute the average reaction time for your class.

Dividing Whole Numbers and Decimals

Practice Worksheet
Workbook S210, Copymaster S210, or Duplicating Master S210

WORKSHEET 42 (Use after page 97.)

RIDDLE TIME
To find the answer to the riddle:
1. Divide.
2. Cross out each box below that contains an answer.
3. Read the answer using the letters in the remaining boxes.

What do you break by naming it?

1. 2)16.04 → 8.02
2. 3)9.36 → 3.12
3. 5)22.85 → 4.57
4. 7)1.47 → 0.21
5. 6)2.076 → 0.346

6. 4)4.56 → 1.14
7. 8)67.60 → 8.45
8. 9)42.39 → 4.71
9. 11)38.5 → 3.5
10. 52)71.24 → 1.37

11. 24)6.504 → 0.271
12. 45)54.45 → 1.21
13. 36)147.6 → 4.1
14. 81)72.09 → 0.89

F	S	A	R	I	I	L
1.14	0.83	1.21	8.02	10.6	3.5	8.49
E	I	N	T	C	O	C
0.37	0.89	1.78	0.21	0.346	4.71	2.1
M	F	A	L	E	E	S
1.37	4.57	4.1	8.45	9.65	3.12	0.271

Answer: SILENCE

© D. C. Heath and Company   S210   Dividing a decimal by a whole number

Challenge Problem

Place the digits 0, 1, 2, 3, 4, and 5 in the boxes so that the quotient will be 4.45.

```
         4.45
  1 2 ) 5 3 . 4 0
```

Copymaster S444

Class Starter Quiz 39
on previous lesson

Divide.

1. 43.8 ÷ 6   7.3
2. 13.95 ÷ 9   1.55
3. 2.480 ÷ 20 0.124
4. 165.33 ÷ 9 18.37
5. 22 ÷ 5   4.4

Copymaster S394

Problem-Solving Skills
Choosing the correct operation
Solving a two-step problem

Skills Reviewed
Adding, subtracting, and multiplying whole numbers and decimals
Comparing decimals
Writing standard numerals for decimals
Simplifying expressions

Starting the Lesson
Problem Solving Direct the students' attention to the comments at the top of page 98. Ask the students whose calculations should be used to compute these:

(457 × 4) + 825   (Matt and Anne's)
(150 − 125) ÷ 5   (Susan and David's)
20 − (5.75 × 3)   (Matt and Susan's)

Cumulative Skill Practice Challenge the students to an estimation hunt by saying, "Pick the largest difference in the first row of exercises." (Exercise 4) Then have the students pick the largest difference in the second row of exercises. (Exercise 6)

98

Problem solving Choosing the operation

This calculator only does addition. *This one only does subtraction.* *Multiplication is the only operation this one will do.* *Division only on this one.*

Whose calculator solved it? Name the two people who worked together to solve each problem. Decide when a calculator would be useful.

1. Names ?Matt ?Anne
My friend delivers 35 papers 6 days a week. She also delivers 48 Sunday papers. How many papers does she deliver each week?
Answer: 258 papers

2. Names Susan David ? ?
We bought some $3 records. We gave the cashier $20 and got $8 in change. How many records did we buy?
Answer: 4 records

3. Names Anne David ? ?
My friend and I earned $27.50 yesterday and $15.40 today. We shared the money equally. How much money did I get?
Answer: $21.45

4. Names Matt Susan ? ?
I bought 4 frozen pizzas at $2.90 each. How much change should I get from a $20 bill?
Answer: $8.40

Name the two people who should work together to solve each problem. Then solve each problem.

5. Names Matt Anne ? ?
After 5 friends each ate 4 tacos, there were 3 tacos left. How many tacos were there to start with?
Answer: ? 23

6. Names Susan David ? ?
I bought 5 hockey tickets. I gave the cashier $20 and got $2.50 in change. What was the price of each ticket?
Answer: ? $3.50

7. Names Anne David ? ?
At my school, 106 girls and 92 boys signed up to play volleyball. A total of 18 teams were formed. How many players were on each team?
Answer: ? 11

8. Names Matt Anne ? ?
We bought a 14.5-pound watermelon at $.08 per pound and a $2.98 bag of oranges. What was the total cost?
Answer: ? $4.14

98 Chapter 4

Cumulative Skill Practice

Subtract. *(page 36)*

1. 3902 − 1755 = 2147
2. 7301 − 879 = 6422
3. 5026 − 3477 = 1549
4. 6085 − 392 = 5693
5. 5004 − 1679 = 3325
6. 86,294 − 7,381 = 78,913
7. 72,855 − 6,974 = 65,881
8. 68,302 − 9,756 = 58,546
9. 29,001 − 17,362 = 11,639
10. 30,047 − 18,492 = 11,555

Less than (<) or greater than (>)? *(page 42)*

11. 16.1 > 16.0
12. 7.86 > 7.68
13. 0.38 > 0.3
14. 42.79 < 42.8
15. 3.1 > 2.97
16. 6.347 < 63.4
17. 0.852 < 1.2
18. 1.3 > 1.03
19. 529.6 < 530
20. 14.006 < 14.06
21. 73.02 > 72.03
22. 0.021 < 0.21

Give the difference. *(page 46)*

23. 72.3 − 2.59 = 69.71
24. 29.7 − 15.92 = 13.78
25. 74.36 − 2.1 = 72.26
26. 26.083 − 7.461 = 18.622
27. 74.4 − 3.88 = 70.52
28. 65.0 − 42.5 = 22.5
29. 12 − 3.7 = 8.3
30. 52 − 9.64 = 42.36
31. 29.4 − 2.94 = 26.46
32. 56.7 − 8.821 = 47.879
33. 37.4 − 18 = 19.4
34. 45.6 − 2.735 = 42.865

Multiply. *(page 64)*

35. 538 × 132 = 71,016
36. 603 × 115 = 69,345
37. 492 × 236 = 116,112
38. 871 × 382 = 332,722
39. 729 × 533 = 388,557
40. 821 × 364 = 298,844

MIXED PRACTICE
Complete.

41. The standard numeral for twelve thousandths is __?__ 0.012
42. The standard numeral for 31 and six hundredths is __?__ 31.06
43. 3521 + 8731 = __?__ 12,252
44. 6.34 + 9.83 = __?__ 16.17
45. 4.8 × 5.2 = __?__ 24.96
46. 12.94 × 6.7 = __?__ 86.698
47. 928 + 74 + 7 = __?__ 1009
48. 9.6 + 0.57 + 18 = __?__ 28.17
49. (9.1 + 2.54) × 3 = __?__ 34.92
50. 10.5 + (4.8 × 5) = __?__ 34.5
51. (12.3 − 7.6) + 11.7 = __?__ 16.4

Dividing Whole Numbers and Decimals

Problem-Solving Worksheet
Workbook S211, Copymaster S211, or Duplicating Master S211

Challenge Problems

Suppose the 5 key does not work on your calculator. Tell how you might use your broken calculator to do these multiplication problems.

1. 375 × 34   (374 × 34) + 34
2. 52 × 47   (42 × 47) + (10 × 47)

Answers will vary.

Copymaster S444

Class Starter Quiz 40
on previous lesson

Solve.

1. I bought 6 movie tickets. I gave the cashier $20 and got back $5 in change. What was the price of each ticket? **$2.50**
2. Two friends bought a 12-pound watermelon at $.15 per pound. They shared the cost equally. How much money was each person's share? **$.90**

Copymaster S394

Lesson Objective
To divide mentally by 10, 100, or 1000

Problem-Solving Skills
Choosing appropriate information
Choosing the correct operation
Following directions

Starting the Lesson
Mental Math The comment from the banker at the top of page 100 is also on ■ **Visual Aid 15** (copymaster or transparency S123). Use the banker's comment. Tell the students that the banker can fill in the missing numbers in five seconds. Have the students determine how fast they can do it. Have them describe the method they use to get the missing numbers.

Exercise Note
Mental Math Point out to the students that division by 10, 100, or 1000 should always be done mentally rather than by the division algorithm. Thus, for exercises 1–60 they need record only the quotients on their papers.

Mental math—dividing by 10, 100, or 1000

The banker can fill in these missing numbers in 5 seconds. How fast can you do it?

8000 $1 bills equals **800** $10 bills, **80** $100 bills, or **8** $1000 bills.

The banker knows that if you divide a number by 10, 100, or 1000, the quotient is less than the number. So, she moves the decimal point to the left.

Here's how *to divide by 10, 100, or 1000.*

$8000 \div 10 = 800.0$ or 800 — Dividing by 10 moves the decimal point 1 place to the left.

$8000 \div 100 = 80.00$ or 80 — Dividing by 100 moves the decimal point 2 places to the left.

$8000 \div 1000 = 8.000$ or 8 — Dividing by 1000 moves the decimal point 3 places to the left.

Other examples:
Dividing by

10	$23.25 \div 10 = 2.325$
100	$6.1 \div 100 = 0.061$
1000	$9.75 \div 1000 = 0.00975$

Some zeros had to be written before the decimal point could be placed in the quotient.

EXERCISES
MENTAL MATH Divide. Write answers only. Here are scrambled answers for the next two rows of exercises: 0.536 6.7 0.067 3.62 36.2 5.36 0.0536 0.67

1. $36.2 \div 10$ **3.62**
2. $362 \div 10$ **36.2**
3. $53.6 \div 100$ **0.536**
4. $53.6 \div 1000$ **0.0536**
5. $536 \div 100$ **5.36**
6. $67 \div 1000$ **0.067**
7. $6700 \div 1000$ **6.7**
8. $67 \div 100$ **0.67**
9. $712.2 \div 10$ **71.22**
10. $712.2 \div 100$ **7.122**
11. $712.2 \div 1000$ **0.7122**
12. $71.22 \div 10$ **7.122**
13. $18.6 \div 1000$ **0.0186**
14. $18.6 \div 100$ **0.186**
15. $18.6 \div 10$ **1.86**
16. $0.186 \div 10$ **0.0186**

17. 60 ÷ 10   6
18. 60 ÷ 100   0.6
19. 60 ÷ 1000   0.06
20. 600 ÷ 10   60
21. 14.32 ÷ 10   1.432
22. 14.32 ÷ 100   0.1432
23. 14.32 ÷ 1000   0.01432
24. 143.2 ÷ 100   1.432
25. 486 ÷ 1000   0.486
26. 486 ÷ 100   4.86
27. 486 ÷ 10   48.6
28. 4.86 ÷ 10   0.486
29. 421.9 ÷ 1000   0.4219
30. 421.9 ÷ 100   4.219
31. 421.9 ÷ 10   42.19
32. 42.19 ÷ 1000   0.04219
33. 242 ÷ 10   24.2
34. 242 ÷ 1000   0.242
35. 242 ÷ 100   2.42
36. 24.2 ÷ 1000   0.0242
37. 81.5 ÷ 100   0.815
38. 81.5 ÷ 10   8.15
39. 81.5 ÷ 1000   0.0815
40. 815 ÷ 10   81.5
41. 7 ÷ 1000   0.007
42. 7 ÷ 10   0.7
43. 7 ÷ 100   0.07
44. 70 ÷ 100   0.7
45. 9.3 ÷ 100   0.093
46. 9.3 ÷ 1000   0.0093
47. 9.3 ÷ 10   0.93
48. 93 ÷ 1000   0.093
49. 2.03 ÷ 10   0.203
50. 2.03 ÷ 1000   0.00203
51. 2.03 ÷ 100   0.0203
52. 20.3 ÷ 10   2.03
53. 765 ÷ 1000   0.765
54. 765 ÷ 10   76.5
55. 765 ÷ 100   7.65
56. 76.5 ÷ 10   7.65
57. 0.6 ÷ 100   0.006
58. 0.6 ÷ 1000   0.0006
59. 0.6 ÷ 10   0.06
60. 60 ÷ 1000   0.06

Solve. Use the money facts.

61. What is the height of $50,000 in
    a. $10 bills?   20 inches
    b. $100 bills?   2 inches
    c. $1000 bills?   0.2 inches
    d. $1 bills?   200 inch

62. What is the weight of $50,000 in
    a. $10 bills?   11 pounds
    b. $100 bills?   1.1 pounds
    c. $1000 bills?   0.11 pounds
    d. $1 bills?   110 pound

MONEY FACTS

$50,000 in $10 bills is 20 inches high.

$50,000 in $10 bills weighs 11 pounds.

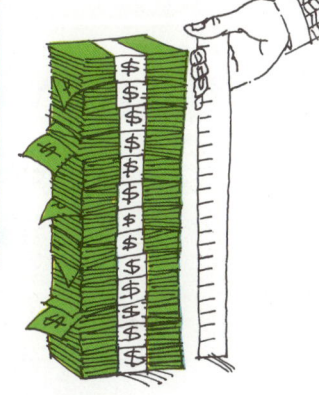

Make your money stretch Following directions

63. a. Suppose you have 1 million dollars in $1 bills. If you laid the bills end-to-end, how many miles of money would you have? To find the number of miles, write a   97.85

   5 in the hundredths place
   7 in the ones place
   8 in the tenths place
   9 in the tens place

b. Suppose that you have 1 billion dollars in $1 bills. If you laid the bills end-to-end, how many miles of money would you have? *Hints: 1000 million is 1 billion. Use your answer from part a.*   97,850

Dividing Whole Numbers and Decimals **101**

Extra Practice
Page 478   Skill 16

Practice Worksheet
Workbook S212, Copymaster S212, or Duplicating Master S212

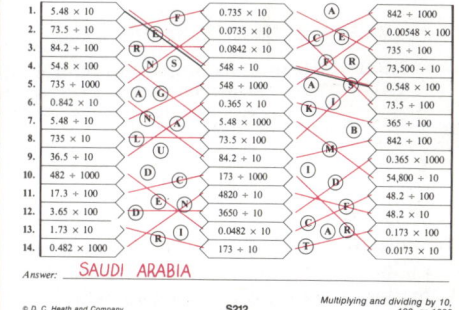

Challenge Problems
Find the missing numbers.

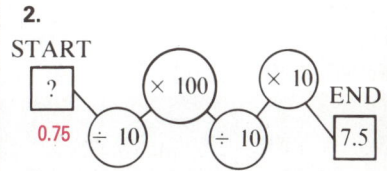

Copymaster S445

101

Class Starter Quiz 41
on previous lesson

Divide.

1. 42.3 ÷ 10   4.23
2. 0.15 ÷ 100   0.0015
3. 67 ÷ 1000   0.067
4. 258 ÷ 1000   0.258
5. 7 ÷ 10   0.7
6. 93 ÷ 100   0.93
7. 8.5 ÷ 100   0.085
8. 175 ÷ 10   17.5
9. 0.6 ÷ 10   0.06
10. 76.5 ÷ 100   0.765

Copymaster S395

Lesson Objective
To divide a decimal by a decimal

Problem-Solving Skills
Finding information in a newspaper article
Choosing the correct operation

Starting the Lesson
What are the facts? Have the students study the article at the top of page 102 for 60 seconds. Then tell the students to close their books and decide whether these statements are true or false:

- Skip Gilligan's attempt at setting an onion-peeling record ended after 3.5 minutes. (True)
- He peeled more than 30 pounds of onions. (True)
- He was forced to stop due to sore fingers. (False)
- Skip is a 28-year-old construction worker. (False)

Here's How *Note*
After discussing exercises 1–5, it may be helpful to have several of exercises 6–57 demonstrated at the chalkboard before the students begin their independent work.

102

Dividing a decimal by a decimal

ONION TEARS
Skip Gilligan's attempt at setting a new onion-peeling record ended in tears. After 3.5 minutes and 31.5 pounds of peeled onions, Gilligan was forced to stop due to eye discomfort. With tears in his eyes, the 28-year-old cafeteria worker said he would be back for another shot at the record.

1. Read the newspaper article. How many pounds of onions did Gilligan peel in 3.5 minutes? 31.5

2. To compute how many pounds of onions he peeled per minute, you would divide 31.5 by what number? 3.5

Here's how to divide a decimal by a decimal.   31.5 ÷ 3.5 = ?

Hard problem!

3.5) 31.5
↑   ↑
divisor   dividend

Multiply both divisor and dividend by 10 to get a whole-number divisor.

3.5) 31.5

Divide.

      9.
3.5) 31.5
     −315
        0

3. Look at the *Here's how*. To get a whole-number divisor, both decimal points were moved [?] place(s) to the right. one

4. How many pounds of onions did Gilligan peel per minute? 9

5. Check these examples. Are the answers correct? Yes

a. 8.32 ÷ 0.32

        26.
0.32) 8.32
     −64
      192
    −192
        0   Yes

Move both decimal points 2 places to the right.

b. 0.9856 ÷ 0.004

         246.4
0.004) 0.9856
     −8
      18
    −16
      25
    −24
      16
    −16
       0   Yes

Move both decimal points 3 places to the right.

EXERCISES

Divide. Here are scrambled answers for the next row of exercises: 3.6 6.8 6.5 38

6. 0.2)1.36 = 6.8
7. 0.5)3.25 = 6.5
8. 0.4)1.44 = 3.6
9. 0.02)0.76 = 38
10. 0.07)4.62 = 66
11. 0.02)2.12 = 106
12. 0.005)0.515 = 103
13. 0.007)7.280 = 1040
14. 0.6)1.5 = 2.5
15. 0.8)2.56 = 3.2
16. 0.03)1.26 = 42
17. 0.005)0.485 = 97
18. 0.2)5.46 = 27.3
19. 4.5)22.5 = 5
20. 0.05)0.075 = 1.5
21. 0.08)1.016 = 12.7
22. 0.4)2.488 = 6.22
23. 0.3)0.06 = 0.2
24. 0.05)0.75 = 15
25. 0.008)0.512 = 64
26. 1.2)31.2 = 26
27. 0.15)0.57 = 3.8
28. 3.4)3.57 = 1.05
29. 0.26)0.884 = 3.4
30. 7.4)40.7 = 5.5
31. 5.9)21.24 = 3.6
32. 0.81)49.41 = 61
33. 0.94)10.528 = 11.2
34. 9.3)6.045 = 0.65
35. 0.66)0.3366 = 0.51
36. 5.2)21.164 = 4.07
37. 0.48)1.5216 = 3.17
38. 4.5 ÷ 0.3 15
39. 12.6 ÷ 0.2 63
40. 56.84 ÷ 0.7 81.2
41. 4.32 ÷ 0.4 10.8
42. 1.684 ÷ 0.02 84.2
43. 0.7085 ÷ 0.05 14.17
44. 0.1128 ÷ 0.08 1.41
45. 0.567 ÷ 0.09 6.3
46. 0.341 ÷ 0.11 3.1
47. 1.875 ÷ 0.25 7.5
48. 26.22 ÷ 5.7 4.6
49. 23.24 ÷ 2.8 8.3
50. 4.14 ÷ 1.8 2.3
51. 8.4 ÷ 2.1 4
52. 1.56 ÷ 0.12 13
53. 1.260 ÷ 0.12 10.5
54. 43.2 ÷ 0.24 180
55. 0.522 ÷ 0.036 14.5
56. 1.6254 ÷ 3.01 0.54
57. 43.594 ÷ 0.71 61.4

Solve.

58. Calvin set out to break the onion-peeling record. He peeled 12.6 pounds in 1.5 minutes. Did he peel more or less than 9 pounds per minute? *Hint: Divide to get the answer.* Less

59. In 1979, a team of 57 men pushed a baby carriage 345.25 miles in 24 hours. How many miles did they average per hour? Round the answer to the nearest hundredth of a mile. 14.39

60. Fourteen students leapfrogged for 148 hours at the average rate of 3.75 miles per hour. How many miles did they leapfrog? 555

61. The record for treading water is 98.5 hours. How many minutes is that? 5910

Just for the record

62. In 1981, Norman Johnson sliced a 12-inch cucumber into 22 slices per inch. It took him 19.11 seconds. What was his average time per slice? Round the answer to the nearest hundredth.
0.07 second per slice

Dividing Whole Numbers and Decimals **103**

Practice Worksheet
Workbook S213, Copymaster S213, or Duplicating Master S213

Challenge Problems

Use the digits. Fill in the boxes to get the answer.

1. [1 4 2 8]
   ☐.☐ ÷ ☐.☐ = 7
   8.4 ÷ 1.2 = 7

2. [8 4 1 2 6]
   ☐☐.☐ ÷ ☐.☐ = 7
   16.8 ÷ 2.4 = 7

Copymaster S445

103

Class Starter Quiz 42
on previous lesson

Divide.
1. 6.25 ÷ 0.5   12.5
2. 0.84 ÷ 0.02   42
3. 0.423 ÷ 0.003   141
4. 2.4 ÷ 1.2   2
5. 1.68 ÷ 21   0.08

Copymaster S395

Lesson Objectives

To divide a decimal by a decimal
To write a remainder as a decimal

Problem-Solving Skills

Reading a chart
Selecting data from a display
Choosing the correct operation
Making a drawing

Starting the Lesson

The chart and drawing at the top of page 104 are also on ■ **Visual Aid 16** (copymaster or transparency S123). Use the chart and drawing and ask questions like these before discussing exercises 1–5.

- What is this type of race called? (BMX race)
- On this track, is the distance from start to finish more or less than half a mile? (Less)
- Did the two BMX riders complete the race in more or less than half a minute? (More)

Here's How Note

After discussing exercises 1–5, you may wish to use several of exercises 6–37 as chalkboard examples before the students begin their independent work.

104

More on dividing a decimal by a decimal

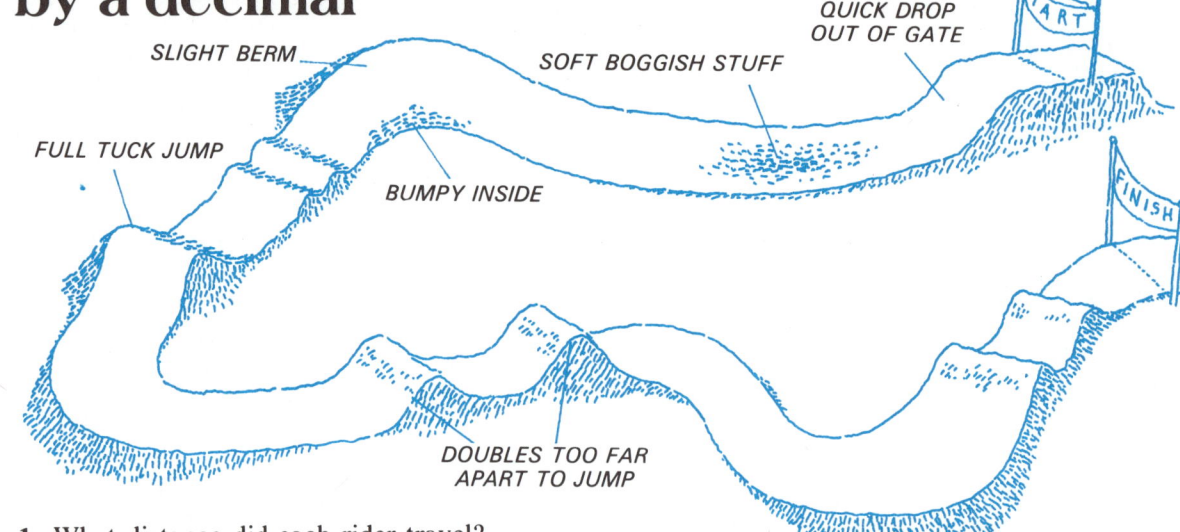

1. What distance did each rider travel? 0.2 mile
2. Which rider completed the race in 32.4 seconds? LeDuc
3. What was LeDuc's time in hours? 0.009
4. To compute LeDuc's average speed in miles per hour, you would divide 0.2 (miles traveled) by what number? 0.009

	BMX RACING		
Rider	Distance in miles	Time in seconds	in hours
LeDuc	0.2	32.4	0.009
Elliott	0.2	39.6	0.011

Here's how to divide a decimal by a decimal.

0.2 ÷ 0.009 = ?

You have to write some 0's in the dividend to place the decimal point.

5. Look at the *Here's how*. Notice that the division will never come out even. What is LeDuc's average speed rounded to the nearest tenth? 22.2 miles per hour

104 Chapter 4

EXERCISES

Divide. *Hint: You will need to write some zeros.*
Here are scrambled answers for the next row of exercises: 5.5 5 7.2 7.5

6. 0.5)3.6 = 7.2
7. 0.2)1.1 = 5.5
8. 0.06)0.3 = 5
9. 0.4)3 = 7.5
10. 0.8)56 = 70
11. 0.4)1.4 = 3.5
12. 0.02)7 = 350
13. 0.005)0.047 = 9.4
14. 0.2).01 = 0.05
15. 0.4)0.26 = 0.65
16. 0.08)1.24 = 15.5
17. 0.006)0.285 = 47.5
18. 1.2)4.2 = 3.5
19. 2.5)0.55 = 0.22
20. 3.1)124 = 40
21. 4.2)6.3 = 1.5

Divide. Round each quotient to the nearest tenth.
Here are scrambled answers for the next row of exercises: 0.4 2.3 0.2 4.7

22. 0.3)1.4 = 4.7
23. 0.6)0.14 = 0.2
24. 1.4)0.528 = 0.4
25. 2.2)4.98 = 2.3
26. 2.5)6.1 = 2.4
27. 3.1)7.5 = 2.4
28. 0.06)0.5 = 8.3
29. 1.2)0.54 = 0.5
30. 7.5)62.91 = 8.4
31. 3.9)7.503 = 1.9
32. 2.6)7.4 = 2.8
33. 0.31)0.09 = 0.3
34. 12.1)156.3 = 12.9
35. 1.84)192.6 = 104.7
36. 23.5)6.843 = 0.3
37. 3.41)18.5 = 5.4

Solve. *Refer to the chart and drawing on page 104.*

38. Was Elliott's average speed more or less than 20 miles per hour? *Hint: Divide to answer the question.* Less

39. How many seconds faster did LeDuc ride the 0.2-mile track than Elliott? 7.2

40. From the starting line to the full tuck jump is 0.09 mile. How far is it from the full tuck jump to the finish line? 0.11 mile

41. LeDuc's riding time from start to the full tuck jump was 14.7 seconds. How many seconds did it take LeDuc to ride from the full tuck jump to the finish line? 17.7 seconds

Who's ahead?

As the winner crosses the finish line, Cherie is 4 feet ahead of Eric. Kenny is 2 feet behind Justin. Eric is 6 feet ahead of Kenny.

42. Who is the winner? Cherie
43. Who is last? Kenny
44. How many feet separate the first and last racers? 10

Draw a picture like this one. It will help you solve the problems.

Dividing Whole Numbers and Decimals **105**

Extra Practice
Page 479 Skill 17

Practice Worksheet
Workbook S214, Copymaster S214, or Duplicating Master S214

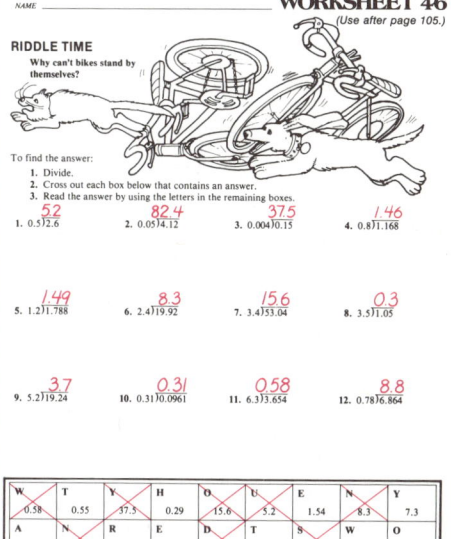

Project

Researching information
Find how long it takes you to walk, taking normal steps, a distance of 30 feet. Then compute your walking speed in feet per second and in miles per hour.

Copymaster S477

Class Starter Quiz 43
on previous lesson

Divide. Round each quotient to the nearest tenth.

1. 1.6 ÷ 0.3   5.3
2. 2.48 ÷ 0.6   4.1
3. 0.102 ÷ 0.09   1.1
4. 7.5 ÷ 2.2   3.4

Copymaster S395

Lesson Objective
To interpret remainders in the solution to a problem

Starting the Lesson
Problem-Solving Cover-up Use the chalkboard or mask ■ *Visual Aid 17* (copymaster or transparency S124).

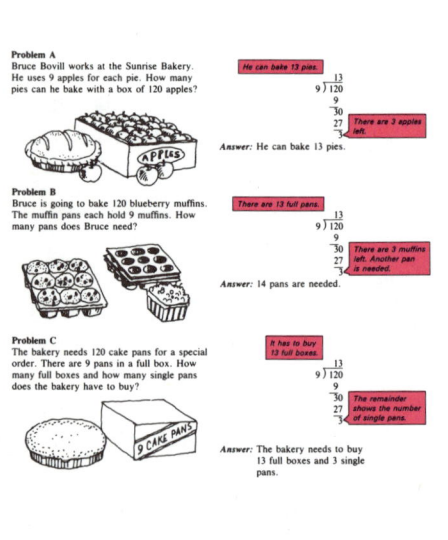

Have the students, working in small groups, study the problems on page 106 for several minutes. Then have them close their books, look at the visual aid, and tell what has been covered up.

106

PROBLEM SOLVING *interpreting remainders*

Sometimes when you use division to solve a problem, you have to think about how to use the remainder.

Problem A
Bruce Bovill works at the Sunrise Bakery. He uses 9 apples for each pie. How many pies can he bake with a box of 120 apples?

He can bake 13 pies.

$$9\overline{)120}$$ gives 13 remainder 3

There are 3 apples left.

Answer: He can bake 13 pies.

Problem B
Bruce is going to bake 120 blueberry muffins. The muffin pans each hold 9 muffins. How many pans does Bruce need?

There are 13 full pans.

$$9\overline{)120}$$ gives 13 remainder 3

There are 3 muffins left. Another pan is needed.

Answer: 14 pans are needed.

Problem C
The bakery needs 120 cake pans for a special order. There are 9 pans in a full box. How many full boxes and how many single pans does the bakery have to buy?

It has to buy 13 full boxes.

$$9\overline{)120}$$ gives 13 remainder 3

The remainder shows the number of single pans.

Answer: The bakery needs to buy 13 full boxes and 3 single pans.

106   Chapter 4

PROBLEMS

Solve. Think about the remainder. Then choose the correct answer.

1. A muffin recipe uses 4 cups of flour. How many full recipes can be made from 22 cups of flour?
   (a.) 5   b. 6

2. You can bake 2 dozen cookies on one cookie sheet. How many cookie sheets are needed to bake 15 dozen cookies?
   a. 7   (b.) 8

3. The oven can bake 18 loaves of bread at the same time. How many times would the oven have to be used to bake 60 loaves of bread?
   a. 3   (b.) 4

4. A carrot cake recipe calls for 15 eggs. There are 170 eggs on hand in the bakery. How many full recipes can be made?
   (a.) 11   b. 12

Solve. Think about the remainder before you write the answer.

5. A donut recipe calls for 12 cups of milk. How many full recipes can be made from 160 cups of milk? 13

6. The oven can bake 8 pans of dinner rolls at one time. How many times would the oven be used to bake 35 pans of rolls? 5

7. Adela has $4.80 and wants to buy as many croissants as she can. Croissants sell for $.65 each. How may can she buy? 7

8. Holiday cookies sell for $1.95 a dozen. Mark has $10 and wants to buy as many cookies as he can. How many dozens of cookies can he buy? 5

Dough photo   Visual thinking

9. In what order were the photographs taken?   E, D, B, A, F, C

A   B   C

D   E   F

Dividing Whole Numbers and Decimals   **107**

Practice Worksheet

Workbook S215, Copymaster S215, or Duplicating Master S215

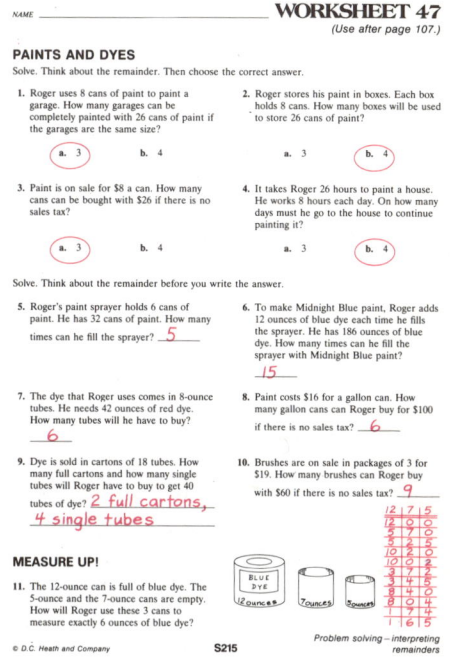

Challenge Problem

Use a calculator. Write a question that fits the answer.

A 1-pound can of coffee costs $2.79.

A 3-pound can of coffee costs $8.07.

How much is saved by buying a 3-pound can instead of a 1-pound can?

Answer: $.30

Copymaster S445

Class Starter Quiz 44
on previous lesson

Solve. Think about the remainder before you write the answer.

1. A recipe calls for 8 cups of milk. How many full recipes can be made from 30 cups of milk? **3**
2. You can put 12 doughnuts in a box. How many boxes are needed for 52 doughnuts? **5**

Copymaster S395

Problem-Solving Skill
Finding information on a computer printout

Skills Reviewed
Multiplying whole numbers and decimals
Dividing whole numbers and decimals
Simplifying expressions
Rounding decimals
Adding and subtracting whole numbers and decimals

Starting the Lesson
Problem Solving Write this student identification number on the chalkboard:

185-76-335

Tell the students that part of the ID number contains a student's month, day of month, and year of birth. Challenge them to figure out the student's year of birth. (1976) Then tell them to read the top of page 108 to find out how to break the code to find the student's birthday.

Cumulative Skill Practice Challenge the students to an estimation hunt by saying, "Pick the largest product in the first row of exercises." (Exercise 3) Then have students use their estimation skills to pick the largest product in the second, third, fourth, fifth and sixth rows of exercises. (Exercises 7, 12, 15, 19, and 23.)

108

Problem solving
COMPUTERS IN SCHOOLS

Schools use computers to store student data. In the Franklin School District, a computer assigns each student an identification number.

Look at Carol's identification number. The middle two digits of the number tell Carol's year of birth. The last three digits tell her birthday.

To find her birthday, divide the last 3-digit number by 32. The whole number in the quotient is her month of birth, and the remainder is the day of the month.

FRANKLIN SCHOOL DISTRICT

Carol Hanson
Student

16
Homeroom

A. Collier
Advisor

185-76-335

Year of birth | Code for birthday

$$32 \overline{\smash{)}335} \quad \begin{array}{r} 10 \leftarrow \text{month} \\ -320 \\ \hline 15 \leftarrow \text{day of month} \end{array}$$

Carol was born on October 15, 1976.

Use the information on the computer printout to answer these questions.

1. My name is Chris Harris. What is my date of birth?

July 22, 1978

		HOME	
NAME	I.D. NUMBER	ROOM	ADVISOR
Hanson, Carol	185-76-335	16	Collier
Harris, Chris	204-78-246	16	Vandegrift
Holmes, Brian	133-77-133	29	Collier
Holway, Amy	218-78-336	34	Vandegrift
Hull, Jack	199-77-190	34	Haver
Hurst, Mary	306-79-230	16	Collier
Heinbrough, Karen	156-80-121	16	Haver
Hernandez, Ralph	254-77-144	29	Collier
Hirai, George	314-80-299	34	Vandegrift

2. Give the birthday for each student.
   a. Amy Holway   October 16, 1978
   b. Mary Hurst   July 6, 1979
   c. Jack Hull   May 30, 1977

3. Mr. Haver's birthday is March 25. Which student on the list has the same birthday as Mr. Haver? **Karen Heinbrough**

4. Who is the youngest student on the list? **George Hirai**

Solve.

5. How many students on the list have birthdays during the school year? **7**

6. Brian Ogel was born on March 22, 1980. What are the last 5 digits of his student ID number? **80-118**

7. Suppose you registered in the Franklin School District. What would be the last 5 digits of your student ID number? *Answers will vary.*

108 Chapter 4

Cumulative Skill Practice

Give the product. (page 68)

1. 2.6 × 0.9 2.34
2. 3.98 × 0.6 2.388
3. 27.5 × 0.8 22
4. 2.9 × 0.7 2.03
5. 6.2 × 0.59 3.658
6. 0.38 × 1.5 0.57
7. 36 × 0.42 15.12
8. 3.1 × 0.2 0.62
9. 206 × 4.3 885.8
10. 8.36 × 19 158.84
11. 0.35 × 50 17.5
12. 14.6 × 82 1197.2

Divide. If the division does not come out even, round the quotient to the nearest whole number. (pages 88, 92)

13. 28)1932 69
14. 39)3042 78
15. 15)1470 98
16. 62)1426 23
17. 17)1003 59
18. 12)10,446 871
19. 56)69,454 1240
20. 36)20,882 580
21. 41)35,867 875
22. 48)31,200 650
23. 117)72,594 620
24. 117)70,551 603
25. 117)24,687 211
26. 400)14,400 36
27. 504)16,128 32

Simplify. (page 94)

28. 24 ÷ 4 + 2 8
29. 30 − 5 + 7 32
30. 8 × 4 − 1 31
31. 36 ÷ 6 ÷ 3 2
32. 14 × 5 ÷ 5 14
33. 18 + 8 ÷ 2 22
34. (25 + 25) ÷ 10 5
35. 16 × (7 − 3) 64
36. 20 × (10 − 4) 120
37. 36 + 12 ÷ 4 + 2 41
38. (36 + 12) ÷ 4 + 2 14
39. 36 + 12 ÷ (4 + 2) 38
40. 32 × 8 − 8 + 4 252
41. 32 × (8 − 8) + 4 4
42. 32 × 8 − (8 + 4) 244

Give the quotient. Round the quotient to the nearest tenth. (pages 102, 104)

43. 16.4 ÷ 0.6 27.3
44. 3.74 ÷ 0.2 18.7
45. 9.05 ÷ 0.05 181
46. 28.7 ÷ 0.3 95.7
47. 4.38 ÷ 0.04 109.5
48. 8.4 ÷ 0.08 105
49. 6.15 ÷ 1.2 5.1
50. 2.038 ÷ 2.4 0.8
51. 4.06 ÷ 0.25 16.2

MIXED PRACTICE

Complete.

52. 35.082 rounded to the nearest tenth is ? 35.1
53. 7.465 rounded to the nearest hundredth is ? 7.47
54. 19 + 758 + 65 = ? 842
55. 357 × 8 = ? 2856
56. 46.38 − 21.67 = ? 24.71
57. 248 × 106 = ? 26,288
58. 1702 − 485 = ? 1217
59. 5.783 + 12 + 8.2 = ? 25.983
60. (5.9 + 3.5) × 2 = ? 18.8
61. 0.8 × (6.7 − 2.5) = ? 3.36
62. 12.6 − (4.3 + 2.17) = ? 6.13

Dividing Whole Numbers and Decimals

Problem-Solving Worksheet
Workbook S216, Copymaster S216, or Duplicating Master S216

NAME _____ **WORKSHEET 48**
(Use after page 108.)

LOOSE CHANGE
One nickel, some dimes, and three pennies total 48¢. How many dimes are there? 4

BE A SCOREKEEPER!

SCORE BOOK		
PLAYER	SHOTS	POINTS
Reed	XXOXO	7
Davis	X●O●X	6
Boyce	XXX●●	8
Gilbert	XO●	3
Lewis	XXXXX	10

How many points does each ● represent?

HIGHS AND LOWS

TEMPERATURE (°F)		
CITY	HIGH	LOW
Boston	77°	53°
Chicago	78°	62°
Cleveland	80°	54°
Denver	85°	67°
Detroit	83°	65°

Which city had a high temperature of more than 80° and a low of less than 66°? Detroit

WATCH OUT!
The watch shows the same time upside down as right side up. What is the next time this will happen again? 11:11

DOLLARS AND SENSE
Joanne had less than 5 one-dollar bills. She traded her dollar bills for an equal number of nickels and dimes. How many dollar bills did Joanne have? 3

WORK BACKWARD
Find the Start number.

START 6 → +2 → ×4 → END 10, +2, −4

ALL IN THE FAMILY
A mechanic's brother won the car race. But the man who won had no brother. How is this possible? The mechanic was a woman.

© D. C. Heath and Company   S216   Problem solving

Project

Using library resources
Use an almanac to find which two presidents of the United States had their birthdays on the same month and day of the month.

Warren Harding and James Polk both had a November 2 birthday.

Copymaster S477

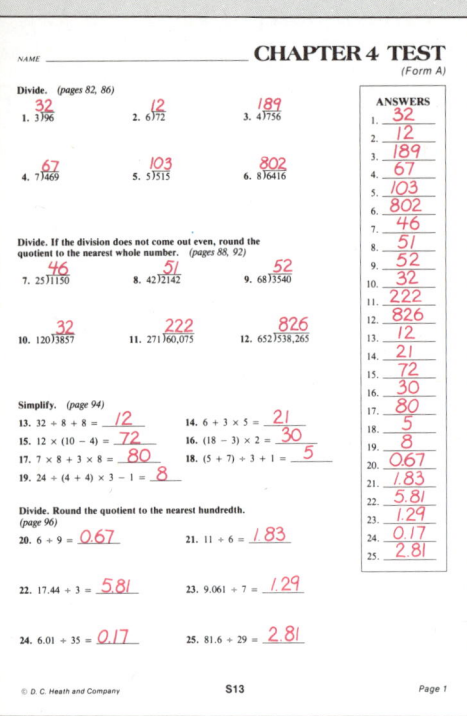
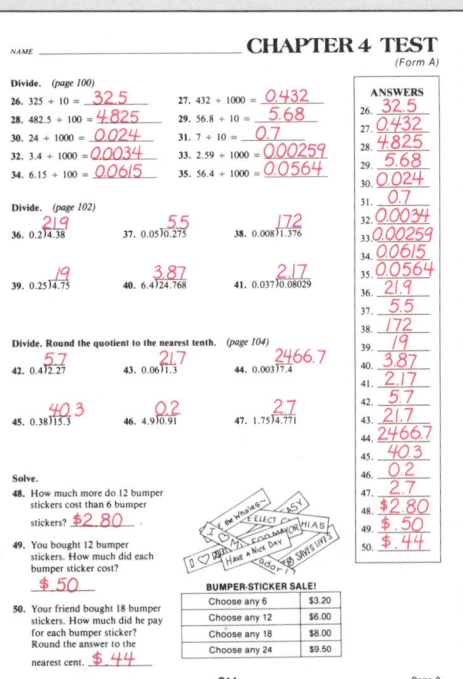

Chapter REVIEW

Here are scrambled answers for the review exercises:

| 1 | 3 | 6 | 26 | divide | left | quotient | zeros |
| 2 | 5 | 7 | 140 | add | multiply | subtract | |

1. 6, quotient **2.** zeros, 26 **3.** 5, multiply, add

1. To complete this division exercise, divide 24 by [?]. The answer is called the [?]. *(page 82)*

$$6\overline{)324}$$
$$-30$$
$$24$$

2. To estimate this quotient, find the first digit in the quotient and write [?] in the remaining places. To find the first digit of this quotient, think about dividing [?] by 6. *(pages 86, 88)*

$$62\overline{)2609}$$

3. In this division exercise, the quotient is 42 and the remainder is [?]. To check the division, [?] 42 by 62 and [?] 5. *(page 88)*

$$62\overline{)2609}\ \ 42\ R5$$
$$-248$$
$$129$$
$$-124$$
$$5$$

4. 7 **5.** divide, subtract **6.** 140

4. A zero was written in the tenths place and the division was carried out to that place. The quotient rounded to the nearest whole number is [?]. *(page 92)*

$$3\overline{)22.0}\ \ 7.3$$
$$-21$$
$$10$$
$$-9$$
$$1$$

5. To simplify this expression, first [?] and then [?]. *(page 94)*

$$6 - 4 \div 2$$

6. To carry out the division to the next place, you would divide [?] by 34. *(page 96)*

$$34\overline{)198.6}\ \ 5.8$$
$$-170$$
$$286$$
$$-272$$
$$14$$

7. 3, left **8.** 1 **9.** 2

7. Dividing by 1000 moves the decimal point [?] places to the [?]. *(page 100)*

$$16.5 \div 1000 = 0.165$$

8. Before dividing, move both decimal points [?] place(s) to the right. *(page 102)*

$$1.4\overline{)2.884}$$

9. To place the decimal point in the dividend, you first write [?] zeros. *(page 104)*

$$0.007\overline{)1.6}$$

Chapter 4

Chapter TEST

Divide. (pages 82, 86)

1. 3)93̄ → 31
2. 5)85̄ → 17
3. 9)198̄ → 22
4. 6)738̄ → 123
5. 8)816̄ → 102
6. 2)608̄ → 304
7. 4)1624̄ → 406
8. 7)4235̄ → 605
9. 8)6424̄ → 803
10. 5)4195̄ → 839

Divide. If the division does not come out even, round the quotient to the nearest whole number. (pages 88, 92)

11. 24)888̄ → 37
12. 18)1098̄ → 61
13. 30)2734̄ → 91
14. 56)8365̄ → 149
15. 53,216 ÷ 500   106
16. 38,296 ÷ 213   180
17. 71,025 ÷ 432   164
18. 67,055 ÷ 621   108

Simplify. (page 94)

19. 8 − 4 + 2   6
20. 7 × 5 − 3   32
21. 20 + 8 ÷ 4   22
22. 3 + 9 ÷ 3   6
23. 5 + 8 ÷ 4 + 9   16
24. (16 + 8) ÷ 4 + 2   8
25. 16 + 8 ÷ 4 + 2   20
26. 8 + 4 × 2 − 1   15

Divide. (pages 96, 100)

27. 19.2 ÷ 8   2.4
28. 118.4 ÷ 4   29.6
29. 109.62 ÷ 9   12.18
30. 6.45 ÷ 5   1.29
31. 558.7 ÷ 37   15.1
32. 99.84 ÷ 48   2.08
33. 1483.2 ÷ 72   20.6
34. 205.44 ÷ 32   6.42
35. 364 ÷ 10   36.4
36. 528 ÷ 100   5.28
37. 327 ÷ 1000   0.327
38. 326 ÷ 100   3.26
39. 123.9 ÷ 100   1.239
40. 682.3 ÷ 10   68.23
41. 529.64 ÷ 100   5.2964
42. 65.2 ÷ 100   0.652

Divide. Round each quotient to the nearest tenth. (pages 102, 104)

43. 0.5)4.8̄ → 9.6
44. 0.6)31.26̄ → 52.1
45. 0.9)4.32̄ → 4.8
46. 1.6)35̄ → 21.9
47. 1.1)5.73̄ → 5.2
48. 3.6)7.039̄ → 2.0
49. 0.64)15.3̄ → 23.9
50. 3.2)6.84̄ → 2.1

Solve.

51. How much will it cost to have a roll of 12 and a roll of 24 printed? $6.10
52. How much will it cost to have 3 rolls of 15 printed? $8.16
53. How much does each print cost for a roll of 36? Round your answer to the nearest cent. $.14
54. You had 2 rolls of 12 and 1 roll of 24 printed. You gave the clerk $20. How much change should you have received? $11.51

PHOTO FINISHING	
12 EXPOSURES	$2.39
15 EXPOSURES	2.72
24 EXPOSURES	3.71
36 EXPOSURES	5.03

Dividing Whole Numbers and Decimals **111**

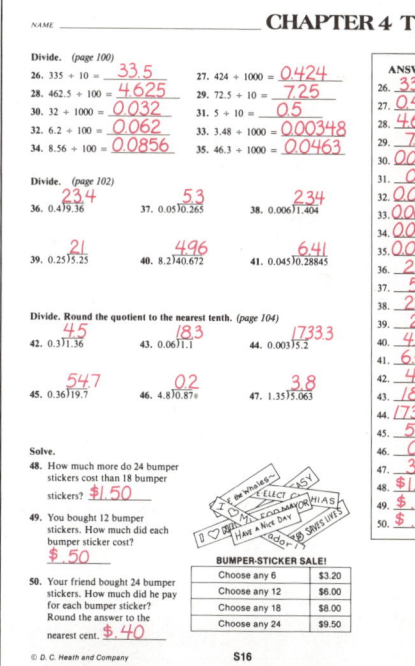

Cumulative Test
(Chapters 1–4)

Use Copymaster S109 to provide the students with an answer sheet in standardized test format.

Answers for Cumulative Test, Chapters 1–4

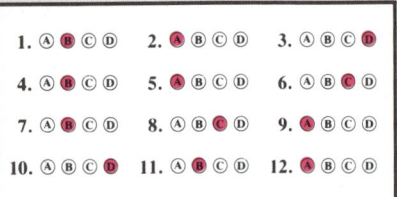

The table below correlates test items with student text pages.

Test Item	Page(s) Taught	Skill Practice
1	4	p. 91, exercises 1–16
2	6	p. 91, exercises 17–31
3	22	p. 91, exercises 32–46
4	38	p. 99, exercises 1–10
5	42	p. 99, exercises 11–22
6	46	p. 99, exercises 23–34
7	64	p. 99, exercises 35–40
8	68	p. 109, exercises 1–12
9	88, 92	p. 109, exercises 13–27
10	94	p. 109, exercises 28–42
11	102, 104	p. 109, exercises 43–51
12	64	

112

Cumulative TEST — Standardized Format

Choose the correct letter.

1. The short word-name for 8,020,000 is
   - A. 8 billion, 20 million
   - **B. 8 million, 20 thousand**
   - C. 8 million, 2 thousand
   - D. none of these

2. Add.   56,394
         + 26,748
   - **A. 83,142**
   - B. 72,032
   - C. 72,042
   - D. none of these

3. Give the sum.
   14.6 + 31.5 + 2.06
   - A. 66.7
   - B. 47.16
   - C. 481.6
   - **D. none of these**

4. Subtract.   35,064
              − 13,297
   - A. 22,233
   - **B. 21,767**
   - C. 22,767
   - D. none of these

5. Which number is greater than 0.06?
   - **A. 0.5**
   - B. 0.04
   - C. 0.059
   - D. none of these

6. Give the difference.
   424.7 − 38.46
   - A. 40.1
   - B. 382.36
   - **C. 386.24**
   - D. none of these

7. Multiply.   629
              × 240
   - A. 15,096
   - **B. 150,960**
   - C. 149,660
   - D. none of these

8. Give the product.
   3.62 × 1.7
   - A. 5.774
   - B. 61.54
   - **C. 6.154**
   - D. none of these

9. Divide.   68) 7140
   - **A. 105**
   - B. 15
   - C. 106
   - D. none of these

10. Simplify.
    20 − 8 ÷ 4 × 2
    - A. 6
    - B. 19
    - C. 36
    - **D. none of these**

11. Give the quotient rounded to the nearest tenth.
    4.615 ÷ 0.3
    - A. 15.3
    - **B. 15.4**
    - C. 1.5
    - D. none of these

12. You can buy a car for $325 down and 12 payments of $48.75 each. What is the total cost?
    - **A. $910**
    - B. $373.75
    - C. $260
    - D. none of these

112   Chapter 4

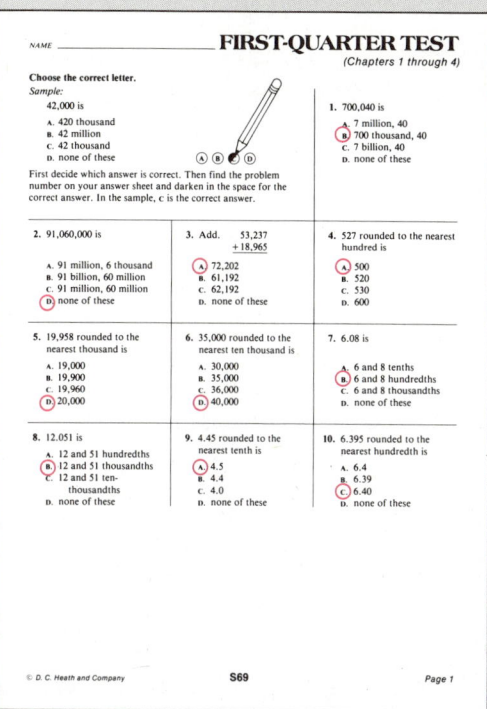

First-Quarter Test

The first-quarter test shown on these two pages is in standardized format so that the students can become accustomed to taking standardized tests.

Use Copymaster S92 or Duplicating Master S92 to provide the students with an answer sheet in standardized test format.

Copymaster S106 has a quick-score answer key for the first-quarter test.

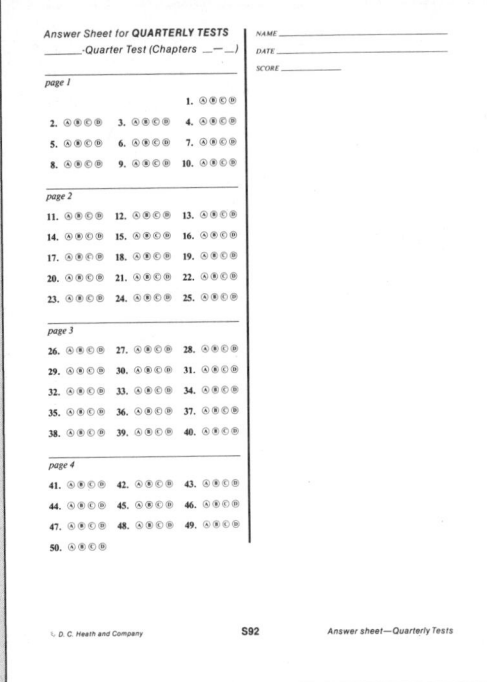

The table below correlates test items with student text pages.

Test Item	Text Page
1	p. 4
2	p. 4
3	p. 6
4	p. 10
5	p. 10
6	p. 10
7	p. 14
8	p. 14
9	p. 18
10	p. 18
11	p. 20
12	p. 22
13	p. 20

Test Item	Text Page
14	p. 20
15	p. 32
16	p. 32
17	p. 38
18	p. 38
19	p. 42
20	p. 42
21	p. 44
22	p. 46
23	p. 46
24	p. 44
25	p. 46
26	p. 56

Test Item	Text Page
27	p. 56
28	p. 58
29	p. 62
30	p. 64
31	p. 68
32	p. 68
33	p. 70
34	p. 70
35	p. 72
36	p. 72
37	p. 68
38	p. 68
39	p. 86

Test Item	Text Page
40	p. 88
41	p. 92
42	p. 94
43	p. 94
44	p. 96
45	p. 100
46	p. 100
47	p. 102
48	p. 104
49	p. 98
50	p. 98

Graphs and Statistics

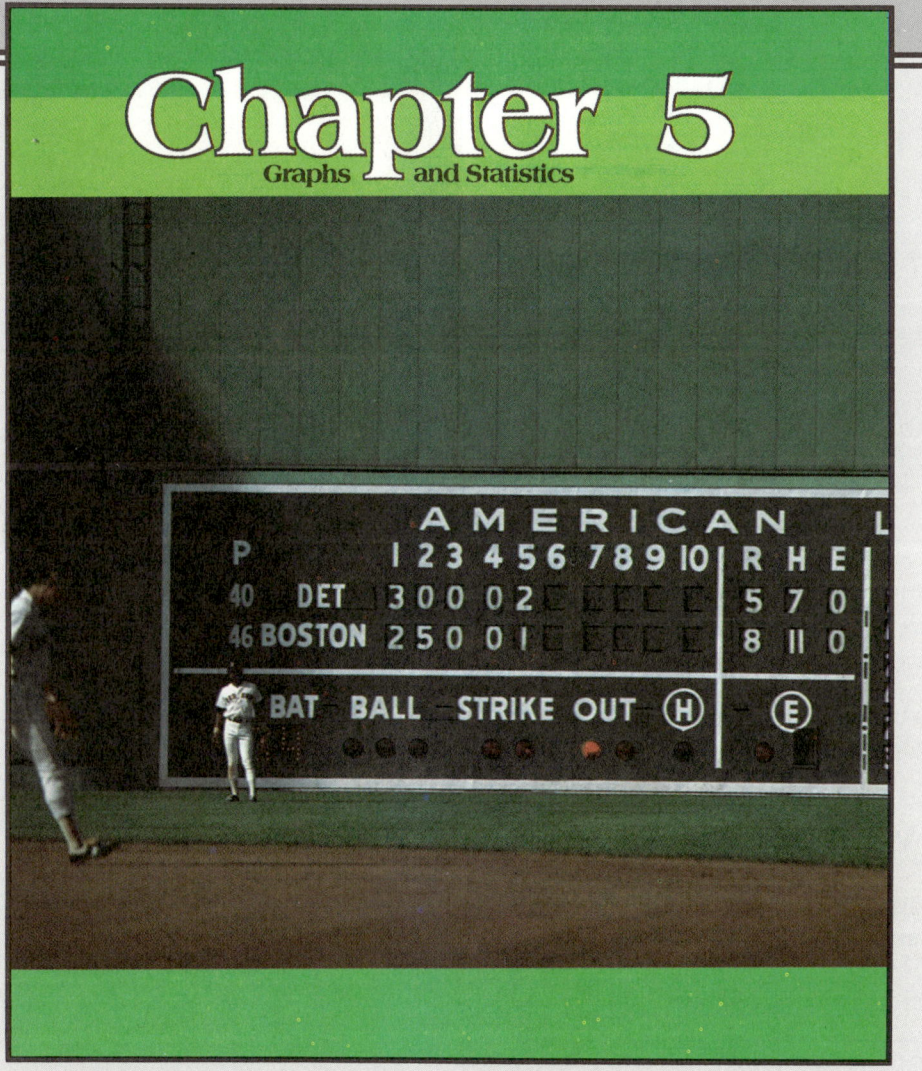

Resources

- **Class Starter Quizzes 45-53** *(Copymasters S396-S398)*
- **Visual Aids 18-23** *(Copymasters or Transparencies S125-S130)*
- **Worksheets 49-58** *(Copymasters, Duplicating Masters, or Workbook pages S217-S226)*
- **Challenge Problems** for pages 123, 131, 133 *(Copymasters S445-S446)*
- **Projects** for pages 121, 129 *(Copymaster S477)*
- **Tests** *(Copymasters or Duplicating Masters S17-S20)*

Lesson Objectives
To read data from frequency tables
To make a frequency table from a set of data

Problem-Solving Skills
Reading and interpreting a table
Organizing and tabulating information

Starting the Lesson
The National Road-Sign Quiz from page 114 and a blank frequency table are on ■ **Visual Aid 18** (copymaster or transparency S125). Have the students take the National Road-Sign Quiz. Correct and save their quiz papers. After discussing questions 1–7, use the students' papers to make a frequency table of their scores. Then have the students answer questions 3–7 from the information on their own frequency table.

Answer for page 114.

19.

Age	Number of Occurrences	
	Tally	Frequency
14	⊮ II	7
15	⊮ ⊮ I	11
15½	I	1
16	⊮ ⊮ ⊮ ⊮ ⊮ ⊮	30
16½	I	1
18	I	1

Organizing data

A group of students took the road-sign quiz. Their scores are listed on this card.

National Road-Sign Quiz
Choose the correct answer.

(a.) Slippery when wet
b. Road curves
c. Steep hill

a. Housing
b. Youth hostel
(c.) Hospital

a. No passing
(b.) No parking
c. No pedestrians

a. Zoo
b. Hunting area
(c.) Deer crossing

NAME	SCORE	NAME	SCORE
B. Barrett	3	H. Lee	1
M. Cataldo	4	B. Lynch	3
A. Collier	3	P. Perez	4
C. DeBold	4	A. Riccio	3
J. Dombrowski	2	A. Sargent	2
D. Dori	4	R. Smith	1
J. Goldman	2	G. Summers	3
V. Hawkins	3	J. Vandegrift	3
A. Jacobson	1	S. Vlahach	4
K. Kennedy	2	C. Werner	0

1. Look at the score card. How many students had a score of 4? 5

2. How many students had a score of 3? 7

To help answer questions like these, you can make a frequency table.

Here's how *to organize the data (quiz scores) in a frequency table.*

A tally mark shows each time a score occurred. The frequency column shows the total number of times each score occurred.

Frequency Table for Quiz Scores

SCORE	NUMBER OF TIMES SCORE OCCURRED	
	TALLY	FREQUENCY
4	⊮	5
3	⊮ II	7
2	IIII	4
1	III	3
0	I	1

3. How many students had a score less than 4? 15

4. How many had a score greater than 2? 12

5. What was the highest score? lowest score? 4, 0

6. Which score occurred most often? least often? 3, 0

7. How many students took the quiz? 20

EXERCISES

Use the frequency tables to answer the questions.

8. What was the highest score on the engine-repair exam? 15
9. Which score occurred most often? 13
10. How many students scored 13 or less? 20
11. How many students took the exam? 27
12. What was the difference between the highest and lowest scores? 5
13. A score of 14 or better is an A. How many students received an A? 7

14. How many families in the survey have 3 cars? 7
15. How many families have 2 or more cars? 25
16. How many families have at least 1 car? 47
17. What is the most common number of cars per family? 1
18. How many families were in the survey? 50

Frequency Table for Engine-Repair Exam Scores

SCORE	NUMBER OF OCCURRENCES	
	TALLY	FREQUENCY
15					3
14						4
13	||||-|||| ||	12
12							6
11		0
10				2

Frequency Table for Cars per Family

CARS PER FAMILY	NUMBER OF FAMILIES	
	TALLY	FREQUENCY
4				2
3								7
2															16
1																				22
0					3

Try it yourself! Organizing information

19. Make a frequency table for the driver's-license information on the map.
20. In how many states can you obtain a driver's license before age 16? 27

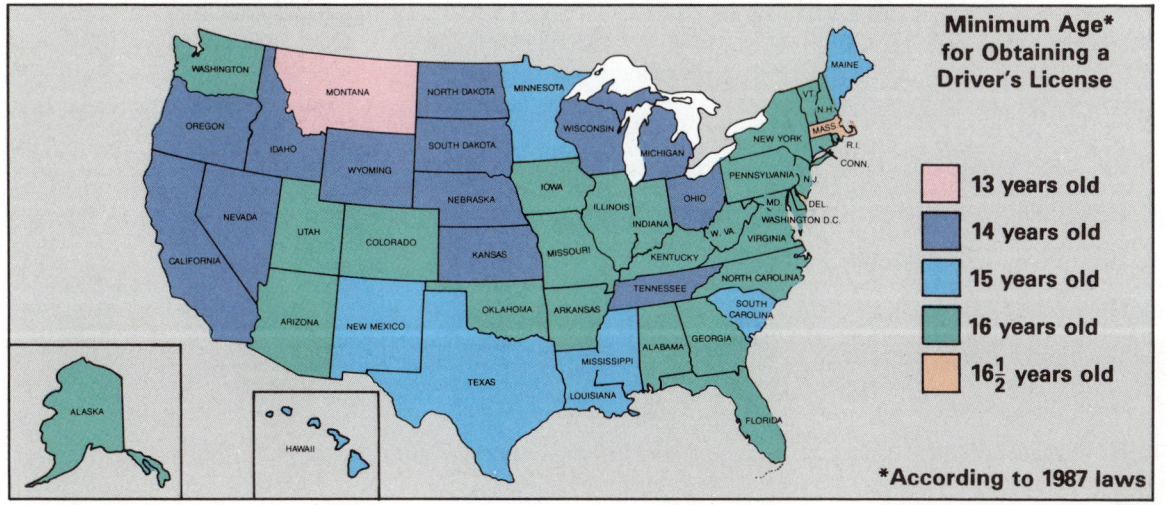

Minimum Age* for Obtaining a Driver's License
- 13 years old
- 14 years old
- 15 years old
- 16 years old
- 16½ years old

*According to 1987 laws

Graphs and Statistics **115**

Practice Worksheet

Workbook S217, Copymaster S217, or Duplicating Master S217

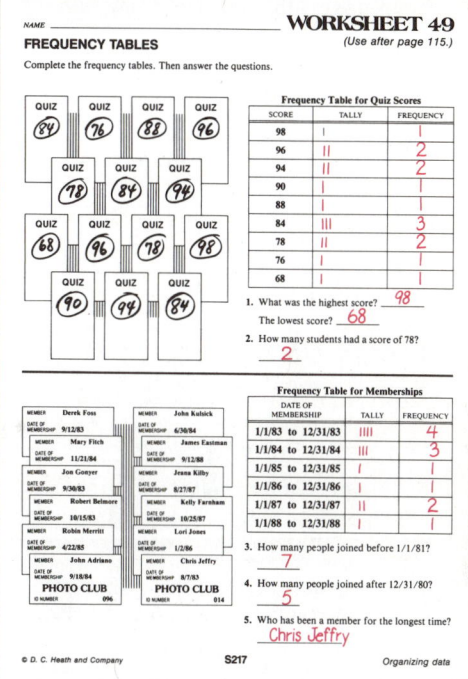

Group Project
Collecting data

Have the students work in small groups and conduct a survey of teenagers to find out the kind of car they like best. Then have them use the data to show their findings on a frequency chart.

115

Using bar graphs

Fifty people took part in a TV trivia survey. In Part 1 of the survey, each person was asked to name these characters.

WHO ARE THESE CHARACTERS?

The results of Part 1 of the survey are shown in this **bar graph.**

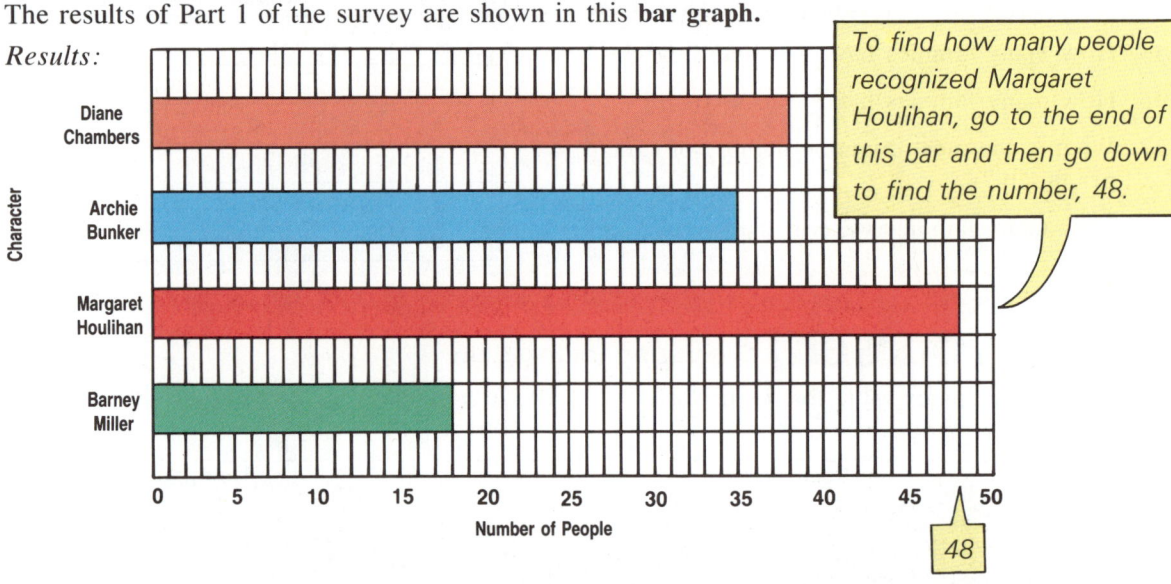

To find how many people recognized Margaret Houlihan, go to the end of this bar and then go down to find the number, 48.

1. Read the bar graph. How many people recognized Margaret Houlihan? 48
2. How many recognized Barney Miller? 18
3. How many knew Archie Bunker? 35
4. How many knew Diane Chambers? 38
5. Which character was recognized by the most people? the fewest people? Margaret Houlihan, Barney Miller
6. How many more people recognized Margaret Houlihan than Barney Miller? 30
7. How many of the 50 people surveyed did not recognize Archie Bunker? 15

EXERCISES

Use the bar graphs to answer the questions.

In Part 2 of the survey, each person was asked to name the TV show in which the character appeared.

8. How many people knew that Diane appeared in *Cheers*? **24**

9. How many knew that Archie appeared in *All in the Family*? **25**

10. How may more people knew the name of the TV show Barney appeared in than knew the TV show Archie appeared in? **22**

Results:

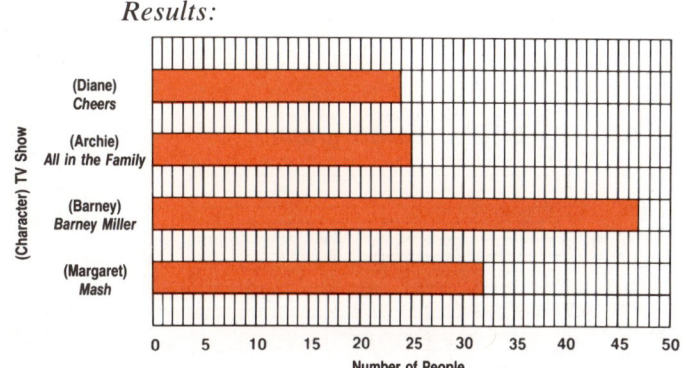

In Part 3 each person was asked to name the actor or actress who played the character.

11. Which actor or actress was identified by the most people? the fewest people? **Hal Linden, Loretta Swit**

12. How many people identified Shelley Long? **15**

13. How many people identified Hal Linden? **25**

14. How many more people identified Carroll O'Conner than identified Loretta Swit? **4**

Results:

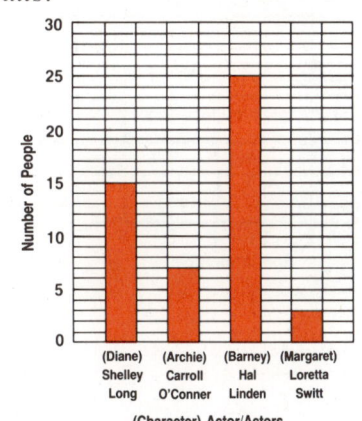

You're the reporter

Use the bar graph. Complete the story.

15. The price of a movie ticket in **?** was $1.50. A movie ticket in 1980 cost **?**. **1975, $2**

16. Between 1975 and **?**, the price of a ticket increased $1.00, from $1.50 to **?**. **1983, $2.50**

17. In 1975, 4 tickets cost **?**. The cost of **?** tickets in 1980 was $10.00. **$6, 5**

Graphs and Statistics 117

Extra Practice

Page 480 Skill 19

Practice Worksheet

Workbook S218, Copymaster S218, or Duplicating Master S218

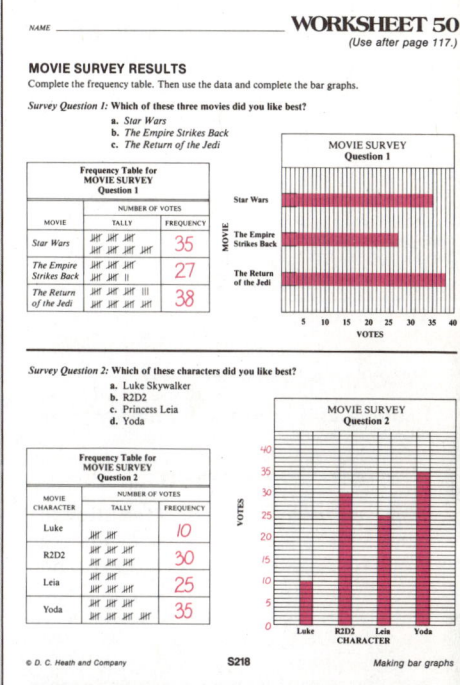

Group Project

Collecting and presenting data

Have the students work in small groups and conduct a survey of teenagers to find out the kind of music they like best.

MUSIC	TALLY
Popular	
Rock	
Country	
Jazz	
Classical	
Other	

Then have them use the data to show their findings on a bar graph.

Class Starter Quiz 46
on previous lesson

Solve. Use the bar graphs on page 117.

1. How many people knew that Archie appeared in *All in the Family*? 25
2. How many people knew that Hal Linden played Barney Miller? 25
3. How many more people named Shelley Long than named Loretta Swit? 12

Copymaster S396

Lesson Objective
To read data from line graphs

Problem-Solving Skill
Reading and interpreting line graphs

Starting the Lesson
What are the facts? Have the students study the newspaper article and line graph on page 118 for 60 seconds. Then have them close their books and decide whether these statements are true or false:

- Zella was this year's Coos County Grand Champion lamb. (True)
- Zella is 7 months old and weighs 50 pounds. (False)
- The line graph shows Zella's weight increase. (True)
- When Zella was 1 month old, she weighed 14 pounds. (True)
- Zella won a red ribbon for her owner, Julie Eastman. (False)

Exercise Note
The line graphs on pages 118 and 119 are also on ■ **Visual Aids 20 and 21** (copymasters or transparencies S127–S128). You might wish to use these visual aids when discussing exercises 1–13.

Using line graphs

Livestock that are raised to be shown at fairs receive special care and attention. Feed is controlled and growth is measured monthly. The **line graph** shows Zella's weight increase.

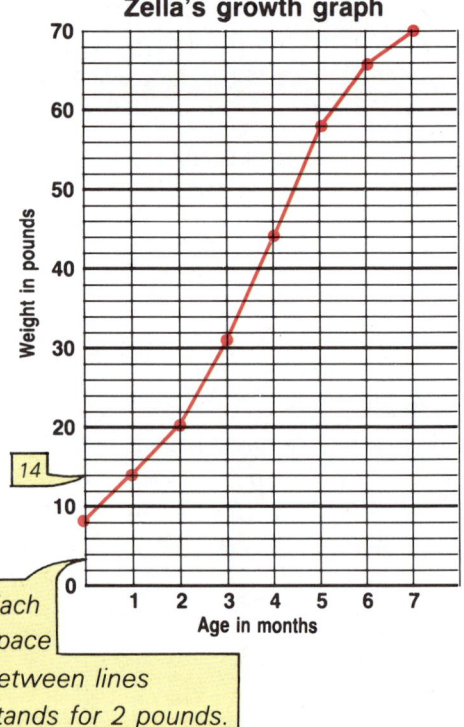

Each space between lines stands for 2 pounds.

ZELLA WINS THE BLUE!
LANCASTER, N.H. This year's Grand Champion Lamb is Zella. Zella is 7 months old and weighs 70 pounds.

This year's Coos County Grand Champion Lamb with her owner, Julie Eastman.

To find Zella's weight when she was 1 month old, look up along the 1-month line, then look left to find her weight, 14 pounds.

1. Read the line graph. How much did Zella weigh at
   a. 4 months? 44 lb
   b. 5 months? 58 lb
   c. 7 months? 70 lb
2. How old was Zella when she weighed
   a. 20 pounds? 2 mo
   b. 31 pounds? 3 mo
   c. 66 pounds? 6 mo
3. How many pounds did Zella gain during the first two months? 12
4. How much more did Zella weigh at 3 months than she did at 1 month? 17 lb
5. When did Zella gain more, in the third month or in the sixth month? Third
6. During which 3-month period did Zella gain the most weight?
   Between the 2nd and 5th months

Chapter 5

EXERCISES

Use the line graphs to answer the questions.

7. What was the fair attendance at
   a. 10 A.M.? 600
   b. noon? 800
   c. 6 P.M.? 1800
   d. 10 P.M.? 2700

8. What was the approximate time when the attendance reached
   a. 1200 people? 2 P.M.
   b. 1800 people? 6 P.M.
   c. 2500 people? 8 P.M.
   d. 2700 people? 10 P.M.

9. Fair officials predicted that 3000 people would be in attendance opening day, September 1. How far off was their prediction? 200 people

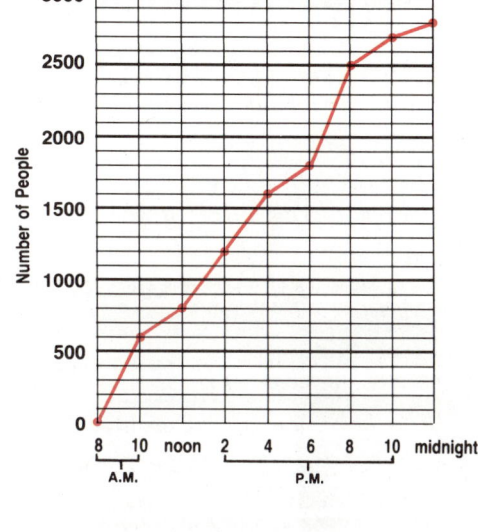

Lancaster Fair Attendance for September 1

10. Workers at the booth worked 4-hour shifts. The first shift took in $150. How much did the second shift take in? $250

11. How much money did the Lancaster Booster Club take in between 8 P.M. and midnight? $300

12. The shift Carmen worked took in $500. What time did she get off work? 8 P.M.

13. The Booster Club charged $5 for each hat. About how many hats did the club sell on September 1? 240

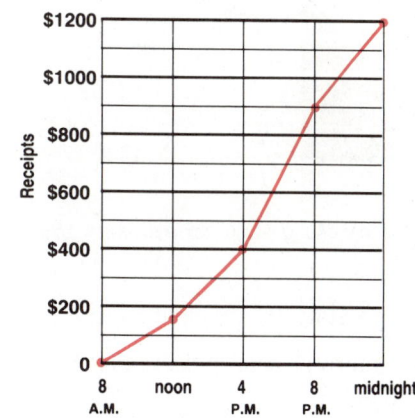

Lancaster Booster Club Hat Sales Booth Receipts for September 1

Don't be fooled! Interpreting data

14. Each bar graph compares attendance at the Lancaster Fair. One of the two graphs has been drawn to mislead people. Which graph is misleading? How is it misleading? b
    The scale begins at 2500. Therefore it appears that many more people attended the fair on Sept. 1 than on Sept. 2.

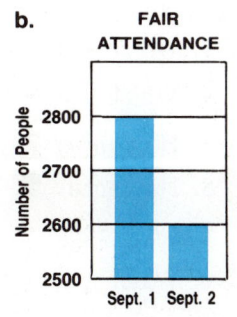

Graphs and Statistics **119**

Extra Practice
Page 480   Skill 20

Practice Worksheet
Workbook S219, Copymaster S219, or Duplicating Master S219

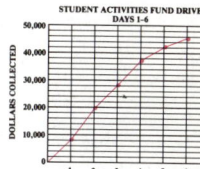

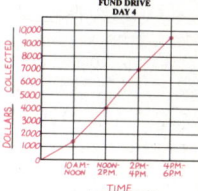

Group Project

Collecting and presenting data
Have the students work in small groups to make a line graph that will show the daily high temperatures in the community for a week.

119

Using circle graphs and picture graphs

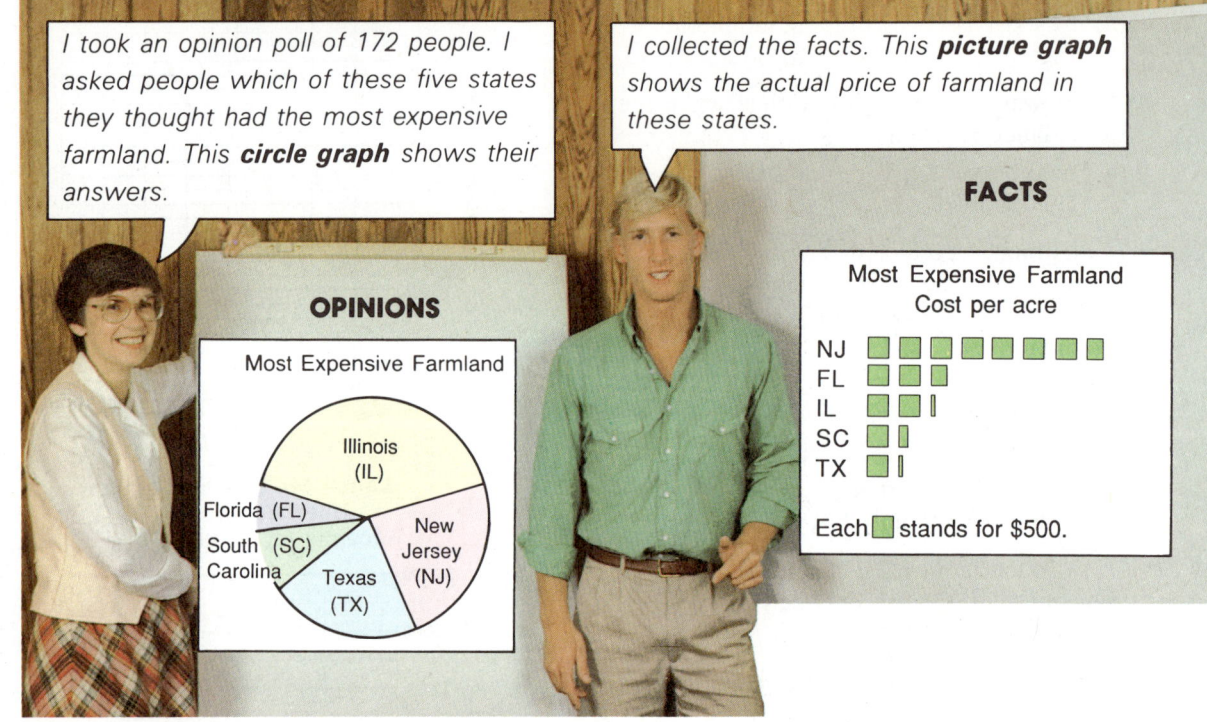

*I took an opinion poll of 172 people. I asked people which of these five states they thought had the most expensive farmland. This **circle graph** shows their answers.*

*I collected the facts. This **picture graph** shows the actual price of farmland in these states.*

Use the circle graph to answer these questions.

1. Which state was picked by the most people? Illinois
2. Which state was picked by the fewest people? Florida
3. Did more people pick Texas than picked South Carolina? Yes
4. Did more people pick New Jersey than picked Illinois? No

Use the picture graph to answer these questions.

5. Which state has the most expensive farmland? New Jersey
6. Which state's farmland averages about $1400 an acre? About $3900 an acre? Florida, New Jersey
7. How many 💲 would be used to show $1500-an-acre farmland? 3
8. How many 💲 would be used to show $1250-an-acre farmland? $2\frac{1}{2}$
9. Does South Carolina's farmland average more or less than $2000 an acre? less
10. Which state's farmland costs about $300 per acre more than Illinois's? Florida

120   Chapter 5

EXERCISES

Use the circle graph to answer the questions.

11. How many people thought Alaska had the least expensive farmland? 86

12. Which state was picked by 42 people? Wyoming

13. Which state was picked by the fewest people? Nevada

14. Which two states were picked by a total of 128 people? Alaska, Wyoming

15. Which three states were picked by a total of 75 people? New Mexico, Arizona, Wyoming

Use the picture graph to answer the questions.

16. Which state's farmland costs about
    a. $109 per acre? Alaska
    b. $191 per acre? Nevada
    c. $195 per acre? Arizona
    d. $146 per acre? New Mexico

17. Which state's farmland costs about $90 per acre more than Alaska's? Arizona

18. Does Wyoming's farmland cost more or less than $150 per acre? More

19. How many 💲 would be used to show $260-per-acre land? 13

Make a picture graph

20. Use the information in the chart to make a picture graph. You may choose any symbol to represent pounds of fruit. Draw one symbol for each 4 pounds. Be sure to label your graph.

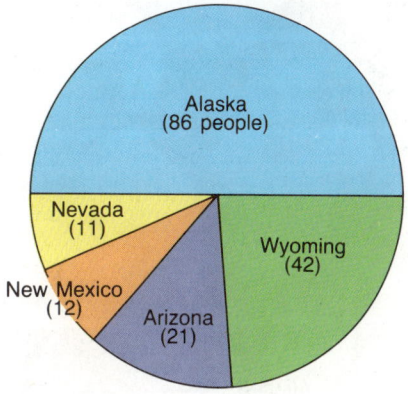

Least Expensive Farmland According to an Opinion Poll

- Alaska (86 people)
- Wyoming (42)
- Arizona (21)
- New Mexico (12)
- Nevada (11)

Least Expensive Farmland
(based on cost per acre
U.S. Census Bureau averages)

Alaska	💲💲💲💲💲🌗
Wyoming	💲💲💲💲💲💲💲🌗
New Mexico	💲💲💲💲💲💲💲🌗
Nevada	💲💲💲💲💲💲💲💲💲🌗
Arizona	💲💲💲💲💲💲💲💲💲🌗

Each 💲 stands for $20.

Yearly Consumption of Fruit by the Average American

TYPE	NUMBER OF POUNDS
Fresh	80
Canned	36
Frozen	12
Dried	4

Graphs and Statistics

Extra Practice
Page 481   Skill 21

Practice Worksheet
Workbook S220, Copymaster S220, or Duplicating Master S220

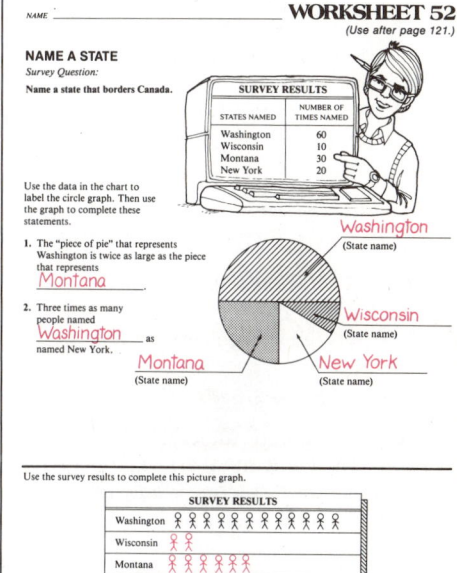

Project

Using library resources

Use a book of world records to find the cost per square foot of the most expensive land in the world.

Answers will vary. Records may change from year to year.

Copymaster S477

Class Starter Quiz 48
on previous lesson

Solve. Use the circle graph and picture graph on page 121.

1. How many people thought Wyoming had the least expensive farmland? **42**
2. Which two states were picked by a total of 33 people? **New Mexico and Arizona**
3. Which state's farmland costs about $80 per acre more than Alaska's? **Nevada's**

Copymaster S396

Problem-Solving Skills
Reading a graph
Choosing the correct operation

Skills Reviewed
Writing standard numerals
Rounding whole numbers and decimals
Adding and subtracting whole numbers and decimals
Simplifying expressions

Starting the Lesson
Problem Solving The line graph on page 122 is also on ■ **Visual Aid 22** (copymaster or transparency S129). Have the students read the introductory paragraph on page 122 and then use the graph to answer questions like these:

- What is the title of the graph? (Total Miles Traveled)
- On which day did Andy and Sarah travel 400 miles? (3rd day)

Cumulative Skill Practice Write these six answers on the chalkboard:

    3,375,000   540,000   74.018
    0.07   26.08   41.81

Challenge the students to an answer hunt by saying, "Look at exercises 1–43 on page 123. Find the six exercises that have these answers. You have five minutes to find as many of the exercises as you can." (Exercises 6, 17, 22, 26, 34, and 40)

122

Problem solving Reading a graph

Andy and Sarah went on a 5-day motorcycle trip. They started in Dallas, rode to Houston, and spent the night. They spent the next three nights in San Antonio, Lubbock, and Amarillo, and returned to Dallas on the fifth day.

Sarah made the line graph using the information from their log book. Use the graph to answer these questions. Decide when a calculator would be useful.

1. The log book says they were in San Antonio the second night. How many miles is it from Dallas to San Antonio? **450**

2. How many miles is it from San Antonio to Lubbock? **400**

3. How far is it from Lubbock to Amarillo? **150 miles**

4. On which day did they travel 150 miles? **4th**

5. On which day did they travel the most miles? How many miles did they travel that day? **3rd, 400**

6. The log book shows that they traveled 4 hours on the second day. How many miles per hour did they average the second day? **50 mph**

7. During which day did their trip mileage reach a total of 1000 miles? **4th**

8. They were between what two cities when their trip was half over? *Hint: Divide the total distance by 2 and use the graph.* **San Antonio and Lubbock**

9. One day they drove 10 hours and averaged 40 miles per hour. They were between what cities? **San Antonio and Lubbock**

10. Could they have completed the trip in 4 days if they had driven 50 miles per hour for 6 hours each day? **No**

122 Chapter 5

Cumulative Skill Practice

Write the standard numeral. *(page 4)*

1. 26 thousand, 429   26,429
2. 18 thousand, 92   18,092
3. 238 thousand, 164   238,164
4. 37 million   37,000,000
5. 259 million   259,000,000
6. 3 million, 375 thousand   3,375,000
7. two hundred twenty-five thousand, four hundred seven   225,407
8. thirty-three million, eight hundred seventy-three thousand   33,873,000
9. six billion, one hundred ninety-three million, four hundred thousand   6,193,400,000

Round to the nearest ten thousand. *(page 10)*

10. 76,501   80,000
11. 48,400   50,000
12. 74,391   70,000
13. 93,866   90,000
14. 356,200   360,000
15. 635,000   640,000
16. 824,999   820,000
17. 544,099   540,000

Write the standard numeral. *(page 14)*

18. 5 hundredths   0.05
19. 36 thousandths   0.036
20. 12 and 5 tenths   12.5
21. 59 and 74 hundredths   59.74
22. 74 and 18 thousandths   74.018
23. 3951 ten-thousandths   0.3951

Round to the nearest hundredth. *(page 18)*

24. 74.573   74.57
25. 6.3048   6.30
26. 0.0659   0.07
27. 5.3625   5.36
28. 9.3511   9.35
29. 0.005   0.01
30. 13.949   13.95
31. 372.678   372.68

Give the sum. *(page 22)*

32. 2.793 + 3.899   6.692
33. 5.466 + 1.805   7.271
34. 18.05 + 8.03   26.08
35. 0.84 + 5.96 + 7.6   14.4
36. 56.4 + 1.82 + 6.6   64.82
37. 7 + 3.85 + 6.8   17.65

Give the difference. *(page 46)*

38. 54.5 − 4.56   49.94
39. 32.6 − 19.38   13.22
40. 45.61 − 3.8   41.81
41. 22 − 5.9   16.1
42. 9 − 0.748   8.252
43. 6.1 − 3.045   3.055

MIXED PRACTICE
Complete.

44. 3.6 × 0.8 = __?__   2.88
45. 0.3048 ÷ 0.8 = __?__   0.381
46. 5.643 + 2.817 = __?__   8.46
47. 5.544 ÷ 1.2 = __?__   4.62
48. 5.3 + 2.89 + 6 = __?__   14.19
49. 19.23 × 1.6 = __?__   30.768
50. 35.2 − 14.9 = __?__   20.3
51. 0.85 ÷ 50 = __?__   0.017
52. 23 − 9.74 = __?__   13.26
53. 24 ÷ 6 ÷ 3 = __?__   1.3
54. 20 + 16 ÷ 2 = __?__   28
55. (30 + 20) ÷ 5 = __?__   10
56. 18 × (5 − 2) = __?__   54
57. 24 + 12 ÷ 4 + 2 = __?__   29
58. (24 + 12) ÷ 4 + 2 = __?__   11

Graphs and Statistics

Problem-Solving Worksheet
Workbook S221, Copymaster S221, or Duplicating Master S221

WORKSHEET 53 (Use after page 122.)

NAME _____

MAKE A LIST
What scores can you get with 3 darts? Make a list.

SCRAMBLED MATH
Unscramble the letters to get the answer.
NINE   ELEVEN   FORTY-FIVE
EINN times LEVNEE minus TORFY-EIVF plus REETH equals   57 .
THREE

9   11   13
15   17   19
21   23   27

How many ways can you score more than 20 points? 3

GUESS AND CHECK
Put plus (+) or minus (−) in each ☐ to get the answer shown on the calculator.
165 [+] 13 [−] 88 [−] 55 [=]

35

AND THE WINNER IS...
Three horses finished a race. Dazzler came in 100 seconds after Carrot Top. Royal Flush came in 75 seconds before Dazzler. Which horse won the race?   Carrot Top

BUY NOW!
How much money do you save if you buy 4 cans of tennis balls at the sale price? $1.20

HOW OLD?
Maria's age is twice Tom's age. The sum of their ages is 30. How old is Maria?   20

© D. C. Heath and Company   S221   Problem solving

Challenge Problems

Look at page 123. Estimate which two decimals in exercises 24–31 have a sum of about:

1. 20.2   6.3048 + 13.949
2. 15.6   6.3048 + 9.3511
3. 80.0   74.573 + 5.3625
4. 379.0   372.678 + 6.3048

If you have a calculator, use it to see whether you made the correct choices.

Copymaster S445

Class Starter Quiz 49

on previous lesson

Solve. Use the line graph on page 122.

1. How many miles is it from Dallas to San Antonio? 450
2. On which day did Andy and Sarah travel 200 miles? 2nd
3. They traveled 8 hours on the third day. How many miles per hour did they average the third day? 50

Copymaster S397

Lesson Objective

To compute the mean of a set of data

Problem-Solving Skills

Reading a chart
Choosing the correct operation
Analyzing data

Starting the Lesson

Have the students read about the coin-stacking experiment at the top of page 124. Then have a few students stack 30 pennies to see how long it takes. Each coin must be flat on a table to start and must be picked up (not slid off the edge).

Analyzing data— finding the mean

Nine students in a psychology class conducted an experiment on hand-eye coordination. Each student was asked to stack 30 pennies as quickly as possible.

The results are shown in the chart.

1. What was Dan Holway's time? 43.3 seconds
2. Who stacked the coins in 44.2 seconds?
   Judy Conrey
3. Who had the shortest time? the longest time?
   Polly Smith, Charlie Allen
4. The average of a set of numbers is called the **mean**. Whose time do you think is closer to the mean time, Ed Dorr's or Polly Smith's?
   Ed Dorr's

STUDENT	TIME (SECONDS)
Dan Holway	43.3
Julia Belmore	38.2
Charlie Allen	56.1
Ed Dorr	45.4
Judy Conrey	44.2
Tracy McDonald	45.2
Polly Smith	37.1
José Rivera	52.8
Angelo Robinson	46.3

Here's how *to find the mean.*

Find the sum of the times.

```
  43.3
  38.2
  56.1
  45.4
  44.2
  45.2
  37.1
  52.8
+ 46.3
─────
 408.6
```

Divide the sum by the number of students.

```
       45.4
   9 ) 408.6
      − 36
        48
      − 45
         36
       − 36
          0
```

The mean time is 45.4 seconds.

5. Look at the *Here's how*. Who had a time that was the same as the mean time? Ed Dorr
6. Which students had a time longer than the mean time? Charlie Allen, José Rivera, Angelo Robinson
7. Which students had a time shorter than the mean time? Dan Holway, Julia Belmore, Judy Conrey, Tracy McDonald, Polly Smith

EXERCISES

Find the mean. Round each answer to the nearest tenth.
Here are scrambled answers for the next two rows of exercises: 182.6 32.3 9.8 264.6

8. 12 4 14 9 *9.8*
9. 35 39 27 28 *32.3*
10. 195 176 183 178 181 *182.6*
11. 253 276 248 281 265 *264.6*
12. 361 375 392 386 379 *378.6*
13. 432 481 467 429 450 *451.8*
14. 18.6 19.4 21.6 *19.9*
15. 13.5 14.8 14.2 *14.2*
16. 29.6 32.7 31.9 28.7 *30.7*
17. 46.8 48.9 43.7 41.2 *45.2*
18. 112.2 114.7 116.3 *114.4*
19. 129.6 128.7 126.3 *128.2*
20. 18 28 16 32 14 26 10 *20.6*
21. 15 21 18 24 31 27 19 *22.1*
22. 28 31 19 42 37 16 12 *26.4*
23. 19 29 17 33 15 27 11 *21.6*

Use the chart on page 124 to answer the questions.

24. Who stacked the coins in a shorter time, Ed Dorr or Julia Belmore? *Julia Belmore*
25. Which student had a time that was 1.2 seconds less than the mean time? *Judy Conrey*
26. Who stacked the coins 2.1 seconds faster than Ed Dorr? *Dan Holway*
27. Who stacked the coins 1.1 seconds slower than Tracy McDonald? *Angelo Robinson*

Tee time! Analyzing data

To qualify for the finals in the Heavy Hitters Golf Ball Driving Contest, golfers must have a mean distance of 205 yards on 4 drives.

GOLFER	LENGTH OF DRIVE (YARDS)			
	1	2	3	4
Ann	195	210	200	?
Jan	205	195	200	?
Nan	198	205	218	?

Use the clues to name each golfer.

28. I hit my fourth drive 215 yards and I just barely made the finals! *Ann*

29. I made the finals with a fourth drive of 200 yards. *Nan*

30. I never made the finals. *Jan*

Hint: To make the finals, what must the four drives total?

Graphs and Statistics **125**

Practice Worksheet

Workbook S222, Copymaster S222, or Duplicating Master S222

NAME _____ **WORKSHEET 54**
(Use after page 125.)

PRESIDENTIAL TRIVIA
Do you know the middle names of these presidents?

- John F. Kennedy (JFK)
- Lyndon B. Johnson (LBJ)
- Dwight D. Eisenhower (IKE)
- Franklin D. Roosevelt (FDR)

To find the middle names:
Find the *mean* of each set of numbers.
Then use the DECODER to get the answer.

1. 14, 6, 16, 12 *12* A	2. 472, 486, 493, 497, 480 *485.6* D	3. 223.3, 225.8, 227.4 *225.5* F
4. 39, 42, 34, 53, 48, 27, 23 *38* I	5. 230.7, 239.8, 237.5 *236* N	6. 32.8, 34.4, 39.7, 41.5 *37.1* R
7. 506, 493, 498, 502, 501 *500* T	8. 18, 23, 27, 32, 19, 21, 30, 36, 29, 32, 23, 19, 16 *25* Z	9. 197, 178, 185, 182, 183 *185*
10. 29.7, 30.5, 31.3 *30.5* E	11. 29, 39, 37, 43, 25, 37, 21 *33* G	12. 24.6, 25.9, 25.4 *25.3* L
13. 29, 32, 29, 39, 41, 38, 30 *34* O	14. 118.2, 123.8, 131.6, 126.4 *125* S	15. 49.6, 45.3, 45.1, 48.4 *47.1* V

DECODER

1. *F I T Z G E R A L D* (JFK)
   225.5 38 500 25 33 30.5 37.1 12 25.3 485.6
2. *B A I N E S* (LBJ)
   185 12 38 236 30.5 125
3. *D A V I D* (IKE)
   485.6 12 47.1 38 485.6
4. *D E L A N O* (FDR)
   485.6 30.5 25.3 12 236 34

© D. C. Heath and Company   S222   Finding the mean

Group Project

Collecting data

Have each student try the coin-stacking experiment described on page 124. Each coin must be flat on a table to start and must be picked up (not slid off the edge). Compile the data for the entire class, and have the students use a calculator to find the mean.

125

Class Starter Quiz 50
on previous lesson
Find the mean.

1. 7 9 12 8   9
2. 35 40 39   38
3. 6.4 3.2 4.4 2.8   4.2
4. 6 8 7 3 9 9   7

Copymaster S397

Lesson Objective
To compute the median, mode, and range of a set of data

Problem-Solving Skills
Reading and interpreting a chart
Collecting, organizing, and analyzing data

Starting the Lesson
Have the students take the Famous-Faces Quiz at the top of page 126. Correct and score their papers. After discussing exercises 1–3, have the students find the median, mode, and range of their Famous-Faces Quiz scores.

Analyzing data—finding the median, mode, and range

FAMOUS-FACES QUIZ   Whose famous face is in the picture?

Cartoonist (10 points)
a. Johnny Hart
b. Walt Disney
c. Charles Schulz

Tennis player (8 points)
a. Tracy Austin
b. Chris Evert-Lloyd
c. Pam Shriver

U.S. President (5 points)
a. Harry Truman
b. John F. Kennedy
c. Abraham Lincoln

Singer (7 points)
a. Juice Newton
b. Barbra Streisand
c. Tina Turner

Fifteen high school students took the Famous-Faces Quiz. Their scores are shown in the chart.

NAME	SCORE	NAME	SCORE	NAME	SCORE
T. Coyle	5	A. Milan	20	C. Sanford	13
L. Cummings	13	M. Olsen	18	A. Travers	17
F. Epstein	15	K. Panetta	7	L. Vita	5
J. Kenney	12	J. Perry	13	K. Wong	23
J. Lubell	15	J. Quiroga	22	M. Young	25

1. What would be your score if you knew the cartoonist and the singer?   17
2. What is the lowest score shown on the chart? the highest score?   5, 25

Here's how *to find the* **median,** *the* **mode,** *and the* **range** *of a collection of data.*

Rank the numbers from least to greatest.

The **median** is the score in the middle.

The **mode** is the number that appears most often.

The **range** is the difference between the largest number and the smallest number.

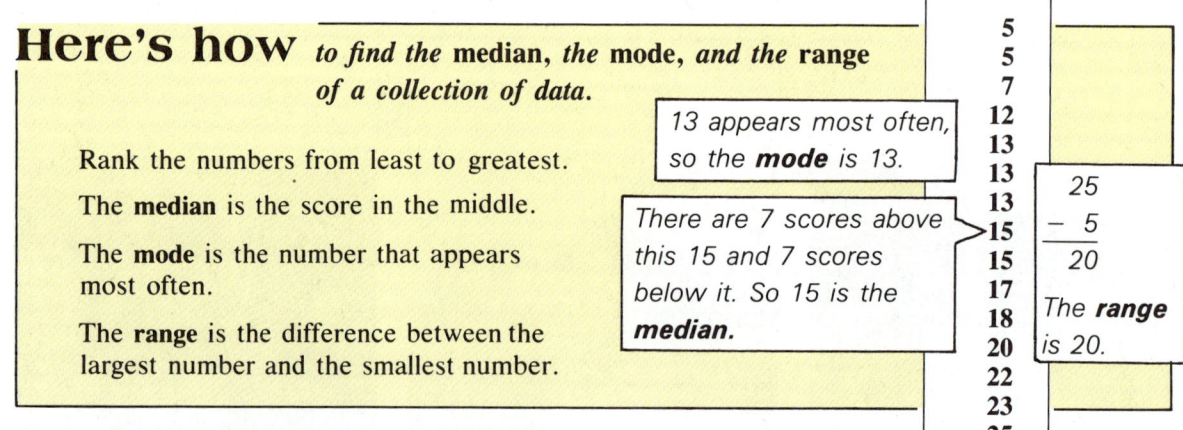

13 appears most often, so the **mode** is 13.

There are 7 scores above this 15 and 7 scores below it. So 15 is the **median**.

25
− 5
20
The **range** is 20.

3. Look at the *Here's how*. What are the median, the mode, and the range of the Famous-Faces Quiz scores?   Median 15, mode 13, range 20

EXERCISES

Find the median.

4. 13, 12, 11, 6, 7, 12, 9   11
5. 12, 9, 8, 14, 10, 11   10.5

   Hint: Average the two middle numbers, 10 and 11.

6. 14, 7, 12, 7, 7, 10, 13   10
7. 13, 14, 9, 11, 13, 10, 9   11
8. 183, 182, 183, 180, 187, 182   182.5
9. 12, 15, 13, 19, 13, 15, 13   13
10. 1243, 1245, 1301, 1248, 1256, 1276, 1287, 1308, 1299   1276
11. 6797, 6785, 6897, 6579, 6057, 6896, 6570, 6075, 6597, 6895, 6579, 6719, 6852   6719

Find the mode and range. (It is possible for a list to have more than one mode.)

12. 5, 7, 9, 7, 5, 9, 7   7, 4
13. 11, 15, 12, 15, 10, 14, 15   15, 5
14. 20, 27, 22, 23, 21, 24, 22   22, 7
15. 66, 82, 69, 73, 68, 66, 76, 66   66, 16
16. 16, 19, 21, 16, 18, 19, 23   16, 19; 7
17. 43, 39, 47, 43, 48, 41, 38, 43   43, 10
18. 104, 108, 103, 106, 103, 107, 101, 103, 108, 104, 100, 110   103, 10
19. 506, 511, 508, 503, 500, 503, 505, 508, 505, 503, 514, 501, 505, 509   503, 505; 14

Solve. Refer to the data on page 126.

20. Whose scores were greater than the median?   A. Milan, M. Olsen, J. Quiroga, K. Wong, A. Travers, M. Young
21. Did any student know all four famous people?   No
22. What is the mean rounded to the nearest tenth?   14.9
23. Is the mean greater than or less than the median?   Less than

Project — Collecting, organizing, and analyzing data

Answers will vary.

24. Ask your classmates to estimate to the nearest 30 minutes how many minutes that they listened to the radio yesterday.
25. Order the times from least to greatest.
26. Find the mode and the median.
27. Compute the mean.

Graphs and Statistics

Class Starter Quiz 51
on previous lesson

Find the median.
1. 12, 11, 10, 5, 7, 11, 8   10
2. 16, 13, 14, 14, 11, 17   14

Find the mode.
3. 6, 8, 10, 8, 6, 10, 8   8
4. 1, 5, 2, 5, 10, 4, 5   5

Find the range.
5. 6, 3, 10, 13, 6, 2, 14   12
6. 16, 19, 21, 23   7

Copymaster S397

Lesson Objectives
To interpret bar, line, circle, and picture graphs
To construct bar, line, circle, and picture graphs

Problem-Solving Skills
Finding information from graphs
Making a graph

Starting the Lesson
Write these five bumper-sticker slogans on the chalkboard:

DRAGON POWER!
Big Green Machine
Hayden High is #1
MEAN AND GREEN
Go for it, Dragons!

Before the students open their books, tell them that the Hayden High School Band sold these five bumper stickers to raise money for new uniforms. Have the students guess which bumper sticker sold the most. Then have them turn to page 128 to check their guesses.

Exercise Note
The graphs on pages 128 and 129 are also on ■ **Visual Aid 23** (copymaster or transparency S130). You may wish to use the visual aid when discussing exercises 1–11.

128

Reading and constructing graphs

The Haydon High School Band sold bumper stickers to raise money for new uniforms. Five different bumper stickers were offered. The table shows the distribution of the first 100 bumper stickers sold.

Frequency Table for Bumper-Sticker Sales (First 100)

Bumper sticker	Number sold	
	TALLY	FREQUENCY
DRAGON POWER!	|||| |||| |||| |||| |||| |||| |||| ||||	40
Big Green Machine	|||| |||| |||| ||||	20
Haydon High is #1	||||	5
MEAN AND GREEN	|||| ||||	10
Go for it, Dragons	|||| |||| |||| |||| ||||	25

Answer these questions to see how you would fill in the missing information on the graphs.

Look at the bar graph and the frequency table.

1. What bumper sticker does the longest bar represent?
    Dragon Power
2. What number is represented by H? A? B? J?
    35, 0, 5, 45
3. The shortest bar represents what bumper sticker?
    Haydon High is #1
4. What bumper sticker is represented by a? b? c? d? e?
    Dragon Power; Big Green Machine; Haydon High is #1; Mean and Green; Go for it, Dragons

Look at the circle graph and the frequency table.

5. What number does each color represent?
    Green, 40; Blue, 25; Yellow, 10; Red, 5; Orange, 20
6. The blue sector represents sales of which bumper sticker?   *Go for it, Dragons*
7. Which graph do you think allows you to compare the sales more accurately?   *Bar graph*

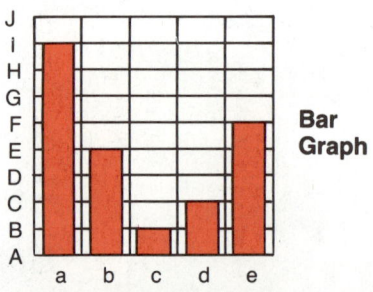

Bar Graph

BUMPER-STICKER SALES (First 100)

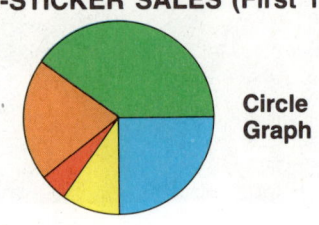

Circle Graph

128   Chapter 5

EXERCISES

Solve.

8. Look at the sales chart and the picture graph.
   a. How many 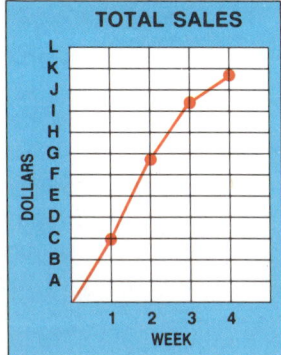 were used to show the first week's sales? 10
   b. What does each 💵 represent? $15

9. Look at the sales chart and the line graph.
   a. Sales for the first week were $150. What number is represented by C? A? L? $150, $50, $600
   b. The total sales goal was $500. Did the band reach this goal? Yes

10. Look at both graphs. Which one do you think shows total sales better? Why?
    Line graph. Less counting is involved.

11. Which of the two graphs do you think looks more interesting? Why?
    Answers will vary.

Join the band
Making a graph

12. The Haydon High School Band has 66 members. Complete the frequency table and make a bar or picture graph showing the distribution of instruments in the band.

INSTRUMENT	TALLY	FREQUENCY
Baritone	II	?
Bassoon	II	?
Clarinet	|||| |||| ||||	?
Flute	|||| ||||	?
French horn	||||	?
Oboe	II	?
Percussion	|||| I	?
Piccolo	I	?
Saxophone	||||	?
Trombone	||||	?
Trumpet	|||| |||| II	?
Tuba	II	?

SALES— FIRST 4 WEEKS

Sales Chart

Week	Sales
1	$150
2	$185
3	$135
4	$ 70

Line Graph

TOTAL SALES

Picture Graph

WEEKLY SALES

Week 1
Week 2
Week 3
Week 4

Graphs and Statistics **129**

Extra Practice
Page 481   Skill 22

Practice Worksheet
Workbook S224, Copymaster S224, or Duplicating Master S224

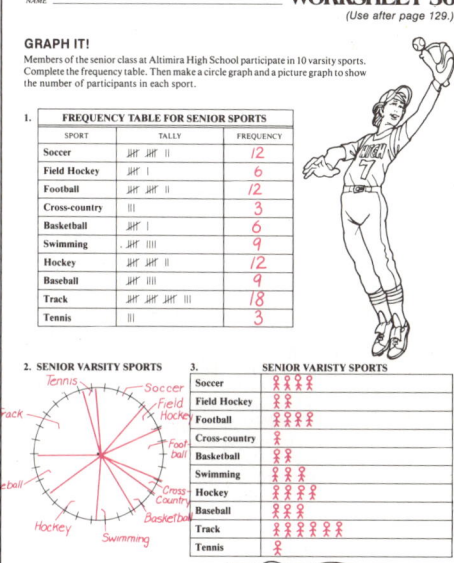

Project

Making bar graphs

Find a line graph or a circle graph in a newspaper or magazine. Then show the same information in a bar graph.

Copymaster S477

Answer for page 129.

12.

Haydon High School Band	
Baritone	♩
Bassoon	♩
Clarinet	♩♩♩♩♩♩
Flute	♩♩♩♩
French Horn	♩♩
Oboe	♩
Percussion	♩♩♩
Piccolo	•
Saxophone	♩♩♩
Trombone	♩♩♩
Trumpet	♩♩♩♩♩♩
Tuba	♩

Each ♩ represents 2 players.

129

Class Starter Quiz 52
on previous lesson

Complete the picture graph of weekly sales.

Week	Sales
1	$40
2	$70
3	$55
4	$30

```
          WEEKLY SALES
Week 1 □□□□
Week 2 □□□□□□□
Week 3 □□□□□▯
Week 4 □□□
Each □ represents $10.
```

Copymaster S397

Lesson Objective
To choose the most sensible answer when selecting the answer to a problem

Problem-Solving Skill
Making up a problem

Starting the Lesson
Group Activity Read the problem on page 130 to the students. Have the students, working in small groups, decide which answer makes sense. Then have the students open their books to page 130 to check their thinking. After discussing the PLAN and SOLVE step with the class, you may want to have the students do exercises 1–11 in small groups.

130

PROBLEM SOLVING *choosing sensible answers*

When you solve a problem, be sure your answer makes sense.

Problem
Every player on the Eagles high school basketball team went to watch the Rockets play the Utah Jazz. How many of them went to the game?

       2   15   52

Here's how *to decide if an answer makes sense.*

STUDY Read the problem carefully.
What facts are given?
What is the question?

PLAN and SOLVE How will you use the facts to answer the question?

Decide which answer makes sense.

- 2 **is not** a sensible answer because it takes more than 2 players to make up a basketball team.

- 15 **is** a sensible answer because there are enough players to make up a team of 5 players and have a reasonable number of extra players.

- 52 **is not** a sensible answer because there would be too many extra players.

Answer: 15 players went to the game.

CHECK Make sure you check each answer.

130 Chapter 5

PROBLEMS

Solve. Choose the most sensible answer.

1. The basketball team rode to the game in vans. How many vans were needed?
   <u>2</u>   5   10

2. The players shared seats in the vans. How many players sat in each seat?
   1   <u>3</u>   8

3. The team had to travel 100 miles to get to the Summit Arena in Houston. How many hours did it travel?
   1   <u>3</u>   9

4. The van drivers parked the vans in a special parking lot reserved for vans and buses. How many vans were in the lot?
   2   <u>20</u>   4000

5. Every seat in the Summit Arena was filled by game time. How many people were at the game?
   16   160   <u>16,016</u>

6. Hugo bought a hamburger and juice drink for $3. How much did the juice drink cost?
   <u>$1</u> $3 $5

7. Rick bought a Rocket's pennant. How much money did he get back from a $10 bill?
   <u>$5</u> $10 $20

8. At halftime what was the total number of points scored by both teams?
   5   <u>105</u>   205

9. With 2 seconds left in the game, it was the Utah Jazz 111, the Rockets 109. The Rockets scored a basket and won the game by 1 point. How many points was their final basket worth?
   1   2   <u>3</u>

10. Before the game started, the announcer introduced the starting players from both teams. How many players were introduced?
    10   15   <u>20</u>

Win some/lose some   Making up a problem

11. The chart shows the final standings for a recent year.

    Make up a problem that has this answer: 6 games.   *Answers will vary.*

National Basketball Association Midwest Division		
Club	Won	Lost
Houston	51	31
Denver	47	35
Dallas	44	38
Utah	42	40
Sacramento	37	45
San Antonio	35	47

Graphs and Statistics

Practice Worksheet

Workbook S225, Copymaster S225, or Duplicating Master S225

WORKSHEET 57
(Use after page 131.)

NAME _____

SOFTBALL

Solve. Choose the most sensible answer.

1. Corey's softball team practices every day after school. How many minutes does the team practice each week?
   15   45   (240)

2. The team has at least one substitute for each of the 9 players on the field. How many players are on the team?
   12   (22)   72

3. Each of the games played last week was 7 innings in length. What was the total number of innings played last week?
   9   (21)   51

4. The team had to travel 80 miles to play in a tournament. How many hours did the team travel?
   1   (2)   8

5. The smallest crowd at a softball game this year was 300. What was the total attendance at last week's 3 games?
   200   900   (9000)

6. The final score of today's game was 8 to 5. What was the total number of runs scored during the first 5 innings.
   (9)   15   20

7. Corey scored 3 runs in today's game. That was more runs than she scored in any of the other 8 games played this season. What is the total number of runs scored by Corey so far this season?
   (12)   24   50

8. Corey's friend Heidi has had 2 or more hits in each of the 9 games. How many hits has Heidi so far this season?
   15   (21)   56

9. Jill bought a bag of popcorn and a juice drink for $1.75. How much did the popcorn cost?
   ($0.75) $1.80 $3.50

10. Ken bought a hamburger and a juice drink for $2.25. The hamburger cost twice as much as the juice drink. How much did the juice drink cost?
    $0.50   ($0.75) $1.00

11. Brian bought a hamburger. How much change did he get back from a $5-bill?
    $3.50   ($5) $7.50

12. Week-day games start right after school and always finish before dark. Each game is about how many minutes long?
    15   (90)   300

RUN THAT BY ME AGAIN!

13. The scores of last week's games were 8 to 7, 15 to 1, and 3 to 2. Corey's team won 2 of the games but the team's opponents scored twice as many runs! How many runs did Corey's team score?   <u>8+1+3 or 12</u>

© D.C. Heath and Company   S225   *Problem solving — choosing sensible answers*

Challenge Problem

Write a question that fits the answer.

Tina and 3 friends earned $25 on Tuesday, $23 on Wednesday, and $32 on Thursday?

<u>If they share the money equally, how much does each person get?</u>

Answer: $20

Copymaster S446

Class Starter Quiz 53
on previous lesson

Solve. Choose the most sensible answer.

1. The Eagles won most of their high school basketball games. How many games did they win?
      3   13   130
2. Andy bought a box of popcorn and a juice drink for $1.75. How much did the popcorn cost?
      $.75 $1.75 $2.00

Copymaster S398

Problem-Solving Skill
Reading and interpreting graphs

Skills Reviewed
Multiplying decimals
Multiplying and dividing by 10, 100, or 1000
Dividing decimals
Computing the mean
Rounding decimals
Adding and subtracting whole numbers and decimals
Simplifying expressions

Starting the Lesson
Problem Solving Discuss the computer graphics on page 132. Use questions such as "What is the title of the circle graph? What data are shown on the vertical scale of the bar graph? How many students are enrolled in the fine-arts activities for the current year?"

Cumulative Skill Practice Challenge the students to an estimation hunt by saying, "Pick the smallest product in the first row of exercises." (Exercise 1) Then have the students pick the smallest product in the second and third rows of exercises. (Exercises 9 and 15)

132

Problem solving
USING COMPUTER GRAPHICS

Lori Adriano is the Fine Arts Coordinator for Lincoln High School. She uses a personal computer to store data about students who participate in school-sponsored fine-arts activities. A special **software** package helped her produce these graphs.

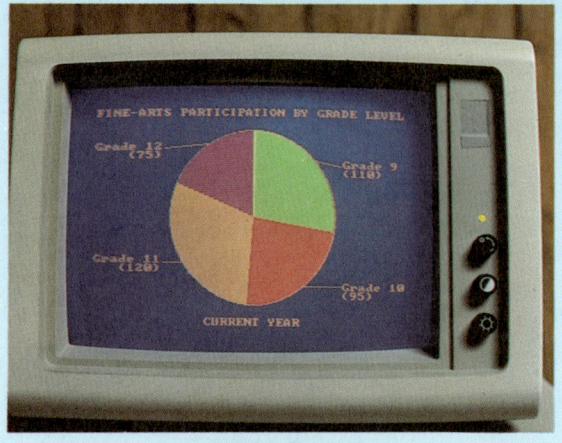
GRAPH 1

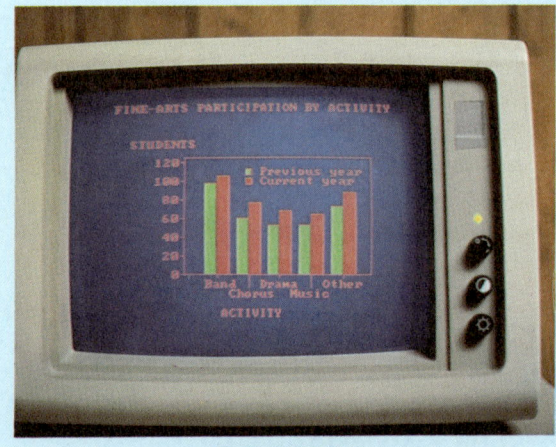
GRAPH 2

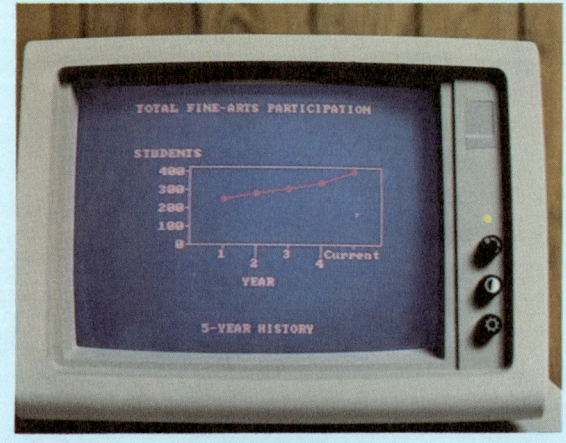
GRAPH 3

Decide which graph or graphs Ms. Adriano could use to answer the question. Then write a reasonable answer.

1. Parent: "Which fine-arts activity is the most popular?"
   Graph 2, Band

2. School Board Member: "Are the school's fine-arts activities as popular with seniors as they are with freshmen?" Graph 1, No

3. Taxpayer: "How many students are enrolled in fine-arts activities anyway?"
   Graph 1 or 3, 400

4. Parent: "Is participation in fine-arts activities increasing or decreasing?"
   Graph 3, Increasing

5. School Board Member: "Our costs have gone up nearly 25% over the past 5 years. How do you account for this?"
   Graph 3, Student participation has been increasing over the past 5 years.

6. Taxpayer: "I read in the paper that 50 players tried out for football this year. Why doesn't the school do more for non-athletes?" Graph 2 or 3, Four hundred students participate in the fine-arts program. They participate in band, chorus, drama, music, and other activities.

132 Chapter 5

Cumulative Skill Practice

Give the product. *(page 68)*

1. 3.4 × 0.8 2.72
2. 5.61 × 0.9 5.049
3. 32.4 × 0.4 12.96
4. 305 × 5.8 1769
5. 9.63 × 17 163.71
6. 0.33 × 40 13.2
7. 3.51 × 2.08 7.3008
8. 4.07 × 71 288.97
9. 0.02 × 1.03 0.0206

Give the product. *(page 72)*

10. 39 × 10 390
11. 39 × 100 3900
12. 39 × 1000 39,000
13. 2.07 × 100 207
14. 2.07 × 1000 2070
15. 2.07 × 10 20.7
16. 0.007 × 10 0.07
17. 0.007 × 1000 7
18. 0.007 × 100 0.7

Give the quotient. *(page 100)*

19. 36.4 ÷ 10 3.64
20. 36.4 ÷ 100 0.364
21. 36.4 ÷ 1000 0.0364
22. 8 ÷ 100 0.08
23. 8 ÷ 10 0.8
24. 8 ÷ 1000 0.008
25. 83.07 ÷ 1000 0.08307
26. 83.07 ÷ 10 8.307
27. 83.07 ÷ 100 0.8307

Give the quotient rounded to the nearest tenth. *(pages 102, 104)*

28. 13.5 ÷ 0.6 22.5
29. 2.81 ÷ 0.3 9.4
30. 5.08 ÷ 0.05 101.6
31. 6.34 ÷ 1.4 4.5
32. 2.059 ÷ 2.5 0.8
33. 5.07 ÷ 0.12 42.3
34. 7.49 ÷ 5.11 1.5
35. 3.716 ÷ 2.9 1.3
36. 38.2 ÷ 6.05 6.3

Find the mean. *(page 124)*

37. 124, 117, 131 124
38. 243, 257, 232 244
39. 706, 599, 639 648
40. 18, 27, 15, 24 21
41. 58, 40, 63, 46 51.75
42. 163, 174, 150, 157 161
43. 16, 23, 15, 20, 18 18.4
44. 43, 81, 49, 58, 74 61
45. 93, 85, 90, 93, 87 89.6

MIXED PRACTICE

Complete.

46. 8.8352 rounded to the nearest hundredth is ?   8.84
47. 16.071 rounded to the nearest tenth is ?   16.1
48. 5.42 + 3.61 + 2 = ?   11.03
49. 28.4 − 13.95 = ?   14.45
50. 4.8 + 19.63 + 11 = ?   35.43
51. (26.2 + 18.3) − 2.7 ?   41.8
52. 26.2 + (18.3 − 2.7) = ?   41.8
53. (9.74 − 3.5) − 2.6 = ?   3.64
54. 9.74 − (3.5 − 2.6) = ?   8.84
55. (6.4 + 3.7) + 1.9 = ?   12
56. 6.4 + (3.7 + 1.9) = ?   12

Graphs and Statistics **133**

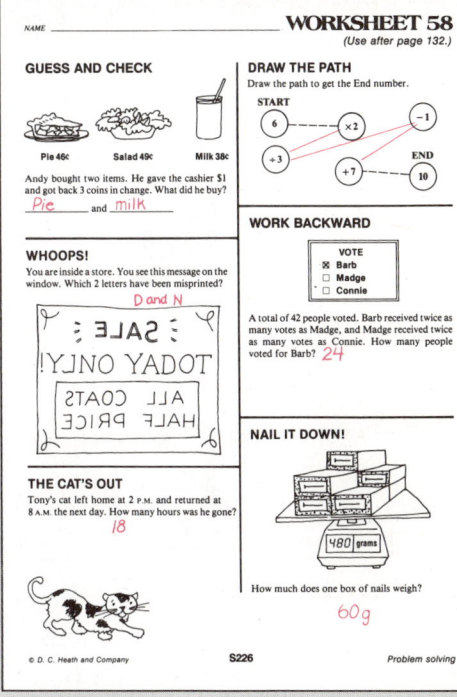

Chapter REVIEW

Here are scrambled answers for the review exercises:

1	200	500	circle	mean	mode	range
4	300		bar	line	median	picture

1. This type of graph is a [?] graph. The game at which the most popcorn was sold was Game [?]. At Game 3, [?] boxes of popcorn were sold. *(page 116)*

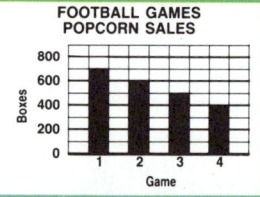

2. This type of graph is called a [?] graph. About [?] boxes of popcorn were sold during the first quarter. *(page 118)*

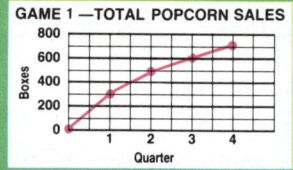

3. This type of graph is called a [?] graph. Each picture on this graph represents [?] people. *(page 120)*

4. This graph is called a [?] graph. The attendance at Games 2 and [?] made up one half of the total attendance. *(page 120)*

5. Add the numbers on the temperature chart, then divide by 7 to find the [?] temperature.
   Subtract 10 from 20 to find the [?] of the temperatures.
   18° is the temperature that appears most often in the chart. It is called the [?].
   Rank the temperatures. The temperature in the middle is called the [?]. *(pages 124, 126)*

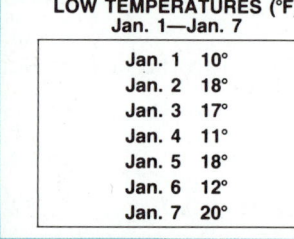

LOW TEMPERATURES (°F)
Jan. 1 — Jan. 7

Jan. 1	10°
Jan. 2	18°
Jan. 3	17°
Jan. 4	11°
Jan. 5	18°
Jan. 6	12°
Jan. 7	20°

1. bar; 1; 500   2. line; 300   3. picture; 200   4. circle; 4   5. mean; range; mode; median

134 *Chapter 5*

Chapter TEST

Use the frequency table to answer each question. *(page 114)*

1. How many students saw 3 movies last month? 6
2. How many didn't go to a movie? 3
3. How many saw 2 or fewer movies? 17
4. How many students were in the survey? 26

Use the bar graph to answer each question. *(page 116)*

5. Who spent the most time doing homework? Kim
6. How many hours did David spend doing homework? 12
7. How many hours did Jan spend doing homework? 18
8. How many more hours did Kim spend doing homework than Gayle? 4

Use the line graph to answer each question. *(page 118)*

9. By the end of the second week, Anne had saved $8. How much had she saved by the end of the third week? $14
10. During which week did Anne save the most? 4
11. How much did Anne save during the six weeks? $30

Use the picture graph to answer each question. *(page 120)*

12. Who has the most records? the fewest? Bob, Randy
13. How many records does Bob have? 28
14. How many records does Loni have? 21
15. How many records do Carl and Randy have together? 33

Use the list of scores to answer each question. *(pages 124, 126)*

16. What is the mean of Kathleen's math test scores? 80
17. What is the median? 79
18. What is the mode? 77
19. What is the range of her math test scores? 7

NUMBER OF MOVIES ATTENDED LAST MONTH

Number of Movies	Number of Students	
	TALLY	FREQUENCY
4					3
3							6
2												12
1				2
0					3

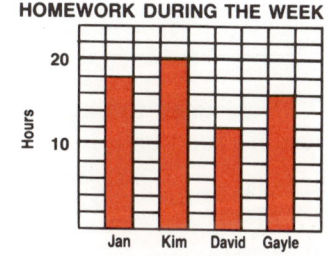

HOMEWORK DURING THE WEEK

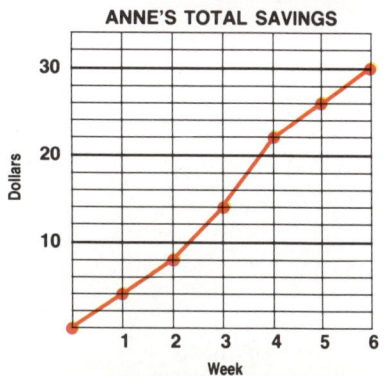

ANNE'S TOTAL SAVINGS

NUMBER OF RECORDS IN COLLECTION

Bob ● ● ● ● ● ● ●
Carl ● ● ● ● ◐
Loni ● ● ● ● ● ◔
Randy ● ● ● ◔

Each ● stands for 4 records.

KATHLEEN'S MATH TEST SCORES

83 77 79 77 84

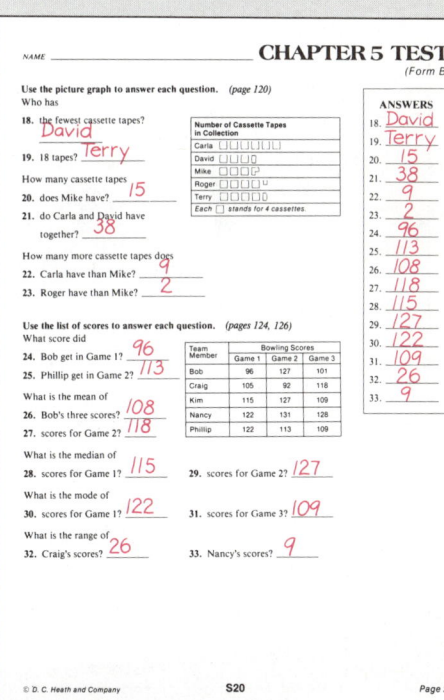

Graphs and Statistics **135**

Cumulative Test
(Chapters 1–5)

Use copymaster S109 to provide the students with an answer sheet in standardized test format.

Answers for Cumulative Test, Chapters 1–5

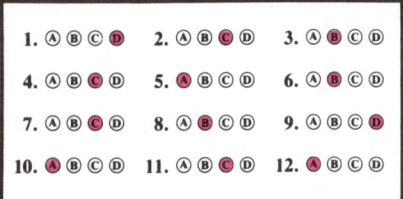

The table below correlates test items with student text pages.

Test Item	Page(s) Taught	Skill Practice
1	4	p. 123, exercises 1–9
2	10	p. 123, exercises 10–17
3	14	p. 123, exercises 18–23
4	18	p. 123, exercises 24–31
5	22	p. 123, exercises 32–37
6	46	p. 123, exercises 38–43
7	68	p. 133, exercises 1–9
8	72	p. 133, exercises 10–18
9	100	p. 133, exercises 19–27
10	102, 104	p. 133, exercises 28–36
11	124	p. 133, exercises 37–45
12	68	

136

Cumulative TEST — Standardized Format

Choose the correct letter.

1. The standard numeral for 3 billion, 32 thousand is
   A. 3,032,000
   B. 3,320,000
   C. 3,032,000,000
   D. none of these

2. 253,599 rounded to the nearest ten thousand is
   A. 300,000
   B. 254,000
   C. 250,000
   D. none of these

3. The standard numeral for 16 and 64 thousandths is
   A. 16.64
   B. 16.064
   C. 16.0064
   D. none of these

4. 7.695 rounded to the nearest hundredth is
   A. 7.60
   B. 7.69
   C. 7.70
   D. none of these

5. Give the sum.
   $6.4 + 0.65 + 12.9$
   A. 19.95
   B. 25.8
   C. 18.95
   D. none of these

6. Give the difference.
   $19.7 - 1.93$
   A. 0.04
   B. 17.77
   C. 17.83
   D. none of these

7. Give the product.
   23.6×1.09
   A. 4.484
   B. 25.474
   C. 25.724
   D. none of these

8. Give the product.
   5.772×1000
   A. 577.2
   B. 5772
   C. 57,720
   D. none of these

9. Give the quotient.
   $2.41 \div 100$
   A. 241
   B. 24.1
   C. 0.241
   D. none of these

10. Give the quotient rounded to the nearest tenth.
    $3.609 \div 3.4$
    A. 1.1
    B. 1.6
    C. 1.06
    D. none of these

11. The mean of 80, 83, 87, 88, and 92 is
    A. 9
    B. 87
    C. 86
    D. none of these

12. You bought 4 adult tickets for $2.75 each and 3 children's tickets for $1.25 each. How much did you spend?
    A. $14.75
    B. $13.25
    C. $14.25
    D. none of these

136 *Chapter 5*

Number Theory, Fractions, and Decimals

Resources

- **Class Starter Quizzes 54-70** *(Copymasters S398-S402)*
- **Visual Aids 24-31** *(Copymasters or Transparencies S131-S136)*
- **Manipulatives**
  Manipulative Activities 6-10 *(Copymaster S516-S517)*
  Fraction pieces *(Copymasters or Transparencies S530-S531)*
- **Worksheets 59-76** *(Copymasters, Duplicating Masters, or Workbook pages S227-S244)*
- **Challenge Problems** for pages 139, 141, 143, 145, 147, 151, 153, 155, 161, 163, 167, 169, 171, 173 *(Copymasters S446-S449)*
- **Projects** for pages 149, 157, 159, 165 *(Copymaster S478)*
- **Mental Math Extensions** for Skills 23-31 *(Copymasters S497-S501)*
- **Tests** *(Copymasters or Duplicating Masters S21-S24)*

Lesson Objectives

To write numbers with exponents as standard numerals

To use exponents to write products

Problem-Solving Skill

Finding a pattern

Starting the Lesson

Use the arrangement of blocks at the top of the page and exercises 1 and 2 to show the students that instead of writing the same factor several times, we can use an exponent. Next, put these drawings on the chalkboard.

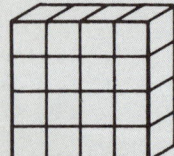 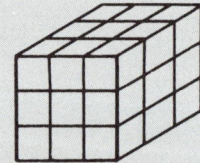

Have the students tell how they would use exponents to write the number of blocks in each arrangement. (4^2 and 3^3)

Exercise Note

Calculator Some calculators have a built-in constant- or K-function. (Check the owner's manual that came with the calculator.) The constant multiplication function can be used to compute powers of numbers when there is no power key on the calculator. For example, to evaluate 5^4, you might key 5 ⊗ K (or 5 ⊗ or 5 ⊗ ⊗ depending on the calculator) to get into the constant function. Then press = 3 times to get 5^4.

Keying sequence:
5 ⊗ K = = =

Display:
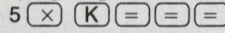

You might want to let students who have calculators with constant- or K-functions experiment with them to evaluate numbers like 5^5 or 6^6.

5 ⊗ K = = = =
6 ⊗ K = = = = =

138

Exponents

Instead of writing the same factor several times, you can write the factor once and use an **exponent**.

The exponent tells you how many times the **base** is used as a factor.

Notice the three ways to write the number of blocks.

1. Look at the blue blocks. How do you write the number of blocks using an exponent? 3^2

2. Look at the red blocks. How do you write the number of blocks using an exponent? 4^3

9
3×3

3^2 ← exponent
↑
base

Read "3^2" as "3 squared" or "3 to the second power."

64
$4 \times 4 \times 4$

4^3

Read "4^3" as "4 cubed" or "4 to the third power."

Here are some more examples.

Written using an exponent	Written as a product of factors	Written as a standard numeral
4^2 4 to the second power	4×4	16
2^1 2 to the first power	2	2
3^4 3 to the fourth power	$3 \times 3 \times 3 \times 3$	81
2^5 2 to the fifth power	$2 \times 2 \times 2 \times 2 \times 2$	32
1^6 1 to the sixth power	$1 \times 1 \times 1 \times 1 \times 1 \times 1$	1
10^3 10 to the third power	$10 \times 10 \times 10$	1000

3. Look at the examples on page 138.
   a. How do you write 3^4 as a product of factors? $3 \times 3 \times 3 \times 3$
   b. How do you write 2^5 as a standard numeral? 32

EXERCISES

Write as a product of factors.
Here are scrambled answers for the next row of exercises:
$1 \times 1 \times 1 \times 1 \quad 4 \quad 2 \times 2 \times 2 \quad 3 \times 3 \quad 5 \times 5 \times 5 \quad 5 \times 5 \times 5 \times 5 \times 5$

4. 2^3 $\quad 2 \times 2 \times 2$
5. 3^2 $\quad 3 \times 3$
6. 1^4 $\quad 1 \times 1 \times 1 \times 1$
7. 4^1 $\quad 4$
8. 5^3 $\quad 5 \times 5 \times 5$
9. 5^5 $\quad 5 \times 5 \times 5 \times 5 \times 5$
10. 1^3 $\quad 1 \times 1 \times 1$
11. 2^1 $\quad 2$
12. 3^3 $\quad 3 \times 3 \times 3$
13. 5^2 $\quad 5 \times 5$
14. 3^1 $\quad 3$
15. 5^1 $\quad 5$
16. 3^4 $\quad 3 \times 3 \times 3 \times 3$
17. 6^2 $\quad 6 \times 6$
18. 6^3 $\quad 6 \times 6 \times 6$
19. 3^5 $\quad 3 \times 3 \times 3 \times 3 \times 3$
20. 4^2 $\quad 4 \times 4$
21. 2^4 $\quad 2 \times 2 \times 2 \times 2$
22. 2^2 $\quad 2 \times 2$
23. 2^6 $\quad 2 \times 2 \times 2 \times 2 \times 2 \times 2$
24. 1^4 $\quad 1 \times 1 \times 1 \times 1$
25. 4^4 $\quad 4 \times 4 \times 4 \times 4$
26. 2^5 $\quad 2 \times 2 \times 2 \times 2 \times 2$
27. 4^3 $\quad 4 \times 4 \times 4$

Write as a standard numeral.
Here are scrambled answers for the next row of exercises: $\quad 5 \quad 16 \quad 81 \quad 100 \quad 125 \quad 256$

28. 10^2 100
29. 2^4 16
30. 3^4 81
31. 5^1 5
32. 4^4 256
33. 5^3 125
34. 2^2 4
35. 3^3 27
36. 10^5 100,000
37. 2^5 32
38. 5^2 25
39. 4^3 64
40. 6^3 216
41. 2^3 8
42. 1^6 1
43. 3^2 9
44. 10^4 10,000
45. 2^7 128
46. 2^6 64
47. 10^3 1000
48. 4^5 1024
49. 2^1 2
50. 4^2 16
51. 5^4 625

Write using an exponent.
Here are scrambled answers for the next row of exercises: $\quad 5 \quad 2^3 \quad 3^4$

52. $3 \times 3 \times 3 \times 3 \quad 3^4$
53. $2 \times 2 \times 2 \quad 2^3$
54. $5 \quad 5^1$
55. $4 \times 4 \quad 4^2$
56. $1 \times 1 \times 1 \times 1 \times 1 \times 1 \quad 1^6$
57. $2 \times 2 \times 2 \times 2 \times 2 \quad 2^5$
58. $10 \quad 10^1$
59. $8 \times 8 \times 8 \quad 8^3$
60. $3 \times 3 \times 3 \times 3 \times 3 \quad 3^5$
61. $10 \times 10 \times 10 \quad 10^3$
62. $6 \times 6 \times 6 \times 6 \times 6 \quad 6^5$
63. $10 \times 10 \times 10 \times 10 \quad 10^4$

Pattern search

Solve. [Hint: Look for a pattern.]

64. $2^1 = 2$
    $2^2 = 4$
    $2^3 = 8$
    $2^4 = 16$
    $2^5 = 32$
    .
    .
    .
    $2^{40} = 1,099,511,627,77\underline{\ ?\ }$ 6
    What is the missing digit?

65. $1^3 = \underline{\ ?\ }$ 1
    $1^3 + 2^3 = \underline{\ ?\ }$ 9
    $1^3 + 2^3 + 3^3 = \underline{\ ?\ }$ 36
    $1^3 + 2^3 + 3^3 + 4^3 = \underline{\ ?\ }$ 100
    $1^3 + 2^3 + 3^3 + 4^3 + 5^3 = \underline{\ ?\ }$ 225
    $1^3 + 2^3 + 3^3 + 4^3 + 5^3 + 6^3 = \underline{\ ?\ }$ 1521

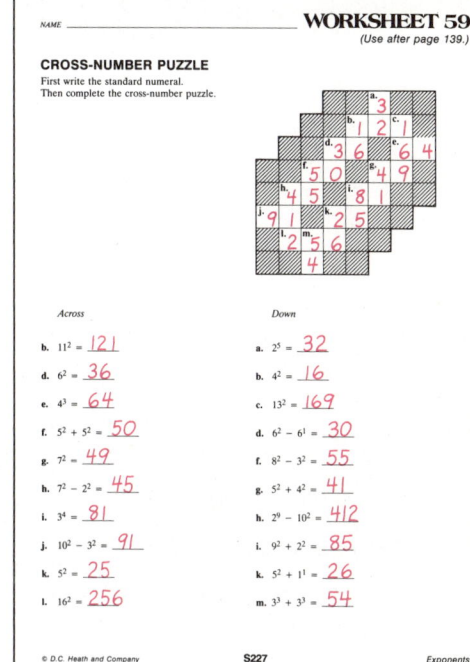

Practice Worksheet
Workbook S227, Copymaster S227, or Duplicating Master S227

WORKSHEET 59 (Use after page 139.)

CROSS-NUMBER PUZZLE
First write the standard numeral. Then complete the cross-number puzzle.

Across
b. $11^2 =$ 121
d. $6^2 =$ 36
e. $4^3 =$ 64
f. $5^2 + 5^2 =$ 50
g. $7^2 =$ 49
h. $7^2 - 2^2 =$ 45
i. $3^4 =$ 81
j. $10^2 - 3^2 =$ 91
k. $5^2 =$ 25
l. $16^2 =$ 256

Down
a. $2^5 =$ 32
b. $4^2 =$ 16
c. $13^2 =$ 169
d. $6^2 - 6^1 =$ 30
e. $8^2 - 3^2 =$ 55
g. $5^2 + 4^2 =$ 41
h. $2^9 - 10^2 =$ 412
i. $9^2 + 2^2 =$ 85
k. $5^2 + 1^1 =$ 26
m. $3^3 + 3^3 =$ 54

Challenge Problem
Continue the pattern.
$1^3 = 1^2 - 0^2$
$2^3 = 3^2 - 1^2$
$3^3 = 6^2 - 3^2$
$4^3 = \boxed{10}^2 - \boxed{6}^2$
$5^3 = \boxed{15}^2 - \boxed{10}^2$
$6^3 = \boxed{21}^2 - \boxed{15}^2$

Copymaster S446

Class Starter Quiz 54
on previous lesson

Write as a standard numeral.

1. 2^5 32   2. 10^1 10
3. 3^4 81   4. 6^3 216

Write using an exponent.

5. $4 \times 4 \times 4$ 4^3
6. $10 \times 10 \times 10 \times 10$ 10^4
7. 8×8 8^2
8. $3 \times 3 \times 3 \times 3 \times 3$ 3^5

Copymaster S398

Lesson Objective
To use rules to test for divisibility by 2, 3, 4, 5, 9, and 10

Problem-Solving Skill
Estimating quantity

Starting the Lesson
Put these numbers on the chalkboard:

    35   15
    60   26
    14   40

Encourage the students to discover the divisibility rules for 2, 5, and 10 by asking these questions:

- Which of the numbers listed are divisible by 2? (60, 26, 14, and 40) How can you tell by looking at the number? (Its last digit is divisible by 2.)
- Which of the numbers are divisible by 5? (35, 15, 60, and 40) How can you tell by looking at the number? (When the last digit is 0 or 5, it is divisible by 5.)
- Which of the numbers are divisible by 10? (60 and 40) How can you tell by looking at the number? (When the last digit is 0, it is divisible by 10.)

Then carefully go over the rules for divisibility given on page 140.

Divisibility

There are 288 football cards in the stack. If you divide 288 by 6, you get a quotient of 48 and a remainder of 0. Since the remainder is 0, we say that 288 is **divisible** by 6.

```
      48
  6)288
     24
     48
     48
      0
```

1. Is 288 divisible by 2? Yes
2. Is 288 divisible by 3? Yes

You can tell whether a whole number is divisible by certain other numbers without actually dividing.

Here's how *to use rules to test for divisibility.*

A whole number is divisible by

2 if its last digit is divisible by 2.

Example:

288 is divisible by 2.

A number that is divisible by 2 is called an **even number**.
A number that is not divisible by 2 is called an **odd number**.

4 if its last two digits are divisible by 4.

Example:

7024 is divisible by 4.

9 if the sum of its digits is divisible by 9.

Example:

3951 is divisible by 9.
$3 + 9 + 5 + 1 = 18$

3 if the sum of the digits is divisible by 3.

Example:

288 is divisible by 3.

$2 + 8 + 8 = 18$

5 if the last digit is divisible by 5, that is, if the last digit is 0 or 5.

Example:

3865 is divisible by 5.

10 if its last digit is divisible by 10, that is, if the last digit is 0.

Example:

5820 is divisible by 10.

3. Look at the *Here's how*. Is 7024 divisible by 4? Yes

Chapter 6

EXERCISES

Is the number divisible by 2?

4. 127 No	5. 854 Yes	6. 208 Yes	7. 357 No	8. 621 No	9. 960 Yes
10. 2351 No	11. 5632 Yes	12. 3904 Yes	13. 1756 Yes	14. 8340 Yes	15. 5735 No

Is the number divisible by 3?

16. 120 Yes	17. 252 Yes	18. 641 No	19. 433 No	20. 828 Yes	21. 720 Yes
22. 2583 Yes	23. 9792 Yes	24. 5533 No	25. 8766 Yes	26. 4638 Yes	27. 7101 Yes

Is the number divisible by 4?

28. 624 Yes	29. 427 No	30. 522 No	31. 336 Yes	32. 715 No	33. 212 Yes
34. 2138 No	35. 5240 Yes	36. 3732 Yes	37. 6308 Yes	38. 7145 No	39. 3236 Yes

Is the number divisible by 5?

40. 340 Yes	41. 515 Yes	42. 444 No	43. 813 No	44. 629 No	45. 950 Yes
46. 3285 Yes	47. 7696 No	48. 6280 Yes	49. 3758 No	50. 6100 Yes	51. 4791 No

Is the number divisible by 9?

52. 531 Yes	53. 635 No	54. 414 Yes	55. 783 Yes	56. 925 No	57. 809 No
58. 5580 Yes	59. 3274 No	60. 6921 Yes	61. 9783 Yes	62. 7011 Yes	63. 6975 Yes

Is the number divisible by 10?

64. 570 Yes	65. 360 Yes	66. 465 No	67. 832 No	68. 990 Yes	69. 750 Yes
70. 3520 Yes	71. 7292 No	72. 8630 Yes	73. 5440 Yes	74. 9715 No	75. 6134 No

How many cards? Estimating quantity

76. Estimate the number of football cards. Do not count past 20.
    Answers will vary.

77. Study the clues to find the number of football cards. 135

    Clues:
    - There are more than 120.
    - There are fewer than 140.
    - There is an odd number of cards.
    - The number is divisible by 9.

Number Theory, Fractions, and Decimals

Class Starter Quiz 55
on previous lesson

True or false?

1. 1246 is divisible by 2? True
2. 7214 is divisible by 3? False
3. 9224 is divisible by 4? True
4. 9285 is divisible by 5? True
5. 6820 is divisible by 10? True

Copymaster S398

Lesson Objective
To write the prime factorization of a composite number

Problem-Solving Skill
Making a list

Starting the Lesson
Write the following on the chalkboard:

Prime numbers:
   2, 3, 5, ?, ?, 13
Composite numbers:
   4, 6, 8, ?, ?, 12

Before the students open their books, challenge them to find the missing numbers in each list. Then have the students check their answers with the list at the top of page 142.

Answers for page 143.
26. 3×5
27. $2 \times 3 \times 5$
28. 3×7
29. 2×19
30. 3^3
31. $2 \times 3 \times 7$
32. 7^2
33. 2×5^2
34. $2^2 \times 7$
35. $2^4 \times 3$
36. $2^4 \times 5$
37. $2^5 \times 3$
38. 2^5
39. $2^2 \times 3 \times 7$
40. 3×5^2
41. $2 \times 3 \times 11$
42. 2×3^3
43. 3^4
44. $4 = 2 + 2$
    $6 = 3 + 3$
    $8 = 3 + 5$
    $10 = 5 + 5$
    $12 = 5 + 7$
    $14 = 7 + 7$
    $16 = 5 + 11$
    $18 = 7 + 11$
    $20 = 7 + 13$
    $22 = 11 + 11$
    $24 = 11 + 13$
45. $7 = 2 + 2 + 3$
    $9 = 3 + 3 + 3$
    $11 = 3 + 3 + 5$
    $13 = 3 + 3 + 7$
    $15 = 2 + 2 + 11$
    $17 = 2 + 2 + 13$
    $19 = 3 + 3 + 13$
    $21 = 7 + 7 + 7$
    $23 = 5 + 7 + 11$
    $25 = 5 + 7 + 13$
    $27 = 7 + 7 + 13$

Prime and composite numbers

The table shows all the divisors or factors of the numbers 1 through 12.

A whole number that has exactly two factors is called a **prime number**. You can also think of a prime number as a number greater than 1 that is only divisible by 1 and itself.

A whole number (other than 0) that has more than two factors is called a **composite number**.

Whole Numbers	Divisors or Factors
1	1
2	1, 2
3	1, 3
4	1, 2, 4
5	1, 5
6	1, 2, 3, 6
7	1, 7
8	1, 2, 4, 8
9	1, 3, 9
10	1, 2, 5, 10
11	1, 11
12	1, 2, 3, 4, 6, 12

1. Which of the numbers in the chart are prime numbers?

2. Which of the numbers in the chart are composite numbers?

Every composite number can be expressed as a product of prime numbers. To express a composite number as a product of prime numbers is to give the **prime factorization** of the number.

Here's how *to use a factor tree to find the prime factorization of a composite number.*

Factor 60.   60
Factor 30.   2 × 30
Factor 6.   2 × 6 × 5
Product of prime numbers → 2 × 2 × 3 × 5
Prime factorization → 60 = 2 × 2 × 3 × 5
Prime factorization → 60 = $2^2 \times 3 \times 5$
using exponents

Factor 72.   72
Factor 36.   2 × 36
Factor 18.   2 × 2 × 18
Factor 9.   2 × 2 × 2 × 9
              2 × 2 × 2 × 3 × 3
72 = 2 × 2 × 2 × 3 × 3
72 = $2^3 \times 3^2$

3. Look at the *Here's how*. What is the prime factorization of 60?

4. What is the prime factorization of 72 using exponents?

Chapter 6

EXERCISES

Tell whether the number is prime or composite. The prime numbers are underlined.

5. 10   6. 12   7. <u>13</u>   8. <u>19</u>   9. 16   10. 25
11. 33   12. 42   13. 51   14. 69   15. <u>73</u>   16. <u>83</u>

Copy and complete each factor tree.

17.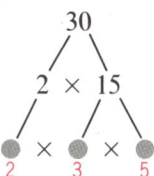
18.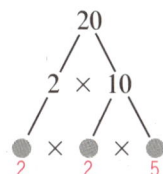
19.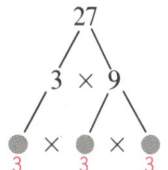

20.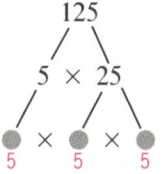
21.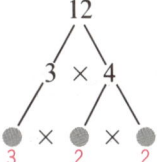
22.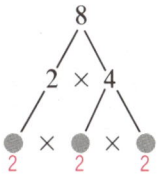

23.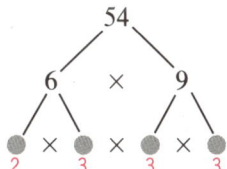
24.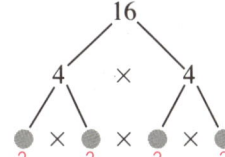
25.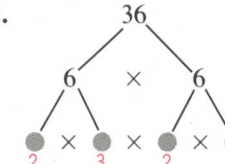

Give the prime factorization. Use exponents when possible.

26. 15   27. 30   28. 21   29. 38   30. 27   31. 42
32. 49   33. 50   34. 28   35. 48   36. 80   37. 96
38. 32   39. 84   40. 75   41. 66   42. 54   43. 81

Prime time

Making a list

44. Start with 4 and write the next ten even numbers as the sum of two prime numbers.

45. Start with 7 and write the next ten odd numbers as the sum of three prime numbers.

Number Theory, Fractions, and Decimals 143

Practice Worksheet

Workbook S229, Copymaster S229, or Duplicating Master S229

WORKSHEET 61
(Use after page 143.)

SKILL DRILL
Complete each factor tree. Do not use 1 as a factor. There is more than one way to complete each factor tree.

1. 12: 4×3, 2×2×3
2. 18: ?×?, 2×3×3
   FACTORS: 2 7 3 5 5 7 3
3. 8: ?×?, 2×2×2
4. 42: ?×?, 2×3×7
5. 27: ?×?, 3×3×3
6. 36: ?×?, 2×2×3×3
7. 24: ?×?, 2×2×2×3
8. 16: ?×?, 2×2×2×2
9. 40: ?×?, 2×2×2×5
10. 54: ?×?, 2×3×3×3
11. 60: ?×?, 2×2×3×5

© D.C. Heath and Company   S229   Prime factorization

Challenge Problems

Use a calculator to give the prime factorization.

1. 1800 $2^3 \times 3^2 \times 5^2$
2. 1134 $2 \times 3^4 \times 7$
3. 1287 $3^2 \times 11 \times 13$
4. 1768 $2^3 \times 13 \times 17$

Copymaster S447

Class Starter Quiz 56
on previous lesson

Give the prime factorization.

1. 6 2×3
2. 15 3×5
3. 33 3×11
4. 24 $2^3 \times 3$
5. 30 $2 \times 3 \times 5$
6. 49 7^2

Copymaster S398

Lesson Objective
To find the greatest common factor of a pair of numbers

Problem-Solving Skill
Using a guess-and-check strategy

Starting the Lesson
Sketch these factor cards on the chalkboard:

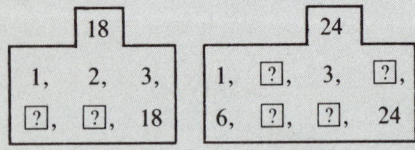

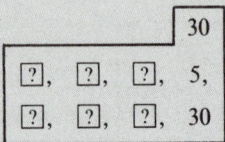

Before the students open their books, challenge them to find the missing numbers on each card. Then have the students check their answers with the factor cards at the top of page 144.

Greatest common factor

1. Look at the factor card for 12. What are the factors of 12?
   1, 2, 3, 4, 6, 12

2. What are the factors of 18?
   1, 2, 3, 6, 9, 18

3. What are the factors of both 12 and 18? They are called **common factors** of 12 and 18.
   1, 2, 3, 6

4. What is the largest number that is a common factor of 12 and 18? 6
   This number is called the **greatest common factor (GCF)** of 12 and 18.

FACTOR CARDS

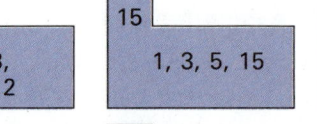

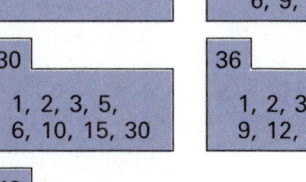

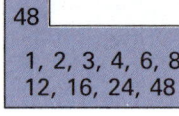

Here's how *to find the GCF.*

Find the GCF of 18 and 24.

Method 1. Find the GCF by listing the factors.
Factors of 18: 1, 2, 3, 6, 9, 18
Factors of 24: 1, 2, 3, 4, 6, 8, 12, 24
The GCF of 18 and 24 is 6.

Method 2. Find the GCF by using prime factorization.
$18 = 2 \times 3 \times 3$
$24 = 2 \times 2 \times 2 \times 3$
To find the GCF, multiply the factors that are common to both 18 and 24.
$GCF = 2 \times 3 = 6$

5. Look at the *Here's how*. What is the GCF of 18 and 24? 6

6. Check these examples. Give each GCF.

   a. $15 = 3 \times 5$
      $30 = 2 \times 3 \times 5$
      GCF = ? 15

   b. $36 = 2 \times 2 \times 3 \times 3$
      $40 = 2 \times 2 \times 2 \times 5$
      GCF = ? 4

   c. $18 = 2 \times 3 \times 3$
      $48 = 2 \times 2 \times 2 \times 2 \times 3$
      GCF = ? 6

EXERCISES

Use the factor cards on page 144.
Give the greatest common factor.
Here are scrambled answers for the next row of exercises: 3 5 6 12 18

7. 12, 15 3
8. 15, 40 5
9. 12, 24 12
10. 18, 48 6
11. 18, 36 18

12. 24, 36 12
13. 12, 30 6
14. 15, 24 3
15. 12, 40 4
16. 30, 48 6

17. 15, 48 3
18. 24, 30 6
19. 18, 40 6
20. 24, 48 24
21. 30, 40 10

Use either method.
Give the greatest common factor.
Here are scrambled answers for the next row of exercises: 3 4 5 6 10

22. 4, 8 4
23. 5, 10 5
24. 6, 9 3
25. 18, 6 6
26. 10, 20 10

27. 18, 27 9
28. 8, 12 4
29. 12, 4 4
30. 20, 12 4
31. 6, 42 6

32. 5, 20 5
33. 16, 8 8
34. 25, 10 5
35. 32, 24 8
36. 5, 40 5

37. 20, 15 5
38. 20, 4 4
39. 32, 16 16
40. 5, 15 5
41. 35, 25 5

42. 32, 8 8
43. 35, 10 5
44. 6, 24 6
45. 36, 4 4
46. 40, 32 8

47. 36, 42 6
48. 50, 35 5
49. 48, 32 16
50. 60, 45 15
51. 44, 55 11

True or false?

52. The sum of two even numbers is an even number. True
53. The product of two odd numbers is an odd number. False
54. 2 is a factor of any even number. True
55. 3 is a factor of any odd number. False
56. 1 is a common factor of any two numbers. True
57. The greatest common factor of two prime numbers is 1. True

Multiplication triangles Guess and check

The product is written between each pair of factors.
Copy and complete each multiplication triangle.

58. (triangle: top 2, left circle 6, right circle ?=8, bottom 3—?=12—4)

59. (triangle: top 8, ?, circles 32 and ?=48, bottom 4—?=24—6)

60. (triangle: top ?=4, circles 12 and 20, bottom ?=3—15—?=5)

Practice Worksheet
Workbook S230, Copymaster S230, or Duplicating Master S230

WORKSHEET 62 (Use after page 145.)

SKILL DRILL
First list the factors of each number. Then circle the greatest common factor.

Challenge Problem

Two slot cars run on a circular track. Car A can circle the track in 4 seconds and car B can circle the track in 6 seconds. Car A is alongside car B just before they start to run, but car A immediately runs ahead of car B. How many seconds will it be before car A is again alongside car B? 12 seconds

Copymaster S447

Number Theory, Fractions, and Decimals

Class Starter Quiz 57
on previous lesson

Give the greatest common factor.

1. 3, 4	1	2. 10, 20	10
3. 5, 6	1	4. 8, 6	2
5. 25, 10	5	6. 20, 12	4

Copymaster S399

Lesson Objective

To find equivalent fractions for a given fraction

Problem-Solving Skills

Finding information in a display
Using logical reasoning

Starting the Lesson

Before discussing exercises 1–6, sketch these shapes on the chalkboard:

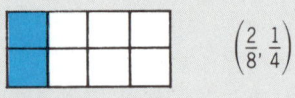 $\left(\frac{2}{8}, \frac{1}{4}\right)$

 $\left(\frac{4}{8}, \frac{1}{2}\right)$

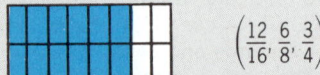 $\left(\frac{12}{16}, \frac{6}{8}, \frac{3}{4}\right)$

Ask the students to write as many fractions as they can to tell how much of each drawing is shaded. Then use exercises 1–6 on page 146 and the *Here's how* to acquaint the students with how to change a fraction to an equivalent fraction.

Here's How Note

Use of Concrete Materials You may wish to use the fraction pieces from copymasters or transparencies S530–S531 to demonstrate equivalent fractions. See ■ *Manipulative Activity 6* on copymaster S516 in the Teacher's Resource Binder.

146

Equivalent fractions

Jan traded school pictures with her close friends. Here is her picture collection.

1. How many of Jan's friends are wearing glasses? 3
2. How many friends are pictured? 8
3. What fraction of her close friends are wearing glasses? $\frac{3}{8}$
4. $\frac{2}{8}$, or $\frac{1}{4}$, of Jan's close friends are boys. Do you agree or disagree with this statement? Agree

Here's how *to change a fraction to an equivalent fraction.*

To change a fraction to an equivalent fraction, multiply or divide both numerator and denominator by the same whole number (not 0).

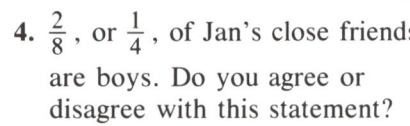

5. The *Here's how* shows that $\frac{1}{4}$ and ? are equivalent fractions. $\frac{2}{8}$

6. Complete these equivalent fractions.

a. $\frac{6}{5} = \frac{?}{15}$ 18 — You multiply 5 by 3 to get 15. So multiply 6 by 3 to find the missing numerator.

b. $\frac{15}{10} = \frac{?}{2}$ 3 — You have to divide 10 by 5 to get 2. So divide 15 by 5 to find the missing numerator.

146 Chapter 6

EXERCISES

Complete to get an equivalent fraction.

7. $\frac{2}{5} = \frac{?}{15}$ [×3] [×3] [×5] 6
8. $\frac{4}{3} = \frac{?}{6}$ [×2] [×2] [×2] 8
9. $\frac{10}{12} = \frac{?}{6}$ [÷2] [÷2] [÷4] 5
10. $\frac{9}{15} = \frac{?}{5}$ [÷3] [÷3] [÷3] 3
11. $\frac{6}{24} = \frac{?}{4}$ [÷6] [÷6] [×3] 1

12. $\frac{1}{2} = \frac{?}{10}$ 5
13. $\frac{3}{8} = \frac{?}{16}$ 6
14. $\frac{12}{16} = \frac{?}{4}$ 3
15. $\frac{6}{9} = \frac{?}{3}$ 2
16. $\frac{7}{8} = \frac{?}{24}$ 21

17. $\frac{1}{8} = \frac{?}{16}$ 2
18. $\frac{5}{2} = \frac{?}{10}$ 25
19. $\frac{4}{9} = \frac{?}{27}$ 12
20. $\frac{6}{8} = \frac{?}{4}$ 3
21. $\frac{5}{6} = \frac{?}{18}$ 15

22. $\frac{16}{10} = \frac{?}{5}$ 8
23. $\frac{9}{12} = \frac{?}{4}$ 3
24. $\frac{8}{12} = \frac{?}{3}$ 2
25. $\frac{4}{7} = \frac{?}{21}$ 12
26. $\frac{3}{18} = \frac{?}{6}$ 1

27. $\frac{10}{9} = \frac{?}{90}$ 100
28. $\frac{24}{30} = \frac{?}{15}$ 12
29. $\frac{25}{75} = \frac{?}{15}$ 5
30. $\frac{2}{9} = \frac{?}{45}$ 10
31. $\frac{30}{50} = \frac{?}{5}$ 3

Give the "next" three equivalent fractions.

32. $\frac{1}{2}, \frac{2}{4}, \frac{3}{6}, ?, ?, ?$ $\frac{4}{8}, \frac{5}{10}, \frac{6}{12}$
33. $\frac{2}{3}, \frac{4}{6}, \frac{6}{9}, ?, ?, ?$ $\frac{8}{12}, \frac{10}{15}, \frac{12}{18}$
34. $\frac{1}{4}, \frac{2}{8}, \frac{3}{12}, ?, ?, ?$ $\frac{4}{16}, \frac{5}{20}, \frac{6}{24}$

35. $\frac{1}{5}, \frac{2}{10}, ?, ?, ?$ $\frac{3}{15}, \frac{4}{20}, \frac{5}{25}$
36. $\frac{3}{4}, \frac{6}{8}, ?, ?, ?$ $\frac{9}{12}, \frac{12}{16}, \frac{15}{20}$
37. $\frac{4}{5}, \frac{8}{10}, ?, ?, ?$ $\frac{12}{15}, \frac{16}{20}, \frac{20}{25}$

38. $\frac{1}{6}, ?, ?, ?$ $\frac{2}{12}, \frac{3}{18}, \frac{4}{24}$
39. $\frac{3}{8}, ?, ?, ?$ $\frac{6}{16}, \frac{9}{24}, \frac{12}{32}$
40. $\frac{6}{5}, ?, ?, ?$ $\frac{12}{10}, \frac{18}{15}, \frac{24}{20}$

Solve. Use the collection of pictures on page 146.

41. What fraction of the pictures have been autographed? $\frac{5}{8}$
42. What fraction of the pictures have not been autographed? $\frac{3}{8}$
43. What fraction of those pictured are girls? Give two equivalent fractions. $\frac{6}{8}, \frac{3}{4}$
44. What fraction of the girls pictured wear glasses? Give two equivalent fractions. $\frac{2}{6}, \frac{1}{3}$

Smile!

45. Study the clues to find in what year this photograph was taken. 1902

    Clues:
    - The year rounded to the nearest ten is 1900.
    - The sum of the digits is 12.

Number Theory, Fractions, and Decimals 147

Extra Practice
Page 482   Skill 23

Practice Worksheet
Workbook S231, Copymaster S231, or Duplicating Master S231

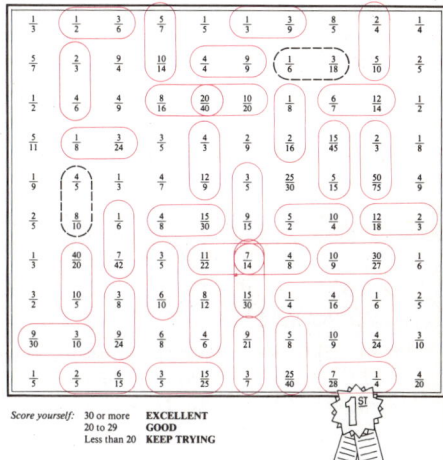

Challenge Problem

Find 18 ways to use any of the digits ①②③④⑥ to make a fraction equivalent to $\frac{1}{2}$. *Hint: You can use the digits more than once.*

$\frac{1}{2}, \frac{2}{4}, \frac{3}{6}, \frac{6}{12}, \frac{11}{22}, \frac{12}{24},$
$\frac{13}{26}, \frac{16}{32}, \frac{21}{42}, \frac{22}{44}, \frac{23}{46}, \frac{31}{62},$
$\frac{32}{64}, \frac{33}{66}, \frac{61}{122}, \frac{62}{124}, \frac{63}{126}, \frac{66}{132}$

Copymaster S447

Class Starter Quiz 58
on previous lesson

Complete to get an equivalent fraction.

1. $\frac{2}{3} = \frac{?}{12}$ 8
2. $\frac{8}{10} = \frac{?}{5}$ 4
3. $\frac{10}{7} = \frac{?}{70}$ 100
4. $\frac{18}{30} = \frac{?}{15}$ 9
5. $\frac{2}{3} = \frac{?}{15}$ 10
6. $\frac{15}{20} = \frac{?}{4}$ 3

Copymaster S399

Lesson Objective
To write fractions in lowest terms

Problem-Solving Skill
Using logical reasoning

Starting the Lesson
Before the students open their textbooks, ask them to write the prime factorizations of 12 and 18. Then have them use this information to answer exercises 1–3. Next, go over the examples in the *Here's how* and exercises 6 and 7 to acquaint the students with how to write a fraction in lowest terms.

Here's How Note
Use of Concrete Materials You may wish to use the fraction pieces from copymasters or transparencies S530–S531 to demonstrate writing fractions in lowest terms. See ■ **Manipulative Activity 7** on copymaster S516 in the Teacher's Resource Binder.

148

Writing fractions in lowest terms

The numerator and denominator of a fraction are sometimes called the **terms** of the fraction.

1. What are the terms of $\frac{12}{18}$? 12 and 18

2. What are the common factors of 12 and 18? 1, 2, 3, 6

3. What is the greatest common factor (GCF) of 12 and 18? 6

Here's how *to write a fraction in lower terms.*

To write a fraction in **lower terms**, divide both terms by a common factor greater than 1.

$\boxed{\div 2}$ $\boxed{\div 3}$ $\boxed{\div 3}$

$\frac{12}{18} = \frac{6}{9}$ ← lower terms $\frac{30}{36} = \frac{10}{12}$ $\frac{24}{15} = \frac{8}{5}$

$\boxed{\div 2}$ $\boxed{\div 3}$ $\boxed{\div 3}$

4. Which resulting fraction above cannot be written in lower terms? $\frac{8}{5}$

Here's how *to write a fraction in lowest terms.*

To write a fraction in **lowest terms**, divide both terms by the greatest common factor.

$\frac{24}{36}$ The GCF is 12.

Divide both terms by the GCF.

$\boxed{\div 12}$

$\frac{24}{36} = \frac{2}{3}$ ← lowest terms

$\boxed{\div 12}$

A fraction is in **lowest terms** if 1 is the GCF of its terms.

5. Look at the *Here's how*. $\frac{24}{36}$ written in lowest terms is ? . $\frac{2}{3}$

148   *Chapter 6*

EXERCISES

Which fractions cannot be written in lower terms?

6. $\frac{1}{4}$ 7. $\frac{8}{2}$ 8. $\frac{2}{3}$ 9. $\frac{8}{10}$ 10. $\frac{2}{6}$ 11. $\frac{3}{2}$ 12. $\frac{6}{10}$

13. $\frac{9}{15}$ 14. $\frac{5}{12}$ 15. $\frac{6}{9}$ 16. $\frac{7}{21}$ 17. $\frac{10}{6}$ 18. $\frac{5}{4}$ 19. $\frac{7}{8}$

20. $\frac{24}{48}$ 21. $\frac{36}{18}$ 22. $\frac{9}{10}$ 23. $\frac{18}{30}$ 24. $\frac{9}{20}$ 25. $\frac{18}{27}$ 26. $\frac{13}{24}$

(Circled: 6, 8, 11, 14, 18, 19, 24, 26)

Write each fraction in lowest terms.

Here are scrambled answers for the next row of exercises: $\frac{3}{5}$ $\frac{4}{5}$ $\frac{1}{2}$ $\frac{3}{4}$ $\frac{3}{1}$ $\frac{5}{8}$ $\frac{4}{9}$

27. $\frac{9}{12}$ $\frac{3}{4}$ 28. $\frac{36}{12}$ $\frac{3}{1}$ 29. $\frac{18}{30}$ $\frac{3}{5}$ 30. $\frac{24}{30}$ $\frac{4}{5}$ 31. $\frac{15}{24}$ $\frac{5}{8}$ 32. $\frac{15}{30}$ $\frac{1}{2}$ 33. $\frac{20}{45}$ $\frac{4}{9}$

34. $\frac{6}{8}$ $\frac{3}{4}$ 35. $\frac{3}{9}$ $\frac{1}{3}$ 36. $\frac{2}{6}$ $\frac{1}{3}$ 37. $\frac{3}{12}$ $\frac{1}{4}$ 38. $\frac{10}{30}$ $\frac{1}{3}$ 39. $\frac{30}{36}$ $\frac{5}{6}$ 40. $\frac{14}{42}$ $\frac{1}{3}$

41. $\frac{5}{15}$ $\frac{1}{3}$ 42. $\frac{4}{6}$ $\frac{2}{3}$ 43. $\frac{18}{15}$ $\frac{6}{5}$ 44. $\frac{9}{6}$ $\frac{3}{2}$ 45. $\frac{4}{12}$ $\frac{1}{3}$ 46. $\frac{10}{15}$ $\frac{2}{3}$ 47. $\frac{25}{10}$ $\frac{5}{2}$

48. $\frac{24}{8}$ $\frac{3}{1}$ 49. $\frac{15}{10}$ $\frac{3}{2}$ 50. $\frac{20}{50}$ $\frac{2}{5}$ 51. $\frac{40}{50}$ $\frac{4}{5}$ 52. $\frac{14}{24}$ $\frac{7}{12}$ 53. $\frac{15}{25}$ $\frac{3}{5}$ 54. $\frac{9}{24}$ $\frac{3}{8}$

55. $\frac{18}{14}$ $\frac{9}{7}$ 56. $\frac{18}{6}$ $\frac{3}{1}$ 57. $\frac{15}{45}$ $\frac{1}{3}$ 58. $\frac{40}{30}$ $\frac{4}{3}$ 59. $\frac{11}{22}$ $\frac{1}{2}$ 60. $\frac{20}{24}$ $\frac{5}{6}$ 61. $\frac{22}{33}$ $\frac{2}{3}$

62. $\frac{24}{32}$ $\frac{3}{4}$ 63. $\frac{14}{16}$ $\frac{7}{8}$ 64. $\frac{25}{15}$ $\frac{5}{3}$ 65. $\frac{14}{6}$ $\frac{7}{3}$ 66. $\frac{10}{40}$ $\frac{1}{4}$ 67. $\frac{15}{20}$ $\frac{3}{4}$ 68. $\frac{20}{40}$ $\frac{1}{2}$

69. $\frac{10}{24}$ $\frac{5}{12}$ 70. $\frac{12}{18}$ $\frac{2}{3}$ 71. $\frac{20}{32}$ $\frac{5}{8}$ 72. $\frac{25}{20}$ $\frac{5}{4}$ 73. $\frac{12}{24}$ $\frac{1}{2}$ 74. $\frac{30}{20}$ $\frac{3}{2}$ 75. $\frac{18}{21}$ $\frac{6}{7}$

76. $\frac{24}{18}$ $\frac{4}{3}$ 77. $\frac{18}{36}$ $\frac{1}{2}$ 78. $\frac{8}{24}$ $\frac{1}{3}$ 79. $\frac{30}{18}$ $\frac{5}{3}$ 80. $\frac{24}{36}$ $\frac{2}{3}$ 81. $\frac{30}{24}$ $\frac{5}{4}$ 82. $\frac{75}{100}$ $\frac{3}{4}$

No bones about it

Did you know there is only one bone in your skull that can move? It is the one in your lower jaw, which permits you to talk, laugh, and chew food.

83. How many bones are in your skull? 22
Clues:
- There are more than 12 but less than 30 bones.
- 11 is a divisor of the number of bones.

Number Theory, Fractions, and Decimals **149**

Extra Practice

Page 482 Skill 24

Practice Worksheet

Workbook S232, Copymaster S232, or Duplicating Master S232

Project

Reading a yardstick

Find each measure on a yardstick and compare it to the total length of the stick. Then tell what fractional part of a yard, in lowest terms, each measure is.

1. 12 inches $\frac{1}{3}$
2. 8 inches $\frac{2}{9}$
3. 9 inches $\frac{1}{4}$
4. 16 inches $\frac{4}{9}$
5. 18 inches $\frac{1}{2}$
6. 2 feet $\frac{2}{3}$
7. 27 inches $\frac{3}{4}$
8. 4 inches $\frac{1}{9}$

Copymaster S478

Class Starter Quiz 59
on previous lesson

Write each fraction in lowest terms.

1. $\frac{3}{6}$ $\frac{1}{2}$
2. $\frac{9}{15}$ $\frac{3}{5}$
3. $\frac{24}{30}$ $\frac{4}{5}$
4. $\frac{32}{8}$ $\frac{4}{1}$
5. $\frac{5}{15}$ $\frac{1}{3}$
6. $\frac{14}{24}$ $\frac{7}{12}$
7. $\frac{30}{18}$ $\frac{5}{3}$
8. $\frac{50}{20}$ $\frac{5}{2}$

Copymaster S399

Lesson Objectives
To identify the least common multiple of a pair of numbers
To identify the least common denominator of a pair of fractions

Problem-Solving Skills
Drawing a diagram
Using logical reasoning

Starting the Lesson
Use of Concrete Materials The multiple strips at the top of page 150 are also on ■ **Visual Aid 24** (copymaster or transparency S131). Before discussing exercises 1–4, challenge the students to look for the pattern and identify what the next four numbers on each strip would be. You may wish to cut the visual aid and use only the appropriate strips for discussing the first four exercises.

Here's How Note
Point out that the least common multiple of the denominators is the *least common denominator*. You may wish to go over several exercises from 5–59 orally before assigning the independent work.

Least Common Multiple

1. **a.** Look at multiple strip for 3. Are 3, 6, 9, 12, 15, 18, 21, and 24 multiples of 3? Yes

   b. Look at the multiple strip for 4. List the given multiples. 4, 8, 12, 16, 20, 24, 28, 32

   Notice that 12 and 24 are multiples of both 3 and 4. They are called **common multiples** of 3 and 4.

2. What is the smallest number that is a common multiple of 3 and 4? 12
   This number is called the **least common multiple (LCM)** of 3 and 4.

Multiple Strips

2	2	4	6	8	10	12	14	16	18
3	3	6	9	12	15	18	21	24	
4	4	8	12	16	20	24	28	32	
5	5	10	15	20	25	30	35		
6	6	12	18	24	30	36	42		
7	7	14	21	28	35	42	49		
8	8	16	24	32	40	48	56		

Here's how *to find the LCM.*   Find the LCM of 4 and 6.

Method 1.

List the multiples of 4 and 6. Then find the first multiple of the smaller number (4) that is also a multiple of the larger number (6).

4: 4 8 **12** 16 20 24 28 32
6: 6 **12** 18 24 30 36 42

The LCM of 4 and 6 is 12.

Method 2.

Find the prime factors of 4 and 6.

⌐—common
4 = 2 × 2
6 = 2 × 3
          ⌐—not common

Multiply the factors that are common to both 4 and 6 by the factors that are not common to both 4 and 6.

LCM = 2 × 2 × 3 = 12

3. Look at the *Here's how*. What is the LCM of 4 and 6? 12

Here's how *to find the least common denominator of two fractions.*

To find the **least common denominator** of two fractions, find the LCM of the denominator.

$\frac{3}{4}, \frac{5}{8}$ The least common denominator is 8.

$\frac{5}{6}, \frac{4}{9}$ The least common denominator is 18.

---

4. Look at the *Here's how*. What is the least common denominator of $\frac{5}{6}$ and $\frac{4}{9}$? 18

EXERCISES

Give the least common multiple. Use either method.
Here are scrambled answers for the next row of exercises: 9 18 20 4 24

5. 2, 4 **4**
6. 9, 3 **9**
7. 4, 10 **20**
8. 2, 9 **18**
9. 8, 6 **24**
10. 5, 6 **30**
11. 5, 10 **10**
12. 3, 9 **9**
13. 3, 8 **24**
14. 6, 3 **6**
15. 6, 4 **12**
16. 4, 12 **12**
17. 7, 2 **14**
18. 3, 12 **12**
19. 5, 10 **10**
20. 7, 4 **28**
21. 9, 5 **45**
22. 5, 4 **20**
23. 8, 5 **40**
24. 6, 7 **42**

Give the least common denominator.

25. $\frac{1}{3}, \frac{1}{6}$ **6**
26. $\frac{2}{5}, \frac{3}{2}$ **10**
27. $\frac{1}{3}, \frac{5}{12}$ **12**
28. $\frac{4}{3}, \frac{3}{8}$ **24**
29. $\frac{1}{8}, \frac{1}{12}$ **24**
30. $\frac{1}{2}, \frac{3}{20}$ **20**
31. $\frac{1}{12}, \frac{7}{8}$ **24**
32. $\frac{3}{4}, \frac{7}{6}$ **12**
33. $\frac{5}{9}, \frac{7}{6}$ **18**
34. $\frac{1}{10}, \frac{2}{15}$ **30**
35. $\frac{4}{5}, \frac{1}{6}$ **30**
36. $\frac{3}{8}, \frac{1}{6}$ **24**
37. $\frac{3}{4}, \frac{9}{10}$ **20**
38. $\frac{1}{8}, \frac{7}{10}$ **40**
39. $\frac{1}{7}, \frac{1}{3}$ **21**
40. $\frac{1}{7}, \frac{3}{8}$ **56**
41. $\frac{2}{5}, \frac{3}{4}$ **20**
42. $\frac{3}{7}, \frac{1}{5}$ **35**
43. $\frac{1}{9}, \frac{3}{8}$ **72**
44. $\frac{2}{15}, \frac{1}{30}$ **30**
45. $\frac{1}{20}, \frac{3}{40}$ **40**
46. $\frac{2}{11}, \frac{1}{33}$ **33**
47. $\frac{1}{15}, \frac{5}{6}$ **30**
48. $\frac{7}{3}, \frac{2}{11}$ **33**
49. $\frac{4}{9}, \frac{1}{8}$ **72**
50. $\frac{4}{5}, \frac{1}{3}$ **15**
51. $\frac{5}{4}, \frac{2}{9}$ **36**
52. $\frac{3}{4}, \frac{7}{12}$ **12**
53. $\frac{1}{3}, \frac{7}{24}$ **24**
54. $\frac{6}{7}, \frac{1}{4}$ **28**
55. $\frac{2}{3}, \frac{1}{36}$ **36**
56. $\frac{1}{10}, \frac{1}{100}$ **100**
57. $\frac{3}{10}, \frac{1}{25}$ **50**
58. $\frac{1}{8}, \frac{1}{80}$ **80**
59. $\frac{2}{25}, \frac{7}{30}$ **150**

Credit cutting Logical reasoning

Credit-card companies suggest that expired cards be destroyed. With 3 straight cuts, this credit card was cut into 7 pieces.

60. What is the greatest number of pieces you can get with 5 straight cuts? *Hints: Draw a diagram. Use the clues to check your answer.* **16**

   Clues:
   - There are more than 10 but less than 20 pieces.
   - The number of pieces is a multiple of 8.

Extra Practice
Page 483   Skill 25

Practice Worksheet
Workbook S233, Copymaster S233, or Duplicating Master S233

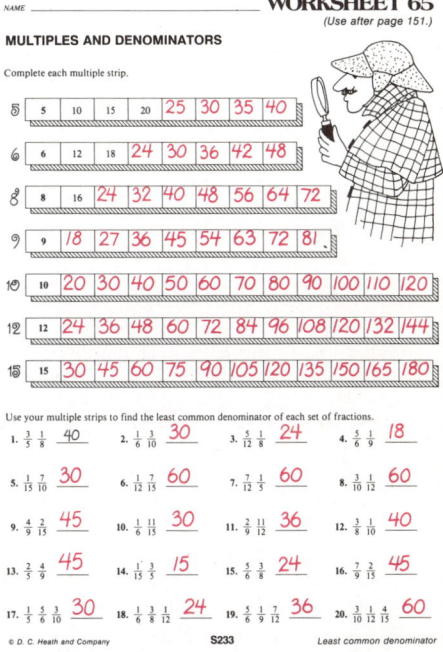

Challenge Problem
Jim's age this year is a multiple of 7. Next year his age will be a multiple of 5. His older sister is now 20. How old is Jim now?
14 years old

Copymaster S447

Class Starter Quiz 60
on previous lesson

Find the least common denominator.

1. $\frac{1}{2}, \frac{3}{5}$ 10
2. $\frac{1}{3}, \frac{2}{9}$ 9
3. $\frac{3}{10}, \frac{1}{4}$ 20
4. $\frac{3}{7}, \frac{1}{6}$ 42
5. $\frac{7}{6}, \frac{2}{15}$ 30
6. $\frac{6}{7}, \frac{3}{4}$ 28

Copymaster S399

Lesson Objective
To compare two fractions.

Problem-Solving Skills
Using data from a circle graph
Using information from a display

Starting the Lesson
The picture of Mount Rushmore at the top of page 152 is also on ■ **Visual Aid 25** (copymaster or transparency S131). Use the picture of Mount Rushmore and take a student poll. Ask the students to name *one* of the presidents whose face is carved into Mount Rushmore. Use a frequency table to record the results of the poll.

Presidents	Tally
George Washington	
Thomas Jefferson	
Abraham Lincoln	
Theodore Roosevelt	

Have the students determine what fraction of the class named each president. Then have the students compare their polling results with the results in the graph on page 152.

Here's How Note
Use of Concrete Materials You may wish to use the fraction pieces from copymasters or transparencies S530–S531 to demonstrate comparing fractions. See ■ **Manipulative Activity 8** on copymaster S516 in the Teacher's Resource Binder.

152

Comparing fractions
MOUNT RUSHMORE RECALL

A group of high school students were asked to name one of the presidents whose face is carved into Mount Rushmore. The circle graph shows the results of the poll.

1. What fraction of the students named Abraham Lincoln? $\frac{3}{8}$
2. Thomas Jefferson was named by what fraction of the students? $\frac{1}{8}$
3. What two fractions would you compare to decide whether more students named Lincoln than named Jefferson? $\frac{3}{8}, \frac{1}{8}$

Here's how to compare two fractions with a common denominator. $\frac{3}{8} \bullet \frac{1}{8}$

To compare fractions with a common denominator, compare the numerators.

$\frac{3}{8} > \frac{1}{8}$ 3 is greater than 1. So $\frac{3}{8}$ is greater than $\frac{1}{8}$.

4. Which president, Abraham Lincoln or Thomas Jefferson, did more students name?
Abraham Lincoln

Here's how to compare two fractions with different denominators. $\frac{3}{8} \bullet \frac{1}{2}$

To compare fractions with different denominators, compare equivalent fractions with the same denominator.

Find the least common denominator.	Write equivalent fractions.	Compare.
$\frac{3}{8} \bullet \frac{1}{2}$	$\frac{3}{8} \bullet \frac{1}{2}$	$\frac{3}{8} \bullet \frac{1}{2}$
8	$\frac{3}{8} \quad \frac{4}{8}$	$\frac{3}{8} < \frac{4}{8}$ So $\frac{3}{8}$ is less than $\frac{1}{2}$.

5. Which president, Abraham Lincoln or George Washington, did more students name?
George Washington

152   Chapter 6

EXERCISES

< or >?

6. $\dfrac{2}{5} < \dfrac{3}{5}$
7. $\dfrac{5}{4} < \dfrac{7}{4}$
8. $\dfrac{5}{7} > \dfrac{3}{7}$
9. $\dfrac{5}{9} > \dfrac{4}{9}$
10. $\dfrac{3}{8} > \dfrac{0}{8}$

11. $\dfrac{4}{4} < \dfrac{5}{4}$
12. $\dfrac{6}{5} < \dfrac{7}{5}$
13. $\dfrac{7}{8} > \dfrac{5}{8}$
14. $\dfrac{0}{6} < \dfrac{1}{6}$
15. $\dfrac{7}{4} < \dfrac{8}{4}$

16. $\dfrac{5}{3} > \dfrac{3}{3}$
17. $\dfrac{3}{8} > \dfrac{2}{8}$
18. $\dfrac{7}{9} > \dfrac{4}{9}$
19. $\dfrac{11}{10} < \dfrac{13}{10}$
20. $\dfrac{0}{5} < \dfrac{1}{5}$

<, >, or =? *Hint: First write equivalent fractions with the same denominator.*

21. $\dfrac{1}{4} \; \bullet \; \dfrac{3}{8}$ < $\;\;\boxed{\dfrac{2}{8}}\;\boxed{\dfrac{3}{8}}$
22. $\dfrac{5}{6} \; \bullet \; \dfrac{2}{3}$ > $\;\;\boxed{\dfrac{5}{6}}\;\boxed{\dfrac{4}{6}}$
23. $\dfrac{1}{3} \; \bullet \; \dfrac{2}{7}$ > $\;\;\boxed{\dfrac{7}{21}}\;\boxed{\dfrac{6}{21}}$
24. $\dfrac{1}{3} \; \bullet \; \dfrac{1}{4}$ > $\;\;\boxed{\dfrac{?}{12}}\;\boxed{\dfrac{?}{12}}$
25. $\dfrac{3}{2} \; \bullet \; \dfrac{5}{4}$ > $\;\;\boxed{\dfrac{?}{4}}\;\boxed{\dfrac{?}{4}}$

26. $\dfrac{1}{3} > \dfrac{3}{10}$
27. $\dfrac{3}{8} < \dfrac{3}{4}$
28. $\dfrac{1}{6} > \dfrac{1}{8}$
29. $\dfrac{2}{9} = \dfrac{4}{18}$
30. $\dfrac{2}{5} > \dfrac{1}{4}$

31. $\dfrac{3}{4} > \dfrac{2}{3}$
32. $\dfrac{5}{6} > \dfrac{3}{4}$
33. $\dfrac{3}{4} > \dfrac{3}{5}$
34. $\dfrac{2}{3} < \dfrac{7}{9}$
35. $\dfrac{0}{3} = \dfrac{0}{7}$

36. $\dfrac{3}{7} = \dfrac{9}{21}$
37. $\dfrac{9}{16} < \dfrac{5}{8}$
38. $\dfrac{4}{7} < \dfrac{5}{8}$
39. $\dfrac{7}{8} < \dfrac{8}{9}$
40. $\dfrac{15}{12} = \dfrac{5}{4}$

41. $\dfrac{7}{10} > \dfrac{69}{100}$
42. $\dfrac{6}{100} > \dfrac{55}{1000}$
43. $\dfrac{9}{1000} < \dfrac{1}{10}$
44. $\dfrac{49}{1000} > \dfrac{4}{100}$
45. $\dfrac{7}{10} < \dfrac{73}{100}$

Solve.

46. On the day that Marcia visited Mount Rushmore, $\dfrac{1}{12}$ of the visitors were from Texas and $\dfrac{1}{8}$ were from California. From which state were there more visitors? **California**

47. On another day, $\dfrac{1}{4}$ of the visitors were from North Dakota and $\dfrac{7}{20}$ of the visitors were from Nebraska. Were there more visitors from North Dakota or Nebraska? **Nebraska**

You're the statistical clerk!

Statistical clerks gather information from surveys and records. Business people depend on statistical clerks to help them make decisions.

48. Which statement is correct?
    a. A greater fraction of girls knew the location of Mount Rushmore.
    b. A greater fraction of boys knew the location of Mount Rushmore.

SURVEY FINDINGS
8 out of 15 boys knew that Mount Rushmore was located in South Dakota.
7 out of 12 girls knew its location.

Number Theory, Fractions, and Decimals

Extra Practice
Page 483 Skill 26

Practice Worksheet
Workbook S234, Copymaster S234, or Duplicating Master S234

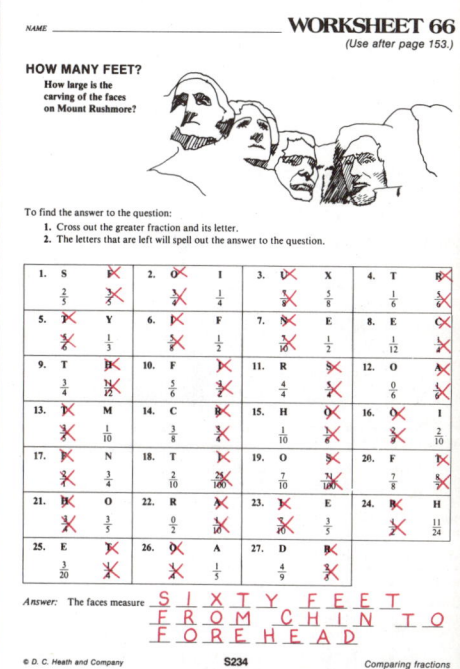

Challenge Problem
Match each fraction to a tag on the number line.

$\dfrac{5}{3} \quad \dfrac{2}{3} \quad \dfrac{11}{10} \quad \dfrac{1}{2} \quad \dfrac{6}{4} \quad \dfrac{3}{4}$

$\dfrac{1}{2} \quad \dfrac{2}{3} \quad \dfrac{3}{4} \quad \dfrac{11}{10} \quad \dfrac{6}{4} \quad \dfrac{5}{3}$

Copymaster S448

Class Starter Quiz 61
on previous lesson

<, >, or = ?

1. $\frac{1}{4}$ ● $\frac{1}{8}$ >   2. $\frac{2}{6}$ ● $\frac{1}{3}$ =

3. $\frac{3}{5}$ ● $\frac{3}{4}$ <   4. $\frac{3}{5}$ ● $\frac{7}{10}$ <

5. $\frac{5}{6}$ ● $\frac{7}{8}$ <   6. $\frac{3}{10}$ ● $\frac{29}{100}$ >

Copymaster S400

Problem-Solving Skills
Finding information in a newspaper ad
Solving a multistep problem

Skills Reviewed
Writing the word-name for a decimal
Adding and subtracting decimals
Comparing decimals
Multiplying and dividing decimals
Simplifying expressions

Starting the Lesson

Problem Solving The newspaper ad at the top of page 154 is also on ■ **Visual Aid 26** (copymaster or transparency S132). Use the ad and ask questions like these:

- What is the cost per month for the one-bedroom apartment? ($200)
- Is $200 per month enough money to rent an efficiency apartment in the new building? (No)
- Would it cost more or less than $2500 to rent the furnished three-bedroom apartment for one year? (More)

Cumulative Skill Practice Write these four answers on the chalkboard:

308 thousandths
27.9   84.39   97.54

Challenge the students to an answer hunt by saying, "Look at exercises 1–51 on page 155. Find the four exercises that have these answers. You have four minutes to find as many of the exercises as you can." (Exercises 9, 20, 26, and 48)

154

Problem solving   Renting an apartment

Use the ad to answer these questions.
Decide when a calculator would be useful.

1. How much would I spend per year for the first apartment listed? $2340

2. My "take-home" pay is $830 each month. How much would I have left after paying the rent on the second apartment listed? $480

Apartments— Furnished

EFFICIENCY APARTMENT
New carpet, spotless. Bus at door, all shopping half block away. ALL UTILITIES PAID BY LANDLORD. $195 per month. Call 999-1668.

2 or 3 bedrooms, $350 per month, all utilities paid. 104 E. Green. 999-4457. 999-4688.

Quiet 2-bedroom second-floor apartment in private home. Close to downtown. References, deposit, and lease. $275 per month plus electricity. After 5 P.M.: 999-8923.

NEW BUILDING
Now Leasing
New furniture & appliances. 3-bedroom apartments have dishwashers.
3 bedrooms—$395 per month
Efficiencies—$215 per month
999-5046 or 999-3689

Unfurnished 2-bedroom apartment. Maximum 3 persons. $250 per month. Phone: Hodges Real Estate, 999-5126.

One-bedroom furnished apartment. Available immediately. $200 per month. 1-888-3355 days.

3. Look at the third apartment listed. If the average cost of electricity is $74.50 per month, how much would you spend a year for rent and electricity? $4194

4. Suppose that you and 2 of your friends decided to rent the next-to-the-last apartment listed. What would your share of the rent be each month? Round your answer to the nearest cent. $83.33

5. Suppose that your "take-home" pay is $724 per month and that you rented the third apartment listed. If the electric bill was $89.76 the first month, how much would you have left after paying the rent and electric bill? $359.24

6. You decided that you could pay a maximum of $340 a month for rent and utilities. The monthly utilities for the efficiency apartment in the new building are estimated to be $76 for heat and $68 for electricity. Should you rent the apartment? No

7. According to a tenant of the last apartment listed, the total amount paid for the apartment and utilities for last year was $3762. How much did the utilities cost if the rent was the same last year as this year? $1362

8. Suppose that you and 4 of your friends rented a 3-bedroom apartment in the new building. How much would each of you pay in a year for rent? $948

9. The estimated total rent and utilities for a 3-bedroom apartment in the new building is $6480 per year. What is the estimated average monthly utility bill? $145

154   Chapter 6

Cumulative Skill Practice

Write the short word-name. *(page 14)*

1. 8.4
2. 16.3
3. 0.451
4. 9.86
5. 0.03
6. 5.036
7. 0.29
8. 3.8352
9. 0.308
10. 10.61
11. 112.74 — 112 and 74 hundredths
12. 11.274 — 11 and 274 thousandths
13. 1.1274 — 1 and 274 ten thousandths
14. 16.02 — 16 and 2 hundredths
15. 31.301 — 31 and 301 thousandths

Give the sum. *(page 22)*

16. 8.24 + 6.59   14.83
17. 59.2 + 36.4   95.6
18. 6.095 + 4.968   11.063
19. 6.09 + 4.196   10.286
20. 15 + 12.9   27.9
21. 8.6 + 3.36   11.96
22. 2.3 + 4 + 5.8   12.1
23. 18 + 5.6 + 10   33.6
24. 8.4 + 6 + 5.9   20.3
25. 16.34 + 21.7 + 32.5   70.54
26. 45.6 + 38 + 0.79   84.39
27. 56.7 + 42.3 + 87   186

<, >, or = ? *(page 42)*

28. 0.6 > 0.4
29. 0.05 < 0.06
30. 0.008 > 0.007
31. 24.3 > 24.2
32. 9.63 < 9.635
33. 0.32 > 0.3
34. 17.1 > 17.08
35. 0.034 < 0.34
36. 6.804 < 6.84
37. 28.24 > 2.842
38. 2.0 > 1.99
39. 3.008 < 3.8

Give the difference. *(page 46)*

40. 9.8 − 3.4   6.4
41. 6.54 − 2.39   4.15
42. 12.346 − 7.591   4.755
43. 17 − 9.04   7.96
44. 18.3 − 2.67   15.63
45. 33.4 − 1.839   31.561
46. 5.43 − 2.976   2.454
47. 42.3 − 9.9   32.4
48. 100 − 2.46   97.54
49. 36.8 − 8.37   28.43
50. 59 − 8.75   50.25
51. 1 − 0.399   0.601

MIXED PRACTICE

Complete.

52. The standard numeral for eight million, five hundred twenty thousand is ?   8,520,000

53. The standard numeral for sixteen billion, forty-seven million is ?   16,047,000,000

54. 7.4 × 0.5 = ?   3.7
55. 1.248 ÷ 0.6   ?   2.08
56. 1.26 × 0.35 = ?   0.441
57. 0.51 ÷ 0.3 = ?   1.7
58. 8.03 × 0.7 = ?   5.621
59. 4.764 ÷ 1.2 = ?   3.97
60. (1.8 + 3.5) − 2.7 = ?   2.6
61. (5.72 − 3) − 0.54 = ?   2.18
62. 5.72 − (3 − 0.54) = ?   3.26

1. 8 and 4 tenths
2. 16 and 3 tenths
3. 451 thousandths
4. 9 and 86 hundredths
5. 3 hundredths
6. 5 and 36 thousandths
7. 29 hundredths
8. 3 and 8352 ten-thousandths
9. 308 thousandths
10. 10 and 61 hundredths

Number Theory, Fractions, and Decimals **155**

Problem-Solving Worksheet
Workbook S235, Copymaster S235, or Duplicating Master S235

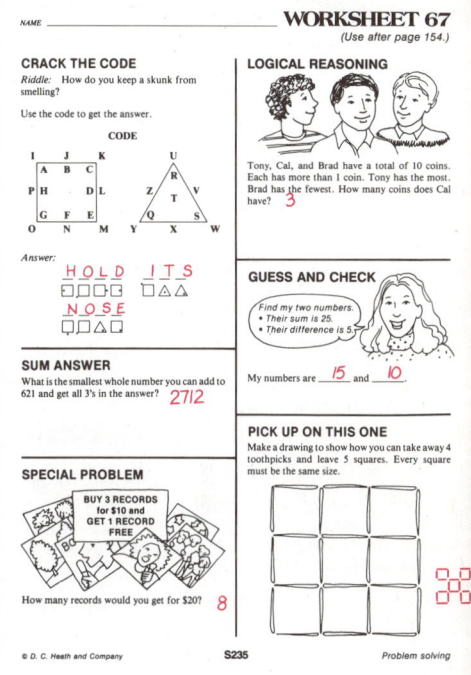

Challenge Problems

Look at page 155. Estimate which two decimals in exercises 1–15 have a sum of about:

1. 144   112.74 + 31.301
2. 42.6   11.274 + 31.301
3. 1.4   1.1274 + 0.308
4. 24.4   8.4 + 16.02

If you have a calculator, use it to see whether you made the correct choices.

Copymaster S448

Writing whole numbers and mixed numbers as fractions

1. Look at a whole pizza. Does $1 = \frac{4}{4}$?   Yes
2. Look at 2 whole pizzas. Does $2 = \frac{8}{4}$?   Yes

Here's how *to change a whole number to a fraction.* $2 = \frac{?}{4}$

Write the whole number over the denominator 1.	Multiply the numerator and denominator by the same whole number (not 0).
$\frac{2}{1}$	$\frac{2}{1} = \frac{8}{4}$ (× 4) There are 8 fourths in 2.

3. What would you have to multiply the numerator and denominator by to change $\frac{2}{1}$ to eighths?   8

4. Look at the pizzas. There are 2 whole pizzas plus ? fourths.   3
   The mixed number $2\frac{3}{4}$ is read as "2 and $\frac{3}{4}$."

Here's how *to change a mixed number to a fraction.* $2\frac{3}{4} = ?$

Multiply the denominator by the whole number. (This gives the number of fourths in 2.)	Add the numerator. (This gives the number of fourths in $2\frac{3}{4}$.)
$2\frac{3}{4}$	$2\frac{3}{4} = \frac{11}{4}$ There are 11 fourths in $2\frac{3}{4}$.

5. Look at the pizzas. Does $2\frac{3}{4} = \frac{11}{4}$?   Yes

6. Check these examples. Are they correct?   Yes

   a. $3 = \frac{6}{2}$ b. $4 = \frac{12}{3}$ c. $1\frac{1}{2} = \frac{3}{2}$ d. $2\frac{2}{3} = \frac{8}{3}$

156 *Chapter 6*

EXERCISES

Change to thirds.

Here are scrambled answers for the next row of exercises: $\frac{21}{3}$ $\frac{27}{3}$ $\frac{30}{3}$ $\frac{3}{3}$ $\frac{9}{3}$ $\frac{6}{3}$ $\frac{42}{3}$

7. 2 $\frac{6}{3}$
8. 9 $\frac{27}{3}$
9. 3 $\frac{9}{3}$
10. 1 $\frac{3}{3}$
11. 7 $\frac{21}{3}$
12. 10 $\frac{30}{3}$
13. 14 $\frac{42}{3}$

14. 6 $\frac{18}{3}$
15. 12 $\frac{36}{3}$
16. 4 $\frac{12}{3}$
17. 11 $\frac{33}{3}$
18. 5 $\frac{15}{3}$
19. 8 $\frac{24}{3}$
20. 15 $\frac{45}{3}$

Change to fifths.

21. 5 $\frac{25}{5}$
22. 1 $\frac{5}{5}$
23. 6 $\frac{30}{5}$
24. 2 $\frac{10}{5}$
25. 10 $\frac{50}{5}$
26. 7 $\frac{35}{5}$
27. 13 $\frac{65}{5}$

28. 4 $\frac{20}{5}$
29. 12 $\frac{60}{5}$
30. 9 $\frac{45}{5}$
31. 8 $\frac{40}{5}$
32. 3 $\frac{15}{5}$
33. 11 $\frac{55}{5}$
34. 15 $\frac{75}{5}$

Change each mixed number to a fraction.

35. $1\frac{1}{3}$ $\frac{4}{3}$
36. $1\frac{1}{2}$ $\frac{3}{2}$
37. $2\frac{1}{2}$ $\frac{5}{2}$
38. $2\frac{1}{3}$ $\frac{7}{3}$
39. $1\frac{1}{4}$ $\frac{5}{4}$
40. $3\frac{1}{4}$ $\frac{13}{4}$
41. $7\frac{3}{4}$ $\frac{31}{4}$

42. $1\frac{2}{3}$ $\frac{5}{3}$
43. $2\frac{3}{4}$ $\frac{11}{4}$
44. $4\frac{1}{3}$ $\frac{13}{3}$
45. $3\frac{2}{3}$ $\frac{11}{3}$
46. $2\frac{2}{5}$ $\frac{12}{5}$
47. $3\frac{4}{5}$ $\frac{19}{5}$
48. $9\frac{1}{6}$ $\frac{55}{6}$

49. $4\frac{3}{4}$ $\frac{19}{4}$
50. $4\frac{1}{6}$ $\frac{25}{6}$
51. $5\frac{3}{4}$ $\frac{23}{4}$
52. $4\frac{3}{5}$ $\frac{23}{5}$
53. $5\frac{5}{6}$ $\frac{35}{6}$
54. $2\frac{3}{8}$ $\frac{19}{8}$
55. $11\frac{1}{8}$ $\frac{89}{8}$

56. $6\frac{3}{8}$ $\frac{51}{8}$
57. $3\frac{5}{8}$ $\frac{29}{8}$
58. $6\frac{3}{10}$ $\frac{63}{10}$
59. $8\frac{7}{8}$ $\frac{71}{8}$
60. $4\frac{9}{10}$ $\frac{49}{10}$
61. $5\frac{5}{8}$ $\frac{45}{8}$
62. $7\frac{4}{5}$ $\frac{39}{5}$

63. $6\frac{1}{2}$ $\frac{13}{2}$
64. $3\frac{5}{6}$ $\frac{23}{6}$
65. $4\frac{7}{8}$ $\frac{39}{8}$
66. $4\frac{2}{3}$ $\frac{14}{3}$
67. $5\frac{4}{5}$ $\frac{29}{5}$
68. $2\frac{3}{5}$ $\frac{13}{5}$
69. $9\frac{2}{3}$ $\frac{29}{3}$

70. $2\frac{7}{8}$ $\frac{23}{8}$
71. $2\frac{1}{6}$ $\frac{13}{6}$
72. $6\frac{2}{3}$ $\frac{20}{3}$
73. $7\frac{5}{6}$ $\frac{47}{6}$
74. $3\frac{3}{4}$ $\frac{15}{4}$
75. $9\frac{3}{8}$ $\frac{75}{8}$
76. $13\frac{1}{4}$ $\frac{53}{4}$

77. $10\frac{3}{5}$ $\frac{53}{5}$
78. $12\frac{1}{2}$ $\frac{25}{2}$
79. $11\frac{2}{3}$ $\frac{35}{3}$
80. $15\frac{1}{2}$ $\frac{31}{2}$
81. $12\frac{3}{4}$ $\frac{51}{4}$
82. $10\frac{3}{8}$ $\frac{83}{8}$
83. $11\frac{1}{4}$ $\frac{45}{4}$

Pizza puzzle Logical reasoning

84. Beth, Maria, and John ate one sausage and one bacon pizza. Study the clues to find what each person ate.

Clues:
- Each pizza was cut into fifths.
- Maria didn't eat bacon pizza.
- Maria ate 3 pieces.
- Beth didn't eat sausage pizza.
- Beth ate 1 more piece than John.

Hint: Draw a picture.

Beth—4 pieces of bacon pizza
Maria—3 pieces of sausage pizza
John—2 pieces of sausage and 1 piece of bacon

Number Theory, Fractions, and Decimals **157**

Extra Practice
Page 484 Skill 27

Practice Worksheet
Workbook S236, Copymaster S236, or Duplicating Master S236

WORKSHEET 68 (Use after page 157.)

HOW MANY POUNDS?

A pizza baked at the Oma Pizza Restaurant in Glen Falls, New York, held a world record as the largest pizza ever baked. What was the weight of this pizza?

To find the answer:
1. Write each mixed number as a fraction.
2. Write the letter under its matching number in the DECODER.

[Worksheet puzzle grid with exercises 1–34]

DECODER answer: EIGHTEEN THOUSAND SIX HUNDRED SIXTY-FOUR pounds

Project

Reading a yardstick

Measure your height and round it to the nearest inch. Then use a whole or mixed number to express your height in feet.

Copymaster S478

Class Starter Quiz 63
on previous lesson

Change each mixed number to a fraction.

1. $1\frac{1}{4}$ $\frac{5}{4}$ 2. $2\frac{1}{3}$ $\frac{7}{3}$ 3. $1\frac{2}{5}$ $\frac{7}{5}$
4. $2\frac{3}{4}$ $\frac{11}{4}$ 5. $3\frac{7}{8}$ $\frac{31}{8}$ 6. $6\frac{7}{10}$ $\frac{67}{10}$
7. $2\frac{5}{8}$ $\frac{21}{8}$ 8. $10\frac{3}{8}$ $\frac{83}{8}$ 9. $4\frac{2}{3}$ $\frac{14}{3}$

Copymaster S400

Lesson Objective
To write fractions as whole numbers or mixed numbers

Problem-Solving Skill
Finding information in an ad

Starting the Lesson
What are the facts? Have the students study the newspaper ad at the top of page 158 for 30 seconds. Then tell them to close their books and decide whether these statements are true or false:

- For each album you buy at the Stereo Connection, you get one fourth of a coupon. (True)
- If you have one half of a coupon, you can buy an album for half price. (False)
- When you buy four albums, you can use your coupon to get another album free. (True)

Here's How Note
Use of Concrete Materials You may wish to use the fraction pieces from copymasters or transparencies S530–S531 to demonstrate writing fractions as whole numbers or as mixed numbers. See ■ **Manipulative Activity 10** on copymaster S517 in the Teacher's Resource Binder.

The coupons at the top of page 158 are also on ■ **Visual Aid 28** (copymaster or transparency S133). You may wish to cut out and rearrange the coupon fourths when discussing the *Here's how* and exercises 1–5. Additional coupon fourths are provided. Use these to illustrate other examples.

158

Writing fractions as whole numbers or mixed numbers

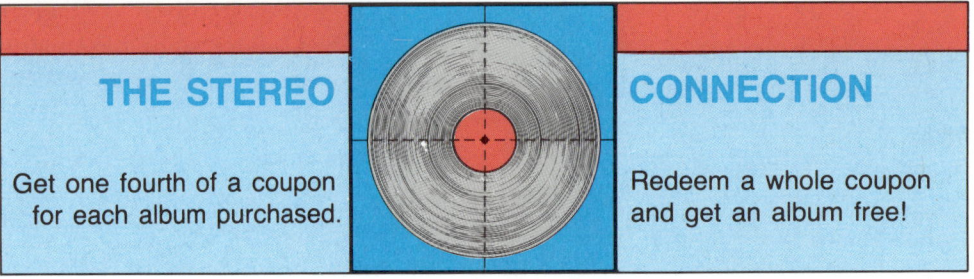

THE STEREO CONNECTION

Get one fourth of a coupon for each album purchased.

Redeem a whole coupon and get an album free!

1. How many fourths do you need to make a whole coupon? 4
2. If you have 8 fourths, how many whole coupons do you have? 2

Here's how to change $\frac{8}{4}$ to a whole number.

To change a fraction to a whole number, divide the numerator by the denominator.

number of fourths in one → $4\overline{)8}$ ← number of fourths in all

So $\frac{8}{4} = 2$

3. $\frac{20}{4}$ is equal to what whole number? 5

Here's how to change $\frac{11}{4}$ to a mixed number.

To change a fraction to a mixed number, divide the numerator by the denominator.

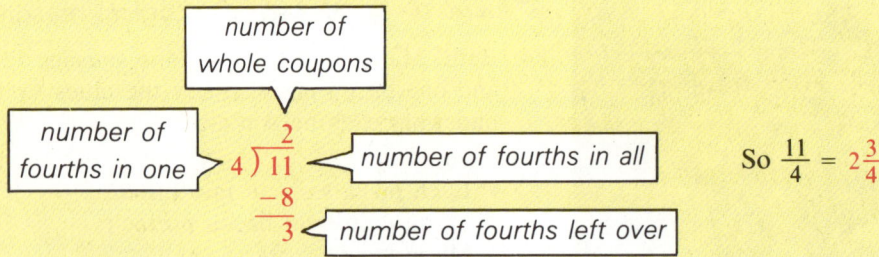

So $\frac{11}{4} = 2\frac{3}{4}$

4. $\frac{21}{4}$ is equal to what mixed number? $5\frac{1}{4}$

5. A fraction can be changed to a whole number or mixed number if the denominator is ___?___ the numerator. less than
   (less than/greater than)

158   Chapter 6

EXERCISES

Change each fraction to a whole number.
Here are scrambled answers for the next row of exercises: 5 4 2 1 8 6 3

6. $\frac{9}{3}$ 3
7. $\frac{10}{5}$ 2
8. $\frac{16}{2}$ 8
9. $\frac{6}{6}$ 1
10. $\frac{16}{4}$ 4
11. $\frac{18}{3}$ 6
12. $\frac{25}{5}$ 5

13. $\frac{15}{5}$ 3
14. $\frac{10}{2}$ 5
15. $\frac{3}{3}$ 1
16. $\frac{12}{4}$ 3
17. $\frac{16}{8}$ 2
18. $\frac{12}{6}$ 2
19. $\frac{24}{3}$ 8

20. $\frac{20}{10}$ 2
21. $\frac{24}{6}$ 4
22. $\frac{18}{2}$ 9
23. $\frac{24}{8}$ 3
24. $\frac{5}{5}$ 1
25. $\frac{12}{3}$ 4
26. $\frac{50}{25}$ 2

Change each fraction to a whole number or a mixed number.
Here are scrambled answers for the next row of exercises: 3 $1\frac{5}{6}$ $1\frac{1}{2}$ 2 $1\frac{1}{4}$ $2\frac{3}{5}$ 9

27. $\frac{3}{2}$ $1\frac{1}{2}$
28. $\frac{5}{4}$ $1\frac{1}{4}$
29. $\frac{9}{3}$ 3
30. $\frac{13}{5}$ $2\frac{3}{5}$
31. $\frac{11}{6}$ $1\frac{5}{6}$
32. $\frac{16}{8}$ 2
33. $\frac{81}{9}$ 9

34. $\frac{13}{10}$ $1\frac{3}{10}$
35. $\frac{16}{4}$ 4
36. $\frac{5}{2}$ $2\frac{1}{2}$
37. $\frac{7}{4}$ $1\frac{3}{4}$
38. $\frac{17}{8}$ $2\frac{1}{8}$
39. $\frac{10}{3}$ $3\frac{1}{3}$
40. $\frac{11}{9}$ $1\frac{2}{9}$

41. $\frac{19}{5}$ $3\frac{4}{5}$
42. $\frac{14}{3}$ $4\frac{2}{3}$
43. $\frac{15}{4}$ $3\frac{3}{4}$
44. $\frac{27}{10}$ $2\frac{7}{10}$
45. $\frac{11}{2}$ $5\frac{1}{2}$
46. $\frac{18}{3}$ 6
47. $\frac{27}{7}$ $3\frac{6}{7}$

48. $\frac{35}{5}$ 7
49. $\frac{25}{3}$ $8\frac{1}{3}$
50. $\frac{30}{6}$ 5
51. $\frac{35}{2}$ $17\frac{1}{2}$
52. $\frac{29}{6}$ $4\frac{5}{6}$
53. $\frac{36}{4}$ 9
54. $\frac{36}{6}$ 6

55. $\frac{20}{5}$ 4
56. $\frac{13}{6}$ $2\frac{1}{6}$
57. $\frac{37}{10}$ $3\frac{7}{10}$
58. $\frac{20}{3}$ $6\frac{2}{3}$
59. $\frac{19}{6}$ $3\frac{1}{6}$
60. $\frac{42}{5}$ $8\frac{2}{5}$
61. $\frac{21}{7}$ 3

62. $\frac{28}{3}$ $9\frac{1}{3}$
63. $\frac{14}{14}$ 1
64. $\frac{8}{5}$ $1\frac{3}{5}$
65. $\frac{50}{5}$ 10
66. $\frac{29}{3}$ $9\frac{2}{3}$
67. $\frac{15}{3}$ 5
68. $\frac{19}{8}$ $2\frac{3}{8}$

A stack of singles! Following directions

69. The Beatles are considered the most successful recording group. If you stacked the single records that they sold between 1963 and 1973 in one stack, it would be ? feet high. 3,493,590

To find ?, write a
- 0 in the ones place.
- 5 in the hundreds place.
- 4 in the hundred thousands place.
- 9 in both the tens place and the ten thousands place.
- 3 in both the millions place and the thousands place.

70. About how many miles high would the stack be? *Hint: There are 5280 feet in a mile.* About 662 miles

Extra Practice
Page 484 Skill 28

Practice Worksheet
Workbook S237, Copymasters S237, or Duplicating Masters S237

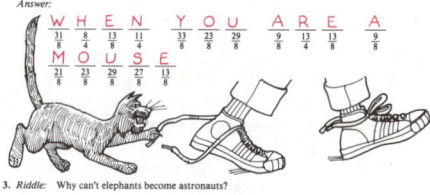

Project

Researching information
Find how mixed numbers are used in stock quotations and how to read stock quotations.

Copymaster S478

Number Theory, Fractions, and Decimals

Class Starter Quiz 64

on previous lesson

Change each fraction to a whole number or a mixed number.

1. $\frac{5}{2}$ $2\frac{1}{2}$ 2. $\frac{12}{3}$ 4 3. $\frac{15}{8}$ $1\frac{7}{8}$
4. $\frac{9}{4}$ $2\frac{1}{4}$ 5. $\frac{13}{9}$ $1\frac{4}{9}$ 6. $\frac{21}{7}$ 3
7. $\frac{27}{10}$ $2\frac{7}{10}$ 8. $\frac{43}{5}$ $8\frac{3}{5}$ 9. $\frac{20}{3}$ $6\frac{2}{3}$

Copymaster S400

Lesson Objective

To write fractions and mixed numbers in simplest form

Problem-Solving Skill

Using logical reasoning

Starting the Lesson

Write these fractions and mixed numbers on the chalkboard:

$\frac{7}{10}$	$4\frac{1}{2}$	$1\frac{1}{3}$
$\frac{1}{2}$	$6\frac{2}{5}$	$4\frac{5}{6}$
$\frac{5}{8}$	$2\frac{4}{8}$	$7\frac{1}{3}$
$\frac{3}{12}$	$8\frac{3}{4}$	$\frac{10}{7}$

Ask the students which fraction or mixed number does not belong in each column of numbers. (The numbers $\frac{3}{12}$, $2\frac{4}{8}$, and $\frac{10}{7}$ do not belong, since they are not written in simplest form.) Then discuss exercises 1–6 to acquaint the students with the rules for writing fractions and mixed numbers in simplest form.

160

Writing fractions and mixed numbers in simplest form

1. Look at the yellow cards. Which two fractions have numerators less than their denominators? $\frac{6}{8}, \frac{8}{12}$ Are these fractions less than or greater than 1? Less than

Here's how to write $\frac{6}{8}$ in simplest form.

Write fractions less than 1 in lowest terms.

$\frac{6}{8} = \frac{3}{4}$ ← simplest form

2. How would you write $\frac{8}{12}$ in simplest form? $\frac{2}{3}$

3. What two mixed numbers are written on the cards? $2\frac{2}{4}, 3\frac{4}{6}$

Here's how to write $2\frac{2}{4}$ in simplest form.

Write mixed numbers with the fraction part less than 1 *and* in lowest terms.

$2\frac{2}{4} = 2\frac{1}{2}$ ← simplest form

4. How would you write $3\frac{4}{6}$ in simplest form? $3\frac{2}{3}$

5. Look at the yellow cards. Which fractions are greater than or equal to 1? $\frac{6}{6}, \frac{18}{6}, \frac{14}{3}, \frac{20}{5}, \frac{18}{4}$

Here's how to write $\frac{20}{5}$ and $\frac{14}{3}$ in simplest form.

Write fractions that are greater than or equal to 1 as a whole number or as a mixed number in simplest form.

$\frac{20}{5} = 4$ ← simplest form $\frac{14}{3} = 4\frac{2}{3}$ ← simplest form

6. Write each fraction in simplest form.

  a. $\frac{18}{6}$ 3 b. $\frac{18}{4}$ $4\frac{1}{2}$ c. $\frac{6}{6}$ 1

160 Chapter 6

EXERCISES

Write in simplest form.

Here are scrambled answers for the next row of exercises: $\frac{3}{5}$ $\frac{1}{2}$ $\frac{3}{4}$ $\frac{5}{6}$ $\frac{1}{6}$ $\frac{2}{3}$ $\frac{4}{5}$

7. $\frac{6}{10}$ $\frac{3}{5}$
8. $\frac{2}{12}$ $\frac{1}{6}$
9. $\frac{5}{10}$ $\frac{1}{2}$
10. $\frac{6}{8}$ $\frac{3}{4}$
11. $\frac{4}{6}$ $\frac{2}{3}$
12. $\frac{15}{18}$ $\frac{5}{6}$
13. $\frac{20}{25}$ $\frac{4}{5}$

14. $\frac{6}{18}$ $\frac{1}{3}$
15. $\frac{8}{14}$ $\frac{4}{7}$
16. $\frac{10}{12}$ $\frac{5}{6}$
17. $\frac{5}{20}$ $\frac{1}{4}$
18. $\frac{16}{24}$ $\frac{2}{3}$
19. $\frac{14}{16}$ $\frac{7}{8}$
20. $\frac{6}{9}$ $\frac{2}{3}$

Write in simplest form.

21. $2\frac{2}{4}$ $2\frac{1}{2}$
22. $4\frac{2}{8}$ $4\frac{1}{4}$
23. $3\frac{4}{6}$ $3\frac{2}{3}$
24. $5\frac{2}{6}$ $5\frac{1}{3}$
25. $4\frac{6}{8}$ $4\frac{3}{4}$
26. $6\frac{10}{12}$ $6\frac{5}{6}$
27. $7\frac{4}{16}$ $7\frac{1}{4}$

28. $4\frac{3}{12}$ $4\frac{1}{4}$
29. $8\frac{5}{10}$ $8\frac{1}{2}$
30. $3\frac{3}{9}$ $3\frac{1}{3}$
31. $12\frac{10}{15}$ $12\frac{2}{3}$
32. $5\frac{9}{12}$ $5\frac{3}{4}$
33. $10\frac{8}{10}$ $10\frac{4}{5}$
34. $1\frac{9}{15}$ $1\frac{3}{5}$

35. $6\frac{4}{8}$ $6\frac{1}{2}$
36. $3\frac{8}{24}$ $3\frac{1}{3}$
37. $2\frac{15}{30}$ $2\frac{1}{2}$
38. $5\frac{10}{18}$ $5\frac{5}{9}$
39. $4\frac{5}{15}$ $4\frac{1}{3}$
40. $1\frac{7}{14}$ $1\frac{1}{2}$
41. $10\frac{8}{64}$ $10\frac{1}{8}$

Write in simplest form.

42. $\frac{6}{3}$ 2
43. $\frac{10}{2}$ 5
44. $\frac{12}{4}$ 3
45. $\frac{20}{4}$ 5
46. $\frac{36}{3}$ 12
47. $\frac{24}{8}$ 3
48. $\frac{14}{7}$ 2

49. $\frac{9}{2}$ $4\frac{1}{2}$
50. $\frac{8}{3}$ $2\frac{2}{3}$
51. $\frac{7}{4}$ $1\frac{3}{4}$
52. $\frac{9}{5}$ $1\frac{4}{5}$
53. $\frac{10}{3}$ $3\frac{1}{3}$
54. $\frac{11}{4}$ $2\frac{3}{4}$
55. $\frac{6}{1}$ 6

56. $\frac{17}{3}$ $5\frac{2}{3}$
57. $\frac{15}{4}$ $3\frac{3}{4}$
58. $\frac{12}{8}$ $1\frac{1}{2}$
59. $\frac{15}{10}$ $1\frac{1}{2}$
60. $\frac{16}{12}$ $1\frac{1}{3}$
61. $\frac{18}{12}$ $1\frac{1}{2}$
62. $\frac{34}{11}$ $3\frac{1}{11}$

63. $\frac{8}{10}$ $\frac{4}{5}$
64. $\frac{10}{8}$ $1\frac{1}{4}$
65. $\frac{16}{3}$ $5\frac{1}{3}$
66. $\frac{3}{6}$ $\frac{1}{2}$
67. $\frac{9}{8}$ $1\frac{1}{8}$
68. $\frac{16}{14}$ $1\frac{1}{7}$
69. $\frac{35}{30}$ $1\frac{1}{6}$

70. $\frac{33}{6}$ $5\frac{1}{2}$
71. $\frac{36}{5}$ $7\frac{1}{5}$
72. $\frac{24}{36}$ $\frac{2}{3}$
73. $\frac{18}{5}$ $3\frac{3}{5}$
74. $\frac{22}{16}$ $1\frac{3}{8}$
75. $\frac{28}{6}$ $4\frac{2}{3}$
76. $\frac{31}{7}$ $4\frac{3}{7}$

77. $\frac{8}{12}$ $\frac{2}{3}$
78. $\frac{6}{24}$ $\frac{1}{4}$
79. $\frac{8}{1}$ 8
80. $\frac{10}{25}$ $\frac{2}{5}$
81. $\frac{25}{10}$ $2\frac{1}{2}$
82. $\frac{18}{36}$ $\frac{1}{2}$
83. $\frac{35}{15}$ $2\frac{1}{3}$

84. $\frac{42}{6}$ 7
85. $\frac{6}{42}$ $\frac{1}{7}$
86. $\frac{25}{8}$ $3\frac{1}{8}$
87. $\frac{16}{18}$ $\frac{8}{9}$
88. $\frac{24}{32}$ $\frac{3}{4}$
89. $\frac{32}{24}$ $1\frac{1}{3}$
90. $\frac{13}{11}$ $1\frac{2}{11}$

Face fact Logical reasoning

91. Study the clues to find how many muscles you use to smile and how many muscles you use to frown.

Clues: 43 muscles to frown
17 muscles to smile
- It takes more muscles to frown.
- If you add the numbers, you get 60.
- If you subtract the numbers, you get 26.

Number Theory, Fractions, and Decimals

Extra Practice
Page 485 Skill 29

Practice Worksheet
Workbook S238, Copymaster S238, or Duplicating Master S238

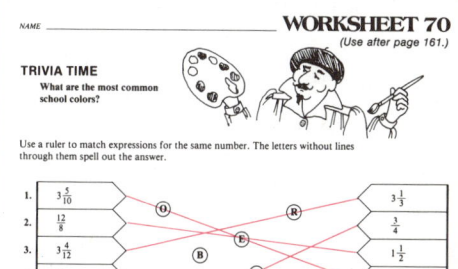

Challenge Problems
Use a calculator to write each of these fractions in lowest terms:

1. $\frac{1920}{2560}$ $\frac{3}{4}$

2. $\frac{9702}{14{,}553}$ $\frac{2}{3}$

Copymaster S448

Class Starter Quiz 65
on previous lesson

Write in simplest form.

1. $\dfrac{4}{10}$ $\dfrac{2}{5}$ 2. $\dfrac{16}{20}$ $\dfrac{4}{5}$ 3. $\dfrac{15}{18}$ $\dfrac{5}{6}$

4. $2\dfrac{3}{6}$ $2\dfrac{1}{2}$ 5. $8\dfrac{5}{10}$ $8\dfrac{1}{2}$ 6. $4\dfrac{8}{12}$ $4\dfrac{2}{3}$

7. $\dfrac{11}{2}$ $5\dfrac{1}{2}$ 8. $\dfrac{25}{8}$ $3\dfrac{1}{8}$ 9. $\dfrac{20}{5}$ 4

Copymaster S401

Lesson Objective
To write quotients as mixed numbers

Problem-Solving Skills
Finding information in a display
Choosing the correct operation
Using logical reasoning

Starting the Lesson
Write these division exercises on the chalkboard:

3)$\overline{42}$ 7)$\overline{84}$ 4)$\overline{51}$ 5)$\overline{75}$

Ask the students which quotient does not belong in the group of division exercises. (4)$\overline{51}$ does not belong, since its quotient has a remainder.) Then use exercises 1–5 and the *Here's how* to show the students how to write quotients as mixed numbers.

Writing quotients as mixed numbers

1. How many photos can be mounted on one page of the album? 6

2. To find how many pages would be needed to mount 100 photos, you would divide 100 by what number? 6

Here's how to write a quotient as a mixed number. $100 \div 6 = ?$

Step 1. Divide.

```
      16   ← number of
6)100        full pages
   -6
   40
  -36   ← number of
    4     photos
          remaining
```

Step 2. Write the quotient as a mixed number.

```
   16 4/6   ← Write the
6)100       remainder
   -6       over the
   40       divisor.
  -36
    4
```

Step 3. Write the mixed number in simplest form.

```
   16 4/6 = 16 2/3
6)100
   -6
   40
  -36
    4
```

3. Look at the first step in the *Here's how*. After 16 pages of the album were filled, how many photos would be left to be mounted? 4

4. Look at the last step of the *Here's how*. It would take 16 and ⬜ pages to mount all the photos. $\dfrac{2}{3}$

5. Check these examples. Are they correct? Yes

a.
```
     24 3/5
  5)123
    -10
     23
    -20
      3
```

b.
```
     17 6/8 = 17 3/4
  8)142
    -8
    62
   -56
     6
```

c.
```
     16 7/18
 18)295
    -18
    115
   -108
      7
```

d.
```
     25 7/21 = 25 1/3
 21)532
    -42
    112
   -105
      7
```

EXERCISES

Divide. Write each quotient as a mixed number in simplest form.

Here are scrambled answers for the next row of exercises: $14\frac{2}{3}$ $5\frac{1}{3}$ $8\frac{1}{3}$ $4\frac{1}{2}$ $8\frac{2}{5}$

6. $4\overline{)18}$ — $4\frac{1}{2}$
7. $3\overline{)44}$ — $14\frac{2}{3}$
8. $6\overline{)50}$ — $8\frac{1}{3}$
9. $5\overline{)42}$ — $8\frac{2}{5}$
10. $9\overline{)48}$ — $5\frac{1}{3}$

11. $7\overline{)45}$ — $6\frac{3}{7}$
12. $8\overline{)75}$ — $9\frac{3}{8}$
13. $6\overline{)74}$ — $12\frac{1}{3}$
14. $4\overline{)90}$ — $22\frac{1}{2}$
15. $9\overline{)78}$ — $8\frac{2}{3}$

16. $5\overline{)162}$ — $32\frac{2}{5}$
17. $7\overline{)253}$ — $36\frac{1}{7}$
18. $8\overline{)153}$ — $19\frac{1}{8}$
19. $9\overline{)124}$ — $13\frac{7}{9}$
20. $3\overline{)124}$ — $41\frac{1}{3}$

21. $12\overline{)283}$ — $23\frac{7}{12}$
22. $15\overline{)406}$ — $27\frac{1}{15}$
23. $21\overline{)592}$ — $28\frac{4}{21}$
24. $25\overline{)685}$ — $27\frac{2}{5}$
25. $18\overline{)665}$ — $36\frac{17}{18}$

26. $24\overline{)862}$ — $35\frac{11}{12}$
27. $28\overline{)906}$ — $32\frac{5}{14}$
28. $32\overline{)900}$ — $28\frac{1}{8}$
29. $30\overline{)820}$ — $27\frac{1}{3}$
30. $36\overline{)912}$ — $25\frac{1}{3}$

31. $44\overline{)2688}$ — $61\frac{1}{11}$
32. $42\overline{)2324}$ — $55\frac{1}{3}$
33. $48\overline{)3996}$ — $83\frac{1}{4}$
34. $60\overline{)2565}$ — $42\frac{3}{4}$
35. $40\overline{)2420}$ — $60\frac{1}{2}$

36. $1450 \div 30$ — $48\frac{1}{3}$
37. $1480 \div 16$ — $92\frac{1}{2}$
38. $1850 \div 26$ — $71\frac{2}{13}$
39. $2187 \div 36$ — $60\frac{3}{4}$

40. $3136 \div 42$ — $74\frac{2}{3}$
41. $2440 \div 60$ — $40\frac{2}{3}$
42. $2691 \div 72$ — $37\frac{3}{8}$
43. $3570 \div 64$ — $55\frac{25}{32}$

44. $3475 \div 50$ — $69\frac{1}{2}$
45. $3396 \div 48$ — $70\frac{3}{4}$
46. $2282 \div 84$ — $27\frac{1}{6}$
47. $3660 \div 80$ — $45\frac{3}{4}$

48. $4526 \div 30$ — $150\frac{13}{15}$
49. $4586 \div 15$ — $305\frac{11}{15}$
50. $8452 \div 12$ — $704\frac{1}{3}$
51. $6881 \div 17$ — $404\frac{13}{17}$

52. $3146 \div 27$ — $116\frac{14}{27}$
53. $9107 \div 23$ — $395\frac{22}{23}$
54. $3106 \div 94$ — $33\frac{2}{47}$
55. $3524 \div 15$ — $234\frac{14}{15}$

Solve.

56. You bought 2 rolls of film for $2.48 a roll. How much did you spend for the film? **$4.96**

57. You had 2 rolls of 24 developed for $6.79 a roll and 1 roll of 36 developed for $8.11. What was the total cost? **$21.69**

58. You can take 36 pictures on a large roll of film. If a large roll costs $3.69, how much does the film cost for each picture? Round the answer to the nearest cent. **$.10**

59. You have 68 photos to put in an album. If you put 6 photos on each page, how many pages will you need? Give the answer as a mixed number in simplest form. **$11\frac{1}{3}$**

Photo count!

60. Study these clues to find how many pictures are in the pile. **42**

    Clues:
    - There are fewer than 50.
    - If you put 8 on a page, you will have 2 left over.
    - If you put 9 on a page, you will have 6 left over.

Number Theory, Fractions, and Decimals

Class Starter Quiz 66
on previous lesson

Divide. Write each quotient as a mixed number in simplest form.

1. $44 \div 5$ $8\frac{4}{5}$ 2. $50 \div 8$ $6\frac{1}{4}$
3. $65 \div 7$ $9\frac{2}{7}$ 4. $90 \div 4$ $22\frac{1}{2}$
5. $283 \div 10$ $28\frac{3}{10}$ 6. $144 \div 50$ $2\frac{22}{25}$

Copymaster S401

Problem-Solving Skills
Reading a map
Choosing the correct operation
Solving a multistep problem

Skills Reviewed
Multiplying and dividing decimals
Finding the median of a set of numbers
Writing numbers with exponents as standard numerals and vice versa
Finding the prime factorization of a number
Finding the GCF and the LCM of pairs of numbers
Multiplying by 10, 100, or 1000

Starting the Lesson
Problem Solving The route marked on the map on page 164 is on ■ **Visual Aid 29** (copymaster or transparency S134). Use the map and ask questions like these:

- How many miles is it from Los Angeles to San Bernardino? (65)
- What is the driving time from Santa Barbara to Los Angeles? (1 h 56 min)
- Is 3 hours enough driving time to travel from Barstow to Death Valley? (No)

Cumulative Skill Practice Challenge the students to an estimation hunt by saying, "Pick the largest product in the first row of exercises." (Exercise 3) Then have the students pick the largest product in each of the next three rows of exercises. (Exercises 4, 7, and 10)

164

Problem solving

The red numbers show the miles. The blue numbers show the driving time in hours and minutes. You decide to take the trip shown in yellow.

Solve.
Decide when a calculator would be useful.

1. How many miles is it from Los Angeles to Barstow? 138

2. What is the driving time from Barstow to Death Valley? 3 hours 38 minutes

3. You leave Death Valley at 8:00 A.M. If you plan two 15-minute rest stops and an hour for lunch, at what time should you arrive in Yosemite? 4:18 P.M.

4. Your fuel tank will hold 14.7 gallons. In San Francisco it takes 10.8 gallons to fill the tank. How many gallons were in the tank when you reached San Francisco? 3.9

5. If the fuel costs $1.32 per gallon, how much should you have paid for 10.8 gallons? $14.26

6. If you want to arrive at Monterey by noon, by what time should you leave San Francisco? 9:19 A.M.

7. On your way to Sequoia National Park, you fill the fuel tank. If the tank holds 14.7 gallons and your car gets 28.5 miles per gallon, how far can you drive before running out of fuel? 418.95 miles

8. You leave Sequoia and average 52.5 miles per hour for the first 2.5 hours. How far do you travel during this time? 131.25 miles

9. You plan to leave Sequoia and drive no more than 6 hours today. Can you reach Los Angeles today? No

10. Suppose that it takes 16.5¢ per mile to operate your car. What would be the total car expense for your trip? $268.29

164 Chapter 6

DRIVING DISTANCES

Cumulative Skill Practice

Give the product. (page 68)

1. 2.74 × 7 19.18
2. 3.82 × 6 22.92
3. 7.23 × 9 65.07
4. 6.84 × 0.6 4.104
5. 31.5 × 0.004 0.126
6. 0.98 × 0.09 0.0882
7. 24.96 × 12 299.52
8. 35.8 × 2.4 85.92
9. 2.694 × 5.1 13.7394
10. 5.07 × 38 192.66
11. 3.81 × 5.31 20.2311
12. 0.064 × 4.5 0.288

Give the product. (page 72)

13. 125 × 100 12,500
14. 74 × 10 740
15. 52 × 1000 52,000
16. 0.563 × 100 56.3
17. 0.563 × 10 5.63
18. 0.563 × 1000 563
19. 0.64 × 100 64
20. 0.64 × 10 6.4
21. 0.64 × 1000 640
22. 12.87 × 100 1287
23. 12.87 × 10 128.7
24. 12.87 × 1000 12,870

Give the quotient rounded to the nearest tenth. (page 104)

25. 9.4 ÷ 0.6 15.7
26. 8.3 ÷ 0.3 27.7
27. 2.73 ÷ 0.7 3.9
28. 5.64 ÷ 0.5 11.3
29. 8.422 ÷ 0.04 210.6
30. 5.75 ÷ 0.07 82.1
31. 42 ÷ 1.1 38.2
32. 6.38 ÷ 2.4 2.7
33. 74.26 ÷ 0.36 206.3
34. 96.32 ÷ 9.3 10.4
35. 6.389 ÷ 0.56 11.4
36. 9.62 ÷ 0.45 21.4

Find the median. (page 126)

37. 93, 97, 58, 83, 86 86
38. 59, 63, 57, 55, 61 59
39. 39, 40, 40, 37, 35 39
40. 28, 27, 26, 30 27.5
41. 92, 96, 90, 89 91
42. 215, 217, 218, 216 216.5
43. 74, 70, 68, 75, 76 74
44. 23, 27, 24, 18 23.5
45. 53, 51, 47, 59 52
46. 308, 306, 312, 304 307
47. 66, 65, 68, 72 67
48. 71, 71, 75, 78, 73 73

MIXED PRACTICE

Complete.

49. The standard numeral for 2^3 is ?   8
50. 5 × 5 written using an exponent is ? 5^2
51. The prime factorization of 12 is ? $2 × 2 × 3$ or $2^2 × 3$
52. The greatest common factor of 9 and 12 is ?   3
53. The least common multiple of 10 and 15 is ?   30

Number Theory, Fractions, and Decimals **165**

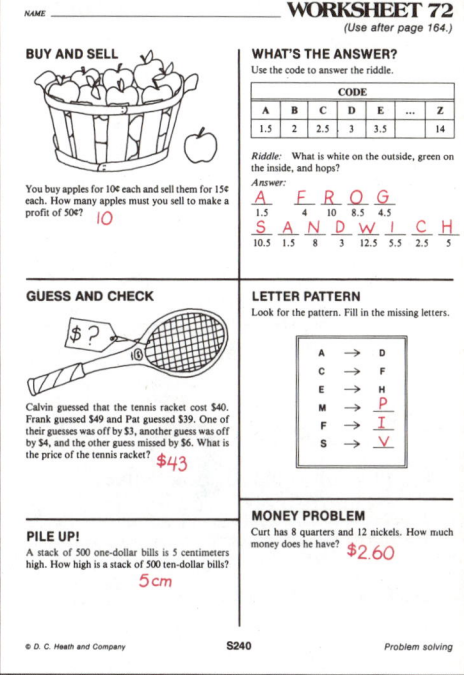

Problem-Solving Worksheet
Workbook S240, Copymaster S240, or Duplicating Master S240

Project

Reading a map

Assume you can average 50 miles per hour on interstate highways, 45 miles per hour on numbered highways, and 35 miles per hour on other roads. Use a state road map and choose two cities. Then use the distance information on the map to compute the shortest driving time between the two cities.

Copymaster S478

Class Starter Quiz 67
on previous lesson

Solve. Use the map on page 164.

1. How many miles is it from San Francisco to Monterey? 125
2. If you want to arrive at San Luis Obispo at 10 P.M., by what time should you leave Santa Barbara? 7:58 P.M.

Copymaster S401

Lesson Objective
To write fractions and mixed numbers as decimals

Problem-Solving Skills
Finding information in a recipe
Choosing the correct operation
Looking for patterns

Starting the Lesson
Before the students open their books, challenge them to name the meal. Read the recipe on page 166 and have them write what type of meal they think the recipe makes. Then have them open their books to page 166 and check their answers.

Exercise Note
Calculator You may want to show the students how to use the calculator to change a mixed number to a decimal.

Example: Change $12\frac{5}{16}$ to a decimal.

Step 1. Divide the numerator by the denominator.

⑤ ÷ ⑯ = 0.3125

Step 2. Add the whole number part.

⊕ 12 = 12.3125

166

Writing fractions and mixed numbers as decimals

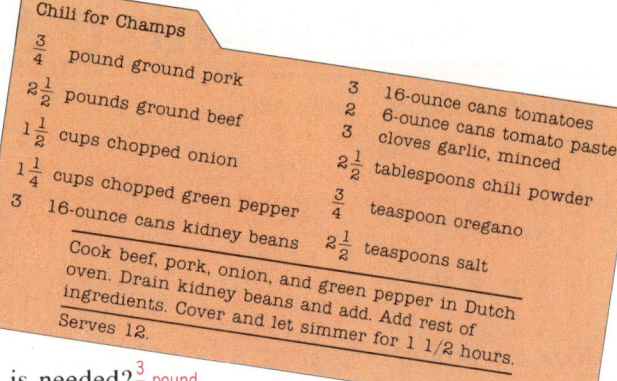

1. Look at the recipe. How much ground pork is needed? $\frac{3}{4}$ pound

2. How much ground pork is in the package? To decide whether the package contains the proper amount of ground pork for the recipe, you can change $\frac{3}{4}$ to a decimal. 0.75 pound

Here's how to change $\frac{3}{4}$ to a decimal.

To change a fraction to a decimal, divide the numerator by the denominator.

```
    0.75
4 ) 3.00
   -28
    ‾‾
    20
   -20
    ‾‾
     0
```
So $\frac{3}{4} = 0.75$

3. Look at the *Here's how*. Is $\frac{3}{4}$ pound the same as 0.75 pound? Yes

4. Does the package contain the proper amount of ground pork? Yes

5. Look at the recipe. How much ground beef is needed? $2\frac{1}{2}$ pounds

Study this example.

```
     0.666
3 ) 2.000
   -18
    ‾‾
    20
   -18
    ‾‾
    20
   -18
    ‾‾
     2
```

If the division does not end, round the quotient.

So $\frac{2}{3} \approx 0.67$

Read ≈ as "is approximately equal to."

6. To what place was the decimal rounded? Hundredths

166 Chapter 6

EXERCISES

Change each fraction to a decimal. Here are scrambled answers for the next row of exercises: 0.2 0.25 0.4 0.75 0.3 0.625 0.375

7. $\frac{1}{5}$ 0.2
8. $\frac{3}{10}$ 0.3
9. $\frac{1}{4}$ 0.25
10. $\frac{3}{4}$ 0.75
11. $\frac{2}{5}$ 0.4
12. $\frac{5}{8}$ 0.625
13. $\frac{3}{8}$ 0.375

14. $\frac{1}{16}$ 0.0625
15. $\frac{3}{16}$ 0.1875
16. $\frac{9}{20}$ 0.45
17. $\frac{7}{16}$ 0.4375
18. $\frac{5}{16}$ 0.3125
19. $\frac{12}{25}$ 0.48
20. $\frac{15}{5}$ 3

21. $\frac{9}{8}$ 1.125
22. $\frac{12}{5}$ 2.4
23. $\frac{15}{4}$ 3.75
24. $\frac{25}{8}$ 3.125
25. $\frac{9}{16}$ 0.5625
26. $\frac{13}{10}$ 1.3
27. $\frac{3}{20}$ 0.15

Change each mixed number to a decimal.

28. $2\frac{1}{2}$ $2\frac{1}{2} = 2 + \frac{1}{2}$; $\frac{1}{2} = 0.5$; $2\frac{1}{2} = 2.5$
29. $2\frac{3}{8}$ 2.375
30. $3\frac{1}{2}$ 3.5
31. $1\frac{1}{8}$ 1.125
32. $4\frac{3}{5}$ 4.6
33. $9\frac{1}{4}$ 9.25

34. $2\frac{1}{4}$ 2.25
35. $3\frac{4}{5}$ 3.8
36. $4\frac{2}{5}$ 4.4
37. $2\frac{5}{8}$ 2.625
38. $3\frac{3}{16}$ 3.1875

39. $7\frac{7}{8}$ 7.875
40. $5\frac{3}{8}$ 5.375
41. $4\frac{3}{4}$ 4.75
42. $3\frac{3}{8}$ 3.375
43. $6\frac{1}{5}$ 6.2
44. $4\frac{3}{10}$ 4.3
45. $5\frac{5}{8}$ 5.625

Change to a decimal rounded to the nearest hundredth.

46. $\frac{1}{3}$ 0.33
47. $\frac{1}{6}$ 0.17
48. $\frac{2}{3}$ 0.67
49. $\frac{1}{9}$ 0.11
50. $\frac{5}{6}$ 0.83
51. $\frac{5}{9}$ 0.56
52. $\frac{3}{14}$ 0.21

53. $\frac{5}{7}$ 0.71
54. $\frac{13}{3}$ 4.33
55. $\frac{1}{12}$ 0.08
56. $\frac{11}{6}$ 1.83
57. $\frac{17}{12}$ 1.42
58. $\frac{15}{6}$ 2.50
59. $\frac{7}{22}$ 0.32

60. $\frac{4}{9}$ 0.44
61. $\frac{5}{3}$ 1.67
62. $\frac{13}{6}$ 2.17
63. $\frac{5}{12}$ 0.42
64. $\frac{20}{3}$ 6.67
65. $\frac{17}{6}$ 2.83
66. $\frac{4}{7}$ 0.57

Solve. Use the recipe on page 166.

67. Ground beef costs $1.50 a pound. How much will the ground beef cost for the recipe? $3.75

68. A 6-ounce can of tomato paste costs $.48 and a 16-ounce can of tomatoes costs $.59. What will be the total cost of these ingredients for the recipe? $2.73

Can you spot it?

Divide. Find the fraction that does not belong in each group.

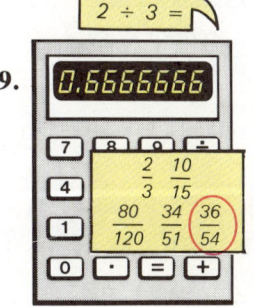

69. 0.6666666 ; $2 \div 3 =$; $\frac{2}{3}$ $\frac{10}{15}$ $\frac{80}{120}$ $\frac{34}{51}$ ⊙$\frac{36}{54}$

70. 0.64 ; $\frac{64}{100}$ $\frac{96}{150}$ $\frac{144}{225}$ ⊙$\frac{9}{25}$ $\frac{224}{350}$

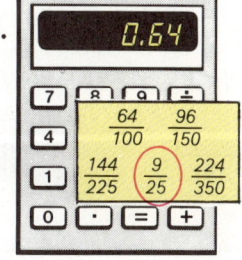

71. 1.2222222 ; $\frac{11}{9}$ $\frac{44}{36}$ ⊙$\frac{77}{54}$ $\frac{132}{108}$ $\frac{286}{234}$

Extra Practice
Page 485 Skill 30

Practice Worksheet
Workbook S241, Copymaster S241, or Duplicating Master S241

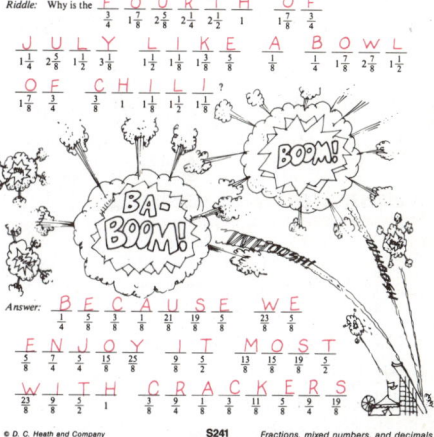

Challenge Problems

1. Find the decimal name for each of these fractions.

   $\frac{1}{9} = ?$ 0.111… $\frac{2}{9} = ?$ 0.222…

   $\frac{3}{9} = ?$ 0.333… $\frac{4}{9} = ?$ 0.444…

2. Look for the pattern. What is the decimal name for each of these fractions?

   $\frac{5}{9} = ?$ 0.555… $\frac{6}{9} = ?$ 0.666…

   $\frac{7}{9} = ?$ 0.777… $\frac{8}{9} = ?$ 0.888…

Copymaster S449

Number Theory, Fractions, and Decimals

Class Starter Quiz 68
on previous lesson

Change to a decimal.

1. $\frac{1}{4}$ 0.25 **2.** $\frac{1}{2}$ 0.5 **3.** $\frac{5}{4}$ 1.25

4. $\frac{9}{20}$ 0.45 **5.** $\frac{1}{8}$ 0.125 **6.** $\frac{13}{10}$ 1.3

7. $5\frac{1}{2}$ 5.5 **8.** $2\frac{3}{4}$ 2.75 **9.** $5\frac{3}{8}$ 5.375

Copymaster S401

Lesson Objective
To write decimals as fractions or mixed numbers

Problem-Solving Skill
Using a guess-and-check strategy

Starting the Lesson
Use of Concrete Materials Prepare a copy of ■ *Visual Aid 30* (dot grid, copymaster or transparency S135) or draw 10 × 10 grids on the chalkboard. Have the students use a decimal and a fraction in simplest form to tell what part of each square is shaded.

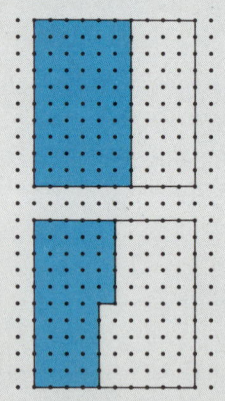

Record the answers on the chalkboard.

0.6 $\frac{6}{10} = \frac{3}{5}$

0.45 $\frac{45}{100} = \frac{9}{20}$

Then go over the exercises and the *Here's how* examples on page 168.

168

Writing decimals as fractions or mixed numbers

1. What decimal is shown on this calculator? 0.75

2. Is the decimal less than or greater than 1? Less than

Here's how to change a decimal (less than 1) to a fraction in simplest form.   0.75 = ?

Read the decimal.	Write as a fraction.	Write in simplest form.
0.75	$0.75 = \frac{75}{100}$	$0.75 = \frac{75}{100}$
75 hundredths		$= \frac{3}{4}$

3. What decimal is shown on this calculator? 2.6

4. When 2.6 is changed to a mixed number, what should the whole-number part be? 2

Here's how to change a decimal (greater than 1) to a mixed number in simplest form.   2.6 = ?

Read the decimal.	Write as a mixed number.	Write in simplest form.
2.6	$2.6 = 2\frac{6}{10}$	$2.6 = 2\frac{6}{10}$
2 and 6 tenths		$= 2\frac{3}{5}$

168 *Chapter 6*

EXERCISES

Change to a fraction in simplest form. Here are scrambled answers for the next row of exercises: $\frac{3}{4}$ $\frac{1}{4}$ $\frac{1}{8}$ $\frac{3}{5}$ $\frac{1}{2}$

5. 0.6 $\frac{3}{5}$
6. 0.25 $\frac{1}{4}$
7. 0.5 $\frac{1}{2}$
8. 0.75 $\frac{3}{4}$
9. 0.125 $\frac{1}{8}$
10. 0.8 $\frac{4}{5}$
11. 0.24 $\frac{6}{25}$
12. 0.48 $\frac{12}{25}$
13. 0.9 $\frac{9}{10}$
14. 0.150 $\frac{3}{20}$
15. 0.35 $\frac{7}{20}$
16. 0.375 $\frac{3}{8}$
17. 0.72 $\frac{18}{25}$
18. 0.4 $\frac{2}{5}$
19. 0.16 $\frac{4}{25}$
20. 0.36 $\frac{9}{25}$
21. 0.65 $\frac{13}{20}$
22. 0.875 $\frac{7}{8}$
23. 0.45 $\frac{9}{20}$
24. 0.05 $\frac{1}{20}$

Change to a mixed number in simplest form.

25. 2.25 $2\frac{1}{4}$
26. 1.4 $1\frac{2}{5}$
27. 2.400 $2\frac{2}{5}$
28. 5.5 $5\frac{1}{2}$
29. 9.35 $9\frac{7}{20}$
30. 7.8 $7\frac{4}{5}$
31. 3.75 $3\frac{3}{4}$
32. 6.08 $6\frac{2}{25}$
33. 12.375 $12\frac{3}{8}$
34. 4.04 $4\frac{1}{25}$
35. 6.28 $6\frac{7}{25}$
36. 8.44 $8\frac{11}{25}$
37. 6.85 $6\frac{17}{20}$
38. 4.50 $4\frac{1}{2}$
39. 3.6 $3\frac{3}{5}$
40. 8.52 $8\frac{13}{25}$
41. 3.875 $3\frac{7}{8}$
42. 10.350 $10\frac{7}{20}$
43. 6.15 $6\frac{3}{20}$
44. 5.625 $5\frac{5}{8}$

<, =, or >?

45. $\frac{1}{4}$ ● 0.2   **>**
46. 0.3 ● $\frac{1}{4}$ **>**
47. 0.1 ● $\frac{1}{10}$ **=**
48. $\frac{2}{3}$ ● 0.3   **>**
49. $\frac{1}{5}$ ● 0.25   **<**
50. $\frac{2}{5}$ ● 0.4   **=**
51. $\frac{1}{2}$ ● 0.6   **<**
52. $1\frac{3}{8}$ ● 1.38   **<**
53. 0.375 ● $\frac{2}{5}$ **<**
54. $\frac{3}{5}$ ● 0.625   **<**
55. $\frac{3}{4}$ ● 0.80   **<**
56. 0.62 ● $\frac{31}{50}$ **=**
57. 1.5 ● $1\frac{1}{2}$ **=**
58. $2\frac{3}{4}$ ● 2.7   **>**
59. $1\frac{7}{8}$ ● 1.85   **>**
60. $\frac{3}{4}$ ● 0.075   **>**
61. 3.4 ● $3\frac{3}{8}$ **>**
62. 2.08 ● $2\frac{1}{10}$ **<**
63. 3.625 ● $3\frac{5}{8}$ **=**
64. 1.05 ● $1\frac{1}{20}$ **=**
65. 2.5 ● $2\frac{1}{2}$ **=**
66. 3.15 ● $3\frac{3}{20}$ **=**
67. 2.66 ● $2\frac{2}{5}$ **>**
68. 4.77 ● $4\frac{3}{4}$ **>**

Four fun Guess and check

69. Use the digit 4 four times to build each of the numbers from 1 through 10. You may add, subtract, multiply, and/or divide. Here are two examples:

    $44 \div 44 = 1$
    $4 \times (4 - 4) + 4 = 4$

Extra Practice
Page 486   Skill 31

Practice Worksheet
Workbook S242, Copymaster S242, or Duplicating Master S242

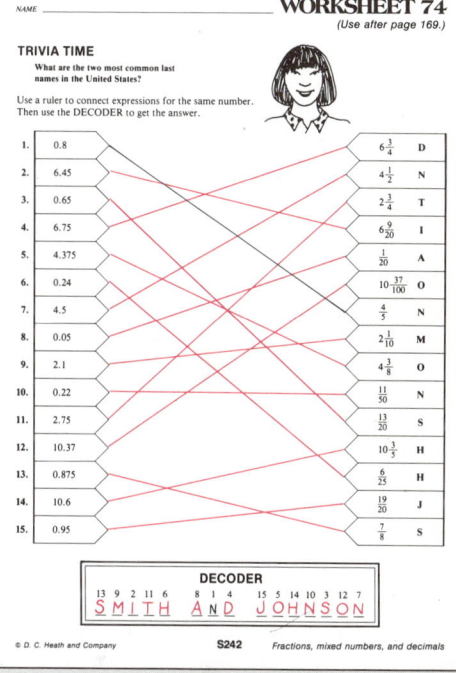

Challenge Problems

Unscramble the letters of each of these math words:

1. IARFCTNO — Fraction
2. ATENURMOR — Numerator
3. ALCDIEM — Decimal
4. EMXID UNBMRE — Mixed number

Copymaster S449

Answer for page 169.
69. Answers will vary.
Sample answers are given.
$4 \div 4 + 4 \div 4 = 2$
$(4 + 4 + 4) \div 4 = 3$
$(4 + 4 \times 4) \div 4 = 5$
$(4 + 4) \div 4 + 4 = 6$
$44 \div 4 - 4 = 7$
$4 + 4 \times 4 \div 4 = 8$
$4 + 4 + 4 \div 4 = 9$
$(44 - 4) \div 4 = 10$

Class Starter Quiz 69
on previous lesson

Change to a fraction or a mixed number in simplest form.

1. 0.4 $\frac{2}{5}$ 2. 0.75 $\frac{3}{4}$ 3. 0.125 $\frac{1}{8}$
4. 0.7 $\frac{7}{10}$ 5. 0.24 $\frac{6}{25}$ 6. 0.375 $\frac{3}{8}$
7. 2.5 $2\frac{1}{2}$ 8. 6.25 $6\frac{1}{4}$ 9. 3.625 $3\frac{5}{8}$

Copymaster S402

Lesson Objective
To solve problems using a guess-and-check strategy

Starting the Lesson
Problem-Solving Cover-up Use the chalkboard or mask ■ **Visual Aid 31** (copymaster or transparency S136).

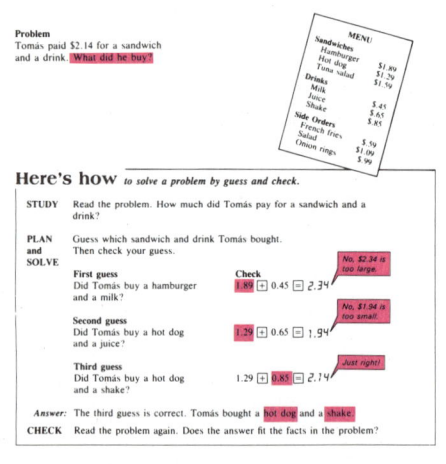

Have the students, working in small groups, study the problems and the problem-solving steps on page 170 for several minutes. Then have them close their books, look at the visual aid and tell what has been covered up.

PROBLEM SOLVING *guess and check*

You can solve some problems by making careful guesses and then checking them.

Problem
Tomás paid $2.14 for a sandwich and a drink. What did he buy?

Here's how *to solve a problem by guess and check.*

STUDY Read the problem. How much did Tomás pay for a sandwich and a drink?

PLAN and SOLVE Guess which sandwich and drink Tomás bought. Then check your guess.

First guess
Did Tomás buy a hamburger and a milk?
Check: 1.89 + 0.45 = 2.34
No, $2.34 is too large.

Second guess
Did Tomás buy a hot dog and a juice?
1.29 + 0.65 = 1.94
No, $1.94 is too small.

Third guess
Did Tomás buy a hot dog and a shake?
1.29 + 0.85 = 2.14
Just right!

Answer: The third guess is correct. Tomás bought a hot dog and a shake.

CHECK Read the problem again. Does the answer fit the facts in the problem?

PROBLEMS

Check each first guess. Keep guessing and checking until you get the right answer. Use the menu prices on page 170.

1. Lisa paid $2.88 for a sandwich and a side order. What did she buy? *Hamburger and onion rings*

    First guess: Did Lisa buy a hamburger and french fries? *No*

2. Chen bought a drink and a side order. He spent $1.64. What did he buy? *Juice and onion rings*

    First guess: Did Chen buy a juice and a salad? *No*

Solve using guess and check. Use the menu prices on page 170.

3. Which two sandwiches cost a total of $3.48? *Hamburger and tuna salad*

4. Which two side orders cost a total of $1.68? *French fries and salad*

5. Bien paid $2.24 for a sandwich and a drink. What did he buy? *Tuna salad and juice*

6. Rosa bought a sandwich and a side order. She spent $2.68. What did she buy? *Tuna salad and salad*

7. Cindy bought a sandwich, a drink, and a salad. She spent $3.83. What kind of sandwich did she buy? *Hamburger*

8. Matt paid $2.63 for a tuna salad, a drink, and a side order. What kind of drink did he buy? *Milk*

9. Kim paid less than $2.33 for a sandwich, a drink, and a side order. What did she buy? *Hot dog, milk, and french fries*

10. Luis paid more than $3.83 for a sandwich, a drink, and a side order. What kind of sandwich and drink did he have? *Hamburger and shake*

What's the price? Logical reasoning

11. Study the clues. What is the price of the eggs? the pancakes? the bacon?

Eggs and bacon cost $2.50

Pancakes and bacon cost $2.75

Pancakes and eggs cost $2.25

Eggs $1.00, Pancakes $1.25, Bacon $1.50

Number Theory, Fractions, and Decimals **171**

Practice Worksheet

Workbook S243, Copymaster S243, or Duplicating Master S243

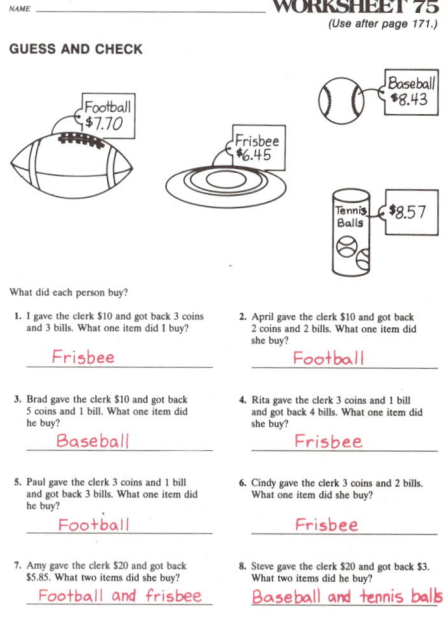

Challenge Problem

Write a question that fits the answer.

John spent $10 of his $25 weekly income on movies.

What fraction of his weekly income did John spend on movies?

Answer: $\frac{2}{5}$

Copymaster S449

171

Class Starter Quiz 70
on previous lesson

Solve using the guess-and-check strategy. Use the menu prices on page 170.

1. Which two sandwiches cost $3.78? 2 hamburgers
2. Brad paid $2.38 for a sandwich and a side order. What did he buy? A hot dog and a salad

Copymaster S402

Problem-Solving Skills
Reading a cash-register receipt
Choosing the correct operation

Skills Reviewed
Comparing fractions
Writing fractions and mixed numbers in simplest form
Changing fractions and mixed numbers to decimals
Adding and subtracting whole numbers and decimals
Multiplying and dividing whole numbers and decimals
Simplifying expressions

Starting the Lesson
Problem Solving Have the students use the information on the cash-register receipt to answer questions like these:

- Which food item costs $.59? (Celery)
- Which food item costs $1.19? (Cereal)
- Which item costs 5 cents more than the mushroom soup? (Carrots)

Starting the Lesson
Cumulative Skill Practice Write these four answers on the chalkboard:

$1\frac{4}{7}$ $\frac{4}{5}$ 1.3   4.5

Challenge the students to an answer hunt by saying, "Look at exercises 1–72 on page 173. Find the four exercises that have these answers. You have four minutes to find as many of the exercises as you can." (Exercises 24, 30, 43, and 47)

Problem solving
COMPUTERS IN STORES

Most items sold in stores are marked with light and dark bands called the Universal Product Code (UPC).

An optical scanner reads and sends the code to a computer. The computer uses the code to search its memory for the price of the item. The price is then printed on a cash-register receipt.

Solve. Decide when a calculator would be useful.

```
      **TRIPLE S**
       STORE #315

GRND BEEF           7.17
MM FROZEN OJ        1.01
MILK                 .97
POTATOES            1.07
CARROTS              .44
DIET P COLA          .99
CELERY               .59
SUGAR               1.29
PNUT BUTTER         2.38
SLTN CRACKERS       1.35
MUSH SOUP            .39
CEREAL              1.19

     TOTAL         18.84

     CASH          20.00

     CHANGE         1.16

     THANK YOU
# 20416 C013 R 06 T12:40
```

7. A—MM FROZEN OJ
   and
   B—SUGAR
   or
   A—SLTN CRACKERS
   and
   B—CEREAL

C—PNUT BUTTER
D—CEREAL

1. What was the total cost of the vegetables? $2.10
2. Did the meat item cost more than 3 times as much as the vegetables? Yes
3. This purchase was made by 3 friends who share an apartment. What was each person's share of the cost? $6.28
4. The amount of this purchase is about $\frac{1}{5}$ of their weekly food allowance. About how much do they spend each week for food? $94.20
5. The ground beef will be used to prepare a meat loaf to serve 5 people. What is the average cost per serving? (Round to the nearest cent.) $1.43
6. What is the least number of coins that could be given in change? 3
7. Each of these UPCs is from an item printed on the receipt. Decide which item each code is on.

 costs 28¢ more than

 costs twice as much as

Cumulative Skill Practice

<, =, or >? *(page 152)*

1. $\frac{2}{5}$ < $\frac{3}{5}$
2. $\frac{5}{8}$ > $\frac{3}{8}$
3. $\frac{7}{10}$ < $\frac{9}{10}$
4. $\frac{5}{9}$ > $\frac{4}{9}$
5. $\frac{1}{6}$ < $\frac{5}{6}$
6. $\frac{1}{2}$ < $\frac{3}{4}$
7. $\frac{2}{3}$ < $\frac{5}{6}$
8. $\frac{2}{4}$ = $\frac{1}{2}$
9. $\frac{1}{4}$ < $\frac{1}{3}$
10. $\frac{3}{8}$ = $\frac{6}{16}$

Write in simplest form. *(page 160)*

11. $\frac{6}{8}$ $\frac{3}{4}$
12. $\frac{9}{3}$ 3
13. $5\frac{4}{6}$ $5\frac{2}{3}$
14. $\frac{10}{3}$ $3\frac{1}{3}$
15. $\frac{6}{36}$ $\frac{1}{6}$
16. $3\frac{2}{4}$ $3\frac{1}{2}$
17. $\frac{17}{34}$ $\frac{1}{2}$
18. $\frac{12}{5}$ $2\frac{2}{5}$
19. $\frac{5}{20}$ $\frac{1}{4}$
20. $4\frac{5}{10}$ $4\frac{1}{2}$
21. $\frac{18}{24}$ $\frac{3}{4}$
22. $3\frac{6}{10}$ $3\frac{3}{5}$
23. $1\frac{10}{12}$ $1\frac{5}{6}$
24. $1\frac{8}{14}$ $1\frac{4}{7}$
25. $\frac{33}{5}$ $6\frac{3}{5}$
26. $\frac{12}{16}$ $\frac{3}{4}$
27. $\frac{18}{7}$ $2\frac{4}{7}$
28. $\frac{10}{2}$ 5
29. $2\frac{3}{9}$ $2\frac{1}{3}$
30. $\frac{8}{10}$ $\frac{4}{5}$
31. $3\frac{7}{14}$ $3\frac{1}{2}$

Change to a decimal. *(page 166)*

32. $\frac{1}{2}$ 0.5
33. $1\frac{1}{5}$ 1.2
34. $\frac{9}{10}$ 0.9
35. $\frac{1}{4}$ 0.25
36. $\frac{1}{16}$ 0.0625
37. $1\frac{1}{8}$ 1.125
38. $1\frac{3}{10}$ 1.3
39. $2\frac{2}{5}$ 2.4
40. $\frac{3}{8}$ 0.375
41. $\frac{3}{4}$ 0.75
42. $\frac{5}{2}$ 2.5
43. $\frac{13}{10}$ 1.3
44. $3\frac{3}{5}$ 3.6
45. $\frac{25}{16}$ 1.5625
46. $1\frac{3}{16}$ 1.1875
47. $\frac{9}{2}$ 4.5
48. $\frac{7}{10}$ 0.7
49. $4\frac{7}{8}$ 4.875
50. $\frac{9}{4}$ 2.25
51. $\frac{5}{16}$ 0.3125
52. $\frac{9}{8}$ 1.125

Change to a decimal rounded to the nearest hundredth. *(page 166)*

53. $\frac{1}{3}$ 0.33
54. $\frac{1}{6}$ 0.17
55. $\frac{1}{12}$ 0.08
56. $\frac{2}{3}$ 0.67
57. $\frac{1}{9}$ 0.11
58. $\frac{5}{3}$ 1.67
59. $\frac{4}{7}$ 0.57
60. $\frac{7}{12}$ 0.58
61. $\frac{11}{6}$ 1.83
62. $\frac{4}{9}$ 0.44
63. $\frac{5}{12}$ 0.42
64. $\frac{10}{9}$ 1.11
65. $\frac{7}{6}$ 1.17
66. $\frac{15}{9}$ 1.67
67. $\frac{5}{6}$ 0.83
68. $\frac{4}{3}$ 1.33
69. $\frac{2}{9}$ 0.22
70. $\frac{13}{6}$ 2.17
71. $\frac{7}{3}$ 2.33
72. $\frac{11}{12}$ 0.92
73. $\frac{3}{11}$ 0.27

MIXED PRACTICE
Complete.

74. $53 \times 100 =$? 5300
75. $15.8 + 2.34 =$? 18.14
76. $9.2 \times 0.07 =$? 0.644
77. $14.63 - 2.87 =$? 11.76
78. $61.2 \div 0.7 =$? 87.429
79. $19 - 6.91 =$? 12.09
80. $6.9 + 8 + 5.37 =$? 20.27
81. $97.6 \times 10 =$? 976
82. $0.2106 \div 2.6 =$? 0.081
83. $65.2 \div 10 =$? 6.52
84. $0.08 \times 1000 =$? 80
85. $3.25 \div 1000 =$? 0.00325
86. $(3.4 \times 10) \div 2 =$? 17
87. $3.4 \times (10 \div 2) =$? 17
88. $5.74 - (3.1 + 2.5) =$? 0.14

Number Theory, Fractions, and Decimals

Problem-Solving Worksheet
Workbook S244, Copymaster S244, or Duplicating Master S244

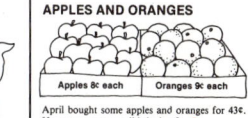

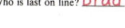

Challenge Problem

Use the code to answer the riddle.

CODE

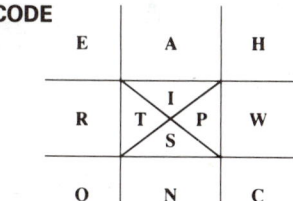

Riddle: What did Baby Corn say to Mother Corn?

Answer:

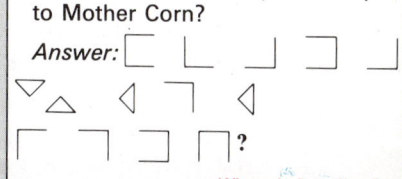

Where is Pop Corn?

Copymaster S449

Chapter REVIEW

Here are scrambled answers for the review exercises:

| 1 | 3^2 | 5 | 100 | divisor | GCF | lowest | mixed | tenth |
| 2 | 4 | 8 | common | greater | LCM | prime | remainder | |

1. A whole number that has exactly two factors is called a ⟨?⟩ number. The prime factorization of 18 is $2 \times$ ⟨?⟩. *(page 142)*

2. To write a fraction in lowest terms, divide both terms by the ⟨?⟩. This fraction is in lowest terms because ⟨?⟩ is the GCF of its terms. *(page 148)*

$$\frac{2}{3}$$

3. To find the least common denominator of two fractions, you find the ⟨?⟩ of the denominators. The least common denominator of these fractions is ⟨?⟩. *(page 150)*

$$\frac{3}{4} \text{ and } \frac{3}{8}$$

4. To compare these fractions, compare equivalent fractions with a ⟨?⟩ denominator. *(page 152)*

$$\frac{2}{3} \bullet \frac{3}{4}$$

5. To change this mixed number to a fraction, multiply 5 by 4 and add ⟨?⟩ to find the numerator and use ⟨?⟩ for the denominator. *(page 156)*

$$4\frac{2}{5} = ?$$

6. To change this fraction to a mixed number, divide 13 by ⟨?⟩. *(page 158)*

$$\frac{13}{4} = ?$$

7. To write this fraction in simplest form, write it in ⟨?⟩ terms. *(page 160)*

$$\frac{10}{15} = ?$$

8. This mixed number is **not** in simplest form, since the fraction part is ⟨?⟩ than 1. *(page 160)*

$$3\frac{6}{5}$$

9. To write this fraction in simplest form, you would change it to a ⟨?⟩ number in simplest form. *(page 160)*

$$\frac{17}{3}$$

10. The last step in writing this quotient as a mixed number is to write the ⟨?⟩ over the ⟨?⟩. *(page 162)*

$$7\overline{)255}$$
$$-21$$
$$45$$
$$-42$$
$$3$$
(quotient 36)

11. The fraction $\frac{5}{6}$ written as a decimal rounded to the nearest ⟨?⟩ is 0.8. *(page 166)*

$$\frac{5}{6} = ?$$

$$6\overline{)5.00}$$
$$-48$$
$$20$$
$$-18$$
$$2$$
(quotient 0.83)

12. To change this decimal to a fraction in simplest form, write 25 as the numerator and ⟨?⟩ as the denominator. Then write the fraction in simplest form. *(page 168)*

$$0.25 = ?$$

---

1. prime, 3^2 2. GCF, 1 3. LCM, 8 4. common 5. 2, 5
6. 4 7. lowest 8. greater 9. mixed
10. remainder, divisor 11. tenth 12. 100

Chapter TEST

Give the prime factorization. Use exponents when possible. *(page 142)*

1. 10 2×5
2. 12 $2^2 \times 3$
3. 28 $2^2 \times 7$
4. 40 $2^3 \times 5$
5. 48 $2^4 \times 3$
6. 66 $2 \times 3 \times 11$
7. 72 $2^3 \times 3^2$

Write each fraction in lowest terms. *(page 148)*

8. $\frac{6}{12}$ $\frac{1}{2}$
9. $\frac{6}{9}$ $\frac{2}{3}$
10. $\frac{9}{6}$ $\frac{3}{2}$
11. $\frac{6}{18}$ $\frac{1}{3}$
12. $\frac{20}{15}$ $\frac{4}{3}$
13. $\frac{21}{24}$ $\frac{7}{8}$
14. $\frac{50}{25}$ 2

Find the least common denominator. *(page 150)*

15. $\frac{3}{4}$ $\frac{1}{2}$ 4
16. $\frac{3}{10}$ $\frac{2}{5}$ 10
17. $\frac{1}{2}$ $\frac{1}{3}$ 6
18. $\frac{1}{3}$ $\frac{3}{8}$ 24
19. $\frac{5}{6}$ $\frac{2}{9}$ 18

<, =, or >? *(page 152)*

20. $\frac{3}{8}$ ● $\frac{1}{4}$ >
21. $\frac{2}{3}$ ● $\frac{5}{6}$ <
22. $\frac{1}{3}$ ● $\frac{1}{4}$ >
23. $\frac{1}{3}$ ● $\frac{5}{12}$ <
24. $\frac{4}{5}$ ● $\frac{5}{6}$ <

Change each mixed number to a fraction. *(page 156)*

25. $1\frac{1}{2}$ $\frac{3}{2}$
26. $3\frac{1}{4}$ $\frac{13}{4}$
27. $2\frac{2}{3}$ $\frac{8}{3}$
28. $4\frac{3}{4}$ $\frac{19}{4}$
29. $2\frac{7}{8}$ $\frac{23}{8}$
30. $3\frac{5}{6}$ $\frac{23}{6}$
31. $7\frac{4}{5}$ $\frac{39}{5}$

Change each fraction to a whole number or mixed number. *(page 158)*

32. $\frac{5}{2}$ $2\frac{1}{2}$
33. $\frac{8}{2}$ 4
34. $\frac{11}{4}$ $2\frac{3}{4}$
35. $\frac{12}{5}$ $2\frac{2}{5}$
36. $\frac{15}{3}$ 5
37. $\frac{23}{6}$ $3\frac{5}{6}$
38. $\frac{25}{8}$ $3\frac{1}{8}$

Write in simplest form. *(page 160)*

39. $\frac{6}{8}$ $\frac{3}{4}$
40. $\frac{8}{2}$ 4
41. $\frac{3}{2}$ $1\frac{1}{2}$
42. $\frac{5}{10}$ $\frac{1}{2}$
43. $2\frac{4}{8}$ $2\frac{1}{2}$
44. $\frac{16}{4}$ 4
45. $11\frac{11}{22}$ $11\frac{1}{2}$
46. $4\frac{2}{6}$ $4\frac{1}{3}$
47. $\frac{8}{12}$ $\frac{2}{3}$
48. $\frac{20}{6}$ $3\frac{1}{3}$
49. $\frac{9}{3}$ 3
50. $\frac{10}{15}$ $\frac{2}{3}$
51. $3\frac{6}{9}$ $3\frac{2}{3}$
52. $\frac{13}{4}$ $3\frac{1}{4}$

Divide. Write each quotient as a mixed number in simplest form. *(page 162)*

53. $3\overline{)127}$ $42\frac{1}{3}$
54. $4\overline{)925}$ $231\frac{1}{4}$
55. $9\overline{)384}$ $42\frac{2}{3}$
56. $12\overline{)7023}$ $585\frac{1}{4}$
57. $15\overline{)4316}$ $287\frac{11}{15}$
58. $28\overline{)2970}$ $106\frac{1}{14}$

Change to a decimal. *(page 166)*

59. $\frac{1}{2}$ 0.5
60. $\frac{3}{4}$ 0.75
61. $\frac{7}{8}$ 0.875
62. $\frac{1}{16}$ 0.0625
63. $4\frac{9}{10}$ 4.9
64. $2\frac{7}{8}$ 2.875
65. $3\frac{4}{5}$ 3.8

Change to a fraction or mixed number in simplest form. *(page 168)*

66. 0.6 $\frac{3}{5}$
67. 0.25 $\frac{1}{4}$
68. 0.2 $\frac{1}{5}$
69. 0.75 $\frac{3}{4}$
70. 1.4 $1\frac{2}{5}$
71. 3.25 $3\frac{1}{4}$
72. 2.375 $2\frac{3}{8}$

Number Theory, Fractions, and Decimals

Cumulative Test
(Chapters 1–6)

Use Copymaster S109 to provide the students with an answer sheet in standardized test format.

Answers for Cumulative Test, Chapters 1–6

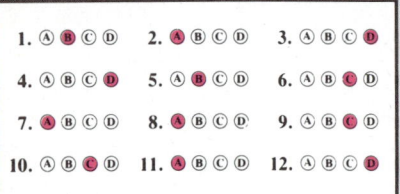

The table below correlates test items with student text pages.

Test Item	Page Taught	Skill Practice
1	14	p. 155, exercises 1–15
2	22	p. 155, exercises 16–27
3	42	p. 155, exercises 28–39
4	46	p. 155, exercises 40–51
5	68	p. 165, exercises 1–12
6	72	p. 165, exercises 13–24
7	104	p. 165, exercises 25–36
8	126	p. 165, exercises 37–48
9	152	p. 173, exercises 1–10
10	160	p. 173, exercises 11–31
11	166	p. 173, exercises 53–73
12	116	

176

Cumulative TEST — Standardized Format

Choose the correct letter.

1. The short word-name for 50.012 is
   A. 50 and 12 hundredths
   B. 50 and 12 thousandths
   C. 50 and 12 tenths
   D. none of these

2. Give the sum.
   $36.09 + 14.8 + 321.7$
   A. 372.59
   B. 69.74
   C. 83.06
   D. none of these

3. Which number is less than 0.04?
   A. 0.3
   B. 0.05
   C. 0.041
   D. none of these

4. Give the difference.
   $52.46 - 3.521$
   A. 17.25
   B. 48.941
   C. 48.938
   D. none of these

5. Give the product.
   4.03×4.6
   A. 185.38
   B. 18.538
   C. 18.428
   D. none of these

6. Give the product.
   10.463×100
   A. 0.10463
   B. 104.63
   C. 1046.3
   D. none of these

7. Give the quotient rounded to the nearest tenth.
   $3.742 \div 1.9$
   A. 2.0
   B. 1.9
   C. 0.2
   D. none of these

8. The median of 38, 42, 36, 36, and 43 is
   A. 38
   B. 36
   C. 39
   D. none of these

9. $\frac{1}{3} < ?$
   A. $\frac{1}{4}$
   B. $\frac{3}{10}$
   C. $\frac{3}{8}$
   D. none of these

10. $\frac{24}{16}$ in simplest form is
   A. $\frac{3}{2}$
   B. $1\frac{8}{16}$
   C. $1\frac{1}{2}$
   D. none of these

11. Change to a decimal rounded to the nearest hundredth.
   $\frac{5}{6} = ?$
   A. 0.83
   B. 0.84
   C. 1.20
   D. none of these

12.
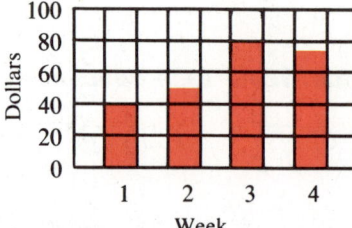
SAL'S EARNINGS

Sal's total earnings were
   A. $80
   B. $70
   C. $130
   D. none of these

176   Chapter 6

Adding and Subtracting Fractions and Mixed Numbers

Chapter 7
Adding and Subtracting Fractions and Mixed Numbers

Resources

- **Class Starter Quizzes 71-79** *(Copymasters S402-S404)*
- **Visual Aids 32-35** *(Copymasters or Transparencies S137-S140)*
- **Manipulatives**
  Manipulative Activities 11-14 *(Copymasters S517-S518)*
  Fraction pieces *(Copymasters or Transparencies S530-S531)*
- **Worksheets 77-86** *(Copymasters, Duplicating Masters, or Workbook pages S245-S254)*
- **Challenge Problems** for pages 179, 181, 189, 191, 193, 195, 197 *(Copymasters S450-S451)*
- **Projects** for pages 183, 185, 187 *(Copymaster S479)*
- **Mental Math Extensions** for Skills 32-35 *(Copymasters S502-S503)*
- **Tests** *(Copymasters or Duplicating Masters S25-S28)*

Lesson Objective
To add fractions with common denominators

Problem-Solving Skills
Choosing information from a display
Making a list

Starting the Lesson
Write these coins and weights on the chalkboard:

penny $\frac{2}{28}$ ounce

nickel $\frac{3}{28}$ ounce

dime $\frac{5}{28}$ ounce

quarter $\frac{6}{28}$ ounce

Ask the students to use their estimation skills to match each coin with its weight. Then have the students open their books to page 178 to check their answers.

Here's How Note
Use of Concrete Materials You may wish to use the fraction pieces from copymasters or transparencies S530–S531 to demonstrate adding fractions with common denominators. See ■ *Manipulative Activity 11* on copymaster S517 in the Teacher's Resource Binder.

Exercise Note
Problem Solving Encourage the students to make a list for each of exercises 42 and 43.

COINS		
dimes	nickels	pennies
✓✓	✓	
✓✓		✓✓✓✓✓
✓	✓✓✓	

Adding fractions with common denominators

1. What is the weight of a nickel? $\frac{5}{28}$ ounce
2. Which coin weighs $\frac{2}{28}$ ounce? Dime
3. You have 3 coins that are worth 35¢. What three fractions would you add to find how many ounces they weigh?
 $\frac{5}{28}, \frac{5}{28},$ and $\frac{6}{28}$

$\frac{3}{28}$ ounce $\frac{6}{28}$ ounce

$\frac{5}{28}$ ounce $\frac{2}{28}$ ounce

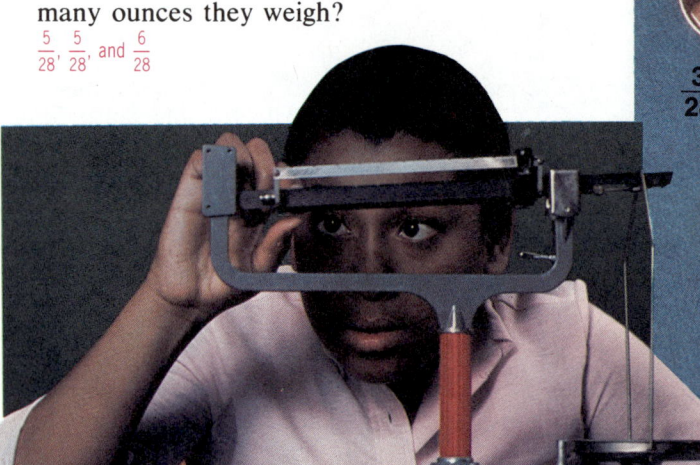

Here's how to add fractions with common denominators. $\frac{5}{28} + \frac{5}{28} + \frac{6}{28} = ?$

MENTAL MATH Add the numerators and use the common denominator.

nickel   nickel   quarter
↓   ↓   ↓
$\frac{5}{28} + \frac{5}{28} + \frac{6}{28} = \frac{16}{28}$
$= \frac{4}{7}$

4. Look at the *Here's how*. What is the total weight of the 3 coins? $\frac{4}{7}$

5. Check these examples. Give each sum in simplest form.

a. $\frac{3}{28} + \frac{5}{28} = \frac{8}{28}$
   $= ?$ $\frac{2}{7}$

b. $\frac{2}{3} + \frac{1}{3} = \frac{3}{3}$
   $= ?$ 1

c. $\frac{4}{5} + \frac{2}{5} + \frac{3}{5} = \frac{9}{5}$
   $= ?$ $1\frac{4}{5}$

Chapter 7

EXERCISES

MENTAL MATH Add. Write the sum in simplest form.
Here are scrambled answers for the next row of exercises: $1\frac{1}{4}$ $\frac{5}{7}$ $\frac{2}{5}$ $\frac{1}{2}$ 1

6. $\frac{3}{8} + \frac{1}{8}$ $\frac{1}{2}$
7. $\frac{3}{10} + \frac{1}{10}$ $\frac{2}{5}$
8. $\frac{2}{7} + \frac{3}{7}$ $\frac{5}{7}$
9. $\frac{3}{8} + \frac{7}{8}$ $1\frac{1}{4}$
10. $\frac{1}{3} + \frac{2}{3}$ 1

11. $\frac{5}{9} + \frac{1}{9}$ $\frac{2}{3}$
12. $\frac{1}{5} + \frac{2}{5}$ $\frac{3}{5}$
13. $\frac{5}{12} + \frac{3}{12}$ $\frac{2}{3}$
14. $\frac{2}{5} + \frac{3}{5}$ 1
15. $\frac{1}{6} + \frac{5}{6}$ 1

16. $\frac{3}{10} + \frac{2}{10}$ $\frac{1}{2}$
17. $\frac{7}{12} + \frac{7}{12}$ $1\frac{1}{6}$
18. $\frac{5}{8} + \frac{7}{8}$ $1\frac{1}{2}$
19. $\frac{1}{4} + \frac{3}{4}$ 1
20. $\frac{3}{5} + \frac{4}{5}$ $1\frac{2}{5}$

21. $\frac{4}{15} + \frac{1}{15}$ $\frac{1}{3}$
22. $\frac{31}{50} + \frac{9}{50}$ $\frac{4}{5}$
23. $\frac{17}{100} + \frac{33}{100}$ $\frac{1}{2}$
24. $\frac{5}{16} + \frac{7}{16}$ $\frac{3}{4}$
25. $\frac{7}{10} + \frac{9}{10}$ $1\frac{3}{5}$

26. $\frac{3}{10} + \frac{1}{10} + \frac{2}{10}$ $\frac{3}{5}$
27. $\frac{1}{6} + \frac{1}{6} + \frac{5}{6}$ $1\frac{1}{6}$
28. $\frac{1}{12} + \frac{5}{12} + \frac{3}{12}$ $\frac{3}{4}$
29. $\frac{1}{8} + \frac{3}{8} + \frac{5}{8}$ $1\frac{1}{8}$

30. $\frac{1}{5} + \frac{2}{5} + \frac{4}{5}$ $1\frac{2}{5}$
31. $\frac{2}{5} + \frac{1}{5} + \frac{1}{5}$ $\frac{4}{5}$
32. $\frac{6}{7} + \frac{3}{7} + \frac{2}{7}$ $1\frac{4}{7}$
33. $\frac{5}{9} + \frac{2}{9} + \frac{5}{9}$ $1\frac{1}{3}$

Solve. Use the coin facts on page 178.

34. What coins are worth 40¢ and weigh $\frac{13}{28}$ ounce? 1 quarter, 1 dime 1 nickel

35. What coins are worth 15¢ and weigh $\frac{15}{28}$ ounce? 3 nickels

36. What coins are worth 15¢ and weigh $\frac{17}{28}$ ounce? 1 dime, 5 pennies

37. What coins are worth 50¢ and weigh $\frac{15}{28}$ ounce? 1 quarter, 2 dimes, 1 nickel

38. What coins are worth 60¢ and weigh $\frac{1}{2}$ ounce? 2 quarters, 1 dime

39. What coins are worth 30¢ and weigh $\frac{3}{4}$ ounce? 1 quarter, 5 pennies

Change, please Making a list

40. How many ways can you make change for a nickel? 1

41. How many ways can you make change for a dime? *Hint: The answer is not 2.* 3

42. How many ways can you make change for a quarter? *Hint: Make a list.* 12

43. How many ways can you make change for a half-dollar? 49

Adding and Subtracting Fractions and Mixed Numbers

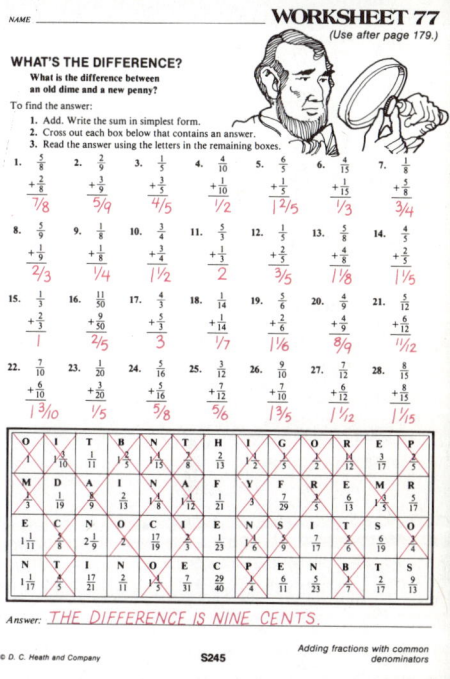

Practice Worksheet
Workbook S245, Copymaster S245, or Duplicating Master S245

Challenge Problem
Use this code and find the three-letter word on page 179 that has a sum of $1\frac{7}{26}$. the

a	b	c	.....	z
$\frac{1}{26}$	$\frac{2}{26}$	$\frac{3}{26}$	.....	$\frac{26}{26}$

Copymaster S450

Class Starter Quiz 71
on previous lesson

Add. Write the sum in simplest form.

1. $\frac{4}{9} + \frac{2}{9}$ $\frac{2}{3}$
2. $\frac{3}{8} + \frac{7}{8}$ $1\frac{1}{4}$
3. $\frac{1}{4} + \frac{3}{4}$ 1
4. $\frac{7}{10} + \frac{5}{10}$ $1\frac{1}{5}$
5. $\frac{1}{8} + \frac{3}{8} + \frac{5}{8}$ $1\frac{1}{8}$

Copymaster S402

Lesson Objective
To add fractions with different denominators

Problem-Solving Skill
Reading a map

Starting the Lesson
The map at the top of page 180 is also on ■ **Visual Aid 32** (copymaster or transparency S137). Use the map and ask these questions:

- Which scenic point is between Dry Gulch and Deer Pond? (Rainbow Falls)
- Which scenic point is seven eighths of a mile from Roaring Rapids? (Lost Mine)
- Is it more or less than one-half mile from Ranger's Tower to Fox Ridge? (More)

Then use the map to discuss exercises 1–3.

Point out that sometimes, as in the first example in exercise 4, only one of the fractions will have to be renamed, since the other one has a denominator that can be the common denominator.

Here's How Note
Use of Concrete Materials You may wish to use the fraction pieces from copymasters or transparencies S530–S531 to demonstrate adding fractions with different denominators. See ■ **Manipulative Activity 12** on copymaster S518 in the Teacher's Resource Binder.

180

Adding fractions with different denominators

This map shows the distance between scenic points on a hiking trail.

1. What is the shortest hiking distance from the Trail Entrance to Dry Gulch? $\frac{1}{4}$ mile

2. What two fractions would you add to compute the distance in miles from Rainbow Falls through Deer Pond to Lost Mine? $\frac{1}{2}$ and $\frac{2}{3}$

Here's how to add fractions with different denominators. $\frac{1}{2} + \frac{2}{3} = ?$

Find the least common denominator.	Change to equivalent fractions.	Add. Write the sum in simplest form.
$\frac{1}{2}$ $+\frac{2}{3}$ } 6	$\frac{1}{2} = \frac{3}{6}$ $+\frac{2}{3} = +\frac{4}{6}$	$\frac{1}{2} = \frac{3}{6}$ $+\frac{2}{3} = +\frac{4}{6}$ $\frac{7}{6} = 1\frac{1}{6}$

3. Look at the *Here's how*. How far is it from Rainbow Falls to Lost Mine? $1\frac{1}{6}$

4. Check these examples. Give each sum in simplest form.

   a. $\frac{1}{2} + \frac{3}{4} = \frac{2}{4} + \frac{3}{4}$
      $= \frac{5}{4}$
      $= ?$ $1\frac{1}{4}$

   b. $\frac{1}{4} + \frac{1}{3} + \frac{1}{2} = \frac{3}{12} + \frac{4}{12} + \frac{6}{12}$
      $= \frac{13}{12}$
      $= ?$ $1\frac{1}{12}$

180 Chapter 7

EXERCISES

Add. Give the sum in simplest form.

Here are scrambled answers for the next row of exercises: $\frac{5}{8}$ $1\frac{1}{8}$ $\frac{7}{24}$ $1\frac{5}{24}$ $\frac{3}{5}$ $\frac{3}{4}$ 1

5. $\frac{1}{2} + \frac{1}{4} = \frac{3}{4}$
6. $\frac{1}{6} + \frac{1}{8} = \frac{7}{24}$
7. $\frac{3}{8} + \frac{1}{4} = \frac{5}{8}$
8. $\frac{3}{10} + \frac{7}{10} = 1$
9. $\frac{2}{5} + \frac{1}{5} = \frac{3}{5}$
10. $\frac{5}{8} + \frac{1}{2} = 1\frac{1}{8}$
11. $\frac{5}{6} + \frac{3}{8} = 1\frac{5}{24}$

12. $\frac{7}{16} + \frac{1}{4} = \frac{11}{16}$
13. $\frac{1}{3} + \frac{5}{9} = \frac{8}{9}$
14. $\frac{5}{12} + \frac{2}{3} = 1\frac{1}{12}$
15. $\frac{5}{6} + \frac{1}{4} = 1\frac{1}{12}$
16. $\frac{3}{10} + \frac{1}{2} = \frac{4}{5}$
17. $\frac{2}{3} + \frac{3}{4} = 1\frac{5}{12}$
18. $\frac{5}{9} + \frac{5}{6} = 1\frac{7}{18}$

19. $\frac{1}{3} + \frac{1}{6} = \frac{1}{2}$
20. $\frac{3}{4} + \frac{1}{8} = \frac{7}{8}$
21. $\frac{3}{5} + \frac{7}{10} = 1\frac{3}{10}$
22. $\frac{5}{8} + \frac{1}{2} = 1\frac{1}{8}$
23. $\frac{1}{3} + \frac{1}{4} = \frac{7}{12}$
24. $\frac{1}{5} + \frac{3}{10} = \frac{1}{2}$
25. $\frac{2}{3} + \frac{2}{3} = 1\frac{1}{3}$

26. $\frac{2}{3} + \frac{1}{5} = \frac{13}{15}$
27. $\frac{1}{2} + \frac{11}{16} = 1\frac{3}{16}$
28. $\frac{2}{5} + \frac{1}{4} = \frac{13}{20}$
29. $\frac{5}{8} + \frac{1}{6} = \frac{19}{24}$
30. $\frac{9}{16} + \frac{1}{2} = 1\frac{1}{16}$
31. $\frac{2}{5} + \frac{3}{10} = \frac{7}{10}$
32. $\frac{5}{9} + \frac{1}{6} = \frac{13}{18}$

33. $\frac{1}{2} + \frac{1}{4} + \frac{1}{8} = \frac{7}{8}$
34. $\frac{3}{4} + \frac{3}{8} + \frac{1}{2} = 1\frac{5}{8}$
35. $\frac{1}{16} + \frac{3}{8} + \frac{1}{4} = \frac{11}{16}$
36. $\frac{1}{8} + \frac{5}{16} + \frac{3}{4} = 1\frac{3}{16}$

Solve. Use the map on page 180.

37. What is the shortest hiking distance from the Trail Entrance to Roaring Rapids? $\frac{3}{4}$ mi
38. What is the shortest hiking distance from Lost Mine to the Ranger's Tower? $1\frac{5}{8}$ mi
39. If you hiked at 1 mile per hour, could you hike from the Trail Entrance to Lost Mine in less than $1\frac{1}{2}$ hours? No
40. Which is the shorter route from the Trail Entrance to Ranger's Tower, over Fox Ridge or through Dry Gulch and past Roaring Rapids? Over Fox Ridge

Where are you? Reading a map

Use the map. At which scenic points would you find these trail signs?

41.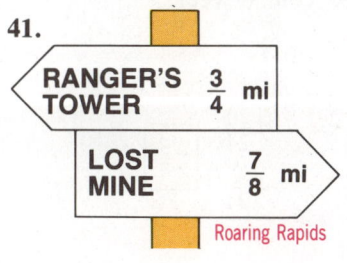
Roaring Rapids

42.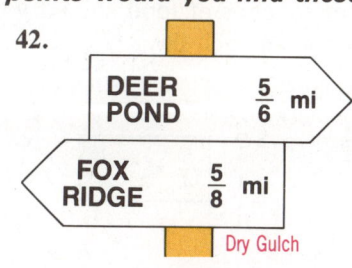
Dry Gulch

43.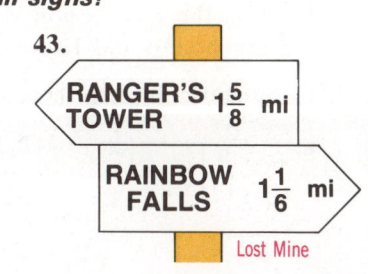
Lost Mine

Adding and Subtracting Fractions and Mixed Numbers

Extra Practice
Page 486 Skill 32

Practice Worksheet
Workbook S246, Copymaster S246, or Duplicating Master S246

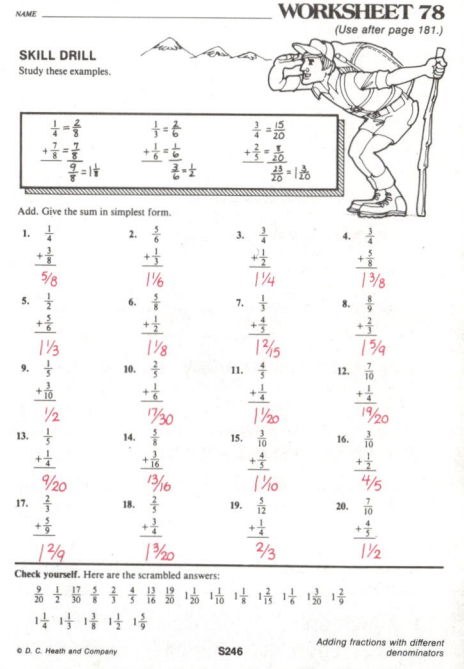

Challenge Problems
Find the missing digit.

1. $\frac{1}{3} + \frac{1}{\boxed{2}} = \frac{5}{6}$
2. $\frac{1}{\boxed{5}} + \frac{3}{4} = \frac{19}{20}$
3. $\frac{\boxed{3}}{8} + \frac{1}{3} = \frac{17}{24}$
4. $\frac{1}{4} + \frac{\boxed{1}}{6} = \frac{5}{12}$

Copymaster S450

Class Starter Quiz 72
on previous lesson

Add. Write the sum in simplest form.

1. $\frac{1}{2} + \frac{1}{3}$ $\frac{5}{6}$
2. $\frac{3}{8} + \frac{1}{4}$ $\frac{5}{8}$
3. $\frac{3}{10} + \frac{1}{2}$ $\frac{4}{5}$
4. $\frac{1}{5} + \frac{7}{20}$ $\frac{11}{20}$
5. $\frac{3}{4} + \frac{1}{8}$ $\frac{7}{8}$
6. $\frac{1}{3} + \frac{2}{5}$ $\frac{11}{15}$

Copymaster S402

Lesson Objective
To add mixed numbers

Problem-Solving Skills
Selecting information from a chart
Following instructions and checking answers

Starting the Lesson
Estimation Before the students open their books, have them guess how many cups of dog food a 12-week-old chihuahua should be fed each day. Have the students write their guesses. Record the high and low guesses on the chalkboard. Then say, "Open your book to page 182. Read the chart. What answer does the book give?" $\left(2\frac{1}{3} \text{ cups}\right)$

Adding mixed numbers

RECOMMENDED DAILY AMOUNTS OF FOOD		
DOG SIZE	AGE 5-9 WEEKS	AGE 10-15 WEEKS
SMALL Scottish Terrier Chihuahua	$1\frac{1}{4}$ cups	$2\frac{1}{3}$ cups
MEDIUM Poodle Welsh Terrier	$2\frac{1}{2}$ cups	$4\frac{1}{2}$ cups
LARGE Greyhound Schnauzer	$3\frac{2}{3}$ cups	$6\frac{3}{4}$ cups

1. How many cups of food should you feed an 8-week-old poodle each day? $2\frac{1}{2}$

2. How much daily food is recommended for a 13-week-old schnauzer? $6\frac{3}{4}$

3. You have a 6-week-old greyhound and a 12-week-old poodle. What two mixed numbers would you add to find the total number of cups of food you should feed them each day? $3\frac{2}{3}$ and $4\frac{1}{2}$

Here's how *to add mixed numbers.* $3\frac{2}{3} + 4\frac{1}{2} = ?$

Write equivalent fractions with a common denominator.

$$3\frac{2}{3} = 3\frac{4}{6}$$
$$+4\frac{1}{2} = +4\frac{3}{6}$$

Add the fractions. Since the sum is greater than 1, regroup.

$$3\frac{2}{3} = 3\frac{4}{6}$$
$$+4\frac{1}{2} = +4\frac{3}{6}$$
$$\frac{7}{6} = 1\frac{1}{6}$$
$$\frac{7}{6}$$

Add the whole numbers.

$$3\frac{2}{3} = 3\frac{4}{6}$$
$$+4\frac{1}{2} = +4\frac{3}{6}$$
$$8\frac{1}{6}$$

4. Look at the *Here's how*. How many cups of food should you feed your 6-week-old greyhound and 12-week-old poodle each day? $8\frac{1}{6}$

5. Copy and complete these examples.

a. $3\frac{2}{3}$
   $+2\frac{1}{3}$
   $\overline{6}$ $\frac{3}{3} = 1$

b. $2\frac{1}{2} = 2\frac{2}{4}$
   $+6\frac{3}{4} = +6\frac{3}{4}$
   $\overline{9\frac{1}{4}}$ $\frac{5}{4} = 1\frac{1}{4}$

c. $4\frac{5}{6} = 4\frac{20}{24}$
   $+4\frac{3}{8} = +4\frac{9}{24}$
   $\overline{9\frac{5}{24}}$ $\frac{29}{24}$

Chapter 7

EXERCISES

Add. Write the sum in simplest form.
Here are scrambled answers for the next row of exercises: $8\frac{7}{8}$ $10\frac{1}{8}$ $8\frac{1}{4}$ $7\frac{3}{4}$ $9\frac{1}{8}$ $6\frac{1}{6}$

6. $3\frac{1}{2} + 4\frac{1}{4} = 7\frac{3}{4}$
7. $5\frac{3}{4} + 2\frac{1}{2} = 8\frac{1}{4}$
8. $3\frac{1}{8} + 5\frac{3}{4} = 8\frac{7}{8}$
9. $6\frac{3}{4} + 2\frac{3}{8} = 9\frac{1}{8}$
10. $3\frac{1}{2} + 6\frac{5}{8} = 10\frac{1}{8}$
11. $4\frac{2}{3} + 1\frac{1}{2} = 6\frac{1}{6}$

12. $7\frac{3}{4} + 3\frac{1}{4} = 11$
13. $8\frac{1}{6} + 4\frac{1}{9} = 12\frac{5}{18}$
14. $9\frac{5}{6} + 5\frac{1}{4} = 15\frac{1}{12}$
15. $8 + 8\frac{3}{4} = 16\frac{3}{4}$
16. $5\frac{1}{2} + 6\frac{7}{8} = 12\frac{3}{8}$
17. $4\frac{2}{5} + 1\frac{1}{2} = 5\frac{9}{10}$

18. $4\frac{2}{3} + 9\frac{5}{6} = 14\frac{1}{2}$
19. $8\frac{5}{12} + 6\frac{7}{8} = 15\frac{7}{24}$
20. $5 + 5\frac{3}{8} = 10\frac{3}{8}$
21. $7\frac{3}{8} + 9\frac{7}{12} = 16\frac{23}{24}$
22. $9\frac{3}{5} + 8\frac{2}{5} = 18$
23. $6\frac{1}{3} + 4\frac{2}{3} = 11$

24. $9\frac{4}{5} + 5\frac{3}{10}$ $15\frac{1}{10}$
25. $7\frac{5}{12} + 4\frac{1}{3}$ $11\frac{3}{4}$
26. $6\frac{1}{2} + 5\frac{3}{10}$ $11\frac{4}{5}$
27. $9\frac{1}{10} + 3\frac{1}{2}$ $12\frac{3}{5}$
28. $1\frac{1}{5} + 3\frac{3}{10}$ $4\frac{1}{2}$

29. $3\frac{5}{12} + 2\frac{1}{2}$ $5\frac{11}{12}$
30. $6\frac{1}{2} + 2\frac{1}{3}$ $8\frac{5}{6}$
31. $1\frac{1}{4} + 3\frac{2}{5}$ $4\frac{13}{20}$
32. $2\frac{7}{8} + 1\frac{1}{2}$ $4\frac{3}{8}$
33. $4\frac{1}{6} + 2\frac{2}{3}$ $6\frac{5}{6}$

Solve. Use the chart on page 182.

34. Is $4\frac{5}{6}$ cups of food enough food for a 6-week-old Welsh terrier and a 14-week-old Chihuahua?   Yes

35. Is $9\frac{5}{12}$ cups of food enough food for a 7-week-old greyhound and a 14-week-old schnauzer?   No

Check the sums Checking answers

36. Find and correct the two wrong answers.

a. Two tenths plus one and three tenths plus six tenths equals   [2.1]

b. Fifty-one hundredths plus six tenths plus fourteen hundredths equals   [1.75]   1.25

c. One and eight tenths plus sixty-two hundredths plus three and two tenths equals   [5.42]   5.62

Adding and Subtracting Fractions and Mixed Numbers **183**

Extra Practice
Page 487 Skill 33

Practice Worksheet
Workbook S247, Copymaster S247,
Duplicating Master S247

Project

Researching information

The recommended daily amount of food for a 5-week-old to 9-week-old schnauzer is $3\frac{2}{3}$ cups. Look in a grocery store for the costs of different-sized packages of dry dog food. Then compute the cost of feeding a 5-week-old schnauzer for 4 weeks.

Copymaster S479

Class Starter Quiz 73
on previous lesson

Add. Write the sum in simplest form.

1. $4\frac{1}{2} + 2\frac{3}{8}$ $6\frac{7}{8}$
2. $2\frac{1}{4} + 3\frac{1}{2}$ $5\frac{3}{4}$
3. $3\frac{1}{3} + 1\frac{2}{3}$ 5
4. $4\frac{1}{5} + 2\frac{3}{10}$ $6\frac{1}{2}$
5. $8\frac{1}{6} + 4\frac{1}{5}$ $12\frac{11}{30}$
6. $4\frac{2}{5} + 5\frac{3}{4}$ $10\frac{3}{20}$

Copymaster S403

Problem-Solving Skills
Making a picture
Choosing the correct operation

Skills Reviewed
Adding and subtracting decimals
Multiplying and dividing decimals
Rounding decimals
Simplifying expressions

Starting the Lesson
Problem Solving

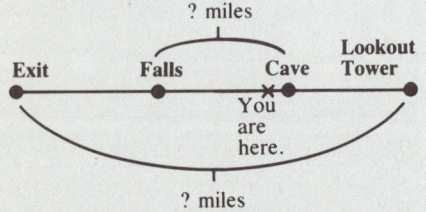

Have the students use the information on the signpost at the top of page 184 to find the missing number of miles in the drawing. (Falls to cave, $3\frac{1}{4}$ miles; exit to lookout tower, $9\frac{3}{4}$ miles)

Cumulative Skill Practice Write these four answers on the chalkboard:

    43.1   1.782
    6.9   5.342

Challenge the students to an answer hunt by saying, "Look at exercises 1–48 on page 185. Find the four exercises that have these answers. You have four minutes to find as many of the exercises as you can." (Exercises 6, 11, 20, and 29)

184

Problem solving
Making a drawing

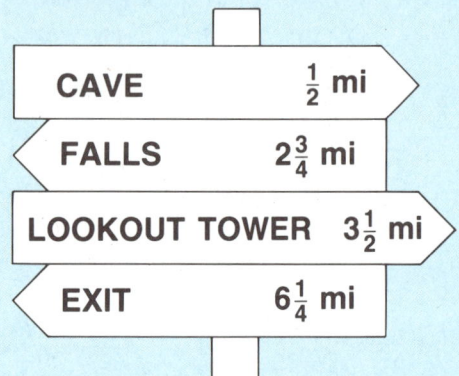

Use the picture to answer these hikers' questions.

1. How many miles is it from the cave to the falls? $3\frac{1}{4}$

2. When I get to the lookout tower, how far will I be from the exit? $9\frac{3}{4}$ miles

3. "How many miles is it from the falls to the lookout tower?" $6\frac{1}{4}$

4. "How far is it from the cave to the exit?" $6\frac{3}{4}$ miles

5. "When I'm at the cave, how far is it to the lookout tower?" 3 miles

6. "Which is closer to the cave, the falls or the lookout tower?" The lookout tower

Draw another picture to solve these problems.

7. How many miles is the round trip from the entrance to the scenic view and back again? 20

8. Suppose you started at the entrance and hiked 7 miles toward the pond. How far are you from the scenic view? 3 miles

9. When you are halfway between the scenic view and the pond, how far are you from the entrance? 11 miles

10. Is the ledge closer to the pond or to the scenic view? How much closer? Scenic view, 2 miles

184 Chapter 7

Cumulative Skill Practice

Give the sum. *(page 22)*

1. 3.4 + 4.56   7.96
2. 6.87 + 3.9   10.77
3. 7.24 + 45.6   52.84
4. 3.24 + 6.2 + 8.4   17.84
5. 25.6 + 52.5 + 2.85   80.95
6. 36 + 5.3 + 1.8   43.1
7. 0.25 + 0.6 + 3   3.85
8. 6.7 + 5 + 0.92   12.62
9. 13 + 2.5 + 0.35   15.85

Give the difference. *(page 46)*

10. 63.5 − 49.7   13.8
11. 20.7 − 13.8   6.9
12. 8.00 − 4.47   3.53
13. 8.7 − 5.99   2.71
14. 30 − 7.24   22.76
15. 43.2 − 3.64   39.56
16. 8.6 − 2.73   5.87
17. 13 − 4.53   8.47
18. 10.18 − 8.45   1.73

Give the product. *(page 68)*

19. 3.7 × 0.8   2.96
20. 2.97 × 0.6   1.782
21. 47.5 × 0.5   23.75
22. 5.2 × 0.68   3.536
23. 0.28 × 1.3   0.364
24. 24 × 0.41   9.84
25. 3.62 × 3.06   11.0772
26. 2.05 × 61   125.05
27. 7.03 × 4.41   31.0023

Give the quotient. *(page 100)*

28. 63 ÷ 10   6.3
29. 534.2 ÷ 100   5.342
30. 629.8 ÷ 1000   0.6298
31. 242 ÷ 100   2.42
32. 71.5 ÷ 1000   0.0715
33. 6 ÷ 100   0.06
34. 3.03 ÷ 10   0.303
35. 4.29 ÷ 100   0.0429
36. 7042 ÷ 1000   7.042

Give the quotient rounded to the nearest hundredth. *(page 104)*

37. 0.3) 1.4   4.67
38. 0.7) 0.52   0.74
39. 1.2) 6.4   5.33
40. 2.6) 3.24   1.25
41. 0.06) 0.5   8.33
42. 2.2) 4.96   2.25
43. 0.006) 0.04   6.67
44. 0.12) 0.5   4.17
45. 3.2) 7.5   2.34
46. 0.04) 2.93   73.25
47. 0.7) 3.4   4.86
48. 2.1) 3.43   1.63

MIXED PRACTICE

Complete.

49. 5.051 rounded to the nearest tenth is   ?   5.1
50. 32.163 rounded to the nearest hundredth is   ?   32.16
51. 4.8790 rounded to the nearest whole number is   ?   5
52. 8 × 3 + 2 × 3 =   ?   30
53. (8 × 2) × 3 =   ?   48
54. 8 × (3 + 2) × 3 =   ?   120
55. 18 − (10 + 2) ÷ 2 =   ?   12
56. 18 − 10 + 2 ÷ 2 =   ?   9
57. (18 − 10) + 2 ÷ 2 =   ?   9

Adding and Subtracting Fractions and Mixed Numbers

Problem-Solving Worksheet
Workbook S248, Copymaster S248, or Duplicating Master S248

WORKSHEET 80 (Use after page 184.)

NAME _____

PEANUT PROBLEM

SPECIAL PEANUTS $2.40/lb

Pounds	1/4	1/2	3/4	1
Cost	$.60	$1.20	$1.80	$2.40

Amy bought 1½ pounds of peanuts. She spent $ 3.60

Brian gave the clerk $5 and got $.20 in change. He bought 2 pounds of peanuts.

TRIANGLE TANGLE

How many triangles are there? *Hint: There are more than 10.* 13

LOGICAL REASONING

Sidney has some pennies. When he counts them by fives, there are three left over. When he counts them by tens, there are eight left over. Sidney has almost 50¢ in pennies. How many pennies does he have? 48

MISSING DIGITS

Fill in the missing digits.

```
   2 5 8
 ×   2 3
   7 7 4
 5 1 6 0
 5 9 3 4
```

```
 1 5 4 8
 -  8 0 ?
 1 6 0 7 3 4
```

TIME IT

TIME CARD

Name Jan Baker
Week Ending May 15

DAY	A.M. HOURS	P.M. HOURS	TOTAL HOURS
M	3½	4	7½
T	2¼	3	5½
W	4	2¼	6½
Th	3	4	7
F	4½	1½	6
		Total	32½

Fill in the time card. On which day did Jan earn $24 if she was paid $4 an hour? Friday

COUNT YOUR CHANGE!

You bought this record. You gave the clerk $2. What 5 coins did you get back in change?

quarter quarter
nickel   penny
penny

© D. C. Heath and Company   S248   Problem solving

Project

Reading a map

Use the map on page 8 to find the missing mileages on each of these road signs:

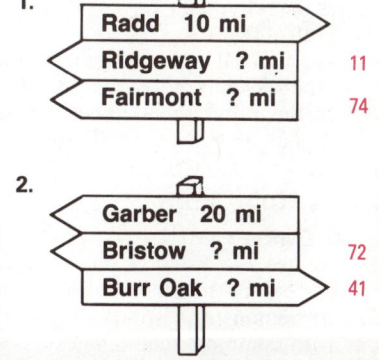

1. Radd 10 mi; Ridgeway ? mi 11; Fairmont ? mi 74
2. Garber 20 mi; Bristow ? mi 72; Burr Oak ? mi 41

Copymaster S479

Class Starter Quiz 74
on previous lesson

Solve. Use the signpost at the top of page 184.

1. Is the cave more or less than 7 miles from the exit? Less
2. How many miles is the round trip from the exit to the lookout tower and back again?
   $19\frac{1}{2}$

Copymaster S403

Lesson Objective

To subtract fractions with common denominators

Problem-Solving Skills

Finding information in a display
Choosing the correct operation
Making a drawing

Starting the Lesson

The dashboard gauges at the top of page 186 are also on ■ **Visual Aid 33** (copymaster or transparency S138).

What are the facts? Have the students study the dashboard displays for 30 seconds and then tell them to close their books. Challenge them to answer these true-or-false questions from memory:

- Before the rally, the gasoline tank was full. (False)
- After the rally, the tank was less than half full. (True)
- Before the rally, the odometer read less than 39,000 miles. (True)
- After the rally, the odometer read more than 39,000 miles. (False)

Here's How Note

Use of Concrete Materials You may wish to use the fraction pieces from copymasters or transparencies S530–S531 to demonstrate subtracting fractions with common denominators. See ■ **Manipulative Activity 13** on copymaster S518 in the Teacher's Resource Binder.

186

Subtracting fractions with common denominators

You won first prize at a sports-car rally.

BEFORE RALLY

AFTER RALLY

1. Fuel check! What fraction of a tank did you have before the rally? $\frac{3}{4}$
2. Fuel check! What fraction of a tank did you have after the rally? $\frac{1}{4}$
3. Would you add or subtract to compute what fraction of a tank you used? Subtract

Here's how to subtract fractions with a common denominator.

$$\frac{3}{4} - \frac{1}{4} = ?$$

MENTAL MATH
Subtract the numerators and use the common denominator.

$$\frac{3}{4} - \frac{1}{4} = \frac{2}{4}$$
$$= \frac{1}{2}$$

4. Look at the *Here's how*. What fraction of a tank did you use on the rally? $\frac{1}{2}$
5. Check these examples. Give each difference in simplest form.

   a. $\frac{7}{8} - \frac{3}{8} = \frac{4}{8}$
      $= ?\ \frac{1}{2}$

   b. $\frac{5}{12} - \frac{1}{12} = \frac{4}{12}$
      $= ?\ \frac{1}{3}$

   c. $\frac{5}{2} - \frac{2}{2} = \frac{3}{2}$
      $= ?\ 1\frac{1}{2}$

186 Chapter 7

EXERCISES

MENTAL MATH Subtract. Write the difference in simplest form.

Here are scrambled answers for the next row of exercises: $\frac{5}{6} \quad \frac{1}{4} \quad \frac{1}{6} \quad \frac{1}{2} \quad 0 \quad \frac{2}{5}$

6. $\frac{3}{5} - \frac{1}{5}$ $\frac{2}{5}$
7. $\frac{2}{4} - \frac{1}{4}$ $\frac{1}{4}$
8. $\frac{6}{6} - \frac{1}{6}$ $\frac{5}{6}$
9. $\frac{4}{6} - \frac{3}{6}$ $\frac{1}{6}$
10. $\frac{5}{8} - \frac{1}{8}$ $\frac{1}{2}$
11. $\frac{3}{8} - \frac{3}{8}$ 0

12. $\frac{5}{8} - \frac{3}{8}$ $\frac{1}{4}$
13. $\frac{5}{9} - \frac{0}{9}$ $\frac{5}{9}$
14. $\frac{5}{4} - \frac{1}{4}$ 1
15. $\frac{7}{6} - \frac{3}{6}$ $\frac{2}{3}$
16. $\frac{5}{6} - \frac{1}{6}$ $\frac{2}{3}$
17. $\frac{6}{4} - \frac{2}{4}$ 1

18. $\frac{3}{5} - \frac{0}{5}$ $\frac{3}{5}$
19. $\frac{11}{8} - \frac{5}{8}$ $\frac{3}{4}$
20. $\frac{5}{9} - \frac{2}{9}$ $\frac{1}{3}$
21. $\frac{12}{8} - \frac{6}{8}$ $\frac{3}{4}$
22. $\frac{11}{4} - \frac{3}{4}$ 2
23. $\frac{7}{4} - \frac{4}{4}$ $\frac{3}{4}$

24. $\frac{9}{4} - \frac{3}{4}$ $1\frac{1}{2}$
25. $\frac{5}{8} - \frac{4}{8}$ $\frac{1}{8}$
26. $\frac{8}{6} - \frac{2}{6}$ 1
27. $\frac{5}{3} - \frac{2}{3}$ 1
28. $\frac{7}{6} - \frac{2}{6}$ $\frac{5}{6}$
29. $\frac{10}{9} - \frac{4}{9}$ $\frac{2}{3}$

30. $\frac{12}{4} - \frac{3}{4}$ $2\frac{1}{4}$
31. $\frac{7}{8} - \frac{3}{8}$ $\frac{1}{2}$
32. $\frac{13}{4} - \frac{3}{4}$ $2\frac{1}{2}$
33. $\frac{10}{8} - \frac{2}{8}$ 1
34. $\frac{9}{6} - \frac{6}{6}$ $\frac{1}{2}$
35. $\frac{10}{4} - \frac{2}{4}$ 2

36. $\frac{2}{3} - \frac{2}{3}$ 0
37. $\frac{7}{4} - \frac{3}{4}$ 1
38. $\frac{8}{9} - \frac{3}{9}$ $\frac{5}{9}$
39. $\frac{4}{5} - \frac{2}{5}$ $\frac{2}{5}$
40. $\frac{9}{8} - \frac{3}{8}$ $\frac{3}{4}$
41. $\frac{13}{7} - \frac{8}{7}$ $\frac{5}{7}$

42. $\frac{12}{10} - \frac{7}{10}$ $\frac{1}{2}$
43. $\frac{7}{8} - \frac{4}{8}$ $\frac{3}{8}$
44. $\frac{6}{4} - \frac{2}{4}$ 1
45. $\frac{5}{6} - \frac{5}{6}$ 0
46. $\frac{6}{5} - \frac{3}{5}$ $\frac{3}{5}$
47. $\frac{5}{6} - \frac{3}{6}$ $\frac{1}{3}$

Solve. Look at the top of page 186.

48. How many miles had your car been driven before the rally? 38,016

49. How many miles had your car been driven after the rally? 38,186

50. How many miles was the rally? *Hint: Use your answers to problems 48 and 49.* 170

51. If you used 7.8 gallons of gasoline on the rally, how many miles did you average per gallon? Round the answer to the nearest tenth. 21.8

Where are they? Making a sketch

52. Study the clues to find out how far Carol and Joe are from the checkpoint at Clara's Corner. *Hint: Make a sketch.*

    Clues: 9.9 miles
    - They left Bruskville and drove 18.6 miles east to Clara's Corner.
    - They turned right on Highway 26 and drove 13.9 miles to the second checkpoint at Harold's Hollow.
    - They drove 6.4 miles beyond Harold's Hollow.
    - They turned around and started back for Clara's Corner. They drove 10.4 miles toward Clara's Corner.

Adding and Subtracting Fractions and Mixed Numbers

Practice Worksheet

Workbook S249, Copymaster S249, or Duplicating Master S249

WORKSHEET 81 (Use after page 187.)

RIDDLE TIME
What did one tailpipe say to the other?

Use a ruler to connect expressions for the same number. Then use the DECODER to get the answer.

1. $\frac{3}{5} - \frac{1}{5}$		$\frac{5}{8} - \frac{1}{8}$ O
2. $\frac{3}{4} - \frac{1}{4}$		$\frac{4}{4} - \frac{1}{4}$ M
3. $\frac{2}{4} - \frac{1}{4}$		$\frac{3}{16} - \frac{1}{16}$ X
4. $\frac{2}{4} - \frac{1}{4}$		$\frac{5}{6} - \frac{1}{6}$ U
5. $\frac{11}{8} - \frac{5}{8}$		$\frac{7}{10} - \frac{1}{10}$ B
6. $\frac{7}{8} - \frac{3}{8}$		$\frac{9}{8} - \frac{1}{8}$ I
7. $\frac{7}{10} - \frac{4}{10}$		$\frac{5}{8} - \frac{1}{4}$ A
8. $\frac{7}{8} - \frac{6}{8}$		$\frac{9}{4} - \frac{1}{4}$ H
9. $\frac{5}{8} - \frac{1}{8}$		$\frac{7}{6} - \frac{5}{6}$ Y
10. $\frac{10}{9} - \frac{5}{9}$		$\frac{5}{9} - \frac{1}{9}$ S
11. $\frac{5}{3} - \frac{3}{3}$		$\frac{4}{10} - \frac{1}{10}$ E
12. $\frac{4}{9} - \frac{0}{9}$		$\frac{7}{10} - \frac{2}{10}$ T
13. $\frac{3}{3} - \frac{3}{3}$		$\frac{1}{7} - \frac{2}{7}$ D
14. $\frac{7}{8} - \frac{2}{8}$		$\frac{11}{18} - \frac{1}{18}$ A
15. $\frac{4}{7} - \frac{1}{7}$		$\frac{6}{3} - \frac{1}{3}$ E

DECODER

1	2	3	4	5	6	7	8	9	10	11	12	13	14	15
B	O	Y	A	M	I	E	X	H	A	U	S	T	E	D

Answer: BOY, AM I EXHAUSTED.

Subtracting fractions with a common denominator

Project

Using library resources

Use an almanac to find your favorite sports car's gas mileage: the number of miles the car can be driven on one gallon of gas. Use what you found to compute the number of gallons required to drive 170 miles in a sports car rally.

Copymaster S479

Class Starter Quiz 75
on previous lesson

Subtract. Write the difference in simplest form.

1. $\frac{3}{8} - \frac{1}{8}$ $\frac{1}{4}$
2. $\frac{3}{4} - \frac{1}{4}$ $\frac{1}{2}$
3. $\frac{11}{8} - \frac{3}{8}$ 1
4. $\frac{5}{6} - \frac{1}{6}$ $\frac{2}{3}$
5. $\frac{7}{8} - \frac{1}{8}$ $\frac{3}{4}$
6. $\frac{5}{3} - \frac{2}{3}$ 1

Copymaster S403

Lesson Objective
To subtract fractions with different denominators

Problem-Solving Skills
Reading a circle graph
Checking answers

Starting the Lesson
Sketch this circle graph on the chalkboard:

FAVORITE AFTER-SCHOOL SPORTS OF TEENAGERS

Swimming
Jogging
Cycling
Other Sports

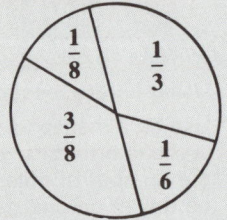

Before the students open their books, have them guess which fraction of the circle graph represents each sport. Then say, "Open your books to page 188 and check your guesses with the graph at the top of the page."

Here's How Note
Use of Concrete Materials You may wish to use the fraction pieces from copymasters or transparencies S530–S531 to demonstrate subtracting fractions with different denominators. See ■ **Manipulative Activity 14** on copymaster S518 in the Teacher's Resource Binder.

188

Subtracting fractions with different denominators

The circle graph shows the results of a teenage survey. Each teenager was asked to name his/her favorite after-school sport.

1. What fraction of those surveyed preferred swimming? $\frac{1}{6}$
2. What fraction preferred cycling? $\frac{3}{8}$
3. What fraction preferred jogging? $\frac{1}{3}$
4. Would you add or subtract to find how much greater the fraction for cycling is than the fraction for jogging? Subtract

FAVORITE AFTER-SCHOOL SPORTS OF TEENAGERS

Here's how *to subtract fractions with different denominators.* $\frac{3}{8} - \frac{1}{3} = ?$

Find the least common denominator.	Change to equivalent fractions.	Subtract.
$\frac{3}{8}$ $-\frac{1}{3}$ 24	$\frac{3}{8} = \frac{9}{24}$ $-\frac{1}{3} = -\frac{8}{24}$	$\frac{3}{8} = \frac{9}{24}$ $-\frac{1}{3} = -\frac{8}{24}$ $\frac{1}{24}$

5. Look at the *Here's how*. How much greater was the fraction of teenagers who preferred cycling than the fraction who preferred jogging? $\frac{1}{24}$

6. Check these examples. Give each difference in simplest form.

a. $\frac{7}{3}$
   $-\frac{1}{3}$
   $\frac{6}{3} = ?$ 2

b. $\frac{5}{6} = \frac{5}{6}$
   $-\frac{1}{3} = -\frac{2}{6}$
   $\frac{3}{6} = ?$ $\frac{1}{2}$

c. $\frac{5}{3} = \frac{25}{15}$
   $-\frac{3}{5} = -\frac{9}{15}$
   $\frac{16}{15} = ?$ $1\frac{1}{15}$

Chapter 7

EXERCISES

Subtract. Give the difference in simplest form.
Here are scrambled answers for the next row of exercises: $\frac{3}{4}$ $\frac{3}{10}$ $\frac{1}{9}$ $\frac{1}{3}$ $\frac{5}{12}$ $\frac{13}{24}$ $\frac{1}{4}$

7. $\frac{2}{3} - \frac{5}{9} = \frac{1}{9}$
8. $\frac{3}{2} - \frac{3}{4} = \frac{3}{4}$
9. $\frac{9}{10} - \frac{3}{5} = \frac{3}{10}$
10. $\frac{5}{6} - \frac{1}{2} = \frac{1}{3}$
11. $\frac{7}{8} - \frac{1}{3} = \frac{13}{24}$
12. $\frac{7}{12} - \frac{1}{3} = \frac{1}{4}$
13. $\frac{3}{4} - \frac{1}{3} = \frac{5}{12}$

14. $\frac{5}{6} - \frac{2}{3} = \frac{1}{6}$
15. $\frac{3}{4} - \frac{3}{8} = \frac{3}{8}$
16. $\frac{2}{3} - \frac{5}{12} = \frac{1}{4}$
17. $\frac{3}{2} - \frac{7}{8} = \frac{5}{8}$
18. $\frac{5}{6} - \frac{0}{3} = \frac{5}{6}$
19. $\frac{3}{4} - \frac{2}{3} = \frac{1}{12}$
20. $\frac{5}{6} - \frac{1}{4} = \frac{7}{12}$

21. $\frac{3}{8} - \frac{1}{3} = \frac{1}{24}$
22. $\frac{5}{9} - \frac{1}{6} = \frac{7}{18}$
23. $\frac{7}{4} - \frac{3}{4} = 1$
24. $\frac{9}{5} - \frac{3}{10} = 1\frac{1}{2}$
25. $\frac{7}{8} - \frac{5}{6} = \frac{1}{24}$
26. $\frac{1}{2} - \frac{1}{3} = \frac{1}{6}$
27. $\frac{7}{10} - \frac{1}{5} = \frac{1}{2}$

Solve. Use the circle graph on page 188.

28. How much greater was the fraction of teens who preferred jogging than the fraction of teens who preferred swimming? $\frac{1}{6}$

29. How much greater was the fraction of teens who preferred cycling than the fraction of teens who preferred swimming? $\frac{5}{24}$

30. What fraction of those surveyed preferred either jogging or swimming? $\frac{1}{2}$

31. What fraction did not prefer the three most-preferred sports? $\frac{1}{8}$

32. What fraction did not choose jogging? $\frac{2}{3}$

33. What fraction did not choose either cycling or swimming? $\frac{11}{24}$

Check the differences Checking answers

34. Find and correct the two wrong answers.

a. Four and three tenths minus three and six tenths equals
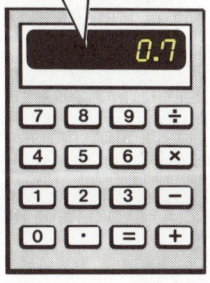
0.86

b. Two and sixteen hundredths minus one and three tenths equals
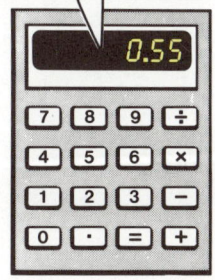

c. Three and seven tenths minus fifty-one hundredths equals
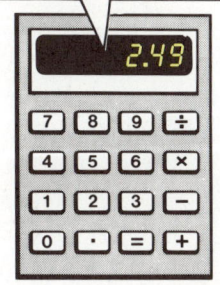
3.19

Adding and Subtracting Fractions and Mixed Numbers

Extra Practice
Page 487   Skill 34

Practice Worksheet
Workbook S250, Copymaster S250, or Duplicating Master S250

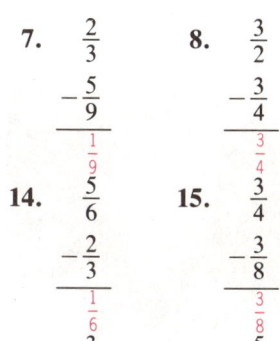

Challenge Problems

Find the missing digits.

1. $\frac{5}{\boxed{8}} - \frac{1}{4} = \frac{3}{8}$
2. $\frac{3}{4} - \frac{1}{\boxed{3}} = \frac{5}{12}$
3. $\frac{\boxed{5}}{9} - \frac{1}{3} = \frac{2}{9}$
4. $\frac{5}{6} - \frac{\boxed{2}}{3} = \frac{1}{6}$

Copymaster S450

Class Starter Quiz 76
on previous lesson

Subtract. Write the difference in simplest form.

1. $\frac{3}{4} - \frac{1}{2}$ $\frac{1}{4}$ 2. $\frac{1}{2} - \frac{2}{5}$ $\frac{1}{10}$
3. $\frac{7}{8} - \frac{1}{2}$ $\frac{3}{8}$ 4. $\frac{3}{2} - \frac{7}{8}$ $\frac{5}{8}$
5. $\frac{7}{10} - \frac{3}{5}$ $\frac{1}{10}$ 6. $\frac{2}{3} - \frac{1}{4}$ $\frac{5}{12}$

Copymaster S403

Lesson Objective
To subtract mixed numbers without regrouping

Problem-Solving Skills
Selecting information from a display
Using logical reasoning

Starting the Lesson
What are the facts? Allow students 30 seconds to study the comments at the top of page 190. Then have them close their books and try to decide from memory whether these statements are true or false:

- Steve said he was $68\frac{3}{4}$ inches tall. (True)
- Arlo said he was shorter than Steve. (True)
- The girl said she was shorter than Steve. (True)
- The girl's name is Polly. (False)

190

Subtracting mixed numbers without regrouping

I'm $3\frac{1}{4}$ inches shorter than Steve.

Arlo

I'm the shortest. The difference between my height and Steve's is $4\frac{1}{2}$ inches.

Holly

I'm $68\frac{3}{4}$ inches tall.

Steve

1. Who is the shortest? Holly
2. Who is the tallest? Steve
3. To find how many inches tall Holly is, you subtract $4\frac{1}{2}$ from what number? $68\frac{3}{4}$

Here's how to subtract mixed numbers. $68\frac{3}{4} - 4\frac{1}{2} = ?$

Write equivalent fractions with a common denominator.

$68\frac{3}{4} = 68\frac{3}{4}$
$- 4\frac{1}{2} = -4\frac{2}{4}$

Subtract the fractions.
Subtract the whole numbers.

$68\frac{3}{4} = 68\frac{3}{4}$
$- 4\frac{1}{2} = -4\frac{2}{4}$
$\phantom{- 4\frac{1}{2} = -}64\frac{1}{4}$

4. Look at the *Here's how*. How tall is Holly? $64\frac{1}{4}$ inches

5. Check these examples. Is Arlo $65\frac{1}{2}$ inches or 72 inches tall? $65\frac{1}{2}$

a. $68\frac{3}{4}$
   $+ 3\frac{1}{4}$
   $\overline{71\frac{4}{4}} = 72$

b. $68\frac{3}{4}$
   $- 3\frac{1}{4}$
   $\overline{65\frac{2}{4}} = 65\frac{1}{2}$

190 Chapter 7

EXERCISES

Subtract. Give each difference in simplest form.

Here are scrambled answers for the next row of exercises: $3\frac{3}{8}$ $4\frac{1}{8}$ $2\frac{1}{4}$ $1\frac{1}{4}$ $5\frac{1}{4}$ $2\frac{3}{8}$

6. $4\frac{3}{8} - 2\frac{1}{8} = 2\frac{1}{4}$
7. $8\frac{1}{2} - 5\frac{1}{8} = 3\frac{3}{8}$
8. $10\frac{3}{4} - 5\frac{1}{2} = 5\frac{1}{4}$
9. $6\frac{7}{8} - 2\frac{3}{4} = 4\frac{1}{8}$
10. $3\frac{1}{2} - 2\frac{1}{4} = 1\frac{1}{4}$
11. $4\frac{5}{8} - 2\frac{1}{4} = 2\frac{3}{8}$

12. $19\frac{4}{5} - 8\frac{3}{10} = 11\frac{1}{2}$
13. $17\frac{5}{12} - 14\frac{1}{3} = 3\frac{1}{12}$
14. $36\frac{1}{2} - 25\frac{3}{10} = 11\frac{1}{5}$
15. $17\frac{3}{8} - 12\frac{1}{8} = 5\frac{1}{4}$
16. $3\frac{4}{5} - \frac{3}{5} = 3\frac{1}{5}$
17. $2\frac{5}{9} - 1\frac{1}{3} = 1\frac{2}{9}$

18. $5\frac{1}{3} - 2\frac{1}{4} = 3\frac{1}{12}$
19. $3\frac{2}{3} - 1\frac{1}{2} = 2\frac{1}{6}$
20. $12\frac{3}{4} - 1\frac{1}{6} = 11\frac{7}{12}$
21. $8\frac{3}{4} - 2\frac{2}{5} = 6\frac{7}{20}$
22. $9\frac{1}{4} - 3\frac{1}{6} = 6\frac{1}{12}$
23. $10\frac{2}{9} - 3\frac{1}{10} = 7\frac{11}{90}$

24. $12\frac{3}{5} - 3\frac{1}{10}$ $9\frac{1}{2}$
25. $16\frac{1}{2} - 8\frac{1}{4}$ $8\frac{1}{4}$
26. $38\frac{1}{2} - 21\frac{3}{10}$ $17\frac{1}{5}$
27. $15\frac{3}{4} - 9\frac{1}{8}$ $6\frac{5}{8}$

28. $9\frac{7}{8} - 3\frac{1}{2}$ $6\frac{3}{8}$
29. $15\frac{1}{4} - 3\frac{1}{6}$ $12\frac{1}{12}$
30. $2\frac{7}{10} - 1\frac{3}{5}$ $1\frac{1}{10}$
31. $8\frac{4}{5} - 6\frac{11}{20}$ $2\frac{1}{4}$

32. $16\frac{1}{2} - 4\frac{1}{5}$ $12\frac{3}{10}$
33. $3\frac{5}{12} - 1\frac{1}{4}$ $2\frac{1}{6}$
34. $6\frac{4}{9} - 2\frac{1}{3}$ $4\frac{1}{9}$
35. $7\frac{5}{8} - 6\frac{11}{24}$ $1\frac{1}{6}$

How many pounds?

Solve. Use the clues.

36. How much does Holly weigh? $110\frac{1}{2}$ pounds
    Clues: • Steve guessed 115 pounds and missed by $4\frac{1}{2}$ pounds.
    • Arlo guessed 100 pounds and missed by $10\frac{1}{2}$ pounds.

37. How much does Arlo weigh? $145\frac{3}{4}$ pounds
    Clues: • Holly guessed 140 pounds and missed by $5\frac{3}{4}$ pounds.
    • Steve guessed 148 pounds and missed by $2\frac{1}{4}$ pounds.

38. How much does Steve weigh? $143\frac{1}{2}$ pounds
    Clues: • Arlo guessed 160 pounds and missed by $16\frac{1}{2}$ pounds.
    • Holly guessed 150 pounds and missed by $6\frac{1}{2}$ pounds.

Logical reasoning

39. Who do you think is the best guesser, Arlo, Holly, or Steve? Why?
    Answers will vary.

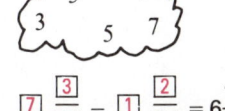

Adding and Subtracting Fractions and Mixed Numbers

Class Starter Quiz 77
on previous lesson

Subtract. Write the difference in simplest form.

1. $12\frac{1}{2} - 5\frac{1}{4}$ $7\frac{1}{4}$
2. $5\frac{7}{8} - 3\frac{1}{4}$ $2\frac{5}{8}$
3. $6\frac{7}{10} - 2\frac{1}{5}$ $4\frac{1}{2}$
4. $7\frac{4}{5} - 2\frac{3}{10}$ $5\frac{1}{2}$
5. $5\frac{1}{4} - 3\frac{1}{6}$ $2\frac{1}{12}$
6. $6\frac{1}{2} - 3\frac{2}{5}$ $3\frac{1}{10}$

Copymaster S404

Lesson Objective
To subtract mixed numbers with regrouping

Problem-Solving Skills
Selecting information from a chart
Reading a map and a map scale

Starting the Lesson
Before discussing exercises 1–3, have the students use the chart at the top of page 192 to answer these questions:

- On a clear day, how far can you see from a hot-air balloon at a height of 50 feet? $\left(8\frac{1}{2}\text{ miles}\right)$
- Can you see more or less than 60 miles at a height of 3000 feet? (More)
- Are you above or below 100 feet if you can see a distance of 10 miles? (Below)

Exercise Note
Use of Concrete Materials The map at the bottom of page 193 is also on ■ **Visual Aid 34** (copymaster or transparency S139). Use the map when discussing exercises 38–40. You may want to make duplicate copies of ■ **Visual Aid 34** for students to use. They can cut out the scale of miles and use it to measure distances between cities.

192

Subtracting mixed numbers with regrouping

At 300 feet I can see 22 miles!

HEIGHT OF BALLOON (IN FEET)	DISTANCE SEEN FROM BALLOON (IN MILES)
10	$3\frac{7}{8}$
50	$8\frac{1}{2}$
100	$12\frac{3}{8}$
200	$17\frac{3}{8}$
300	22
1000	$38\frac{3}{4}$
3000	$67\frac{1}{8}$

1. Your hot-air balloon is 3000 feet above the ground. Your friend's balloon is at 100 feet. Look at the chart. What two numbers would you use to compute how many more miles you can see than your friend? $67\frac{1}{8}$ and $12\frac{3}{8}$

Here's how *to subtract mixed numbers.* $67\frac{1}{8} - 12\frac{3}{8} = ?$

Not enough eighths!	Regroup 1 for $\frac{8}{8}$.	Subtract.
$67\frac{1}{8}$ $-12\frac{3}{8}$	$\overset{66\ 9}{\cancel{67}\frac{\cancel{1}}{8}}$ $-12\frac{3}{8}$	$\overset{66\ 9}{\cancel{67}\frac{\cancel{1}}{8}}$ $-12\frac{3}{8}$ $54\frac{6}{8} = 54\frac{3}{4}$

2. Look at the *Here's how*. How many miles farther can you see? $54\frac{3}{4}$

3. Study these examples. Is each difference in simplest form? Yes

a. $22 = 21\frac{2}{2}$ ◁ Regrouped 1 for $\frac{2}{2}$.
    $-8\frac{1}{2} = -8\frac{1}{2}$
    _____
    $13\frac{1}{2}$

b. $8\frac{1}{2} = \overset{7\ 12}{\cancel{8}\frac{\cancel{4}}{8}}$ ◁ Changed to a common denominator. Regrouped 1 for $\frac{8}{8}$.
    $-3\frac{7}{8} = -3\frac{7}{8}$
    _____
    $4\frac{5}{8}$

192 Chapter 7

EXERCISES

Subtract. Write the difference in simplest form.
Here are scrambled answers for the next row of exercises: $3\frac{1}{4}$ $4\frac{3}{5}$ $7\frac{2}{9}$ $3\frac{3}{4}$ $2\frac{6}{7}$ $3\frac{4}{5}$

4. $6\frac{1}{8} - 2\frac{3}{8} = 3\frac{3}{4}$
5. $8\frac{7}{9} - 1\frac{5}{9} = 7\frac{2}{9}$
6. $9\frac{1}{8} - 5\frac{7}{8} = 3\frac{1}{4}$
7. $6\frac{5}{7} - 3\frac{6}{7} = 2\frac{6}{7}$
8. $8\frac{2}{5} - 3\frac{4}{5} = 4\frac{3}{5}$
9. $6 - 2\frac{1}{5} = 3\frac{4}{5}$

10. $4\frac{3}{7} - 1\frac{5}{7} = 2\frac{5}{7}$
11. $6\frac{5}{9} - 3\frac{8}{9} = 2\frac{2}{3}$
12. $7 - 3\frac{1}{2} = 3\frac{1}{2}$
13. $12\frac{1}{2} - 6\frac{1}{4} = 6\frac{1}{4}$
14. $20 - 8\frac{5}{8} = 11\frac{3}{8}$
15. $2\frac{2}{9} - 1\frac{1}{3} = \frac{8}{9}$

16. $4\frac{1}{8} - 2\frac{1}{2} = 1\frac{5}{8}$
17. $5\frac{3}{4} - 3\frac{1}{2} = 2\frac{1}{4}$
18. $9\frac{1}{10} - 3\frac{1}{5} = 5\frac{9}{10}$
19. $5\frac{3}{8} - 2\frac{1}{4} = 3\frac{1}{8}$
20. $14\frac{1}{8} - 6\frac{3}{4} = 7\frac{3}{8}$
21. $13\frac{1}{4} - 2\frac{1}{2} = 10\frac{3}{4}$

22. $8\frac{1}{3} - 2\frac{1}{2} = 5\frac{5}{6}$
23. $10\frac{1}{4} - 2\frac{2}{3} = 7\frac{7}{12}$
24. $6\frac{3}{8} - 4\frac{5}{6} = 1\frac{13}{24}$
25. $7\frac{7}{8} - 1\frac{3}{4} = 6\frac{1}{8}$
26. $15\frac{7}{8} - 12\frac{1}{4} = 3\frac{5}{8}$
27. $14\frac{1}{6} - 3\frac{2}{3} = 10\frac{1}{2}$

28. $6\frac{3}{4} - 1\frac{3}{5}$ $5\frac{3}{20}$
29. $8\frac{1}{2} - 4\frac{2}{3}$ $3\frac{5}{6}$
30. $7\frac{1}{5} - 2\frac{1}{10}$ $5\frac{1}{10}$
31. $10\frac{3}{4} - 3\frac{7}{8}$ $6\frac{7}{8}$
32. $10 - 6\frac{1}{2}$ $3\frac{1}{2}$

33. $29\frac{2}{3} - 8\frac{1}{2}$ $21\frac{1}{6}$
34. $24\frac{1}{2} - 1\frac{3}{4}$ $22\frac{3}{4}$
35. $14\frac{2}{3} - 3\frac{1}{2}$ $11\frac{1}{6}$
36. $10\frac{5}{8} - 4\frac{3}{4}$ $5\frac{7}{8}$
37. $30 - 8\frac{3}{4}$ $21\frac{1}{4}$

Oh, say can you see? Reading a map

Use the map scale and the chart on page 192 to solve these problems.

38. You are in a hot-air balloon 100 feet above Boone, Nebraska. What 4 cities can you see? *Boone, St. Edward, Albion, Newman Grove*
39. Now you are 300 feet above Petersburg. What 9 cities can you see? *Petersburg, Neligh, Tilden, Elgin, Newman Grove, Albion, Spalding, Boone, St. Edward*
40. When you are 3000 feet above O'Neill, can you see as far as Greeley? *Yes*

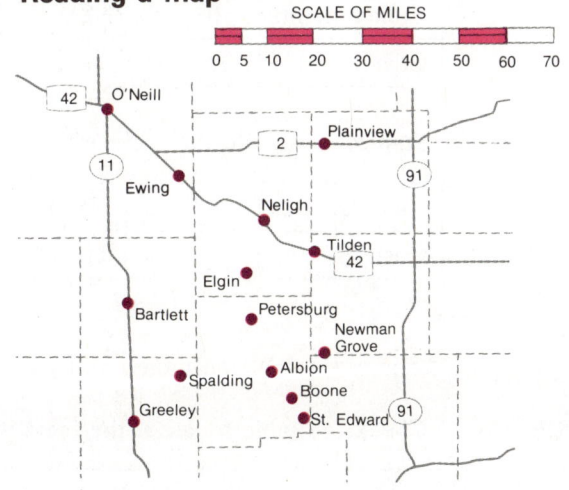

Extra Practice
Page 488 Skill 35

Practice Worksheet
Workbook S252, Copymaster S252, or Duplicating Master S252

Challenge Problem

Complete this Magic Square so that the sums along every row, column, and diagonal are $16\frac{7}{8}$.

9	$1\frac{1}{8}$	$6\frac{3}{4}$
$3\frac{3}{8}$	$5\frac{5}{8}$	$7\frac{7}{8}$
$4\frac{1}{2}$	$10\frac{1}{8}$	$2\frac{1}{4}$

Copymaster S451

PROBLEM SOLVING *making a table*

Some problems can be solved by making a table.

Problem
How much would you pay for 9 enlargements?

Here's how *to solve the problem by making a table and finding a pattern.*

STUDY   Read the ad. What are the facts?

PLAN and SOLVE   **Step 1.** Make rows in a table for the number of enlargements and the cost.

Enlargements										
Cost										

Step 2. Read the ad for buying enlargements. Fill in the facts.

Enlargements	0	1	2	3						
Cost	$0	$3	$6	$6						

Step 3. Complete the table. Look for a pattern.

Enlargements	0	1	2	3	4	5	6	7	8	9
Cost	$0	$3	$6	$6	$9	$12	$12	$15	$18	$18

+3 +3 +0 +3 +3 +0 +3 +3 +0

Answer: Look at the table. You would pay $18 for 9 enlargements.

CHECK   Check the table to see if the facts were filled in correctly.

194   *Chapter 7*

PROBLEMS

Solve by completing the table and finding a pattern.

1.
   36 Exposures
   135 Film — $2.50
   BUY 1 — GET 1 FREE!

   How much will 6 rolls cost?

Rolls of 135 film	0	1	2	3	4	5	6
Cost	$0	$2.50	$2.50	$5.00	?	?	?

   $5.00, $7.50, $7.50

2.
   Flashcubes
   $1.60 a pack
   BUY 2 — GET 1 FREE!

   How much will 6 packs cost?

Packs of flashcubes	0	1	2	3	4	5	6
Cost	$0	$1.60	$3.20	$3.20	?	?	?

   $4.80, $6.40, $6.40

Solve by making a table and finding a pattern.

3. 20-exposure, 135 film costs $2.15 a roll. For every 3 rolls you buy, you get 1 free roll. How much will 8 rolls cost? $12.90

4. 5 × 7 enlargements cost $2.25 each. For every 4 enlargements you buy, you get 1 free. How much will 12 enlargements cost? $22.50

5. 24-exposure, 35-mm film costs $8.50 to develop. For every 3 rolls you develop, you get 1 roll developed free. How much will it cost to develop 15 rolls? $102

6. Photo-album pages cost $.25 each. For every 5 photo-album pages you buy, you get 1 free page. How much will 12 pages cost? $2.50

7. 12-exposure, 110 film costs $1.75 a roll. For every 5 rolls you buy, you get 2 free rolls. How much will 14 rolls cost? $17.50

8. Color prints cost $.25 each. For every 8 prints you buy, you get 1 free print. How many color prints can you get for $4? 18

Say cheese! Visual thinking

9. Which photo will you see when the film is developed?

Adding and Subtracting Fractions and Mixed Numbers

Class Starter Quiz 79
on previous lesson

Solve by making a table.

Posters cost $1.50 each. For every 3 posters that you buy, you get 1 free poster. How much will 12 posters cost? $13.50

Copymaster S404

Problem-Solving Skills

Using information on a display
Choosing the correct operation

Skills Reviewed

Computing the mean
Writing fractions as decimals or decimals as fractions
Adding and subtracting mixed numbers
Testing for divisibility by 3
Using prime factorization to find the GCF and the LCM of pairs of numbers

Starting the Lesson

Problem Solving Have the students study the Fuel Data Panel displays. Then ask the students which buttons were pushed to display:

- the miles per gallon the car is getting at this instant. (Button on the left)
- the number of miles you can travel on the fuel remaining in the tank. (Button on the right)
- the number of gallons of fuel used since the last fill-up. (Push the button on the right twice.)

Cumulative Skill Practice Write these four answers on the chalkboard:

$0.8 \quad \frac{9}{20} \quad 11\frac{7}{8} \quad 16\frac{1}{2}$

Challenge the students to an answer hunt by saying, "Look at exercises 1–47 on page 197. Find the four exercises that have these answers. You have four minutes to find as many of the exercises as you can." (Exercises 18, 29, 39, and 47)

196

Problem solving
COMPUTERS IN AUTOMOBILES

You recently purchased a new car that has an advanced electronics system and a built-in computer. You are in your car driving up a steep hill and decide to use the Fuel Data Panel.

Solve. Decide when a calculator would be useful.

There are three buttons on the panel. You press the one on the left. The digital display shows the miles per gallon your car is producing at this instant.

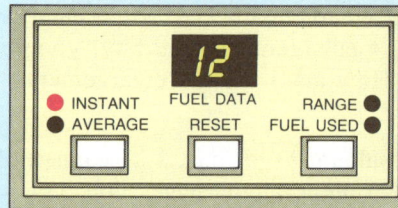

1. a. At this rate, how far will your car travel on 22 gallons of fuel? 264 miles

   b. If the rate improves to 18 miles per gallon, how much further will your car travel at this new rate? 132 miles

You want to know how many miles you can travel on the fuel remaining in the tank. You press the button on the right.

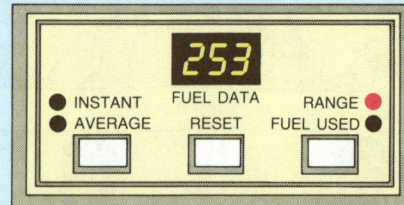

3. You plan to drive 185 miles to Lancaster and then drive on to Colebrook, another 91 miles. If the computer's estimate is accurate, can you make the trip to Colebrook without stopping for fuel? No

You want to know your car's average miles per gallon since the system was last reset. (To reset system, press middle button.) You press the button on the left a second time.

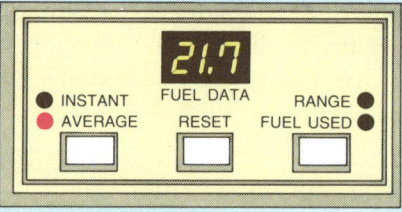

2. a. At this rate, how many gallons have been used if the car was driven 247 miles? (Round the answer to the nearest tenth of a gallon.) 11.4

   b. At this rate, how many miles would the car have been driven if 17.5 gallons of fuel had been used? 379.75

You want to know how many gallons of fuel you have used since your last fill-up. You press the button on the right a second time.

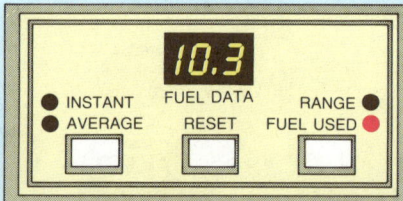

4. a. Your fuel tank holds 22 gallons of fuel. How many gallons are left in the tank? 11.7

   b. Gasoline costs $1.39 per gallon. Can you fill your tank for $15.00? Yes

196 Chapter 7

Cumulative Skill Practice

Find the mean. Round the answer to the nearest tenth. (page 124)

1. 19, 35, 28   27.3
2. 13, 42, 27   27.3
3. 28, 36, 1   21.7
4. 2.8, 4.7, 1.8   3.1
5. 1.8, 2.8, 1.9, 10.8   4.3
6. 28, 56, 32, 45   40.3
7. 13, 20, 17, 19, 15   16.8
8. 28, 14, 25, 13, 20   20
9. 3.7, 4.3, 4.4, 5.6   4.5

Write as a decimal. (page 166)

10. $\frac{1}{2}$ 0.5
11. $\frac{7}{8}$ 0.875
12. $\frac{3}{4}$ 0.75
13. $\frac{2}{5}$ 0.4
14. $\frac{5}{8}$ 0.625
15. $3\frac{1}{2}$ 3.5
16. $4\frac{3}{8}$ 4.375
17. $\frac{3}{10}$ 0.3
18. $\frac{4}{5}$ 0.8
19. $1\frac{3}{5}$ 1.6
20. $4\frac{1}{2}$ 4.5
21. $6\frac{1}{8}$ 6.125
22. $6\frac{2}{5}$ 6.4
23. $3\frac{1}{10}$ 3.1

Change to a fraction or mixed number in simplest form. (page 168)

24. 0.6 $\frac{3}{5}$
25. 0.08 $\frac{2}{25}$
26. 0.25 $\frac{1}{4}$
27. 0.8 $\frac{4}{5}$
28. 0.5 $\frac{1}{2}$
29. 0.45 $\frac{9}{20}$
30. 0.75 $\frac{3}{4}$
31. 0.625 $\frac{5}{8}$
32. 1.75 $1\frac{3}{4}$
33. 2.5 $2\frac{1}{2}$
34. 6.125 $6\frac{1}{8}$
35. 5.6 $5\frac{3}{5}$
36. 9.45 $9\frac{9}{20}$
37. 6.375 $6\frac{3}{8}$

Add. Give the sum in simplest form. (page 182)

38. $8\frac{1}{3} + 4\frac{1}{6} = 12\frac{1}{2}$
39. $5\frac{1}{8} + 6\frac{3}{4} = 11\frac{7}{8}$
40. $6\frac{5}{8} + 3\frac{1}{4} = 9\frac{7}{8}$
41. $3\frac{1}{6} + 5\frac{5}{12} = 8\frac{7}{12}$
42. $2\frac{1}{6} + 7\frac{2}{3} = 9\frac{5}{6}$

Subtract. Give the difference in simplest form. (pages 190, 192)

43. $6\frac{7}{8} - 3\frac{1}{8} = 3\frac{3}{4}$
44. $12\frac{5}{6} - 9\frac{1}{6} = 3\frac{2}{3}$
45. $23\frac{7}{8} - 17\frac{1}{4} = 6\frac{5}{8}$
46. $29\frac{2}{3} - 18\frac{1}{6} = 11\frac{1}{2}$
47. $24\frac{1}{3} - 7\frac{5}{6} = 16\frac{1}{2}$

MIXED PRACTICE
Complete.

48. The standard numeral for 10^3 is __?__   1000
49. A whole number is divisible by 3 if the sum of its digits is divisible by __?__   3
50. The prime factorization of 20 is __?__ $2 \times 2 \times 5$ or $2^2 \times 5$
51. The GCF of 18 and 24 is __?__   6
52. The LCM of 12 and 20 is __?__   60

Adding and Subtracting Fractions and Mixed Numbers

Problem-Solving Worksheet
Workbook S254, Copymaster S254, or Duplicating Master S254

WORKSHEET 86 (Use after page 196.)

CAN YOU PRICE IT?
How much does one baseball cost? $4
How much does one football cost? $7

WHAT'S THE ANSWER?
Use the code to answer the riddle.

CODE						
A	B	C	D	E	...	Z
$\frac{1}{3}$	$\frac{2}{3}$	1	$1\frac{1}{3}$	$1\frac{2}{3}$	...	$8\frac{1}{3}$

Riddle: How do you catch a squirrel?
Answer: CLIMB A TREE AND ACT LIKE A NUT

FRACTION ACTION
Here's how to use 1, 2, 3, and 4 to get $5\frac{1}{2}$.
$\frac{4}{1} + \frac{3}{2} = 5\frac{1}{2}$

Find a way to use 1, 2, 3, and 4 to get 5.
$\frac{4}{2} + \frac{3}{1} = 5$

FIND THE PATTERN
Look for the pattern. How many beads are inside the bag? 39

YOU'RE THE CLERK
What 5 coins would you give as change to a customer who bought a 19¢ pencil and gave you a dollar bill?
quarter quarter quarter nickel penny

© D. C. Heath and Company   S254   Problem solving

Challenge Problems

Look at page 197. Estimate which two decimals in exercises 24–37 have a sum of:

1. 2.75   2.5 + 0.25
2. 0.95   0.5 + 0.45
3. 6.2   0.6 + 5.6
4. 6.925   0.8 + 6.125

If you have a calculator, use it to see whether you made the correct choices.

Copymaster S451

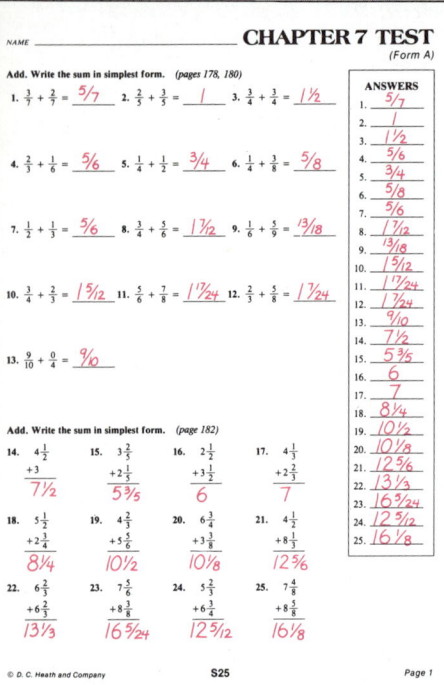
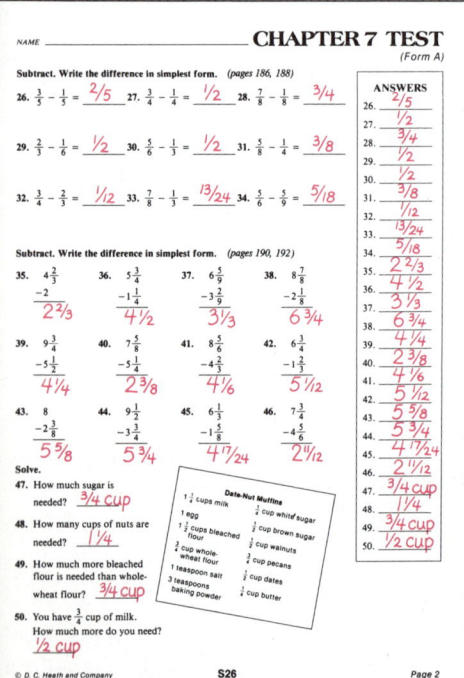

Chapter REVIEW

Here are scrambled answers for the review exercises:

8	add	equivalent	numerators	subtract
12	common	fractions	regroup	whole
20	denominator	numbers	simplest	

1. add   2. common, 12

1. To complete this sum, you would [?] the numerators and use the common denominator. *(page 178)*

$$\frac{3}{5} + \frac{1}{5} = \frac{?}{?}$$

2. To complete this sum, you would first change to equivalent fractions with a [?] denominator. The least common denominator is [?]. *(page 180)*

$$\frac{3}{4} + \frac{2}{3} = \frac{?}{?} + \frac{?}{?}$$

3. fractions, whole, simplest   4. subtract, denominator

3. To complete this addition exercise, you would add the [?] and then add the [?] numbers. The last step would be to write the sum in [?] form. *(page 182)*

$$2\frac{2}{3} = 2\frac{4}{6}$$
$$+1\frac{1}{2} = +1\frac{3}{6}$$

4. To complete this difference, you would [?] the numerators and use the common [?]. *(page 186)*

$$\frac{5}{9} - \frac{4}{9} = \frac{?}{?}$$

5. equivalent, numerators   6. 20, numbers

5. To complete this difference, you would first change to [?] fractions with a common denominator. Then you would subtract the [?] and use the common denominator. *(page 188)*

$$\frac{5}{6} - \frac{2}{3} = \frac{?}{?} - \frac{?}{?}$$

6. To complete this subtraction exercise, you would first change to equivalent fractions using the least common denominator, [?]. Then you would subtract the fractions and whole [?]. *(page 190)*

$$4\frac{4}{5} = 4\frac{?}{?}$$
$$-2\frac{1}{4} = -2\frac{?}{?}$$

7. regroup, 8

7. The next step in this subtraction exercise would be to [?] one for [?] eighths. *(page 192)*

$$5\frac{1}{8} = 5\frac{1}{8}$$
$$-3\frac{1}{2} = -3\frac{4}{8}$$

198 *Chapter 7*

Chapter TEST

Add. Write the sum in simplest form. *(pages 178, 180)*

1. $\frac{5}{9} + \frac{2}{9}$ = $\frac{7}{9}$
2. $\frac{1}{8} + \frac{3}{8}$ = $\frac{1}{2}$
3. $\frac{5}{6} + \frac{3}{6}$ = $1\frac{1}{3}$
4. $\frac{2}{3} + \frac{1}{3}$ = 1
5. $\frac{7}{9} + \frac{4}{9}$ = $1\frac{2}{9}$
6. $\frac{1}{6} + \frac{2}{3}$ = $\frac{5}{6}$
7. $\frac{1}{2} + \frac{1}{3}$ = $\frac{5}{6}$
8. $\frac{5}{8} + \frac{3}{4}$ = $1\frac{3}{8}$
9. $\frac{3}{4} + \frac{2}{3}$ = $1\frac{5}{12}$
10. $\frac{3}{4} + \frac{1}{2}$ = $1\frac{1}{4}$

Add. Write the sum in simplest form. *(page 182)*

11. $3\frac{1}{3} + 2\frac{1}{3} = 5\frac{2}{3}$
12. $2\frac{1}{2} + 1\frac{1}{2} = 4$
13. $5\frac{3}{5} + 3\frac{1}{5} = 8\frac{4}{5}$
14. $4\frac{3}{8} + 4\frac{1}{8} = 8\frac{1}{2}$
15. $2\frac{4}{9} + 4\frac{5}{9} = 7$
16. $4\frac{3}{7} + 8\frac{5}{7} = 13\frac{1}{7}$
17. $5\frac{2}{5} + 2\frac{1}{2} = 7\frac{9}{10}$
18. $3\frac{1}{2} + 6\frac{1}{3} = 9\frac{5}{6}$
19. $4\frac{2}{3} + 5\frac{5}{6} = 10\frac{1}{2}$
20. $6\frac{3}{4} + 2\frac{5}{6} = 9\frac{7}{12}$
21. $5\frac{7}{8} + 5\frac{1}{2} = 11\frac{3}{8}$
22. $4\frac{1}{5} + 3\frac{1}{10} = 7\frac{3}{10}$

Subtract. Write the difference in simplest form. *(pages 186, 188)*

23. $\frac{4}{5} - \frac{1}{5} = \frac{3}{5}$
24. $\frac{5}{6} - \frac{1}{6} = \frac{2}{3}$
25. $\frac{5}{8} - \frac{1}{8} = \frac{1}{2}$
26. $\frac{8}{9} - \frac{2}{9} = \frac{2}{3}$
27. $\frac{5}{7} - \frac{1}{7} = \frac{4}{7}$
28. $\frac{1}{2} - \frac{1}{4} = \frac{1}{4}$
29. $\frac{3}{4} - \frac{1}{3} = \frac{5}{12}$
30. $\frac{5}{9} - \frac{1}{6} = \frac{7}{18}$
31. $\frac{7}{8} - \frac{1}{3} = \frac{13}{24}$
32. $\frac{1}{2} - \frac{3}{8} = \frac{1}{8}$

Subtract. Write the difference in simplest form. *(pages 190, 192)*

33. $5\frac{3}{5} - 2\frac{1}{5} = 3\frac{2}{5}$
34. $6\frac{3}{4} - 4\frac{1}{4} = 2\frac{1}{2}$
35. $7\frac{2}{3} - 1\frac{1}{6} = 6\frac{1}{2}$
36. $8\frac{4}{5} - 3\frac{1}{4} = 5\frac{11}{20}$
37. $6\frac{7}{8} - 2\frac{2}{3} = 4\frac{5}{24}$
38. $9\frac{4}{5} - 3\frac{1}{2} = 6\frac{3}{10}$
39. $9\frac{1}{4} - 6\frac{1}{2} = 2\frac{3}{4}$
40. $7\frac{1}{2} - 4\frac{3}{5} = 2\frac{9}{10}$
41. $8 - 5\frac{5}{6} = 2\frac{1}{6}$
42. $5 - 2\frac{7}{8} = 2\frac{1}{8}$
43. $6\frac{2}{3} - 3\frac{3}{4} = 2\frac{11}{12}$
44. $8\frac{1}{3} - 2\frac{4}{9} = 5\frac{8}{9}$

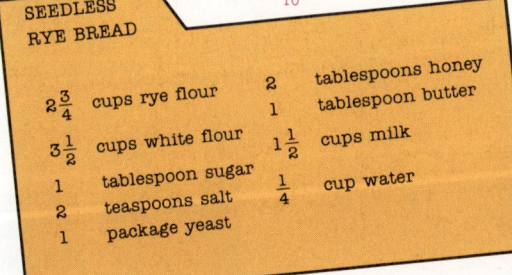

SEEDLESS RYE BREAD

$2\frac{3}{4}$ cups rye flour
$3\frac{1}{2}$ cups white flour
1 tablespoon sugar
2 teaspoons salt
1 package yeast
2 tablespoons honey
1 tablespoon butter
$1\frac{1}{2}$ cups milk
$\frac{1}{4}$ cup water

Solve. Write the answer in simplest form.

45. How many cups of flour are needed? $6\frac{1}{4}$

46. You have $\frac{3}{4}$ cup of milk. How much more do you need? $\frac{3}{4}$

Adding and Subtracting Fractions and Mixed Numbers **199**

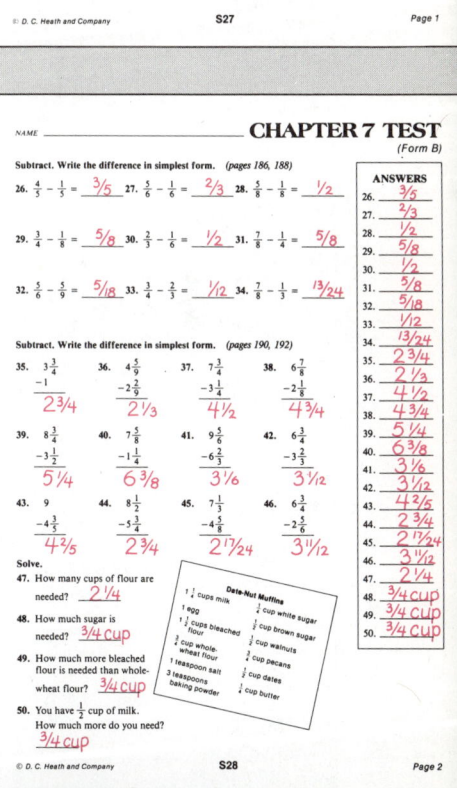

Cumulative Test
(Chapters 1–7)

Use Copymaster S109 to provide the students with an answer sheet in standardized test format.

Answers for Cumulative Test, Chapters 1–7

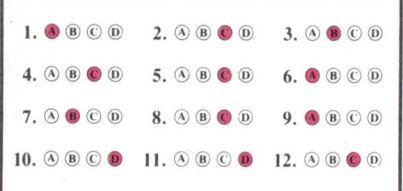

The table below correlates test items with student text pages.

Test Item	Page(s) Taught	Skill Practice
1	22	p. 185, exercises 1–9
2	46	p. 185, exercises 10–18
3	68	p. 185, exercises 19–27
4	100	p. 185, exercises 28–36
5	104	p. 185, exercises 37–48
6	124	p. 197, exercises 1–9
7	166	p. 197, exercises 10–23
8	168	p. 197, exercises 24–37
9	182	p. 197, exercises 38–42
10	190, 192	p. 197, exercises 43–47
11	118	
12	192	

200

Cumulative TEST — Standardized Format

Choose the correct letter.

1. Give the sum.
$81.6 + 42.53 + 231.5$
A. 355.63
B. 73.84
C. 354.63
D. none of these

2. Give the difference.
$64.03 - 2.946$
A. 34.57
B. 3.457
C. 61.084
D. none of these

3. Give the product.
20.6×1.28
A. 26.268
B. 26.368
C. 263.68
D. none of these

4. Give the quotient.
$516.3 \div 100$
A. 51,630
B. 0.5163
C. 5.163
D. none of these

5. Give the quotient rounded to the nearest hundredth.
$21.57 \div 3.7$
A. 5.82
B. 0.58
C. 5.83
D. none of these

6. The mean of 48, 52, 46, 46, and 53 is
A. 49
B. 46
C. 48
D. none of these

7. $2\frac{7}{8} = ?$
A. 0.875
B. 2.875
C. 2.125
D. none of these

8. $1.75 = ?$
A. $\frac{3}{4}$
B. $1\frac{1}{4}$
C. $1\frac{3}{4}$
D. none of these

9. Add. $3\frac{3}{4} + 2\frac{2}{3}$
A. $6\frac{5}{12}$ C. $5\frac{5}{7}$
B. $5\frac{5}{12}$ D. none of these

10. Subtract. $5\frac{1}{3} - 3\frac{1}{2}$
A. $2\frac{1}{6}$
B. $2\frac{5}{6}$
C. $1\frac{1}{6}$
D. none of these

11. QUIZ SCORES

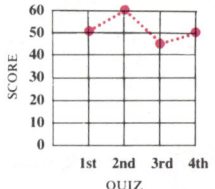

The score on the third quiz was
A. 60   B. 55
C. 50   D. 45

12. You volunteer to work 12 hours on a class project. You work $2\frac{1}{2}$ hours on Friday and $6\frac{3}{4}$ hours on Saturday. How many hours do you have left to work?
A. $7\frac{3}{4}$ B. $3\frac{3}{4}$
C. $2\frac{3}{4}$ D. none of these

Multiplying and Dividing Fractions and Mixed Numbers

Resources

- **Class Starter Quizzes 80-87** *(Copymasters S404-S406)*
- **Visual Aids 36-38** *(Copymasters or Transparencies S141-S143)*
- **Manipulatives**
  Manipulative Activities 15-17 *(Copymaster S519)*
  Fraction pieces, area tiles *(Copymasters or Transparencies S530-S531, S533)*
- **Worksheets 87-95** *(Copymasters, Duplicating Masters, or Workbook pages S255-S263)*
- **Challenge Problems** for pages 203, 207, 209, 211, 213, 217, 219 *(Copymasters S451-S453)*
- **Projects** for pages 205, 215 *(Copymaster S480)*
- **Mental Math Extensions** for Skills 36-41 *(Copymasters S504-S506)*
- **Tests** *(Copymasters or Duplicating Masters S29-S31, S73-S76)*

Lesson Objective
To multiply a fraction by a fraction

Problem-Solving Skills
Finding information in a display
Solving a multistep problem

Starting the Lesson
Before discussing exercises 1–6, draw a square on the chalkboard. Ask a student to draw a line that divides your square into halves. Ask another student to draw a line that divides each half into halves. Shade in one section.

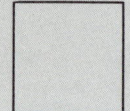 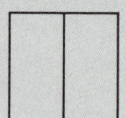 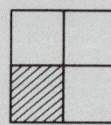

Write $\frac{1}{2}$ of $\frac{1}{2}$ = ? on the chalkboard. Point out to the students that their diagram shows that $\frac{1}{2}$ of $\frac{1}{2}$ is $\frac{1}{4}$. Repeat with thirds. Write $\frac{1}{3}$ of $\frac{1}{3}$ = $\frac{1}{9}$ on the chalkboard.

Here's How Note
Use of Concrete Materials You may wish to use the fraction pieces from copymasters or transparencies S530–S531 to demonstrate multiplying fractions. See ■ **Manipulative Activity 15** on copymaster S519 in the Teacher's Resource Binder.

202

Multiplying fractions

$\frac{2}{3}$

$\frac{3}{4}$

1. How many stamps are on the page? 6
2. How many stamps can be mounted on a full page? 12
3. What fraction of the page is covered with stamps? $\frac{6}{12}$ or $\frac{1}{2}$
4. To find what fraction of the page is covered with stamps, you could multiply $\frac{2}{3}$ by what fraction? $\frac{3}{4}$

Here's how to multiply fractions. $\frac{2}{3} \times \frac{3}{4} = ?$

Multiply the numerators to get the numerator of the product and multiply the denominators to get the denominator of the product.

Multiply numerators.
Multiply denominators.

$\frac{2}{3} \times \frac{3}{4} = \frac{6}{12}$

Write the product in simplest form.

$\frac{2}{3} \times \frac{3}{4} = \frac{6}{12}$
$= \frac{1}{2}$

5. Look at the *Here's how*. What fraction of the page is covered with stamps? $\frac{1}{2}$
6. Check these examples. Give each product in simplest form.

   a. $\frac{2}{3} \times \frac{3}{8} = \frac{6}{24}$
   $= ?$ $\frac{1}{4}$

   b. $2 \times \frac{3}{4} = \frac{2}{1} \times \frac{3}{4}$
   $= ?$ $1\frac{1}{2}$

   $\frac{2}{1}$

202   Chapter 8

EXERCISES

Multiply. Write the product in simplest form.
Here are scrambled answers for the next row of exercises: $\frac{5}{8}$ 1 $1\frac{1}{5}$ $2\frac{1}{4}$ $\frac{1}{16}$ $\frac{2}{7}$

7. $\frac{1}{2} \times \frac{2}{5}$ $\frac{1}{5}$
8. $3 \times \frac{3}{4}$ $2\frac{1}{4}$
9. $\frac{5}{6} \times \frac{3}{4}$ $\frac{5}{8}$
10. $\frac{2}{3} \times \frac{3}{2}$ 1
11. $\frac{1}{4} \times \frac{1}{4}$ $\frac{1}{16}$
12. $\frac{2}{3} \times \frac{3}{7}$ $\frac{2}{7}$

13. $\frac{2}{3} \times 2$ $1\frac{1}{3}$
14. $\frac{1}{2} \times \frac{2}{3}$ $\frac{1}{3}$
15. $\frac{1}{2} \times \frac{1}{3}$ $\frac{1}{6}$
16. $\frac{1}{4} \times \frac{5}{8}$ $\frac{5}{32}$
17. $3 \times \frac{1}{3}$ 1
18. $\frac{1}{5} \times \frac{1}{5}$ $\frac{1}{25}$

19. $\frac{4}{3} \times \frac{2}{5}$ $\frac{8}{15}$
20. $\frac{3}{2} \times \frac{3}{4}$ $1\frac{1}{8}$
21. $4 \times \frac{3}{4}$ 3
22. $\frac{3}{8} \times \frac{2}{3}$ $\frac{1}{4}$
23. $\frac{0}{3} \times \frac{4}{5}$ 0
24. $\frac{2}{7} \times \frac{7}{3}$ $\frac{2}{3}$

25. $\frac{2}{5} \times \frac{3}{2}$ $\frac{3}{5}$
26. $\frac{3}{4} \times \frac{2}{3}$ $\frac{1}{2}$
27. $4 \times \frac{2}{3}$ $2\frac{2}{3}$
28. $\frac{5}{6} \times \frac{1}{3}$ $\frac{5}{18}$
29. $\frac{3}{4} \times \frac{3}{4}$ $\frac{9}{16}$
30. $10 \times \frac{4}{5}$ 8

31. $\frac{3}{4} \times 3$ $2\frac{1}{4}$
32. $\frac{3}{2} \times \frac{1}{4}$ $\frac{3}{8}$
33. $\frac{5}{2} \times \frac{2}{5}$ 1
34. $\frac{2}{9} \times \frac{3}{4}$ $\frac{1}{6}$
35. $\frac{3}{2} \times \frac{0}{2}$ 0
36. $\frac{3}{5} \times \frac{5}{8}$ $\frac{3}{8}$

37. $\frac{1}{3} \times \frac{3}{4}$ $\frac{1}{4}$
38. $\frac{5}{8} \times \frac{4}{5}$ $\frac{1}{2}$
39. $3 \times \frac{5}{6}$ $2\frac{1}{2}$
40. $\frac{9}{2} \times \frac{4}{3}$ 6
41. $\frac{3}{8} \times \frac{3}{8}$ $\frac{9}{64}$
42. $\frac{1}{2} \times \frac{1}{2}$ $\frac{1}{4}$

43. $\frac{5}{8} \times \frac{2}{5}$ $\frac{1}{4}$
44. $\frac{6}{5} \times \frac{15}{2}$ 9
45. $\frac{7}{10} \times \frac{5}{4}$ $\frac{7}{8}$
46. $6 \times \frac{3}{10}$ $1\frac{4}{5}$
47. $\frac{5}{6} \times \frac{6}{5}$ 1
48. $\frac{3}{5} \times \frac{5}{3}$ 1

49. $2 \times \frac{1}{2}$ 1
50. $\frac{5}{12} \times \frac{3}{2}$ $\frac{5}{8}$
51. $\frac{4}{3} \times \frac{3}{4}$ 1
52. $\frac{4}{3} \times \frac{7}{12}$ $\frac{7}{9}$
53. $\frac{15}{16} \times \frac{4}{5}$ $\frac{3}{4}$
54. $4 \times \frac{3}{4}$ 3

55. $5 \times \frac{2}{5}$ 2
56. $\frac{3}{5} \times \frac{1}{3}$ $\frac{1}{5}$
57. $\frac{1}{5} \times \frac{1}{4}$ $\frac{1}{20}$
58. $8 \times \frac{9}{8}$ 9
59. $\frac{2}{7} \times \frac{1}{2}$ $\frac{1}{7}$
60. $\frac{12}{13} \times \frac{1}{2}$ $\frac{6}{13}$

61. $\frac{1}{6} \times \frac{2}{3}$ $\frac{1}{9}$
62. $\frac{1}{2} \times \frac{4}{5}$ $\frac{2}{5}$
63. $6 \times \frac{1}{2}$ 3
64. $\frac{1}{3} \times \frac{2}{5}$ $\frac{2}{15}$
65. $6 \times \frac{1}{6}$ 1
66. $9 \times \frac{4}{3}$ 12

Solve.

67. Two thirds of the stamps in a collection are U.S. stamps. One fourth of the U.S. stamps are airmail stamps. What fraction of the stamps are U.S. airmail stamps? $\frac{1}{6}$

68. One eighth of the stamps in an album were from Germany and one third of the stamps were from France. What fraction of the stamps were not from Germany or France? $\frac{13}{24}$

Airmail error

69. One of the most valuable United States stamps was printed in 1918. Its value resulted from a printer's error (the airplane was upside down). In 1979, 4 of these 24¢ stamps sold for a total of $500,000. The sale price was how many times the face value? Round the answer to the nearest tenth. 520,833.3

Multiplying and Dividing Fractions and Mixed Numbers

Extra Practice
Page 488 Skill 36

Practice Worksheet
Workbook S255, Copymaster S255, or Duplicating Master S255

Challenge Problem
Find a way to fold a piece of notebook paper to show that $\frac{3}{4} \times \frac{1}{2} = \frac{3}{8}$.

Answers will vary.

Copymaster S451

A fraction of a whole number

1. What is the regular price of the women's jogging shoes? $36
2. The sale price of the women's jogging shoes is what fraction of the regular price? $\frac{2}{3}$
3. To find the sale price, you would find $\frac{2}{3}$ of what price? $36

Here's how to find a fraction of a number. $\frac{2}{3}$ of $36 = ?

MENTAL MATH METHOD!
When the denominator is a divisor of the whole number, you can divide the whole number by the denominator and then multiply the result by the numerator.

$$36 \div 3 \times 2 = 24$$

$\frac{2}{3}$ of $36 = $24

REGULAR METHOD
Change the whole number to a fraction and multiply.

$\frac{2}{3}$ of $36 = $\frac{2}{3} \times \frac{$36}{1}$

$= \frac{$72}{3}$

$= $24

Note:

Dividing a number by 3 gives $\frac{1}{3}$ of the number.

Multiplying that result by 2 gives $\frac{2}{3}$ of the number.

4. Look at the *Here's how*. What is the sale price of the women's jogging shoes? $24
5. In which method was the computing easier? Shortcut method
6. You can use the shortcut when the denominator of the fraction is a divisor of the whole number.

EXERCISES

MENTAL MATH Use the mental math method to complete each exercise.
Here are scrambled answers for the next row of exercises: 4 12 3 27

7. $\frac{1}{4}$ of 12 = ? 3
8. $\frac{2}{3}$ of 18 = ? 12
9. $\frac{3}{4}$ of 36 = ? 27
10. $\frac{2}{5}$ of 10 = ? 4

11. $\frac{3}{5}$ of 20 = ? 12
12. $\frac{3}{8}$ of 48 = ? 18
13. $\frac{1}{5}$ of 15 = ? 3
14. $\frac{5}{6}$ of 30 = ? 25

15. $\frac{3}{10}$ of 60 = ? 18
16. $\frac{1}{3}$ of 33 = ? 11
17. $\frac{5}{8}$ of 48 = ? 30
18. $\frac{3}{4}$ of 48 = ? 36

19. $\frac{2}{3}$ of $30 = ? $20
20. $\frac{3}{5}$ of $30 = ? $18
21. $\frac{1}{2}$ of $32 = ? $16
22. $\frac{9}{10}$ of $100 = ? $90

23. $\frac{1}{8}$ of $48 = ? $6
24. $\frac{7}{10}$ of $60 = ? $42
25. $\frac{4}{5}$ of $45 = ? $36
26. $\frac{7}{8}$ of $56 = ? $49

You decide which method to use. Give the answer in simplest form.
Here are scrambled answers for the next row of exercises: $10\frac{1}{2}$ 21 15 $15\frac{1}{3}$

27. $\frac{1}{2}$ of 21 = ? $10\frac{1}{2}$
28. $\frac{3}{4}$ of 28 = ? 21
29. $\frac{2}{3}$ of 23 = ? $15\frac{1}{3}$
30. $\frac{5}{6}$ of 18 = ? 15

31. $\frac{4}{5}$ of 25 = ? 20
32. $\frac{2}{3}$ of 16 = ? $10\frac{2}{3}$
33. $\frac{1}{6}$ of 24 = ? 4
34. $\frac{3}{8}$ of 31 = ? $11\frac{5}{8}$

35. $\frac{3}{4}$ of 15 = ? $11\frac{1}{4}$
36. $\frac{1}{3}$ of 28 = ? $9\frac{1}{3}$
37. $\frac{7}{10}$ of 20 = ? 14
38. $\frac{7}{8}$ of 27 = ? $23\frac{5}{8}$

39. $\frac{11}{12}$ of 12 = ? 11
40. $\frac{2}{5}$ of 12 = ? $4\frac{4}{5}$
41. $\frac{1}{5}$ of 12 = ? $2\frac{2}{5}$
42. $\frac{3}{4}$ of 12 = ? 9

43. $\frac{3}{8}$ of $40 = ? $15
44. $\frac{4}{5}$ of $35 = ? $28
45. $\frac{5}{6}$ of $48 = ? $40
46. $\frac{2}{3}$ of $60 = ? $40

Solve. Use the ad on page 204.

47. What is the sale price of the men's tennis shoes? $20
48. What is the sale price of the men's jogging shoes? $33
49. What is the difference in the sale price of the two men's styles? $13
50. How much would one save by buying the men's jogging shoes on sale? $11

Pick-a-pair Predicting

51. You have 4 pairs of brown socks and 5 pairs of blue socks. It is dark and the lights go out in your room! How many socks must you pick to be sure that you have a pair that matches? 3

Multiplying and Dividing Fractions and Mixed Numbers

Extra Practice
Page 489 Skill 37

Practice Worksheet
Workbook S256, Copymaster S256, or Duplicating Master S256

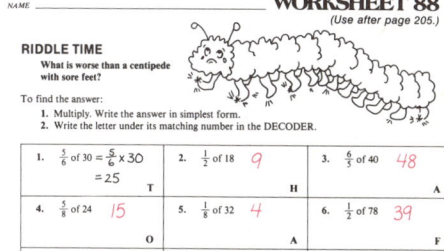

Project

Using a glossary

Look for these words in the Glossary of your math book:

1. One half of the letters of a 4-letter math word are "a." What is the word? area
2. One fifth of the letters of a 10-letter math word are "f." What is the word? difference
3. Five eighths of the letters of an 8-letter math word are vowels. What is the word? equation or evaluate

Copymaster S480

Class Starter Quiz 81
on previous lesson

Give the answer in simplest form.

1. $\frac{1}{3}$ of 21   7
2. $\frac{2}{3}$ of 15   10
3. $\frac{3}{10}$ of 40   12
4. $\frac{4}{5}$ of 30   24
5. $\frac{1}{2}$ of 15 $7\frac{1}{2}$
6. $\frac{3}{4}$ of 5 $3\frac{3}{4}$
7. $\frac{3}{8}$ of 8   3
8. $\frac{5}{6}$ of 48   40

Copymaster S405

Lesson Objective
To multiply mixed numbers

Problem-Solving Skills
Finding information in a recipe
Choosing the correct operation
Following directions and checking answers

Starting the Lesson
Have the students look at the recipe card on page 206. "How many bars does the recipe make?" (48) Ask the students to find the amount of each ingredient that would be needed to double the recipe, that is, to make 8 dozen, or 96, bars.

Multiplying mixed numbers

SPICE BARS
Makes 4 dozen bars

$1\frac{1}{2}$ cups all-purpose flour
$1\frac{1}{4}$ cups sugar
$\frac{1}{2}$ cup milk
$\frac{1}{2}$ cup vegetable oil
2 eggs
$\frac{3}{4}$ teaspoon salt
1 teaspoon baking soda
$1\frac{1}{4}$ teaspoons cinnamon
1 teaspoon cloves
$\frac{1}{2}$ cup chopped nuts
$\frac{1}{2}$ cup raisins

Preheat oven to 350°F. Grease 9" x 10" jelly-roll pan. Combine all ingredients in large bowl and mix well. Turn batter into prepared pan, spreading evenly. Bake until golden, about 20 minutes. Cool in pan on rack 10 minutes. Cut into bars.

1. How many bars does this recipe make?   48

2. How much sugar is needed to make the recipe? $1\frac{1}{4}$ cups

3. You want to make $2\frac{1}{2}$ times the recipe. What two mixed numbers should you multiply to find how many cups of sugar you will need? $2\frac{1}{2}$ and $1\frac{1}{4}$

Here's how to multiply mixed numbers. $2\frac{1}{2} \times 1\frac{1}{4} = ?$

Change each mixed number to a fraction and multiply.

Change to fractions.

$2\frac{1}{2} \times 1\frac{1}{4} = \frac{5}{2} \times \frac{5}{4}$

Multiply.

$2\frac{1}{2} \times 1\frac{1}{4} = \frac{5}{2} \times \frac{5}{4}$
$= \frac{25}{8}$

Write the product in simplest form.

$2\frac{1}{2} \times 1\frac{1}{4} = \frac{5}{2} \times \frac{5}{4}$
$= \frac{25}{8}$
$= 3\frac{1}{8}$

4. Look at the *Here's how*. How many cups of sugar will you need? Does the answer seem reasonable? $3\frac{1}{8}$; Yes

5. Check these examples. Give each product in simplest form.

    a. $3 \times 2\frac{2}{3} = \frac{3}{1} \times \frac{8}{3}$
    $= \frac{24}{3}$
    $= ?$ 8

    b. $2\frac{3}{4} \times 4\frac{2}{5} = \frac{11}{4} \times \frac{22}{5}$
    $= \frac{242}{20}$
    $= ?$ $12\frac{1}{10}$

206   Chapter 8

EXERCISES

Multiply. Write the product in simplest form.
Here are scrambled answers for the next row of exercises: $5\frac{1}{2}$ $4\frac{1}{6}$ 7 2 $6\frac{3}{5}$

6. $2 \times 3\frac{1}{2}$ *7*
7. $1\frac{1}{2} \times 1\frac{1}{3}$ *2*
8. $2\frac{3}{4} \times 2$ *$5\frac{1}{2}$*
9. $1\frac{2}{3} \times 2\frac{1}{2}$ *$4\frac{1}{6}$*
10. $2\frac{1}{5} \times 3$ *$6\frac{3}{5}$*

11. $2\frac{1}{4} \times 3\frac{1}{2}$ *$7\frac{7}{8}$*
12. $2\frac{4}{5} \times 3$ *$8\frac{2}{5}$*
13. $3\frac{1}{2} \times 1\frac{3}{4}$ *$6\frac{1}{8}$*
14. $2\frac{1}{2} \times 3\frac{1}{2}$ *$8\frac{3}{4}$*
15. $2\frac{1}{2} \times 2\frac{1}{2}$ *$6\frac{1}{4}$*

16. $2\frac{1}{3} \times 4\frac{1}{2}$ *$10\frac{1}{2}$*
17. $3\frac{1}{6} \times 2\frac{3}{4}$ *$8\frac{17}{24}$*
18. $3\frac{3}{4} \times 6$ *$22\frac{1}{2}$*
19. $5\frac{1}{2} \times 4\frac{3}{4}$ *$26\frac{1}{8}$*
20. $4\frac{1}{3} \times 3\frac{1}{2}$ *$15\frac{1}{6}$*

21. $2\frac{3}{8} \times 4$ *$9\frac{1}{2}$*
22. $3\frac{2}{3} \times 4\frac{1}{3}$ *$15\frac{8}{9}$*
23. $5\frac{3}{4} \times 2\frac{1}{2}$ *$14\frac{3}{8}$*
24. $2\frac{2}{3} \times 3\frac{1}{6}$ *$8\frac{4}{9}$*
25. $2\frac{1}{5} \times 1\frac{1}{2}$ *$3\frac{3}{10}$*

26. $4\frac{1}{5} \times 5\frac{3}{8}$ *$22\frac{23}{40}$*
27. $3 \times 5\frac{2}{3}$ *17*
28. $1\frac{5}{8} \times 4\frac{1}{2}$ *$7\frac{5}{16}$*
29. $3\frac{3}{8} \times 6\frac{3}{4}$ *$22\frac{25}{32}$*
30. $1\frac{1}{3} \times 6$ *8*

31. $4\frac{1}{2} \times 3$ *$13\frac{1}{2}$*
32. $1\frac{1}{2} \times 2\frac{1}{3}$ *$3\frac{1}{2}$*
33. $4 \times 2\frac{1}{2}$ *10*
34. $3\frac{1}{3} \times 4\frac{1}{3}$ *$14\frac{4}{9}$*
35. $1\frac{2}{3} \times 6$ *10*

Solve. Use the recipe on page 206.

36. You make $1\frac{1}{2}$ times the recipe. How many bars do you make? *72*

37. How much more baking soda than salt is used in the recipe? *$\frac{1}{4}$ tsp*

38. How much flour is needed to double the recipe? *3 cups*

39. How many teaspoons of spices (cinnamon and cloves) are used in the recipe? *$2\frac{1}{4}$*

40. You want to make $3\frac{1}{2}$ times the recipe. How much sugar will you need? *$4\frac{3}{8}$ cups*

41. You have $1\frac{3}{4}$ cups of raisins. Do you have enough to make $3\frac{1}{2}$ times the recipe? *Yes*

Check the products Checking answers

42. Find and correct the two wrong answers.

a. Four and one tenth times two and six tenths equals

10.66

b. Nine tenths times seventy-two hundredths equals

c. Two and eight tenths times one and sixteen hundredths equals

3.248

Multiplying and Dividing Fractions and Mixed Numbers 207

Class Starter Quiz 82
on previous lesson

Multiply. Write the product in simplest form.

1. $2 \times 1\frac{1}{2}$ 3
2. $2\frac{1}{4} \times 1\frac{1}{3}$ 3
3. $3\frac{2}{5} \times 3$ $10\frac{1}{5}$
4. $4\frac{1}{2} \times 2\frac{3}{4}$ $12\frac{3}{8}$
5. $2\frac{1}{5} \times 3\frac{1}{8}$ $6\frac{7}{8}$
6. $1\frac{2}{3} \times 6$ 10

Copymaster S405

Lesson Objective
To relate fractions to customary units of measure

Problem-Solving Skill
Analyzing a sequence of events

Starting the Lesson
Number Sense Write these measurement facts and numbers on the chalkboard:

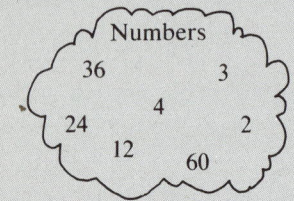

Numbers: 36, 3, 24, 4, 2, 12, 60

1 day = _?_ hours
1 minute = _?_ seconds
1 yard = _?_ feet
1 foot = _?_ inches
1 yard = _?_ inches
1 gallon = _?_ quarts
1 pint = _?_ cups

Before the students open their books, have them choose the numbers to complete the measurement facts. Then tell them to open their books to pages 208 and 209 and check their answers.

208

More on a fraction of a whole number

1. How many hours are there in 2 days? 48

2. How many hours are there in $\frac{1}{2}$ of a day? 12

3. If you know the number of hours in 2 days and the number of hours in $\frac{1}{2}$ of a day, how could you find the number of hours in $2\frac{1}{2}$ days? add

1 day = 24 hours (h)
1 h = 60 minutes (min)
1 min = 60 seconds (s)

Here's how *to find the number of hours in $2\frac{1}{2}$ days.*

MENTAL MATH METHOD!
First find the hours in 2 days and in $\frac{1}{2}$ of a day. Then add.

$2\frac{1}{2}$ days = 48 hours + 12 hours
= 60 hours

[24 hours]

REGULAR METHOD
Change the mixed number to a fraction and multiply.

$2\frac{1}{2}$ days = $2\frac{1}{2} \times 24$ hours
= $\frac{5}{2} \times \frac{24}{1}$ hours
= $\frac{120}{2}$ hours
= 60 hours

[24 hours]

4. Look at the *Here's how*.

   a. How many hours are there in $2\frac{1}{2}$ days? 60

   b. Does the answer make sense? Is it between the number of hours in 2 days (48) and the number of hours in 3 days (72)? Yes, Yes

EXERCISES
Complete. Here are scrambled answers for the next row of exercises: 100   66   90

5. $2\frac{3}{4}$ days = _?_ h   66
6. $1\frac{1}{2}$ h = _?_ min   90
7. $1\frac{2}{3}$ h = _?_ min   100
8. $1\frac{1}{3}$ min = _?_ s   80
9. $1\frac{3}{4}$ h = _?_ min   105
10. $1\frac{2}{3}$ days = _?_ h   40
11. $1\frac{5}{6}$ min = _?_ s   110
12. $2\frac{3}{8}$ days = _?_ h   57
13. $3\frac{2}{3}$ h = _?_ min   220

208   Chapter 8

1 yard (yd) = 3 feet (ft)
1 ft = 12 inches (in.)
1 yd = 36 in.

14. $1\frac{1}{3}$ yd = _?_ ft   4
15. $1\frac{1}{2}$ ft = _?_ in.   18
16. $1\frac{1}{3}$ yd = _?_ in.   48
17. $1\frac{3}{4}$ ft = _?_ in.   21

18. $2\frac{1}{3}$ yd = _?_ ft   7
19. $1\frac{3}{4}$ yd = _?_ in.   63
20. $2\frac{2}{3}$ yd = _?_ ft.   8
21. $2\frac{1}{2}$ yd = _?_ in.   90

22. $2\frac{3}{4}$ ft = _?_ in.   33
23. $3\frac{2}{3}$ yd = _?_ ft   11
24. $3\frac{2}{3}$ ft = _?_ in.   44
25. $3\frac{3}{4}$ yd = _?_ in.   135

1 gallon (gal) = 4 quarts (qt)
1 qt = 2 pints (pt)
1 pt = 2 cups (c)

26. $1\frac{1}{2}$ gal = _?_ qt   6
27. $1\frac{1}{2}$ qt = _?_ pt   3
28. $2\frac{1}{2}$ pt = _?_ c   5
29. $4\frac{1}{2}$ qt = _?_ pt   9
30. $2\frac{3}{4}$ gal = _?_ qt   11
31. $4\frac{1}{2}$ pt = _?_ c   9
32. $3\frac{1}{2}$ gal = _?_ qt   14
33. $3\frac{1}{2}$ pt = _?_ c   7
34. $2\frac{1}{2}$ qt = _?_ pt   5
35. $2\frac{1}{4}$ gal = _?_ qt   9
36. $3\frac{1}{2}$ qt = _?_ pt   7
37. $5\frac{1}{2}$ pt = _?_ c   11

What's the time?

Elapsed time

38. A thunderstorm caused a power failure at 10:05 P.M. The next morning at 7:55 the electric clock showed 6:05.   1 h 50 min
    a. How long was the electricity off?
    b. At what time did the electricity come on again?   11:55 P.M.

Multiplying and Dividing Fractions and Mixed Numbers

Extra Practice
Page 490   Skill 39

Practice Worksheet
Workbook S258, Copymaster S258, or Duplicating Master S258

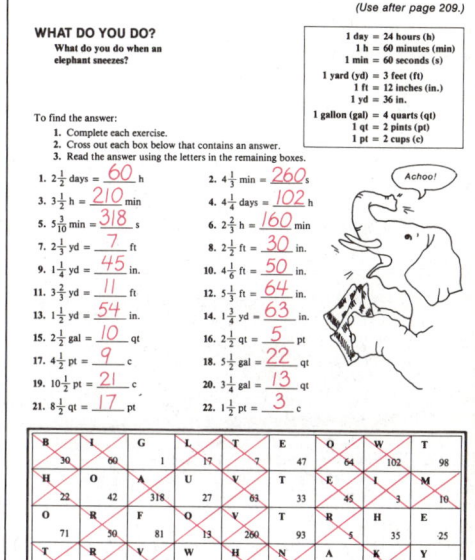

Challenge Problem
Use a calculator to compute how many years old you will be when you have lived 10 million minutes.

19 years old

Copymaster S452

Class Starter Quiz 83
on previous lesson

Complete.

1. $1\frac{1}{2}$ days = _?_ hours   36
2. $1\frac{3}{4}$ hours = _?_ minutes   105
3. $1\frac{1}{3}$ yards = _?_ feet   4
4. $6\frac{1}{2}$ feet = _?_ inches   78
5. $2\frac{3}{4}$ gallons = _?_ quarts   11

Copymaster S405

Problem-Solving Skills
Using a table
Solving a multistep problem

Skills Reviewed
Adding and subtracting decimals
Multiplying and dividing decimals
Comparing fractions
Simplifying expressions
Finding the GCF and LCM of pairs of numbers

Starting the Lesson

Problem Solving The chart at the top of page 210 is also on ■ **Visual Aid 36** (copymaster or transparency S141). Use the chart and ask questions like these:

- Which cost more per pound, Delicious apples or McIntosh apples? (McIntosh apples)
- Is one dollar enough money to buy two pounds of Delicious apples? (No)
- What is the cost of $1\frac{3}{4}$ pounds of Granny Smith apples? ($1.75)

Cumulative Skill Practice Challenge the students to an estimation hunt by saying, "Look at the first row of exercises on page 211. Which exercise has the largest product?" (Exercise 1) Then have the students find the exercise with the largest product in each of the second and third rows. (Exercises 5 and 12)

210

Problem solving
Reading a table
You're the clerk

| | POUNDS |||||
APPLES	$\frac{1}{4}$	$\frac{1}{2}$	$\frac{3}{4}$	1	2
Granny Smith	$.25	$.50	$.75	$1.00	$2.00
Delicious	$.15	$.30	$.45	$.60	$1.20
McIntosh	$.18	$.36	$.54	$.72	$1.44

Use the information in the table to answer each customer's question.

1. What is the cost of 2 pounds of McIntosh apples? $1.44

2. What is the cost of $1\frac{1}{2}$ pounds of Granny Smith apples? $1.50

3. How many pounds of Delicious apples can I buy for $1.80? 3

4. "What is the cost of $2\frac{1}{2}$ pounds of McIntosh apples?" $1.80

5. "How many pounds of Delicious apples can I buy for $1.50?" $2\frac{1}{2}$

6. "What is the cost of $2\frac{3}{4}$ pounds of Granny Smith apples?" $2.75

7. "How many pounds of McIntosh apples can I buy for $.90?" $1\frac{1}{4}$

Solve. Use the apple prices.

8. Mike gave the clerk $5 for $2\frac{3}{4}$ pounds of Delicious apples. How much change did he get? $3.35

9. Karen spent $3.60 for a 5-pound bag of apples. What kind of apples did she buy? McIntosh

10. There are 4 pounds of apples in a bag. The total cost is $2.40. What kind of apples are in the bag? Delicious

11. How much more do 10 pounds of McIntosh apples cost than 10 pounds of Delicious apples? $1.20

12. A customer bought $1\frac{1}{2}$ pounds of Granny Smith apples and some Delicious apples. She spent a total of $3.30. How many pounds of Delicious apples did the customer buy? 3

13. How much does a 6-pound bag of apples cost if it is $\frac{1}{3}$ Granny Smith apples and $\frac{2}{3}$ Delicious apples? $4.40

Chapter 8

Cumulative Skill Practice

Give the product. *(page 68)*

1. 2.3 × 0.4 0.92
2. 2.6 × 0.21 0.546
3. 0.44 × 0.6 0.264
4. 0.2 × 0.35 0.070
5. 14 × 1.2 16.8
6. 0.56 × 2.8 1.568
7. 3.21 × 1.1 3.531
8. 2.11 × 3.8 8.018
9. 9.24 × 3 27.72
10. 14.5 × 2.6 37.70
11. 125 × 1.4 175.0
12. 286 × 3.1 886.6

Give the quotient rounded to the nearest tenth. *(page 104)*

13. 4.64 ÷ 0.5 9.3
14. 8.43 ÷ 0.02 421.5
15. 56.5 ÷ 3 18.8
16. 8.65 ÷ 2 4.3
17. 42 ÷ 0.7 60
18. 3.8 ÷ 0.2 19
19. 6.38 ÷ 1.2 5.3
20. 3.78 ÷ 2.4 1.6
21. 0.8465 ÷ 7.1 0.1
22. 60.04 ÷ 5.2 11.5
23. 0.072 ÷ 0.36 0.2
24. 0.54 ÷ 0.21 2.6

Give the greatest common factor. *(page 144)*

25. 4, 6 2
26. 8, 10 2
27. 12, 16 4
28. 9, 18 9
29. 15, 20 5
30. 16, 24 8
31. 14, 21 7
32. 24, 30 6
33. 24, 36 12
34. 42, 56 14

Give the least common multiple. *(page 150)*

35. 3, 5 15
36. 4, 6 12
37. 5, 10 10
38. 6, 8 24
39. 6, 9 18
40. 10, 15 30
41. 8, 12 24
42. 15, 20 60
43. 12, 18 36
44. 16, 24 48

Less than (<), equal to (=), or greater than (>)? *(page 152)*

45. $\frac{4}{5}$ ● $\frac{3}{5}$ >
46. $\frac{7}{5}$ ● $\frac{3}{5}$ >
47. $\frac{5}{9}$ ● $\frac{4}{9}$ >
48. $\frac{1}{7}$ ● $\frac{3}{7}$ <
49. $\frac{4}{3}$ ● $\frac{7}{3}$ <
50. $\frac{1}{8}$ ● $\frac{1}{6}$ <
51. $\frac{5}{8}$ ● $\frac{1}{2}$ >
52. $\frac{1}{4}$ ● $\frac{1}{5}$ >
53. $\frac{3}{4}$ ● $\frac{2}{3}$ >
54. $\frac{0}{5}$ ● $\frac{0}{4}$ =
55. $\frac{2}{5}$ ● $\frac{1}{2}$ <
56. $\frac{7}{3}$ ● $\frac{5}{2}$ <
57. $\frac{2}{3}$ ● $\frac{7}{10}$ <
58. $\frac{1}{9}$ ● $\frac{1}{8}$ <
59. $\frac{4}{5}$ ● $\frac{2}{3}$ >

MIXED PRACTICE
Complete.

60. 3.4 + 2.36 = ? 5.76
61. 16 − 12.5 = ? 3.5
62. 3.42 × 100 = ? 342
63. 46.74 ÷ 10 = ? 4.674
64. 8.6 × 1000 = ? 8600
65. 9.7 − 2.74 = ? 6.96
66. 6.00 − 2.73 = ? 3.27
67. 0.2 + 0.24 + 0.3 = ? 0.74
68. 0.6 ÷ 100 = ? 0.006
69. (4.6 − 2.3) − 1.4 = ? 0.9
70. 4.6 − (2.3 − 1.4) = ? 3.7
71. (8 − 3.6) + 2.7 = ? 7.1
72. 8 − (3.6 + 2.7) = ? 1.7
73. 15.6 + (5.4 × 10) = ? 69.6
74. (15.6 + 5.4) × 10 = ? 210

Multiplying and Dividing Fractions and Mixed Numbers

Problem-Solving Worksheet
Workbook S259, Copymaster S259, or Duplicating Master S259

WORKSHEET 91 (Use after page 210.)

NAME _____

FRACTION TIME

The clock shows the time now. What time will it be 1¾ hours from now? **4:45**

PASS THE PIZZA
Brian made 3 pizzas and cut them into eighths. How many pieces can he serve if each person eats 4 pieces? **6**

WHAT'S THE ANSWER?
Use the code to answer the riddle.

CODE

A B C	D E F	G H I
J K L	M N O	P Q R
S T U	V W X	Y Z

Riddle: How do you know an elephant will stay for a long time when it comes to visit?
Answer: **IT BRINGS ITS TRUNK**

FAIR SHARE
Check (✓) the two coins you would move from one pile to the other so that both piles have the same amount of money.

WHAT'S THE PATTERN?
Look for the pattern. Fill in the blanks.

(1 × 9) + 2 = 11
(12 × 9) + 3 = 111
(123 × 9) + 4 = 1111
(**1234** × 9) + **5** = 11,111
(**12,345** × 9) + **6** = 111,111

FRACTION ACTION
Fill in the boxes to get the answer.

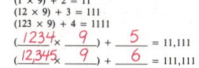

$\frac{3}{4} - \frac{2}{4} = \frac{1}{4}$

© D. C. Heath and Company S259 Problem solving

Challenge Problems

Unscramble the letters of these math words.

1. MUS **Sum**
2. CROPDUT **Product**
3. RAGHP **Graph**
4. EDMO **Mode**
5. NUOIQTET **Quotient**

Copymaster S452

Class Starter Quiz 84
on previous lesson

Solve. Use the price chart on page 210.

1. What is the cost of $2\frac{3}{4}$ pounds of Delicious apples?
   $1.65

2. How much pounds of Granny Smith apples can you buy for $1.50?
   $1\frac{1}{2}$

3. How much more do 3 pounds of McIntosh apples cost than 3 pounds of Delicious apples?
   $.36

Copymaster S405

Lesson Objectives
To identify reciprocals
To divide by a fraction
To divide a fraction by a whole number

Problem-Solving Skills
Using a picture to solve a problem
Reading a chart

Starting the Lesson
Use exercises 1 and 2 to show the students that dividing by $\frac{1}{4}$ is the same as multiplying by 4.

Here's How Note
Use of Concrete Materials You may wish to use the fraction pieces from copymasters or transparencies S530–S531 to demonstrate dividing fractions. See ■ **Manipulative Activity 17** on copymaster S519 in the Teacher's Resource Binder.

212

Dividing fractions

1. a. How far is it around the track? $\frac{1}{4}$ mile
   b. How many times would you have to run around the track to run $\frac{1}{2}$ mile? 2

2. a. How many times would you have to run around the track to run 2 miles? 8
   b. Answer part **a** by dividing.

total miles		miles in each lap		laps
2	÷	$\frac{1}{4}$	=	? 8

It's $\frac{1}{4}$ mile around this track.

   c. Answer part **a** by multiplying.

total miles		laps for each mile		laps
2	×	4	=	? 8

   d. Look at parts **b** and **c**. Dividing by $\frac{1}{4}$ is the same as multiplying by what number? 4

TIME OUT! $\frac{3}{8} \times \frac{8}{3} = 1$

$\frac{8}{3}$ is the reciprocal of $\frac{3}{8}$.

$\frac{3}{8}$ is the reciprocal of $\frac{8}{3}$.

Two numbers are **reciprocals** if their product is 1.

For a fraction not equal to 0, you can find the reciprocal by inverting the fraction.

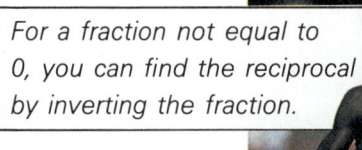

Here's how *to divide by a fraction.*

To divide by a fraction, multiply by its reciprocal.

$\frac{1}{2} \div \frac{1}{4} = \frac{1}{2} \times \frac{4}{1}$ $2 \div \frac{1}{4} = \frac{2}{1} \times \frac{4}{1}$

$\phantom{\frac{1}{2} \div \frac{1}{4}} = \frac{4}{2}$ $\phantom{2 \div \frac{1}{4}} = \frac{8}{1}$

$\phantom{\frac{1}{2} \div \frac{1}{4}} = 2$ $\phantom{2 \div \frac{1}{4}} = 8$

3. Check each division exercise. Then complete the statement.

   a. $\frac{5}{2} \div \frac{2}{3} = \frac{5}{2} \times \frac{3}{2}$ To divide by $\frac{2}{3}$, multiply by ? $\frac{3}{2}$

   $\phantom{\frac{5}{2} \div \frac{2}{3}} = \frac{15}{4}$

   $\phantom{\frac{5}{2} \div \frac{2}{3}} = 3\frac{3}{4}$

   b. $\frac{4}{5} \div 4 = \frac{4}{5} \times \frac{1}{4}$ To divide by 4, multiply by ? $\frac{1}{4}$

   $\phantom{\frac{4}{5} \div 4} = \frac{4}{20}$

   $\phantom{\frac{4}{5} \div 4} = \frac{1}{5}$

212 *Chapter 8*

EXERCISES

Give the reciprocal of each number.

4. 5 $\frac{1}{5}$
5. $\frac{1}{3}$ 3
6. $\frac{1}{8}$ 8
7. $\frac{3}{8}$ $\frac{8}{3}$
8. $\frac{5}{7}$ $\frac{7}{5}$
9. $\frac{2}{5}$ $\frac{5}{2}$
10. 3 $\frac{1}{3}$

11. 6 $\frac{1}{6}$
12. $\frac{3}{5}$ $\frac{5}{3}$
13. $\frac{6}{5}$ $\frac{5}{6}$
14. $\frac{3}{2}$ $\frac{2}{3}$
15. $\frac{9}{10}$ $\frac{10}{9}$
16. 8 $\frac{1}{8}$
17. $\frac{14}{3}$ $\frac{3}{14}$

Divide. Write the quotient in simplest form.

Here are scrambled answers for the next row of exercises: 6   1 $\frac{1}{6}$ $1\frac{1}{8}$ $\frac{9}{16}$ $\frac{2}{3}$

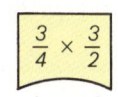 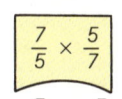 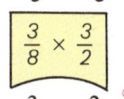 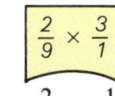

18. $\frac{2}{3} \div 4$ $\frac{1}{6}$
19. $\frac{3}{2} \div \frac{1}{4}$ 6
20. $\frac{3}{4} \div \frac{2}{3}$ $1\frac{1}{8}$
21. $\frac{7}{5} \div \frac{7}{5}$ 1
22. $\frac{3}{8} \div \frac{2}{3}$ $\frac{9}{16}$
23. $\frac{2}{9} \div \frac{1}{3}$ $\frac{2}{3}$

24. $\frac{5}{9} \div \frac{1}{3}$ $1\frac{2}{3}$
25. $\frac{5}{8} \div 2$ $\frac{5}{16}$
26. $\frac{4}{5} \div \frac{3}{3}$ $\frac{4}{5}$
27. $\frac{2}{3} \div \frac{5}{9}$ $1\frac{1}{5}$
28. $5 \div \frac{2}{5}$ $12\frac{1}{2}$
29. $\frac{2}{7} \div \frac{2}{3}$ $\frac{4}{5}$ $\frac{5}{14}$

30. $\frac{3}{5} \div \frac{2}{5}$ $1\frac{1}{2}$
31. $\frac{3}{2} \div \frac{2}{3}$ $2\frac{1}{4}$
32. $\frac{7}{9} \div \frac{4}{3}$ $\frac{7}{12}$
33. $\frac{2}{5} \div 5$ $\frac{2}{25}$
34. $\frac{7}{8} \div \frac{3}{4}$ $1\frac{1}{6}$
35. $\frac{3}{7} \div \frac{27}{49}$ $\frac{7}{9}$

36. $\frac{4}{3} \div \frac{4}{3}$ 1
37. $\frac{3}{8} \div \frac{3}{2}$ $\frac{1}{4}$
38. $\frac{3}{2} \div \frac{3}{8}$ 4
39. $\frac{2}{5} \div \frac{1}{4}$ $1\frac{3}{5}$
40. $\frac{9}{10} \div \frac{4}{5}$ $1\frac{1}{8}$
41. $\frac{1}{5} \div \frac{1}{25}$ 5

42. $4 \div \frac{5}{8}$ $6\frac{2}{5}$
43. $\frac{9}{4} \div \frac{7}{8}$ $2\frac{4}{7}$
44. $6 \div \frac{3}{2}$ 4
45. $\frac{7}{2} \div \frac{9}{4}$ $1\frac{5}{9}$
46. $\frac{7}{8} \div \frac{5}{16}$ $2\frac{4}{5}$
47. $8 \div \frac{1}{2}$ 16

48. $\frac{0}{5} \div \frac{5}{6}$ 0
49. $8 \div \frac{2}{5}$ 20
50. $\frac{5}{9} \div \frac{4}{3}$ $\frac{5}{12}$
51. $\frac{5}{4} \div 3$ $\frac{5}{12}$
52. $\frac{5}{12} \div \frac{5}{12}$ 1
53. $\frac{2}{3} \div \frac{2}{3}$ 1

Solve.

54. How many $\frac{1}{4}$-mile laps must you run to run 4 miles? Hint: $4 \div \frac{1}{4} = ?$ 16

55. You are on a $\frac{1}{2}$-mile relay team. Each runner runs $\frac{1}{8}$ mile. How many runners are on your team? Hint: $\frac{1}{2} \div \frac{1}{8} = ?$ 4

You're the coach Reading a chart

Use the 100-yard-dash times.

56. Which two runners would make the fastest 2-person 200-yard relay team?   Kim, Sandy

57. Which two runners would make the slowest 200-yard relay team?   Jenny, Tom

58. Which of these teams should win a 200-yard relay, Jenny and Dan or Tom and Sandy?   Tom and Sandy

100-yard Dash	
NAME	TIME
Jenny	15.7 seconds
Kim	13.9 seconds
Tom	16.1 seconds
Dan	14.3 seconds
Sandy	13.2 seconds

Multiplying and Dividing Fractions and Mixed Numbers

Extra Practice

Page 490   Skill 40

Practice Worksheet

Workbook S260, Copymaster S260, or Duplicating Master S260

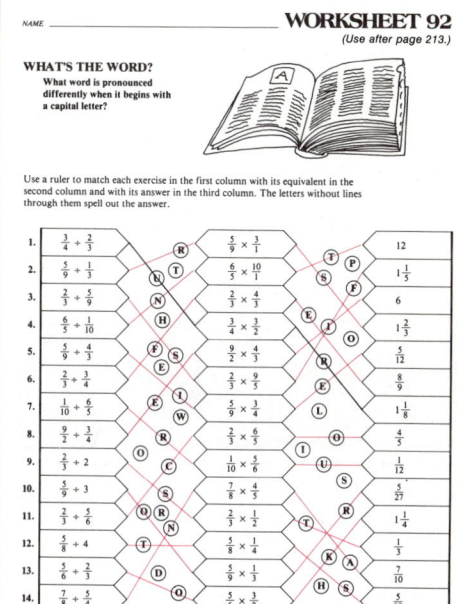

Challenge Problem

Find the missing number.

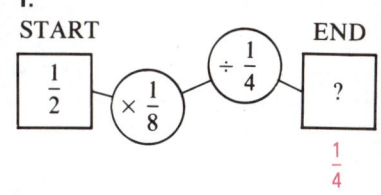

$\frac{1}{4}$

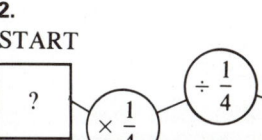

8

Copymaster S452

Class Starter Quiz 85
on previous lesson

Divide. Write the quotient in simplest form.

1. $\frac{2}{3} \div \frac{1}{5}$ $3\frac{1}{3}$
2. $\frac{3}{2} \div \frac{1}{4}$ 6
3. $\frac{2}{5} \div 2$ $\frac{1}{5}$
4. $\frac{4}{3} \div \frac{4}{3}$ 1
5. $8 \div \frac{2}{5}$ 20
6. $\frac{5}{9} \div \frac{4}{3}$ $\frac{5}{12}$
7. $\frac{10}{3} \div \frac{5}{3}$ 2
8. $\frac{9}{2} \div \frac{3}{4}$ 6

Copymaster S406

Lesson Objective
To divide mixed numbers

Problem-Solving Skills
Selecting information from a display
Following directions and checking answers

Starting the Lesson
What are the facts? Have the students study the auto-repair chart at the top of page 214 for 30 seconds. Then tell them to close their books and answer these questions from memory:

- Which job takes longer, a minor tune-up or a major tune-up? (Major tune-up)
- Which job takes $1\frac{1}{4}$ hours? (Minor tune-up)
- Does it take more or less than half an hour to do an oil change? (Less)
- How many cars can have wheel alignments in one hour? (2)

214

Dividing mixed numbers

1. How much time is needed for an oil change? $\frac{1}{3}$ hour
2. How many hours does a minor tune-up take? $1\frac{1}{4}$
3. To find how many minor tune-ups can be completed in $3\frac{3}{4}$ hours, the service manager would divide $3\frac{3}{4}$ by what number? $1\frac{1}{4}$

H & W AUTO REPAIR

JOB	TIME
Oil Change	$\frac{1}{3}$ hour
Wheel Alignment	$\frac{1}{2}$ hour
Minor Tune-up	$1\frac{1}{4}$ hours
Major Tune-up	$2\frac{1}{3}$ hours

Here's how *to divide mixed numbers.*

Change to fractions.

$$3\frac{3}{4} \div 1\frac{1}{4} = \frac{15}{4} \div \frac{5}{4}$$

Divide. Write the quotient in simplest form.

$$3\frac{3}{4} \div 1\frac{1}{4} = \frac{15}{4} \div \frac{5}{4}$$
$$= \frac{15}{4} \times \frac{4}{5}$$
$$= \frac{60}{20}$$
$$= 3$$

4. Look at the *Here's how*. How many minor tune-ups can be completed in $3\frac{3}{4}$ hours? 3

5. Check these examples. Then answer the questions.

$$2\frac{1}{2} \div \frac{1}{2} = \frac{5}{2} \div \frac{1}{2}$$
$$= \frac{5}{2} \times \frac{2}{1}$$
$$= \frac{10}{2}$$
$$= 5$$

$$7 \div 2\frac{1}{3} = \frac{7}{1} \div \frac{7}{3}$$
$$= \frac{7}{1} \times \frac{3}{7}$$
$$= \frac{21}{7}$$
$$= 3$$

a. How many cars can have wheel alignments in $2\frac{1}{2}$ hours? 5

b. How many major tune-ups can be done in 7 hours? 3

214   Chapter 8

EXERCISES

Divide. Write each quotient in simplest form.

Here are scrambled answers for the next row of exercises: $1\frac{1}{4}$ $4\frac{1}{8}$ $1\frac{4}{5}$ $2\frac{1}{7}$ $2\frac{1}{4}$

6. $2\frac{1}{4} \div 1\frac{1}{4}$ $1\frac{4}{5}$
7. $5\frac{1}{2} \div 1\frac{1}{3}$ $4\frac{1}{8}$
8. $5 \div 2\frac{1}{3}$ $2\frac{1}{7}$
9. $4\frac{1}{2} \div 2$ $2\frac{1}{4}$
10. $2\frac{1}{2} \div 2$ $1\frac{1}{4}$

11. $6\frac{1}{2} \div 2\frac{2}{3}$ $2\frac{7}{16}$
12. $6\frac{1}{2} \div 2\frac{1}{4}$ $2\frac{8}{9}$
13. $8 \div 2\frac{1}{4}$ $3\frac{5}{9}$
14. $6\frac{1}{4} \div 1\frac{1}{4}$ 5
15. $1\frac{1}{6} \div 1\frac{1}{2}$ $\frac{7}{9}$

16. $5\frac{7}{8} \div 1\frac{3}{4}$ $3\frac{5}{14}$
17. $7 \div 2\frac{1}{3}$ 3
18. $4\frac{1}{3} \div 2\frac{1}{2}$ $1\frac{11}{15}$
19. $2 \div 1\frac{1}{2}$ $1\frac{1}{3}$
20. $4\frac{1}{2} \div 1\frac{1}{8}$ 4

21. $2\frac{3}{8} \div 1\frac{1}{3}$ $1\frac{25}{32}$
22. $6\frac{3}{4} \div 3$ $2\frac{1}{4}$
23. $1\frac{1}{5} \div 5$ $\frac{6}{25}$
24. $8\frac{1}{2} \div 1\frac{3}{4}$ $4\frac{6}{7}$
25. $6 \div 1\frac{1}{2}$ 4

26. $3\frac{3}{5} \div 1\frac{1}{5}$ 3
27. $6\frac{2}{3} \div 2$ $3\frac{1}{3}$
28. $9\frac{1}{4} \div 2\frac{1}{4}$ $4\frac{1}{9}$
29. $8 \div 2\frac{1}{2}$ $3\frac{1}{5}$
30. $2 \div \frac{2}{3}$ 3

31. $6 \div 4\frac{1}{2}$ $1\frac{1}{3}$
32. $9 \div 2\frac{1}{4}$ 4
33. $2\frac{1}{3} \div 2\frac{1}{3}$ 1
34. $2\frac{1}{2} \div 1\frac{1}{4}$ 2
35. $3\frac{1}{2} \div 3\frac{1}{2}$ 1

36. $1\frac{1}{2} \div 2$ $\frac{3}{4}$
37. $7\frac{1}{2} \div 1\frac{1}{2}$ 5
38. $3\frac{2}{3} \div 1\frac{1}{3}$ $2\frac{3}{4}$
39. $6 \div 1\frac{1}{3}$ $4\frac{1}{2}$
40. $1\frac{1}{2} \div \frac{1}{2}$ 3

41. $4\frac{4}{5} \div 4$ $1\frac{1}{5}$
42. $3\frac{1}{3} \div 6$ $\frac{5}{9}$
43. $7\frac{1}{2} \div \frac{1}{2}$ 15
44. $5\frac{1}{2} \div 10$ $\frac{11}{20}$
45. $10 \div \frac{1}{2}$ 20

Solve. Use the auto-repair information on page 214.

46. How many oil changes can be done in $2\frac{2}{3}$ hours? 8

47. How many hours should it take for a major tune-up and an oil change? $2\frac{2}{3}$

48. How many hours should it take to do wheel alignments on 7 cars? $3\frac{1}{2}$

49. A mechanic starts a minor tune-up at 10:20. At what time should the job be done? $11:35$

Check the quotients Checking answers

50. Find and correct the two wrong answers.

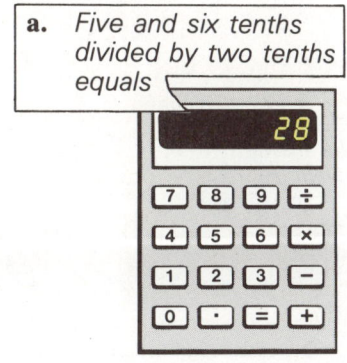

a. Five and six tenths divided by two tenths equals
28
0.32

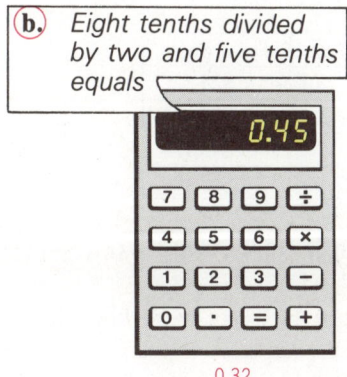

b. Eight tenths divided by two and five tenths equals
0.45

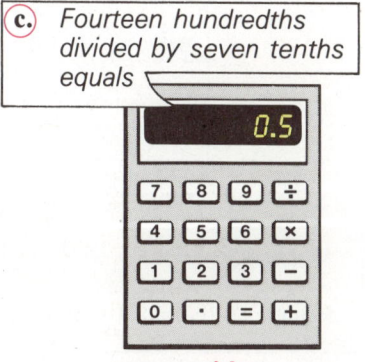

c. Fourteen hundredths divided by seven tenths equals
0.5
0.2

Multiplying and Dividing Fractions and Mixed Numbers

Extra Practice

Page 491 Skill 41

Practice Worksheet

Workbook S261, Copymaster S261, or Duplicating Master S261

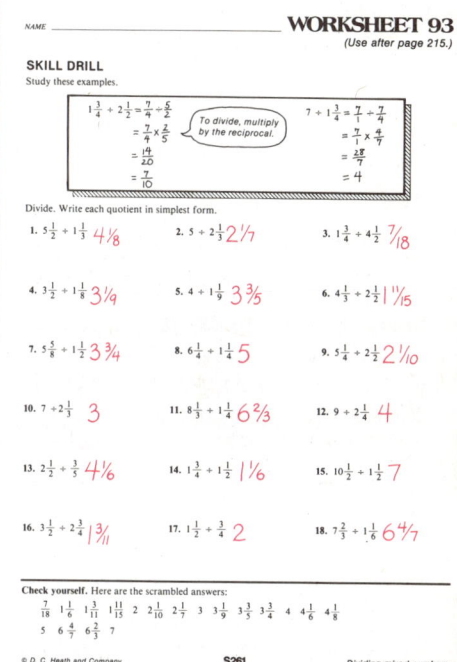

Project

Researching information

Count the number of sheets in a $\frac{1}{2}$-inch-high stack of paper. Then use that number of sheets to compute the thickness of a single sheet of paper.

Copymaster S480

Class Starter Quiz 86
on previous lesson

Divide. Write the quotient in simplest form.

1. $4\frac{1}{4} \div 2\frac{1}{4}$ $1\frac{8}{9}$ 2. $3\frac{1}{2} \div 1\frac{1}{3}$ $2\frac{5}{8}$
3. $5 \div 1\frac{1}{3}$ $3\frac{3}{4}$ 4. $4\frac{1}{2} \div 2$ $2\frac{1}{4}$
5. $4\frac{1}{3} \div 2\frac{1}{2}$ $1\frac{11}{15}$ 6. $8 \div 2\frac{1}{2}$ $3\frac{1}{5}$

Copymaster S406

Lesson Objective
To solve problems using a making-a-list strategy

Starting the Lesson
Problem-Solving Cover-up Use the chalkboard or mask ■ **Visual Aid 37** (copymaster or transparency S142).

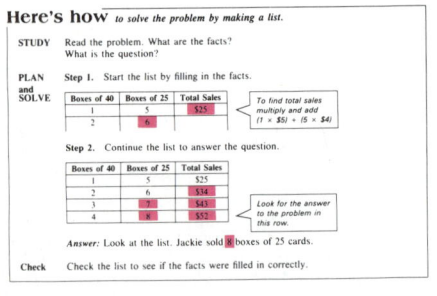

Have the students, working in small groups, study the problem on page 216 and the problem-solving steps for several minutes. Then have them close their books, look at the visual aid, and tell what has been covered up.

216

PROBLEM SOLVING *making a list*

Some problems can be solved by making a list.

Problem
At a recent fund raiser, Martin L. King High School sold assorted cards. When Jackie was at the booth, she sold 4 more boxes of 25 than boxes of 40. Her total sales for these were $52. How many boxes of 25 did she sell?

Here's how *to solve the problem by making a list.*

STUDY   Read the problem. What are the facts?
What is the question?

PLAN and SOLVE   **Step 1.** Start the list by filling in the facts.

Boxes of 40	Boxes of 25	Total Sales
1	5	$25
2	6	

To find total sales multiply and add (1 × $5) + (5 × $4)

Step 2. Continue the list to answer the question.

Boxes of 40	Boxes of 25	Total Sales
1	5	$25
2	6	$34
3	7	$43
4	8	$52

Look for the answer to the problem in this row.

Answer: Look at the list. Jackie sold 8 boxes of 25 cards.

Check   Check the list to see if the facts were filled in correctly.

216   Chapter 8

PROBLEMS

Solve by completing the list. Use the card prices on page 216.

1. Jody sold 3 more boxes of 10 than boxes of 15. Her total for these sales were $24. How many boxes of 10 did she sell? 3

Number of boxes of 10	Number of boxes of 15	Total Sales
1	4	$9
2	5	

2. Wade sold twice as many boxes of 25 as boxes of 15. His total sales for these were $55. How many boxes of 25 did he sell? 10

Number of boxes of 15	Number of boxes of 25	Total Sales
1	2	$11
2	4	

Solve by making a list. Use the card prices on page 216.

3. Lana sold 4 more boxes of 25 than boxes of 10. Her total sales for these were $46. How many boxes of 25 did she sell? 9

4. Ty sold 3 times as many boxes of 15 as boxes of 40. His total sales for these were $70. How many boxes of 15 did he sell? 15

5. Cedric sold the same number of boxes of 10 as boxes of 40. His total sales for these were $35. How many boxes did he sell in all? 10

6. Sara sold 4 times as many boxes of 15 as boxes of 10. Her total sales for these were $56. How many boxes did she sell in all? 20

7. Tyne sold 6 more boxes of 40 than boxes of 10. She sold a total of 28 boxes of these. How much were her total sales? $107

8. Dana sold twice as many boxes of 40 than boxes of 25. He sold a total of 15 boxes of these. How much were his total sales? $70

What's the color? Finding a pattern

9. Here are the first 7 cards in a box of 40. What type is the 40th card? Hint: Look for a pattern. Get Well

Multiplying and Dividing Fractions and Mixed Numbers

Practice Worksheet
Workbook S262, Copymaster S262, or Duplicating Master S262

Challenge Problem

Write a question that fits the answer.

A recipe calls for $4\frac{1}{2}$ cups of flour and $1\frac{1}{4}$ cups of milk. Suppose you wanted to double the recipe.

How many cups of milk would you need?

Answer: $2\frac{1}{2}$

Copymaster S453

Class Starter Quiz 87
on previous lesson

Solve by making a list. Use the card prices on page 216.

Lisa sold 5 more boxes of 15 cards than boxes of 40 cards. Her total sales were $47. How many boxes of 40 cards did she sell? 4

Copymaster S406

Problem-Solving Skills
Using a drawing
Following instructions

Skills Reviewed
Changing fractions to decimals
Adding and subtracting fractions
Multiplying and dividing mixed numbers
Writing standard numerals for decimals
Adding and subtracting decimals
Comparing decimals
Multiplying and dividing decimals

Starting the Lesson

Problem Solving The diagram showing the 24 electrodes is also on ■ **Visual Aid 38** (copymaster or transparency S143). Shade in the appropriate electrodes so that *6:23* is displayed. Have the students tell which electrodes were charged to display:

- the *6*. (Electrodes D, C, B, Q, P, and O)
- the colon. (Electrode R)
- the *2*. (Electrodes H, I, F, S, and T)
- the *3*. (Electrodes L, M, J, X, and W)

Use the visual aid to display other times and ask similar questions.

Cumulative Skill Practice Challenge the students to an estimation hunt by saying, "Find the two sums in exercises 15–26 that are greater than 1." (Exercises 18 and 23) Then have the students find the product in exercises 39–48 that is greater than 10. (Exercise 42)

218

Problem solving

COMPUTERS IN WATCHES

A tiny, battery-powered computer controls the electronic circuits used to display the time on a digital watch.

The diagram shows 24 electrodes. Each electrode controls a part of the display. When the electrode is charged, the part controlled by the computer turns black.

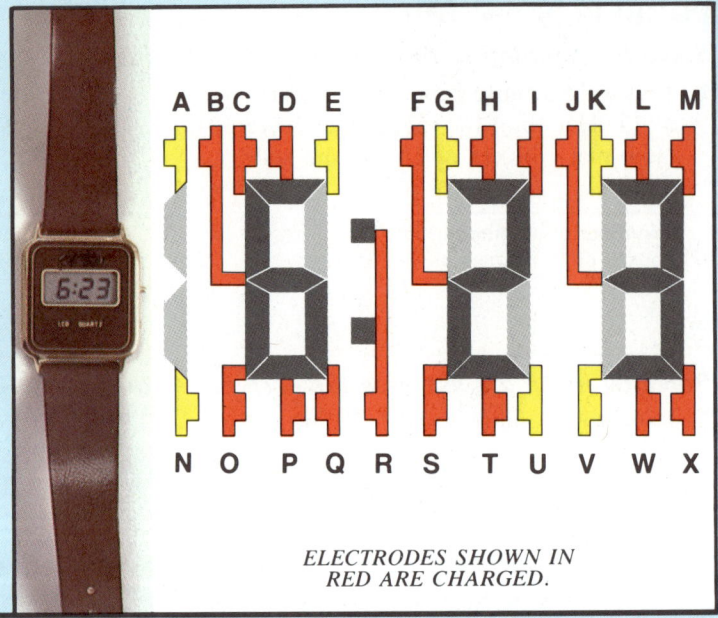

ELECTRODES SHOWN IN RED ARE CHARGED.

Solve. Use the diagram.

1. In the diagram, electrodes B, C, D, F, H, I, J, L, M, O, P, Q, R, S, T, W, and X are charged.
   a. What time is displayed? 6:23
   b. If electrode E was also charged, what time would be displayed? 8:23

2. After one minute passes, the computer will change the *3* to a *4*. To make this change, the computer
   a. removes the charge from electrodes L and ?. W
   b. charges electrode ?. K

3. For each set of charged electrodes, give the time that would be displayed.
   a. E, F, G, I, J, K, L, M, Q, R, U, W, and X   1:49
   b. A, E, I, M, N, Q, R, U, and X   11:11
   c. A, B, D, E, F, H, I, K, L, M, N, O, P, R, T, U, V, W, and X   12:30

4. When *10:00* is displayed, all but 3 of the electrodes are charged. 8:08, 10:06, 10:09, 10:28, 10:38, 10:58, 12:08
   a. Can you find seven more times that use all but 3 of the electrodes?
   b. Can you find a time that uses all but 2 of the electrodes? 10:08

5. A computer can also make it possible for a watch to include a calendar, an alarm, and a calculator. What is the sale price of this watch? $12

218   Chapter 8

Cumulative Skill Practice

Change to a decimal rounded to the nearest hundredth. *(page 166)*

1. $\frac{1}{6}$ 0.17
2. $\frac{1}{3}$ 0.33
3. $\frac{13}{3}$ 4.33
4. $\frac{1}{9}$ 0.11
5. $\frac{5}{3}$ 1.67
6. $\frac{7}{3}$ 2.33
7. $\frac{10}{9}$ 1.11

8. $\frac{1}{12}$ 0.08
9. $\frac{4}{9}$ 0.44
10. $\frac{2}{3}$ 0.67
11. $\frac{6}{11}$ 0.55
12. $\frac{5}{6}$ 0.83
13. $\frac{5}{7}$ 0.71
14. $\frac{3}{7}$ 0.43

Give the sum in simplest form. *(page 180)*

15. $\frac{3}{5} + \frac{1}{5}$ $\frac{4}{5}$
16. $\frac{1}{3} + \frac{1}{2}$ $\frac{5}{6}$
17. $\frac{1}{6} + \frac{1}{3}$ $\frac{1}{2}$
18. $\frac{3}{5} + \frac{7}{10}$ $1\frac{3}{10}$
19. $\frac{1}{8} + \frac{3}{4}$ $\frac{7}{8}$
20. $\frac{2}{5} + \frac{1}{10}$ $\frac{1}{2}$

21. $\frac{1}{4} + \frac{1}{3}$ $\frac{7}{12}$
22. $\frac{3}{10} + \frac{1}{5}$ $\frac{1}{2}$
23. $\frac{4}{9} + \frac{2}{3}$ $1\frac{1}{9}$
24. $\frac{1}{8} + \frac{5}{16}$ $\frac{7}{16}$
25. $\frac{1}{2} + \frac{1}{4}$ $\frac{3}{4}$
26. $\frac{1}{9} + \frac{2}{3}$ $\frac{7}{9}$

Give the difference in simplest form. *(pages 186, 188)*

27. $\frac{3}{4} - \frac{1}{4}$ $\frac{1}{2}$
28. $\frac{7}{10} - \frac{1}{10}$ $\frac{3}{5}$
29. $\frac{3}{2} - \frac{3}{4}$ $\frac{3}{4}$
30. $\frac{2}{3} - \frac{4}{9}$ $\frac{2}{9}$
31. $\frac{9}{10} - \frac{2}{5}$ $\frac{1}{2}$
32. $\frac{4}{5} - \frac{1}{2}$ $\frac{3}{10}$

33. $\frac{7}{8} - \frac{1}{3}$ $\frac{13}{24}$
34. $\frac{5}{12} - \frac{1}{3}$ $\frac{1}{12}$
35. $\frac{3}{4} - \frac{2}{3}$ $\frac{1}{12}$
36. $\frac{3}{2} - \frac{5}{8}$ $\frac{7}{8}$
37. $\frac{5}{6} - \frac{0}{4}$ $\frac{5}{6}$
38. $\frac{2}{3} - \frac{1}{9}$ $\frac{5}{9}$

Give the product in simplest form. *(page 206)*

39. $2 \times 1\frac{1}{2}$ 3
40. $3 \times 2\frac{1}{3}$ 7
41. $4 \times 1\frac{2}{3}$ $6\frac{2}{3}$
42. $3\frac{1}{4} \times 5$ $16\frac{1}{4}$
43. $4 \times 2\frac{1}{2}$ 10

44. $1\frac{1}{3} \times 1\frac{1}{2}$ 2
45. $1\frac{1}{2} \times 1\frac{2}{3}$ $2\frac{1}{2}$
46. $2\frac{1}{4} \times 3\frac{1}{2}$ $7\frac{7}{8}$
47. $3 \times 2\frac{2}{5}$ $7\frac{1}{5}$
48. $3\frac{1}{2} \times 1\frac{1}{4}$ $4\frac{3}{8}$

Give the quotient in simplest form. *(page 214)*

49. $1\frac{1}{2} \div 2$ $\frac{3}{4}$
50. $3\frac{3}{4} \div 3$ $1\frac{1}{4}$
51. $2 \div 1\frac{1}{3}$ $1\frac{1}{2}$
52. $4 \div 2\frac{3}{4}$ $1\frac{5}{11}$
53. $12\frac{1}{2} \div 6$ $2\frac{1}{12}$

54. $5\frac{1}{2} \div 1\frac{1}{3}$ $4\frac{1}{8}$
55. $8 \div 2\frac{1}{2}$ $3\frac{1}{5}$
56. $6\frac{1}{4} \div 1\frac{1}{4}$ 5
57. $5\frac{7}{8} \div 1\frac{3}{4}$ $3\frac{5}{14}$
58. $6\frac{1}{2} \div 2$ $3\frac{1}{4}$

MIXED PRACTICE

Complete.

59. The standard numeral for 8 hundredths is   ?   0.08

60. The standard numeral for 11 and 19 thousandths is   ?   11.019

61. $8.094 + 2.177 =$?   10.271
62. $23 - 14.5 =$?   8.5
63. $4 - 0.125 =$?   3.875

64. 15.1 ? 15.09 >
    (> or <)
65. 0.058 ? 0.58 <
    (> or <)
66. 5.806 ? 5.86 <
    (> or <)

67. $2.546 \times 100 =$?   254.6
68. $0.2158 \div 2.6 =$?   0.083
69. $6.3 \times 0.04 =$?   0.252

Multiplying and Dividing Fractions and Mixed Numbers

Problem-Solving Worksheet
Workbook S263, Copymaster S263, or Duplicating Master S263

WORKSHEET 95 (Use after page 218.)

NAME

POCKET MONEY
There are 15 bills in this wallet. One third of the bills are $10 bills. The rest are $5 bills. How much money is in the wallet? **$100**

GUESS AND CHECK
Which two numbers would you switch so that the product along each side would be 72?

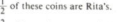
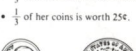 3 and 9

HOW MUCH MONEY?
Which coins are Rita's?
Clues:
- $\frac{1}{2}$ of these coins are Rita's.
- $\frac{2}{3}$ of her coins are worth 10¢.
- $\frac{1}{3}$ of her coins is worth 25¢.

DOLLARS AND SENSE
Sonya bought some $3 records. She gave the clerk $20 and got $2 back in change. How many records did she buy? **6**

Rita's coins:
1 quarter, 2 nickels

MISSING DIGITS
Fill in the missing digits.

PIECE OF PIE
An eighth of a pie sells for $.75. What is the cost of one pie? **$6.00**

© D. C. Heath and Company   S263   Problem solving

Challenge Problem

Use the code to answer the riddle.

CODE

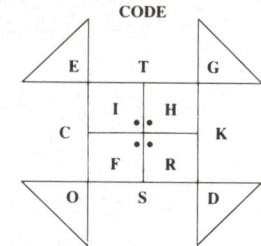

Riddle: Why would you put bug spray on your watch?

Answer:

To get rid of the ticks.

Copymaster S453

Chapter REVIEW

Here are scrambled answers for the review exercises:

1	3	add	fraction	reciprocal
2	4	denominators	inverting	simplest
		divide	numerators	whole

1. numerators, denominators   **2.** 4, 3   **3.** whole

1. To multiply these two fractions, you would multiply the ? to get the numerator of the product and multiply the ? to get the denominator of the product. *(page 202)*

$$\frac{3}{4} \times \frac{1}{2} = ?$$

2. To compute this fraction of a number, you could first divide 36 by ? and then multiply the result by ?. *(page 204)*

$$\frac{3}{4} \text{ of } 36 = ?$$

3. To compute this fraction of a number, you would change the ? number to a fraction and multiply. *(page 204)*

$$\frac{2}{5} \text{ of } 17 = ?$$

4. fraction   **5.** 2, add   **6.** 1, inverting

4. To multiply these mixed numbers, change each mixed number to a ? and multiply. *(page 206)*

$$2\frac{1}{2} \times 1\frac{1}{4} = ?$$

5. To change $2\frac{1}{2}$ feet to inches, you could first find how many inches in ? feet and in $\frac{1}{2}$ of a foot and then ?. *(page 208)*

$$2\frac{1}{2} \text{ feet} = ? \text{ inches}$$

6. Two numbers are reciprocals if their product is ?. For a fraction not equal to 0, you can find the reciprocal by ? the fraction. *(page 212)*

7. reciprocal, divide   **8.** simplest

7. To divide by a fraction, you would multiply by its ?. To divide mixed numbers, change each mixed number to a fraction and ?. *(pages 212, 214)*

8. The last step in this division exercise is to write the quotient in ? form. *(page 214)*

$$2\frac{1}{2} \div 1\frac{3}{4} = \frac{5}{2} \div \frac{7}{4}$$
$$= \frac{5}{2} \times \frac{4}{7}$$
$$= \frac{20}{14}$$
$$= ?$$

Chapter TEST

Multiply. Write the product in simplest form. (page 202)

1. $2 \times \frac{1}{2}$ **1**
2. $\frac{1}{3} \times \frac{1}{3}$ **$\frac{1}{9}$**
3. $\frac{2}{5} \times \frac{1}{3}$ **$\frac{2}{15}$**
4. $\frac{3}{4} \times \frac{1}{2}$ **$\frac{3}{8}$**
5. $\frac{2}{3} \times \frac{3}{5}$ **$\frac{2}{5}$**
6. $\frac{2}{5} \times \frac{1}{2}$ **$\frac{1}{5}$**
7. $4 \times \frac{1}{2}$ **2**
8. $4 \times \frac{1}{8}$ **$\frac{1}{2}$**
9. $\frac{5}{2} \times \frac{4}{3}$ **$3\frac{1}{3}$**
10. $\frac{4}{5} \times \frac{5}{4}$ **1**
11. $\frac{8}{3} \times \frac{5}{2}$ **$6\frac{2}{3}$**
12. $4 \times \frac{5}{4}$ **5**

Complete. (page 204)

13. $\frac{1}{2}$ of $12 = \underline{?}$ **6**
14. $\frac{1}{3}$ of $24 = \underline{?}$ **8**
15. $\frac{2}{3}$ of $18 = \underline{?}$ **12**
16. $\frac{3}{4}$ of $\$20 = \underline{?}$ **\$15**
17. $\frac{3}{4}$ of $\$8 = \underline{?}$ **\$6**

Multiply. Write the product in simplest form. (page 206)

18. $2 \times 1\frac{1}{2}$ **3**
19. $2\frac{1}{3} \times 3$ **7**
20. $3 \times 1\frac{1}{4}$ **$3\frac{3}{4}$**
21. $2\frac{2}{3} \times 4$ **$10\frac{2}{3}$**
22. $5 \times 1\frac{1}{2}$ **$7\frac{1}{2}$**
23. $1\frac{2}{3} \times 2\frac{1}{2}$ **$4\frac{1}{6}$**
24. $2\frac{3}{4} \times 2\frac{3}{4}$ **$7\frac{9}{16}$**
25. $2\frac{5}{6} \times 1\frac{1}{4}$ **$3\frac{13}{24}$**
26. $1\frac{3}{8} \times 3\frac{3}{5}$ **$4\frac{19}{20}$**
27. $1\frac{1}{3} \times 1\frac{2}{3}$ **$2\frac{2}{9}$**

Complete. (page 208)

28. $2\frac{1}{2}$ days $= \underline{?}$ h **60**
29. $1\frac{3}{4}$ h $= \underline{?}$ min **105**
30. $2\frac{1}{2}$ min $= \underline{?}$ sec **150**
31. $1\frac{1}{3}$ yd $= \underline{?}$ ft **4**

Divide. Write each quotient in simplest form. (page 212)

32. $\frac{3}{4} \div 2$ **$\frac{3}{8}$**
33. $3 \div \frac{3}{8}$ **8**
34. $\frac{2}{3} \div \frac{1}{3}$ **2**
35. $\frac{1}{2} \div \frac{1}{3}$ **$1\frac{1}{2}$**
36. $\frac{1}{3} \div \frac{1}{2}$ **$\frac{2}{3}$**
37. $4 \div \frac{1}{2}$ **8**
38. $\frac{5}{6} \div \frac{3}{4}$ **$1\frac{1}{9}$**
39. $6 \div \frac{3}{5}$ **10**
40. $\frac{3}{5} \div 6$ **$\frac{1}{10}$**
41. $\frac{5}{8} \div \frac{3}{4}$ **$\frac{5}{6}$**
42. $\frac{3}{10} \div \frac{2}{5}$ **$\frac{3}{4}$**
43. $\frac{1}{2} \div \frac{3}{4}$ **$\frac{2}{3}$**

Divide. Write each quotient in simplest form. (page 214)

44. $2\frac{1}{2} \div 1\frac{1}{4}$ **2**
45. $3 \div 1\frac{1}{2}$ **2**
46. $2\frac{1}{2} \div 2$ **$1\frac{1}{4}$**
47. $8 \div 2\frac{1}{3}$ **$3\frac{3}{7}$**
48. $5 \div \frac{1}{2}$ **10**
49. $3\frac{1}{4} \div 6$ **$\frac{13}{24}$**
50. $4\frac{3}{4} \div 1\frac{1}{2}$ **$3\frac{1}{6}$**
51. $9 \div 3\frac{3}{8}$ **$2\frac{2}{3}$**
52. $5\frac{5}{6} \div 3\frac{1}{2}$ **$1\frac{2}{3}$**
53. $2\frac{1}{3} \div 3$ **$\frac{7}{9}$**

Solve.

54. How many miles is it from Round Lake to Clear Falls if you hike by Lookout Point? **$5\frac{3}{4}$**

55. Which is the shorter route from Round Lake to Clear Falls? **By Lookout Point**

56. How many miles long is the hiking trail? **12**

57. You hiked the trail in $5\frac{1}{4}$ hours. How many miles per hour did you average? **$2\frac{2}{7}$**

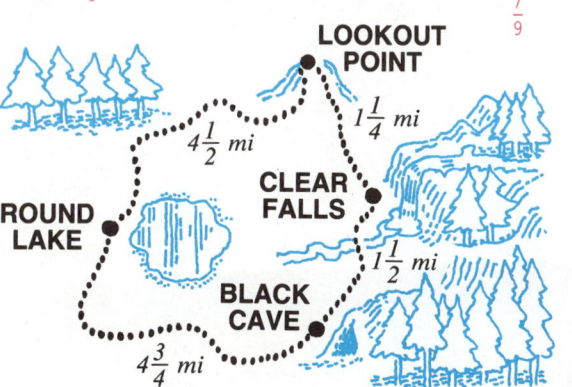

Multiplying and Dividing Fractions and Mixed Numbers **221**

Cumulative TEST — Standardized Format

Choose the correct letter.

1. Give the product.

3.57×2.70

- **A.** 9.6390
- **B.** 8.199
- **C.** 963.90
- **D.** none of these

2. Give quotient rounded to nearest tenth.

$423.8 \div 1.62$

- **A.** 2.6
- **B.** 0.3
- **C.** 261.6
- **D.** none of these

3. The greatest common factor of 12 and 18 is

- **A.** 36
- **B.** 18
- **C.** 6
- **D.** none of these

4. The least common multiple of 9 and 12 is

- **A.** 3
- **B.** 36
- **C.** 9
- **D.** none of these

5. $\frac{7}{8} < ?$

- **A.** $\frac{1}{2}$
- **B.** $\frac{4}{5}$
- **C.** $\frac{5}{6}$
- **D.** none of these

6. Change to a decimal rounded to the nearest hundredth.

$\frac{5}{6} = ?$

- **A.** 0.83
- **B.** 0.84
- **C.** 11.2
- **D.** none of these

7. Give the sum in simplest form.

$\frac{2}{3} + \frac{3}{4}$

- **A.** $\frac{5}{7}$
- **B.** $\frac{1}{2}$
- **C.** $1\frac{5}{12}$
- **D.** none of these

8. Give the difference in simplest form.

$\frac{5}{6} - \frac{3}{8}$

- **A.** 1
- **B.** $\frac{11}{24}$
- **C.** $\frac{1}{12}$
- **D.** none of these

9. Give the product in simplest form.

$2\frac{1}{2} \times 1\frac{1}{4}$

- **A.** 2
- **B.** $6\frac{1}{4}$
- **C.** $12\frac{1}{2}$
- **D.** none of these

10. Give the quotient in simplest form.

$6 \div 3\frac{3}{4}$

- **A.** $1\frac{3}{5}$
- **B.** $\frac{3}{8}$
- **C.** $22\frac{1}{2}$
- **D.** none of these

11. STUDENT TRYOUTS

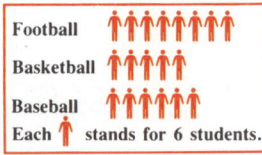

How many more students tried out for football than baseball?

- **A.** 2
- **B.** 12
- **C.** 18
- **D.** none of these

12. Sara jogged 3 miles one day and $4\frac{1}{2}$ miles on each of the next two days. How many miles did she average per day?

- **A.** $3\frac{3}{7}$
- **B.** $3\frac{3}{4}$
- **C.** $4\frac{1}{14}$
- **D.** none of these

Second-Quarter Test

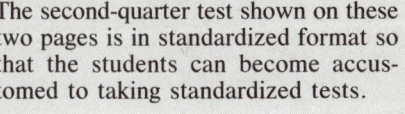

The second-quarter test shown on these two pages is in standardized format so that the students can become accustomed to taking standardized tests.

222A

Use Copymaster S92 or Duplicating Master S92 to provide the students with an answer sheet in standardized test format.

Copymaster S106 has a quick-score answer key for the second-quarter test.

Copymaster S92 or Duplicating Master S92

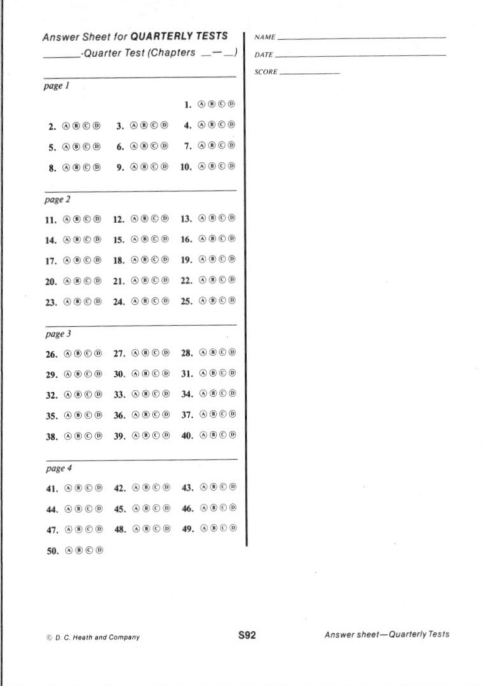

Copymaster S106

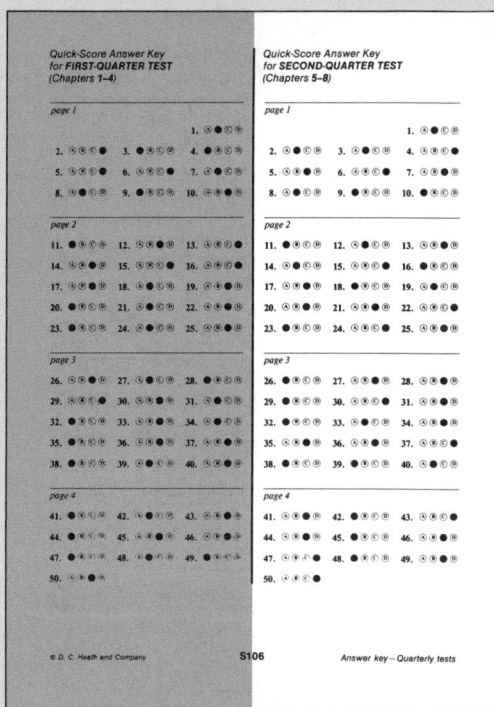

The table below correlates test items with student text pages.

Test Item	Text Page
1	p. 114
2	p. 116
3	p. 116
4	p. 118
5	p. 118
6	p. 120
7	p. 120
8	p. 124
9	p. 126
10	p. 126
11	p. 126
12	p. 142
13	p. 148

Test Item	Text Page
14	p. 150
15	p. 152
16	p. 152
17	p. 156
18	p. 158
19	p. 158
20	p. 160
21	p. 160
22	p. 162
23	p. 166
24	p. 167
25	p. 168
26	p. 168

Test Item	Text Page
27	p. 178
28	p. 180
29	p. 182
30	p. 182
31	p. 186
32	p. 188
33	p. 190
34	p. 190
35	p. 192
36	p. 182
37	p. 192
38	p. 202
39	p. 204

Test Item	Text Page
40	p. 204
41	p. 206
42	p. 206
43	p. 208
44	p. 208
45	p. 208
46	p. 212
47	p. 214
48	p. 214
49	p. 206
50	p. 214

Measurements

Chapter 9
Measurement

Resources

- **Class Starter Quizzes 88-101** *(Copymasters S406-S410)*
- **Visual Aids 39-44** *(Copymasters or Transparencies S143-S147)*
- **Worksheets 96-110** *(Copymasters, Duplicating Masters, or Workbook pages S264-S278)*
- **Challenge Problems** for pages 243, 249, 251, 253 *(Copymasters S453-S454)*
- **Projects** for pages 229, 231, 235, 245, 247 *(Copymasters S480-S481)*
- **Tests** *(Copymasters or Duplicating Masters S33-S36)*

Lesson Objective

To measure lengths with centimeters and millimeters

Problem-Solving Skills

Finding information in a display
Reading a metric ruler
Choosing the correct operation

Starting the Lesson

Use of Concrete Materials Choose a student to read the first paragraph orally. Discuss exercises 1–4. You may wish to use ■ **Visual Aid 39** (centimeter and millimeter rulers on copymaster or transparency S224) for demonstration purposes during the *Here's how* discussion.

Using a metric ruler

The **centimeter** (cm) is a unit of length in the metric system. The length of the 18K (18-carat) gold chain measured to the nearest centimeter is 6 centimeters.

1. What is the length of the 14K gold chain measured to the nearest centimeter? 10 cm

Here's how *to measure to the nearest millimeter*.

Notice that one tenth of a centimeter is 1 millimeter (mm).

First, line up one end of the chain with this end of the ruler.

Then, read the mark on the ruler nearest this end of the chain.

one centimeter   one millimeter   93 mm (9.3 cm)

2. Look at the *Here's how*. What is the length of the chain measured to the nearest millimeter? 93 mm

3. What is the length of the chain to the nearest tenth of a centimeter? to the nearest centimeter? 9.3 cm, 9 cm

4. Look at the ruler. How many millimeters are in 1 centimeter? 10

EXERCISES

Use a metric ruler. Measure each chain to the nearest centimeter.

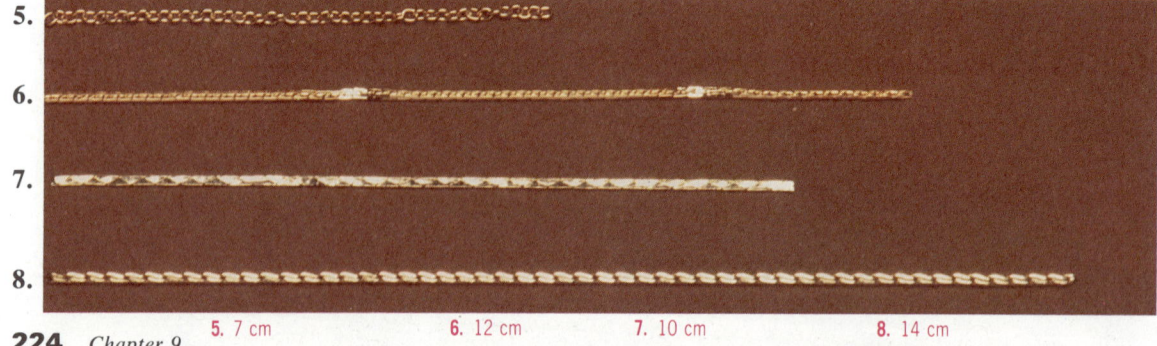

5. 7 cm   6. 12 cm   7. 10 cm   8. 14 cm

9. 61 mm **10.** 91 mm **11.** 96 mm **12.** 104 mm **13.** 124 mm

Measure each chain to the nearest millimeter. Answers will vary slightly. (± 2 mm)

Draw "chains" of these lengths.

14. 3 cm   **15.** 8 cm   **16.** 12 cm   **17.** 90 mm   **18.** 78 mm
19. 58 mm   **20.** 132 mm   **21.** 8.6 cm   **22.** 10.4 cm   **23.** 12.5 cm

Solve. Use the chains pictured at the top of page 224.

24. The 14K gold chain costs $6 per centimeter. What is the cost of the 14K gold chain? $60

25. The 18K gold chain costs $48. What is the cost per centimeter? $8

26. Which costs more, the 14K gold chain or the 18K gold chain? The 14K gold chain

27. What would be the total cost for two 18K chains and one 14K chain? $156

Decoding

28. Use the code to get the answer to the riddle.

CODE: H O E T C W I D A J S L F N R

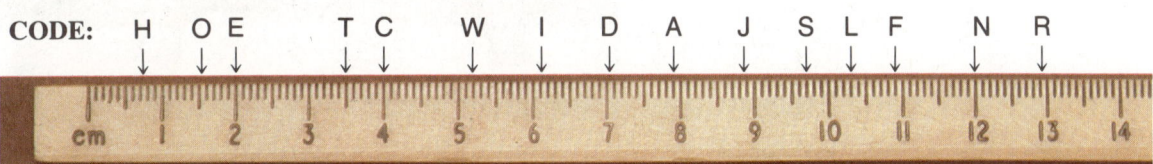

RIDDLE: What's the difference between a jeweler and a jailer?

ANSWER: 1.5 cm * 12 cm * 2 cm
9.7 cm * 20 mm * 10.3 cm * 103 mm * 97 mm
5.2 cm * 79 mm * 35 mm * 4 cm * 7 mm * 20 mm * 9.7 cm
52 mm * 0.7 cm * 6.1 cm * 103 mm * 2 cm
3.5 cm * 7 mm * 20 mm
15 mm * 3.5 cm * 7 mm * 2 cm * 129 mm
52 mm * 7.9 cm * 3.5 cm * 40 mm * 0.7 cm * 2 cm * 97 mm
4 cm * 20 mm * 10.3 cm * 103 mm * 9.7 cm.

One sells watches while the other watches cells.

Measurement 225

Practice Worksheet

Workbook S264, Copymaster S264, or Duplicating Master S264

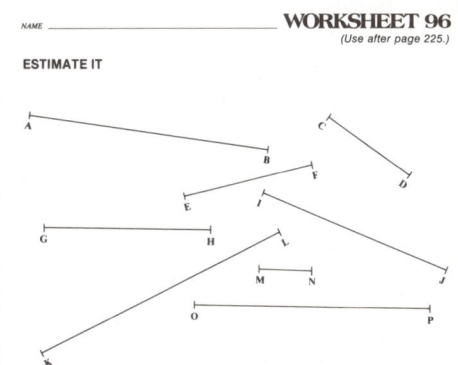

Group Project
Collecting data

Have each student measure to the nearest centimeter his/her height, shoe size, and arm span. Then have the students use their measurements to find the class averages. (If centimeter tapes are not available, the students can use a string and then measure the string using a meterstick or centimeter ruler.)

Class Starter Quiz 88
on previous lesson

Draw "chains" having these lengths.

1. 4 cm	2. 70 mm
3. 9 cm	4. 125 mm
5. 10.5 cm	6. 55 mm

Copymaster S406

Lesson Objective
To become familiar with metric units of length

Problem-Solving Skills
Reading a map
Collecting, organizing, and analyzing data

Starting the Lesson
What are the facts? Have the students study the first two paragraphs and the *Here's how* on page 226 for one minute. Then tell them to close their books and choose *millimeter, centimeter, meter,* or *kilometer* to complete each of these statements:

- The width of a fingernail is about one __?__. (cm)
- The length of a baseball bat is about one __?__. (m)
- The thickness of an eyeglass lens is about one __?__. (mm)
- One thousand long steps measure about one __?__. (km)

The list of metric equivalences at the top of the page is also on ■ **Visual Aid 40** (copymaster or transparency S144). Use the list to help the students see that the metric system is based on powers of 10.

Exercise Note
The map on page 227 is also on ■ **Visual Aid 41** (copymaster or transparency S145). Use the map when discussing exercises 17 and 18.

226

Metric units of length

Could you hit a baseball 100 meters? *Hint: The longest home run ever hit went 188.4 meters. It was hit by Roy "Dizzy" Carlyle.*

The basic unit for measuring length in the metric system is the **meter** (m). A baseball bat is about 1 meter long.

These units are used to measure length in the metric system:

1 kilometer (km) = 1000 meters
1 hectometer (hm) = 100 meters
1 dekameter (dam) = 10 meters
1 meter (m) = 1 meter
1 decimeter (dm) = 0.1 meter
1 centimeter (cm) = 0.01 meter
1 millimeter (mm) = 0.001 meter

Note: The units listed in red are used most often.

Records may change from year to year.

Here's how
to estimate metric units of length.

width of a fingernail— about 1 cm

thickness of an eyeglass lens—about 1 mm

I'd like to hit one a kilometer— that's about 1000 long steps!

one long step— about 1 meter

EXERCISES
Choose mm, cm, m, or km.

1. The distance from home plate to first base is 27.4 __?__. m

2. The height of a first baseman is 182 __?__. cm

3. The thickness of a dime is 1 __?__. mm

4. The length of a river is 450 __?__. km

5. The length of a tennis court is 20 __?__. m

6. The length of a paper clip is 3 __?__. cm

7. The width of a door is 0.6 __?__. m

8. The length of a new pencil is 190 __?__. mm

Which measurement is reasonable?

9. Length of a dollar bill:
   a. 16 mm   (b.) 16 cm   c. 16 m

10. Length of an automobile:
    a. 4.75 cm   b. 4.75 mm   (c.) 4.75 m

11. Length of a baseball bat:
    a. 92 mm   (b.) 92 cm   c. 92 m

12. Height of a nine-story building:
    a. 30 cm   (b.) 30 m   c. 30 km

13. Height of a soup can:
    a. 10 mm   (b.) 10 cm   c. 10 m

14. Width of a thumb:
    (a.) 20 mm   b. 20 cm   c. 20 m

15. Thickness of a nickel:
    (a.) 2 mm   b. 2 cm   c. 2 m

16. Thickness of a dollar bill:
    (a.) 0.1 mm   b. 0.1 cm   c. 0.1 m

AIR DISTANCES

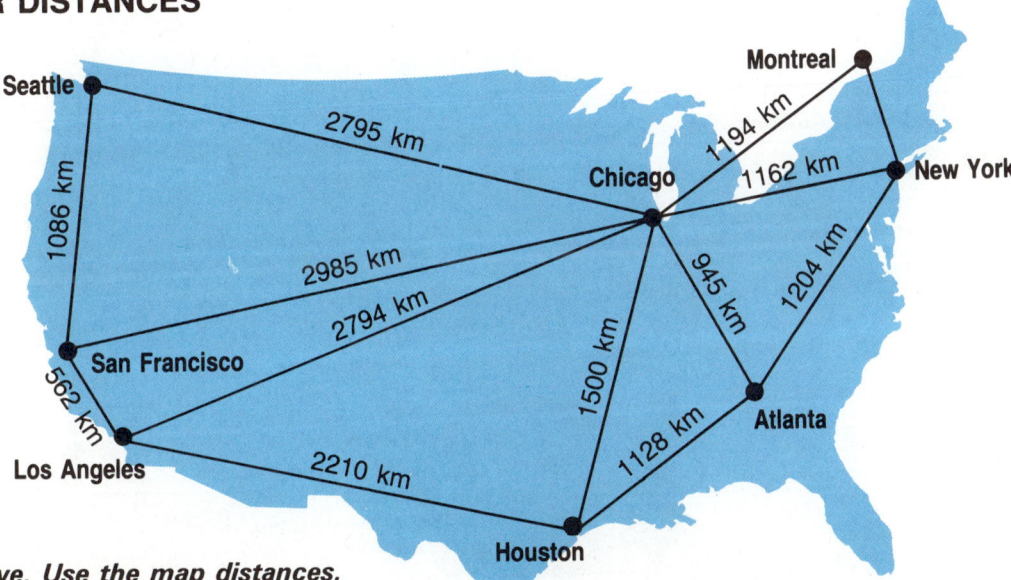

Solve. Use the map distances.

17. How far is it from
    a. San Francisco to New York through Chicago? **4147 km**
    b. New York to Houston through Atlanta? **2332 km**

18. How much farther is it from
    a. Chicago to Los Angeles than from Chicago to Houston? **1294 km**
    b. San Francisco to Chicago than from Los Angeles to Chicago? **191 km**

Project   Collecting, organizing, and analyzing data   Answers will vary.

19. Ask each of your classmates to tell you their arm span to the nearest centimeter.

20. Compute the mean arm span of your classmates. Round to the nearest centimeter.

21. Using the mean arm span, find out how many students it would take to reach from New York City to Los Angeles. Hint: The driving distance is 4684 kilometers.

Measurement **227**

Practice Worksheet
Workbook S265, Copymaster S265, or Duplicating Master S265

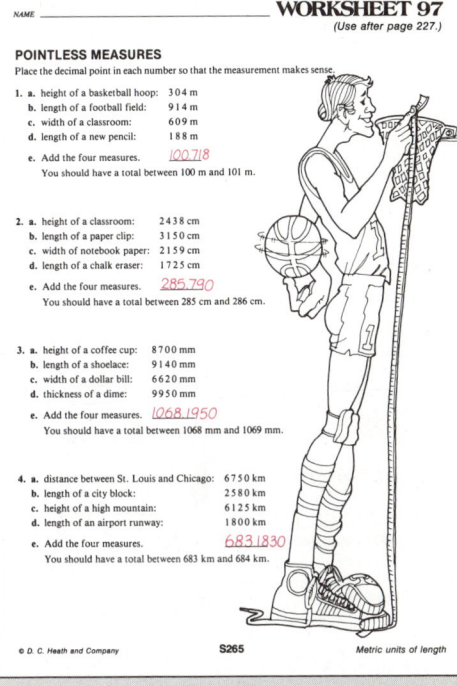

WORKSHEET 97 (Use after page 227.)

POINTLESS MEASURES
Place the decimal point in each number so that the measurement makes sense.

1. a. height of a basketball hoop: 304 m
   b. length of a football field: 914 m
   c. width of a classroom: 609 m
   d. length of a new pencil: 188 m
   e. Add the four measures. **100.718**
      You should have a total between 100 m and 101 m.

2. a. height of a classroom: 2438 cm
   b. length of a paper clip: 3150 cm
   c. width of notebook paper: 2159 cm
   d. length of a chalk eraser: 1725 cm
   e. Add the four measures. **285.790**
      You should have a total between 285 cm and 286 cm.

3. a. height of a coffee cup: 8700 mm
   b. length of a shoelace: 9140 mm
   c. width of a dollar bill: 6620 mm
   d. thickness of a dime: 9950 mm
   e. Add the four measures. **1068.1950**
      You should have a total between 1068 mm and 1069 mm.

4. a. distance between St. Louis and Chicago: 6750 km
   b. length of a city block: 2580 km
   c. height of a high mountain: 6125 km
   d. length of an airport runway: 1800 km
   e. Add the four measures. **683.1830**
      You should have a total between 683 km and 684 km.

© D. C. Heath and Company   S265   Metric units of length

Group Project
Estimating length

Have the students work in pairs. One student holds a metric ruler and the other estimates 8.5 centimeters by pointing to the back of the ruler. The other member of the pair tells his/her partner how close he/she was to the given measurement. Let the students take turns, one naming the length and the other estimating.

227

Class Starter Quiz 89
on previous lesson

Which measurement is reasonable?

1. Length of a new pencil: a
   a. 175 mm b. 175 cm c. 175 m
2. Height of a basketball hoop: c
   a. 3 mm b. 3 cm c. 3 m
3. Width of a dollar bill: b
   a. 6.6 mm b. 6.6 cm c. 6.6 m
4. Cruising altitude of a jet: c
   a. 10 cm b. 10 m c. 10 km

Copymaster S407

Lesson Objective
To make conversions between metric units of length

Problem-Solving Skills
Utilizing metric relationships
Using logical reasoning

Starting the Lesson
Have the students read the information on the entry blanks at the top of the page. Have them record who they think caught the bigger fish. Tell them they will be able to check their answers after completing exercises 1–4.

Exercise Note
Number Sense In exercises 5–37, encourage the students to first decide whether they are converting to a smaller or larger unit. If the conversion is to a smaller unit, multiplication is used. If the conversion is to a larger unit, division is used. Remind the students that the list of metric equivalences on page 226 is helpful when converting from one metric unit to another.

228

Changing units in the metric system

WHO CAUGHT THE BIGGER FISH?

To change units in the metric system, multiply or divide by 10, 100, or 1000.

FISHING CONTEST
Prizes:
1st: $100.00 CASH
2nd: PICKWICK ULTRA LIGHT FISHING POLE
3rd: STAY-DRY HIP WADERS

Name: Mel Criser
Address: 2347 South Vine
Wichita, Kansas
Length: 0.38 m

Name: Nancy Perkins
Address: 42 Falls Road
Salt Lake City, Utah
Length: 475 mm

Here's how *to change from one unit of length to another.*

Mel's fish
0.38 m = ? cm

Since I'm changing to a smaller unit, I should get a larger number. So I should multiply.

Remember: 1 m = 100 cm

0.38 m = 38 cm
    ⌊— × 100 —⌋

Nancy's fish
475 mm = ? cm

Now I'm changing to a larger unit, so I should get a smaller number. Therefore I should divide.

Remember: 10 mm = 1 cm

475 mm = 47.5 cm
    ⌊— ÷ 10 —⌋

1. Look at the *Here's how*. To change from meters to centimeters, multiply by ?. **100**

2. To change from millimeters to centimeters, divide by ?. **10**

3. Who caught the larger fish, Mel or Nancy? **Nancy**

4. Check these examples. Have the units been changed correctly? **Yes**

    a. 1395 m = ? km

      Think: Changing smaller to larger units, so divide.

      Remember: 1000 m = 1 km
               1395 m = **1.395** km
                   ⌊— ÷ 1000 —⌋

    b. 8.53 km = ? m

      Think: Changing larger to smaller units, so multiply.

      Remember: 1 km = 1000 m
               8.53 km = **8530** m
                   ⌊— × 1000 —⌋

228 Chapter 9

EXERCISES

Copy and complete.

5. 7 cm = ? mm 70
6. 3 m = ? cm 300
7. 8 km = ? m 8000
8. 45 cm = ? m 0.45
9. 63 mm = ? cm 6.3
10. 18 km = ? m 18,000
11. 4265 m = ? km 4.265
12. 7.3 km = ? m 7300
13. 4.2 cm = ? mm 42
14. 95 m = ? cm 9500
15. 4.8 m = ? cm 480
16. 420 cm = ? m 4.20
17. 68 mm = ? cm 6.8
18. 3.25 km = ? m 3250
19. 1575 m = ? km 1.575
20. 15 m = ? cm 1500
21. 14 mm = ? cm 1.4
22. 7.8 cm = ? mm 78
23. 250 km = ? m 250,000
24. 300 mm = ? m 0.300
25. 0.4 m = ? mm 400
26. 5 cm + 4 mm = ? mm 54
27. 20 cm + 4 mm = ? mm 204
28. 30 cm + 5 mm = ? cm 30.5
29. 10 cm + 12 mm = ? cm 11.2
30. 8 m + 75 cm = ? cm 875
31. 5 m + 125 cm = ? cm 625
32. 4 m + 50 cm = ? m 4.5
33. 7 m + 200 cm = ? m 9

Solve.

34. The largest salt water fish caught was a 1852-centimeter whale shark. How many meters was that? 18.52

35. The smallest fresh water fish caught was a dwarf pygmy goby. It was only 0.7 centimeter long. How many millimeters was that? 7

36. The greatest depth at which a fish was caught was 8299 meters in the Puerto Rico Trough. How many kilometers was that? 8.299

37. The fastest fish is the cosmopolitan sailfish. It can swim 110 kilometers per hour. How many meters per hour is that? 110,000

You be the judge — Logical reasoning

38. Who caught the biggest fish? Rita
    How big was it? 61 cm

    Clues:
    - Susan's fish was 4 cm longer than Wilda's fish.
    - Wilda's fish was 5 cm shorter than Rita's fish.
    - Rita's fish was 1 cm longer than Monica's fish.
    - Monica caught a 60-cm fish.

Practice Worksheet

Workbook S266, Copymaster S266, or Duplicating Master S266

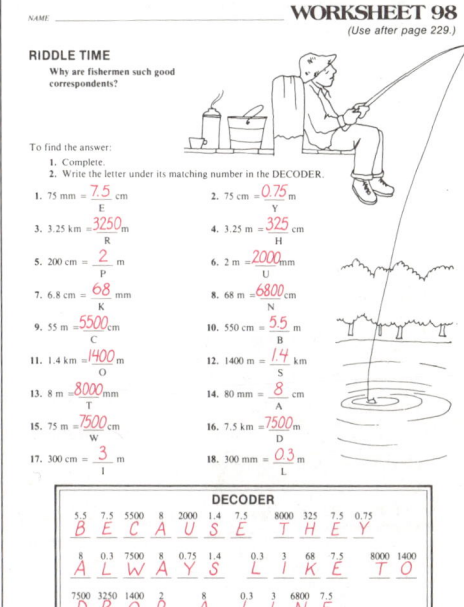

Project

Estimating length

Use your normal step as a unit of measure. Then estimate, by counting steps, the number of kilometers you walk while you are at school.

Copymaster S480

Measurement 229

Class Starter Quiz 90
on previous lesson

Complete.

1. 9 cm = ? mm	90
2. 8 m = ? cm	800
3. 4 km = ? m	4000
4. 2125 m = ? km	2.125
5. 475 cm = ? m	4.75
6. 644 mm = ? cm	64.4
7. 0.6 m = ? mm	600
8. 1.5 km = ? m	1500

Copymaster S407

Lesson Objectives
To become familiar with metric units of liquid volume
To make conversions between metric units of liquid volume

Problem-Solving Skill
Selecting information in an ad

Starting the Lesson
What are the facts? Have the students study the information at the top of page 230 for 60 seconds. Then have them close their books and answer these questions:

- An eyedropper holds about how many milliliters? (1)
- Four glasses of perfume is about how many liters? (1)
- One liter equals how many milliliters? (1000)
- About how much would one liter of the world's most expensive perfume cost? ($15,000)

230

Liquid volume—metric system

WOULD YOU BELIEVE IT!

One drop of the world's most expensive perfume costs $1.50. One liter of the same perfume would cost about $15,000!

A unit for measuring liquid volume in the metric system is the **liter** (L). The **milliliter** (mL) is used to measure small liquid volumes.

1 L = 1000 mL

About 10 drops
1 milliliter of perfume

10 milliliters of perfume

About 4 glasses
1 liter of perfume

1. Which is more, 999 milliliters or 1 liter? 1 liter
2. Choose mL or L.
   a. A 250-[?] bottle of shampoo costs $1.99. mL
   b. A 0.5-[?] bottle of liquid soap costs $1.89. L

Here's how *to change from one unit of liquid volume to another.*

250 mL = ? L	0.5 L = ? mL
Think: Changing from smaller units to larger units, so divide.	*Think:* Changing from larger units to smaller units, so multiply.
Remember: 1000 mL = 1 L	*Remember:* 1 L = 1000 mL
250 mL = 0.250 L ↳ ÷ 1000 ↲	0.5 L = 500 mL ↳ × 1000 ↲

3. Look at the *Here's how*. To change from milliliters to liters, divide by [?]. 1000
4. To change from liters to milliliters, multiply by [?]. 1000

Chapter 9

EXERCISES

Which liquid volume seems reasonable?

5. A coffee cup:
   a. 300 mL   b. 30 mL   c. 3 mL

6. A soft drink can:
   a. 4 mL   b. 40 mL   c. 400 mL

7. A bathtub:
   a. 3 L   b. 30 L   c. 300 L

8. A tablespoon:
   a. 0.5 mL   b. 5 mL   c. 50 mL

9. A thermos bottle:
   a. 80 mL   b. 800 mL   c. 8000 mL

10. A fruit-juice pitcher:
    a. 0.1 L   b. 1 L   c. 10 L

11. An eyedropper:
    a. 1 mL   b. 10 mL   c. 100 mL

12. An automobile gas tank:
    a. 6 L   b. 60 L   c. 600 L

Copy and complete.

13. 6 L = ? mL *6000*
14. 15 L = ? mL *15,000*
15. 2.7 L = ? mL *2700*
16. 4000 mL = ? L *4*
17. 1725 mL = ? L *1.725*
18. 500 mL = ? L *0.500*
19. 5.75 L = ? mL *5750*
20. 0.756 L = ? mL *756*
21. 0.35 L = ? mL *350*
22. 12,000 mL = ? L *12*
23. 870 mL = ? L *0.870*
24. 25 mL = ? L *0.025*
25. 6.05 L = ? mL *6050*
26. 175 mL = ? L *0.175*
27. 100 mL = ? L *0.100*
28. 986 mL = ? L *0.986*
29. 790 L = ? mL *790,000*
30. 2800 L = ? mL *2,800,000*
31. 210 mL = ? L *0.210*
32. 60 mL = ? L *0.060*
33. 5640 L = ? mL *5,640,000*
34. 2 L + 500 mL = ? mL *2500*
35. 5 L + 125 mL = ? mL *5125*
36. 4 L + 100 mL = ? L *4.1*
37. 3 L + 625 mL = ? L *3.625*
38. 46 L + 18 mL = ? mL *46,018*
39. 16 L + 200 mL = ? L *16.2*
40. 75 L + 500 mL = ? mL *75,500*
41. 35 L + 600 mL = ? L *35.6*

Special smells

Use the information in the advertisement. Complete the sentences.

42. A 50-milliliter bottle of ? costs $5.00. *Sweet Rose*

43. A ?-milliliter bottle of Sweet Rose costs $8.00. *80*

44. A 40-milliliter bottle of Twilight costs ? dollars. *$2.00*

45. A 20-milliliter bottle of Twilight and a ?-milliliter bottle of New Spice cost a total of $10.00. *30*

PERFUME $PECIAL$

Twilight	$.05 per mL
Sweet Rose	$.10 per mL
Always Yours	$.20 per mL
New Spice	$.30 per mL

Extra Practice
Page 491 Skill 42

Practice Worksheet
Workbook S267, Copymaster S267, or Duplicating Master S267

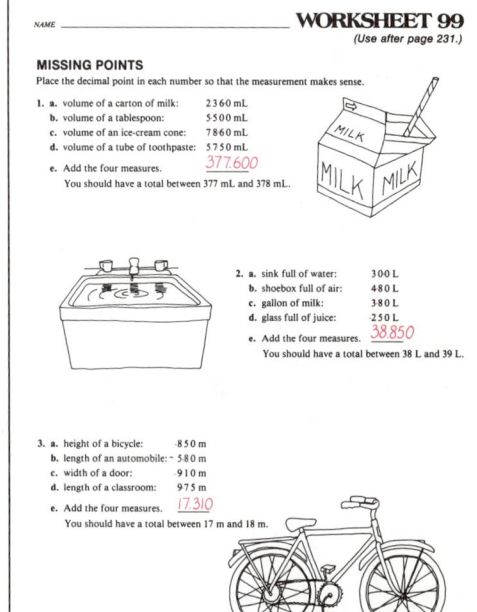

WORKSHEET 99 (Use after page 231.)

MISSING POINTS
Place the decimal point in each number so that the measurement makes sense.

1. a. volume of a carton of milk: 2360 mL
   b. volume of a tablespoon: 5500 mL
   c. volume of an ice-cream cone: 7860 mL
   d. volume of a tube of toothpaste: 5750 mL
   e. Add the four measures. *377.600*
   You should have a total between 377 mL and 378 mL.

2. a. sink full of water: 300 L
   b. shoebox full of air: 480 L
   c. gallon of milk: 380 L
   d. glass full of juice: 250 L
   e. Add the four measures. *38.850*
   You should have a total between 38 L and 39 L.

3. a. height of a bicycle: 850 m
   b. length of an automobile: 580 m
   c. width of a door: 910 m
   d. length of a classroom: 975 m
   e. Add the four measures. *17.310*
   You should have a total between 17 m and 18 m.

Project

Researching information

Estimate and then measure the amount of water that would be lost from a dripping faucet in a day.

Copymaster S480

Class Starter Quiz 91
on previous lesson

Complete.

1. 4 L = _?_ mL   4000
2. 25 L = _?_ mL   25,000
3. 6.5 L = _?_ mL   6500
4. 5000 mL = _?_ L   5
5. 875 mL = _?_ L   0.875
6. 25 mL = _?_ L   0.025

Copymaster S407

Lesson Objectives
To become familiar with metric units of weight (mass)

To make conversions between metric units of weight (mass)

Problem-Solving Skill
Utilizing metric relationships

Starting the Lesson
What are the facts? Have the students study the information at the top of page 232 for 60 seconds. Then have them close their books and answer these questions from memory:

- One kilogram equals how many grams? (1000)
- How many milligrams does it take to equal 1 gram? (1000)
- Does a paper clip weigh 1 gram or 1 kilogram? (1 g)
- Does a wing of a honeybee weigh 1 milligram or 1 gram? (1 mg)
- Can a honeybee carry more or less than 300 times its own weight? (More)

Then discuss exercises 1–3 and the *Here's how*.

232

Weight—metric system

AMAZING FACT!

Did you know that a honeybee can lift more than 300 times its own weight!

A unit for measuring weight (mass) in the metric system is the **gram** (g). The weight of a large paper clip is about 1 gram.

Here are some other units for measuring weight.

$$1 \text{ kilogram (kg)} = 1000 \text{ g}$$
$$1 \text{ milligram (mg)} = 0.001 \text{ g}$$

The weight of this textbook is about 1 kilogram.

Would you believe . . . A honeybee can lift an object as heavy as an egg!

Wing—about 1 mg

Honeybee—about 200 mg

1. Which is heavier, 999 grams or 1 kilogram?   1 kilogram
2. Which is heavier, 999 milligrams or 1 gram?   1 gram

Here's how *to change from one metric unit of weight to another.*

0.06 kg = _?_ g

Changing from larger to smaller units, so multiply.

Remember: 1 kg = 1000 g

    0.06 kg = **60** g
        └× 1000┘

200 mg = _?_ g

Think: Changing from smaller to larger units, so divide.

Remember: 1000 mg = 1 g

    200 mg = **0.2** g
        └÷ 1000┘

3. Check these examples. Have the unit changes been done correctly? **Yes**

    a. 4.6 g = _?_ mg

       Think: Changing from larger to smaller units, so multiply.

       Remember: 1 g = 1000 mg

         4.6 g = **4600** mg
          └× 1000┘

    b. 358 g = _?_ kg

       Think: Changing from smaller to larger units, so divide.

       Remember: 1000 g = 1 kg

         358 g = **0.358** kg
          └÷ 1000┘

EXERCISES

Which weight seems reasonable?

4. A bicycle:
   a. 12 mg   b. 12 g   **c.** 12 kg
5. A dime:
   a. 3 mg   **b.** 3 g   c. 3 kg
6. A straight pin:
   a. 130 mg   b. 130 g   c. 130 kg
7. An automobile:
   a. 2000 mg   b. 2000 g   **c.** 2000 kg
8. A can of tomatoes:
   a. 464 mg   **b.** 464 g   c. 464 kg
9. An apple:
   a. 330 mg   **b.** 330 g   c. 330 kg

Copy and complete.

10. 6000 mg = ? g   *6*
11. 4125 mg = ? g   *4.125*
12. 765 mg = ? g   *0.765*
13. 5000 g = ? kg   *5*
14. 7617 g = ? kg   *7.617*
15. 326 g = ? kg   *0.326*
16. 7 kg = ? g   *7000*
17. 4.2 kg = ? g   *4200*
18. 1.27 kg = ? g   *1270*
19. 4 g = ? mg   *4000*
20. 6.5 g = ? mg   *6500*
21. 0.425 g = ? mg   *425*
22. 3.5 kg = ? g   *3500*
23. 315 g = ? kg   *0.315*
24. 775 mg = ? g   *0.775*
25. 86.3 kg = ? g   *86,300*
26. 489 g = ? kg   *0.489*
27. 9163 g = ? kg   *9.163*
28. 1653 mg = ? g   *1.653*
29. 25.8 kg = ? g   *25,800*
30. 41.3 g = ? mg   *41,300*
31. 6 g + 325 mg = ? mg   *6325*
32. 6 kg + 3500 g = ? g   *9500*
33. 35 g + 1800 mg = ? mg   *36,800*
34. 85 kg + 6000 g = ? kg   *91*
35. 8 g + 435 mg = ? mg   *8435*
36. 25 g + 1500 mg = ? mg   *26,500*
37. 4 g + 666 mg = ? g   *4.666*
38. 12 g + 2545 mg = ? g   *14.545*
39. 6 kg + 825 g = ? g   *6825*
40. 5 kg + 2000 g = ? g   *7000*
41. 9 kg + 125 g = ? kg   *9.125*
42. 3 kg + 3000 g = ? kg   *6*

Are you as strong as an ant?

43. a. An ant weighing 200 mg can lift (with its teeth) a weight of 10 g. How many times its weight can an ant lift? *50*
    b. Suppose that you could lift the same number of times your weight. How much weight would that be? *Answers will vary.*
    c. Are you as strong as an ant? *No*

Measurement **233**

Practice Worksheet

Workbook S268, Copymaster S268, or Duplicating Master S268

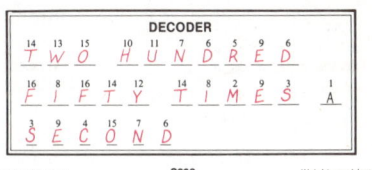

Group Project

Estimating and measuring weight

Provide the students with common classroom objects (pencil, eraser, chalk, book, etc.). Have them work in small groups to complete the chart by first estimating each object's weight and then finding its actual weight using a metric scale.

Object	Estimate	Actual weight

233

Class Starter Quiz 92
on previous lesson

Complete.

1. 7000 mg = __?__ g   7
2. 8000 g = __?__ kg   8
3. 9 kg = __?__ g   9000
4. 4.5 g = __?__ mg   4500
5. 625 mg = __?__ g   0.625
6. 3.5 kg = __?__ g   3500

Copymaster S407

Problem-Solving Skills
Choosing the correct operation
Solving a multistep problem

Skills Reviewed
Multiplying and dividing decimals
Changing fractions to decimals
Changing decimals to mixed numbers
Adding mixed numbers
Adding and subtracting fractions
Multiplying and dividing fractions

Starting the Lesson
Problem Solving Challenge the students to an operation hunt. Have them look at problems 1–5 on page 234 to answer these questions:

- Which problem could you solve by dividing? (Problem 3)
- Which problem could you solve by subtracting? (Problem 1)
- Which problem could you solve by multiplying and subtracting? (Problem 2)

Cumulative Skill Practice Challenge the students to an estimation hunt by saying, "Find the four products in exercises 1–12 that are greater than 30." (Exercises 1, 8, 9, and 10) Then have students find the quotient in exercises 13–22 that is greater than 100. (Exercise 17)

Problem solving
Choosing the operation

YOU'RE THE MANAGER!

Solve.
Decide when a calculator would be useful.

1. You sell a certain brand of plain jeans for $19.98 and the designer jeans for $27.65. How much more do you charge for the designer jeans? $7.67

2. You buy 72 pairs of jeans for $12.45 each and sell them for $18.69 each. How much profit do you make on 72 pairs? $449.28

3. One day 42 customers spent $964.50 in your store. What was the average amount spent per customer? Round the answer to the nearest cent. $22.96

4. One day you sold 60 pairs of jeans. Twenty-four pairs were designer jeans. What fraction of the jeans were not designer jeans? Give the answer in simplest form. $\frac{3}{5}$

5. A denim shirt regularly sells for $18.60. You put it on sale for $\frac{2}{3}$ of the regular price. What is the sale price? $12.40

6. Your store has an area of 996 square feet. Your yearly rent is $14 per square foot. How much is your yearly rent? your monthly rent? $13,944; $1162

7. You pay each employee $5.75 an hour. Your store hours are from 9:00 A.M. to 6:00 P.M. If an employee takes 1 hour off for lunch, how much does it cost per day for each full-time employee? $46

8. One employee works from 9:00 A.M. to 12:45 P.M. How many hours does she work each day? Give the answer as a mixed number in simplest form. $3\frac{3}{4}$

9. a. One employee works from 1:30 P.M. to 6:00 P.M. How many hours is that? Give your answer as a decimal. 4.5
   b. At $5.75 per hour, how much will you pay him each day? Round the answer to the nearest cent. $25.88

Cumulative Skill Practice

Multiply. (page 68)

1. 1.2 × 38 = 45.6
2. 5.9 × 3.2 = 18.88
3. 0.63 × 7.4 = 4.662
4. 0.82 × 23 = 18.86
5. 0.48 × 9.6 = 4.608
6. 6.09 × 3.4 = 20.706
7. 0.61 × 0.55 = 0.3355
8. 0.46 × 72 = 33.12
9. 9.2 × 9.2 = 84.64
10. 5.12 × 35 = 179.20
11. 3.75 × 0.38 = 1.4250
12. 2.15 × 0.2 = 0.430

Divide. Round the quotient to the nearest hundredth. (page 104)

13. 1.7 ÷ 0.3 = 5.67
14. 4.77 ÷ 0.6 = 7.95
15. 2.4 ÷ 0.09 = 26.67
16. 5.8 ÷ 0.06 = 96.67
17. 25 ÷ 0.03 = 833.33
18. 0.5 ÷ 0.6 = 0.83
19. 7.9 ÷ 0.09 = 87.78
20. 0.7 ÷ 0.13 = 5.38
21. 4.26 ÷ 0.27 = 15.78
22. 29.5 ÷ 0.35 = 84.29

Change to a decimal. (page 166)

23. $\frac{1}{4}$ = 0.25
24. $\frac{4}{5}$ = 0.8
25. $\frac{7}{8}$ = 0.875
26. $\frac{3}{5}$ = 0.6
27. $\frac{1}{2}$ = 0.5
28. $\frac{1}{5}$ = 0.2
29. $\frac{3}{4}$ = 0.75
30. $1\frac{1}{2}$ = 1.5
31. $2\frac{3}{10}$ = 2.3
32. $4\frac{3}{8}$ = 4.375
33. $3\frac{5}{8}$ = 3.625
34. $2\frac{3}{4}$ = 2.75
35. $6\frac{2}{5}$ = 6.4
36. $8\frac{7}{10}$ = 8.7

Change to a mixed number in simplest form. (page 168)

37. 2.4 = $2\frac{2}{5}$
38. 3.5 = $3\frac{1}{2}$
39. 4.8 = $4\frac{4}{5}$
40. 9.6 = $9\frac{3}{5}$
41. 5.2 = $5\frac{1}{5}$
42. 6.9 = $6\frac{9}{10}$
43. 4.50 = $4\frac{1}{2}$
44. 6.60 = $6\frac{3}{5}$
45. 2.25 = $2\frac{1}{4}$
46. 7.75 = $7\frac{3}{4}$
47. 8.35 = $8\frac{7}{20}$
48. 3.14 = $3\frac{7}{50}$

Add. Give each sum in simplest form. (page 182)

49. $8\frac{1}{2} + 2\frac{1}{4} = 10\frac{3}{4}$
50. $3\frac{1}{2} + 4\frac{1}{3} = 7\frac{5}{6}$
51. $7\frac{3}{8} + 1\frac{1}{4} = 8\frac{5}{8}$
52. $9\frac{2}{5} + 1\frac{1}{5} = 10\frac{3}{5}$
53. $8 + 2\frac{2}{3} = 10\frac{2}{3}$
54. $6\frac{1}{2} + 3\frac{1}{3} = 9\frac{5}{6}$

MIXED PRACTICE

Complete. Give answers in simplest form.

55. $\frac{1}{2} + \frac{1}{4} = \underline{\ ?\ }$ $\frac{3}{4}$
56. $\frac{1}{4} \times \frac{1}{4} = \underline{\ ?\ }$ $\frac{1}{16}$
57. $\frac{1}{3} + \frac{1}{2} = \underline{\ ?\ }$ $\frac{5}{6}$
58. $\frac{3}{4} - \frac{1}{2} = \underline{\ ?\ }$ $\frac{1}{4}$
59. $\frac{4}{5} \div 4 = \underline{\ ?\ }$ $\frac{1}{5}$
60. $\frac{9}{10} - \frac{2}{5} = \underline{\ ?\ }$ $\frac{1}{2}$
61. $\frac{2}{3} \times \frac{3}{4} = \underline{\ ?\ }$ $\frac{1}{2}$
62. $2 + \frac{4}{5} = \underline{\ ?\ }$ $2\frac{4}{5}$
63. $6 \div \frac{1}{3} = \underline{\ ?\ }$ 18
64. $3 - \frac{1}{4} = \underline{\ ?\ }$ $2\frac{3}{4}$
65. $\frac{7}{8} \div \frac{5}{4} = \underline{\ ?\ }$ $\frac{7}{10}$
66. $\frac{7}{8} \times \frac{3}{2} = \underline{\ ?\ }$ $1\frac{5}{16}$

Measurement

Problem-Solving Worksheet
Workbook S269, Copymaster S269, or Duplicating Master S269

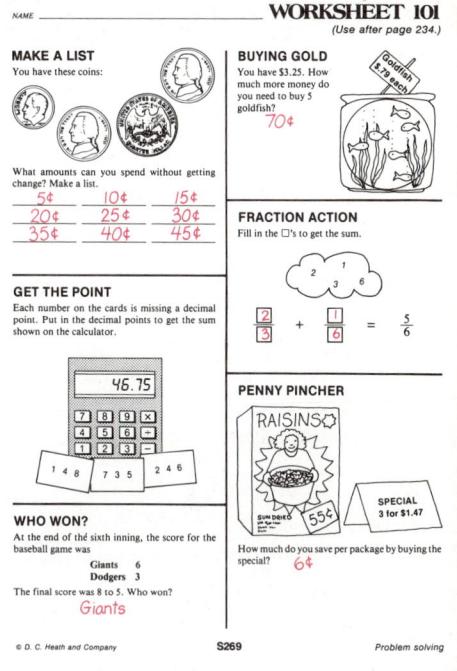

Project

Reading a newspaper ad

Get a newspaper. Use some of the newspaper ads to make up problems similar to problems 1–5 on page 234.

Copymaster S481

Class Starter Quiz 93
on previous lesson

Solve.

1. You buy 24 shirts for $13.80 each and sell them for $19.99 each. How much profit do you make on 24 shirts? **$148.56**
2. A pair of jeans regularly sells for $24.80. You put it on sale for $\frac{3}{4}$ of the regular price. What is the sale price? **$18.60**

Copymaster S408

Lesson Objectives
To read the scale on a Fahrenheit or a Celsius thermometer
To find the difference between two temperatures

Problem-Solving Skills
Reading a graph
Reading a scale

Starting the Lesson
Before the students open their textbooks, ask them whether 70°C, 20°C, or 0°C is the temperature on a nice day. Then say, "Open your books to page 236 and check your guess against the thermometer at the top of the page."

Temperature—degrees Celsius and degrees Fahrenheit

Two common thermometer scales are used to measure temperature.

The metric system uses the Celsius scale.

The customary system use the Fahrenheit scale.

Read the temperature of the "nice day" shown on the thermometer to the right as

25 degrees Celsius (25°C),
or
77 degrees Fahrenheit (77°F).

The "cold day" on the thermometer is 21 degrees Celsius below zero, or −21°C.

1. At what Fahrenheit temperature does water boil? Does water freeze? **212°, 32°**
2. What is body temperature in degrees Celsius? **36°**

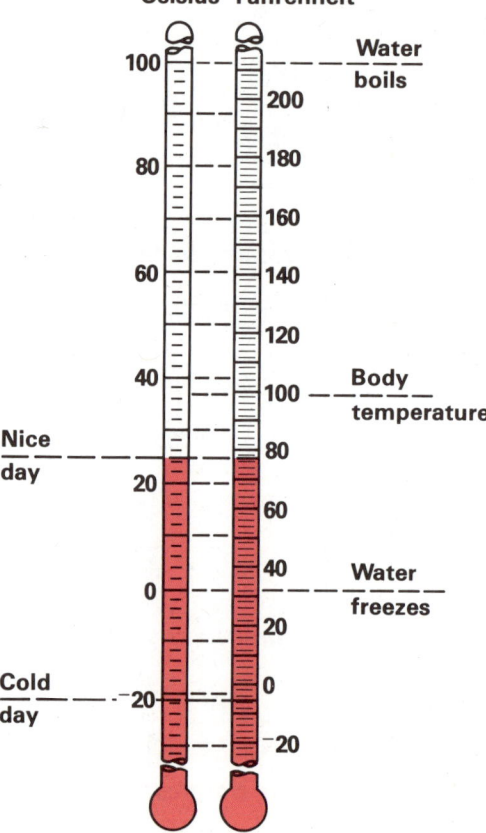

Here's how *to find the difference of two temperatures.*

If both temperatures are above or below zero, subtract the smaller whole number from the larger.

If one temperature is above zero and one is below zero, add the two whole numbers.

Ex. 1 28°F, 15°F

28 − 15 = 13

A difference of 13°F

Ex. 2 −48°C, −16°C

48 − 16 = 32

A difference of 32°C

17°C, −11°C

17 + 11 = 28

A difference of 28°C

3. Look at the *Here's how*. What is the difference of 28°F and 15°F? **13°F** Of 17°C and −11°C? **28°C**

EXERCISES

Give the temperatures shown on these Celsius thermometers.

4. 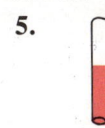 28°C
5.  −4°C
6.  −3°C
7. 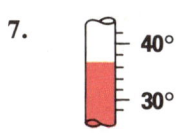 37°C

Give the temperatures shown on these Fahrenheit thermometers.

8.  68°F
9. 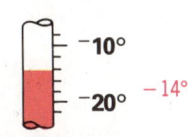 −14°F
10. 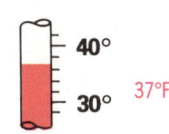 37°F
11. 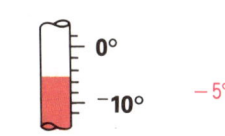 −5°F

Choose the temperature that would be most reasonable for each activity.

12. Skiing   −18°C or 28°C
13. Planting a garden   2°C or 24°C
14. Playing tennis   26°C or 90°C
15. Ice skating   −10°C or −60°C
16. Swimming   86°F or 10°F
17. Water skiing   42°F or 78°F
18. Ice fishing   −16°F or 49°F
19. Riding a bike   62°F or 3°F

Solve.

20. Suppose that a daily high temperature was 74°F and the daily low was 33°F. What was the difference in these two temperatures? 41°

21. It was 20°C when Ron left for school. When he returned home in the afternoon the temperature was 13°C. How many degrees did the temperature fall? 7°

22. Sharon mixed two liquids in science class. She recorded the temperature of the mixture every 30 seconds. The temperature started at 21°C. It went up 2°, up 1°, down 2°, and down 3°. What was the temperature then? 19°C

23. When the skaters arrived at the skating rink the temperature was −5°F. The temperature went up to 0°F. Then it went up to 8°F. How many degrees did it go up between −5°F and 8°F? 13°

Hot spots   Reading a graph

24. What is the highest recorded temperature in the state of Kansas? 121°F

25. In what year was Arizona's highest temperature recorded? 1905

26. How many states have higher recorded temperatures than North Dakota? 3

27. The lowest recorded temperature was about −129°F. The highest recorded temperature was about 136°F. What is the difference between these two temperatures? 265°

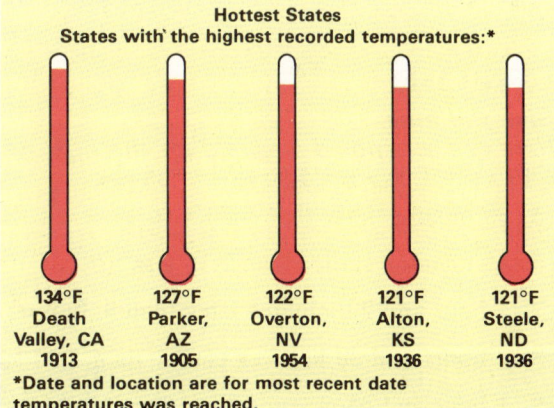

Hottest States
States with the highest recorded temperatures:*

- 134°F Death Valley, CA 1913
- 127°F Parker, AZ 1905
- 122°F Overton, NV 1954
- 121°F Alton, KS 1936
- 121°F Steele, ND 1936

*Date and location are for most recent date temperatures was reached.

Records may change from year to year.

Measurement 237

Practice Worksheet
Workbook S270, Copymaster S270, or Duplicating Master S270

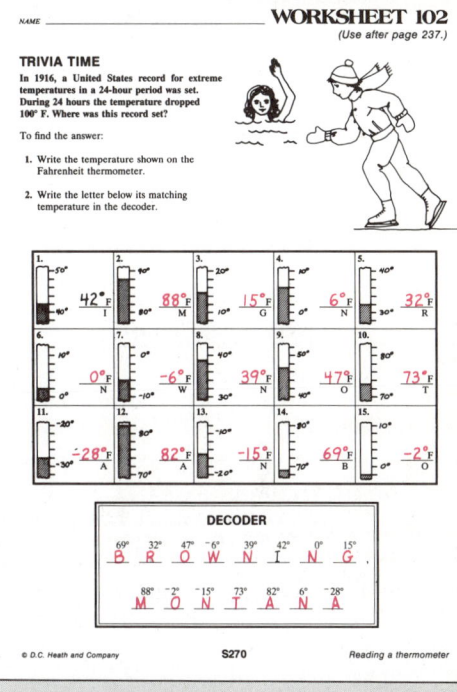

Group Project
Collecting data

Have students work in small groups to use a thermometer and measure the temperature at the same time each day for two weeks. Then have them use the data to make a line graph of the data. (*Note:* Each group could be given a different time of day. Then you could compare the graphs.)

237

Elapsed time

A.M. is used for times after 12:00 midnight and before 12:00 noon.

P.M. is used for times after 12:00 noon and before 12:00 midnight.

1. Look at Paige's time card. At what time did she begin work on Monday? At what time did she finish work? 3:45 P.M., 6:30 P.M.

The amount of time that Paige worked on Monday is the **elapsed time** from 3:45 P.M. to 6:30 P.M.

Johnson, Paige		
Day	Time In	Time Out
Monday	3:45 P.M.	6:30 P.M.
Tuesday	3:30 P.M.	6:30 P.M.
Wednesday	3:45 P.M.	7:00 P.M.
Thursday	4:30 P.M.	7:15 P.M.
Friday	—	—
Saturday	9:00 A.M.	2:15 P.M.
Sunday	—	—

Here's how to find the elapsed time.

Began at 3:45 P.M. → 15 min later → 2h 30min later → Finished at 6:30 P.M.

15min + 2h 30min = 2h 45min

2. Look at the *Here's how*. How long did Paige work on Monday? 2 h 45 min

EXERCISES

Give the time.

Here are scrambled answers for the next two rows of exercises:
6:15 A.M.   7:00 A.M.   1:30 P.M.   1:35 P.M.

3. 30 minutes later than 6:30 A.M. 7:00 A.M.

4. 30 minutes later than 5:45 A.M. 6:15 A.M.

5. 45 minutes earlier than 2:15 P.M. 1:30 P.M.

6. 25 minutes earlier than 2:00 P.M. 1:35 P.M.

7. 1 hour and 20 minutes later than 4:30 P.M. 5:50 P.M.

8. 2 hours and 30 minutes earlier than 1:00 P.M. 10:30 A.M.

9. 1 hour and 45 minutes later than 11:30 A.M. 1:15 P.M.

10. 2 hours and 40 minutes earlier than 1:30 A.M. 10:50 P.M.

Class Starter Quiz 94
on previous lesson

Give the difference of each pair of temperatures.

1. 13°C	9°C	4°C
2. 25°C	17°C	8°C
3. 9°C	−5°C	14°C
4. 20°C	−10°C	30°C

Copymaster S408

Lesson Objective
To find elapsed time

Problem-Solving Skills
Reading a time card
Collecting, organizing, and analyzing data

Starting the Lesson
Write these sentences on the chalkboard:

A.M. is used for times after 12:00 <u>HGIMDNIT</u> and before 12:00 <u>ONON</u>.
P.M. is used for times <u>EFATR</u> 12:00 noon and <u>REBFOE</u> 12:00 midnight.

Before the students open their textbooks, challenge them to unscramble the underlined words so that the sentences make sense. Then have them read the sentences at the top of page 238 to see if they correctly unscrambled the letters.

Give the elapsed time.
Here are scrambled answers for the next three rows of exercises:
15 min   30 min   15 min   1 h   1 h 10 min   2 h

11. from 6:00 A.M. to 7:00 A.M.   1h
12. from 4:30 A.M. to 5:00 A.M.   30 min
13. from 1:15 A.M. to 1:30 A.M.   15 min
14. from 9:45 P.M. to 10:00 P.M.   15 min
15. from 10:30 A.M. to 12:30 A.M.   2 h
16. from 8:05 P.M. to 9:15 P.M.   1 h 10 min
17. from 5:20 A.M. to 5:40 A.M.   20 min
18. from 7:35 P.M. to 8:15 P.M.   40 min
19. from 3:40 A.M. to 5:00 A.M.   1 h 20 min
20. from 11:30 P.M. to noon   12 h 30 min
21. from 6:50 A.M. to 8:15 A.M.   1 h 25 min
22. from 4:10 P.M. to 6:30 P.M.   2 h 20 min
32. from 9:25 P.M. to 11:45 P.M.   2 h 20 min
24. from 8:55 A.M. to 11:30 A.M.   2 h 35 min
25. from 10:45 P.M. to midnight   1 h 15 min

Solve. Use the time card on page 238.

26. How long did Paige work on Tuesday?   3 h
27. How long did Paige work on Wednesday?   3 h 15 min
28. How many hours did she work on Saturday?   5 h 15 min
29. What was the total number of hours worked for the week?   16 h
30. If Paige is paid $4.36 an hour, how much does she earn for working 16 hours? $69.76
31. If Paige is paid $4.36 an hour, how much does she earn for work from 9:00 A.M. to 2:15 P.M.? $22.89

Project Collecting, organizing, and analyzing data

32. Ask your classmates how much time they spent watching TV last night. Have them round their times to the nearest $\frac{1}{4}$ hour. Make a frequency table.   Answers will vary.
33. Show the results on a graph.   Graphs will vary.
34. Write some facts shown by your graph.   Answers will vary.

Measurement **239**

Practice Worksheet
Workbook S271, Copymaster S271, or Duplicating Master S271

NAME _____   **WORKSHEET 103**
(Use after page 239.)

MOVIE MINUTES

MOVIE BEGINS   MOVIE ENDS

Complete this table.

	Time Movie Begins	Time Movie Ends	Length of Movie (in hours and minutes)
1.	2:00 P.M.	3:45 P.M.	1 h 45 min
2.	3:30 P.M.	4:15 P.M.	0 h 45 min
3.	4:20 P.M.	6:20 P.M.	2 h 0 min
4.	7:35 P.M.	9:25 P.M.	1 h 50 min
5.	11:30 A.M.	1:40 P.M.	2 h 10 min
6.	12:50 A.M.	2:45 P.M.	1 h 55 min
7.	4:35 P.M.	6:40 P.M.	2 h 5 min
8.	5:40 P.M.	8:00 P.M.	2 h 20 min
9.	6:15 P.M.	7:55 P.M.	1 h 40 min
10.	8:10 P.M.	10:25 P.M.	2 h 15 min
11.	7:25 P.M.	9:10 P.M.	1 h 45 min
12.	11:30 A.M.	2:05 P.M.	2 h 35 min

© D.C. Heath and Company   S271   Elapsed time

Group Project
Estimating time

Have the students work in small groups. Estimate the time it would take to do these activities:

- Walk a mile.
- Play one side of a 45 RPM record.
- Dial a 7-digit telephone number.
- Count from 1 to 100.

Then have them compare their estimates to determine which estimates are similar.

Length—customary units

GUESS AND CHECK

1. Guess which pencil is longer. (Later, you will check your guess by measuring each pencil with an inch ruler.) Answers will vary.

Here's how to measure with an inch ruler.

2. Look at the *Here's how*. What is the length of the key measured to the nearest $\frac{1}{8}$ inch? (Give the answer in simplest form.) $1\frac{3}{4}$ in.

3. a. Measure the orange pencil to the nearest $\frac{1}{8}$ inch. $3\frac{5}{8}$ in.

   b. Measure the green pencil to the nearest $\frac{1}{8}$ inch. $3\frac{1}{2}$ in.

   c. Which pencil is longer? Orange

4. Was your guess for question 1 correct? Answers will vary.

EXERCISES

Measure each segment to the nearest $\frac{1}{8}$ inch. Give the answer in simplest form.

5. ———————————————— $4\frac{1}{4}$ in.

6. ———— 2 in.

7. ———————————————————— $5\frac{1}{4}$ in.

8. ———————————— $3\frac{1}{2}$ in.

9. ———————————————— $4\frac{3}{4}$ in.

10. ———— $1\frac{3}{8}$ in. or $1\frac{1}{2}$ in.

Draw segments having these lengths.

11. $1\frac{1}{2}$ inches
12. $2\frac{1}{8}$ inches
13. $3\frac{3}{4}$ inches
14. $\frac{7}{8}$ inch
15. $4\frac{5}{8}$ inches
16. $6\frac{1}{4}$ inches
17. $\frac{1}{2}$ inch
18. $5\frac{3}{8}$ inches
19. $2\frac{1}{2}$ inches
20. $1\frac{1}{8}$ inches
21. $5\frac{5}{8}$ inches
22. $3\frac{7}{8}$ inches

23. Draw a segment that is 3 inches long.
    a. How many $\frac{1}{2}$ inches long is it? 6
    b. How many $\frac{1}{4}$ inches long is it? 12

24. Draw a segment that is $4\frac{3}{4}$ inches long.
    a. How many $\frac{1}{4}$ inches long is it? 19
    b. How many $\frac{1}{8}$ inches long is it? 38

Find the treasure Following directions

This treasure map was found in a bottle on a deserted island. Use the map to locate the buried treasure.

25. At the base of which tree would you dig to find the treasure? B

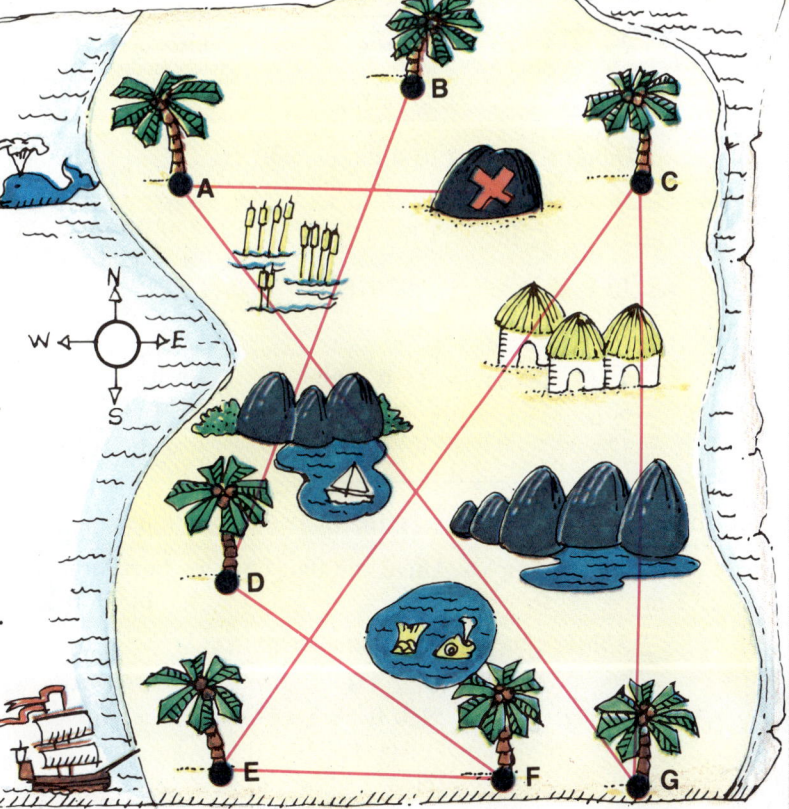

To find the treasure on the map:

Start at "X" on the boulder.
Go west $1\frac{3}{4}$ inches to a tree.
Then go 4 inches to another tree.
Go north $3\frac{1}{8}$ inches to a tree.
Then go $3\frac{7}{8}$ inches to another tree.
Go $1\frac{1}{2}$ inches east to a tree.
Then go $1\frac{3}{4}$ inches northwest to another tree.
Next, go $2\frac{3}{4}$ inches to another tree.
Dig at the base of the last tree.

Measurement **241**

Practice Worksheet
Workbook S272, Copymaster S272, or Duplicating Master S272

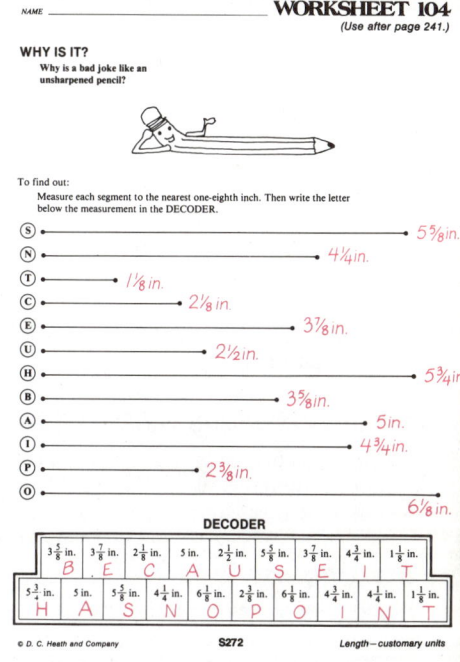

Group Project
Estimating length

Have the students work in pairs. One student holds a ruler and the other estimates $5\frac{1}{2}$ inches by pointing to the back of the ruler. The other member of the pair tells his/her partner how close he/she was to the given measurement. Let students take turns, one naming the length and the other estimating.

241

Class Starter Quiz 96
on previous lesson

Draw line segments having these lengths:

1. $2\frac{1}{2}$ inches
2. $4\frac{3}{4}$ inches
3. $5\frac{7}{8}$ inches
4. $3\frac{3}{8}$ inches
5. $7\frac{1}{4}$ inches
6. $6\frac{5}{8}$ inches

Copymaster S408

Lesson Objective
To make conversions between customary units of length

Problem-Solving Skills
Utilizing relationships among customary units of length
Using logical reasoning

Starting the Lesson
Before discussing exercises 1 and 2, have the students read the information at the top of the page. Ask them to decide, without computing, whether a 185-inch apple peel is a new record.

Exercise Note
Number Sense In exercises 3–32, encourage the students to first decide whether they are converting to a smaller or a larger unit. If the conversion is to a larger unit, they should divide. If the conversion is to a smaller unit, they should multiply.

Changing units of length—customary

RECORD BREAKER?

Is 185 inches of apple peel a new world record? The longest unbroken apple peel is 57 yards 1 foot by Kathy Wafler of Wolcott, New York.

This apple peel is 185 inches long. I bet that's a new world record!

Here are the facts you will need to know to change from one unit of length to another:

12 inches (in.) = 1 foot (ft)
3 ft = 1 yard (yd)
36 in. = 1 yd
5280 ft = 1 mile (mi)

Here's how *to change from one unit of length to another.*

185 in. = _?_ ft _?_ in.

Think: Changing to a larger unit, so divide.

Remember: 12 in. = 1 ft

```
       15
   12)185
      -12
       65
      -60
        5
```
185 in. = **15** ft **5** in.

57 yd 1 ft = _?_ ft

Think: Change to a smaller unit, so multiply.

Remember: 1 yd = 3 ft

```
   57
  × 3
  171
  + 1
  172
```
57 yd 1 ft = **172** ft

1. Look at the *Here's how*. Is 185 inches of apple peel a new world record? **No**

2. Check these examples. Have the unit changes been done correctly? **Yes**

   a. 2 mi = _?_ ft

   Think: Changing to a smaller unit, so multiply.

   Remember: 1 mi = 5280 ft
   2 mi = **10,560** ft
   └─ × 5280 ─┘

   b. 66 ft = _?_ yd

   Think: Changing to a larger unit, so divide.

   Remember: 3 ft = 1 yd
   66 ft = **22** yd
   └─ ÷ 3 ─┘

EXERCISES

Copy and complete.

3. 6 ft = ? in. 72
4. 5 yd = ? ft 15
5. 1 mi = ? ft 5280
6. 48 in. = ? ft 4
7. 12 ft = ? yd 4
8. 3 mi = ? ft 15,840
9. 36 in. = ? yd 1
10. 3 yd = ? in. 108
11. 5 ft = ? in. 60
12. 120 ft = ? yd 40
13. 13 yd = ? ft 39
14. 72 in. = ? yd 2
15. 26 in. = 2 ft ? in. 2
16. 7 ft = 2 yd ? ft 1
17. 40 in. = 1 yd ? in. 4
18. 50 in. = 4 ft ? in. 2
19. 11 ft = 3 yd ? ft 2
20. 49 in. = 1 yd ? in. 13
21. 80 in. = 6 ft ? in. 8
22. 17 ft = 5 yd ? ft 2
23. 68 in. = 1 yd ? in. 32
24. 2 ft 3 in. = ? in. 27
25. 3 yd 1 ft = ? ft 10
26. 2 yd 4 in. = ? in. 76
27. 3 ft 7 in. = ? in. 43
28. 7 yd 2 ft = ? ft 23
29. 3 yd 10 in. = ? in. 118
30. 4 ft 3 in. = ? in. 51
31. 8 yd 3 ft = ? ft 27
32. 7 yd 6 in. = ? in. 258

Solve.

33. The record for spitting watermelon seeds is 65 feet 4 inches. How many inches is that? 784

34. The record for an egg to be thrown and then caught without breaking is 116 yards 2 feet. How many feet is that? 350

35. The record for catching a thrown grape in the mouth is 319 feet 8 inches. Suppose you caught a grape that was thrown 89 yards 10 inches. Would you break the record? No

36. The record for throwing a 2-pound rolling pin is 175 feet 5 inches. Suppose you threw a 2-pound rolling pin 58 yards 1 foot. Would you break the record? No

How many apples?

Records may change from year to year.

Logical reasoning

37. Study the clues to find how many apples are in the bag.

    Clues:
    - There are fewer than 30.
    - If you divided them among 4 people, you would have 3 left over.
    - If you divided them among 5 people, you would have 4 left over. 19

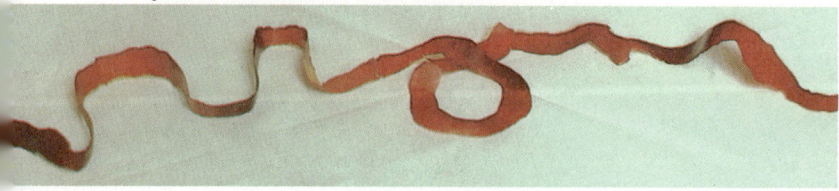

Measurement 243

Practice Worksheet

Workbook S273, Copymaster S273, or Duplicating Master S273

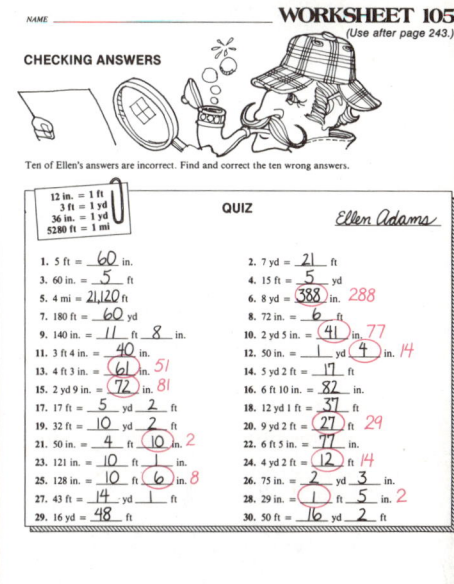

Challenge Problem

Suppose that you were to stack the 40 billion hamburgers sold by a famous fast-food chain one on top of the other. Use a calculator to compute the height of the stack in miles, given that a hamburger is about 1.5 inches high. *Hint:* There are 5280 feet in a mile.

946,969.7 miles high

Copymaster S453

243

Class Starter Quiz 97
on previous lesson

Complete.

1. 5 ft = _?_ in.   60
2. 7 yd = _?_ ft   21
3. 36 in. = _?_ ft   3
4. 15 ft = _?_ yd   5
5. 2 ft 5 in. = _?_ in.   29
6. 9 yd 2 ft = _?_ ft   29

Copymaster S409

Lesson Objective
To make conversions between customary units of liquid volume

Problem-Solving Skills
Selecting information from a recipe
Choosing the correct operation
Using logical reasoning

Starting the Lesson
Have the students read the ingredients on the recipe card at the top of the page. Ask them to write what dessert they think the recipe makes. (Ice cream)

Liquid volume— customary units

CAN YOU NAME THIS DESSERT?
20 cups sugar
5 teaspoons salt
12 quarts milk
48 eggs
2 cups vanilla
10 quarts whipping cream

2 cups (c) = 1 pint (pt)
2 pints = 1 quart (qt)
4 quarts = 1 gallon (gal)

1 cup   1 pint   1 quart   1 gallon

1. Read the recipe. How many quarts of milk are needed to make the dessert? 12
2. How many quarts are in 1 gallon? 4
3. To compute the number of gallons of milk, you would divide 12 by what number? 4

Here's how *to change from one unit of liquid volume to another.*

12 qt = _?_ gal   10 qt = _?_ pt

Think: Changing to a larger unit, so divide.   *Think:* Changing to a smaller unit, so multiply.

Remember: 4 qt = 1 gal   *Remember:* 1 qt = 2 pt
12 qt = 3 gal   10 qt = 20 pt
   ÷ 4   × 2

4. Check these examples. Have the units been changed correctly? Yes

   a. 13 qt = _?_ gal _?_ qt   **b.** 11 pt 1 c = _?_ c

   Remember: 4 qt = 1 gal   *Remember:* 1 pt = 2 c

$$\begin{array}{r}3\\4\overline{)13}\\-12\\\hline 1\end{array}$$ 13 qt = 3 gal 1 qt $$\begin{array}{r}11\\\times 2\\\hline 22\\+1\\\hline 23\end{array}$$ 11 pt 1 c = 23 c

EXERCISES

Copy and complete.

5. 18 pt = _?_ qt   9
6. 12 c = _?_ pt   6
7. 8 qt = _?_ gal   2
8. 6 qt = _?_ pt   12
9. 3 gal = _?_ qt   12
10. 6 pt = _?_ c   12
11. 1 qt = _?_ pt   2
12. 1 qt = _?_ c   4
13. 1 gal = _?_ c   16
14. 4 c = _?_ qt   1
15. 8 pt = _?_ gal   1
16. 16 c = _?_ gal   1
17. 3 pt 1 c = _?_ c   7
18. 5 qt 1 pt = _?_ pt   11
19. 3 gal 2 qt = _?_ qt   14
20. 3 gal 2 pt = _?_ qt   13
21. 8 qt 2 c = _?_ pt   17
22. 1 qt 1 pt = _?_ c   6
23. 15 pt = _?_ qt 1 pt   7
24. 19 qt = _?_ gal 3 qt   4
25. 34 pt = _?_ gal 1 qt   4
26. 17 pt = 8 qt _?_ pt   1
27. 15 qt = 3 gal _?_ qt   3
28. 25 pt = 3 gal _?_ pt   1

Solve. Use the recipe on page 244.

29. a. How many dozen eggs are needed? 4
    b. If eggs cost $.89 a dozen, what will be the total cost of the eggs? $3.56

30. a. How many cups of whipping cream are needed? 40
    b. If whipping cream costs $.98 a cup, what will be the total cost of the whipping cream? $39.20

31. a. How many $\frac{1}{4}$-cup bottles of vanilla are needed? 8
    b. If a $\frac{1}{4}$-cup bottle of vanilla costs $1.98, what will be the total cost for the vanilla? $15.84

32. a. The recipe makes 40 quarts of vanilla ice cream. How many gallons is that? 10
    b. If you bought the same amount of ice cream for $4.89 per gallon, what would be the total cost? $48.90

What's number one?

Logical reasoning

33. What is the most popular flavor of ice cream? Vanilla

    Clues:
    - More people like vanilla than cherry.
    - More people like chocolate than strawberry.
    - Fewer people like chocolate than vanilla.

Measurement **245**

Practice Worksheet

Workbook S274, Copymaster S274, or Duplicating Master S274

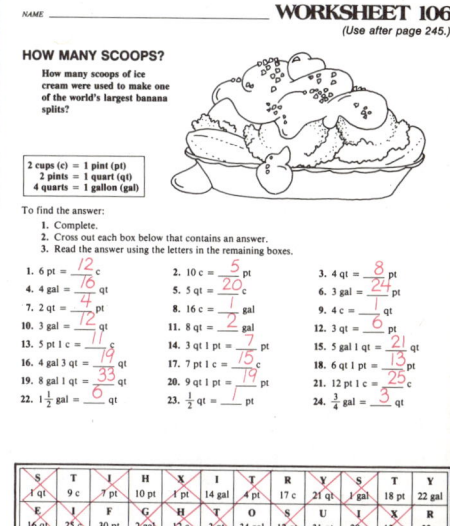

Class Starter Quiz 98
on previous lesson

Complete.

1. 14 pt = _?_ qt   7
2. 10 c = _?_ pt   5
3. 12 qt = _?_ gal   3
4. 3 qt = _?_ pt   6
5. 6 pt = _?_ c   12
6. 4 gal = _?_ qt   16
7. 4 pt 1 c = _?_ c   9
8. 19 qt = _?_ gal 3 qt 4

Copymaster S409

Lesson Objective
To make conversions between customary units of weight

Problem-Solving Skills
Selecting information from a display
Choosing the correct operation
Solving a multistep problem
Using logical reasoning

Starting the Lesson
Before discussing exercises 1–6, have the students study the clues at the top of the page. Ask them to write what fruit they think is in the bag. (Apples)

Weight—customary units

16 ounces (oz) = 1 pound (lb)
2000 lb = 1 ton (T)

1. What is the weight of the bag of fruit? 32 oz
2. How many ounces equal 1 pound? 16
3. To compute the number of pounds of fruit in the bag, you would divide 32 by what number? 16

Here's how to change from one unit of weight to another.

32 oz = _?_ lb

Think: Changing from smaller to larger units, so divide.

Remember: 16 oz = 1 lb
32 oz = 2 lb
└ ÷ 16 ┘

The 2-pound bag of fruit costs $1.60. That's $.80 per pound. There must be apples in the bag!

4. Use the sale prices. How much would 32 ounces of pears cost? $2.40
5. How much would a ton of oranges cost? $2000
6. Check these examples. Have the units been changed correctly? Yes

   a. 6000 lb = _?_ T

   Think: Changing to a larger unit, so divide.

   6000 lb = 3 T
   └ ÷ 2000 ┘

   b. 2 lb 5 oz = _?_ oz

   Think: Changing to a smaller unit, so multiply.

   2 lb 5 oz = 37 oz
   └ × 16 + 5 ┘

EXERCISES

Copy and complete.

7. 3 lb = ? oz 48
8. 48 oz = ? lb 3
9. 4 T = ? lb 8000
10. 8000 lb = ? T 4
11. 10 lb = ? oz 160
12. 160 oz = ? lb 10
13. 80 oz = ? lb 5
14. 10,000 lb = ? T 5
15. 14 lb = ? oz 224
16. 1 lb 8 oz = ? oz 24
17. 2 lb 3 oz = ? oz 35
18. 4 lb 10 oz = ? oz 74
19. 20 oz = ? lb 4 oz 1
20. 52 oz = ? lb 4 oz 3
21. 30 oz = ? lb 14 oz 1
22. 7500 lb = ? T 1500 lb 3
23. 9050 lb = 4 T ? lb 1050
24. 2060 lb = ? T 60 lb 1
25. 18 oz = 1 lb ? oz 2
26. 5000 lb = ? T 1000 lb 2
27. 100 oz = ? lb 4 oz 6
28. 22 lb = ? oz 352
29. 64 oz = ? lb 4
30. 56,000 lb = ? T 28
31. 6016 lb = 3T ? lb 16
32. 40 oz = 2 lb ? oz 8
33. 2500 lb = 1T ? lb 500

Solve.

34. An empty truck weighs 12,500 pounds. When full of apples, it weighs 8 tons. What is the weight of the apples in pounds? 3500

35. A truck contained 3 tons of oranges. After some oranges were unloaded, 1 ton 450 pounds of oranges remained. How many pounds of oranges were unloaded? 3550

36. A 2-pound bag of bananas costs 78¢. A 16-ounce bag of bananas costs 45¢. Which costs less per pound? 2 lb bag

37. A 2-pound 8-ounce bag of peaches costs $4.00. A 1-pound 4-ounce bag of peaches costs $2.25. Which costs less per ounce? 2 lb 8 oz bag

Fruit salad! Logical reasoning

38. Study the clues to find how many apples, oranges, and pears are in the fruit salad.

    Clues:
    * There are 12 oranges.
    * If you add the number of pears and apples, you get 2 more than the number of oranges.
    * There are 4 more pears than apples.

    12 oranges,
    9 pears,
    5 apples

Extra Practice
Page 492 Skill 43

Practice Worksheet
Workbook S275, Copymaster S275, or Duplicating Master S275

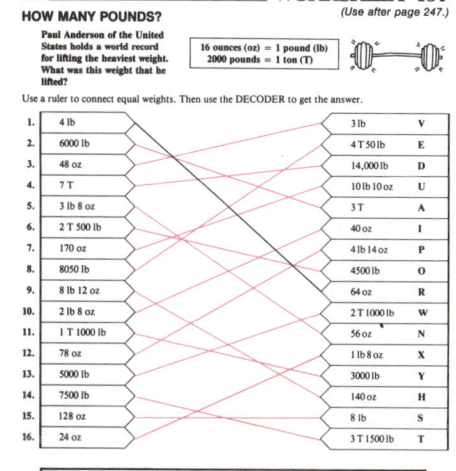

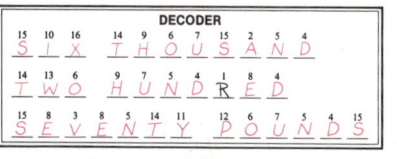

Project

Estimating weight

Estimate the number of apples in a ton. Weigh an apple. Then use what you found to compute the number of apples in a ton. Compare the estimated number of apples with the computed number of apples.

Copymaster S481

Class Starter Quiz 99
on previous lesson

Complete.
1. 5 lb = _?_ oz   80
2. 32 oz = _?_ lb   2
3. 3 T = _?_ lb   6000
4. 1 lb 7 oz = _?_ oz   23
5. 52 oz = _?_ lb 4 oz   3
6. 3050 lb = _?_ T 1050 lb   1

Copymaster S409

Lesson Objective
To compute with customary units

Problem-Solving Skills
Choosing the correct operation
Reading a table

Starting the Lesson
Number Sense Write these measurement facts on the chalkboard:

Numbers: 3, 5280, 16, 2, 12, 36, 4, 2, 2000

1 ft = _?_ in.
1 yd = _?_ ft = _?_ in.
1 mi = _?_ ft   1 pt = _?_ c
1 qt = _?_ pt   1 gal = _?_ qt
1 lb = _?_ oz   1 T = _?_ lb

Before the students open their books, have them choose the numbers to complete the measurement facts. Then tell them to use the chart at the top of page 248 and check their answers.

Here's How Note
Go over the examples carefully. Point out that in computing with customary units, the regrouping is not always 10 for 1 or 1 for 10, but rather varies according to the units used.

248

Computing with customary units

DO YOU REMEMBER THESE MEASUREMENT FACTS?

LENGTH	LIQUID VOLUME	WEIGHT
1 ft = 12 in.	1 pt = 2 c	1 lb = 16 oz
1 yd = 3 ft = 36 in.	1 qt = 2 pt	1 T = 2000 lb
1 mi = 5280 ft	1 gal = 4 qt	

Here's how
to compute with customary units of measurement.

EXAMPLE 1.   15 ft 4 in. + 12 ft 10 in. = ?

Step 1.   Step 2.
Add inches and regroup.   Add feet.
  1 ft   1 ft
  15 ft 4 in.   15 ft 4 in.
+ 12 ft 10 in.   + 12 ft 10 in.
         2 in.   28 ft 2 in.

14 in. = 1 ft 2 in.

EXAMPLE 2.   14 lb 9 oz − 8 lb 12 oz = ?

Step 1.   Step 2.
Regroup.   Subtract.
       16 oz + 9 oz
   13 lb 25 oz   13 lb 25 oz
   ~~14 lb 9 oz~~   ~~14 lb 9 oz~~
 − 8 lb 12 oz   − 8 lb 12 oz
                           5 lb 13 oz

EXERCISES
Add. Here are scrambled answers for the next row of exercises: 6 ft 7 in.   6 yd   6 ft 3 in.   7 yd 1 ft

1.   3 ft 8 in.
   + 2 ft 7 in.
     6 ft 3 in.

2.   2 yd 2 ft
   + 4 yd 2 ft
     7 yd 1 ft

3.   4 ft 10 in.
   + 1 ft 9 in.
     6 ft 7 in.

4.   1 yd 1 ft
   + 4 yd 2 ft
     6 yd

248   Chapter 9

5. 3 gal 3 qt
   +1 gal 2 qt
   _____
   5 gal 1 qt

6. 4 qt 1 pt
   +3 qt 1 pt
   _____
   8 qt

7. 2 gal 2 qt
   +5 gal 1 qt
   _____
   7 gal 3 qt

8. 2 pt 1 c
   +4 pt 1 c
   _____
   7 pt

9. 6 lb 12 oz
   +3 lb 8 oz
   _____
   10 lb 4 oz

10. 2 T 1500 lb
    +3 T 1400 lb
    _____
    6 T 900 lb

11. 8 lb 14 oz
    +9 lb 10 oz
    _____
    18 lb 8 oz

12. 6 T 1000 lb
    +3 T 1800 lb
    _____
    10 T 800 lb

Here are scrambled answers for the next row of exercises: 3 yd 1 ft   2 ft 6 in.   2 ft 11 in.   2 yd 2 ft

13. 5 ft 9 in.
    −2 ft 10 in.
    _____
    2 ft 11 in.

14. 4 yd 1 ft
    −1 yd 2 ft
    _____
    2 yd 2 ft

15. 7 ft 2 in.
    −4 ft 8 in.
    _____
    2 ft 6 in.

16. 6 yd
    −2 yd 2 ft
    _____
    3 yd 1 ft

17. 5 lb 2 oz
    −2 lb 12 oz
    _____
    2 lb 6 oz

18. 6 T 500 lb
    −3 T 1000 lb
    _____
    2 T 1500 lb

19. 4 lb 10 oz
    −1 lb 11 oz
    _____
    2 lb 15 oz

20. 30 T 1800 lb
    − 9 T 1900 lb
    _____
    20 T 1900 lb

21. 4 gal 1 qt
    −1 gal 2 qt
    _____
    2 gal 3 qt

22. 4 qt
    −2 qt 1 pt
    _____
    1 qt 1 pt

23. 5 gal 2 qt
    −2 gal 3 qt
    _____
    2 gal 3 qt

24. 4 gal
    −1 gal 1 qt
    _____
    2 gal 3 qt

Solve.

25. You and a friend cut a watermelon into two pieces. Your piece weighed 14 pounds 9 ounces, and her piece weighed 12 pounds 10 ounces. How much more did your piece weigh? 1 lb 15 oz

26. You bought 3 pounds 7 ounces of green grapes and 2 pounds 9 ounces of red grapes. How many pounds of grapes did you buy altogether? 6

Antique math — Reading a table

Use the table to complete each exercise.

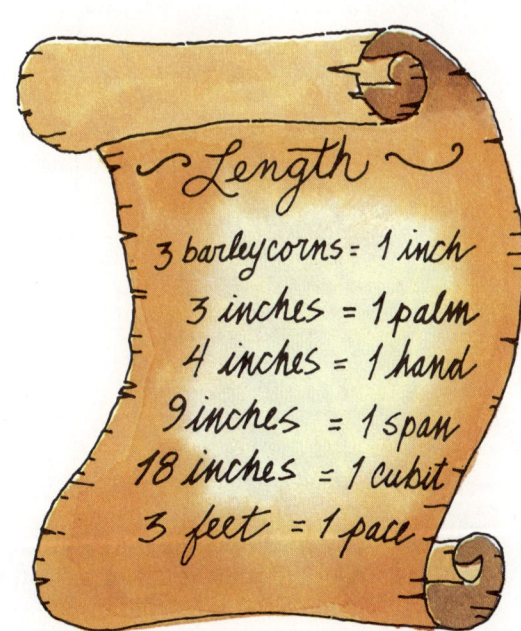

Length
3 barleycorns = 1 inch
3 inches = 1 palm
4 inches = 1 hand
9 inches = 1 span
18 inches = 1 cubit
3 feet = 1 pace

27. 27 barleycorns = __?__ in.   9
28. 15 in. = __?__ palms   5
29. 2 spans + 1 hand = __?__ in.   22
30. 1 pace + 1 cubit = __?__ in.   54
31. 1 pace = __?__ cubits   2
32. 6 palms = __?__ spans   2
33. 2 cubits = __?__ spans   4
34. 3 hands = __?__ palms   4
35. 1 pace − 1 cubit = __?__ spans   2
36. 1 cubit − 1 span = __?__ palms   3
37. 1 hand + 1 palm = __?__ barleycorns   21
38. 1 hand − 1 palm = __?__ barleycorns   3

Measurement

Practice Worksheet
Workbook S276, Copymaster S276, or Duplicating Master S276

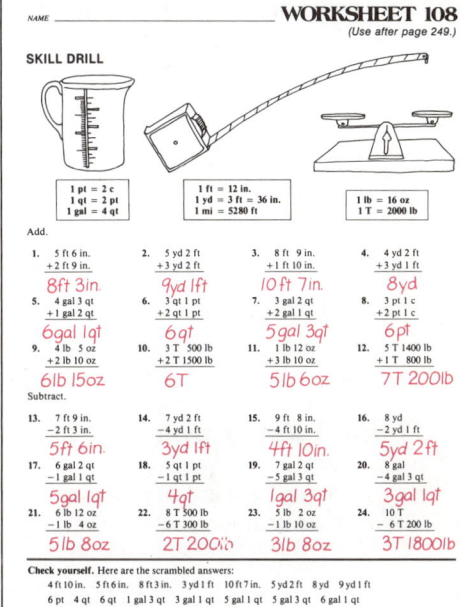

Challenge Problems

Unscramble the letters in these math words:

1. OLAGLN — Gallon
2. DOPUN — Pound
3. RUQAT — Quart
4. EIML — Mile
5. EUONC — Ounce
6. ELIKOMTRE — Kilometer
7. RAGM — Gram
8. RTLIE — Liter

Copymaster S453

Class Starter Quiz 100
on previous lesson

Add.

1. 3 ft 7 in.
    +1 ft 9 in.
    5 ft 4 in.

2. 6 lb 10 oz
    +1 lb 9 oz
    8 lb 3 oz

Subtract.

3. 4 yd 1 ft
    −1 yd 2 ft
    2 yd 2 ft

4. 5 gal 2 qt
    −3 gal 3 qt
    1 gal 3 qt

Copymaster S409

Lesson Objective
To solve problems using a working-backward strategy

Starting the Lesson
Problem-Solving Cover-Up Use the chalkboard or mask ▪ **Visual Aid 44** (copymaster or transparency S147).

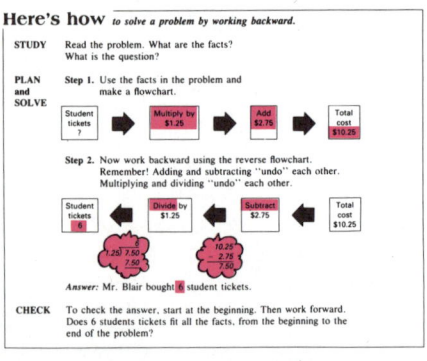

Have the students, working in small groups, study the problem and the problem-solving steps on page 250 for several minutes. Then have them close their books, look at the visual aid, and tell what has been covered up.

250

PROBLEM SOLVING *working backward*

Sometimes it is easier to find the answer to a problem by working backward.

Problem
Mr. Blair bought tickets to the zoo. He bought some $1.25 student tickets and a $2.75 adult ticket. The total cost was $10.25. How many student tickets did he buy?

Here's how *to solve a problem by working backward.*

STUDY   Read the problem. What are the facts? What is the question?

PLAN and SOLVE

Step 1. Use the facts in the problem and make a flowchart.

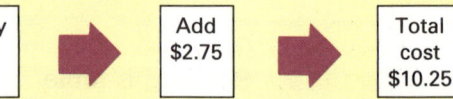

Step 2. Now work backward using the reverse flowchart. Remember! Adding and subtracting "undo" each other. Multiplying and dividing "undo" each other.

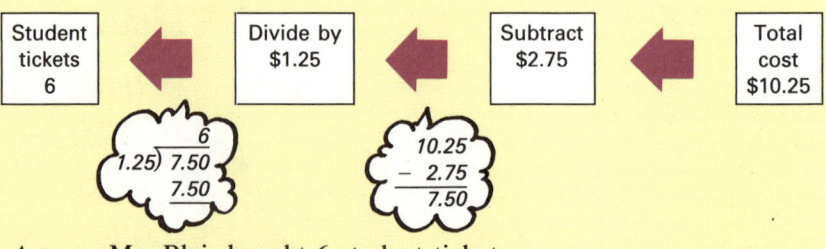

Answer: Mr. Blair bought 6 student tickets.

CHECK   To check the answer, start at the beginning. Then work forward. Does 6 students tickets fit all the facts, from the beginning to the end of the problem?

250   Chapter 9

PROBLEMS

Solve by working backward using a reverse flowchart.

1. Janet bought a bag of peanuts to feed the monkeys. She gave 15 peanuts to her brother and divided the ones that were left among herself and her 3 friends. Janet's final share was 12 peanuts. How many peanuts did she have at the start? 63

2. Lonny spent half of his money at the refreshment stand and $6.75 at the gift shop. Then he had $2.65 left. How much money did Lonny have at the start? $18.80

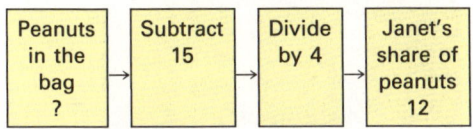

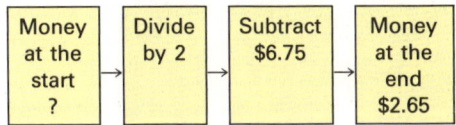

Solve by working backward.

3. Some people got on an empty guided-tour bus at the elephant house. At the monkey house, 8 people got off and 15 people got on. Then there were 21 people on the bus. How many people got on at the elephant house? 14

4. When Noriko entered the seal house, half of the seals were in the water. Then 6 seals got out of the water and 8 got in. Noriko then counted 10 seals in the water. How many seals are there in all? 16

5. The smallest monkey in the zoo is the pygmy marmoset. If you multiply its weight by 3 and then add 2.5, you get 10 ounces. How much does the pygmy marmoset weigh? 2.5 ounces

6. The largest elephant in the zoo is the African elephant. If you divide its weight by 10 and then subtract 500, you get 1000 pounds. How much does it weigh?
15,000 pounds

7. The tallest animal is the giraffe. If you take one half of its height and then add 2, you get 6 feet. How tall is the giraffe? 8 feet

8. The longest snake is the python. If you subtract 3 feet from its length and divide by 6, you get 5 feet. How long is the python? 33 feet

Hippo fact Making up a problem

9. Use the hippo facts to write a problem that can be solved by working backward. Answers will vary.

The hippopotamus' lower canine teeth measure 3 feet long and weigh 8 pounds each!

Class Starter Quiz 101
on previous lesson

Solve by working backward.

> Kevin spent half his money in the morning. He then spent $10.50 in the afternoon. He now has $6.30 left. How much money did Kevin have at the start? $33.60

Copymaster S410

Problem-Solving Skills

Finding information in a computer display
Choosing the correct operation

Skills Reviewed

Subtracting mixed numbers
Multiplying and dividing fractions
Making conversions between metric units
Subtracting customary units
Rounding decimals
Simplifying expressions

Starting the Lesson

Problem Solving Have the students read the paragraph at the top of page 252. Ask the students, "What did Richard Dorr do to get the information displayed on Screen B?" (He selected "4".) "What did he do to get the information displayed on Screen C?" (He selected "10".) "How did he get Screen D?" (He pressed the ENTER key.)

Cumulative Skill Practice Write these five answers on the chalkboard:

$5\frac{2}{5}$ 0 $\frac{1}{10}$
5.8   2 ft 5 in.

Challenge the students to an answer hunt by saying, "Look at exercises 1–46 on page 253. Find the five exercises that have these answers. You have five minutes to find as many of the exercises as you can." (Exercises 5, 13, 24, 33, and 41)

252

Problem solving

GARDENING WITH A COMPUTER

Richard Dorr used his home computer to help his family plan their garden. THE HOME GARDENER program helped them decide how much to plant, how much seed to buy, and how much fertilizer to buy. The steps used to plan one part of the garden are shown below.

A 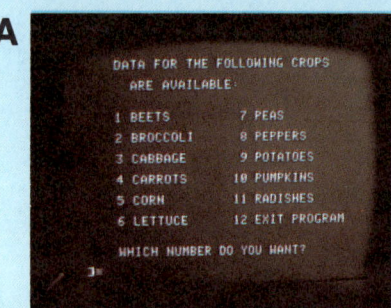 Richard selected "4."

B 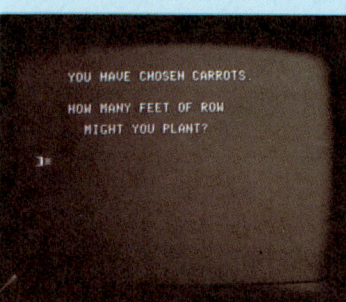 Richard entered "10."

C 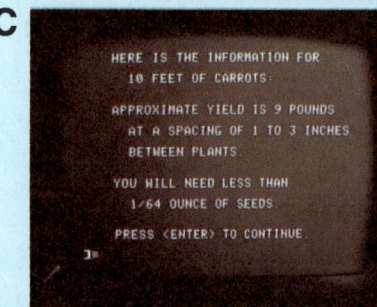 Richard pressed the ENTER key.

D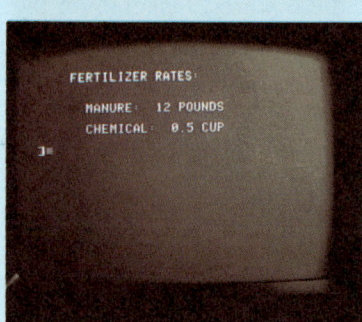

Use the data from the screens to solve these problems.

1. How many different crops does the program provide data for? 11

2. The program assumes that the crops are planted in rows. How long a row of carrots was planned? 10 feet

3. a. What is the suggested spacing for the seeds? 1 to 3 inches

   b. If they used the widest spacing suggested, would they have more than 3 dozen plants in the row? Yes

4. a. How many pounds of carrots should they expect to grow? 9 pounds

   b. A similar row of tomato plants would yield 3 times the weight yielded by the carrot plants. How many pounds of tomatoes could they grow in the row? 27 pounds

5. Richard's father found about 6 ounces of chemical fertilizer left in a bottle. Was this enough fertilizer for the row of carrots? (*Hint: There are 8 ounces in a cup.*) Yes

6. Assume the weight of the seeds used to plant the row of carrots was $\frac{1}{64}$ ounce. How many pounds of carrots could they grow if they planted 1 ounce of seeds? 576 pounds

Cumulative Skill Practice

Subtract. Give the difference in simplest form. (pages 190, 192)

1. $3\frac{3}{4} - 1\frac{1}{4} = 2\frac{1}{2}$
2. $5\frac{5}{6} - 2\frac{1}{6} = 3\frac{2}{3}$
3. $4 - 2\frac{1}{3} = 1\frac{2}{3}$
4. $6 - 4\frac{1}{2} = 1\frac{1}{2}$
5. $9 - 3\frac{3}{5} = 5\frac{2}{5}$
6. $8\frac{1}{2} - 2\frac{3}{4} = 5\frac{3}{4}$

Give the product in simplest form. (page 202)

7. $\frac{1}{3} \times \frac{1}{3} = \frac{1}{9}$
8. $\frac{3}{4} \times \frac{2}{3} = \frac{1}{2}$
9. $\frac{3}{2} \times \frac{3}{2} = 2\frac{1}{4}$
10. $\frac{5}{9} \times \frac{3}{2} = \frac{5}{6}$
11. $\frac{4}{5} \times \frac{7}{2} = 2\frac{4}{5}$
12. $\frac{7}{8} \times \frac{4}{3} = 1\frac{1}{6}$
13. $\frac{0}{2} \times \frac{3}{8} = 0$
14. $\frac{3}{4} \times \frac{4}{3} = 1$
15. $\frac{5}{6} \times \frac{10}{3} = 2\frac{7}{9}$
16. $\frac{5}{9} \times \frac{3}{3} = \frac{5}{9}$

Give the quotient in simplest form. (page 212)

17. $\frac{4}{9} \div \frac{1}{3} = 1\frac{1}{3}$
18. $6 \div \frac{2}{3} = 9$
19. $\frac{5}{6} \div \frac{1}{2} = 1\frac{2}{3}$
20. $4 \div \frac{2}{3} = 6$
21. $\frac{5}{8} \div \frac{1}{4} = 2\frac{1}{2}$
22. $\frac{9}{4} \div \frac{9}{4} = 1$
23. $\frac{3}{2} \div \frac{3}{4} = 2$
24. $\frac{2}{5} \div 4 = \frac{1}{10}$
25. $\frac{3}{5} \div \frac{3}{8} = 1\frac{3}{5}$
26. $\frac{3}{10} \div \frac{2}{5} = \frac{3}{4}$

Complete. (page 228)

27. 9 cm = ? mm   90
28. 4 m = ? cm   400
29. 3 km = ? m   3000
30. 3715 m = ? km   3.715
31. 5.9 km = ? m   5900
32. 6.3 cm = ? mm   63
33. 58 mm = ? cm   5.8
34. 35 cm = ? m   0.35
35. 26 km = ? m   26,000
36. 1368 m = ? km   1.368
37. 5.75 km = ? m   5750
38. 1.6 m = ? mm   1600

Subtract. (page 248)

39. 6 ft 4 in. − 2 ft 8 in. = 3 ft 8 in.
40. 6 yd 1 ft − 3 yd 2 ft = 2 yd 2 ft
41. 8 ft − 5 ft 7 in. = 2 ft 5 in.
42. 9 yd − 2 yd 1 ft = 6 yd 2 ft
43. 8 lb 3 oz − 2 lb 10 oz = 5 lb 9 oz
44. 4 T 200 lb − 2 T 500 lb = 1 T 1700 lb
45. 10 lb 6 oz − 3 lb 12 oz = 6 lb 10 oz
46. 6 T − 2 T 1000 lb = 3 T 1000 lb

MIXED PRACTICE

Complete.

47. 5.3281 rounded to the nearest hundredth is ?   5.33
48. 6.504 rounded to the nearest whole number is ?   7
49. 15.248 rounded to the nearest tenth is ?   15.2
50. $(5.14 + 3.6) \times 2 = $?   17.48
51. $5.14 + (3.6 \times 2) = $?   12.34
52. $(12 - 3.1) + 2.3 = $?   11.2
53. $12 - (3.1 + 2.3) = $?   6.6
54. $(4.8 \div 2) - 1.8 = $?   0.6
55. $4.8 \div (2 - 1.8) = $?   24

Measurement 253

Problem-Solving Worksheet
Workbook S278, Copymaster S278, or Duplicating Master S278

WORKSHEET 110 (Use after page 252.)

NAME _____

WHAT'S THE SALE PRICE?
SAVE $37.29
WAS $219.99
NOW $ 182.70

MYSTERY FRACTIONS
Add the two mystery fractions and you get 1. Subtract them and you get $\frac{1}{2}$. What are the two mystery fractions?
¾ and ¼

SCRAMBLED MATH
Unscramble the letters to get the answer.
NVEES times INNE plus WETVLE equals
SEVEN NINE TWELVE
75

HOW OLD?

Brenda is 28 years old. Calvin is $\frac{1}{2}$ of Brenda's age. Troy is $\frac{1}{2}$ of Calvin's age. How old is Troy? 7

GUESS AND CHECK

Menu	
Hot Cakes	$1.75
Eggs	$1.40
Muffin	$ .85
Milk	$ .45
Juice	$ .65

Tim had hot cakes and milk. He spent a total of $ 2.20
Laura had eggs, muffin, and juice. She spent $2.90.

WIN OR LOSE
The Lions won 3 out of the 8 games they have played. How many of their remaining games must they win in order to win $\frac{1}{2}$ of the 20 games played? 7

© D. C. Heath and Company   S278   Problem solving

Challenge Problem

Use the code to answer the riddle.

CODE

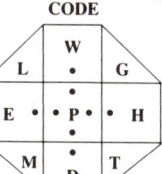
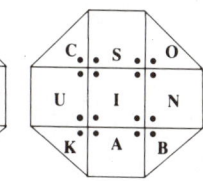

Riddle: When is a gardener like a mystery writer?

Answer: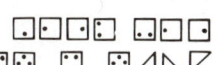

When she digs up a plot.

Copymaster S454

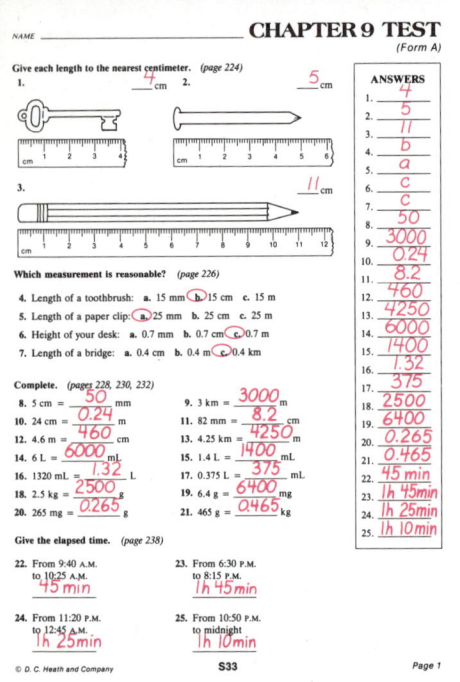

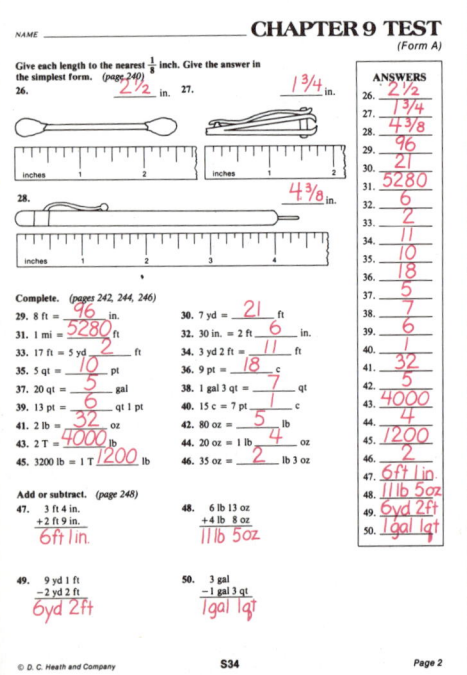

Chapter REVIEW

Here are scrambled answers for the review exercises:

1000	divide	grams	milliliters	ton
5280	Fahrenheit	inch	millimeters	yard
Celsius	feet	kilogram	multiply	
centimeter	gallon	meters	pound	

1. centimeter, millimeters   **2.** meters, multiply, divide   **3.** milliliters, 1000

1. One tenth of a [?] is 1 millimeter. The length of this segment is 23 [?]. (page 224)

2. 1 kilometer = 1000 [?]. To change from kilometers to meters, you would [?] by 1000. To change from millimeters to centimeters, you would [?] by 10. (pages 226, 228)

3. 1 liter = 1000 [?]. To change from milliliters to liters, you would divide by [?]. (page 230)

6000 km = ? m

700 mm = ? cm

8000 mL = ? L

4. kilogram, grams   **5.** inch   **6.** feet, 5280

4. 1 [?] = 1000 grams. To change from kilograms to [?], you would multiply by 1000. (page 232)

5. The length of this segment measured to the nearest $\frac{1}{8}$ [?] is $1\frac{3}{8}$ inches. (page 240)

6. 5280 [?] = 1 mile. To change from miles to feet, you would multiply by [?]. (page 242)

6 kg = ? g

3 mi = ? ft

7. gallon, ton   **9.** pound   **10.** yard

7. 4 quarts = 1 [?]. 2000 pounds = 1 [?]. (pages 244, 246)

8. Water boils at 100 degrees [?]. Water freezes at 32 degrees [?]. (page 236)

9. To do this addition exercise, you would first add the ounces and then regroup 16 ounces for 1 [?]. (page 248)

10. To do this subtraction exercise, you would first need to regroup 1 [?] for 3 feet. (page 248)

5 lb 9 oz
+ 3 lb 11 oz

9 yd 1 ft
− 5 yd 2 ft

8. Celsius, Fahrenheit

254   Chapter 9

Chapter TEST

Measure each length to the nearest centimeter. *(page 224)*

1. ______ 6 cm **2.** ______ 5 cm

3. ______ 5 cm **4.** ______ 7 cm

5. ______ 14 cm

Which measurement is reasonable? *(pages 226, 236)*

6. Length of a key:
(**a.**) 52 mm **b.** 52 cm **c.** 52 m

7. Length of a baseball bat:
a. 91 mm (**b.**) 91 cm **c.** 91 m

8. Height of a door:
a. 2.1 cm (**b.**) 2.1 m **c.** 2.1 km

9. Length of a train:
a. 0.6 cm **b.** 0.6 m (**c.**) 0.6 km

10. Temperature of a cold drink:
a. −20°C (**b.**) 4°C **c.** 80°C

11. Temperature of a summer day:
a. 15°F **b.** 35°F (**c.**) 75°F

Copy and complete. *(pages 228, 230, 232)*

12. 6 cm = ? mm 60
13. 4 m = ? cm 400
14. 9 km ? m 9000
15. 82 mm = ? cm 8.2
16. 3.8 m = ? cm 380
17. 7.4 cm = ? mm 74
18. 8 L = ? mL 8000
19. 1.5 L = ? mL 1500
20. 1250 mL = ? L 1.25
21. 0.475 L = ? mL 475
22. 750 mL = ? L 0.750
23. 65 mL = ? L 0.065
24. 5000 mg = ? g 5
25. 3000 g = ? kg 3
26. 2.1 kg = ? g 2100

Measure each segment to the nearest $\frac{1}{8}$ inch. Give the answer in simplest form. *(page 240)*

27. ______ $2\frac{3}{8}$ in. **28.** ______ $2\frac{1}{8}$ in.

29. ______ $5\frac{3}{4}$ in.

Copy and complete. *(pages 242, 244, 246)*

30. 7 ft = ? in. 84
31. 1 mi = ? ft 5280
32. 33 ft = ? yd 11
33. 72 in. = ? yd 2
34. 2 ft 4 in. = ? in. 28
35. 4 yd 1 ft = ? ft 13
36. 16 pt = ? qt 8
37. 14 c = ? pt 7
38. 16 qt = ? gal 4
39. 8 c = ? qt 2
40. 13 pt = 6 qt ? pt 1
41. 9 qt = 2 gal ? qt 1
42. 4 lb = ? oz 64
43. 64 oz = ? lb 4
44. 3 T = ? lb 6000

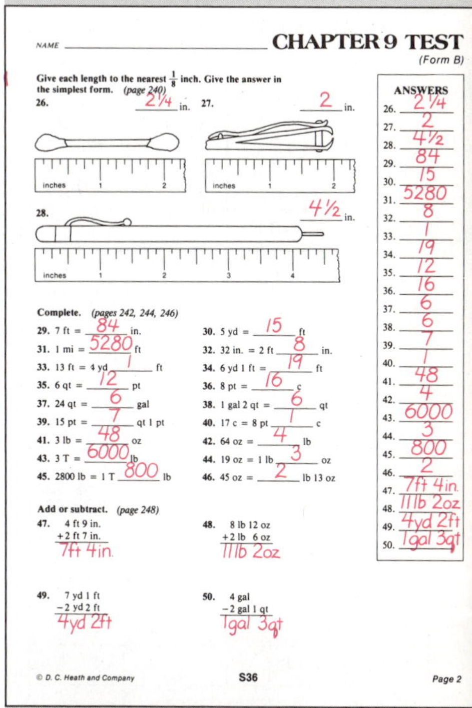

Add or subtract. *(page 248)*

45. 5 ft 9 in. + 3 ft 8 in. = 9 ft 5 in.

46. 8 lb 14 oz + 5 lb 7 oz = 14 lb 5 oz

47. 7 yd 1 ft − 2 yd 2 ft = 4 yd 2 ft

48. 4 gal − 1 gal 3 qt = 2 gal 1 qt

NAME ______ **CHAPTER 9 TEST** (Form B)

Give each length to the nearest centimeter. *(page 224)*

1. 5 cm 2. 5 cm

3. 12 cm

Which measurement is reasonable? *(page 226)*

4. Length of a fork: a. 16 mm (b.) 16 cm c. 16 m
5. Length of a toothpick: (a.) 56 mm b. 56 cm c. 56 m
6. Height of a basketball net: a. 3.1 mm b. 3.1 cm (c.) 3.1 m
7. Length of a tunnel: a. 0.3 cm b. 0.3 m (c.) 0.3 km

Complete. *(pages 228, 230, 232)*

8. 4 cm = 40 mm
9. 5 km = 5000 m
10. 32 cm = 0.32 m
11. 64 mm = 6.4 cm
12. 3.8 m = 380 cm
13. 3.65 km = 3650 m
14. 8 L = 8000 mL
15. 2.3 L = 2300 mL
16. 1560 mL = 1.56 L
17. 0.275 L = 275 mL
18. 3.5 kg = 3500 g
19. 4.8 g = 4800 mg
20. 326 mg = 0.326 g
21. 382 g = 0.382 kg

Give the elapsed time. *(page 238)*

22. From 7:50 A.M. to 8:35 A.M. 45 min
23. From 5:30 P.M. to 7:10 P.M. 1h 40 min
24. From 11:40 A.M. to 12:55 P.M. 1h 15 min
25. From 10:45 P.M. to midnight 1h 15 min

ANSWERS: 1. 5 2. 5 3. 12 4. b 5. a 6. c 7. c 8. 40 9. 5000 10. 0.32 11. 6.4 12. 380 13. 3650 14. 8000 15. 2300 16. 1.56 17. 275 18. 3500 19. 4800 20. 0.326 21. 0.382 22. 45 min 23. 1h 40 min 24. 1h 15 min 25. 1h 15 min

© D. C. Heath and Company S35 Page 1

NAME ______ **CHAPTER 9 TEST** (Form B)

Give each length to the nearest $\frac{1}{8}$ inch. Give the answer in the simplest form. *(page 240)*

26. $2\frac{1}{4}$ in. 27. 2 in.

28. $4\frac{1}{2}$ in.

Complete. *(pages 242, 244, 246)*

29. 7 ft = 84 in.
30. 5 yd = 15 ft
31. 1 mi = 5280 ft
32. 32 in. = 2 ft 8 in.
33. 13 ft = 4 yd 1 ft
34. 6 yd 1 ft = 19 ft
35. 6 qt = 12 pt
36. 8 pt = 16 c
37. 24 qt = 6 gal
38. 1 gal 2 qt = 6 qt
39. 15 pt = 7 qt 1 pt
40. 17 c = 8 pt 1 c
41. 3 lb = 48 oz
42. 64 oz = 4 lb
43. 3 T = 6000 lb
44. 19 oz = 1 lb 3 oz
45. 2800 lb = 1 T 800 lb
46. 45 oz = 2 lb 13 oz

Add or subtract. *(page 248)*

47. 4 ft 9 in. + 2 ft 7 in. = 7 ft 4 in.
48. 8 lb 12 oz + 2 lb 6 oz = 11 lb 2 oz
49. 7 yd 1 ft − 2 yd 2 ft = 4 yd 2 ft
50. 4 gal − 2 gal 1 qt = 1 gal 3 qt

ANSWERS: 26. $2\frac{1}{4}$ 27. 2 28. $4\frac{1}{2}$ 29. 84 30. 15 31. 5280 32. 8 33. 1 34. 19 35. 12 36. 16 37. 6 38. 6 39. 7 40. 1 41. 48 42. 4 43. 6000 44. 3 45. 800 46. 2 47. 7 ft 4 in. 48. 11 lb 2 oz 49. 4 yd 2 ft 50. 1 gal 3 qt

© D. C. Heath and Company S36 Page 2

Copymasters S35 and S36 or Duplicating Masters S35 and S36

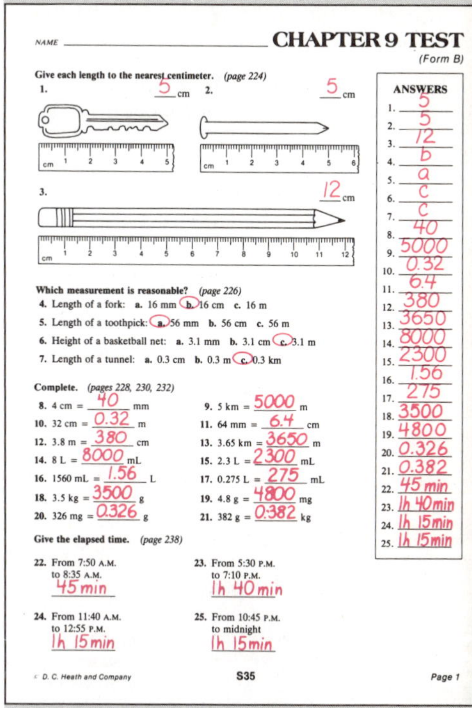

Cumulative Test
(Chapters 1–9)

Use Copymaster S109 to provide the students with an answer sheet in standardized test format.

Answers for Cumulative Test, Chapters 1–9

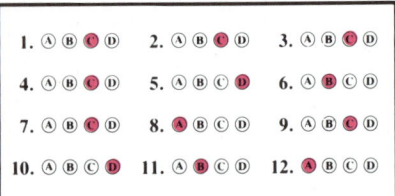

The table below correlates test items with student text pages.

Test Item	Page Taught	Skill Practice
1	68	p. 235, exercises 1–12
2	104	p. 235, exercises 13–22
3	166	p. 235, exercises 23–36
4	168	p. 235, exercises 37–48
5	116	
6	182	p. 235, exercises 49–54
7	192	p. 253, exercises 1–6
8	202	p. 253, exercises 7–16
9	212	p. 253, exercises 17–26
10	228	p. 253, exercises 27–38
11	248	p. 253, exercises 39–46
12	68	

Cumulative TEST — Standardized Format

Choose the correct letter.

1. Multiply. 5.637×2.04
   A. 114.9948
   B. 13.5288
   (C.) 11.49948
   D. none of these

2. Divide. Round the quotient to the nearest hundredth.
   $0.65 \overline{)6.69}$
   A. 1.03
   B. 1.29
   (C.) 10.29
   D. none of these

3. Change to a decimal.
   $2\frac{7}{8} = ?$
   A. 2.78
   B. 3.375
   (C.) 2.875
   D. none of these

4. Change to a mixed number in simplest form.
   3.200
   A. $3\frac{1}{4}$
   B. $3\frac{200}{1000}$
   (C.) $3\frac{1}{5}$
   D. none of these

5.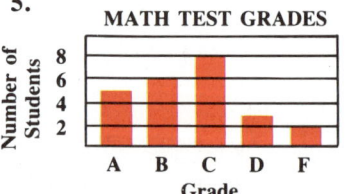
   MATH TEST GRADES
   How many students made a grade of C or better?
   A. 8   B. 11
   C. 13   (D.) none of these

6. Add. $3\frac{2}{3} + 2\frac{3}{4}$
   A. $5\frac{5}{12}$
   (B.) $6\frac{5}{12}$
   C. $5\frac{5}{7}$
   D. none of these

7. Subtract. $4\frac{1}{5} - 2\frac{1}{2}$
   A. $2\frac{3}{10}$ B. $2\frac{7}{10}$
   (C.) $1\frac{7}{10}$ D. none of these

8. Give the product.
   $\frac{3}{8} \times \frac{4}{5} = ?$
   (A.) $\frac{3}{10}$ B. $\frac{7}{40}$
   C. $\frac{12}{13}$ D. none of these

9. Give the quotient.
   $\frac{6}{5} \div \frac{3}{4} = ?$
   A. $1\frac{1}{9}$ B. $\frac{9}{10}$
   (C.) $1\frac{3}{5}$ D. none of these

10. Complete.
   236 cm = ? m
   A. 23.6
   B. 236
   C. 23,600
   (D.) none of these

11. Subtract.   8 ft 2 in. − 3 ft 9 in.
   A. 5 ft 7 in.
   (B.) 4 ft 5 in.
   C. 4 ft 3 in.
   D. none of these

12. You had $27. Then you worked 2.5 hours for $3.80 per hour. How much money did you have then?
   (A.) $36.50   B. $30.80
   C. $71.30   D. none of these

Ratio and Proportion

Chapter 10
Ratio and Proportion

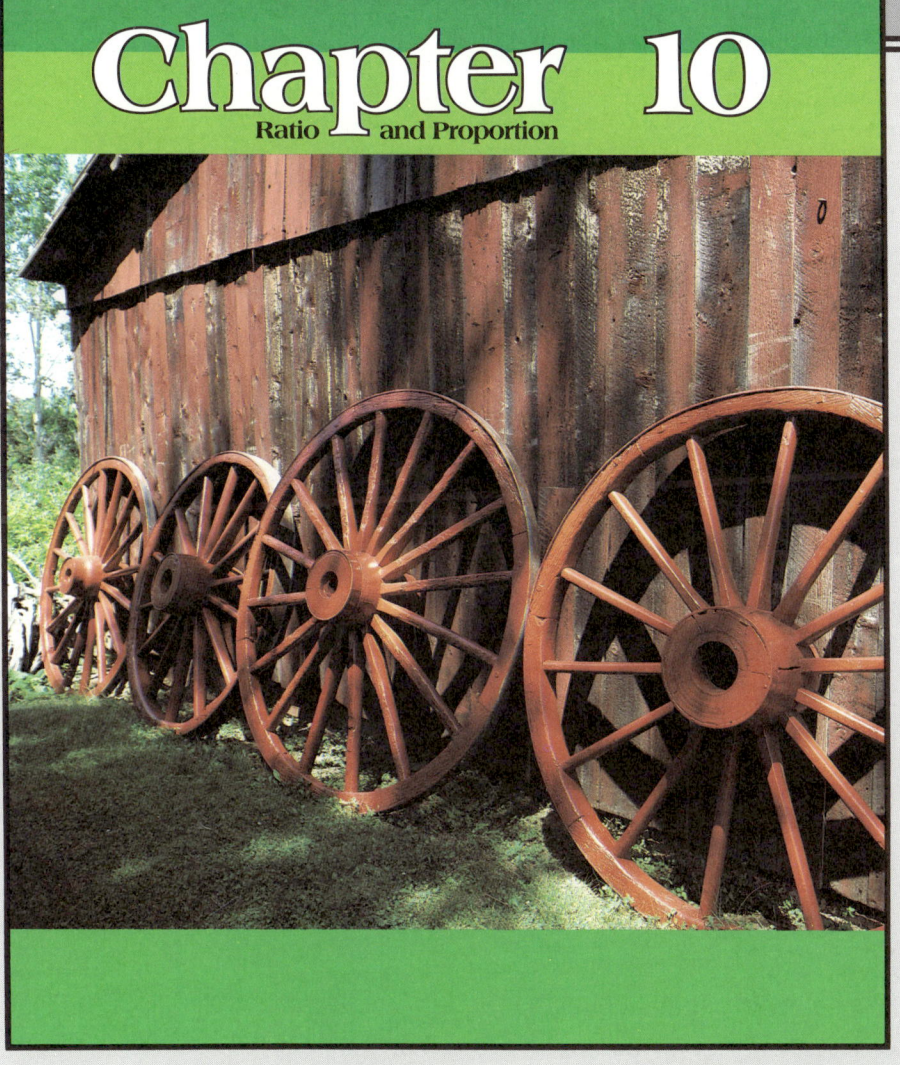

Resources

- **Class Starter Quizzes 102-110** *(Copymasters S410-S412)*
- **Visual Aids 39, 45-49** *(Copymasters or Transparencies S143, S148-S150)*
- **Manipulatives**
  Manipulative Activity 18 *(Copymaster S520)*
  Area tiles *(Copymasters or Transparencies S532-S533)*
- **Worksheets 111-120** *(Copymasters, Duplicating Masters, or Workbook pages S279-S288)*
- **Challenge Problems** for pages 259, 261, 267, 271, 275, 277 *(Copymasters S454-S455)*
- **Projects** for pages 263, 265 *(Copymasters S481-S482)*
- **Mental Math Extension** for Skill 44 *(Copymaster S507)*
- **Tests** *(Copymasters or Duplicating Masters S37-S40)*

Lesson Objectives

To give the ratio of two quantities
To change ratios to higher or lower terms

Problem-Solving Skill

Using information in a display

Starting the Lesson

Have the students read the introductory paragraph at the top of the page. Ask them what color is the result of mixing two jars of yellow paint with three jars of blue paint. (Green) Then use the display of paint jars and discuss exercises 1–5.

Here's How Note

Use of Concrete Materials You may wish to use the triangular and square tiles from copymasters or transparencies S532–S533 to demonstrate equal ratios. See ■ **Manipulative Activity 18** on copymaster S520 in the Teacher's Resource Binder.

Ratios

You can use a **ratio** to compare two numbers. The paint for the birdhouse was a custom mixture of yellow and blue. The ratio of yellow paint to blue paint was 2 to 3. Here are three ways to write the ratio:

2 to 3 $\frac{2}{3}$ 2 : 3

Read each ratio as "2 to 3."

1. Look at the jars of paint shown above. What is the ratio of jars of yellow paint to jars of blue paint? 2 to 3

2. What is the ratio of jars of blue paint to jars of yellow paint? 3 to 2

3. Suppose that you wanted to mix the same color and use 4 jars of yellow paint. How many jars of blue paint should you use? 6

Here's how *to find equal ratios.*

You can find equal ratios by thinking about equivalent fractions.

	1 batch	2 batches	3 batches
jars of yellow paint →	$\frac{2}{3}$ =	$\frac{4}{6}$ =	$\frac{6}{9}$
jars of blue paint →			

4. Look at the *Here's how*.
   a. How many jars of yellow paint would you mix with 6 jars of blue paint? 4
   b. How many jars of blue paint would you mix with 6 jars of yellow paint? 9
   c. You would have to multiply both the numerator and the denominator of $\frac{2}{3}$ by [?] to get $\frac{4}{6}$. 2
   d. You would have to divide both numerator and denominator of $\frac{6}{9}$ by [?] to get $\frac{2}{3}$. 3

5. Suppose you decide to use 12 jars of yellow paint. How many jars of blue paint would you need to mix the custom green? 18

EXERCISES

Give each ratio as a fraction in lowest terms.
Here are scrambled answers for the next row of exercises: $\frac{4}{3}$ $\frac{1}{4}$ $\frac{5}{2}$ $\frac{1}{2}$ $\frac{4}{9}$

6. 4 to 8 $\frac{1}{2}$
7. 3 to 12 $\frac{1}{4}$
8. 15 to 6 $\frac{5}{2}$
9. 8 to 6 $\frac{4}{3}$
10. 16 to 36 $\frac{4}{9}$

11. $\frac{14}{21}$ $\frac{2}{3}$
12. $\frac{10}{4}$ $\frac{5}{2}$
13. $\frac{16}{6}$ $\frac{8}{3}$
14. $\frac{9}{45}$ $\frac{1}{5}$
15. $\frac{18}{27}$ $\frac{2}{3}$

16. 12 : 20 $\frac{3}{5}$
17. 14 : 8 $\frac{7}{4}$
18. 18 : 32 $\frac{9}{16}$
19. 12 : 18 $\frac{2}{3}$
20. 24 : 18 $\frac{4}{3}$

21. 32 : 18 $\frac{16}{9}$
22. 17 : 51 $\frac{1}{3}$
23. 70 : 50 $\frac{7}{5}$
24. 26 : 39 $\frac{2}{3}$
25. 22 : 33 $\frac{2}{3}$

Copy and complete. Hint: Think about equivalent fractions.

26. $\frac{3}{4} = \frac{?}{8}$ 6
27. $\frac{1}{4} = \frac{?}{12}$ 3
28. $\frac{8}{5} = \frac{40}{?}$ 25
29. $\frac{4}{3} = \frac{12}{?}$ 9

30. $\frac{8}{3} = \frac{?}{21}$ 56
31. $\frac{?}{16} = \frac{5}{4}$ 20
32. $\frac{?}{3} = \frac{10}{30}$ 1
33. $\frac{1}{3} = \frac{?}{12}$ 4

34. $\frac{12}{?} = \frac{24}{22}$ 11
35. $\frac{?}{15} = \frac{2}{3}$ 10
36. $\frac{15}{30} = \frac{3}{?}$ 6
37. $\frac{7}{9} = \frac{?}{54}$ 42

38. $\frac{6}{5} = \frac{36}{?}$ 30
39. $\frac{5}{7} = \frac{?}{28}$ 20
40. $\frac{4}{?} = \frac{16}{28}$ 7
41. $\frac{?}{3} = \frac{50}{30}$ 5

42. $\frac{1}{3} = \frac{?}{75}$ 25
43. $\frac{3}{20} = \frac{9}{?}$ 60
44. $\frac{10}{3} = \frac{100}{?}$ 30
45. $\frac{16}{18} = \frac{?}{9}$ 8

46. $\frac{2}{3} = \frac{?}{24}$ 16
47. $\frac{3}{4} = \frac{27}{?}$ 36
48. $\frac{5}{8} = \frac{?}{32}$ 20
49. $\frac{3}{2} = \frac{30}{?}$ 20

50. $\frac{7}{8} = \frac{21}{?}$ 24
51. $\frac{?}{100} = \frac{9}{5}$ 180
52. $\frac{4}{3} = \frac{40}{?}$ 30
53. $\frac{?}{100} = \frac{4}{5}$ 80

54. $\frac{?}{24} = \frac{3}{8}$ 9
55. $\frac{6}{5} = \frac{?}{50}$ 60
56. $\frac{?}{18} = \frac{5}{3}$ 30
57. $\frac{5}{2} = \frac{100}{?}$ 40

You mix it!

To get Sunset Orange, mix 1 part white, 3 parts yellow, and 2 parts red.

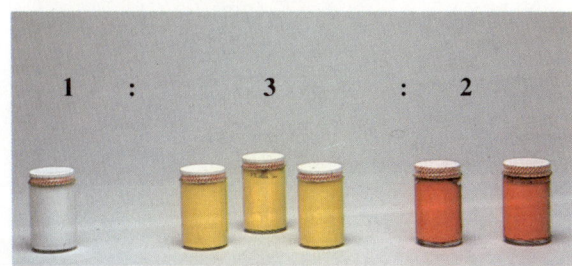

1 : 3 : 2

58. How many jars of white paint would you need if you used 4 jars of red? 2

59. How many jars of red would you need to mix with 9 jars of yellow? 6

60. Suppose you needed 12 jars of Sunset Orange. How many jars of each of the three colors would you need?
2 white, 6 yellow, 4 red

Practice Worksheet
Workbook S279, Copymaster S279, or Duplicating Master S279

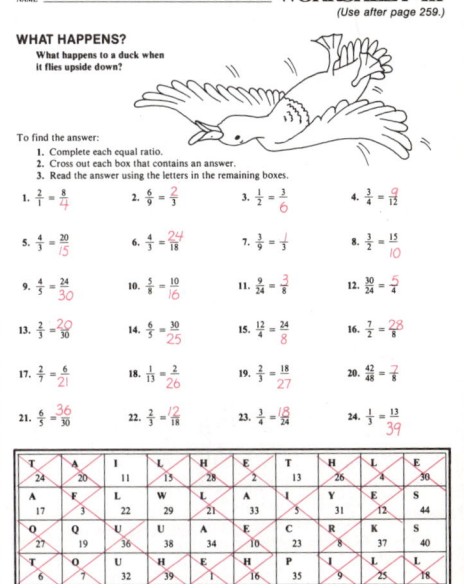

Challenge Problem

Use this month's calendar to write as many ratios as you can. Here are a few examples:

- Number of Saturdays to number of days in the month
- Number of even-numbered days to number of odd-numbered days
- Number of school days to number of non-school days

Answers will vary.

Copymaster S454

Ratio and Proportion

Class Starter Quiz 102
on previous lesson

Complete.

1. $\frac{3}{4} = \frac{?}{12}$ 9
2. $\frac{8}{5} = \frac{24}{?}$ 15
3. $\frac{?}{16} = \frac{5}{4}$ 20
4. $\frac{4}{?} = \frac{16}{28}$ 7
5. $\frac{2}{3} = \frac{?}{24}$ 16
6. $\frac{?}{18} = \frac{5}{3}$ 30

Copymaster S410

Lesson Objective
To compare the cross products to tell whether the ratios are equal or not equal

Problem-Solving Skills
Finding information in a table
Using logical reasoning

Starting the Lesson
Before going over exercises 1–6, have the students study the table at the top of the page. Ask them to write who they think is the better free-throw shooter, Nancy or Ingrid. After completing exercise 6, ask the students again who they think is the better free-throw shooter, Nancy or Ingrid. (Ingrid)

Exercise Note
Mental Math You may wish to use several of exercises 7–16 as oral exercises before making a written assignment. Have the students verbalize the method they are using in comparing the cross products.

Proportions

PLAYER	FREE THROWS MADE	FREE THROWS ATTEMPTED
Nancy B.	|||	++++
Linda D.	||||	++++ |||
Ingrid G.	||||	++++	
Susan L.	|||	++++	
Amanda M.	++++	++++ ++++	
Paige R.	++++	++++ ++++	
Kathy R.	++++		++++ ||||
Maria T.	++++	++++ ++++ ||

1. Look at the table. How many free throws did Nancy make? 3
2. How many free throws did Nancy attempt? 5
3. What is Nancy's ratio of free throws made to free throws attempted? $\frac{3}{5}$
4. What is Kathy's ratio of free throws made to free throws attempted? $\frac{6}{10}$

TIME OUT! An equation stating that two ratios are equal is called a **proportion**. Every proportion has a related multiplication equation.

Proportion

Nancy's ratio $\frac{3}{5} \leftarrow = \rightarrow \frac{6}{10}$ Kathy's ratio

Multiplication Equation

$3 \times 10 = 5 \times 6$

The two products 3×10 and 5×6 are called **cross products**.

Here's how to use the cross products to tell whether or not two ratios are equal.

If the cross products are equal, then the ratios are equal.

Nancy's ratio $\frac{3}{5} \bullet \frac{6}{10}$ Kathy's ratio

Since $3 \times 10 = 5 \times 6$, we know that $\frac{3}{5} = \frac{6}{10}$.

If the cross products are not equal, then the ratios are not equal.

Nancy's ratio $\frac{3}{5} \bullet \frac{4}{6}$ Ingrid's ratio

Since $3 \times 6 \neq 5 \times 4$, we know that $\frac{3}{5} \neq \frac{4}{6}$.

5. Look at the *Here's how*. Is Nancy's ratio equal to Kathy's ratio? Yes
6. Is Nancy's ratio equal to Ingrid's ratio? No

EXERCISES

Tell whether the ratios are equal (=) or not equal (≠). Hint: Compare the cross products.

7. $\frac{1}{2} \bullet \frac{3}{4}$ ≠
8. $\frac{4}{12} \bullet \frac{1}{3}$ =
9. $\frac{5}{9} \bullet \frac{2}{5}$ ≠
10. $\frac{3}{2} \bullet \frac{9}{6}$ =
11. $\frac{4}{7} \bullet \frac{3}{5}$ ≠

12. $\frac{5}{8} \bullet \frac{3}{5}$ ≠
13. $\frac{3}{9} \bullet \frac{4}{12}$ =
14. $\frac{5}{10} \bullet \frac{2}{4}$ =
15. $\frac{5}{4} \bullet \frac{3}{2}$ ≠
16. $\frac{7}{2} \bullet \frac{12}{4}$ ≠

17. $\frac{14}{8} \bullet \frac{21}{12}$ =
18. $\frac{13}{16} \bullet \frac{9}{12}$ ≠
19. $\frac{12}{15} \bullet \frac{8}{10}$ =
20. $\frac{6}{9} \bullet \frac{4}{6}$ =
21. $\frac{15}{9} \bullet \frac{14}{8}$ ≠

22. $\frac{2}{12} \bullet \frac{3}{18}$ =
23. $\frac{8}{10} \bullet \frac{16}{20}$ =
24. $\frac{9}{8} \bullet \frac{6}{5}$ ≠
25. $\frac{13}{15} \bullet \frac{11}{13}$ ≠
26. $\frac{14}{16} \bullet \frac{35}{40}$ =

27. $\frac{3}{4} \bullet \frac{0.5}{1}$ ≠
28. $\frac{2}{0.5} \bullet \frac{8}{2}$ =
29. $\frac{5}{0.75} \bullet \frac{4}{1}$ ≠
30. $\frac{5}{1} \bullet \frac{2}{0.4}$ =
31. $\frac{0.6}{2} \bullet \frac{3}{5}$ ≠

32. $\frac{6}{9} \bullet \frac{0.4}{0.6}$ =
33. $\frac{0.6}{0.5} \bullet \frac{4}{3}$ ≠
34. $\frac{0.3}{0.6} \bullet \frac{0.4}{0.8}$ =
35. $\frac{4.5}{6.0} \bullet \frac{7.5}{10.0}$ =
36. $\frac{1.5}{0.5} \bullet \frac{6.0}{2.0}$ =

37. $\frac{3}{4} \bullet \frac{1\frac{1}{2}}{2}$ =
38. $\frac{3}{2} \bullet \frac{2}{1\frac{1}{4}}$ ≠
39. $\frac{1\frac{1}{3}}{4} \bullet \frac{2}{6}$ =
40. $\frac{2}{1\frac{3}{8}} \bullet \frac{5}{3}$ ≠
41. $\frac{2\frac{1}{2}}{5} \bullet \frac{4}{9}$ ≠

42. $\frac{9}{3} \bullet \frac{7\frac{1}{2}}{2\frac{1}{2}}$ =
43. $\frac{1\frac{1}{4}}{7\frac{1}{2}} \bullet \frac{2}{12}$ =
44. $\frac{1\frac{1}{4}}{2} \bullet \frac{2\frac{1}{8}}{3}$ ≠
45. $\frac{2\frac{2}{3}}{1\frac{1}{2}} \bullet \frac{3}{2\frac{3}{4}}$ ≠
46. $\frac{3\frac{3}{5}}{2\frac{1}{4}} \bullet \frac{1\frac{1}{5}}{1}$ ≠

Solve. Use the table on page 260.

47. What is Linda's ratio of free throws made to free throws attempted? $\frac{4}{8}$

48. What is Susan's ratio of free throws made to free throws attempted? $\frac{3}{6}$

49. Compare the ratios from exercises 47 and 48. Is Linda's ratio equal to Susan's ratio? Yes

50. Which player's ratio (free throws made to free throws attempted) is equal to Ingrid's ratio? Paige's

51. What is Paige's ratio of free throws missed to free throws attempted? $\frac{3}{9}$

52. What is Amanda's ratio of free throws made to free throws missed? $\frac{5}{6}$

Who won?

53. With only two minutes left to play, the Jefferson Tigers were ahead of the Washington Badgers 48 to 47. During the last two minutes, the two teams made a total of 3 field goals (2-point goals) and no free throws. Both teams scored 50 or more points. Which team won, and what was the final score? Badgers, 51 to 50

Ratio and Proportion 261

Practice Worksheet
Workbook S280, Copymaster S280, or Duplicating Master S280

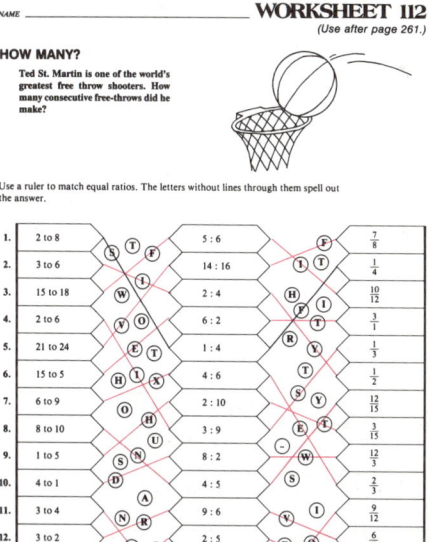

Challenge Problems

Use a clockface to help you answer these questions.

1. What time is it when the ratio of hours past noon to hours until midnight is 7 to 5? 7 P.M.

2. What time is it when the ratio of hours past noon to hours until midnight is 1 to 2? 4 P.M.

Copymaster S454

Solving proportions

Lockheed Vega
Wingspan: **41 feet**   Length: **27½ feet**

In 1932 Amelia Earhart became the first woman to fly across the Atlantic solo and nonstop.

In 1924 a team of U.S. Army pilots were the first to fly around the world.

Douglas Chicago
Wingspan: **50 feet**   Length: **35½ feet**

1. What is the wingspan of the 1924 *Douglas Chicago*? 50 feet

A model of the *Douglas Chicago* is to be $\frac{1}{20}$ the size of the real airplane.

To find how many feet the wingspan of the model should be, you can set up and solve a proportion.

Here's how to set up and solve a proportion.

When setting up a proportion, make sure that the ratios are in the same order.

wingspan of model airplane → $\frac{n}{50} = \frac{1}{20}$ ← model airplane
wingspan of real airplane →

Write the multiplication equation. $20n = 50 \times 1$

20n is a short way to write 20 × n.

Since 20 times *n* equals 50, divide both sides of the equation by 20 to find *n*.

$$\frac{20n}{20} = \frac{50}{20}$$

$$n = 2\frac{1}{2}$$

Check: $\frac{2\frac{1}{2}}{50} = \frac{1}{20}$ $2\frac{1}{2} \times 20 = 50 \times 1$

2. Look at the *Here's how*. If the model is $\frac{1}{20}$ the size of the real airplane, how long should the wingspan of the model be? $2\frac{1}{2}$ feet

Chapter 10

EXERCISES

Copy and complete. Give answers in simplest form.
Here are scrambled answers for the next row of exercises: 20 $1\frac{2}{5}$ $4\frac{4}{5}$ $\frac{3}{4}$ 5

3. $\frac{5}{2} = \frac{n}{8}$
   $2n = 40$
   $\frac{2n}{2} = \frac{40}{?}$
   $n = ?$ 20

4. $\frac{n}{8} = \frac{3}{5}$
   $5n = ?$ 24
   $\frac{5n}{5} = \frac{24}{5}$
   $n = ?$ $4\frac{4}{5}$

5. $\frac{2}{n} = \frac{4}{1\frac{1}{2}}$
   $4n = 2 \times 1\frac{1}{2}$
   $4n = 3$
   $\frac{4n}{4} = \frac{3}{?}$ 4
   $n = ?$ $\frac{3}{4}$

6. $\frac{2\frac{1}{2}}{4} = \frac{n}{8}$
   $4n = 2\frac{1}{2} \times 8$
   $4n = 20$
   $\frac{4n}{4?} = \frac{20}{4?}$
   $n = ?$ 5

7. $\frac{n}{1\frac{3}{4}} = \frac{4}{5}$
   $5n = 1\frac{3}{4} \times 4$
   $5n = 7$
   $\frac{5n}{5?} = \frac{7}{5?}$
   $n = ?$ $1\frac{2}{5}$

Solve each proportion. Give the answer in simplest form.

8. $\frac{1}{2} = \frac{6}{n}$ 12
9. $\frac{1}{4} = \frac{12}{n}$ 3
10. $\frac{n}{5} = \frac{2}{10}$ 1
11. $\frac{5}{n} = \frac{10}{20}$ 10
12. $\frac{1}{6} = \frac{7}{n}$ 42
13. $\frac{3}{10} = \frac{n}{5}$ $1\frac{1}{2}$
14. $\frac{7}{12} = \frac{4}{n}$ $6\frac{6}{7}$
15. $\frac{8}{n} = \frac{10}{3}$ $2\frac{2}{5}$
16. $\frac{n}{6} = \frac{5}{9}$ $3\frac{1}{3}$
17. $\frac{11}{4} = \frac{n}{12}$ 33
18. $\frac{1\frac{1}{2}}{3} = \frac{n}{2}$ 1
19. $\frac{10}{7} = \frac{1\frac{1}{4}}{n}$ $\frac{7}{8}$
20. $\frac{4}{n} = \frac{8}{2\frac{3}{4}}$ $1\frac{3}{8}$
21. $\frac{n}{2\frac{1}{2}} = \frac{4}{6}$ $1\frac{2}{3}$
22. $\frac{5}{4} = \frac{1\frac{1}{4}}{n}$ 1

Solve. Refer to the airplanes pictured on page 262.

23. Suppose that you want to make a model of the *Lockheed Vega* that is $\frac{1}{30}$ the size of the real airplane.

    a. What should the wingspan of the model be? $1\frac{11}{30}$ feet

    b. How long should the model be? $\frac{11}{12}$ foot

Check it out! Checking answers

24. Find and correct the two wrong answers.

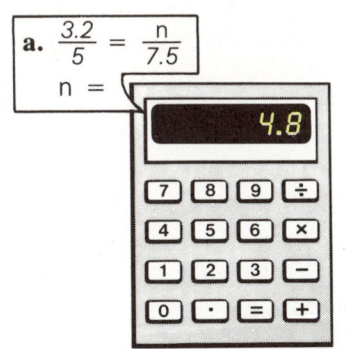
a. $\frac{3.2}{5} = \frac{n}{7.5}$
n = 4.8

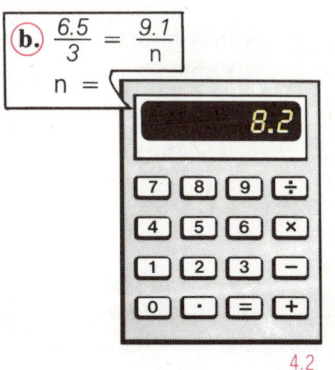
b. $\frac{6.5}{3} = \frac{9.1}{n}$
n = 8.2
4.2

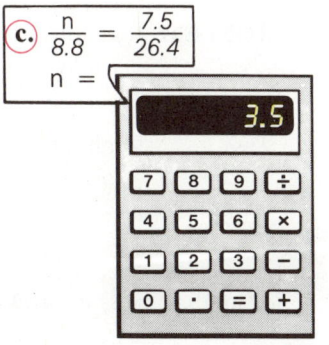
c. $\frac{n}{8.8} = \frac{7.5}{26.4}$
n = 3.5
2.5

Ratio and Proportion **263**

Extra Practice
Page 492 Skill 44

Practice Worksheet
Workbook S281, Copymaster S281, or Duplicating Master S281

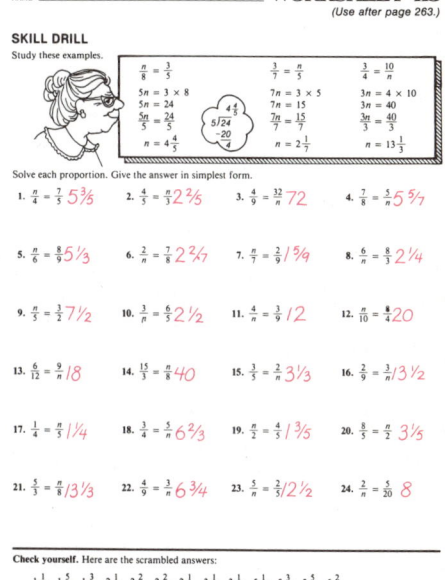

Project
Using proportions

Choose a favorite recipe. Use proportions to alter the recipe so that there is just enough to serve the entire class.

Copymaster S481

Class Starter Quiz 104
on previous lesson

Solve each proportion.

1. $\dfrac{3}{8} = \dfrac{n}{32}$ 12
2. $\dfrac{1}{2} = \dfrac{n}{9}$ $4\dfrac{1}{2}$
3. $\dfrac{1}{6} = \dfrac{5}{n}$ 30
4. $\dfrac{2}{9} = \dfrac{12}{n}$ 54
5. $\dfrac{n}{3} = \dfrac{5}{4}$ $3\dfrac{3}{4}$
6. $\dfrac{2}{3} = \dfrac{7}{n}$ $10\dfrac{1}{2}$

Copymaster S410

Lesson Objective
To solve rate problems using proportions

Problem-Solving Skills
Using a proportion to solve a problem
Reading scales

Starting the Lesson

Estimation Write this problem on the chalkboard:

> You spent $12.50 for 9 gallons of gasoline. At that rate, how much would you spend for 14 gallons?

Before the students open their books, have them estimate the cost of 14 gallons of gasoline. Record the high and low estimates on the chalkboard. Then have the students study problem 1 on page 264 to see if their estimate was close to the exact cost, $19.44.

Rates

A **rate** is a ratio of two quantities.

You spent $12.50 for 9 gallons of gasoline.

Rate: $\dfrac{\$12.50}{9 \text{ gal}}$

Read as "$12.50 per 9 gallons."

Here's how *to use proportions to solve rate problems.*

PROBLEM 1. If you spent $12.50 for 9 gallons of gasoline, how much would you spend for 14 gallons?

dollars → $\dfrac{12.50}{9} = \dfrac{n}{14}$ ← dollars
gallons → ← gallons

Remember: When setting up a proportion, you must be sure that the ratios are in the same order!

$9n = 12.50 \times 14$
$9n = 175$
$n \approx 19.44$

$$\begin{array}{r} 19.444 \\ 9\overline{)175.000} \end{array}$$

Read ≈ as "is approximately equal to."

So, at that rate, you would spend **$19.44** for 14 gallons of gasoline.

PROBLEM 2. You drive 196 miles in 4 hours. At that rate, how many hours will it take you to drive 320 miles?

miles → $\dfrac{196}{4} = \dfrac{320}{n}$ ← miles
hours → ← hours

$196n = 4 \times 320$
$196n = 1280$
$n \approx 6.53$

$$\begin{array}{r} 6.5306 \\ 196\overline{)1280.0000} \end{array}$$

So, at that rate, it would take you about **6.53** hours to drive 320 miles.

EXERCISES

Solve by using proportions. If an answer does not come out evenly, round it to the nearest hundredth.

1. You spend $18 for 12 gallons of gasoline. At that price,
   a. how many gallons could you buy for $15? 10

   Hint: $\dfrac{18}{12} = \dfrac{15}{n}$

   b. how many gallons could you buy for $6.50? 4.33
   c. how much would 9 gallons cost? $13.50
   d. how much would 10.4 gallons cost? $15.60

2. You drive 128 miles in 3 hours. At that speed,
   a. how many miles could you drive in 5 hours? 213.33

   Hint: $\dfrac{128}{3} = \dfrac{n}{5}$

   b. how many miles could you drive in 7 hours? 298.67
   c. how many hours would it take you to drive 200 miles? 4.69
   d. how many hours would it take you to drive 286 miles? 6.70

3. You drive 124 miles and use 4.8 gallons of gasoline. At that rate,
   a. how many miles could you drive on 8 gallons? 206.67
   b. how many miles could you drive on 15 gallons? 387.5
   c. how many gallons would you need for 200 miles? 7.74
   d. how many gallons would you need for 260 miles? 10.06

4. You spend $3.60 to drive 110 miles on a toll road. At that rate,
   a. how many miles could you drive for $1.60? 48.89
   b. how many miles could you drive for $2.00? 61.11
   c. how much would it cost to drive 150 miles? $4.91
   d. how much would it cost to drive 85 miles? $2.78

5. You pay $0.35 tax on a gallon of gasoline. At that rate,
   a. how much tax would you pay on 12 gallons? $4.20
   b. if you pay $5.95 in gasoline tax, how many gallons of gasoline would you purchase? 17

6. You purchase 9 gallons of gasoline and pay $2.61 in tax. At that rate,
   a. how much tax would you pay per gallon of gasoline? $0.29
   b. how much tax would you pay on 15 gallons of gasoline? $4.35

Check your instruments   Reading scales

7. You are 157 miles from Chicago. You want to be in Chicago by 11 o'clock. You look at your speedometer and clock.
   Will you be on time if you keep driving at the same rate? No

Ratio and Proportion   **265**

Problem solving

Logical reasoning

Use the facts and fingerprints to the right. Find the missing information.

Name: Howe   Name: McGrath
Age: 18   Age: 30
Height: 66 in.   Height: 56 in.
Weight: 150 lb   Weight: 100 lb

Name: Delano
Age: 60
Height: 67 in.
Weight: 175 lb

Name: Cleaver
Age: 34
Height: 70 in.
Weight: 190 lb

1.  is 25 pounds heavier than __Howe__ .
   (name)

2.  is 14 inches taller than __McGrath__ .
   (name)

3.  is 4 years younger than __Cleaver__ .
   (name)

4.  is __40__ pounds lighter than Cleaver.
   (number)

5. Two years ago,  was twice as old as __Howe__ .
   (name)

6. The ratio of 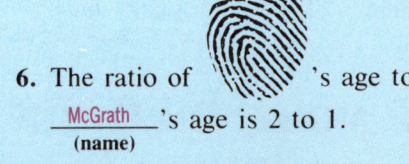's age to __McGrath__'s age is 2 to 1.
   (name)

7. The ratio of 's height to McGrath's height is __5__ to 4.
   (number)

8. The ratio of 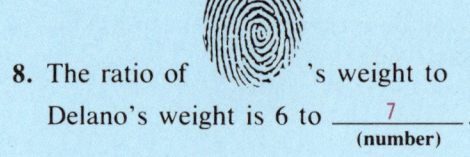's weight to Delano's weight is 6 to __7__ .
   (number)

Cumulative Skill Practice

Multiply. *(page 68)*

1. 1.4 × 52 = 72.8
2. 4.9 × 5.2 = 25.48
3. 0.58 × 6.8 = 3.944
4. 0.94 × 37 = 34.78
5. 0.39 × 9.7 = 3.783
6. 2.04 × 3.1 = 6.324
7. 0.78 × 0.44 = 0.3432
8. 0.53 × 66 = 34.98
9. 8.4 × 8.4 = 70.56
10. 4.18 × 41 = 171.38
11. 2.67 × 0.71 = 1.8957
12. 6.11 × 0.23 = 1.4053

Divide. Round the quotient to the nearest tenth. *(page 104)*

13. $0.6 \overline{)4.3}$ = 7.2
14. $0.3 \overline{)3.91}$ = 13.0
15. $0.09 \overline{)5.2}$ = 57.8
16. $0.03 \overline{)8.7}$ = 290
17. $0.06 \overline{)50}$ = 833.3
18. $1.3 \overline{)5.3}$ = 4.1
19. $0.11 \overline{)37}$ = 336.4
20. $2.7 \overline{)3.7}$ = 1.4
21. $3.7 \overline{)5.04}$ = 1.4
22. $0.65 \overline{)9.1}$ = 14

<, =, or >? *(page 152)*

23. $\frac{1}{2}$ ● $\frac{4}{8}$ =
24. $\frac{3}{8}$ ● $\frac{5}{8}$ <
25. $\frac{1}{2}$ ● $\frac{1}{4}$ >
26. $\frac{3}{9}$ ● $\frac{1}{3}$ =
27. $\frac{1}{3}$ ● $\frac{1}{2}$ <
28. $\frac{5}{8}$ ● $\frac{3}{4}$ <
29. $\frac{2}{3}$ ● $\frac{10}{15}$ =
30. $\frac{3}{4}$ ● $\frac{2}{3}$ >
31. $\frac{5}{6}$ ● $\frac{7}{8}$ <
32. $\frac{5}{6}$ ● $\frac{3}{4}$ >

Change to a decimal rounded to the nearest hundredth. *(page 166)*

33. $\frac{1}{3}$ 0.33
34. $\frac{1}{9}$ 0.11
35. $\frac{1}{6}$ 0.17
36. $\frac{1}{8}$ 0.13
37. $\frac{1}{16}$ 0.06
38. $\frac{3}{4}$ 0.75
39. $\frac{5}{8}$ 0.63
40. $\frac{11}{9}$ 1.22
41. $\frac{15}{6}$ 2.50
42. $\frac{19}{8}$ 2.38
43. $\frac{21}{12}$ 1.75
44. $\frac{25}{16}$ 1.56
45. $\frac{13}{4}$ 3.25
46. $\frac{16}{3}$ 5.33

Give the sum in simplest form. *(page 180)*

47. $\frac{3}{4} + \frac{1}{4}$ = 1
48. $\frac{1}{2} + \frac{1}{4}$ = $\frac{3}{4}$
49. $\frac{5}{9} + \frac{1}{3}$ = $\frac{8}{9}$
50. $4 + \frac{2}{5}$ = $4\frac{2}{5}$
51. $\frac{2}{3} + 3$ = $3\frac{2}{3}$
52. $\frac{7}{8} + \frac{3}{4}$ = $1\frac{5}{8}$
53. $\frac{1}{2} + \frac{3}{8}$ = $\frac{7}{8}$
54. $\frac{3}{10} + \frac{1}{5}$ = $\frac{1}{2}$
55. $\frac{5}{6} + \frac{2}{3}$ = $1\frac{1}{2}$
56. $\frac{2}{3} + \frac{1}{4}$ = $\frac{11}{12}$

MIXED PRACTICE
Complete.

57. The standard numeral for 6^2 is __?__ 36

58. $3 \times 3 \times 3 \times 3$ written using an exponent is __?__ 3^4

59. The prime factorization of 28 is __?__ $2 \times 2 \times 7$ or $2^2 \times 7$

60. The greatest common factor of 15 and 24 is __?__ 3

61. The least common multiple of 9 and 15 is __?__ 45

Ratio and Proportion 267

Problem-Solving Worksheet
Workbook S283, Copymaster S283, or Duplicating Master S283

WORKSHEET 115 (Use after page 266.)

NAME _____

CAN YOU AFFORD IT?

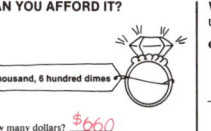

6 thousand, 6 hundred dimes

How many dollars? $660

SPECIAL

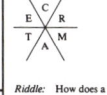

Records .... $7 each
Tapes .... $9 each

Janice bought some records and tapes. She spent a total of $30. How many records did she buy? **3**

DON'T GET LOST!

Use the map. Fill in the missing distances on the sign.

Big Lake 8 km
Pine Grove 4 km
Hill City 24 km

WHAT'S THE ANSWER?
Use the code to answer the riddle.

CODE

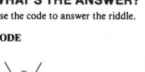

Riddle: How does a witch tell time?
Answer:
S H E
W E A R S
A W I T C H
W A T C H

YOU'RE THE CLERK
What 5 coins would you give a customer who bought a 29¢ pen and gave you a dollar bill?

penny   dime
dime   quarter
quarter

© D. C. Heath and Company   S283   Problem solving

Challenge Problem

Use the code to answer the riddle.

CODE

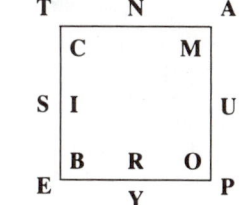

Riddle: What is a store detective called?

Answer:

A counterspy

Copymaster S455

Class Starter Quiz 106
on previous lesson

Complete. Use the information at the top of page 266.

1. <u>Delano</u> is 25 pounds heavier
   (Name)
   than Howe.

2. The ratio of Delano's age to <u>McGrath's</u> age is 2 to 1.
   (Name)

3. The ratio of Cleaver's age to <u>Howe's</u> age is 17 to 9.
   (Name)

Copymaster S411

Lesson Objective
To solve problems that involve scale drawings by using proportions

Problem-Solving Skills
Reading a map
Using a ruler and a map scale
Choosing the correct operation
Solving a multi-step problem

Starting the Lesson
The map at the top of page 268 is also on ■ **Visual Aid 45** (copymaster or transparency S148.) Before going over the *Here's how*, use the map and ask questions like these:

- Which city is closer to Chicago, Pittsburgh or Cincinnati? (Cincinnati)
- According to the scale on the map, 1 centimeter stands for how many kilometers? (150)
- What real distance would 2 centimeters on the map represent? (300 km)

Exercise Note
Use of Concrete Materials You may wish to use the map on ■ **Visual Aid 45** (copymaster or transparency S148) and the millimeter ruler on ■ **Visual Aid 39** (copymaster or transparency S143) when discussing exercises 11–21.

268

Scale drawings

A map is an example of a scale drawing.
On this map, 1 centimeter stands for 150 kilometers.

Here's how to solve a scale-drawing problem.

Since we know the scale, we can measure a distance on the map and solve a proportion to find the actual distance.

PROBLEM. The distance from Minneapolis to Cincinnati on the map is 7.5 centimeters. What is the actual distance from Minneapolis to Cincinnati?

cm on map → $\dfrac{1}{150} = \dfrac{7.5}{n}$ ← cm on map
actual km →   ← actual km

$n = 7.5 \times 150$
$n = 1125$

The actual distance from Minneapolis to Cincinnati is about 1125 kilometers.

268 Chapter 10

EXERCISES

Find the actual distance between the cities. The map distances from page 268 are given.

1. Chicago to Raleigh, 7.8 cm 1170 km
2. Detroit to New York, 5.9 cm 885 km
3. Minneapolis to Toronto, 8.3 cm 1245 km
4. Boston to Washington, 4.8 cm 735 km
5. Green Bay to Buffalo, 5.8 cm 870 km
6. Chicago to New York, 8.6 cm 1290 km
7. St. Louis to Detroit, 5.6 cm 840 km
8. Detroit to Cleveland, 1.2 cm 180 km
9. Minneapolis to Raleigh, 12.3 cm 1845 km
10. Kansas City to Toronto, 10.4 cm 1560 km

Solve. Use a ruler and the map on page 268. Answers for 11–16 may vary slightly.

11. How far is it from Minneapolis to Detroit? 1005 km
12. How far is it from Cleveland to Boston? 990 km
13. How far is it from Minneapolis to Detroit to Washington? 1740 km
14. How far is it from Toronto to Washington to Raleigh? 1170 km
15. How much farther is it from Chicago to Pittsburgh than from Cincinnati to Pittsburgh? 270 km
16. How much farther is it from Kansas City to Pittsburgh than from St. Louis to Pittsburgh? 390 km
17. Which city is about 1350 kilometers east of Des Moines? Pittsburgh
18. Which city is about 700 kilometers northwest of Raleigh? Cincinnati or Pittsburgh

You be the pilot!

19. My airspeed is 600 kilometers per hour. About how many hours is it from Des Moines to Philadelphia? 3
20. I'm flying from Philadelphia to St. Louis. My airspeed is 750 kilometers per hour. About how many hours should the flight take? 2
21. At 600 kilometers per hour, how many hours will it take to fly from St. Louis to Toronto? 2

Extra Practice
Page 493   Skill 45

Practice Worksheet
Workbook S284, Copymaster S284, or Duplicating Master S284

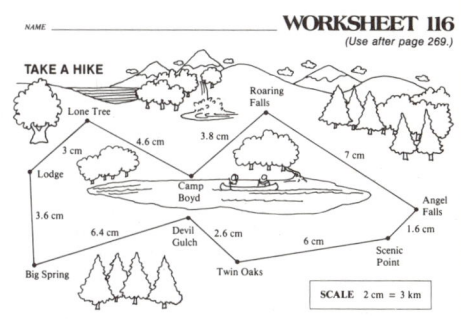

Group Project
Making a scale drawing
Have the students find the outside dimensions of the school building. Tell them to make a scale drawing of the building using the scale

1 centimeter = 1 meter.

Ratio and Proportion

Similar figures

The sails on the box and on the model are the same shape. They are **similar figures**.

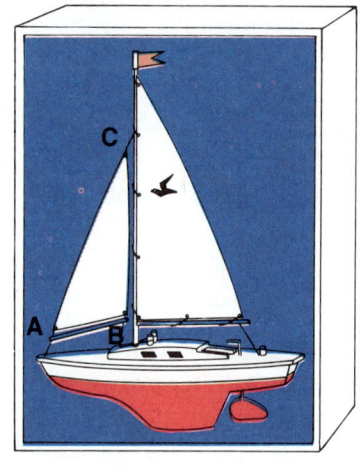

Look at the small sail on the box and on the model.

Side *AB* corresponds to side *RS*.
Side *AC* corresponds to side *RT*.
Side *BC* corresponds to side *ST*.

In similar figures, the ratios of the lengths of corresponding sides are equal.

Here's how to use a proportion to solve a similar-figures problem.

PROBLEM. These two flags are similar. What is the length of side n?

Step 1. Write a proportion.

small flag → $\dfrac{1.5}{2.5} = \dfrac{3}{n}$ ← small flag
large flag →   ← large flag

Step 2. Solve the proportion.

$1.5n = 3 \times 2.5$
$1.5n = 7.5$
$n = 5$

$$1.5 \overline{)7.5} \quad \begin{array}{r} 5. \\ -75 \\ \hline 0 \end{array}$$

The length of side n is 5 centimeters.

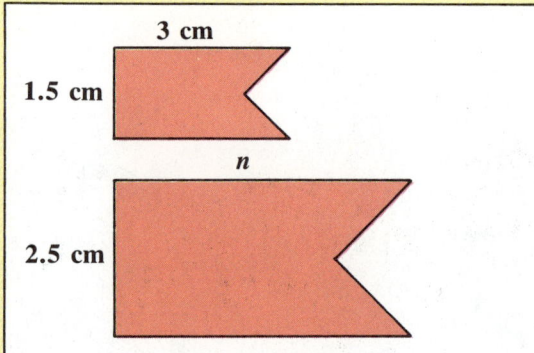

3 cm
1.5 cm
n
2.5 cm

EXERCISES

The two figures are similar. Solve a proportion to find the length of side n.

1.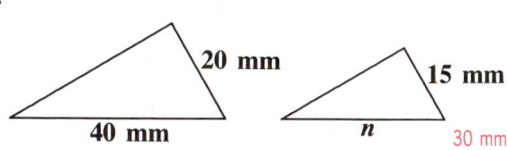
   30 mm

2.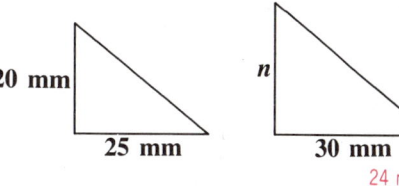
   24 mm

3.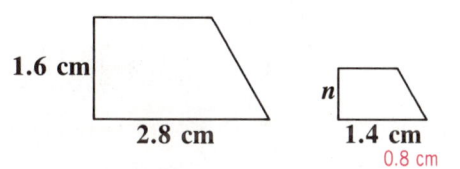
   24 mm

4.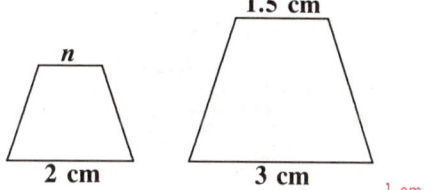
   27 mm

5.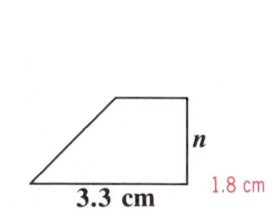
   0.8 cm

6.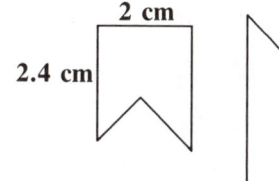
   1 cm

7.
   1.8 cm

8.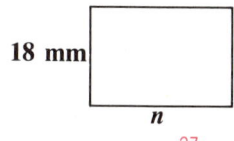
   3 cm

Size it up! Using scale drawings

Solve a proportion to answer the question.

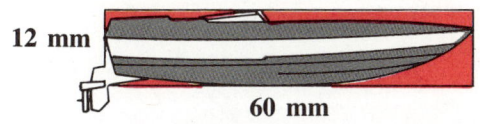

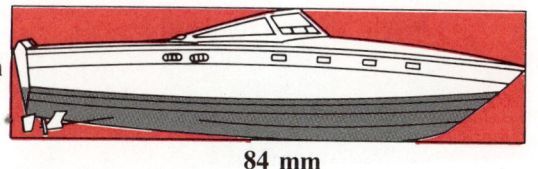

9. The *Starfire* is actually 1.64 meters high. What is its actual length? 8.2 m

10. The *Sport Special* is actually 2.9 meters high. What is its actual length?

Ratio and Proportion **271**

Practice Worksheet
Workbook S285, Copymaster S285, or Duplicating Master S285

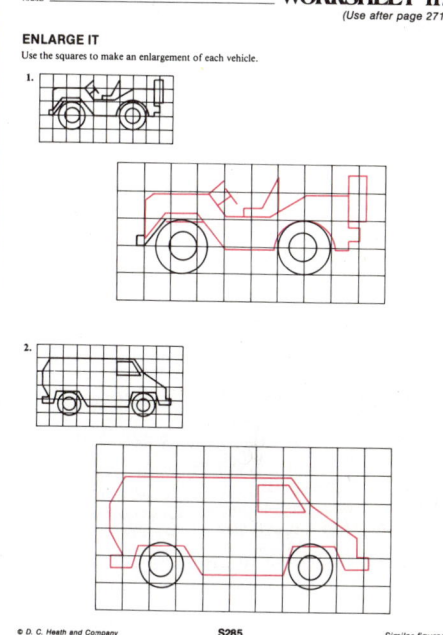

Challenge Problem

Show how to arrange four figures like Figure A to get a larger figure that is similar to Figure A.

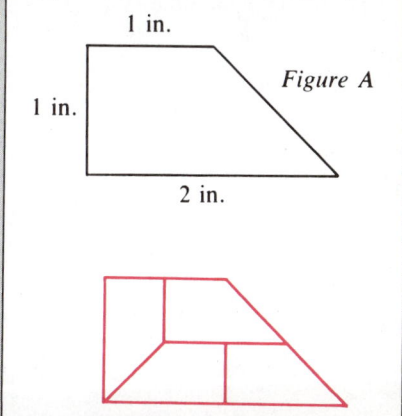

Copymaster S455

271

Class Starter Quiz 108
on previous lesson

The two figures are similar. Solve a proportion to find the length of side *n*.

1. 24 mm / 36 mm ; 12 mm / *n* — 18 mm
2. 10 mm / 20 mm ; *n* / 15 mm — 7.5 mm

Copymaster S411

Lesson Objective
To use similar triangles to solve indirect measurement problems

Problem-Solving Skills
Selecting information from a drawing
Using a drawing and proportion to solve a problem

Starting the Lesson
Estimation The picture at the top of page 272 is on ■ **Visual Aid 47** (copymaster or transparency S149). Before discussing the *Here's how* example, use the picture and have the students guess the height of the loop-the-loop ride. Record the high and low guesses. Then go over the *Here's how* example and have the students check their guesses.

Exercise Note
The pictures at the top of page 273 are also on ■ **Visual Aid 48** (copymaster or transparency S149). Use the pictures when discussing exercises 1–4.

Indirect measurement

MAKE A GUESS!
WHAT IS THE HEIGHT OF THE LOOP-THE-LOOP RIDE?

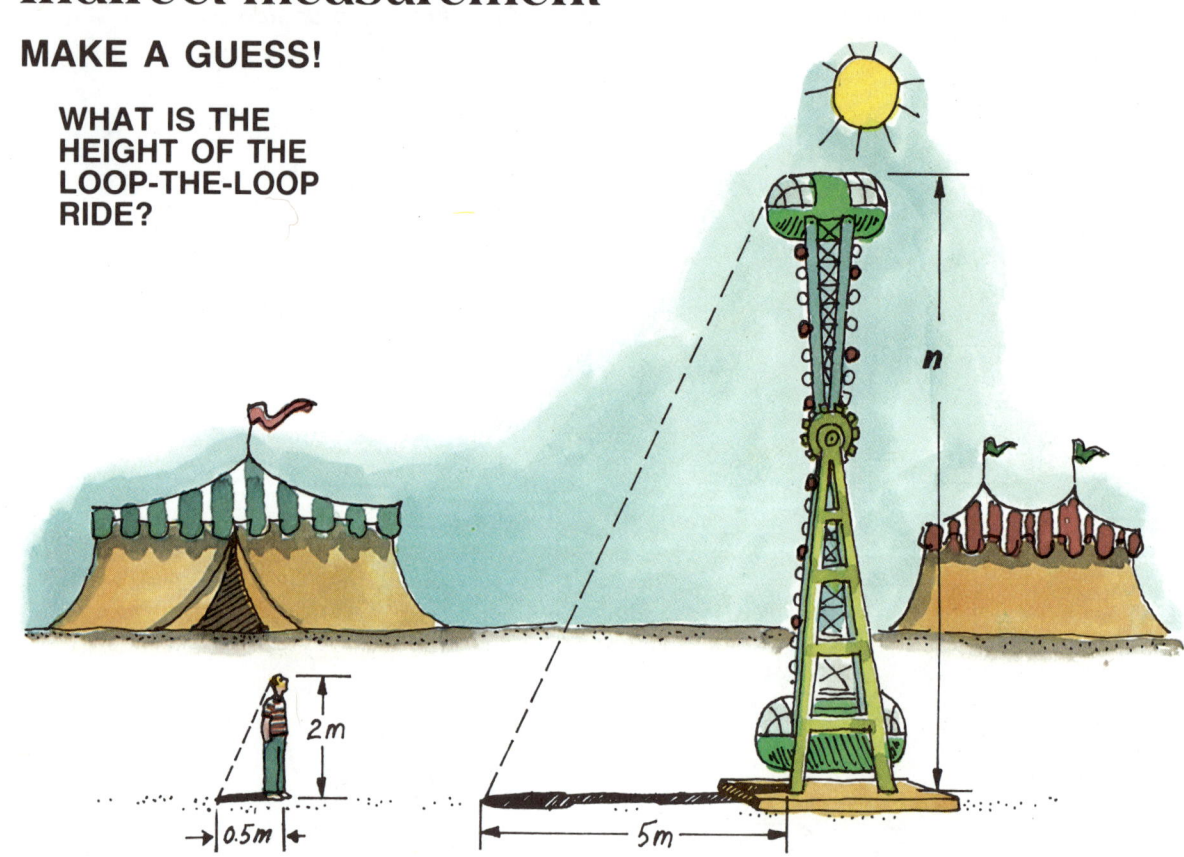

Here's how similar triangles can be used to find lengths that are difficult to measure directly.

The triangle made by the man and his shadow is similar to the triangle made by the loop-the-loop ride and its shadow. We can use this proportion to find the height of the ride.

height of loop-the-loop → $\frac{n}{2} = \frac{5}{0.5}$ ← shadow of loop-the-loop
height of man →   ← shadow of man

$$0.5n = 2 \times 5$$
$$0.5n = 10$$
$$n = 20$$

The height of the loop-the-loop ride is **20 meters**.

EXERCISES

Solve. Round the answer to the nearest hundredth of a meter.

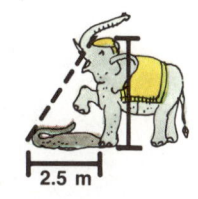
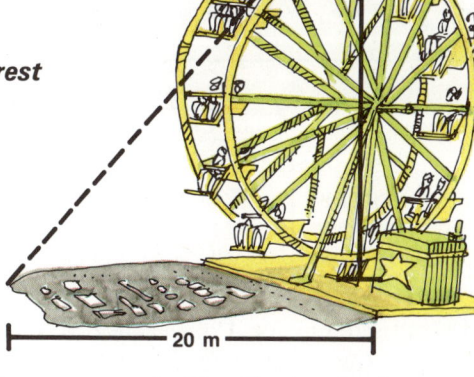

1. The elephant casts a 2.5-meter shadow. A man 2 meters tall casts a 1.5-meter shadow. How tall is the elephant? 3.33 m

2. The Ferris wheel casts a 20-meter shadow. How tall is the Ferris wheel? Round the answer to the nearest hundredth of a meter. *Hint: Use the facts about the man and his shadow.* 26.67 m

Make your own drawings. Then solve the problem.

3. A high-diving pole casts a 21-meter shadow. A man 2 meters tall casts a 1-meter shadow. How tall is the high-diving pole? 42 m

4. A woman 1.5 meters tall casts a 0.75-meter shadow. A diving tank casts a 1.2-meter shadow. How deep is the diving tank? 2.4 m

5. An animal trainer is 2 meters tall and casts a 2.2-meter shadow. A black bear casts a 3.3-meter shadow. How tall is the black bear? 3 m

6. A 2-meter sign casts a 3-meter shadow. A flagpole casts a 45-meter shadow. How tall is the flagpole? 30 m

7. A TV tower is 30 meters high and casts a 10-meter shadow. How tall is a nearby tree that casts a 6-meter shadow? 18 m

8. The fence around a water tower is 2.5 meters high and casts a 4-meter shadow. The water tower casts a 48-meter shadow. How tall is the water tower? 30 m

Animal tracks

9. Unscramble the letters to name each animal.

CBAOBT — BOBCAT

ANOMUTIN INOL — MOUNTAIN LION

10. Study the tracks. The ___(animal)___ takes 100 steps for every 200 steps the ___(animal)___ takes.
    mountain lion ; bobcat

Practice Worksheet

Workbook S286, Copymaster S286, or Duplicating Master S286

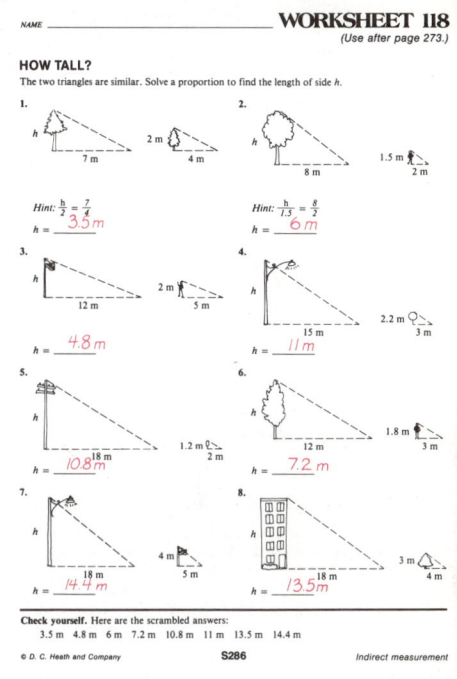

Group Project

Making indirect measurements

Wait for a sunny day. Then have the students, working in small groups, take a meterstick and measuring tape outside. Have them find the height of a flagpole or tree by using the shadow and its relationship to a meterstick and its shadow.

$$\frac{\text{meterstick}}{\text{stick's shadow}} = \frac{\text{flagpole}}{\text{pole's shadow}}$$

Class Starter Quiz 109
on previous lesson

Solve. Draw a sketch.

1. A building casts a 150-meter shadow. At the same time, a man 2 meters tall casts a 3-meter shadow. How tall is the building? 100 m
2. A tree casts a 30-meter shadow. At the same time, a 15-meter stick casts a 2-meter shadow. How tall is the tree? 225 m

Copymaster S412

Lesson Objective
To solve problems using facts from more than one source

Starting the Lesson
The rental charges and newspaper article are on ■ **Visual Aid 49** (copymaster or transparency S150). Use the visual aid when discussing the problem on page 274.

Exercise Note
Many rental agencies consider a week to be 5, 6, or 7 days. Some students may know this and might question the answer to exercise 6 on page 275. A 12 day rental could be considered either a 2-week rental costing $278 or a 1-week, 5-day rental costing $284.

Answer for exercise 9, page 275

Step	8 qt	5 qt	3 qt
1	5	0	3
2	5	3	0
3	2	3	3
4	2	5	1
5	7	0	1
6	7	1	0
7	4	1	3

Step	8 qt	5 qt	3 qt
1	3	5	0
2	3	2	3
3	6	2	0
4	6	0	2
5	1	5	2
6	1	4	3

274

PROBLEM SOLVING *applications*

Rental Charges		
Size	Daily rate	Weekly rate
Compact	$29	$139
Mid-size	35	150
Full-size	39	159

Problem

Andrew rented a full-size car for 4 days. He drove the car 381 miles and spent $17.50 for gasoline. Was his cost per mile higher than the national average?

DRIVING COSTS UP
The average cost of driving a car took an upturn this year, according to a recent study conducted by a national car-rental agency.

Operating costs for compact cars increased from 28.6 cents per mile to 32.7 cents per mile. The costs for mid-size cars increased from 34.6 cents to 38.5 cents per mile, and the costs per mile for full-size cars leaped from 37.3 cents to 43.8 cents.

Here's how *to use the information sources to solve the problem.*

STUDY What facts are given? What is the question?

PLAN and SOLVE Find Andrew's cost per mile. Then compare his cost per mile with the national average in the article.

Step 1 Find the rental charge by multiplying the daily rate ($39 per day) by the number of days (4).

Estimate: Compute:
$4 \times 40 = 160$ $4 \boxed{\times} 39 \boxed{=}$ 156

156 is near the estimate. So 156 seems reasonable.

Step 2. Find the total cost by adding the rental charges ($156) and the gasoline expenses ($17.50).

173.50 is near the estimate. So 173.50 seems reasonable.

Estimate: Compute:
$160 + 20 = 180$ $156 \boxed{+} 17.50 \boxed{=}$ 173.50

Step 3. Find the cost per mile by dividing the total cost ($173.50) by the number of miles driven (381).

Written as cents, the quotient is about 45.5.

Estimate: Compute:
$200 \div 400 = 0.50$ $173.50 \boxed{\div} 381 \boxed{=}$ 0.4553805

Step 4. Compare Andrew's cost per mile to the national average.

Answer: His cost per mile (45.5 ¢) is higher than the national average (43.8 ¢).

CHECK Make sure the correct data was used from the chart and article.

274 Chapter 10

PROBLEMS

Solve. Use the data from page 274.

1. Marsha rented a full-size car for one week. She drove 583 miles and spent $21.75 for gasoline.
   a. What was her total cost? $180.75
   b. What was her cost per mile? $.31
   c. Was her cost per mile higher or lower than the national average? lower

2. Daniel rented a compact car for a week and 3 days. He drove 618 miles and spent $18.50 for gasoline.
   a. How much did Daniel pay for rental charges? $226
   b. What was the total cost of the car for a week and 3 days? $244.50
   c. What was his cost per mile? $.40

3. According to the survey on page 274, how much does it cost to drive a compact car 200 miles? $65.40

4. According to the survey, how much does it cost to drive a full-size car 200 miles? $87.60

5. How much does it cost to rent a full-size car for a week and 2 days? $237

6. How much does it cost to rent a compact car for 12 days? $284 or $278*

7. Which is cheaper, renting a full-size car for 4 days or a compact car for a week? A compact car for a week.
   * $284 on a weekly rate for 1 week and a daily rate for 5 days, $278 on a weekly rate for 2 weeks.

8. Hon is planning a 10-day trip to Florida. She wants to spend less than $250 for a rental car. What size car can she rent? compact

Don't spill the oil! *Acting it out*

9. You have an 8-quart can full of oil. Find a way to use an empty 5-quart can and an empty 3-quart can to measure out 4 quarts of oil.

Ratio and Proportion **275**

Practice Worksheet
Workbook S287, Copymaster S287, or Duplicating Master S287

WORKSHEET 119
(Use after page 275.)

NAME _____

WATER FUN

Rental Charges		
Item	Daily rate	Weekly rate
Canoe	$18	$ 75
Rowboat	15	50
Paddle boat	27	95
Rubber raft	10	50
Motorboat	35	200
Sailboat	25	125

Solve. Write each amount of money correct to the nearest cent.

1. Gerry rented a canoe for 3 days. What was the total cost? $54

2. Sue rented a paddle boat for 2 days and paid with $60. How much change did she get back? $6

3. Which is cheaper, renting a rowboat for 5 days or renting it for a week? How much cheaper? Renting for a week $25

4. Kao rented a motorboat for 6 days. He rented it on the weekly rate. How much did he save by not renting it on the daily rate? $10

5. Sandy rented a paddle boat for 11 days. What was the cheapest possible rental charge? $190

6. Sharon rented a sailboat for a day. She had it out on the lake for 12 hours. How much did it cost her per hour? $2.08

7. Elena and 3 of her friends rented a canoe for a week. They divided the cost evenly. How much did each person pay? $18.75

8. Ed rented a rubber raft for a week. What was the cost per day? How much cheaper is the cost per day than the 1-day rental rate? $7.14 ; $2.86

9. Kevin said, "I spent $50 for rentals over a 2-day period." Which rentals could Kevin have chosen? Sailboat twice, or rowboat and motorboat

10. Carrie rented a different item every day for 3 days. She spent $80. What did she rent? Canoe, paddle boat, and motorboat

© D.C. Heath and Company   S287   Problem solving – applications

---

Challenge Problem

Write a question that fits the answer.

April took 15 shots and made 10 baskets.

What is April's rate of baskets made to shots taken?

Answer: 2 to 3

Copymaster S455

Problem solving

USING COMPUTER SOFTWARE

Maria Cataldo uses a **software** package to draw geometric figures on her computer screen. The software also allows her to enlarge any part of the drawing.

Maria uses a number and a letter to locate each section in her drawing.

When Maria commands the computer to enlarge section 5C, the computer screen looks like this:

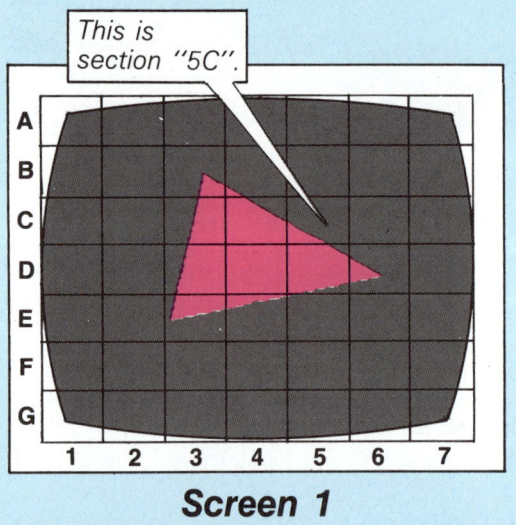

This is section "5C".

Screen 1

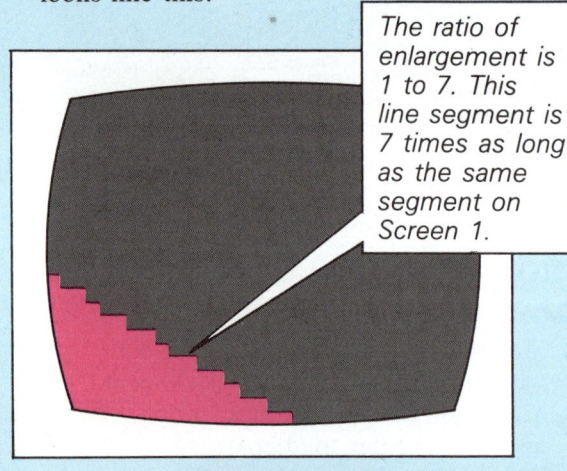

The ratio of enlargement is 1 to 7. This line segment is 7 times as long as the same segment on Screen 1.

Screen 2

Which section from Screen 1 has been enlarged?

1.  3B

2. (image) 5D

3.  3E

Complete.

4. If the ratio of enlargement is 1 to 7, a 2-inch line segment in an original picture would be ? inches in an enlargement. 14

5. If the ratio of enlargement is 1 to 8, a 1.5-inch line segment in an original picture would be ? inches in an enlargement. 12

6. An original picture has a line segment that is 2 inches long. The same line segment is 12 inches long in an enlargement. The ratio of enlargement is 1 to ? . 6

7. A line segment is enlarged from 1.75 inches in the original drawing to 15.75 inches in an enlargement. The ratio of enlargement is 1 to ? . 9

Cumulative Skill Practice

Give the difference in simplest form. *(page 188)*

1. $\frac{3}{5} - \frac{1}{5}$ $\frac{2}{5}$
2. $\frac{3}{4} - \frac{1}{4}$ $\frac{1}{2}$
3. $\frac{1}{2} - \frac{1}{4}$ $\frac{1}{4}$
4. $\frac{1}{3} - \frac{1}{6}$ $\frac{1}{6}$
5. $\frac{3}{8} - \frac{1}{4}$ $\frac{1}{8}$
6. $\frac{7}{8} - \frac{3}{4}$ $\frac{1}{8}$
7. $\frac{1}{2} - \frac{3}{8}$ $\frac{1}{8}$
8. $\frac{3}{10} - \frac{1}{5}$ $\frac{1}{10}$
9. $\frac{5}{6} - \frac{2}{3}$ $\frac{1}{6}$
10. $\frac{2}{3} - \frac{1}{4}$ $\frac{5}{12}$

Complete. *(page 204)*

11. $\frac{1}{2}$ of 12 = ? 6
12. $\frac{1}{3}$ of 15 = ? 5
13. $\frac{2}{3}$ of 18 = ? 12
14. $\frac{3}{4}$ of 24 = ? 18
15. $\frac{7}{8}$ of 16 = ? 14
16. $\frac{3}{5}$ of 25 = ? 15
17. $\frac{5}{8}$ of 32 = ? 20
18. $\frac{3}{4}$ of 48 = ? 36

Give the quotient in simplest form. *(page 214)*

19. $1\frac{1}{2} \div 3$ $\frac{1}{2}$
20. $2\frac{1}{2} \div 2$ $1\frac{1}{4}$
21. $2 \div 1\frac{1}{2}$ $1\frac{1}{3}$
22. $4 \div 2\frac{1}{4}$ $1\frac{7}{9}$
23. $2\frac{2}{3} \div 1\frac{1}{4}$ $2\frac{2}{15}$
24. $5\frac{1}{2} \div 1\frac{3}{8}$ 4
25. $6\frac{1}{4} \div 1\frac{1}{4}$ 5
26. $3\frac{3}{4} \div 2\frac{1}{3}$ $1\frac{17}{28}$

Complete. *(page 230)*

27. 5 L = ? mL 5000
28. 12 L = ? mL 12,000
29. 3.4 L = ? mL 3400
30. 3000 mL = ? L 3
31. 1635 mL = ? L 1.635
32. 400 mL = ? L 0.4

Add. *(page 248)*

33. 4 ft 7 in. + 2 ft 10 in. 7 ft 5 in.
34. 3 yd 2 ft + 1 yd 2 ft 5 yd 1 ft
35. 5 ft 9 in. + 3 ft 8 in. 9 ft 5 in.
36. 4 gal 3 qt + 1 gal 2 qt 6 gal 1 qt

Solve. Give the answer in simplest form. *(page 262)*

37. $\frac{1}{4} = \frac{n}{12}$ 3
38. $\frac{5}{6} = \frac{7}{n}$ $8\frac{2}{5}$
39. $\frac{5}{n} = \frac{3}{15}$ 25
40. $\frac{n}{12} = \frac{5}{8}$ $7\frac{1}{2}$
41. $\frac{9}{n} = \frac{2}{3}$ $13\frac{1}{2}$

MIXED PRACTICE

Complete.

42. 64 × 100 = ? 6400
43. 19.6 + 3.85 = ? 23.45
44. 4.7 × 0.05 = ? 0.235
45. 58.61 − 5.94 = ? 52.67
46. 87.2 ÷ 0.4 = ? 218.5
47. 17 ÷ 4.66 = ? 3.648
48. 3.8 + 12 + 4.25 = ? 20.05
49. 63.8 × 10 = ? 638
50. 20.70 ÷ 1.5 = ? 13.8
51. 83.7 ÷ 10 = ? 8.37
52. 0.09 × 1000 = ? 90
53. 8.97 ÷ 1000 = ? 0.00897
54. (8.6 × 10) ÷ 2 = ? 43
55. 8.6 × (10 ÷ 2) = ? 43
56. 15.4 − (2.6 + 3.07) = ? 9.73

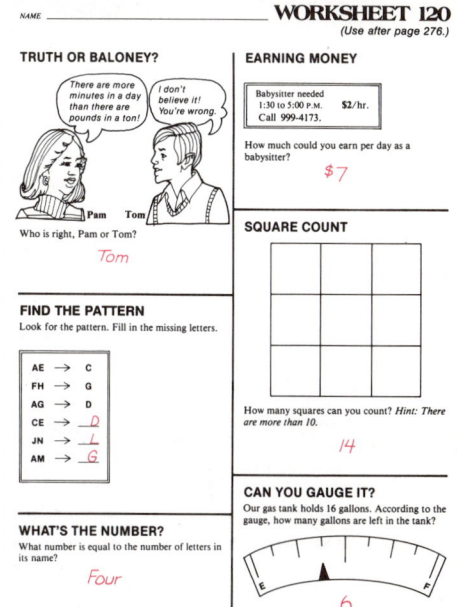

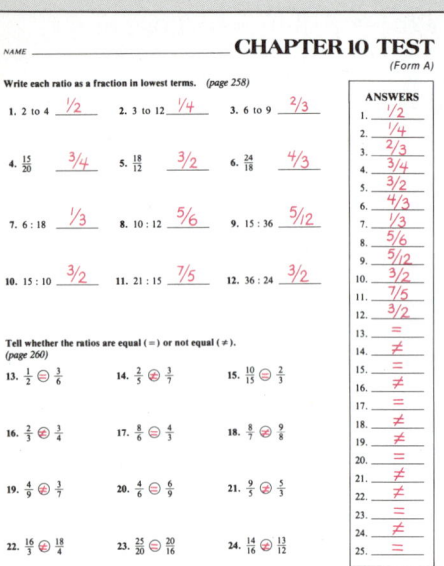
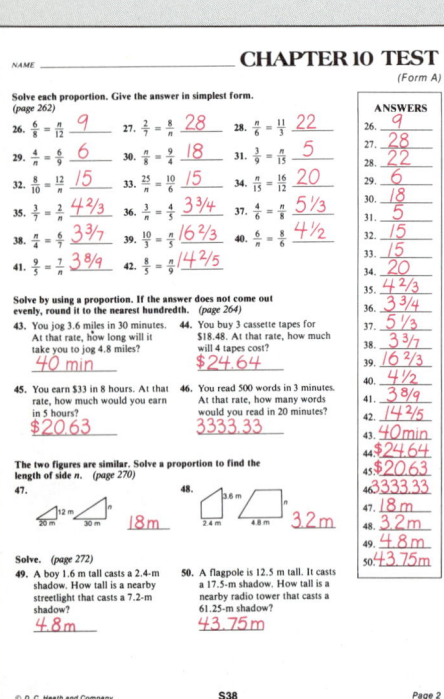

Chapter REVIEW

Here are scrambled answers for the review exercises:

approximately	divide	height	proportion	scale
corresponds	equal	multiplication	rate	similar
cross	fractions	order	ratio	

1. ratio, fractions **2.** proportion, multiplication

1. You can use a [?] to compare two numbers. You can find equal ratios by thinking about equivalent [?]. (page 258)

2. A [?] is an equation that says two ratios are equal. Every proportion has a related [?] equation. (page 260)

3. cross **4.** divide **5.** order, approximately

3. You can use the [?] products to find whether these two ratios are equal. Since 3 × 15 = 5 × 9, you know that the ratios are equal. (page 260)

$$\frac{3}{5} \cdot \frac{9}{15}$$

4. The last step in solving this proportion is to [?] 24 by 5. (page 262)

$$\frac{5}{6} = \frac{4}{n}$$
$$5n = 6 \times 4$$
$$5n = 24$$

5. When setting up a proportion, you must be sure that the ratios are in the same [?]. In this proportion, n is [?] equal to 3.11. (pages 262, 264)

$$\frac{9}{4} = \frac{7}{n}$$
$$9n = 28$$
$$n \approx 3.11$$

6. rate, scale **7.** similar, corresponds, equal **8.** height

6. A [?] is a ratio of two quantities. A map is an example of a [?] drawing. (pages 264, 268)

7. Figures that are the same shape are called [?] figures. In these similar figures, side AB [?] to side RS. In similar figures, the ratios of the lengths of corresponding sides are [?]. (page 270)

8. You can solve this proportion to find the [?] of the flagpole. (page 272)

$$\frac{n}{1.9} = \frac{7.2}{2.4}$$

278 Chapter 10

Chapter TEST

Give each ratio as a fraction in lowest terms. *(page 258)*

1. 3 to 9 $\frac{1}{3}$
2. 8 to 12 $\frac{2}{3}$
3. 16 to 10 $\frac{8}{5}$
4. 18 to 24 $\frac{3}{4}$
5. 7 to 14 $\frac{1}{2}$
6. 12 : 16 $\frac{3}{4}$
7. 15 : 10 $\frac{3}{2}$
8. 21 : 14 $\frac{3}{2}$
9. 32 : 8 $\frac{4}{1}$
10. 56 : 16 $\frac{7}{2}$

Tell whether the ratios are equal (=) or not equal (≠). *(page 260)*

11. $\frac{3}{4}$ ● $\frac{9}{12}$ =
12. $\frac{5}{6}$ ● $\frac{7}{8}$ ≠
13. $\frac{5}{2}$ ● $\frac{7}{4}$ ≠
14. $\frac{4}{6}$ ● $\frac{6}{9}$ =
15. $\frac{4}{8}$ ● $\frac{2}{4}$ =
16. $\frac{4}{3}$ ● $\frac{5}{4}$ ≠
17. $\frac{8}{5}$ ● $\frac{9}{6}$ ≠
18. $\frac{6}{16}$ ● $\frac{9}{24}$ =
19. $\frac{4}{7}$ ● $\frac{7}{12}$ ≠
20. $\frac{5}{6}$ ● $\frac{2}{3}$ ≠

Solve each proportion. Give the answer in simplest form. *(page 262)*

21. $\frac{2}{3} = \frac{n}{9}$ 6
22. $\frac{1}{4} = \frac{10}{n}$ $2\frac{1}{2}$
23. $\frac{5}{8} = \frac{n}{4}$ $2\frac{1}{2}$
24. $\frac{2}{3} = \frac{9}{n}$ $13\frac{1}{2}$
25. $\frac{4}{5} = \frac{8}{n}$ 10
26. $\frac{n}{20} = \frac{3}{4}$ 15
27. $\frac{11}{n} = \frac{5}{4}$ $8\frac{4}{5}$
28. $\frac{3}{n} = \frac{2}{3}$ $4\frac{1}{2}$
29. $\frac{2}{n} = \frac{5}{8}$ $3\frac{1}{5}$
30. $\frac{1}{5} = \frac{2}{n}$ 10
31. $\frac{6}{10} = \frac{n}{8}$ $4\frac{4}{5}$
32. $\frac{4}{7} = \frac{12}{n}$ 21
33. $\frac{7}{n} = \frac{2}{3}$ $10\frac{1}{2}$
34. $\frac{7}{10} = \frac{n}{8}$ $5\frac{3}{5}$
35. $\frac{2}{7} = \frac{n}{5}$ $1\frac{3}{7}$

Solve by using a proportion. *(page 264)*

36. You jog 4.5 miles in 36 minutes. At that rate, how long will it take to jog 6 miles? 48 minutes
37. You buy 3 record albums for $16.50. At that rate, how much will 5 records cost? $27.50
38. You earn $48 in 12 hours. At that rate, how much would you earn in 7 hours? $28
39. You type 129 words in 3 minutes. At that rate, how many words could you type in 10 minutes? 430

The two figures are similar. Solve a proportion to find the length of side n. *(page 270)*

40. 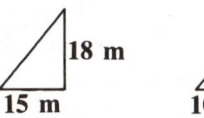   12 m

41. 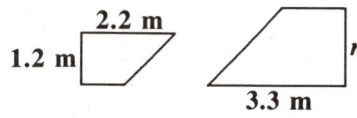 1.8 m

Make a drawing and solve the problem. *(page 272)*

42. A woman 1.8 m tall casts a 1.2-meter shadow. How tall is a nearby building that casts a 14.4-meter shadow? 21.6 m
43. A tower is 60 m high. It casts a 45-meter shadow. How tall is a nearby telephone pole that casts a 6-meter shadow? 8 m

Ratio and Proportion **279**

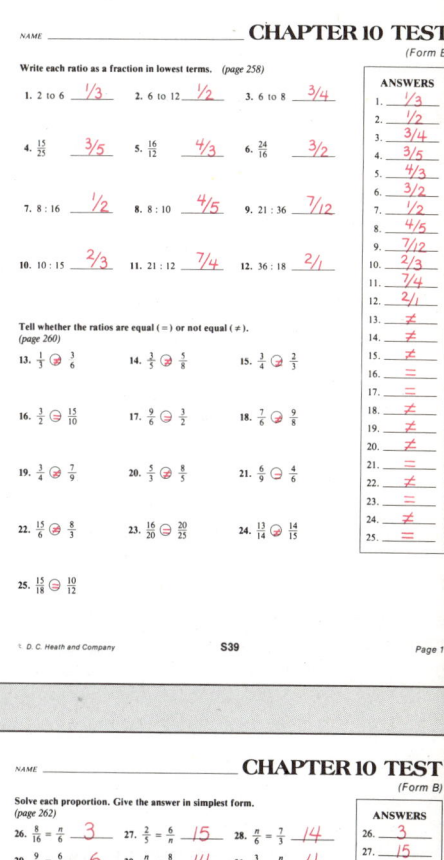

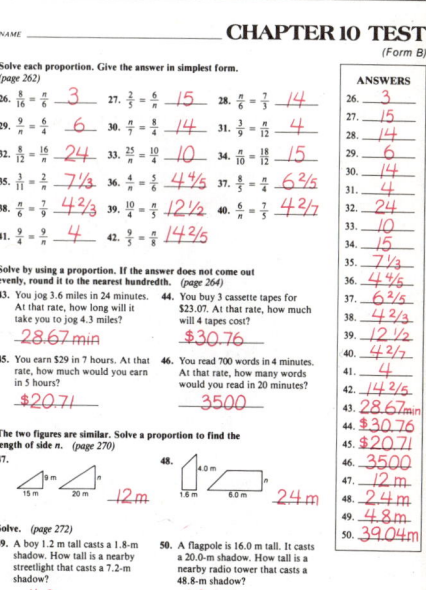

Copymasters S39 and S40
or Duplicating Masters S39 and S40

Cumulative Test
(Chapters 1–10)

Use Copymaster S109 to provide the students with an answer sheet in standardized test format.

Answers for Cumulative Test, Chapters 1–10

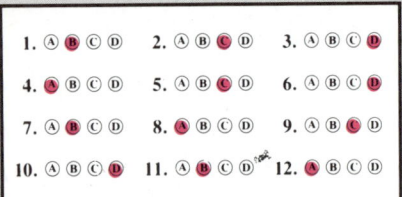

The table below correlates test items with student text pages.

Test Item	Page Taught	Skill Practice
1	68	p. 267, exercises 1–12
2	104	p. 267, exercises 13–22
3	152	p. 267, exercises 23–32
4	166	p. 267, exercises 33–46
5	180	p. 267, exercises 47–56
6	188	p. 277, exercises 1–10
7	204	p. 277, exercises 11–18
8	214	p. 277, exercises 19–26
9	230	p. 277, exercises 27–32
10	248	p. 277, exercises 33–36
11	262	p. 277, exercises 37–41
12	264	

Cumulative TEST — Standardized Format

Choose the correct letter.

1. Multiply. 3.954×36.5
- A. 117.4790
- **B. 144.3210**
- C. 1174.790
- D. none of these

2. Divide. Round the quotient to the nearest tenth.
$$2.03 \overline{)3.959}$$
- A. 1.95
- B. 1.9
- **C. 2.0**
- D. none of these

3. $\frac{5}{8} < ?$
- A. $\frac{1}{2}$
- B. $\frac{4}{9}$
- C. $\frac{3}{5}$
- **D. none of these**

4. Change to a decimal rounded to the nearest hundredth.
$$\frac{2}{3} = ?$$
- **A. 0.67**
- B. 0.66
- C. 0.667
- D. none of these

5. Give the sum. $\frac{5}{6} + \frac{3}{4}$
- A. $\frac{4}{5}$
- B. $\frac{1}{3}$
- **C. $1\frac{7}{12}$**
- D. none of these

6. Give the difference. $\frac{7}{8} - \frac{1}{3}$
- A. $\frac{7}{12}$
- B. $1\frac{1}{5}$
- C. $\frac{1}{4}$
- **D. none of these**

7. Complete. $\frac{3}{4}$ of $24 = ?$
- A. 32
- **B. 18**
- C. 16
- D. none of these

8. Give the quotient. $2\frac{1}{3} \div 1\frac{1}{2}$
- **A. $1\frac{5}{9}$**
- B. $\frac{2}{7}$
- C. $3\frac{1}{2}$
- D. none of these

9. Complete. 1.235 L = ? mL
- A. 123.5
- B. 0.001235
- **C. 1235**
- D. none of these

10. Add. 5 lb 9 oz $+ 2$ lb 11 oz
- A. 9 lb
- B. 8 lb 8 oz
- C. 7 lb 4 oz
- **D. none of these**

11. Solve. $\frac{9}{n} = \frac{2}{3}$
- A. 9
- **B. $13\frac{1}{2}$**
- C. $1\frac{1}{2}$
- D. none of these

12. You work 9 hours and earn $35. At that rate, how much would you earn in 13 hours? Round the answer to the nearest cent.
- **A. $50.56**
- B. $3.34
- C. $24.23
- D. none of these

Percent

Chapter 11
Percent

Resources

- **Class Starter Quizzes 111-120** *(Copymasters S412-S414)*
- **Visual Aids 50-52** *(Copymasters or Transparencies S150-S152)*
- **Manipulatives**
  Manipulative Activities 19-21 *(Copymasters S520-S521)*
  Powers-of-ten tiles, 10-by-10 grid *(Copymasters or Transparencies S529, S535)*
- **Worksheets 121-131** *(Copymasters, Duplicating Masters, or Workbook pages S289-S299)*
- **Challenge Problems** for pages 289, 291, 293, 295, 297, 301, 303 *(Copymasters S456-S457)*
- **Projects** for pages 285, 287, 299 *(Copymaster S482)*
- **Mental Math Extensions** for Skills 46-52 *(Copymasters S507-S510)*
- **Tests** *(Copymasters or Duplicating Masters S41-S44)*

Changing a percent to a fraction

In a survey of high school students, each person was asked to name his or her favorite album. Each album was placed in one of five categories. The graph shows the results of the survey.

Notice that 50% (50 percent) of those surveyed chose a rock album. This means that 50 out of 100, or $\frac{50}{100}$, chose a rock album.

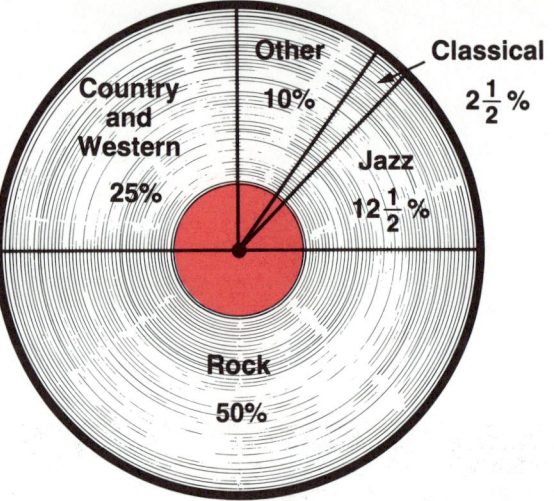

FAVORITE ALBUM SURVEY

1. What two categories of albums were most popular?
2. Were classical music albums more popular than country and western albums?
3. What percent of those surveyed chose a country and western album?
4. What percent of those surveyed chose a jazz album?

Here's how to change a percent to a fraction.

To change a percent to a fraction, first write the percent as a fraction with a denominator of 100. Then write the fraction in simplest form.

Country and Western

$$25\% = \frac{25}{100}$$
$$= \frac{1}{4}$$

Jazz

$$12\tfrac{1}{2}\% = \frac{12\tfrac{1}{2}}{100}$$
$$= 12\tfrac{1}{2} \div 100$$
$$= \frac{25}{2} \div 100$$
$$= \frac{25}{2} \times \frac{1}{100}$$
$$= \frac{25}{200}$$
$$= \frac{1}{8}$$

Divide the numerator by the denominator.

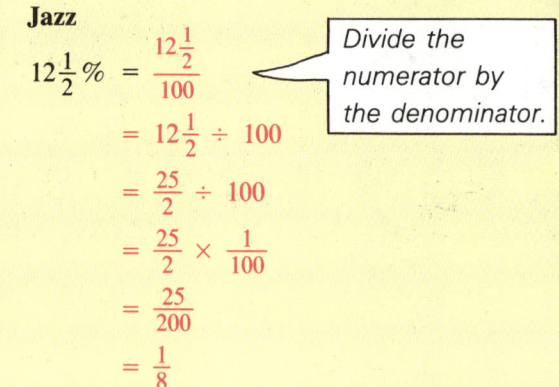

5. Look at the *Here's how*. What fraction of those surveyed chose a country and western album?
6. What fraction of those surveyed chose a jazz album?

EXERCISES

Change to a fraction, whole number, or mixed number.
Give the answer in simplest form.

Here are scrambled answers for the next row of exercises: $\frac{1}{4}$ $\frac{7}{10}$ $\frac{1}{2}$ $\frac{18}{25}$ 1 $1\frac{1}{4}$

7. 50% $\frac{1}{2}$
8. 125% $1\frac{1}{4}$
9. 25% $\frac{1}{4}$
10. 72% $\frac{18}{25}$
11. 100% 1
12. 70% $\frac{7}{10}$
13. 120% $1\frac{1}{5}$
14. 85% $\frac{17}{20}$
15. 96% $\frac{24}{25}$
16. 10% $\frac{1}{10}$
17. 44% $\frac{11}{25}$
18. 45% $\frac{9}{20}$
19. 32% $\frac{8}{25}$
20. 15% $\frac{3}{20}$
21. 75% $\frac{3}{4}$
22. 220% $2\frac{1}{5}$
23. 300% 3
24. 175% $1\frac{3}{4}$
25. 66% $\frac{33}{50}$
26. 48% $\frac{12}{25}$
27. 210% $2\frac{1}{10}$
28. 20% $\frac{1}{5}$
29. 60% $\frac{3}{5}$
30. 200% 2
31. 225% $2\frac{1}{4}$
32. 30% $\frac{3}{10}$
33. 150% $1\frac{1}{2}$
34. 400% 4
35. 74% $\frac{37}{50}$
36. 250% $2\frac{1}{2}$
37. 375% $3\frac{3}{4}$
38. 16% $\frac{4}{25}$
39. 110% $1\frac{1}{10}$
40. 35% $\frac{7}{20}$
41. 275% $2\frac{3}{4}$
42. 160% $1\frac{3}{5}$
43. 40% $\frac{2}{5}$
44. 325% $3\frac{1}{4}$
45. 90% $\frac{9}{10}$
46. 350% $3\frac{1}{2}$
47. 80% $\frac{4}{5}$
48. 5% $\frac{1}{20}$
49. $33\frac{1}{3}$% $\frac{1}{3}$
50. $106\frac{1}{4}$% $1\frac{1}{16}$
51. $137\frac{1}{2}$% $1\frac{3}{8}$
52. $18\frac{3}{4}$% $\frac{3}{16}$
53. $62\frac{1}{2}$% $\frac{5}{8}$
54. $8\frac{1}{3}$% $\frac{1}{12}$
55. $66\frac{2}{3}$% $\frac{2}{3}$
56. $16\frac{2}{3}$% $\frac{1}{6}$
57. $81\frac{1}{4}$% $\frac{13}{16}$
58. $133\frac{1}{3}$% $1\frac{1}{3}$
59. $87\frac{1}{2}$% $\frac{7}{8}$
60. $162\frac{1}{2}$% $1\frac{5}{8}$
61. $166\frac{2}{3}$% $1\frac{2}{3}$
62. $212\frac{1}{2}$% $2\frac{1}{8}$
63. $233\frac{1}{3}$% $2\frac{1}{3}$
64. $206\frac{1}{4}$% $2\frac{1}{16}$
65. $187\frac{1}{2}$% $1\frac{7}{8}$
66. $116\frac{2}{3}$% $1\frac{1}{6}$

Solve. Use the graph on page 282.

67. What fraction of those surveyed chose a classical album? $\frac{1}{40}$
68. What fraction chose either a rock or a country and western album? $\frac{3}{4}$
69. What fraction of those surveyed chose either a jazz or a rock album? $\frac{5}{8}$
70. What fraction of those surveyed did not choose a jazz album? $\frac{7}{8}$

Name that year

71. The first phonograph was made by Thomas Alva Edison. You can find the year by pressing the calculator keys as shown below.

[1][4][4][÷][4][=][×][4][8][=][+][1][6][6][=][−][1][7][=] 1877

Extra Practice
Page 493   Skill 46

Practice Worksheet
Workbook S289, Copymaster S289, or Duplicating Master S289

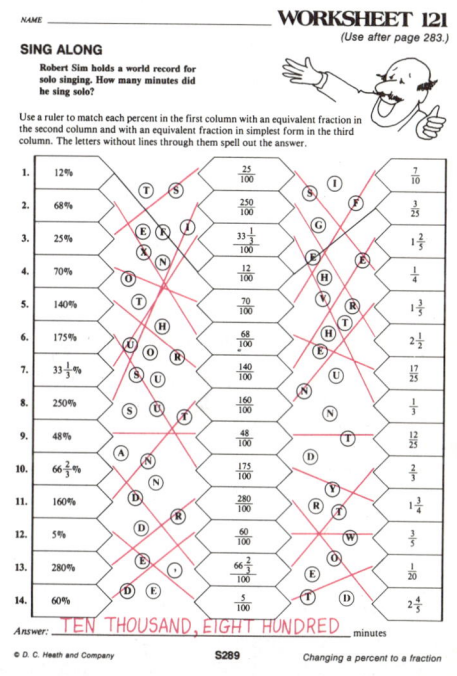

Group Project
Taking a survey

Have a group of students survey 100 students to determine which of these kinds of music they like best.

- Classical
- Country and Western
- Jazz
- Rock
- Other

After they have tallied the results, have them find what percent of the students chose each kind of music.

Changing a fraction to a percent

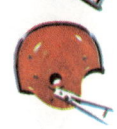

Look at the football helmets. Can you name the team for each helmet?

This question was part of a football survey conducted at Lancaster High School. The results are shown in the table.

NAME	NUMBER OF TEAMS NAMED CORRECTLY
Marty	12
Jill	16
David	18
Robert	14
Ann	20
Terry	17

1. How many teams did Marty name correctly? 12
2. How many teams are there in all? 24
3. What fraction of the teams did Marty name correctly? $\frac{1}{2}$
4. What fraction of the teams did Jill name correctly? $\frac{2}{3}$

Here's how to change a fraction to a percent.

Changing Marty's fraction to a percent

Method 1. Change to an equivalent fraction with a denominator of 100. Then write as a percent.

$$\frac{1}{2} = \frac{50}{100}$$
$$= 50\%$$

Changing Jill's fraction to a percent

Method 2. Since there is no whole number that you can multiply by 3 to get the denominator of 100, solve a proportion.

$$\frac{2}{3} = \frac{n}{100}$$
$$3n = 200$$
$$n = 66\frac{2}{3}$$

So $\frac{2}{3} = \frac{66\frac{2}{3}}{100}$
$= 66\frac{2}{3}\%$

5. Which method would you use to change $\frac{3}{4}$ to a percent? Method 1
6. Which method would you use to change $\frac{1}{6}$ to a percent? Method 2

EXERCISES

Change to a percent. *Hint: First change to an equivalent fraction with a denominator of 100.* Here are scrambled answers for the next row of exercises: 225% 40% 60% 90% 50% 125%

7. $\frac{2}{5}$ 40%
8. $\frac{9}{4}$ 225%
9. $\frac{9}{10}$ 90%
10. $\frac{3}{5}$ 60%
11. $\frac{5}{4}$ 125%
12. $\frac{1}{2}$ 50%

13. $\frac{1}{5}$ 20%
14. $\frac{1}{4}$ 25%
15. $\frac{4}{5}$ 80%
16. 1 100%
17. $\frac{1}{10}$ 10%
18. $\frac{3}{10}$ 30%

19. $\frac{3}{2}$ 150%
20. 2 200%
21. $\frac{7}{5}$ 140%
22. $\frac{3}{4}$ 75%
23. $\frac{5}{2}$ 250%
24. $\frac{7}{4}$ 175%

Change to a percent. *Hint: You may need to solve a proportion.* Here are scrambled answers for the next row of exercises: 175% $16\frac{2}{3}$% $33\frac{1}{3}$% 120% $133\frac{1}{3}$% $83\frac{1}{3}$%

25. $\frac{1}{3}$ $33\frac{1}{3}$%
26. $\frac{1}{6}$ $16\frac{2}{3}$%
27. $\frac{7}{4}$ 175%
28. $\frac{6}{5}$ 120%
29. $\frac{5}{6}$ $83\frac{1}{3}$%
30. $\frac{4}{3}$ $133\frac{1}{3}$%

31. $\frac{9}{16}$ $56\frac{1}{4}$%
32. $\frac{9}{25}$ 36%
33. $\frac{5}{9}$ $55\frac{5}{9}$%
34. $\frac{7}{2}$ 350%
35. $\frac{2}{3}$ $66\frac{2}{3}$%
36. $\frac{3}{8}$ $37\frac{1}{2}$%

37. $\frac{4}{9}$ $44\frac{4}{9}$%
38. $\frac{9}{20}$ 45%
39. $\frac{31}{50}$ 62%
40. $\frac{5}{3}$ $166\frac{2}{3}$%
41. $\frac{5}{12}$ $41\frac{2}{3}$%
42. $\frac{1}{12}$ $8\frac{1}{3}$%

43. $\frac{7}{25}$ 28%
44. $\frac{17}{10}$ 170%
45. $\frac{7}{8}$ $87\frac{1}{2}$%
46. 3 300%
47. $\frac{5}{8}$ $62\frac{1}{2}$%
48. $\frac{11}{25}$ 44%

Solve. Use the survey results on page 284.

49. What percent of the teams did David name correctly? 75%
50. What percent of the teams did Robert name correctly? $58\frac{1}{3}$%
51. What percent of the students surveyed knew more than 15 of the teams? $66\frac{2}{3}$%
52. What percent of the students surveyed knew 15 or fewer of the teams? (*Hint: Use your answer from exercise 51.*) $33\frac{1}{3}$%

How many fans? Solving a simpler problem

The number on the first ticket sold for a ball game was 10394. The number on the last ticket was 19017.

53. How many tickets were sold? *Hint: Look at the chart to do a simpler, but similar problem.*

First Ticket Sold	Last Ticket Sold	Number of Tickets Sold
#10394	#10395	2
#10394	#10396	3

8624

54. Three hundred of the tickets sold were not used. How many people were at the ball game? 8324

Percent **285**

Extra Practice
Page 494   Skill 47

Practice Worksheet
Workbook S290, Copymaster S290, or Duplicating Master S290

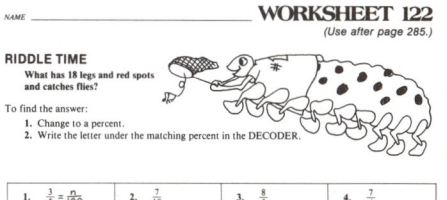

Project

Taking a survey

Survey students to see if they can name the team for each helmet shown on page 284. Find what percent of the teams each student named correctly.

Copymaster S482

Class Starter Quiz 112
on previous lesson

Change to a percent.

1. $\frac{1}{2}$ 50%
2. $\frac{3}{4}$ 75%
3. $\frac{1}{10}$ 10%
4. $\frac{2}{5}$ 40%
5. $\frac{3}{2}$ 150%
6. 2   200%
7. $\frac{5}{4}$ 125%
8. $\frac{1}{3}$ $33\frac{1}{3}$%
9. $\frac{3}{8}$ $37\frac{1}{2}$%

Copymaster S412

Lesson Objectives
To change a percent to a decimal
To change a decimal to a percent

Problem-Solving Skill
Reading a map

Starting the Lesson
Before the students open their books, have them list which states share part of their border with Texas. Then have them use the map on page 286 to check their answers.

Here's How Note
Use of Concrete Materials You may wish to use the powers-of-ten tiles and the 10-by-10 grid from copymasters or transparencies S529 and S535 to demonstrate changing a percent to a decimal and changing a decimal to a percent. See ■ **Manipulative Activity 21** on copymaster S521 in the Teacher's Resource Binder.

Exercise Note
The map at the top of page 286 is also on ■ **Visual Aid 51** (copymaster or transparency S151). You may wish to use the visual aid when discussing exercises 1, 2, and 51–56.

286

Percents and decimals

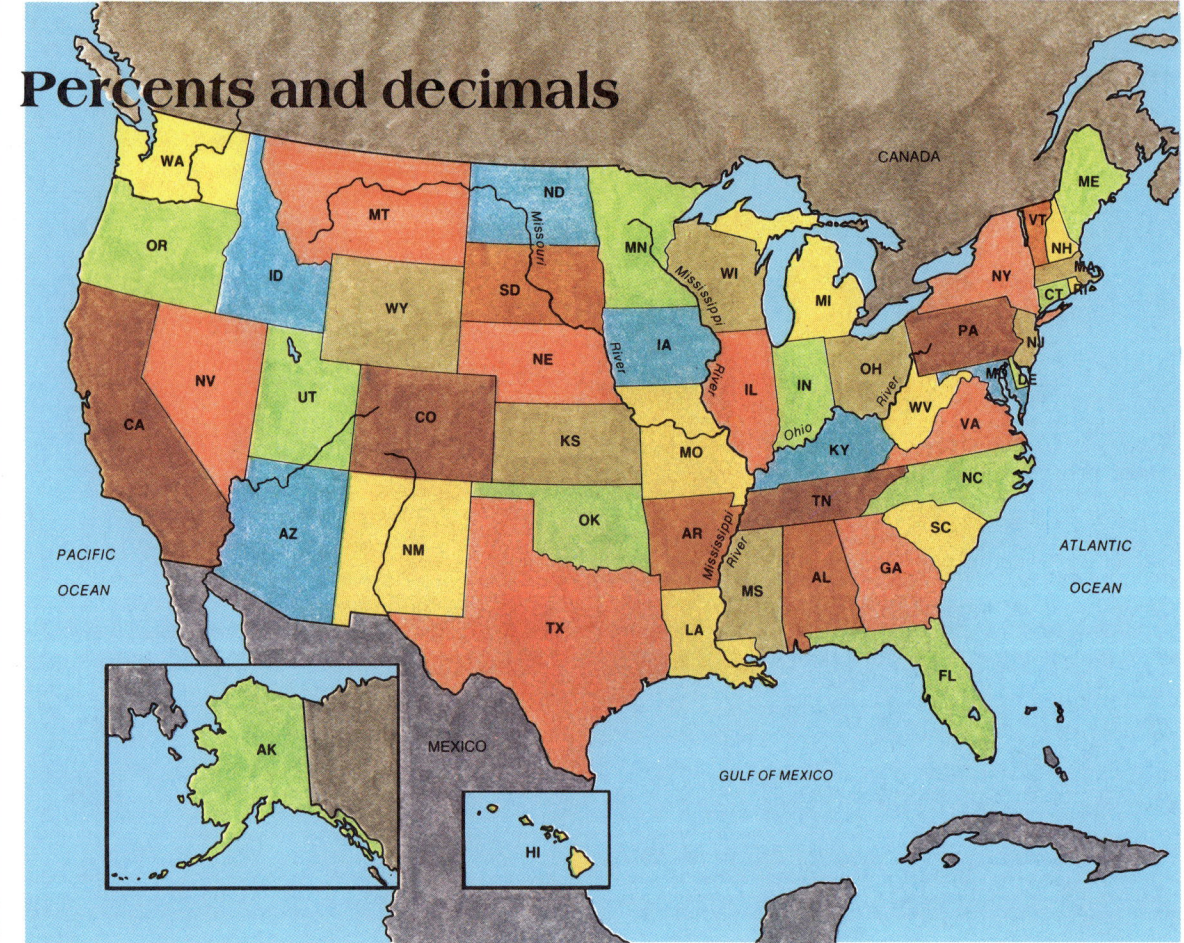

1. What percent of the states share part of their border with Texas? 8%
2. What percent of the states have names beginning with the letter M? 16%

Here's how to change a percent to a decimal.

$8\% = \frac{8}{100}$
$ = 0.08$

$16\% = \frac{16}{100}$
$ = 0.16$

$33\frac{1}{3}\% = \frac{33\frac{1}{3}}{100}$
$\phantom{33\frac{1}{3}\%} = 0.33\frac{1}{3}$

MENTAL MATH
To change a percent to a decimal, move the decimal point two places to the left and remove the percent sign.

$8\% = 0.08$
$16\% = 0.16$
$33\frac{1}{3}\% = 0.33\frac{1}{3}$
$29.5\% = 0.295$
$115\% = 1.15$

286 Chapter 11

Here's how to change a decimal to a percent.

$0.36 = \frac{36}{100} = 36\%$ $0.8 = \frac{80}{100} = 80\%$ $0.66\frac{2}{3} = \frac{66\frac{2}{3}}{100} = 66\frac{2}{3}\%$

MENTAL MATH
To change a decimal to a percent, move the decimal point two places to the right and write the percent sign.

$0.36 = 36\%$ $0.02 = 2\%$
$0.8 = 80\%$ $1.375 = 137.5\%$
$0.66\frac{2}{3} = 66\frac{2}{3}\%$

EXERCISES

Change each percent to a decimal. Hint: Use MENTAL MATH.

3. 5% 0.05
4. 9% 0.09
5. 6% 0.06
6. 25% 0.25
7. 72% 0.72
8. 44% 0.44
9. 125% 1.25
10. 150% 1.5
11. 238% 2.38
12. 282% 2.82
13. 360% 3.6
14. 400% 4
15. 6.25% 0.0625
16. 37.5% 0.375
17. 9.6% 0.096
18. 8.75% 0.0875
19. 62.5% 0.625
20. 4.8% 0.048
21. $33\frac{1}{3}\%$ $0.33\frac{1}{3}$
22. $16\frac{2}{3}\%$ $0.16\frac{2}{3}$
23. $37\frac{1}{2}\%$ $0.37\frac{1}{2}$
24. $166\frac{2}{3}\%$ $1.66\frac{2}{3}$
25. $162\frac{1}{2}\%$ $1.62\frac{1}{2}$
26. $187\frac{1}{2}\%$ $1.87\frac{1}{2}$

Change each decimal to a percent. Hint: Use MENTAL MATH.

27. 0.06 6%
28. 0.08 8%
29. 0.02 2%
30. 0.38 38%
31. 0.65 65%
32. 0.93 93%
33. 0.5 50%
34. 0.8 80%
35. 0.7 70%
36. 0.6 60%
37. 0.2 20%
38. 0.4 40%
39. 1.50 150%
40. 2.5 250%
41. 1.375 137.5%
42. 2.875 287.5%
43. 0.002 0.2%
44. 0.085 8.5%
45. $0.12\frac{1}{2}$ $12\frac{1}{2}\%$
46. $0.33\frac{1}{3}$ $33\frac{1}{3}\%$
47. $0.66\frac{2}{3}$ $66\frac{2}{3}\%$
48. $1.37\frac{1}{2}$ $137\frac{1}{2}\%$
49. $2.16\frac{2}{3}$ $216\frac{2}{3}\%$
50. $2.83\frac{1}{3}$ $283\frac{1}{3}\%$

Solve. Use the map on page 286.

51. What percent of the states are east of the Mississippi River? 52%
52. What percent of the states have names beginning with the letter N? 16%
53. What fraction of the states share part of their border with Mexico? $\frac{2}{25}$
54. What fraction of the states are completely surrounded by water? $\frac{1}{50}$

Map madness! Reading a map

55. Which 4 states share a common corner? Utah, Colorado, Arizona, and New Mexico
56. Name the two states each of which borders on eight other states. Tennessee

Percent **287**

Extra Practice
Pages 494 and 495   Skills 48 and 49

Practice Worksheet
Workbook S291, Copymaster S291, or Duplicating Master S291

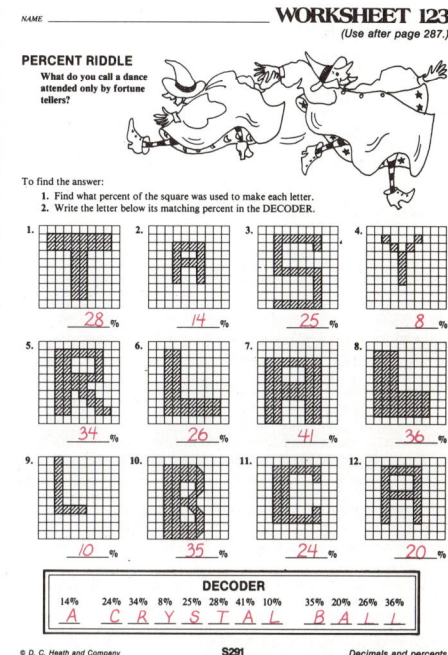

Project

Taking a survey
Survey the members of your family, including yourself. Show them the map on page 286, and find the states visited by at least one family member. Then compute the percent of the states that at least one family member has visited.

Copymaster S482

Class Starter Quiz 113
on previous lesson

Change to a decimal.
1. 15% 0.15   2. 4% 0.04
3. 49% 0.49   4. 175% 1.75
5. 400% 4   6. 37.5% 0.375

Change to a percent.
7. 0.08 8%   8. 0.75 75%
9. 0.65 65%   10. 0.8 80%
11. 2.25 225% 12. $0.66\frac{2}{3}$ $66\frac{2}{3}$%

Copymaster S413

Problem-Solving Skills
Selecting information from a sign
Choosing the correct operation
Solving a multistep problem

Skills Reviewed
Multiplying and dividing decimals
Finding the mean
Adding and subtracting fractions
Finding the prime factorization of a number
Finding the GCF and LCM of pairs of numbers
Writing numbers with exponents as standard numerals

Starting the Lesson
Problem Solving Use the Trail Rules and Information sign and ask questions like these:
- Which trails take less than three hours to hike? (Deer and Pine Ridge)
- If it is 3:30 P.M., is there time to hike Pine Ridge Trail? (Yes)
- Which trail is the hardest trail to hike? (Clear Falls)

Cumulative Skill Practice Challenge the students to an estimation hunt by saying, "Find the four products in exercises 1–12 that are greater than 50." (Exercises 1, 6, 9, and 10) Then have the students find the four quotients in exercises 13–22 that are greater than 100. (Exercises 15, 16, 17, and 19)

288

Problem solving
Reading a chart
YOU'RE THE PARK RANGER

TRAIL RULES AND INFORMATION
All children under 12 years of age must be accompanied by an adult.
All litter is to be carried out by hikers.
All trails are closed at 6:00 P.M.

TRAIL NAME	DISTANCE (miles)	HIKING TIME (hours)
Deer	5.8	$2\frac{1}{2}$
Clear Falls	4.9	$3\frac{3}{4}$
Pine Ridge	5.6	$2\frac{1}{4}$

Use the trail information to answer each hiker's question.

$4\frac{3}{4}$ hours

1. What is the total distance of the 2 shorter trails? 10.5 miles

2. How long would it take us to hike both Deer Trail and Pine Ridge Trail?

3. "How much longer would it take me to hike Clear Falls Trail than Pine Ridge Trail?" $1\frac{1}{2}$ hours

4. "We've hiked 2.8 miles of Pine Ridge Trail. How far is it to the end of the trail?" 2.8 miles

5. "I've hiked $\frac{1}{2}$ of Deer Trail. How far is it to the end of the trail?" 2.9 miles

6. "It is now 8:00 A.M. Would we have time to hike both Deer Trail and Clear Falls Trail by 12:30 P.M.?" No

Solve. Decide when a calculator would be useful.

7. During the first 15 days of July there were 403 hikers. At that rate, how many hikers would there be for the entire month of July? Round your answer to the nearest whole number. 833

8. Small postcards cost $.50 each, large postcards cost $.85 each, and a photo book of all 3 trails costs $5.97. A hiker buys 3 small postcards, 2 large postcards, and a photo book. How much change should he get from a $10 bill? $.83

288   Chapter 11

Cumulative Skill Practice

Multiply. *(page 68)*

1. 64 × 1.3 = 83.2
2. 5.7 × 4.7 = 26.79
3. 0.65 × 9.4 = 6.110
4. 0.86 × 48 = 41.28
5. 0.36 × 7.2 = 2.592
6. 8.3 × 21 = 174.3
7. 0.85 × 0.55 = 0.4675
8. 0.31 × 76 = 23.56
9. 9.3 × 9.3 = 86.49
10. 5.06 × 42 = 212.52
11. 3.74 × 0.52 = 1.9448
12. 2.23 × 0.14 = 0.3122

Divide. Round the quotient to the nearest tenth. *(page 104)*

13. 7.2 ÷ 0.4 = 18
14. 2.81 ÷ 0.3 = 9.4
15. 6.2 ÷ 0.06 = 103.3
16. 6.5 ÷ 0.03 = 216.7
17. 20 ÷ 0.09 = 222.2
18. 5.7 ÷ 2.3 = 2.5
19. 32 ÷ 0.12 = 266.7
20. 5.2 ÷ 3.6 = 1.4
21. 6.01 ÷ 3.9 = 1.5
22. 8.9 ÷ 0.56 = 15.9

Give the mean. Round the answer to the nearest tenth. *(page 124)*

23. 46, 49, 51   48.7
24. 681, 694, 700   691.7
25. 89.6, 90.4, 100.7   93.6
26. 78, 65, 93, 88   81.0
27. 345, 361, 402, 390   374.5
28. 9.82, 7.55, 6.09, 10.32   8.4

Give the sum in simplest form. *(page 180)*

29. $\frac{1}{2} + \frac{1}{4}$ $\frac{3}{4}$
30. $\frac{2}{3} + \frac{5}{6}$ $1\frac{1}{2}$
31. $\frac{1}{4} + \frac{5}{8}$ $\frac{7}{8}$
32. $3 + \frac{2}{3}$ $3\frac{2}{3}$
33. $\frac{4}{5} + \frac{3}{10}$ $1\frac{1}{10}$
34. $\frac{1}{4} + \frac{5}{12}$ $\frac{2}{3}$
35. $\frac{3}{5} + 2$ $2\frac{3}{5}$
36. $\frac{5}{9} + \frac{2}{3}$ $1\frac{2}{9}$
37. $\frac{1}{3} + \frac{1}{2}$ $\frac{5}{6}$
38. $\frac{2}{3} + \frac{4}{5}$ $1\frac{7}{15}$

Give the difference in simplest form. *(page 188)*

39. $\frac{3}{4} - \frac{1}{4}$ $\frac{1}{2}$
40. $\frac{7}{8} - \frac{3}{4}$ $\frac{1}{8}$
41. $1 - \frac{2}{3}$ $\frac{1}{3}$
42. $\frac{1}{2} - \frac{3}{8}$ $\frac{1}{8}$
43. $\frac{3}{10} - \frac{1}{5}$ $\frac{1}{10}$
44. $\frac{2}{3} - \frac{1}{4}$ $\frac{5}{12}$
45. $\frac{3}{4} - \frac{3}{8}$ $\frac{3}{8}$
46. $\frac{1}{2} - \frac{1}{3}$ $\frac{1}{6}$
47. $\frac{5}{6} - \frac{2}{3}$ $\frac{1}{6}$
48. $\frac{3}{4} - \frac{2}{3}$ $\frac{1}{12}$

MIXED PRACTICE
Complete.

49. The prime factorization of 36 is __?__ 2 × 2 × 3 × 3 or $2^2 \times 3^2$
50. The greatest common factor of 21 and 30 is __?__ 3
51. The least common multiple of 9 and 15 is __?__ 45
52. The standard numeral for 2^5 is __?__ 32
53. The standard numeral for 10^4 is __?__ 10,000

Percent   **289**

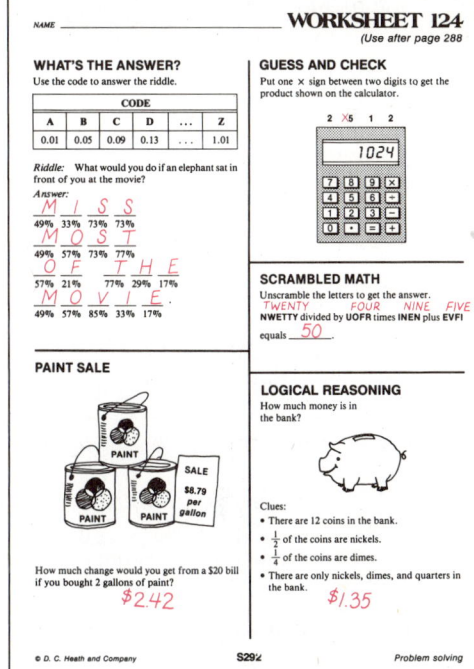

Problem-Solving Worksheet
Workbook S292, Copymaster S292, or Duplicating Master S292

Challenge Problems

Look through exercises 1–12 on page 289 to find which exercise has a product of about:

1. 6.1   Exercise 3
2. 2.6   Exercise 5
3. 0.5   Exercise 7
4. 1.9   Exercise 11

Copymaster S456

Finding a percent of a number

Camping Equipment Sale!

20% off
Reg. $80
Sportsman's Boots

18% off
Reg. $22
Swiss Army Knife

25% off **Reg. $120**
Two-Person Tent

16% off
Reg. $63
Binoculars

50% off
Reg. $29.98
Trail Pack

10% off
Reg. $16.50
Candle Lantern

To find the sale price of an item, you can first compute the discount. The discount is the amount that is subtracted from the regular price.

1. What is the regular price of the tent? $120
2. The discount on the tent is what percent of the regular price? 25%
3. What is the regular price of the binoculars? $63
4. The discount on the binoculars is what percent of the regular price? 16%

Here's how *to find a percent of a number*.

Change the percent to a fraction or decimal and multiply.

Discount on Tent
25% of $120 = n

Change the percent to a fraction and multiply.

25% of $120 = $\frac{1}{4}$ × $120
= $30

Discount on Binoculars
16% of $63 = n

Change the percent to a decimal and multiply.

16% of $63 = 0.16 × $63
= $10.08

```
    63
  ×0.16
  ─────
   3 78
   6 3
  ─────
  10.08
```

5. Look at the *Here's how*. What is the discount on the price of the tent? $30
6. What is the sale price of the tent? $90
7. What is the sale price of the binoculars? $52.92

EXERCISES

Solve by changing the percent to a fraction and multiplying.
Here are scrambled answers for the next row of exercises: 15   18   10

8. 50% of 36 = n   18
9. 25% of 40 = n   10
10. 60% of 25 = n   15
11. 75% of 24 = n   18
12. 30% of 60 = n   18
13. 10% of 80 = n   8
14. 20% of 30 = n   6
15. 40% of 40 = n   16
16. 25% of 44 = n   11
17. 75% of 32 = n   24
18. 100% of 24 = n   24
19. 150% of 18 = n   27

Solve by changing the percent to a decimal and multiplying.
Here are scrambled answers for the next row of exercises: 4.20   4.68   17.94

20. 12% of 35 = n   4.20
21. 9% of 52 = n   4.68
22. 23% of 78 = n   17.94
23. 32% of 156 = n   49.92
24. 54% of 264 = n   142.56
25. 78% of 165 = n   128.7
26. 5.6% of 61 = n   3.416
27. 8.75% of 46 = n   4.025
28. 12.5% of 132 = n   16.5
29. 0.75% of 50 = n   0.375
30. 0.35% of 21.5 = n   0.07525
31. 14.8% of 36.7 = n   5.4316

Solve. Hint: First try to decide which method would be easier.

32. 50% of 18 = n   9
33. 14% of 32 = n   4.48
34. 25% of 48 = n   12
35. 25% of 73 = n   18.25
36. 7.5% of 56 = n   4.2
37. 80% of 20 = n   16
38. 10% of 125 = n   12.5
39. 16.5% of 80 = n   13.2
40. 20% of 50 = n   10

Solve. Look at the items on page 290.

41. a. What is the discount on the price of the boots? $16
    b. What is the sale price of the boots? $64
42. a. What is the discount on the price of the knife? $3.96
    b. What is the sale price of the knife? $18.04
43. What is the sale price of the lantern? $14.85
44. What is the sale price of the small pack? $14.99

Be a super shopper

45. On which camping item can you save the largest amount of money? Two Person Tent

46. Your rich uncle gave you a $100 bill. You bought two of the camping items on sale. You got $21.15 in change. What did you buy? Boots and the lantern

Extra Practice
Page 495   Skill 50

Practice Worksheet
Workbook S293, Copymaster S293, or Duplicating Master S293

WORKSHEET 125 (Use after page 291.)

CAMPING OUT
Graeme Hurry of England holds a world record for camping out. How many consecutive nights did he sleep out?

To find the answer:
1. Solve each problem.
2. Write the letter under its matching number in the DECODER.

1. 50% of 44	2. 75% of 64	3. 80% of 25	4. 25% of 40
22 S	48 R	20 I	10 E
5. 100% of 45	6. 150% of 24	7. 5% of 20	8. 10% of 80
45 D	36 O	1 N	8 T
9. 180% of 35	10. 175% of 20	11. 90% of 150	12. 200% of 7
63 A	35 U	135 N	14 R
13. 120% of 80	14. 60% of 90	15. 40% of 40	16. 250% of 10
96 Y	54 F	16 H	25 X

DECODER
36 135 10   8 16 36 35 22 63 1 45
O N E   T H O U S A N D
54 36 35 48   16 35 135 45 14 10 45
F O U R   H U N D R E D
22 20 25 8 96
S I X T Y

Challenge Problems

Arrange the digits in the boxes to get the answer.

1. [1][2]% of [4] = 0.48

2. [8][3]% of [5] = 4.15

3. [2][9]% of [6] = 1.74

4. [1]% of [7][9] = 0.79

Copymaster S456

Class Starter Quiz 115
on previous lesson

Solve.

1. 75% of 40 = n   30
2. 50% of 84 = n   42
3. 250% of 18 = n   45
4. 64% of 260 = n   166.4
5. 16.5% of 60 = n   9.9

Copymaster S413

Lesson Objective
To estimate a percent of a number

Problem-Solving Skill
Estimating quantity

Starting the Lesson
Provide a 30-second warm-up of fraction–percent equivalents. Have the students give the fraction equivalent for each of these percents.

25% ($\frac{1}{4}$)   20% ($\frac{1}{5}$)

10% ($\frac{1}{10}$) $33\frac{1}{3}$% ($\frac{1}{3}$)

$12\frac{1}{2}$% ($\frac{1}{8}$)   50% ($\frac{1}{2}$)

Then use exercises 1–4 and the example in the *Here's how* to introduce estimating a percent of a number.

Estimating a percent of a number

1. What is the regular price of the concert ticket? $11.50

2. 25% of the ticket price is the same as what fraction of the ticket price? $\frac{1}{4}$

Here's how *to estimate a percent of a number.*

To estimate 25% of $11.50, change 25% to $\frac{1}{4}$ and then find a number close to $11.50 that is easy to multiply.

25% of $11.50 = n

$\frac{1}{4}$ of $12 = $\frac{1}{4}$ × $12

12 is an easy number to take $\frac{1}{4}$ of.   = $3

These will help you estimate.

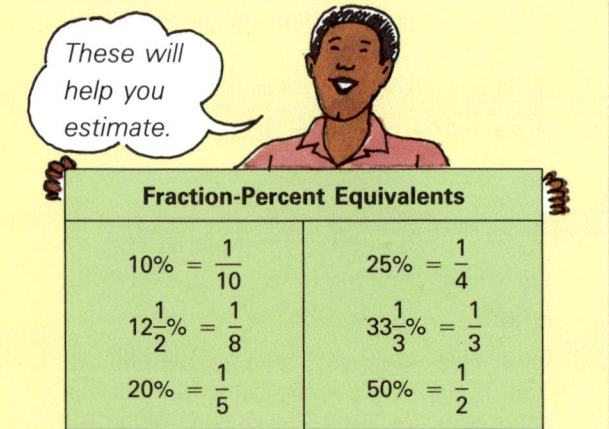

Fraction-Percent Equivalents

10% = $\frac{1}{10}$	25% = $\frac{1}{4}$
$12\frac{1}{2}$% = $\frac{1}{8}$	$33\frac{1}{3}$% = $\frac{1}{3}$
20% = $\frac{1}{5}$	50% = $\frac{1}{2}$

3. Look at the *Here's how*. To estimate 25% of $11.50, take $\frac{1}{4}$ of [?] $12

4. About how much will you save when you buy the ticket at the sale price? $3

292   Chapter 11

EXERCISES

Estimate. Change the percent to a fraction and then find a number that is easy to multiply.
Here are scrambled answers for the next row of exercises: $3, $5, $9

5. 50% of $17.95 $9
   Think: $\frac{1}{2}$ of $18

6. 20% of $24.75 $5
   Think: $\frac{1}{5}$ of $25

7. $33\frac{1}{3}$% of $8.95 $3
   Think: $\frac{1}{3}$ of $9

8. $12\frac{1}{2}$% of $25.50 $3
   Think: $\frac{1}{8}$ of $24

9. 25% of $399 $100
   Think: $\frac{1}{4}$ of $400

10. 10% of $59.99 $6
    Think: $\frac{1}{10}$ of $60

11. 50% of $39.99 $20
12. $33\frac{1}{3}$% of $8.97 $3
13. 25% of $23.77 $6
14. 20% of $15.39 $3
15. $12\frac{1}{2}$% of $16.19 $2
16. 10% of $29.88 $3
17. $33\frac{1}{3}$% of $31.89 $11
18. 25% of $398 $100
19. 50% of $99.99 $50
20. $12\frac{1}{2}$% of $79.29 $10
21. 20% of $497 $100
22. 10% of $89.19 $9
23. 50% of $999 $500
24. 25% of $37 $9
25. $33\frac{1}{3}$% of $35 $12

Choose greater than (>) *or* less than (<) for each ?.

26. 25% of 80 is 20,
    so 25% of 82 is [>] 20.

27. $33\frac{1}{3}$% of 90 is 30,
    so $33\frac{1}{3}$% of 89 is [<] 30.

28. 20% of 50 is 10,
    so 18% of 50 is [<] 10.

29. 50% of 400 is 200,
    so 53% of 400 is [>] 200.

Look it over   Estimating quantity

30. The picture shows some of the people at the concert. Estimate how many people are at the concert if the concert hall has 600 seats. About 300

Percent **293**

Challenge Problem

Rita sold 50% of her plants. Then she gave away 50% of the plants that were left. Now she has only 3 plants. How many plants did she have to begin with?

12

Copymaster S456

Class Starter Quiz 116
on previous lesson

Match. Find the percent of the number by estimating.

1. 28% of $199   c
2. 10% of $69   d
3. $12\frac{1}{2}$% of $23.60   a
4. 20% of $19   e
5. 50% of $49.90   b

a. $2.95
b. $24.95
c. $49.75
d. $6.90
e. $3.80

Copymaster S413

Lesson Objective
To find a percent of a number by solving a proportion

Problem-Solving Skills
Reading a chart
Making a list

Starting the Lesson
Take a Survey Write these foods on the chalkboard:

Chicken   Potatoes
Hamburgers   Steak
Pizza   Tacos

Before the students open their books, take an opinion poll. Ask them which of these foods they think was most often named as a favorite food when 600 teenagers were surveyed. Then have the students use the chart at the top of page 294 to get the answer. (Pizza)

294

More on finding a percent of a number

The "menu" shows the results of a survey of 600 teenagers. They were asked to list some favorite foods and some least-favorite foods.

TEENAGER MENU

FAVORITE FOODS
Pizza 47%
Steak .... $33\frac{1}{3}$%
Hamburgers ... 25%
Chicken .... $18\frac{1}{6}$%
Potatoes .... $12\frac{1}{2}$%
Tacos 10%

LEAST-FAVORITE FOODS
Spinach 20%
Liver .... $16\frac{2}{3}$%
Beans .... $8\frac{1}{3}$%
Broccoli 8%
Peas 6%
Fish 5%

1. What percent of those surveyed listed pizza as a favorite food? 47%
2. How many teenagers were surveyed? 600
3. What decimal would you multiply 600 by to compute the number of teenagers who listed pizza as a favorite food? 0.47
4. What fraction would you multiply 600 by to compute the number of teenagers who listed tacos as a favorite? $\frac{1}{10}$
5. What percent of those surveyed listed liver as a least-favorite food? $16\frac{2}{3}$%

Here's how *to find a percent of a number by solving a proportion.*

To find the number of those surveyed who listed liver as a least-favorite food, you would need to solve the equation $16\frac{2}{3}$% of 600 = n.

When you cannot easily change the percent to a fraction or decimal, find n by solving a proportion.

$16\frac{2}{3}$ out of 100 ... $\dfrac{16\frac{2}{3}}{100} = \dfrac{n}{600}$ ... is how many out of 600?

$$100n = 16\frac{2}{3} \times 600$$
$$= \frac{50}{3} \times 600$$
$$= 10{,}000$$
$$n = 100$$

6. Look at the *Here's how*. How many of those surveyed listed liver as a least-favorite food? 100

294 Chapter 11

EXERCISES

Solve by solving a proportion. Round answers to the nearest hundredth.
Here are scrambled answers for the next row of exercises: 3.75   1.71   3.12

| $6\frac{1}{3}$ out of 100 is equal to how many out of 27? | $8\frac{2}{3}$ out of 100 is equal to how many out of 36? | $12\frac{1}{2}$ out of 100 is equal to how many out of 30? |

7. $6\frac{1}{3}\%$ of $27 = n$ 1.71
8. $8\frac{2}{3}\%$ of $36 = n$ 3.12
9. $12\frac{1}{2}\%$ of $30 = n$ 3.75

10. $33\frac{1}{3}\%$ of $48 = n$ 16
11. $66\frac{2}{3}\%$ of $69 = n$ 46
12. $16\frac{2}{3}\%$ of $39 = n$ 6.5

13. $6\frac{3}{4}\%$ of $45 = n$ 3.04
14. $8\frac{7}{8}\%$ of $56 = n$ 4.97
15. $5\frac{1}{3}\%$ of $18 = n$ 0.96

Solve by multiplying by a fraction, by multiplying by a decimal, or by solving a proportion. *Hint: Look for the easiest method.*

16. 10% of $50 = n$ 5
17. 23% of $125 = n$ 28.75
18. $8\frac{1}{3}\%$ of $21 = n$ 1.75

19. 22.5% of $86 = n$ 19.35
20. 75% of $92 = n$ 69
21. $66\frac{2}{3}\%$ of $33 = n$ 22

22. 80% of $12 = n$ 9.6
23. 8.5% of $120 = n$ 10.2
24. 150% of $29 = n$ 43.5

25. 20% of $144 = n$ 28.8
26. 6.5% of $66 = n$ 4.29
27. 1.9% of $164 = n$ 3.116

28. $4\frac{2}{3}\%$ of $30 = n$ 1.4
29. 50% of $124 = n$ 62
30. 11% of $63 = n$ 6.93

Solve. Use the survey on page 294.

31. How many of those surveyed listed hamburgers as a favorite food? 150

32. How many of those surveyed listed pizza as a favorite food? 282

33. How many listed spinach as a least favorite food? 120

34. How many more listed spinach than broccoli as a least-favorite food? 72

35. How many more listed hamburgers than chicken as a favorite food? 41

36. How many did not list tacos as a favorite food? 540

37. How many did not list the most popular favorite food? 318

38. Which food was chosen as a favorite food by 75 of the teenagers surveyed?
Potatoes

Your order, please

39. A vendor sells only hot dogs and hamburgers. You cannot order more than 3 sandwiches. How many different orders can you place? *Hint: Make a list.* 9

Percent **295**

Class Starter Quiz 117
on previous lesson

Solve.
1. 20% of 60 = n   12
2. 6.5% of 34 = n   2.21
3. $33\frac{1}{3}$% of 66 = n   22
4. 50% of 142 = n   71
5. $8\frac{1}{3}$% of 24 = n   2

Copymaster S414

Lesson Objective

To use a proportion to find a number when a percent of the number is known

Problem-Solving Skills

Choosing the correct operation
Following directions

Starting the Lesson

Before discussing exercises 1–4, have the students study items 1–6 of the driver's test for 60 seconds. Tell them to close their books. Then read test items 1–6 and have the students answer true or false to each item.

Finding the number when a percent is known

RULES OF THE ROAD

Here are the first few items on a driver's test.

1. What percent of the total items did Ann get correct?   80%
2. How many of the total items did she get correct?   52
3. Were there more or fewer than 52 items on the whole test?   More

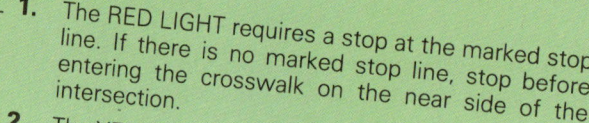

Name Ann Bender
Number correct 52
Percent correct 80%

Part A. True or false?

t 1. The RED LIGHT requires a stop at the marked stop line. If there is no marked stop line, stop before entering the crosswalk on the near side of the intersection.

X f 2. The YELLOW LIGHT warns that the signal is changing from green to red. When the red light appears, you are prohibited from entering the intersection.

t 3. The GREEN LIGHT means you may proceed if it is safe to do so. You must first, however, yield the right-of-way to pedestrians and vehicles that are still within the intersection or an adjacent crosswalk.

t 4. A YELLOW ARROW may appear after a green arrow. It means that the green arrow movement is ending.

t 5. A GREEN ARROW, pointing right or left, means you may make a turn in the direction of the arrow, if you are in the proper lane for such a turn, after yielding the right-of-way to vehicles and pedestrians within the intersection.

f 6. A GREEN ARROW pointing upward means you may turn left or right.

t 7. A GREEN ARROW...

Here's how *to find the number when a percent is known.*

To find the number of items on the test taken by Ann, you would need to solve the equation

$$80\% \text{ of } n = 52$$

You can find *n* by solving a proportion.

$$\boxed{\frac{part}{whole}} \quad \frac{80}{100} = \frac{52}{n} \quad \boxed{\frac{part}{whole}}$$

$$80n = 5200$$
$$n = 65$$

4. Look at the *Here's how*. How many items were on the test Ann took?   65

EXERCISES

Solve by solving a proportion.
Here are scrambled answers for the next row of exercises: 64   48   75

5. 25% of n = 12   48
6. 20% of n = 15   75
7. 75% of n = 48   64
8. 60% of n = 42   70
9. 40% of n = 56   140
10. 5% of n = 8   160
11. 50% of n = 17   34
12. 6% of n = 12   200
13. 80% of n = 20   25
14. 9% of n = 45   500
15. 10% of n = 18   180
16. 30% of n = 15   50

Solve. Round each answer to the nearest tenth.

17. 12.5% of n = 10.2   81.6
18. 6.4% of n = 1.3   20.3
19. 9.3% of n = 4.7   50.5
20. 5.8% of n = 6.4   110.3
21. 0.5% of n = 0.9   180
22. 0.8% of n = 1.3   162.5
23. 1.2% of n = 4.2   3.5
24. 4.8% of n = 1.2   25
25. 5.6% of n = 3.6   64.3
26. 125% of n = 2.3   1.8
27. 175% of n = 12.4   7.1
28. 150% of n = 10.5   7

Solve.

29. You took a test that had 72 questions. You got 18 questions wrong.
    a. What fraction of the questions did you get right? $\frac{3}{4}$
    b. What percent of the questions did you get right?   75%

30. You took a test that had 60 questions. You got 30 questions right.
    a. What fraction of the questions did you get right? $\frac{1}{2}$
    b. What percent of the questions did you get right?   50%

31. You took a test that had 120 questions. You got 80% of the questions right. How many questions did you get right? *Hint: 80% of 120 = n*   96

32. You scored $66\frac{2}{3}$% on a 96-question test. How many questions did you get right?   64

33. You got 60 questions on a test right. You scored 75%. How many questions were on the test? *Hint: 75% of n = 60.*   80

34. You scored 90% on a test. You got 135 of the questions right. How many questions were on the test?   150

What a jam! Following directions

35. The longest traffic jam ever reported was that of February 16, 1980. It stretched northward from Lyon, France, [?] miles toward Paris.   109.3

   To find [?], write a *9* in the ones place, a *3* in the tenths place, a *0* in the tens place, and a *1* in the hundreds place.

Extra Practice
Page 496   Skill 52

Practice Worksheet
Workbook S296, Copymaster S296, or Duplicating Master S296

WORKSHEET 128 (Use after page 297.)

BACKING UP

Charles Creighton and James Hargis hold a world record for driving in reverse. They drove their Ford Model A 1929 roadster in reverse from New York City to [?] without stopping the engine once.

To find the city:
1. Solve each equation by using a proportion.
2. Cross out each box below that contains an answer.
3. Read the name of the city using the letters in the remaining boxes.

1. 30% of n = 15
$\frac{30}{100} = \frac{15}{n}$
$30n = 1500$
$n = 50$

2. 6% of n = 9   150
3. 80% of n = 20   25
4. 25% of n = 44   176
5. 20% of n = 16   80
6. 10% of n = 13   130
7. 40% of n = 36   90
8. 50% of n = 55   110
9. 5% of n = 10   200
10. 75% of n = 225   300
11. 2% of n = 10   500
12. 150% of n = 15   10

Answer: LOS ANGELES

© D. C. Heath and Company   S296   Finding the number when the percent is known

Challenge Problems

Unscramble the letters of each of these math words.

1. CREPETN   Percent
2. UISDCOTN   Discount
3. ELAS CRPIE   Sale price
4. ITROPPORON   Proportion

Copymaster S457

Class Starter Quiz 118
on previous lesson

Solve by using a proportion.

1. 25% of n = 24   96
2. 40% of n = 30   75
3. 50% of n = 17   34
4. 125% of n = 40   32
5. 200% of n = 150   75

Copymaster S414

Lesson Objective
To solve percent problems by using proportions

Problem-Solving Skills
Reading a sale ad
Setting up a proportion to solve any kind of percent problem
Following directions

Starting the Lesson
Number Sense Write this on the chalkboard:

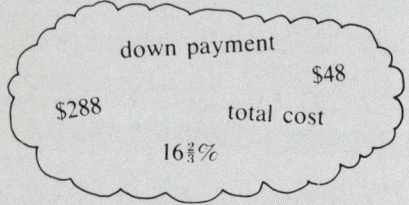

A _?_ down payment on a _?_ stereo system means the _?_ is _?_ of the _?_.

Before the students open their books, tell them to use the words, dollar amounts, and percent to fill in the blanks so the sentence makes sense. Then have them look at the first *Here's how* on page 298 to check on whether they filled in the blanks correctly.

More on percent

SYSTEM A
?% DOWN   **$48 DOWN**

TOTAL COST $288

SYSTEM B
$? DOWN   **$33\frac{1}{3}$% DOWN**

TOTAL COST $220

SYSTEM C
$37\frac{1}{2}$% DOWN   **$92 DOWN**

TOTAL COST $?

Here's how to find the percent of down payment for SYSTEM A.

$n\%$ of $288 = 48$

$\dfrac{part}{whole}$ $\dfrac{48}{288} = \dfrac{n}{100}$ $\dfrac{part}{whole}$

$288n = 4800$

$n = 16\frac{2}{3}$

The percent of down payment is $16\frac{2}{3}\%$

Here's how to find the total cost of SYSTEM C.

$37\frac{1}{2}\%$ of $n = 92$

$\dfrac{part}{whole}$ $\dfrac{37\frac{1}{2}}{100} = \dfrac{92}{n}$ $\dfrac{part}{whole}$

$37\frac{1}{2}n = 9200$

$n = 9200 \div 37\frac{1}{2}$

$n \approx 245.33$

The total cost is $245.33

Here's how to find the down payment for SYSTEM B.

$33\frac{1}{3}\%$ of $220 = n$

$\dfrac{part}{whole}$ $\dfrac{33\frac{1}{3}}{100} = \dfrac{n}{220}$ $\dfrac{part}{whole}$

$100n = 33\frac{1}{3} \times 220$

$= \dfrac{100}{3} \times 220$

$= \dfrac{22{,}000}{3}$

$n = \dfrac{22{,}000}{3} \div 100$

$n \approx 73.33$

The down payment is $73.33

1. The down payment for System A is what percent of the total cost? $16\frac{2}{3}\%$
2. How much is the down payment for System B? $73.33
3. What is the total cost of System C? $245.33

298   Chapter 11

EXERCISES

Solve. Here are scrambled answers for the next row of exercises: 5 125 75

4. $n\%$ of $4 = 3$ 75
5. $n\%$ of $4 = 5$ 125
6. $n\%$ of $20 = 1$ 5
7. $n\%$ of $8 = 4$ 50
8. $n\%$ of $16 = 4$ 25
9. $n\%$ of $1 = 5$ 500

Solve. Here are scrambled answers for the next row of exercises: 0.075 0.32 0.36

10. $2\frac{2}{3}\%$ of $12 = n$ 0.32
11. $1\frac{1}{4}\%$ of $6 = n$ 0.075
12. $4\frac{1}{2}\%$ of $8 = n$ 0.36
13. $16\frac{2}{3}\%$ of $72 = n$ 12
14. $33\frac{1}{3}\%$ of $42 = n$ 14
15. $8\frac{1}{3}\%$ of $96 = n$ 8

Solve. Here are scrambled answers for the next row of exercises: 34 45 48

16. 20% of $n = 9$ 45
17. 25% of $n = 12$ 48
18. 50% of $n = 17$ 34
19. 9% of $n = 36$ 400
20. 1% of $n = 2.56$ 256
21. 24% of $n = 12.48$ 52
22. $12\frac{1}{2}\%$ of $n = 6$ 48
23. $16\frac{2}{3}\%$ of $n = 24$ 144
24. $33\frac{1}{3}\%$ of $n = 17$ 51

Solve. Here are scrambled answers for the next row of exercises: 120 $31\frac{1}{4}$ 6

25. $12\frac{1}{2}\%$ of $n = 15$ 120
26. $16\frac{2}{3}\%$ of $36 = n$ 6
27. $n\%$ of $16 = 5$ $31\frac{1}{4}$
28. $33\frac{1}{3}\%$ of $18 = n$ 6
29. $n\%$ of $8 = 7$ $87\frac{1}{2}$
30. 22% of $n = 11$ 50

Solve.

31. A stereo system that costs $224 has a down payment of $56. What percent of the total cost is the down payment? 25%

32. A tape deck sells for $279. The down payment is $33\frac{1}{3}\%$ of the selling price. How much is the down payment? $93

33. A set of speakers can be purchased for a down payment of $24. The down payment is $16\frac{2}{3}\%$ of the price. What is the price of the speakers? $144

34. A car radio sells for $275. The down payment is 25% of the selling price. How much is the down payment? $68.75

Name that year

35. You can find the year the radio was invented by pressing these calculator keys. 1896

Percent **299**

Extra Practice
Page 497 Skill 53

Practice Worksheet
Workbook S297, Copymaster S297, or Duplicating Master S297

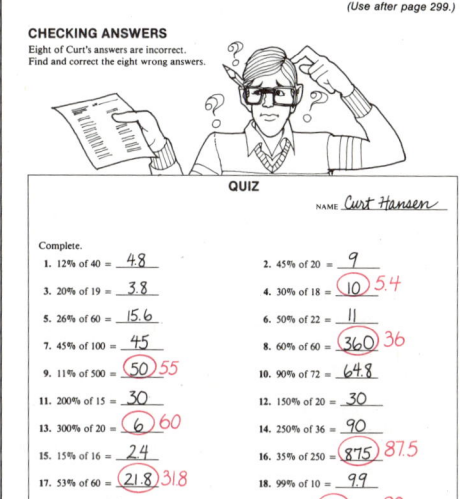

Project

Checking sale ads

Look for percents in ads. Do the computations that involve percent to see if you can find errors or misleading information.

Copymaster S482

299

PROBLEM SOLVING *using estimation*

You can solve some problems by making an estimate.

Problem A
About what fraction of her weekly income does Beth spend on clothes?

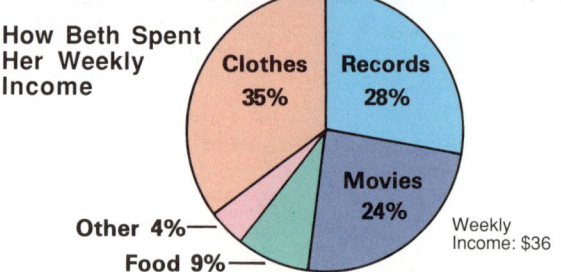

Here's how to solve problem A by estimating.

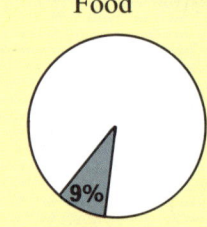 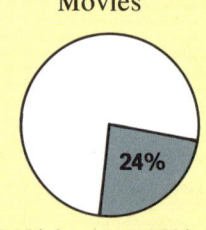 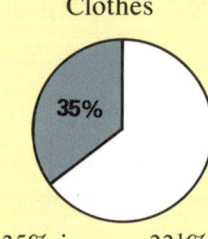 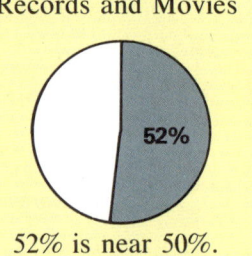

| Food | Movies | Clothes | Records and Movies |

9% is near 10%.　24% is near 25%.　35% is near $33\frac{1}{3}$%.　52% is near 50%.
Round 9% to $\frac{1}{10}$.　Round 24% to $\frac{1}{4}$.　Round 35% to $\frac{1}{3}$.　Round 52% to $\frac{1}{2}$.

Answer: Look at the examples above. Since 35% is near $33\frac{1}{3}$%, Beth spent about $\frac{1}{3}$ of her income on clothes.

Problem B
About how much money did Beth spend on records?

Here's how to solve problem B by estimating.

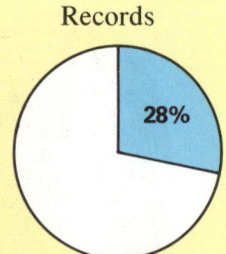

Records

Step 1. *Think:* 28% is about $\frac{1}{4}$.

Step 2. *Think:* $\frac{1}{4}$ of $36 = \frac{1}{4} \times \$36$
$= \$9$

Answer: Beth spent about $9 on records.

PROBLEMS

Estimate. Choose the most reasonable answer.
Use the circle graph on top of page 300.

1. About what fraction of her money did Beth spend on food?
   a. $\frac{1}{10}$ (circled) b. $\frac{1}{4}$ c. $\frac{1}{3}$

2. About what fraction of her money did Beth spend on movies?
   a. $\frac{1}{10}$ **b.** $\frac{1}{4}$ (circled) c. $\frac{1}{3}$

3. About how much money did Beth spend on clothes?
   a. $9 **b.** $12 (circled) c. $18

4. About how much money did Beth spend on movies?
   a. $9 (circled) b. $12 c. $18

Estimate.
Use the circle graph on top of page 300.

5. About what fraction of her money did Beth spend on records and movies? $\frac{1}{2}$

6. About what fraction of her money did Beth spend of food and movies? $\frac{1}{3}$

7. About how much money did Beth spend on records and movies? $18

8. About how much money did Beth spend on food and movies? $12

9. Did Beth spend more or less than half her money on food and clothes? less

10. Did Beth spend more or less than one third of her money on records and food? more

11. On which item did Beth spend about $4? food

12. On which three items did Beth spend about $30? clothes, records, and movies

Where does the money go?   Using estimation

Estimate. Use Tim's circle graph to match the item with the money spent on the item.

13. Records   d   a. $2.50
14. Food   c   b. $5.00
15. Clothes   b   c. $11.00
16. Movies   e   d. $14.00
17. Others   a   e. $17.50

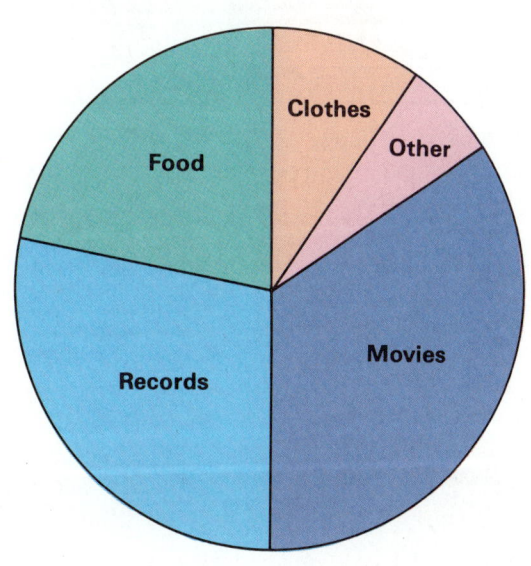

Percent **301**

Extra Practice
Page 497   Skill 54

Practice Worksheet
Workbook S298, Copymaster S298, or Duplicating Master S298

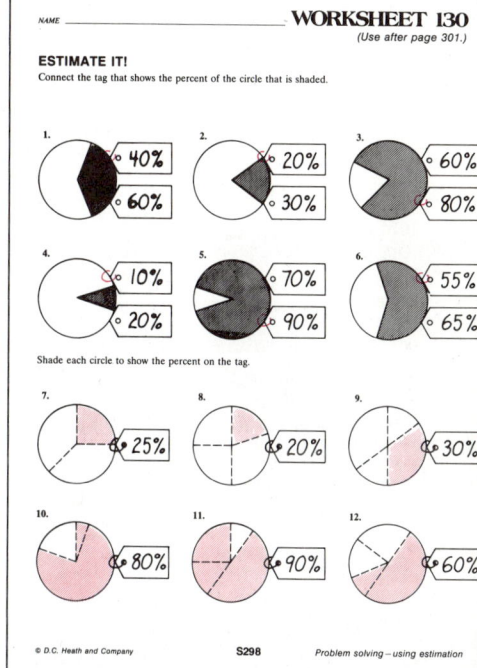

Challenge Problem

Write a question that fits the answer.

You scored 86% on a 50-question test.

How many questions did you miss?

Answer: 7 questions

Copymaster S457

Class Starter Quiz 120
on previous lesson

Estimate. Use the circle graph on page 300.

1. About how much money did Beth spend on movies?
   About $9
2. About how much money did she spend on clothes?
   About $12

Copymaster S414

Problem-Solving Skills

Finding information in a computer display
Choosing the correct operation

Skills Reviewed

Multiplying and dividing fractions
Making conversions between metric units of measure
Subtracting customary units
Solving proportions
Making conversions among fractions, mixed numbers, decimals, and percents
Simplifying fractions

Starting the Lesson

Problem Solving Have the students read the paragraphs at the top of page 302. Ask the students, "What did Eric Jeffrey do to get the information displayed on Screen B?" (Eric selected "2".) "Screen C?" (Eric entered $7.50.) "Screen D?" (Eric entered "4".) "Screen E?" (Eric pushed the ENTER key.) "Screen F?" (Eric entered 1.00.)

Cumulative Skill Practice Write these five answers on the chalkboard:

$\frac{3}{7}$ $\frac{1}{36}$ 238

3 yd 2 ft $\frac{2}{5}$

Challenge the students to an answer hunt by saying, "Look at exercises 1–54 on page 303. Find the five exercises that have these answers. You have five minutes to find as many of the exercises as you can." (Exercises 6, 24, 28, 32, and 46)

302

Problem solving

COMPUTERS AT HOME

Eric Jeffrey's home computer uses the telephone to communicate with a large **main frame** computer. This communication link allows him to use a wide variety of programs stored in the large computer.

Eric has just received a raise, and he uses a HOME MANAGEMENT program to answer some questions about his raise.

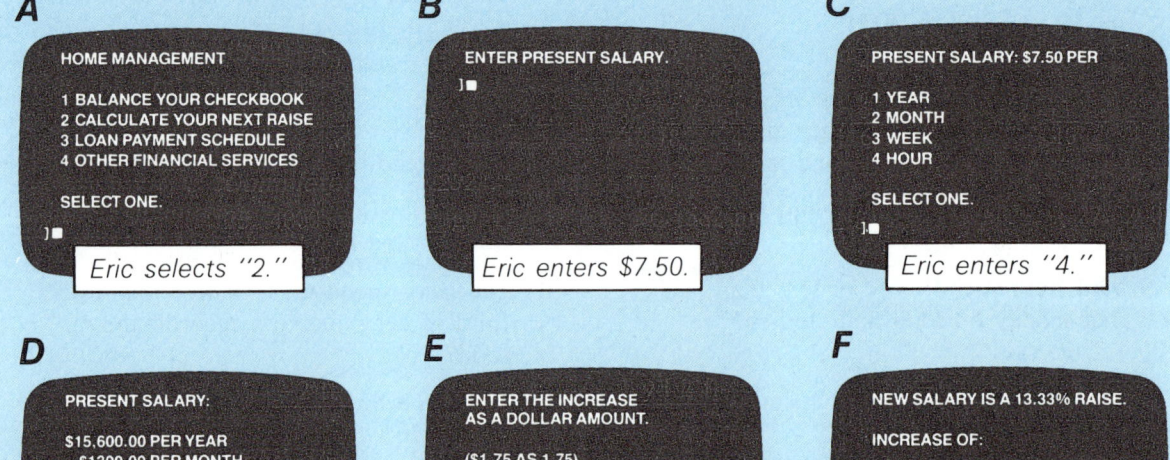

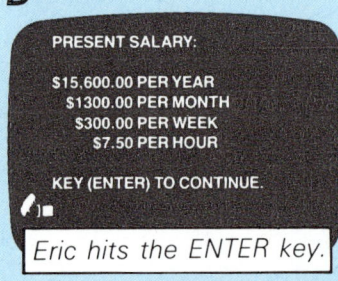

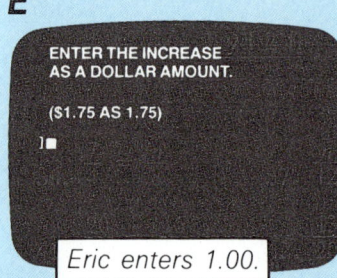

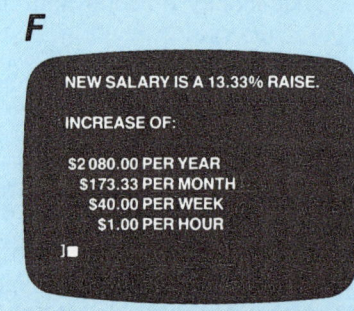

Use Screen D to answer these questions.

1. Is the weekly salary equal to 40 × $7.50? Yes
2. How did the computer calculate Eric's salary per year?
   Multiplied $7.50 × 40 × 52.
3. How did the computer calculate Eric's salary per month?
   Divided $15,600 by 12.

Use Screen F to answer these questions.

4. What is the percent of increase of Eric's raise? 13.33%
5. How much more money will Eric earn per week? $40.

Use Screens D and F to answer these questions.

6. $1.00 is 13.33% of what amount? $7.50
7. What will Eric's new salary per year be? $17,680

302 Chapter 11

Cumulative Skill Practice

Give the product in simplest form. *(page 202)*

1. $\frac{2}{3} \times \frac{4}{4}$ $\frac{2}{3}$
2. $\frac{1}{2} \times \frac{1}{4}$ $\frac{1}{8}$
3. $\frac{1}{6} \times \frac{1}{8}$ $\frac{1}{48}$
4. $\frac{3}{4} \times \frac{2}{3}$ $\frac{1}{2}$
5. $\frac{2}{5} \times \frac{5}{6}$ $\frac{1}{3}$
6. $\frac{3}{4} \times \frac{4}{7}$ $\frac{3}{7}$
7. $2 \times \frac{3}{5}$ $1\frac{1}{5}$
8. $\frac{3}{8} \times \frac{3}{8}$ $\frac{9}{64}$
9. $\frac{0}{2} \times \frac{1}{6}$ 0
10. $\frac{3}{4} \times \frac{2}{5}$ $\frac{3}{10}$
11. $4 \times \frac{2}{9}$ $\frac{8}{9}$
12. $5 \times \frac{1}{2}$ $2\frac{1}{2}$

Give the quotient in simplest form. *(page 212)*

13. $2 \div \frac{1}{2}$ 4
14. $\frac{1}{2} \div \frac{2}{3}$ $\frac{3}{4}$
15. $3 \div \frac{2}{3}$ $4\frac{1}{2}$
16. $\frac{3}{4} \div \frac{3}{8}$ 2
17. $\frac{2}{3} \div \frac{1}{2}$ $1\frac{1}{3}$
18. $4 \div \frac{1}{5}$ 20
19. $\frac{3}{5} \div \frac{3}{8}$ $1\frac{3}{5}$
20. $\frac{3}{10} \div \frac{1}{5}$ $1\frac{1}{2}$
21. $\frac{5}{8} \div \frac{1}{4}$ $2\frac{1}{2}$
22. $\frac{5}{6} \div 2$ $\frac{5}{12}$
23. $\frac{3}{10} \div \frac{3}{4}$ $\frac{2}{5}$
24. $\frac{1}{6} \div 6$ $\frac{1}{36}$

Complete. *(page 232)*

25. 4000 mg = ? g 4
26. 582 mg = ? g 0.582
27. 2000 g = ? kg 2
28. 0.238 kg = ? g 238
29. 8.3 g = ? mg 8300
30. 0.375 g = ? mg 375

Subtract. *(page 248)*

31. 5 ft 6 in. − 2 ft 9 in. **2 ft 9 in.**
32. 5 yd 1 ft − 1 yd 2 ft **3 yd 2 ft**
33. 8 ft 4 in. − 6 ft 9 in. **1 ft 7 in.**
34. 3 gal 2 qt − 1 gal 3 qt **1 gal 3 qt**

Solve. *(page 262)*

35. $\frac{3}{n} = \frac{4}{8}$ 6
36. $\frac{7}{4} = \frac{n}{8}$ 14
37. $\frac{n}{12} = \frac{15}{8}$ $22\frac{1}{2}$
38. $\frac{10}{9} = \frac{16}{n}$ $14\frac{2}{5}$
39. $\frac{8}{n} = \frac{5}{12}$ $19\frac{1}{5}$
40. $\frac{n}{18} = \frac{6}{4}$ 27
41. $\frac{20}{n} = \frac{16}{9}$ $11\frac{1}{4}$
42. $\frac{13}{20} = \frac{n}{18}$ $11\frac{7}{10}$
43. $\frac{6}{8} = \frac{9}{n}$ 12
44. $\frac{6}{n} = \frac{3}{7}$ 14

Change to a fraction in simplest form. *(page 282)*

45. 75% $\frac{3}{4}$
46. 40% $\frac{2}{5}$
47. 100% 1
48. 125% $1\frac{1}{4}$
49. 275% $2\frac{3}{4}$
50. $33\frac{1}{3}$% $\frac{1}{3}$
51. $37\frac{1}{2}$% $\frac{3}{8}$
52. $66\frac{2}{3}$% $\frac{2}{3}$
53. $6\frac{1}{4}$% $\frac{1}{16}$
54. $116\frac{2}{3}$% $1\frac{1}{6}$

MIXED PRACTICE

Complete.

55. $4\frac{2}{3}$ written as a fraction is ? $\frac{14}{3}$
56. $\frac{28}{5}$ written as a mixed number is ? $5\frac{3}{5}$
57. $\frac{20}{6}$ written in simplest form is ? $3\frac{1}{3}$
58. $\frac{1}{16}$ written as a decimal is ? 0.0625
59. 0.8 written as a fraction is ? $\frac{4}{5}$

Percent **303**

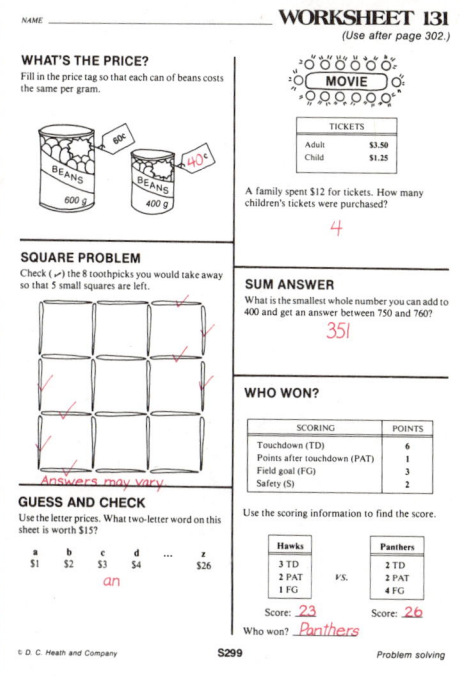

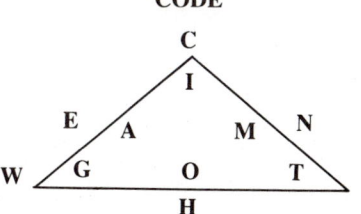

Riddle: Who earns a living without doing a day's work?

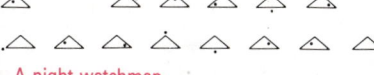

A night watchman

Copymaster S457

303

Chapter REVIEW

Here are scrambled answers for the review exercises:

7	75	25%	denominator	equivalent	multiply	proportion	simplest
12	100	34%	divide	left		percent	right

1. 25% means 25 out of [?]. To change a percent to a fraction, you can first write the percent as a fraction with a [?] of 100. Then you can write the fraction in [?] form. *(page 282)*

   25%

2. To change this fraction to a percent, you could first write the [?] fraction with a denominator of 100. Then you would write that fraction as a [?]. *(page 284)*

   $\frac{3}{4}$

3. To change $\frac{5}{6}$ to a percent, you could solve this [?]. *(page 284)*

   $\frac{5}{6} = \frac{n}{100}$

4. To change a percent to a decimal, you can move the decimal point two places to the [?] and remove the percent sign.
   To change a decimal to a percent, you can move the decimal point two places to the [?] and write a percent sign. *(pages 286, 287)*

5. To find a percent of a number, you can change the percent to a fraction or decimal and [?].
   25% of 48 is [?].
   14% of 50 is [?]. *(page 290)*

6. You can find $8\frac{1}{3}\%$ of 60 by solving this proportion. The next step in solving the proportion would be to [?] both sides of the equation by 100. *(page 294)*

   $8\frac{1}{3}\%$ of $60 = n$

   $\frac{8\frac{1}{3}}{100} = \frac{n}{60}$

   $100n = 8\frac{1}{3} \times 60$
   $= 500$

7. You can solve a proportion to find the number when a percent of it is known. If you solve the proportion, you will find that $n =$ [?] *(page 296)*

   24% of $n = 18$

   $\frac{24}{100} = \frac{18}{n}$

8. To estimate what fraction of Carl's total expenses were used to buy film you would round 23% to [?].

9. To estimate about how much money Carl spent on food you would multiply $48 by [?].

How Carl spent $48 on a weekend trip. *(page 300)*

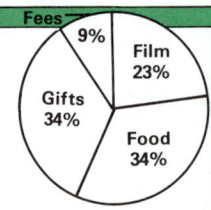

1. 100, denominator, simplest 2. equivalent, percent
3. proportion 4. left, right 5. multiply, 12, 7
6. divide 7. 75 8. 25% 9. 34%

Chapter TEST

Change to a fraction in simplest form. (page 282)

1. 20% $\frac{1}{5}$
2. 25% $\frac{1}{4}$
3. 80% $\frac{4}{5}$
4. 150% $1\frac{1}{2}$
5. 225% $2\frac{1}{4}$
6. $12\frac{1}{2}$% $\frac{1}{8}$

Change to a percent. (page 284)

7. $\frac{1}{2}$ 50%
8. $\frac{2}{5}$ 40%
9. $\frac{3}{4}$ 75%
10. $\frac{7}{5}$ 140%
11. $\frac{9}{4}$ 225%
12. $\frac{1}{3}$ $33\frac{1}{3}$%

Change each percent to a decimal. (page 286)

13. 12% 0.12
14. 9% 0.09
15. 160% 1.6
16. 15.3% 0.153
17. 6.85% 0.0685
18. $16\frac{2}{3}$% $0.16\frac{2}{3}$

Change each decimal to a percent. (page 287)

19. 0.05 5%
20. 0.09 9%
21. 1.375 137.5%
22. 2.326 232.6%
23. 0.004 0.4%
24. $0.12\frac{1}{2}$ $12\frac{1}{2}$%

Solve. (pages 290, 294)

25. 25% of 44 = n 11
26. 6.2% of 32 = n 1.984
27. $8\frac{1}{3}$% of 60 = n 5

Solve. (page 296)

28. 150% of n = 51 34
29. 12.5% of n = 3 24
30. $33\frac{1}{3}$% of n = 21 63

Solve. (page 298)

31. How much is the down payment? $120

$? DOWN
25% DOWN

TOTAL COST $480

32. What is the total cost? $560

$112 DOWN
20% DOWN

TOTAL COST $?

Solve. (page 300)

Sales Results

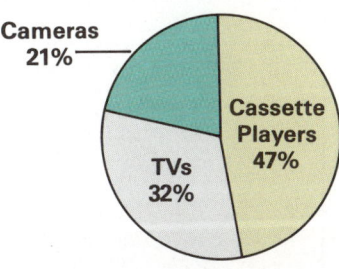

Cameras 21%
Cassette Players 47%
TVs 32%

Total Sales $6000

33. Cassette players were about what fraction of the total sales? $\frac{1}{2}$

34. Estimate the amount that came from TV sales. $2,000

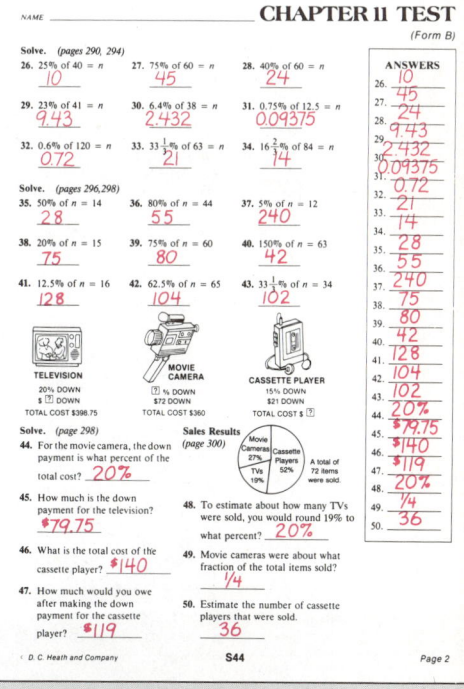

Copymasters S43 and S44
or Duplicating Masters S43 and S44

Percent **305**

Cumulative Test
(Chapters 1–11)

Use Copymaster S109 to provide the students with an answer sheet in standardized test format.

Answers for Cumulative Test, Chapters 1–11

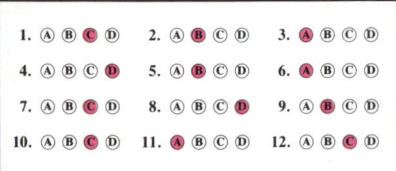

The table below correlates test items with student text pages.

Test Item	Page Taught	Skill Practice
1	68	p. 289, exercises 1–12
2	104	p. 289, exercises 13–22
3	124	p. 289, exercises 23–28
4	180	p. 289, exercises 29–38
5	188	p. 289, exercises 39–48
6	202	p. 303, exercises 1–12
7	212	p. 303, exercises 13–24
8	232	p. 303, exercises 25–30
9	248	p. 303, exercises 31–34
10	262	p. 303, exercises 35–44
11	282	p. 303, exercises 45–54
12	290	

Cumulative TEST — Standardized Format

Choose the correct letter.

1. Multiply.

$$4.239 \times 8.07$$

A. 342.0873
B. 36.8793
C. 34.20873 ✓
D. none of these

2. Divide. Round the quotient to the nearest tenth.

$$62.5 \overline{)191.27}$$

A. 3.06
B. 3.1 ✓
C. 3.6
D. none of these

3. The mean of 81.5, 83.4, 86.9, 87.4, and 90.3 is

A. 85.9 ✓
B. 86.9
C. 87.9
D. none of these

4. Give the sum.

$$\frac{5}{6} + \frac{5}{8}$$

A. $\frac{5}{7}$
B. $\frac{5}{24}$
C. $1\frac{1}{4}$
D. none of these ✓

5. Give the difference.

$$\frac{7}{8} - \frac{2}{3}$$

A. 1
B. $\frac{5}{24}$ ✓
C. $\frac{5}{11}$
D. none of these

6. Give the product.

$$\frac{5}{6} \times \frac{3}{8}$$

A. $\frac{5}{16}$ ✓
B. $2\frac{2}{9}$
C. $\frac{9}{20}$
D. none of these

7. Give the quotient.

$$\frac{7}{8} \div \frac{3}{4}$$

A. $\frac{21}{32}$
B. $\frac{6}{7}$
C. $1\frac{1}{6}$ ✓
D. none of these

8. Complete.

275 mg = __?__ g

A. 2.75
B. 275
C. 27.5
D. none of these ✓

9. Subtract.

$$\begin{array}{r} 9 \text{ gal } 2 \text{ qt} \\ -3 \text{ gal } 3 \text{ qt} \end{array}$$

A. 6 gal 1 qt
B. 5 gal 3 qt ✓
C. 5 gal 1 qt
D. none of these

10. Solve.

$$\frac{7}{n} = \frac{5}{9}$$

A. 3.89
B. 6.43
C. 12.6 ✓
D. none of these

11. Change to a fraction.

$$16\frac{2}{3}\% = ?$$

A. $\frac{1}{6}$ ✓
B. $\frac{1}{16}$
C. $\frac{1}{8}$
D. none of these

12. You took a 72-problem math test. You got 75% of the problems correct. How many was that?

A. 96
B. 36
C. 54 ✓
D. none of these

Consumer Mathematics

Chapter 12
Consumer Mathematics

Resources

- **Class Starter Quizzes 121-130** *(Copymasters S415-S417)*
- **Visual Aids 53-59** *(Copymasters or Transparencies S152-S158)*
- **Worksheets 132-142** *(Copymasters, Duplicating Masters, or Workbook pages S300-S310)*
- **Challenge Problems** for pages 325, 327, 329 *(Copymaster S458)*
- **Projects** for pages 309, 311, 315, 317, 319, 321 *(Copymasters S483-S484)*
- **Tests** *(Copymasters or Duplicating Masters S45-S48, S77-S80)*

Earning money and payroll deductions

Read the want ads to find the jobs.

Pinsetter mechanic

1. I could earn $37.20 a day if I worked as a ___?___.

Waiter/waitress

3. If I worked 10 hours as a ___?___, I could earn $38.

HELP WANTED

WAITER/WAITRESS
THE BURGER BIN
Mon. – Fri.
8-hour day $3.80/hour

PINSETTER MECHANIC
Maple Lanes Bowling Alley
6-hour day $6.20/hour

NEON SIGN REPAIRER
No experience needed.
Learn on the job.
20-hour week $100/week

CHECK-OUT CLERK
BERNIE'S FOOD
30-hour week $120/week
Apply in person.

Neon sign repairer

2. I could earn $5 an hour working as a ___?___.

Check-out clerk

4. In 4 weeks as a ___?___, I could make $480.

EXERCISES
Solve.

5. The Ace Trucking Company pays its drivers 20¢ a mile.
   a. How much would an Ace driver be paid for an 840-mile trip? $168
   b. How many miles would a driver have to drive to make $150? 750

6. The Tip Top Café pays its cooks $5.40 an hour.
   a. How much would a Tip Top cook be paid for a 7.5-hour day? $40.50
   b. How much would a cook be paid for a 37.5-hour week? $202.50

7. The Happy Day Card Company pays its salespeople a commission of 20% on all sales.
   a. How much would you earn for selling $300 worth of Happy Day cards? *Hint: What is 20% of $300?* $60
   b. To earn $100 commission, would you have to sell more or less than $400 worth of Happy Day cards? More

Cindy Davis is a manager at Showtime Videos. Each week she receives with her paycheck a **statement of withholdings and deductions.** This statement lists the amount that her employer deducts from her weekly paycheck.

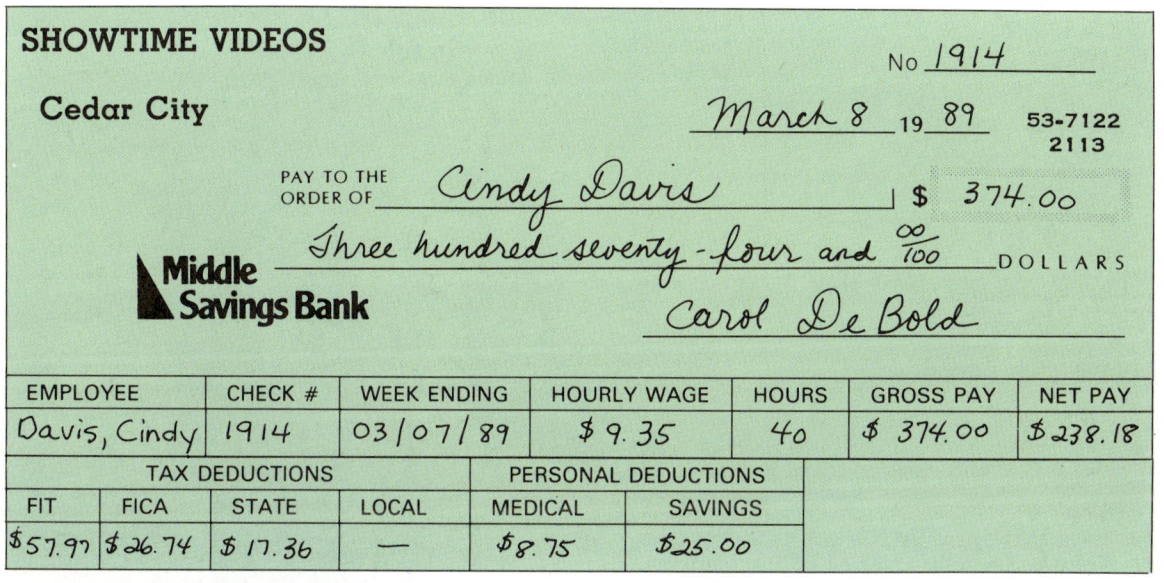

8. How much was Cindy paid per hour? $9.35

9. How many hours did she work? 40

10. To find Cindy's gross pay, $9.35 was multiplied by what number? 40

11. Social Security (FICA) withholding is for retirement income and disability income. How much was withheld for Social Security? $26.74

12. How much did her employer withhold for federal income tax (FIT)? State income tax? $57.97, $17.36

13. How much was subtracted from Cindy's gross pay for personal deductions? $33.75

14. How much was Cindy's net, or "take home," pay? $238.18

You're the boss

15. Complete this statement by computing Cory's gross and net pay.

EMPLOYEE	CHECK #	WEEK ENDING	HOURLY WAGE	HOURS	GROSS PAY	NET PAY
Glinn, Cory	1972	03/15/89	$3.80	29.5	?	?
TAX DEDUCTIONS			PERSONAL DEDUCTIONS		$112.10	$86.43
FIT	FICA	STATE	LOCAL	MEDICAL	UNION DUES	
$15.68	$7.42	$2.57	—	—	—	

Consumer Mathematics **309**

Practice Worksheet
Workbook S300, Copymaster S300, or Duplicating Master S300

Project

Using library resources

Use the *Occupational Outlook Handbook* to find the education requirements, working conditions, and beginning salary of a job that interests you.

Copymaster S483

Class Starter Quiz 121
on previous lesson

Complete.
1. Wage: $4.40 per hour
   Hours: 40
   Weekly pay: _?_ $176
2. Wage: $3.80 per hour
   Hours: 7 hours per day for a 5-day week
   Weekly pay: _?_ $133

Copymaster S415

Lesson Objective
To compute discounts and sale prices

Problem-Solving Skills
Reading a sale ad
Checking information in an ad

Starting the Lesson
Number Sense Write this on the chalkboard:

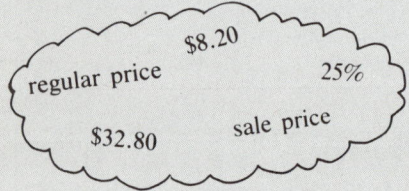

A _?_ pair of jeans on sale at _?_ off the regular price means the _?_ is _?_ less than the _?_.

Before the students open their books, tell them to use the words, dollar amounts, and percent to fill in the blanks so the sentence makes sense. Then have them do exercises 1–4 to check on whether they filled in the blanks correctly.

310

Buying on sale

1. What is the regular price of the jeans? $32.80
2. Will the sale price of the jeans be more or less than $32.80? Less

Here's how to find the sale price of a pair of jeans.

METHOD 1.

```
 $ 3 2.8 0   ← Regular price
 ×   0.2 5
 ─────────
   1 6 4 0 0
   6 5 6 0
 ─────────
 $8.2 0 0 0   ← Discount is $8.20.
```

```
 $32.80   ← Regular price
 − 8.20   ← Discount
 ──────
 $24.60   ← Sale price
```

METHOD 2.

```
 $ 3 2.8 0   ← Regular price
 ×   0.7 5
 ─────────
   1 6 4 0 0
   2 2 9 6 0
 ─────────
 $2 4.6 0 0 0   ← Sale price is $24.60
```

If the discount is 25%, the sale price is 75% of the regular price.

3. Look at the *Here's how*. How much is the savings on a pair of jeans during the sale? $8.20
4. What is the sale price of a pair of jeans? $24.60
5. What is the sale price of a pair of leather boots? $56.61
6. What is the sale price of a flannel shirt? $19.80

EXERCISES

Find the sale price of each item.

7. $170.55

CD PLAYER
Reg. **$189.50**   NOW **10% OFF**

8. $55.92

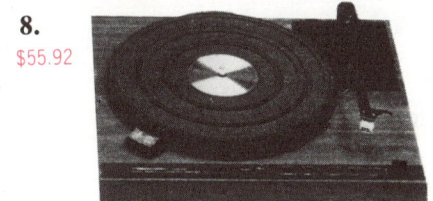

TURNTABLE
Reg. **$69.90**   NOW **20% OFF**

9.

SUPER HEADPHONES
Reg. **$40.40** $30.30

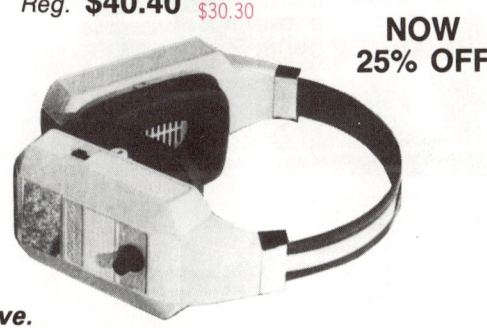

NOW **25% OFF**

10. **HIGH-STYLE STEREO RADIO AND CASSETTE PLAYER**
$62.93   Reg. **$89.90**

NOW **30% OFF**

Solve.

11. A calculator usually costs $9.90. Now it is marked 10% off. What is the sale price? $8.91

12. A television set usually costs $289. Now it is on sale at a 20% discount. What is the sale price? $231.20

13. A radio is on sale at a 25% discount. The regular price is $39.88. How much would you save when buying the radio on sale? $9.97

14. How much would you save by purchasing a stereo set at 30% off the regular price? The regular price is $186. $55.80

Check the ads

15. Two of these ads are incorrect. Find and correct the two wrong sale prices.

a. SAVE $1.99

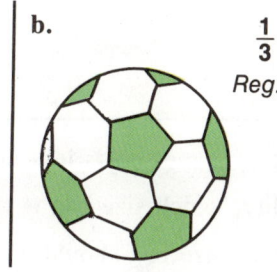

Reg. **$7.29**
SALE ~~$4.30~~
$5.30

b. ⅓ OFF
Reg. **$8.49**

NOW **$5.66**

c. 15% OFF

Reg. **$6.80**
SALE ~~$5.69~~
$5.78

Consumer Mathematics **311**

Practice Worksheet
Workbook S301, Copymaster S301, or Duplicating Master S301

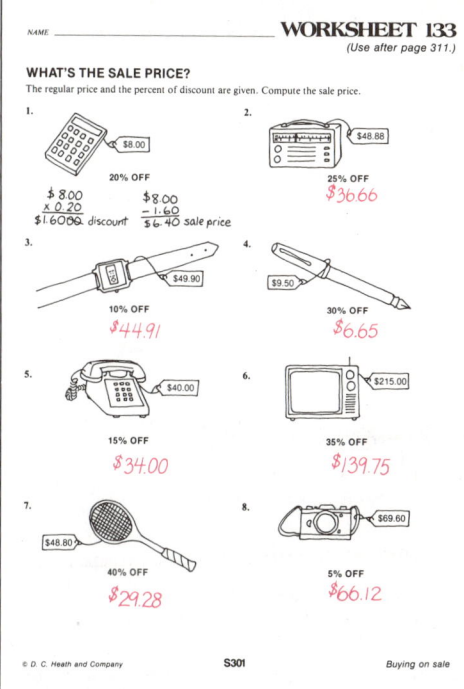

Project

Using a calculator

Look for ads that involve percent discounts. When the regular price and the sale price are given, use a calculator to check whether the correct percent is given.

Copymaster S483

311

Comparison buying

WHO'S RIGHT?

Which is the better buy, the 4.5-pound bag of apples or the 8-pound bag?

If the apples in each bag are the same quality and you could use either amount, you should compute the **unit price** (price per pound) to decide which is the better buy.

Here's how *to find the unit price.*

To compute the unit price of the apples, divide the price by the number of pounds. That will tell you the price per pound.

Small bag   4.5 pounds for $1.79   **Large bag**   8 pounds for $2.99

```
        $ 0.397                     $0.373
    4.5 ) $1.7900               8 ) $2.990
         − 135                      − 24
           440                        59
         − 405                      − 56
           350                        30
         − 315                      − 24
            35                         6
```

Round each unit price to the nearest cent.

Small bag   **Large bag**
$.40 per pound $.37 per pound

1. Look at the *Here's how*. Which size bag of apples costs less per pound? Large bag
2. If both bags of apples are the same quality, which size bag is the better buy? Large bag
3. If you only need 4 pounds of apples, which size bag should you buy? Small bag

EXERCISES

Compute the unit price. Round each answer to the nearest cent.

4. Grapes $.95
   2 pounds for $1.89

5. Pickles $.09
   9 ounces for $.79

   Hint: The unit price is the cost per ounce.

6. Crackers
   10 ounces for 98¢ $.10

7. Ketchup
   8 ounces for 69¢ $.09

8. Tomatoes
   3 pounds for $2.69 $.90

9. Apples
   4 pounds for $1.75 $.44

10. Peanuts
    1.1 pounds for $1.99 $1.81

11. Bread
    14 ounces for $1.10 $.08

12. Sunflower seeds
    6.5 ounces for 89¢ $.14

13. Spaghetti sauce
    12 ounces for 99¢ $.08

Suppose that you could use either amount. Tell which is the better buy.

Hint: Compute and compare the unit prices.

14. Cheese
    a. 3 pounds for $3.19
    b. 2 pounds for $2.29

15. Bananas
    a. 3 pounds for $.89
    b. 5 pounds for $1.39

16. Carrots
    a. 0.5 pound for $.29
    b. 2 pounds for $.99

17. Beans
    a. 8-ounce can for $.49
    b. 12-ounce can for $.89

18. Cereal
    a. 18-ounce box for $1.09
    b. 12-ounce box for $.63

19. Olives
    a. 6-ounce jar for $.99
    b. 10-ounce jar for $1.79

20. Dinner napkins
    a. 100 for $.85
    b. 150 for $1.75

21. Milk
    a. 1 quart for $.99
    b. $\frac{1}{2}$ gallon for $1.79

What's in the bag?   Working backward

Use the ad to find what fruit is in each bag.

22. Cherries

23. Grapes

PRODUCE SALE

PEARS .... $ .88/lb
PEACHES .... $ .75/lb
CHERRIES $1.80/lb
GRAPES $1.20/lb

Consumer Mathematics   313

Practice Worksheet

Workbook S302, Copymaster S302, or Duplicating Master S302

WORKSHEET 134
(Use after page 313.)

NAME _____

COMPARISON SHOPPING
Compute the unit price to answer the questions.

1. a. What does one ounce of the large jar of jelly cost? $.06
   b. What does one ounce of the small jar of jelly cost? $.069
   c. Do you save money by buying the larger size? Yes

2. a. Which size bag of walnuts costs $0.105 per ounce? Small
   b. Which size bag costs $0.095 per ounce? Large
   c. Which size bag of walnuts is the better buy? Large

3. a. How much does one apple in the large bag cost? $.129
   b. How much does one apple in the small bag cost? $.115
   c. Which size bag of apples is the better buy? Small

4. a. Which size box of cereal costs $0.075 per ounce? Giant
   b. Which size box of cereal costs $0.08 per ounce? Medium
   c. Which size box of cereal is the best buy? Giant

© D. C. Heath and Company   S302   Comparison buying

Group Project

Computing unit prices

Have a contest to find the food item in the kitchen that cost the most per ounce. Encourage the students to use a calculator to compute the unit prices.

313

Class Starter Quiz 123
on previous lesson

Which is the better buy?

1. a. 1 pound of cheese for $1.19
   b. 2 pounds for $2.29 b
2. a. 3 pounds of peanuts for $2.55
   b. 5 pounds for $4.35 a

Copymaster S415

Lesson Objective
To solve problems by relating discounts and coupons to real situations

Problem-Solving Skills
Selecting information from sale ads and coupons
Choosing the correct operation

Starting the Lesson
Number Sense Write this on the chalkboard:

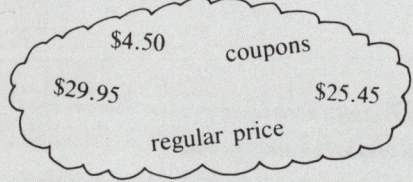

$4.50   coupons
$29.95 $25.45
    regular price

Shoppers who use ? pay less than the ? . For example, with a ? off coupon, a ? camera will cost ? .

Before the students open their books, tell them to use the words and dollar amounts to fill in the blanks so the sentences make sense. Then have them use the item prices and coupon information on page 314 to check on whether they filled in the blanks correctly.

314

Bargain buying

Shoppers who use coupons pay less than the regular price. They get the price reduced by the value of the coupon. For example, with the coupon, the pocket camera will cost $25.45.

POCKET CAMERA
Reg. **$29.95**

$29.95   regular price
− 4.50   value of coupon
$25.45

CAMERA BAG
Reg. **$48.29**

VIDEO CAMERA
$\frac{1}{4}$ OFF *Reg.* **$1240.00**

MOVIE CAMERA
10% OFF
Reg. **$199.00**

$4.50   STORE COUPON $4.50
$4.50 OFF POCKET CAMERA
LIMIT 1 PER CUSTOMER

STORE COUPON
30¢ OFF
ON 1 ROLL
OF 24-EXPOSURE
COLOR FILM
Reg. $2.49
$2.19

STORE COUPON
SAVE $1.99
ON
CAMERA BAG
LIMITED QUANTITY

STORE COUPON
COLOR PRINT PROCESSING
	Reg.	WITH COUPON
12-EXPOSURE	$2.25	$1.99
20-EXPOSURE	$4.35	$3.79
36-EXPOSURE	$6.45	$5.89

BRING THIS COUPON WITH ORDER   NO LIMIT

314   *Chapter 12*

EXERCISES

Solve.

1. What is the regular price of a
   a. pocket camera? $29.95
   b. roll of 24-exposure color film? $2.49

2. With a coupon, how much can you save when you buy a
   a. pocket camera? $4.50
   b. camera bag? $1.99

3. How much more will it cost to process a roll of 36-exposure color film without a coupon? $.56

4. With a coupon, how much would you pay for a camera bag? $46.30

5. You bought a movie camera. What was the sale price? $179.10

6. You are interested in buying a video camera. How much can you save by buying it during the sale? $310

7. Mrs. Kelly bought a roll of 24-exposure color film and a camera bag. How much did she save by using coupons? $2.29

8. A customer paid for 2 rolls of 24-exposure color film with a $10 bill and got $5.62 in change. Did the customer use coupons? Yes

9. Janet has a $15 gift certificate. How much more money does she need to buy a pocket camera using a coupon? $10.45

10. Mr. Harms was charged $13.05 to have 3 rolls of 20-exposure color film processed. Did he have a coupon? No

11. Using only one coupon, what would be the total cost for two pocket cameras? $55.40

12. Which is the better buy, getting 10% off the regular price of a roll of 24-exposure color film or using the coupon? Using the coupon

You decide!

Use the coupons and the prices on page 314 to answer these questions.

13. I bought a roll of 24-exposure color film and a pocket camera. I spent $27.94. Which coupon did I clip out and use? $4.50 off pocket camera

14. I bought a camera bag, a pocket camera, and a roll of 24-exposure film. I spent a total of $78.44. Which coupons did I clip out and use? Coupons for camera bag and film

Practice Worksheet

Workbook S303, Copymaster S303, or Duplicating Master S303

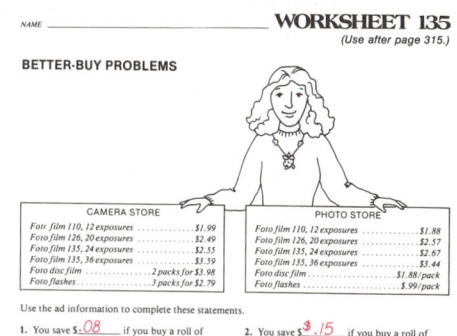

Project

Using coupons

Look in magazines and newspapers for coupons that your family could use. Find how much money your family could save if the coupons were used.

Copymaster S483

Consumer Mathematics

Class Starter Quiz 124
on previous lesson

Solve. Use the sale ad and coupon information on page 314.

1. With a coupon, how much would you pay for a pocket camera? **$25.45**
2. Using only one coupon, what would be the total cost for two rolls of 24-exposure color film? **$4.68**

Copymaster S415

Problem-Solving Skill
Selecting information from a chart

Skills Reviewed
Adding and subtracting mixed numbers
Multiplying and dividing mixed numbers
Making conversions between metric units of measure
Adding and subtracting decimals
Multiplying and dividing decimals
Simplifying expressions

Starting the Lesson
Problem Solving Have the students use the budgets at the top of page 316 to answer questions like these:

- Which person budgets a total of $17 for movies and lunches? (Rhonda)
- Who budgets $\frac{1}{6}$ of total earnings for movies? (Walter)
- Who saves 20% of total earnings? (Kelly)

Cumulative Skill Practice Challenge the students to an estimation hunt by saying, "Find the five sums in exercises 1–6 that are less than 6. (Exercises 1, 2, 3, 4, and 6) Then have the students find the four differences in exercises 7–12 that are less than 4. (Exercises 8, 9, 10, and 12)

Problem solving Making a budget

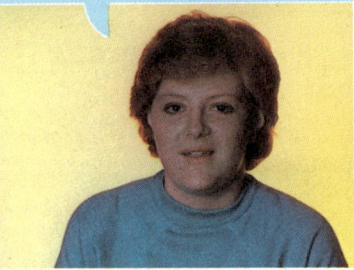

*A **budget** is a plan for using one's money. Here's my weekly budget.*

Kelly
EARNINGS: $30/week
BUDGET
Savings: $6.00
Clothing: 4.50
Movies: 3.50
Records: 3.00
Lunches: 9.00
Other: 4.00

I use a budget to keep track of my spending so that I won't run out of money between paychecks.

Rhonda
EARNINGS: $40/week
BUDGET
Savings: $ 4.00
Clothing: 10.00
Movies: 7.00
Records: 3.00
Lunches: 10.00
Other: 6.00

I keep a budget to make sure I put some money into savings each week.

Walter
EARNINGS: $24/week
BUDGET
Savings: $3.00
Clothing: 6.50
Movies: 4.00
Records: 2.00
Lunches: 6.50
Other: 2.00

Use the budgets to answer the questions.

1. Who budgets a total of $6 for movies and records? **Walter**

2. Who budgets $2.50 more for clothing than for movies? **Walter**

3. Which person budgets $\frac{1}{4}$ of total earnings for clothing? **Rhoda**

4. What fraction of his earnings does Walter save? Give the answer in lowest terms. $\frac{1}{8}$

5. Who spends 10% of total earnings for records? **Kelly**

6. Who spends 50% of total earnings for clothing and lunches? **Rhonda**

Solve. Decide when a calculator would be useful.

7. Holly Moore earns $50 a week. She plans her budget using percents. Find the amount she plans to spend for each category. **Savings—$5.00; Recreation—$12.50; Personal—$7.50; Other—$10.00**

8. How much more money does Holly plan to spend for clothing than for recreation? **$2.50**

9. You earn $65 a week. Using the same categories as Holly, assign a percent to each category that you feel is appropriate for you. Then determine each amount.
Answers will vary.

HOLLY'S BUDGET

CATEGORY	PERCENT	AMOUNT
Clothing	30%	$15.00
Savings	10%	?
Recreation	25%	?
Personal	15%	?
Other	20%	?

316 Chapter 12

Cumulative Skill Practice

Add. Write the sum in simplest form. (page 182)

1. $3\frac{2}{3} + 1\frac{1}{2} = 5\frac{1}{6}$
2. $2\frac{1}{3} + 2\frac{3}{4} = 5\frac{1}{12}$
3. $1\frac{5}{6} + 2\frac{2}{3} = 4\frac{1}{2}$
4. $1\frac{3}{4} + 1\frac{1}{2} = 3\frac{1}{4}$
5. $4\frac{1}{2} + 2\frac{1}{4} = 6\frac{3}{4}$
6. $3\frac{1}{3} + 2\frac{1}{2} = 5\frac{5}{6}$

Subtract. Write the difference in simplest form. (page 192)

7. $6\frac{3}{4} - 1\frac{1}{2} = 5\frac{1}{4}$
8. $5\frac{2}{3} - 3 = 2\frac{2}{3}$
9. $4\frac{1}{2} - 1\frac{3}{4} = 2\frac{3}{4}$
10. $5\frac{1}{8} - 2\frac{1}{2} = 2\frac{5}{8}$
11. $6\frac{3}{5} - 2\frac{1}{2} = 4\frac{1}{10}$
12. $7\frac{2}{7} - 3\frac{1}{2} = 3\frac{11}{14}$

Give the product in simplest form. (page 206)

13. $3\frac{1}{2} \times 2\frac{1}{4} \;\; 7\frac{7}{8}$
14. $1\frac{1}{3} \times 1\frac{1}{2} \;\; 2$
15. $4\frac{1}{2} \times 2\frac{1}{3} \;\; 10\frac{1}{2}$
16. $2\frac{1}{2} \times 1\frac{1}{4} \;\; 3\frac{1}{8}$
17. $1\frac{1}{4} \times 2\frac{1}{3} \;\; 2\frac{11}{12}$
18. $1\frac{2}{3} \times 2\frac{1}{2} \;\; 4\frac{1}{6}$
19. $1\frac{1}{3} \times 1\frac{1}{3} \;\; 1\frac{7}{9}$
20. $2\frac{3}{4} \times 3\frac{2}{3} \;\; 10\frac{1}{12}$
21. $2\frac{3}{8} \times 3 \;\; 7\frac{1}{8}$
22. $2\frac{2}{3} \times 1\frac{3}{4} \;\; 4\frac{2}{3}$

Give the quotient in simplest form. (page 214)

23. $3\frac{3}{4} \div 1\frac{1}{4} \;\; 3$
24. $5 \div 2\frac{1}{2} \;\; 2$
25. $5\frac{5}{6} \div 2\frac{1}{3} \;\; 2\frac{1}{2}$
26. $5\frac{1}{2} \div 2\frac{1}{4} \;\; 2\frac{4}{9}$
27. $6\frac{2}{3} \div 1\frac{1}{6} \;\; 5\frac{5}{7}$
28. $6\frac{1}{4} \div 1\frac{1}{4} \;\; 5$
29. $5\frac{5}{8} \div 1\frac{1}{2} \;\; 3\frac{3}{4}$
30. $5\frac{1}{4} \div 2\frac{1}{2} \;\; 2\frac{1}{10}$
31. $10\frac{1}{2} \div 1\frac{3}{4} \;\; 6$
32. $8 \div 1\frac{2}{3} \;\; 4\frac{4}{5}$

Complete. (page 228)

33. 9 cm = ? mm 90
34. 6 km = ? m 6000
35. 62 mm = ? cm 6.2
36. 3450 m = ? km 3.450
37. 8.4 km = ? m 8400
38. 5.6 cm = ? m 0.056

MIXED PRACTICE

Complete.

39. $5.92 + 11.6 = $? 17.52
40. $57 \times 1000 = $? 57,000
41. $75.81 - 6.8 = $? 69.01
42. $7.4 \times 0.03 = $? 0.222
43. $39 - 7.85 = $? 31.15
44. $85.5 \div 0.9 = $? 95
45. $41.5 \times 10 = $? 415
46. $5.6 + 2.88 + 4 = $? 12.48
47. $50.32 \div 3.7 = $? 13.6
48. $0.26 \times 100 = $? 26
49. $92.3 \div 10 = $? 9.23
50. $0.83 \div 1000 = $? 0.00083
51. $(18.5 - 3) - 1.9 = $? 13.6
52. $18.5 - (3 - 1.9) = $? 17.4
53. $18.5 - (3 + 1.9) = $? 13.6

Consumer Mathematics

Problem-Solving Worksheet

Workbook S304, Copymaster S304, or Duplicating Master S304

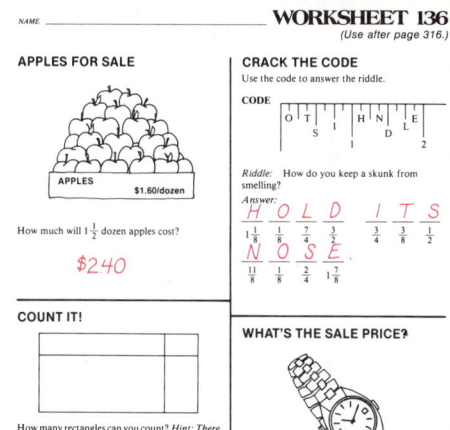

Project

Making a budget

Compute the amount of money you receive in a week. Then make a budget to show how you would like to use that money.

Copymaster S483

Class Starter Quiz 125
on previous lesson

Solve. Use the budgets shown at the top of page 316.

1. How many dollars of her earnings did Rhonda budget for movies? **$7**
2. What fraction of his earnings did Walter budget for savings? **$\frac{1}{8}$**
3. What percent of her earnings did Kelly budget for records? **10%**

Copymaster S416

Lesson Objectives
To understand checking accounts and checks
To balance a checking account

Problem-Solving Skill
Selecting information from checks and a check register

Starting the Lesson
The three checks at the top of page 318 are also on ■ **Visual Aid 54** (copymaster or transparency S153). Have the students use the checks and the clues in exercises 1 and 2 to find out whose checks paid for the stereo and the TV set.

Exercise Note
The check registers on page 319 are on ■ **Visual Aid 55** (copymaster or transparency S154). You may wish to use this in discussing exercises 7-16.

Checking accounts
CHECK IT OUT!

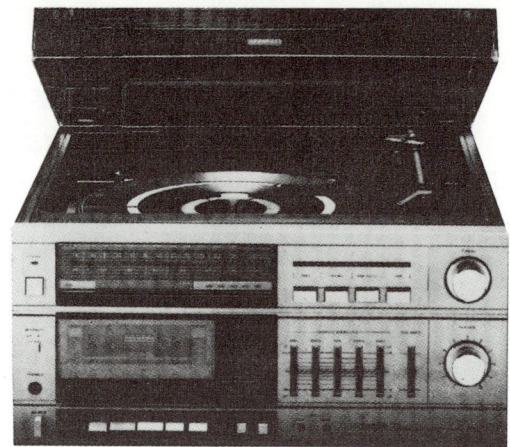

1. Whose check paid for the stereo? **Tony Perez's**
   Clues:
   • The check was written for more than $300.
   • The check was dated in early March.

2. Whose check paid for a TV set? **Frank Horowitz's**
   Clues:
   • The check number is less than 500.
   • The check was written for less than $400.

EXERCISES
Use the checks to answer the questions.

3. Find check No. 175. Who signed the check? What was the amount of the check? **Julie Adams, $426.93**

4. Who signed the check that tells the United Bank of Lexington to pay Ace Electronics $283.17 from his account? **Frank Horowitz**

5. What is the amount of check No. 1215? Why do you think the amount is written in both numerals and words? **$379.86 So that the amount cannot be changed.**

6. Julie Adams keeps her money in the First National Bank. Her checking-account number is 127 415. Whose checking-account number is 741 105? What is Tony Perez's checking-account number? **Frank Horowitz's, 416 913**

Julie's checkbook has a section called a **check register** in which she keeps a record of her checking account.

NUMBER	DATE	DESCRIPTION OF TRANSACTION	AMOUNT OF CHECK	✓	AMOUNT OF DEPOSIT	BALANCE $173	96
173	3/14/89	Top Supervalue	$ 72.14		$	101	82
	3/16/89				500.00	601	82
174	3/21/89	cash	50.00			551	82
175	3/28/89	Ace Electronics	426.93			124	89
176	4/2/89	Reed Bookstore	13.21				

7. On what date was a $500 deposit made? **3/16/85**

8. For how much was the check that was written to Top Supervalue? **$72.14**

9. To whom was a check for $426.93 written? **Ace Electronics**

10. Check No. 174 was written for what amount? **$50.00**

11. What was the balance in Julie's account after check 174 was written? **$551.82**

12. What two numbers would you use to find the balance in her account on April 2? Would you add or subtract to find the balance? **$124.89, $13.21, Subtract**

13. What was the balance in her account after check No. 176 was written? **$111.68**

14. On April 4, Julie wrote check No. 177 for $17.65. What was her balance after that? **$94.03**

Can you figure it?

Find each new balance.

15.

	AMOUNT OF CHECK	✓	AMOUNT OF DEPOSIT	BALANCE 745	32
692.22 a.	53.10			?	
674.94 b.	17.28			?	
779.94 c.			105.00	?	
496.77 d.	283.17			?	
542.52 e.			45.75	?	
484.14 f.	58.38			?	

16.

	AMOUNT OF CHECK	✓	AMOUNT OF DEPOSIT	BALANCE 434	80
371.68 a.	63.12			?	
353.29 b.	18.39			?	
225.36 c.	127.93			?	
475.36 d.			250.00	?	
95.50 e.	379.86			?	
84.43 f.	11.07			?	

Consumer Mathematics

Practice Worksheet

Workbook S305, Copymaster S305, or Duplicating Master S305

WORKSHEET 137 (Use after page 319.)

CHECKS AND BALANCES

Complete each check.
1. Write a check to Midtown Hardware for $53.20.
2. Write a check to EZ Cleaners for $17.89.

Complete each balance.

	CHECK NO.	DATE	PAY TO	AMOUNT	DEPOSIT	BALANCE 623	78
3.	1211	March 1	Crosstown Motors	275.50		348	28
4.	1212	March 1	Bell Telephone	36.52		311	76
5.		March 2			385.16	696	92
6.	1213	March 2	Adams Insurance Co.	59.65		637	27
7.	1214	March 2	Cash	75.00		562	27
8.	1215	March 4	Power and Gas	69.65		492	62
9.	1216	March 7	Midtown Hardware	53.20		439	42
10.	1217	March 7	EZ Cleaners	17.89		421	53
	1218	March 8	Cash	10.00		411	53

© D. C. Heath and Company   S305   Checking accounts

Project

Researching information

Find the answers to these questions:

1. What does "overdrawn" mean?
2. What is a "service charge"?
3. What does it mean to make a check out to "cash"?
4. What is a "cashier's check"?

Copymaster S484

Class Starter Quiz 126
on previous lesson

Solve. Use the check register shown at the top of page 319.

1. What is the number of the check written to Ace Electronics? **175**
2. On what date was a $13.21 check written? **4/2/90**
3. What was the balance in the account after check 173 was written? **$101.82**

Copymaster S416

Lesson Objectives
To understand savings accounts and passbooks
To understand compound interest

Problem-Solving Skills
Selecting information from a savings-account passbook
Reading a graph
Following directions

Starting the Lesson
Have the students take the savings-account quiz shown at the top of page 320. Correct the quizzes before assigning the exercises.

Exercise Note
The passbook on page 320 and the compound-interest graph on page 321 are also on ■ **Visual Aids 56 and 57** (copymasters or transparencies S155–S156).

Savings accounts

CAN YOU PASS THIS SAVINGS ACCOUNT QUIZ?

Word list
added   interest   subtracted
deposit passbook   withdrawal

Use the word list to complete each statement.

1. To open a savings account, you must make a [?]. Each time you make a deposit, it is [?] to the balance of your account. **deposit, added**
2. A [?] is an amount of money you take out of your savings account. When you make a withdrawal, it is [?] from your balance. **withdrawal, subtracted**
3. You may receive a [?] when you open a savings account. Bank tellers use it to record all deposits, withdrawals, [?] earned, and the new balance. **passbook, interest**

EXERCISES

Use James Stickney's passbook to answer the questions.

1. How much money did James deposit on January 19? **$250.00**
2. How much money did he withdraw on March 12? **$65.50**
3. What was the balance in his account on March 12? **$302.44**
4. On what date did James get $2.22 interest? **April 1**
5. What was the balance on June 1? **$489.94**
6. How much interest did James receive for January 19 through June 1? **$9.19**

DEPOSITOR: JAMES STICKNEY ACCOUNT NO. 26-01432				
DATE	WITHDRAWAL	DEPOSIT	INTEREST	BALANCE
01-19		250.00		250.00
02-01			.65	250.65
02-18		115.50		366.15
03-01			1.79	367.94
03-12	65.50			302.44
04-01			2.22	304.66
05-01			2.01	306.67
05-17		180.75		487.42
06-01			2.52	?

The interest on savings accounts is often compounded daily. This means that the interest is added to the account each day. That way, you earn interest on your interest. Answers for 7–9 may vary slightly.

Use the graph to answer the questions.

7. At 6% interest, about how much would a $1000 deposit be worth at the end of 3 years? 5 years? 7 years? $1175, $1325, $1500

8. At 9% interest, about how much would a $1000 deposit be worth at the end of 3 years? 5 years? 7 years? $1300, $1525, $1825

9. At 12% interest, about how much would a $1000 deposit be worth at the end of 3 years? 5 years? 7 years? $1400, $1775, $2225

10. About how many years does it take to double a $1000 deposit at 9% interest? 8

11. About how many years does it take to double a $1000 deposit at 12% interest? 6

12. Suppose you deposited $1000 at 9% interest and a friend deposited $1000 at 6% interest.
    a. About how much more money would you have than your friend at the end of 4 years? $150
    b. About how much more money would you have than your friend at the end of 8 years? $400

Want to be a millionaire?
Here's how to make $1,000,000.

13. Put [?] dollars in a savings account that pays [?]% interest compounded yearly. Then wait [?] years and your account will be worth $1,000,981.

    To find [?], write a 5 in the ones place, a 6 in the hundreds place, and a 9 in the tens place. 695

    To find [?], write a 1 in the tens place and a 6 in the ones place. 16

    To find [?], write a 9 in the ones place and a 4 in the tens place. 49

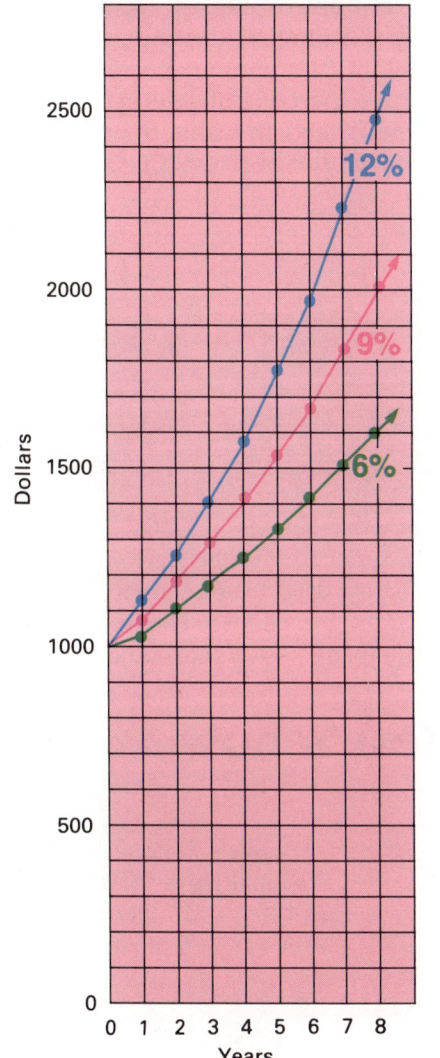

HOW $1000 GROWS
(Interest Compounded Daily)

Consumer Mathematics 321

Practice Worksheet
Workbook S306, Copymaster S306, or Duplicating Master S306

Project
Using library resources

Use an almanac to find which bank is the largest United States commercial bank. Then write the dollar amount of the bank's deposits in words.

Answers will vary. Records may change from year to year.

Copymaster S484

Class Starter Quiz 127
on previous lesson

Solve. Use the compound-interest graph shown on page 321.

1. At 9% interest, about how much would a $1000 deposit be worth at the end of 8 years? **$2000**
2. About how much interest is earned on a $1000 deposit in the first 5 years at 12%? **$750**
3. About how many years does it take a $1000 deposit to earn $500 interest at 6%? **7**

Copymaster S416

Lesson Objective
To compute simple interest

Problem-Solving Skills
Selecting information from a display
Reading an auto-loan application

Starting the Lesson
Write this on the chalkboard:

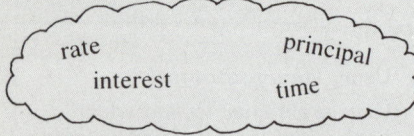

The amount of _?_ (rent for using the money) depends on the _?_ (the amount borrowed), the _?_ (percent of interest charged), and the _?_ for which the money is borrowed.

Before the students open their books, have them use the words to fill in the blanks so that the sentence makes sense. Then have them read the paragraph at the top of page 322 to check their answers.

Borrowing money

The amount of **interest** (rent for using the money) depends on the **principal** (the amount borrowed), the **rate** (percent of interest charged), and the **time** for which the money is borrowed.

PEOPLE'S FINANCE CO.
BORROW UP TO $1000
PAY ONLY **2% PER MONTH**

1. In order to buy this car, you have to borrow $500 for 6 months. Which loan would you choose, People's Finance Company's or Honest Abe's? *Answers will vary.*

Here's how *to use a formula to find the interest (I) for a $500 loan for 6 months.*

People's Finance Co.	Honest Abe
Principal (p) = $500	Principal = $500
Rate (r) = 2% per month	Rate = 18% per year
Time (t) = 6 months	Time = 6 months ($\frac{1}{2}$ year)

FORMULA
$I = p \times r \times t$
$I = \$500 \times 0.02 \times 6$
$= \$60$

$I = p \times r \times t$
$I = \$500 \times 0.18 \times 0.5$
$= \$45$

The time units must be the same. Since the rate is a yearly rate, use 0.5 of a year for t.

2. Look at the *Here's how*. How much interest is People's Finance Company charging? How much interest is Honest Abe charging? **$60, $45**

3. Who is offering the better deal, People's Finance Company or Honest Abe? **Honest Abe**

EXERCISES

Compute the interest.
Here are scrambled answers for the next row of exercises: $4.50 $240 $30

4. Principal = $1000
   Rate = 12% per year
   Time = 2 years $240

5. Principal = $600
   Rate = 10% per year
   Time = 6 months $30

6. Principal = $100
   Rate = 1.5% per month
   Time = 3 months $4.50

7. Principal = $300
   Rate = 16% per year
   Time = 1 year $48

8. Principal = $450
   Rate = 1% per month
   Time = 8 months $36

9. Principal = $700
   Rate = 14% per year
   Time = 9 months $73.50

10. Principal = $200
    Rate = 15% per year
    Time = 1.5 years $45

11. Principal = $4000
    Rate = 12% per year
    Time = 4 years $1920

12. Principal = $300
    Rate = 18.5% per year
    Time = 18 months $83.25

Solve.

13. Brian borrowed $1500 for 1 year to buy a car. The yearly rate was 14%.
    a. How much interest will he owe at the end of the year? $210
    b. What is the total amount he will have to repay at the end of a year? $1710

14. You need $300 to buy a motorcycle.
    a. A bank will lend you the money at 15% for 1 year. How much interest is that? $45
    b. A finance company will loan you the money at 1.5% per month for 12 months. How much interest is that? $54
    c. Which loan is the better deal, the bank's or the finance company's? The bank's

You're the loan officer

15. You work at a bank. Part of your job is to review loan applications. Complete the loan application. Then decide whether or not you would approve the loan.

FIRST NATIONAL BANK — APPLICATION FOR AUTO LOAN

NAME: Amy Higgins   OCCUPATION: editor

MONTHLY INCOME (after deductions): $1400

MONTHLY EXPENSES
- HOUSE PAYMENT: 550
- UTILITIES: 180
- LIVING: 360
- INSURANCE: 80
- OTHER LOAN PAYMENTS: 50
- OTHER: 120
- TOTAL EXPENSES: ? $1340

- COST OF AUTO: $7200
- DOWN PAYMENT: 1000
- BALANCE OWED (cost minus down payment): 6200
- FINANCE CHARGE: 2770
- TOTAL LOAN (balance owed plus finance charge): $8970
- NUMBER OF MONTHLY PAYMENTS: 30
- MONTHLY PAYMENT: ? $299

Do not write below this line

BANK USE ONLY Check one   Loan ( ) approved (X) not approved

Practice Worksheet

Workbook S307, Copymaster S307, Duplicating Master S307

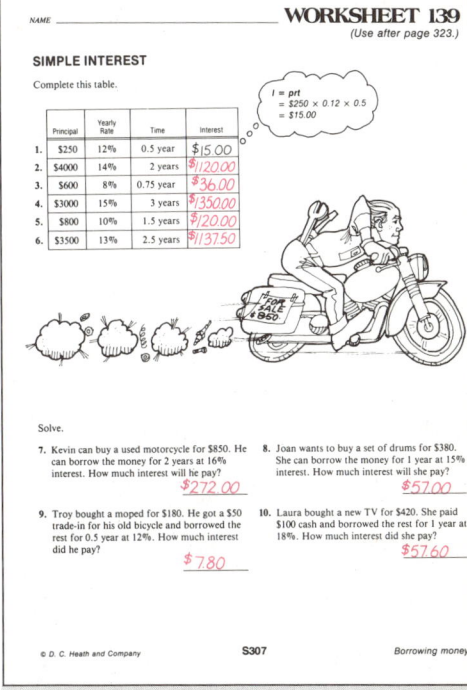

WORKSHEET 139 (Use after page 323.)

SIMPLE INTEREST

Complete this table.

$I = prt$
$= $250 \times 0.12 \times 0.5$
$= 15.00

	Principal	Yearly Rate	Time	Interest
1.	$250	12%	0.5 year	$15.00
2.	$4000	14%	2 years	$120.00
3.	$600	8%	0.75 year	$36.00
4.	$3000	15%	3 years	$1350.00
5.	$800	10%	1.5 years	$120.00
6.	$3500	13%	2.5 years	$1137.50

Solve.

7. Kevin can buy a used motorcycle for $850. He can borrow the money for 2 years at 16% interest. How much interest will he pay? $272.00

8. Joan wants to buy a set of drums for $380. She can borrow the money for 1 year at 15% interest. How much interest will she pay? $57.00

9. Troy bought a moped for $180. He got a $50 trade-in for his old bicycle and borrowed the rest for 0.5 year at 12%. How much interest did he pay? $7.80

10. Laura bought a new TV for $420. She paid $100 cash and borrowed the rest for 1 year at 18%. How much interest did she pay? $57.60

Group Project

Buying a car

Give small groups of students copies of the classified section of a local newspaper listing automobiles for sale. Have each group choose an automobile. Ask them to suppose that they have enough cash to pay 20% of the listed price and that they borrow the rest of the money for 2 years at a yearly rate of 15%. Have each group figure out how much interest they will have to pay and how much they will have to pay each month in order to pay off the loan.

Consumer Mathematics

Paying bills

BILLS, BILLS, BILLS!

At the end of each month, bills are sent to customers. Here are some monthly bills that George Mumby received.

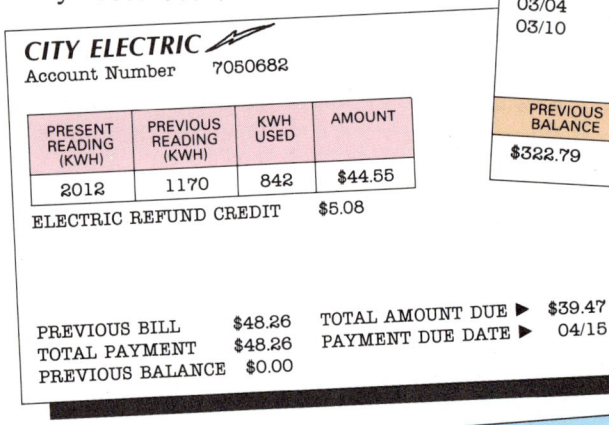

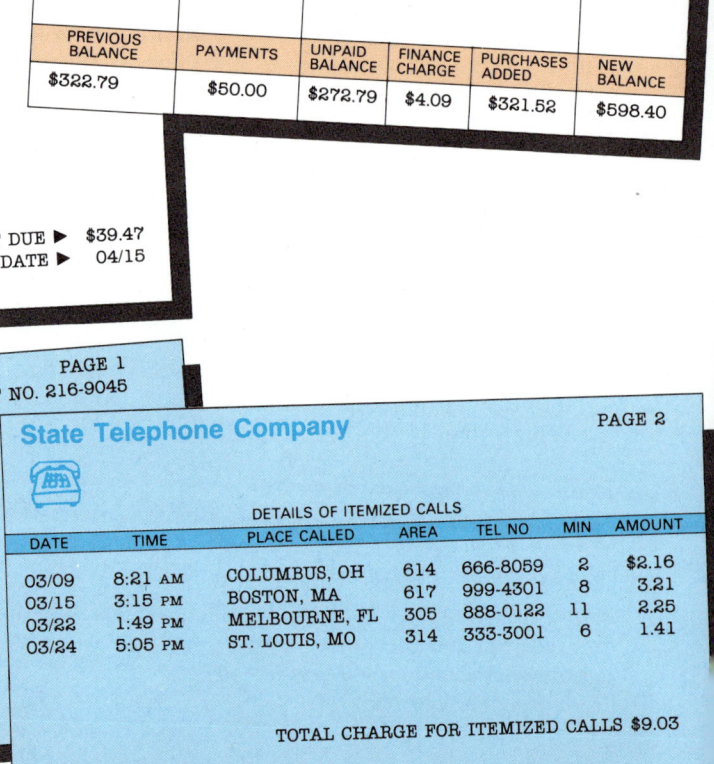

EXERCISES

Use the electric bill to answer these questions.

1. How many kilowatt hours (KWH) of electricity were used during the month? 842

2. What is the total amount George owes the electric company? On what date is the payment due? $39.47, April 15

3. What was the previous month's electricity bill? How much more was it than this month's bill? $48.26, $8.79

Use the Bank Card statement to answer these questions.

4. How much did George charge at the Corner Gift Shop? $12.50

5. What was the amount of the charge at the C & C Muffler Shop? $47.72

6. What was the total amount of new charges? $321.52

7. What two numbers would you use to check the amount of the unpaid balance? Would you add or subtract? $322.79, $50, Subtract

8. Bank Card charges 1.5% per month interest (finance charge) on the unpaid balance. How much was last month's finance charge? $4.09

9. What three numbers would you add to check the amount of the new balance? $272.79, $4.09, $321.52

10. George's new balance is $598.40. If he makes the minimum payment, how much will he still owe? $568.40

11. At 1.5% per month, how much will the finance charge be on an unpaid balance of $570? $8.55

Use the telephone bill to answer these questions.

12. When must the telephone bill be paid? How much does George owe? by April 15, $20.25

13. How much federal tax is included in the bill? $.24

14. What is the monthly charge for service? $10.98

15. On March 15, George made a long-distance call to Boston, Massachusetts. How many minutes long was the call? What was the charge for the call? 8, $3.21

16. How much did it cost per minute for George to call Columbus, Ohio? $1.08

17. If George makes a long-distance call between 11:00 P.M. and 8:00 A.M., he gets a 60% discount. How much would he have saved if he had made the call to Melbourne, Florida, between 11:00 P.M. and 8:00 A.M.? $1.35

Answer the phone

Use the phone rates to answer the questions.

18. Julie called a friend at 10 A.M. on Wednesday. They talked for 5 minutes. How much was she charged for the call? $1.44

19. Julie was charged $1.28 for a call she made on Friday at 9:30 P.M. How long was the call? 6 minutes

LONG DISTANCE IN-STATE PHONE RATES Monday through Friday		
	8 A.M.–5 P.M.	5 P.M.–8 A.M.
First minute	$.44	$.28
Each additional minute	$.25	$.20

Consumer Mathematics

Practice Worksheet

Workbook S308, Copymaster S308, or Duplicating Master S308

WORKSHEET 140
(Use after page 325.)

NAME _____

MONTHLY PAYMENTS

BANK CARD STATEMENT

ACCOUNT NUMBER	CREDIT LIMIT	AVAILABLE CREDIT	STATEMENT DATE	PAYMENT DUE DATE	MINIMUM PAYMENT DUE
431 025 1506	$1000	⑥ $773.10	06/10/85	06/25/85	⑤ $22.69

DATE OF TRANSACTION	REFERENCE NUMBER	CHARGES SINCE LAST STATEMENT	AMOUNT
05/18	612435	UPTOWN MOTORS	$ 41.10
05/23	127067	TED'S CAMERA SHOP	16.95
05/28	427106	PRO SPORTSWEAR	39.65
05/30	372111	HARRY'S RESTAURANT	27.70

PREVIOUS BALANCE	PAYMENTS	UNPAID BALANCE	FINANCE CHARGE	PURCHASES ADDED	NEW BALANCE
175.15	75.15	① $100.00	② $1.50	③ $125.40	④ $226.90

Use the monthly bank card statement to answer the questions. Then use the answers to complete the bank card statement.

1. Subtract the payments from the previous balance to get the unpaid balance. What is the unpaid balance? **$100.00**

2. The finance charge is computed by multiplying 0.015 (1.5%) times the unpaid balance. What is the finance charge? **$1.50**

3. Add the charges since the last payment to get the purchases added. What is the total of the purchases added? **$125.40**

4. Add the unpaid balance, finance charge, and purchases added to get the new balance. What is the new balance? **$226.90**

5. The minimum payment is computed by multiplying 0.10 (10%) times the new balance. What is the minimum payment? **$22.69**

6. Subtract the new balance from the credit limit to get the available credit for the next month. What is the available credit? **$773.10**

© D. C. Heath and Company   S308   Paying bills

Challenge Problems

Unscramble the letters of each of these consumer math words.

1. INUT CPERI — **Unit price**
2. SVNAIGS UCACONT — **Savings account**
3. EDUBGT — **Budget**
4. ETINERST — **Interest**

Copymaster S458

PROBLEM SOLVING *using a sales tax table*

Some problems can be solved by reading information off a table.

Problem
Jeremy bought a set of Talking Teeth. How much sales tax did he pay?

TALKING TEETH	$2.50
MINI SPY CAMERA	$1.95
PHONY ARM CAST	$2.95
BAG OF LAUGHS	$3.95
BALD HEAD WIG	$1.50
VENUS FLY TRAP	$1.75

Here's how *to use a sales tax table*.

STUDY Find the facts in the ad.

PLAN and SOLVE To find the sales tax on the Talking Teeth, find the two amounts in the table that the price is between. Then read across to find the tax, $.10.

Answer: Jeremy paid $.10 sales tax.

CHECK Check to see that the right numbers were used.

SALES TAX TABLE

Amount of Sale	Tax	Amount of Sale	Tax
$0.00 to 0.12	$0.00	3.88 to 4.12	0.16
0.13 to 0.37	0.01	4.13 to 4.37	0.17
0.38 to 0.62	0.02	4.38 to 4.62	0.18
0.63 to 0.87	0.03	4.63 to 4.87	0.19
0.88 to 1.12	0.04	4.88 to 5.12	0.20
1.13 to 1.37	0.05	5.13 to 5.37	0.21
1.38 to 1.62	0.06	5.38 to 5.62	0.22
1.63 to 1.87	0.07	5.63 to 5.87	0.23
1.88 to 2.12	0.08	5.88 to 6.12	0.24
2.13 to 2.37	0.09	6.13 to 6.37	0.25
2.38 to 2.62	0.10	6.38 to 6.62	0.26
2.63 to 2.87	0.11	6.63 to 6.87	0.27
2.88 to 3.12	0.12	6.88 to 7.12	0.28
3.13 to 3.37	0.13	7.13 to 7.37	0.29
3.38 to 3.62	0.14	7.38 to 7.62	0.30
3.63 to 3.87	0.15	7.63 to 7.87	0.31

PROBLEMS

Use the sales tax table to find the sales tax on each item.

1. Mini Spy Camera $.08
2. Phony Arm Cast $.12
3. Bag of Laughs $.16
4. Bald Head Wig $.06

Solve.
Use the ad and sales tax table on page 326.

5. Juan bought a Phony Arm Cast and a Bag of Laughs.
   a. What was the total cost of the items? $6.90
   b. What was the sales tax on the total? $.28
   c. What was the total cost including sales tax? $7.18

6. Diane bought a Mini Spy Camera, a Venus Fly Trap, and Talking Teeth.
   a. What was the total cost of the three items? $6.20
   b. What was the sales tax on the total? $.25
   c. What was the total cost including sales tax? $6.45

7. What is the total cost, including sales tax, for a Phony Arm Cast and Bald Head Wig? $4.63

8. Connie bought Talking Teeth and a Bald Head Wig. What was the total cost, including sales tax? $4.16

9. If you have $5.90, do you have enough money to buy a Bag of Laughs and a Venus Fly Trap? *Hint: Don't forget the sales tax.* No

10. Andrew wants to buy a Mini Spy Camera and a Phony Arm Cast, but he has only $3.00. How much more money does he need? $2.10

11. Megan bought two novelty items. The sales tax was $.13. What two items did she buy? Bald Head Wig and Venus Fly Trap

12. The sales tax on the two novelty items Brad bought was $.28. What two items did he buy? Bag of Laughs and Phony Arm Cast

13. Maria bought three Bags of Laughs and two Phony Arm Casts. She paid with a $20 bill and received $1.54 change. What was the sales tax on the total? $.71

14. Chen bought three of the same novelty item. The sales tax was $.18. The total cost, including sales tax, was $4.68. What novelty item did he buy? Bald Head Wig

Can you sense the heavyweight?

15. One of these 8 coins is counterfeit and is slightly heavier than each of the other 7. Tell how you would use a balance, just two times, to find the counterfeit coin.

Consumer Mathematics **327**

Extra Practice
Page 498   Skill 55

Practice Worksheet
Workbook S309, Copymaster S309, or Duplicating Master S309

Challenge Problem
Write a question that fits the answer.

Amy earns $30 a week. She budgets 10% for savings, 30% for movies, and 50% for clothes.

How much money does Amy budget for movies each week?

Answer: $9

Copymaster S458

Class Starter Quiz 130
on previous lesson

Solve. Use the table on page 326.

1. What is the sales tax on the total cost of 3 wigs? $.18
2. What is the sales tax on the total cost of a flytrap and a camera? $.15

Copymaster S417

Problem-Solving Skills
Reading a computer display
Choosing the correct operation
Using a guess-and-check strategy

Skills Reviewed
Adding with customary units of measure
Changing percents to fractions
Changing fractions to percents
Solving percent problems
Adding and subtracting with fractions and mixed numbers
Multiplying and dividing with fractions and mixed numbers

Starting the Lesson
Problem Solving Have the students use the information on Screens A and B and ask them to answer questions like these:

- Use Screen B. What are the order number and cost of the Friendship Bouquet? (204, $23.95)
- What is the cost of order 219? ($19.50)
- Use Screen A. What number would you select to get the order numbers and the costs of the Happy Birthday! selections? (5)

Cumulative Skill Practice Write these four answers on the chalkboard:

5 ft 5 in. $\frac{1}{8}$ 75%   40

Challenge the students to an answer hunt by saying, "Look at exercises 1–39 on page 329. Find the four exercises that have these answers. You have five minutes to find as many of the exercises as you can." (Exercises 3, 16, 24, and 34)

328

Problem solving

SHOPPING BY COMPUTER

Diane Milotte can use her home computer to order flowers to be sent almost anywhere in the world. Her computer can be linked by telephone to a large computer that will select a florist, place the order, and charge it to Diane's account number.

The screens below show some of the decisions Diane has to make when she places an order.

Screen A
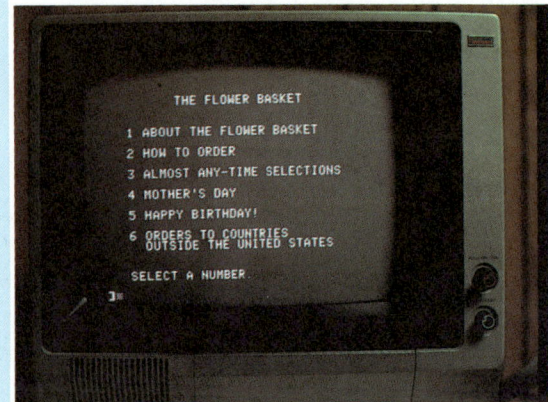

Diane selected "3."
The computer then displayed Screen B.

Screen B
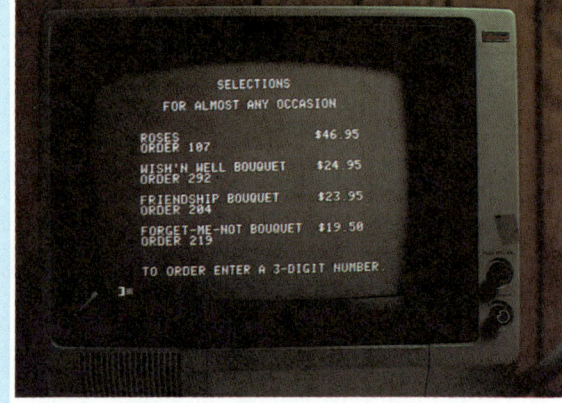

Diane entered "219."
Additional screens requested more information to complete her order.

Solve. Decide when a calculator would be useful.

1. Use Screen A. What information would Diane be asking for if she selected number 2 on Screen A? How to order

2. Use Screen B. What are the order number and the cost of the Wish'n Well Bouquet? 292, $24.95

3. Use Screen A.
   a. What number should Diane select to order a special bouquet for her aunt's birthday? 5
   b. Diane wants to order flowers for her sister in Paris, France. What number should she select? 6

4. Use Screen B.
   a. What 3-digit number should Diane enter to order the least expensive bouquet? 219
   b. If Diane decided to spend $23.95, what 3-digit number should she enter? 204

5. Diane ordered bouquets 107, 292, and 219. How much money was charged to her account? $91.40

6. One month Diane ordered 2 different bouquets. She spent a total of $43.45. Which bouquets did she order?
   Friendship and Forget-Me-Not

328   Chapter 12

Cumulative Skill Practice

Add. *(page 248)*

1. 4 ft 9 in.
   + 1 ft 8 in.
   ___
   6 ft 5 in.

2. 3 yd 2 ft
   + 2 yd 2 ft
   ___
   6 yd 1 ft

3. 3 ft 9 in.
   + 1 ft 8 in.
   ___
   5 ft 5 in.

4. 2 gal 3 qt
   + 1 gal 1 qt
   ___
   4 gal

5. 2 qt 1 pt
   + 1 qt 1 pt
   ___
   4 qt

6. 3 pt 1 c
   + 1 pt 1 c
   ___
   5 pt

7. 7 lb 12 oz
   + 5 lb 8 oz
   ___
   13 lb 4 oz

8. 4 T 1200 lb
   + 1 T 1300 lb
   ___
   6 T 500 lb

Change to a fraction or mixed number in simplest form. *(page 282)*

9. 25% $\frac{1}{4}$
10. 50% $\frac{1}{2}$
11. 60% $\frac{3}{5}$
12. 5% $\frac{1}{20}$
13. 175% $1\frac{3}{4}$
14. 240% $2\frac{2}{5}$
15. $33\frac{1}{3}$% $\frac{1}{3}$
16. $12\frac{1}{2}$% $\frac{1}{8}$
17. $66\frac{2}{3}$% $\frac{2}{3}$
18. $6\frac{1}{4}$% $\frac{1}{16}$
19. $87\frac{1}{2}$% $\frac{7}{8}$
20. $3\frac{1}{8}$% $\frac{1}{32}$

Change to a percent. *(page 284)*

21. $\frac{2}{5}$ 40%
22. $\frac{1}{10}$ 10%
23. $\frac{1}{2}$ 50%
24. $\frac{3}{4}$ 75%
25. $\frac{3}{5}$ 60%
26. $\frac{5}{2}$ 250%
27. 3 300%

Solve. *(page 294)*

28. 12% of 34 = n 4.08
29. 9% of 41.5 = n 3.735
30. 8.5% of 136 = n 11.56
31. 11.4% of 52 = n 5.928
32. $12\frac{1}{2}$% of 56 = n 7
33. $83\frac{1}{3}$% of 72 = n 60

Solve. *(page 296)*

34. 10% of n = 4 40
35. 20% of n = 11 55
36. 25% of n = 16 64
37. 50% of n = 13 26
38. 75% of n = 48 64
39. 150% of n = 39 26

Solve. Round each answer to the nearest tenth. *(page 296)*

40. 4.5% of n = 6 133.3
41. 9.5% of n = 12 126.3
42. 6.4% of n = 15 234.4
43. 1.5% of n = 7.3 486.7
44. 21.4% of n = 10.6 49.5
45. 7.8% of n = 12.4 159.0

MIXED PRACTICE

Complete.

46. $\frac{3}{8} + \frac{1}{2} = $? $\frac{7}{8}$
47. $12 \times 2\frac{1}{2} = $? 30
48. $5 + 4\frac{3}{8} = $? $9\frac{3}{8}$
49. $2\frac{1}{2} \div 1\frac{1}{4} = $? 2
50. $\frac{7}{8} - \frac{1}{4} = $? $\frac{5}{8}$
51. $\frac{2}{3} \div \frac{3}{4} = $? $\frac{8}{9}$
52. $4\frac{1}{3} \times 1\frac{1}{2} = $? $6\frac{1}{2}$
53. $\frac{2}{3} + \frac{3}{4} = $? $1\frac{5}{12}$
54. $\frac{5}{9} \times 0 = $? 0
55. $5\frac{2}{3} + 4\frac{3}{5} = $? $10\frac{4}{15}$
56. $9\frac{1}{3} - 4\frac{5}{8} = $? $4\frac{17}{24}$
57. $\frac{9}{10} \div \frac{3}{5} = $? $1\frac{1}{2}$
58. $8 - 3\frac{4}{5} = $? $4\frac{1}{5}$
59. $4\frac{1}{2} \div 1\frac{2}{3} = $? $2\frac{7}{10}$
60. $\frac{11}{12} - \frac{3}{8} = $? $\frac{13}{24}$

Consumer Mathematics

Problem-Solving Worksheet
Workbook S310, Copymaster S310, or Duplicating Master S310

WORKSHEET 142
(Use after page 328.)

NAME _____

GUESS AND CHECK

2 grams | 5 grams | 6 grams

What coins are worth 50¢ and weigh 15 grams?
1 quarter
2 dimes
1 nickel

EARNING MONEY

WANTED
Newspaper carrier
for the *Register*
Commission Rates:
5¢ per daily paper
10¢ per Sunday paper

You have 30 customers who order the daily paper and 42 who order the Sunday paper. How much commission do you earn in one week?
$13.20

WHAT'S THE COST?

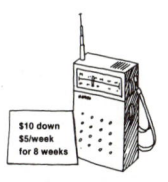

$10 down
$5/week
for 8 weeks

What is the total cost of the radio?
$50

WHAT DAY IS IT?

If March 4th is on a Friday, what day of the week is March 30th?
Wednesday

HEARTBEATS

My pulse rate is 42 heartbeats in 30 seconds. — Sid
My pulse rate is 35 heartbeats in 30 seconds. — Pam

Which person's heart would beat about 5000 times in one hour?
Pam's

LOGICAL REASONING

Each of Dan, Rick, Paul, and Tom is a different height. Dan is shorter than Rick. Only Rick's height is between Dan's and Paul's. At least one has a height that is between Paul's and Tom's.
Who is the shortest? Tom
Who is the tallest? Paul

© D. C. Heath and Company   S310   Problem solving

Challenge Problem

Use this code and find the two-letter word in the first paragraph on page 328 that has a sum of 35. to

CODE

a	b	c	d	e	...	z
1	2	3	4	5	...	26

Copymaster S458

NAME ______________________ **CHAPTER 12 TEST**

(Form A)

Solve. *(pages 308, 309)*

SALES CLERK
Monday-Friday
8-hr day $4.35/hr
Apply in person.
Pedal Power Bike Shop

1. How many hours would you work each week? 40
2. How much does the job pay each week? $174
3. Suppose that $27.40 is withheld for federal income tax and $12.35 is withheld for Social Security. What would your net pay be? $134.25
4. How much less would the net pay be than the gross pay? $39.75

Solve. *(page 310)*

5. What is the regular price of the sander? $54.60
6. How much is the discount on the sander? $8.19
7. What is the sale price of the sander? $46.41

Complete. Round each unit price to the nearest cent. *(page 312)*

8. The unit price of the smaller bag of apples is $.33 per pound.
9. The unit price of the larger bag of apples is $.32 per pound.
10. The larger bag is the better buy.

Solve. *(page 314)*

With a coupon, how much can you save when you buy a

11. small pizza? $.95
12. medium pizza? $1.48
13. large pizza? $1.97
14. How much more do you save on a large pizza than on a small one? $1.02
15. How much would you have to pay for 2 medium pizzas if you used a coupon? $13.32
16. You had $20. You bought a medium pizza with a coupon and paid the regular price for a small pizza. How much money did you have left? $8.13

PIZZA PALACE COUPON
Save on ONE Super Combo pizza.

	Reg.	With coupon
Small	$5.95	$5.00
Medium	$7.40	$5.92
Large	$8.95	$6.98

Limit 1 per customer.

ANSWERS
1. 40
2. $174
3. $134.25
4. $39.75
5. $54.60
6. $8.19
7. $46.41
8. $.33
9. $.32
10. larger
11. $.95
12. $1.48
13. $1.97
14. $1.02
15. $13.32
16. $8.13

 S45 Page 1

NAME ______________________ **CHAPTER 12 TEST**

(Form A)

Solve. *(page 318)*

NO. 106
March 16, 19 89
Pay to the order of Hal Lee $ 34.63
Thirty-four and 63/100 dollars
People's Bank
Lexington, MA
Jo Davis

17. What is the check number? 106
18. To whom was the check written? Hal Lee
19. Who signed the check? Jo Davis
20. Before Jo Davis wrote check no. 106, she had a balance of $406.38. What was the balance after she wrote check no. 106? $371.75

Solve. *(page 320)*

DEPOSITOR: RUTH WEAVER ACCOUNT NO.: 41-0305

DATE	WITHDRAWAL	DEPOSIT	INTEREST	BALANCE
04-03		500.00		500.00
04-23	50.00			450.00
05-01			2.06	452.06
05-15		125.00		577.06

21. How much money was deposited on April 3? $500
22. On what date was $50 withdrawn? April 23
23. How much interest was paid on May 1? $2.06
24. What was the balance after the interest was paid on May 1? $452.06
25. What would the balance be if $39.65 was deposited on May 16? $616.71
26. What would the balance be if $39.65 was deposited on Mav 16 and $50.00 withdrawn on May 17? $566.71

Compute the interest. Round to the nearest cent. *(page 322)*

27. Principal = $100
Rate = 10% per year
Time = 1 year
Interest = $10

28. Principal = $240
Rate = 12% per year
Time = 2 years
Interest = $57.60

29. Principal = $850
Rate = 1.5% per year
Time = 1 year
Interest = $12.75

Solve. *(page 324)*

30. What is the total amount owed to the electric company? $55.36
31. On what date is the payment due? May 15
32. How much more was the previous bill? $13.63
33. How many kilowatt hours (KWH) of electricity were used? 1181

CITY ELECTRIC

PRESENT READING (KWH)	PREVIOUS READING (KWH)	KWH USED	AMOUNT
4806	3625	1181	$62.49

ELECTRIC REFUND CREDIT $7.13
Previous Bill $68.99
Total Payment $68.99 Total Amount Due $55.36
Previous Balance $ 0.00 Payment Due Date 5/15

ANSWERS
17. 106
18. Hal Lee
19. Jo Davis
20. $371.75
21. $500
22. April 23
23. $2.06
24. $452.06
25. $616.71
26. $566.71
27. $10
28. $57.60
29. $12.75
30. $55.36
31. May 15
32. $13.63
33. 1181

 S46 Page 2

Copymasters S45 and S46
or Duplicating Masters S45 and S46

Chapter REVIEW

Here are scrambled answers for the review exercises:

18	40	72	200	interest	net	Robert Thayer	subtracted
38	60	75	added	less	pounds	Security	time

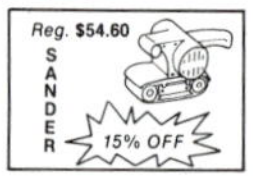

1. Security, net

2. 75, 72

3. pounds, 38

1. An employer is required by law to deduct from a paycheck money for federal income tax and social [?]. "Take home" pay is called an employee's [?] pay. *(page 309)*

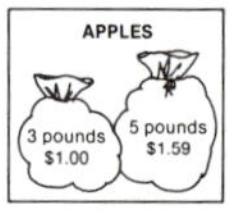

2. If an item is on sale for 25% off the regular price, you can compute the sale price by taking [?]% of the regular price. The sale price of this TV is $[?]. *(page 310)*

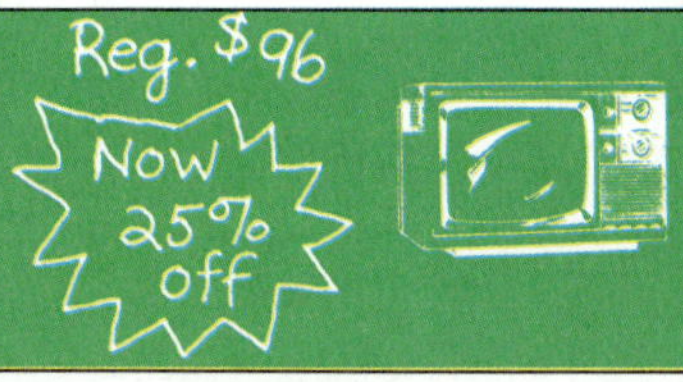

3. To find the price per pound of these oranges, you divide the price by the number of [?]. The unit price of the oranges rounded to the nearest cent is [?]¢. *(page 312)*

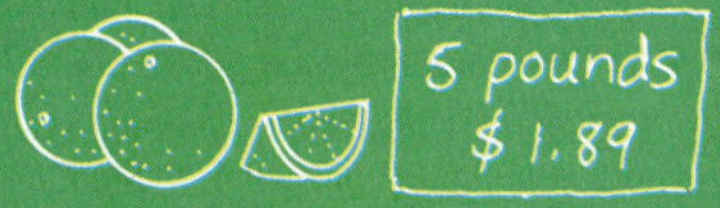

4. 200, 40, Robert Thayer, 60

4. The check number is [?]. The check was written for [?] dollars. [?] signed the check. *(page 280)*
The balance in the checking account before the check was written was $100. The balance after the check was written was [?] dollars. *(page 318)*

200
March 3 89
PAY TO THE ORDER OF Mike's Market 40.00
Forty and 00/100
Robert Thayer
⑆211137122?⑆ 1234025431⑈ 200

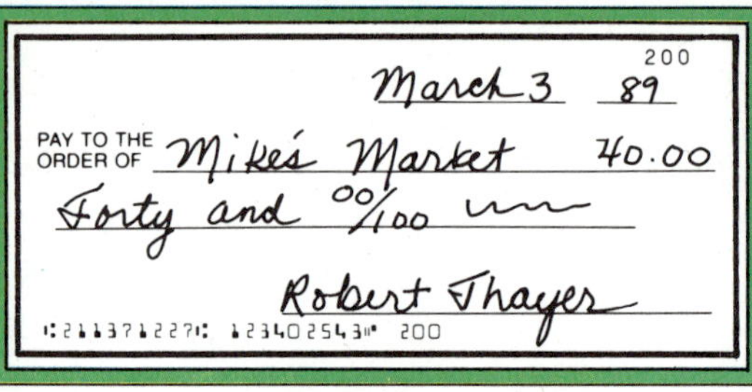

5. added, subtracted

6. time, interest

7. 18

5. Each time you make a deposit in a savings account, the amount is [?] to the balance of your account. Each time you make a withdrawal from a savings account, the amount is [?] from the balance of your account. *(page 320)*

6. The amount of interest charged for a loan depends on the principal, the interest rate, and the [?] for which the money is borrowed. You can use this formula to compute the [?] on a loan. *(page 322)*

$I = p \times r \times t$

7. The interest for this loan would be $[?]. *(page 322)*

Principal = $300
Rate = 12% per year
Time = 6 months

Chapter TEST

Solve. *(pages 308, 309)*

1. a. How much does the job pay each week?
   b. Suppose that Camper's Supply withheld $23.75 for federal income tax and $11.50 for Social Security. What would the net pay be? $134.75
   c. How much less would the net pay be than the gross pay? $35.25

   $170

SALES CLERK
No experience necessary.
Mon—Fri
8-hr day $4.25/hr
Apply in person
CAMPER'S SUPPLY

Compute the sale price. *(page 310)*

2. HIKING BOOTS $51.60
   Reg. $64.50
   20% OFF!

3. TENT $84
   Reg. $112
   25% OFF!

4. BACK PACK $39.20
   Reg. $56
   30% OFF!

Solve. *(pages 312, 314)*

3. What is the unit price of the smaller bag of trail mix? $.16
4. Which is the better buy, the small bag or the large bag? Small bag
5. Camper's Supply Store gave out a 50¢ coupon that could be used on the purchase of a large bag of trail mix. If a customer used a coupon, how much would 3 small bags and 1 large bag of trail mix cost? $7.86

$2.69 16 ounces
$1.89 12 ounces

Solve. *(page 318)*

6. Who signed the check? David Jones
7. What is the number of the check? 2003
8. What is the amount of the check? $48.93
9. To whom was the check written? Camper's Supply
10. David James had a balance of $167.34 in his checking account before he wrote the check. What was his balance after he wrote the check? $118.41

Solve. *(pages 320, 322)*

11. On March 24 you had a balance of $124 in your savings account. On March 25 you withdrew $35, and on March 31 the bank paid you $.82 interest. What was your balance then? $89.82

12. How much interest will you pay if you borrow $240 for 6 months if you are charged 1% per month? $14.40

Consumer Mathematics   331

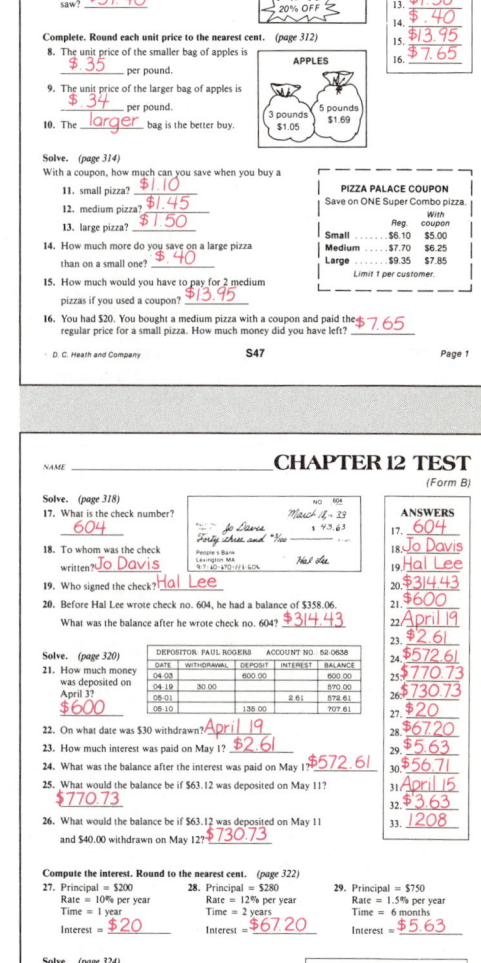

Cumulative Test
(Chapters 1–12)

Use Copymaster S109 to provide the students with an answer sheet in standardized test format.

Answers for Cumulative Test, Chapters 1–12

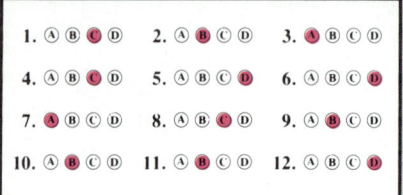

The table below correlates test items with student text pages.

Test Item	Page Taught	Skill Practice
1	182	p. 317, exercises 1–6
2	192	p. 317, exercises 7–12
3	206	p. 317, exercises 13–22
4	214	p. 317, exercises 23–32
5	228	p. 317, exercises 33–38
6	248	p. 329, exercises 1–8
7	282	p. 329, exercises 9–20
8	284	p. 329, exercises 21–27
9	294	p. 329, exercises 28–33
10	296	p. 329, exercises 34–39
11	264	
12	322	

Cumulative TEST — Standardized Format

Choose the correct letter.

1. Add. $2\frac{5}{8} + 1\frac{1}{2}$
- A. $3\frac{1}{8}$
- B. $3\frac{3}{5}$
- (C.) $4\frac{1}{8}$
- D. none of these

2. Subtract. $3\frac{1}{4} - 1\frac{1}{3}$
- A. $2\frac{11}{12}$
- (B.) $1\frac{11}{12}$
- C. $2\frac{1}{12}$
- D. none of these

3. Give the product. $2\frac{1}{2} \times 1\frac{1}{4}$
- (A.) $3\frac{1}{8}$
- B. 2
- C. $\frac{8}{25}$
- D. none of these

4. Give the quotient. $4\frac{2}{3} \div 1\frac{3}{4}$
- A. $8\frac{1}{6}$
- B. $\frac{3}{8}$
- (C.) $2\frac{2}{3}$
- D. none of these

5. 425 mm = ? m
- A. 4.25
- B. 42.5
- C. 4250
- (D.) none of these

6. Add. 4 lb 9 oz + 5 lb 8 oz
- A. 10 lb 7 oz
- B. 10 lb 5 oz
- C. 10 lb 3 oz
- (D.) none of these

7. Change to a fraction. $33\frac{1}{3}\% = ?$
- (A.) $\frac{1}{3}$
- B. $\frac{2}{3}$
- C. 3
- D. none of these

8. Change to a percent. $\frac{5}{6} = ?$
- A. $16\frac{2}{3}\%$
- B. $66\frac{2}{3}\%$
- (C.) $83\frac{1}{3}\%$
- D. none of these

9. Solve. 12% of 42 = n
- A. 350
- (B.) 5.04
- C. 50.4
- D. none of these

10. Solve. 5% of n = 12
- A. 60
- (B.) 240
- C. 0.6
- D. none of these

11. You jog 3 miles in 24 minutes. At that rate, how far can you jog in 36 minutes?
- A. 6 miles
- (B.) 4.5 miles
- C. 1.5 miles
- D. none of these

12. Compute the interest. Principal = $650, Rate = 15% per year, Time = 6 months
- A. $97.50
- B. $195
- C. $585
- (D.) none of these

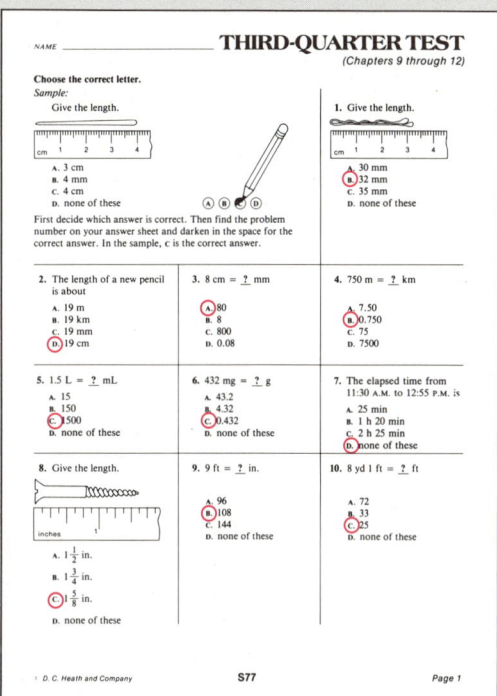

Use Copymaster S92 or Duplicating Master S92 to provide the students with an answer sheet in standardized test format.

Copymaster S107 has a quick-score answer key for the third-quarter test.

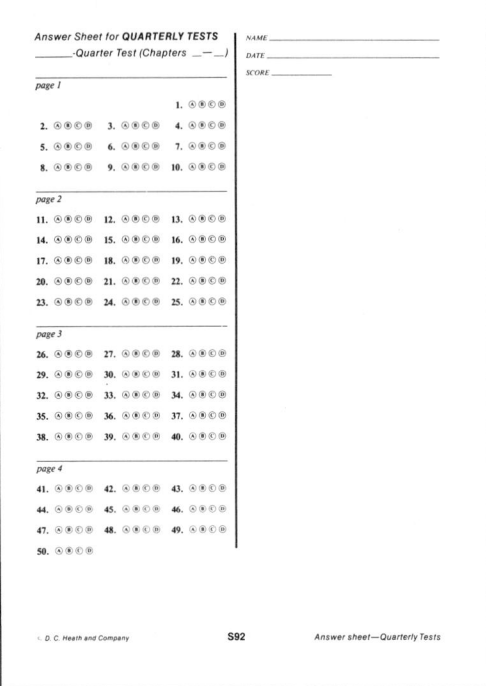

The table below correlates test items with student text pages.

Test Item	Text Page
1	224
2	226
3	228
4	228
5	230
6	232
7	238
8	240
9	242
10	242
11	244
12	246
13	246

Test Item	Text Page
14	248
15	248
16	258
17	260
18	262
19	262
20	264
21	264
22	270
23	270
24	272
25	272
26	282

Test Item	Text Page
27	282
28	284
29	284
30	286
31	286
32	287
33	287
34	290
35	294
36	296
37	296
38	298
39	298

Test Item	Text Page
40	298
41	308
42	309
43	310
44	312
45	314
46	318
47	320
48	322
49	322
50	324

Geometry—Perimeter and Area

Chapter 13
Geometry—Perimeter and Area

Resources

- **Class Starter Quizzes 131-142** *(Copymasters S417-S420)*
- **Visual Aids 30, 60-64** *(Copymasters or Transparencies S135, S159-S162)*
- **Manipulatives**
  Manipulative Activities 22-25 *(Copymasters S522-S523)*
  Area tiles, tangram pieces and shapes, area grid, area-formula pieces *(Copymasters or Transparencies S532-S534, S536-S539, S540, S541)*
- **Worksheets 143-155** *(Copymasters, Duplicating Masters, or Workbook pages S311-S323)*
- **Challenge Problems** for pages 337, 339, 341, 347, 351, 355, 357, 359 *(Copymasters S458-S460)*
- **Projects** for pages 343, 349, 353 *(Copymasters S484-S485)*
- **Tests** *(Copymasters or Duplicating Masters S49-S52)*

Lesson Objectives

To name and draw lines and angles
To use a protractor to measure angles
To classify angles
To use a protractor to draw angles

Problem-Solving Skill

Estimating measurements

Starting the Lesson

Estimation Draw these three angles on the chalkboard:

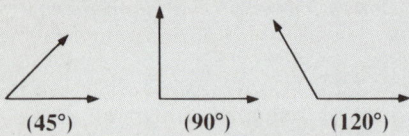

Have the students estimate the degree measure of each angle. To determine who is the best estimator, have each student compute the difference between the real measure and the measure gotten for each angle. For example, getting either 43° or 47° for the 45° angle results in a degree difference of 2°. The student whose "degree-difference" total is lowest is the best estimator.

Here's How Note

Use of Concrete Materials A protractor is on ■ *Visual Aid 60* (copymaster or transparency S159). You may choose to use the visual aid when demonstrating how to use a protractor to measure and draw angles.

Exercise Note

The protractor decoder in exercise 28 is also on ■ *Visual Aid 60* (copymaster or transparency S159).

Lines and Angles

A **straight line,** or simply a **line,** extends indefinitely in two directions. A **ray** is part of a line that starts at a point and extends indefinitely in one direction. A **line segment,** or simply a **segment,** is part of a line between two points of the line.

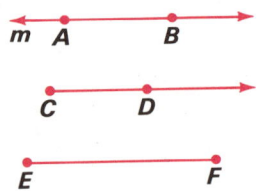

line *m* or line *AB* or $\overleftrightarrow{AB}$

ray *CD* or $\overrightarrow{CD}$

segment *EF* or $\overline{EF}$

An **angle** is formed by two rays with a common endpoint. The common endpoint is called the **vertex** of the angle, and the rays are called the **sides** of the angle.

Here's how to use a protractor to measure an angle.

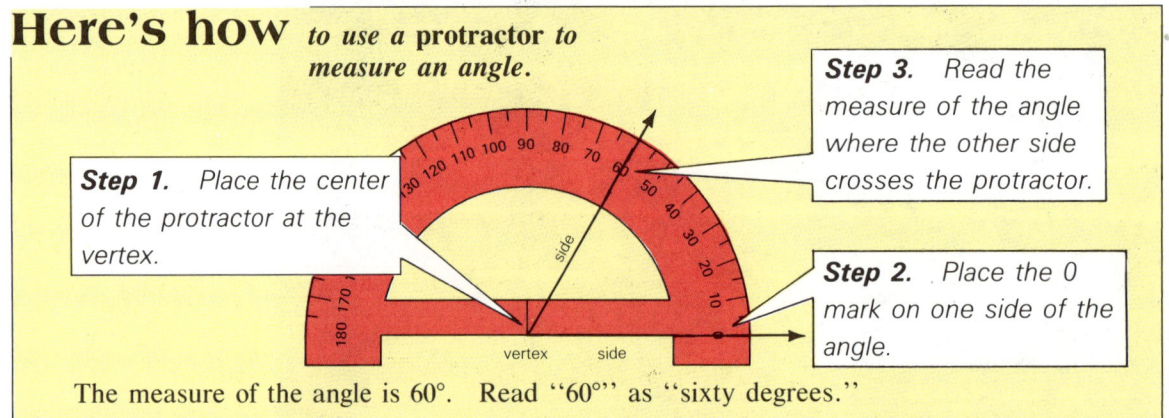

Step 1. Place the center of the protractor at the vertex.

Step 2. Place the 0 mark on one side of the angle.

Step 3. Read the measure of the angle where the other side crosses the protractor.

The measure of the angle is 60°. Read "60°" as "sixty degrees."

An **acute** angle measures between 0° and 90°.

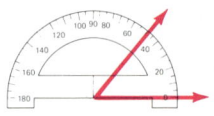

A **right** angle measures 90°.

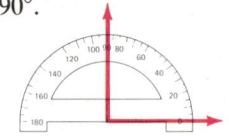

An **obtuse** angle measures between 90° and 180°.

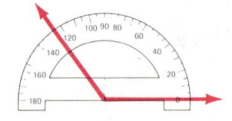

Here's how to draw a 50° angle.

1. Draw one side.

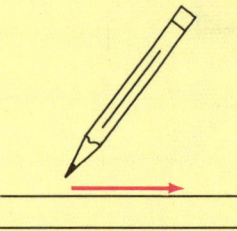

2. Place the protractor as you would for measuring and make a mark at 50°.

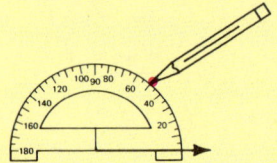

3. Draw the other side.

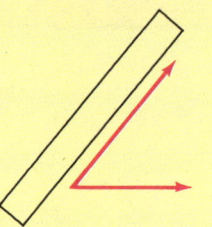

334   Chapter 13

EXERCISES

Name each of the following with words or symbols.

1.
line g

2.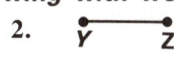
Segment YZ or Segment ZY or $\overline{YZ}$ or $\overline{ZY}$

3.
ray RP or $\overrightarrow{RP}$

4.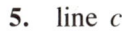
line HB or line BH or $\overleftrightarrow{HB}$ or $\overleftrightarrow{BH}$

Draw a representation of each of the following.

5. line c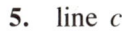
c

6. $\overline{AF}$
A F

7. ray GL
G L

8. $\overrightarrow{BA}$
B A

9. line segment LP
L P

10. $\overleftrightarrow{GK}$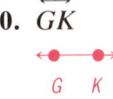
G K

11. $\overline{XY}$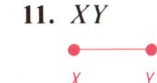
X Y

12. ray RZ
R Z

Measure each angle.

13. 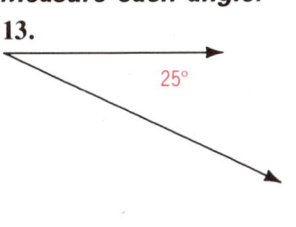 25°

14. 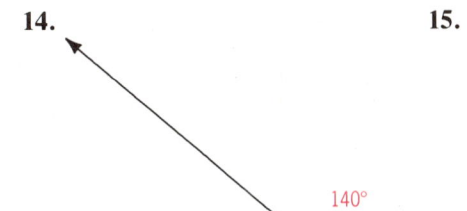 140°

15. 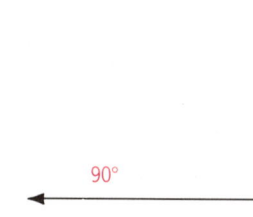 90°

Tell whether each angle is acute, right, or obtuse.

16. 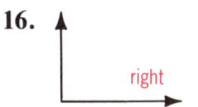 right

17. 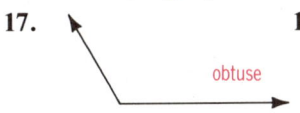 obtuse

18. 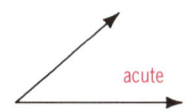 acute

19. 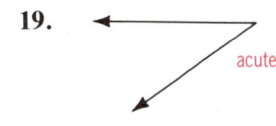 acute

Draw angles having these measures.

20. 45° 21. 120° 22. 90° 23. 75° 24. 150° 25. 32° 26. 135° 27. 88°

Crack the code

28. Use the code to get the answer.

   RIDDLE: What time is it when an elephant stands on your protractor?

   ANSWER: 15° * 53° * 126° * 88°
   12° * 105°
   37° * 84° * 17°
   165°   63° * 82° * 155°
   145° * 45° * 102° * 18° * 43° * 168° * 25° * 11° * 103° * 49°.

   TIME TO GET A NEW PROTRACTOR

Geometry, Perimeter, and Area

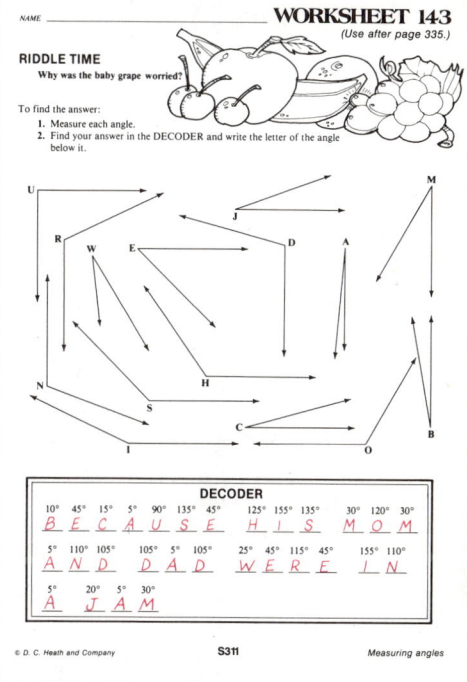

Practice Worksheet
Workbook S311, Copymaster S311, or Duplicating Master S311

Group Project

Measuring angles

Ask the students, working in small groups, to find and measure 15 angles in the classroom. Tell them to use their findings to determine the most common angle in the room.

Perpendicular and parallel lines

Two lines that intersect to form a right angle are **perpendicular**.

line *l* is perpendicular to line *m*
or
line *l* ⊥ line *m*

A square corner means the angle is a right angle.

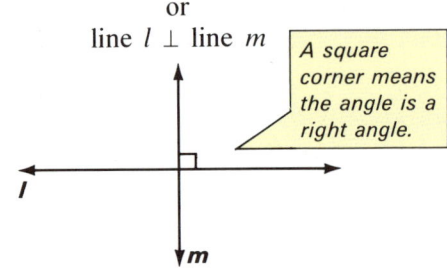

Two segments are perpendicular if the lines that contain them are perpendicular.

segment *AB* is perpendicular to segment *BC*
or
$\overline{AB} \perp \overline{BC}$

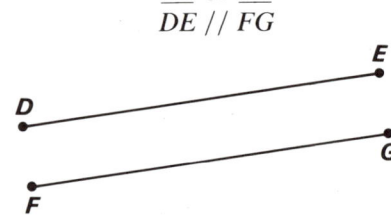

Two lines that do not intersect are **parallel**.

line *u* is parallel to line *v*
or
line *u* // line *v*

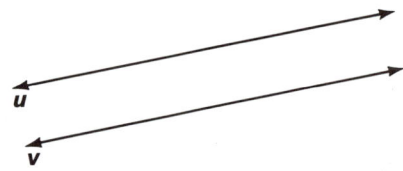

Two segments are parallel if the lines that contain them are parallel.

segment *DE* is parallel to segment *FG*
or
$\overline{DE}$ // $\overline{FG}$

EXERCISES
True or false?

1. line *s* // line *r* True
2. line *r* ⊥ line *s* False
3. line *u* // line *t* True
4. line *u* ⊥ line *r* True
5. $\overline{DC} \perp \overline{BC}$ True
6. $\overline{CB} \perp \overline{BA}$ False
7. $\overline{DA}$ // $\overline{CB}$ True
8. There are more than 10 right angles in the drawing. True
9. There are 4 acute angles in the drawing. True
10. There are more than 4 obtuse angles in the drawing. False

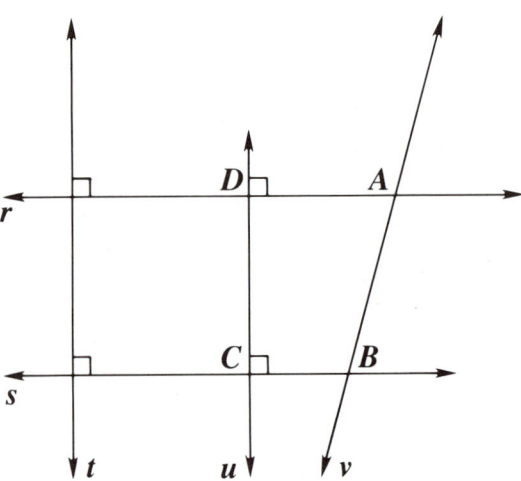

11. Line *l* is perpendicular to line *m*. How many right angles are there? 4

12. Line *p* and line *q* are not perpendicular.
    a. How many acute angles are there? 2
    b. Do all the acute angles have the same measure? yes
    c. How many obtuse angles are there? 2
    d. Do all the obtuse angles have the same measure? yes

13. Line *s* is parallel to line *t*.
    Line *u* is not perpendicular to line *s* or line *t*.
    a. How many acute angles are there? 4
    b. Do all the acute angles have the same measure? yes
    c. How many obtuse angles are there? 4
    d. Do all the obtuse angles have the same measure? yes

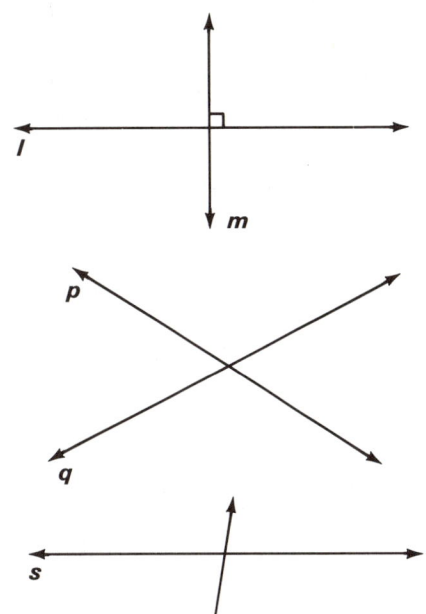

Street wise Logical reasoning

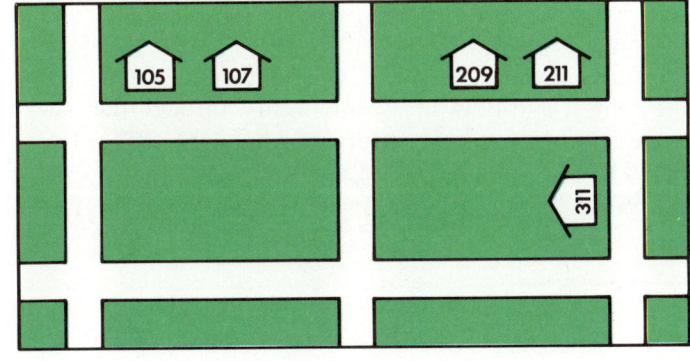

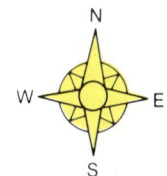

Clues:
- Paul lives east of Cindy.
- Darla lives west of Cindy.
- Paul lives between Cindy and Tim.
- Jan's street is perpendicular to Cindy's street.

Use the clues to answer the questions.

14. Who lives in house number 105? Darla
15. Who lives in house number 209? Paul
16. What is Cindy's house number? 107
17. What is Jan's house number? 311

Geometry, Perimeter, and Area   **337**

Practice Worksheet

Worksheet S312, Copymaster S312, or Duplicating Master S312

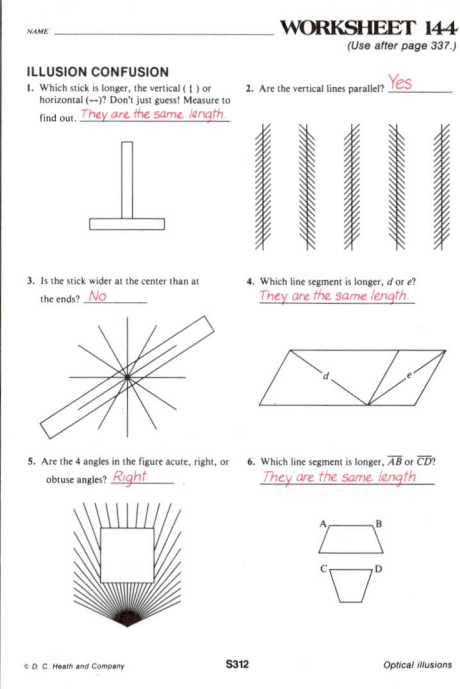

Challenge Problem

Draw the capital letters of the alphabet so that as many as possible have all the lines either parallel or perpendicular.

Copymaster S458

Class Starter Quiz 132
on previous lesson

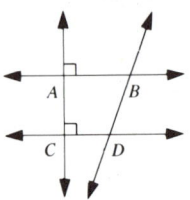

True or false?

1. Line AB is parallel to line CD. True
2. Line BD is perpendicular to line AB. False
3. Line AC is perpendicular to line CD. True
4. Line BD is parallel to line AC. False
5. Line CD is perpendicular to line AC. True

Copymaster S418

Lesson Objectives
To use a straightedge and compass to copy segments
To use a straightedge and compass to bisect segments and angles

Problem-Solving Skill
Following directions

Starting the Lesson
Write this sentence on the chalkboard:

In geometry, a <u>APOCMSS</u> and <u>EDTGITSRAHEG</u> are used to construct drawings.

Before the students open their books challenge them to unscramble the underlined words so that the sentence makes sense. Then have them open their books and read the top of page 336 to see if they correctly unscrambled the letters.

338

Geometric constructions

In geometry, a compass and straightedge are used to construct drawings.

Here's how *to construct a line segment that is the same length as another line segment.*

Step 1.
Use a straightedge to draw a ray and label its endpoint R.

Step 2.
Set the compass legs on points A and B.

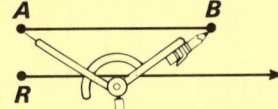

Step 3.
Use this setting to mark off the same length on the ray.

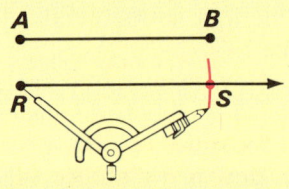

Segment RS is the same length as segment AB.

1. Look at the *Here's how*. Which segment is the same length as segment AB? RS

To **bisect a segment** is to divide it into two segments that are the same length.

Here's how *to construct the perpendicular bisector of a segment.*

Step 1.
Open the compass so that the setting is more than half the length of $\overline{CD}$. Use C as the center and draw arcs above and below $\overline{CD}$.

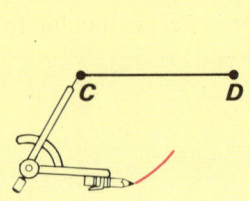

Step 2.
With the same setting, use D as the center, and draw arcs above and below segment CD.

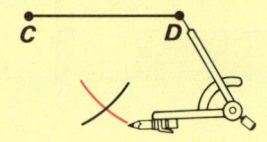

Step 3.
Use a straightedge and draw a line through the points where the arcs cross.

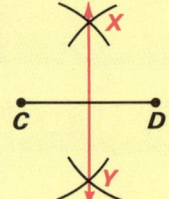

Line XY is the perpendicular bisector of segment CD.

2. Line XY is the perpendicular bisector of which segment? $\overline{CD}$

338   Chapter 13

To **bisect an angle** is to divide the angle into two angles that have the same measure.

Here's how *to bisect an angle.*

Step 1.

Use *V* as the center and draw an arc that intersects both sides.

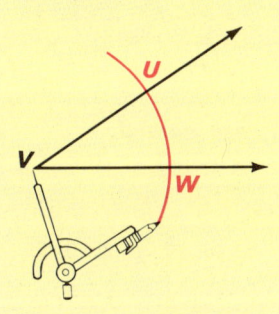

Step 2.

Use *U* and *W* as centers and draw intersecting arcs with the same compass setting.

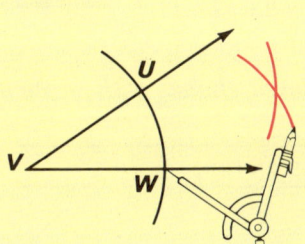

Step 3.

Use a straightedge and draw the ray from *V* through the point where the arcs intersect.

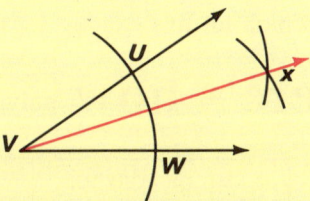

Ray *VX* is the angle bisector of angle *V*.

EXERCISES

Use a straightedge and compass.

Construct segments that are the same length.

3. 4. 5.

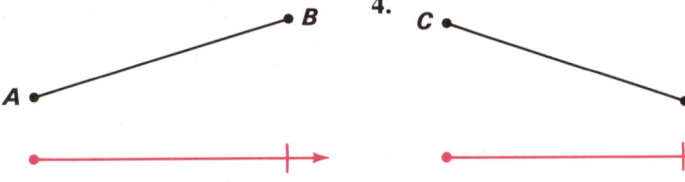

6. Draw a segment. Construct the perpendicular bisector.

7. Draw an angle and bisect it.

8. Draw a large triangle. Construct the bisector of each angle. If you are very careful, the three bisectors will all cross at the same point.

9. Draw a large triangle. Construct the perpendicular bisector of each side. If you are careful, the three perpendicular bisectors will all cross at the same point.

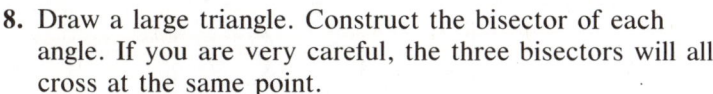

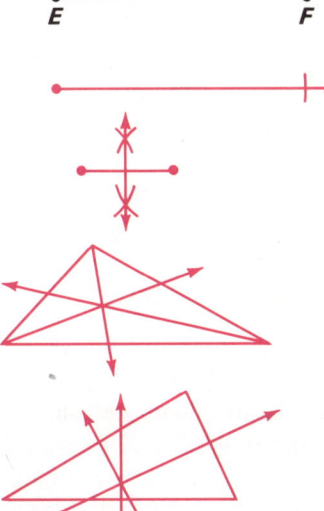

Geometry, Perimeter, and Area **339**

Practice Worksheet

Workbook S313, Copymaster S313, or Duplicating Master S313

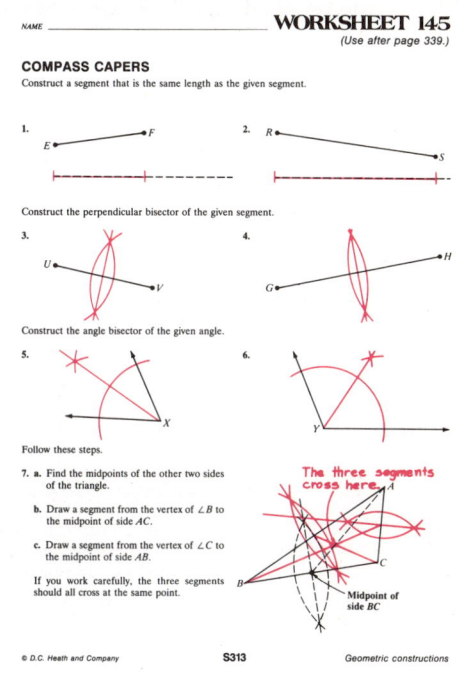

Challenge Problems

Experiment. Use your compass to make these designs.

1.
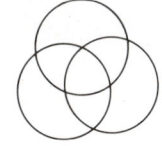

2.

Copymaster S459

339

Polygons

1. Which lake lot is mine? My lot has 2 pairs of parallel sides. It has no right angles.   Lot F

   LOT A
   LOT B
   LOT C
   LOT D
   LOT E
   LOT F
   CASS LAKE

2. Which lot has 4 sides all the same length?   Lot B
3. Which lot has exactly 1 pair of parallel sides?   Lot C

Here's how *polygons (closed shapes with segments as sides) are named.*

NAME OF POLYGON	DESCRIPTION	EXAMPLES
Triangle	3 sides	
Square	4 sides the same length 4 right angles	
Rectangle	4 sides 4 right angles	A square is also a rectangle.
Parallelogram	4 sides 2 pairs of parallel sides	A rectangle is also a parallelogram.
Trapezoid	4 sides Exactly 1 pair of parallel sides	
Pentagon	5 sides	
Hexagon	6 sides	

4. Use the map and the *Here's how* chart to answer these questions.
   a. Which lot is a square?   Lot B
   b. Which 2 lots are rectangles?   Lots A, B
   c. Which 3 lots are parallelograms?   Lots A, B, F
   d. Which lot is a triangle?   Lot E
   e. Which lot is a trapezoid?   Lot C
   f. Which lot is a pentagon?   Lot D

340   Chapter 13

EXERCISES

Name these polygons. Some shapes have more than one name.

5. 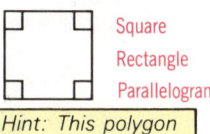 Square, Rectangle, Parallelogram

   Hint: This polygon has 3 names.

6. 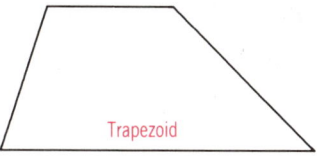 Trapezoid

7. 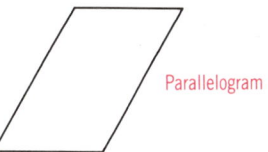 Parallelogram

8. 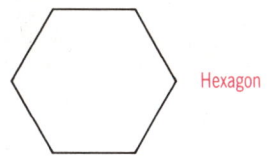 Hexagon

9.  Pentagon

10. 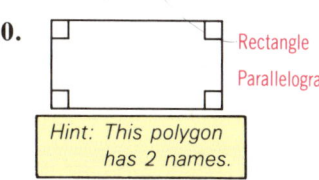 Rectangle, Parallelogram

    Hint: This polygon has 2 names.

11. 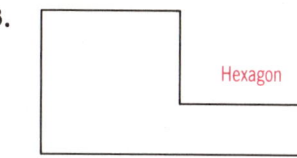 Triangle

12.  Square, Rectangle, Parallelogram

13. Hexagon

Use the clues. Draw and name each polygon.

14. Clues:
    - This polygon has 4 sides.
    - It has no right angles.
    - It has 2 pairs of parallel sides.

    Parallelogram

15. Clues:
    - This polygon has 2 right angles.
    - It has 4 sides.
    - It has 1 acute angle.
    - It has 1 obtuse angle.

    Trapezoid

Angles of a triangle

Here's how to show that the sum of the measures of the angles of a triangle is 180°.

Step 1. Cut out a large triangle. Tear off the corners.

Step 2. Arrange the corners to show that the angle sum is 180°.

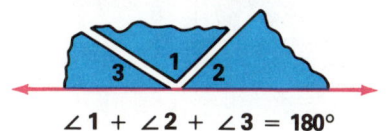

∠1 + ∠2 + ∠3 = 180°

The measures of two angles are given. What is the size of the third angle? Do not use a protractor.

16. 60° 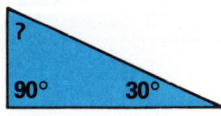 90°, 30°

17. 35° 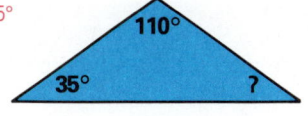 110°, 35°, ?

18. 130° 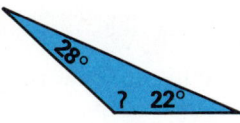 28°, ?, 22°

Geometry, Perimeter, and Area

Class Starter Quiz 134
on previous lesson

Make a sketch of each polygon.
1. square
2. triangle
3. rectangle
4. hexagon
5. pentagon
6. trapezoid

Copymaster S418

Lesson Objective
To compute the perimeter of a polygon, given the length of its sides

Problem-Solving Skills
Finding information in a drawing
Using a formula

Starting the Lesson
Have the students look at the paintings at the top of the page. Ask them which painting needed 96 centimeters of framing. (24 cm × 24 cm painting) Then go over the *Here's how* and discuss questions 1 and 2.

Here's How Note
After going over questions 1 and 2, let the students practice using the formulas on these squares and rectangles.

12 cm × 12 cm
P = 48 cm

1.2 km × 4.5 km
P = 11.4 km

2.3 m × 1.75 m
P = 8.1 cm

Perimeter

FRAMED?

Jeff used narrow rope to frame these paintings. Which one needed 96 centimeters of rope? the boat

24 cm

24 cm

40 cm

32 cm

The **perimeter** of the figure is the distance around the figure.

Here's how *to use formulas to find the perimeters of squares and rectangles.*

Square

24 cm

The perimeter (*P*) of a square is 4 times the length of one side (*s*).

FORMULA $P = 4 \times s$
$P = 4 \times 24$ cm
$= 96$ cm

The perimeter is 96 cm.

Rectangle

40 cm
32 cm

The perimeter (*P*) of a rectangle is 2 times the sum of the length (*l*) and the width (*w*).

FORMULA $P = 2 \times (l + w)$
$P = 2 \times (40 \text{ cm} + 32 \text{ cm})$
$= 2 \times 72$ cm
$= 144$ cm

The perimeter is 144 cm.

1. Look at the *Here's how*. What is the formula for the perimeter of a square? $P = 4 \times s$ What does the letter *s* stand for? Length of a side

2. What is the formula for the perimeter of a rectangle? What does each letter stand for? $P = 2 \times (l + w)$ $P =$ perimeter, $l =$ length, $w =$ width

EXERCISES

Find each perimeter.

3. 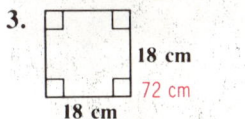 18 cm, 18 cm — 72 cm

4. 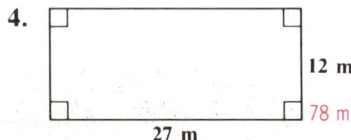 12 m, 27 m — 78 m

5. 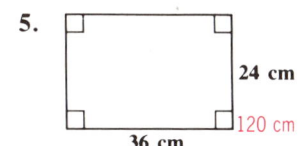 24 cm, 36 cm — 120 cm

6. 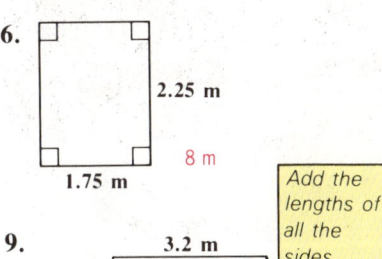 2.25 m, 1.75 m — 8 m

7. 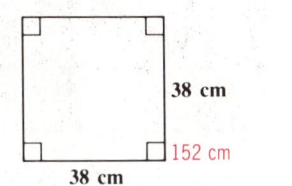 38 cm, 38 cm — 152 cm

8. 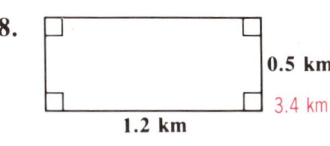 0.5 km, 1.2 km — 3.4 km

Add the lengths of all the sides.

9. 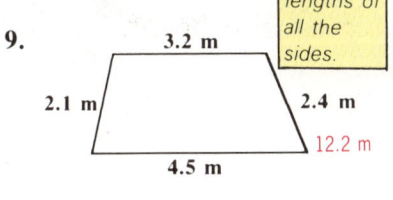 3.2 m, 2.1 m, 2.4 m, 4.5 m — 12.2 m

10. 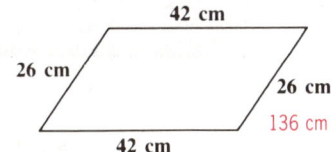 42 cm, 26 cm, 26 cm, 42 cm — 136 cm

11. 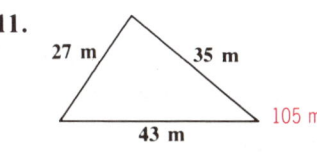 27 m, 35 m, 43 m — 105 m

12. 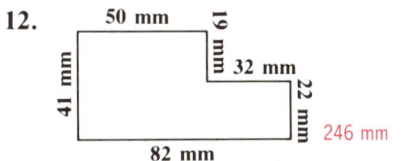 50 mm, 19 mm, 32 mm, 22 mm, 82 mm, 41 mm — 246 mm

13. 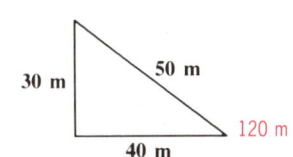 30 m, 50 m, 40 m — 120 m

14. 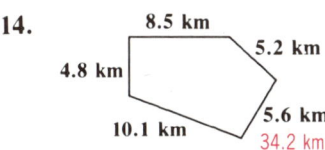 8.5 km, 5.2 km, 5.6 km, 10.1 km, 4.8 km — 34.2 km

Solve.

15. How many centimeters of rope are needed to frame a painting that is 65 centimeters long and 40 centimeters wide? 210 cm

16. A square painting, 25 centimeters on each side, is to be framed. How many centimeters of framing are needed? 100 cm

17. How much fencing is needed to enclose a rectangular yard that is 15 meters by 24 meters? 78 m

18. A square pen is built. The pen is 15.5 meters on each side. How many meters of fencing are needed? 62 m

Triangle tangle

19. Find the triangle that has the same perimeter as triangle *AEI*. Triangle CGK

20. Find 5 triangles that have the same perimeter as triangle *ABL*. Triangles BCD, DEF, FGH, HIJ, JKL

21. Find the 2 parallelograms that have the same perimeter as parallelogram *LCFI*. Parallelograms ADGJ, BEHK

Geometry, Perimeter, and Area **343**

Practice Worksheet

Workbook S315, Copymaster S315, or Duplicating Master S315

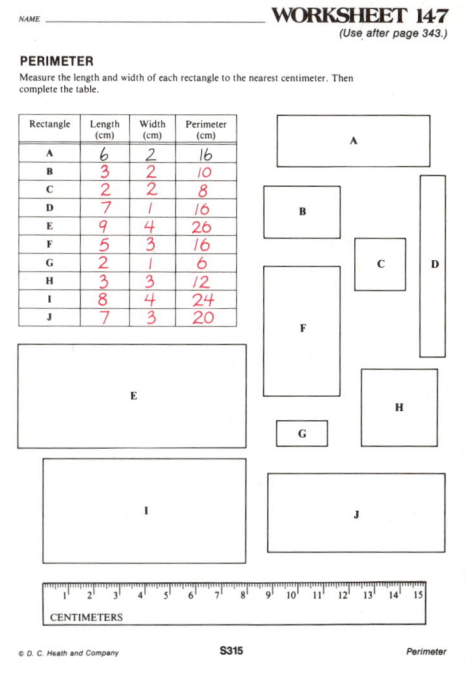

Project

Measuring perimeter

Find the perimeter of the classroom, of a table top, and of your desk top. In each case you will have to decide on an appropriate unit of measurement.

Copymaster S484

Class Starter Quiz 135
on previous lesson

Give the perimeter of each figure.

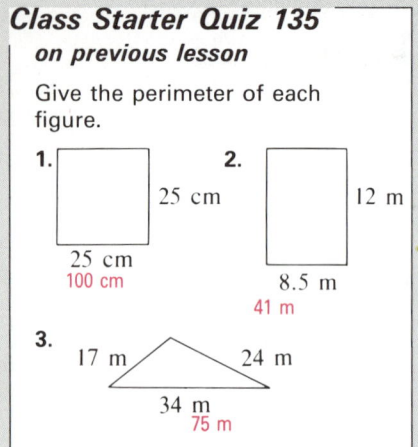

1. 25 cm square, 25 cm — **100 cm**
2. 12 m by 8.5 m — **41 m**
3. Triangle 17 m, 24 m, 34 m — **75 m**

Copymaster S419

Lesson Objective
To compute the circumference of a circle, given its diameter or radius.

Problem-Solving Skills
Finding information in a drawing
Using a formula

Starting the Lesson
What are the facts? Have the students study the information at the top of page 344 for 30 seconds. Then tell them to close their books and answer these questions from memory:

- Which has the greater radius, the front or back wheel? (Front wheel)
- What is the radius of the front wheel? (32 inches)
- Which wheel has a diameter of 20 inches? (Back wheel)
- Is the diameter of the front wheel twice its radius? (Yes)

Here's How Note
After going over questions 1–4, have the students practice by finding the circumference of these circles.

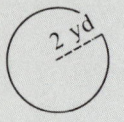 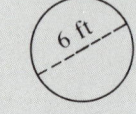 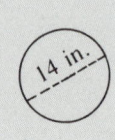

12.56 yd   18.84 ft   43.96 in.

344

Circumference

The radius of the front wheel is 32 inches, and the radius of the back wheel is 10 inches.

1. Which wheel has a diameter of 64 inches, the front or the back? **Front**
2. Make a guess! When the front wheel goes around once, will the bicycle travel more or less than 150 inches? **More**

Notice that the *diameter* is twice the *radius*.

Here's how to use a formula to find the distance around a circle.

The distance around a circle is called the **circumference**. The circumference of a circle is a little more than 3 times the length of its diameter.

To find the circumference (*C*), multiply π (read as "pi") by the diameter (*d*). We'll use 3.14 as a decimal approximation for π.

Front wheel

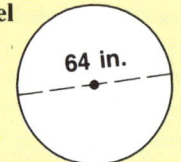

FORMULA $C = \pi \times d$
$C \approx 3.14 \times 64$ in.
≈ 200.96 in.

≈ means "is approximately equal to."

3. Look at the *Here's how*. About how far does the bicycle travel when the front wheel goes around once? **200.96 in.**
4. To compute the circumference of the back wheel, you would multiply 3.14 times what number? **20**

EXERCISES
Find the circumference. Use 3.14 for π. Here are scrambled answers for the next row of exercises: 18.84 in.   31.4 in.   25.12 in.   12.56 in.

5. 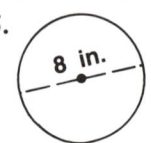 8 in. — 25.12 in.
6. 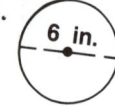 6 in. — 18.84 in.
7. 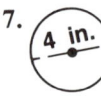 4 in. — 12.56 in.
8. 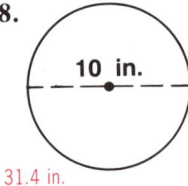 10 in. — 31.4 in.

Chapter 13

9.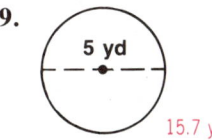
15.7 yd

10.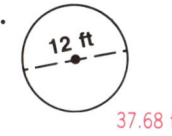
37.68 ft

11.
7.85 ft

12.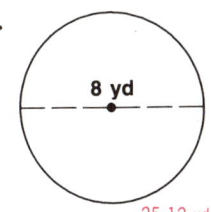
25.12 yd

13. 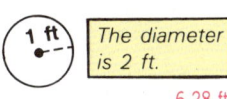 The diameter is 2 ft.
6.28 ft

14.
62.8 in.

15.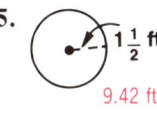
9.42 ft

16.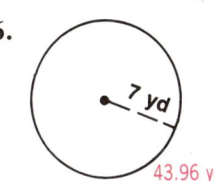
43.96 yd

17. Find and correct the three wrong answers.

   a. *Question:* What is the diameter of a wheel that has a radius of $1\frac{1}{2}$ feet?
      Answer: The diameter is 3 feet.
   b. *Question:* What is the radius of a circle that has a diameter of $6\frac{1}{2}$ feet?
      Answer: The radius is 13 feet. $3\frac{1}{4}$ feet
   c. *Question:* Which has the larger circumference, a circle with a diameter of 3 feet or a circle with a radius of 2 feet?
      Answer: The circle with a diameter of 3 feet. The circle with a radius of 2 feet.
   d. *Question:* What is the circumference of a wheel that has a radius of 26 inches?
      Answer: Approximately 81.64 inches. 163.28 inches
   e. *Question:* What do you get when you divide the circumference of a circle by π (3.14)?
      Answer: You get the diameter of the circle.

Seeing is not believing

Are the red sides straight or curved? *Hint: Check with an edge of a paper.*

18.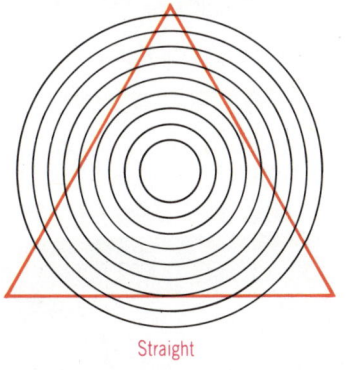
Straight

19.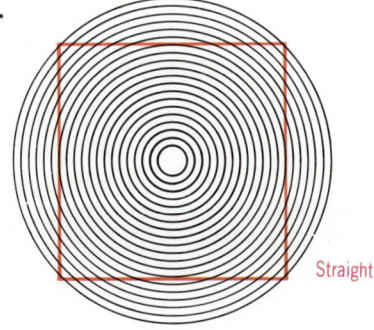
Straight

Geometry, Perimeter, and Area **345**

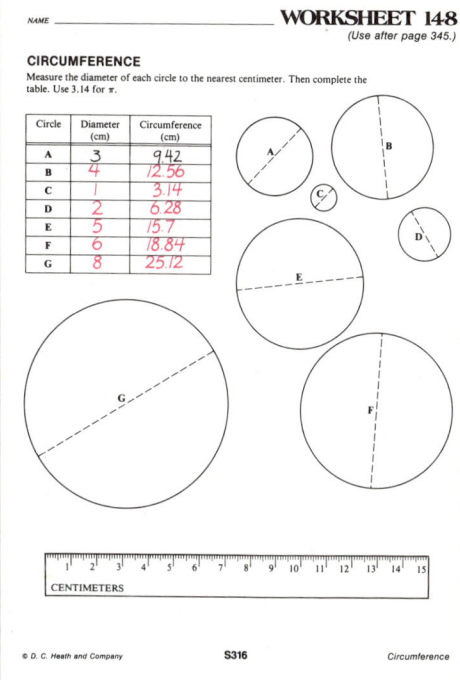

Class Starter Quiz 136
on previous lesson

Give the circumference of each circle. Use 3.14 for π.

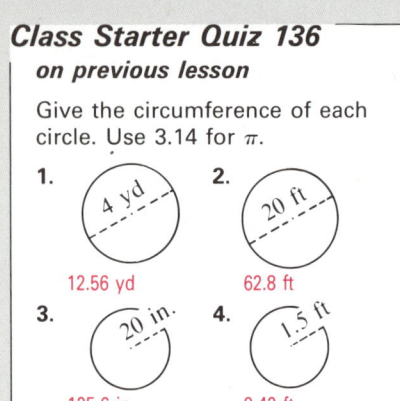

1. 4 yd — 12.56 yd
2. 20 ft — 62.8 ft
3. 20 in. — 125.6 in.
4. 1.5 ft — 9.42 ft

Copymaster S419

Problem-Solving Skills
Finding information in an ad
Using a drawing

Skills Reviewed
Changing mixed numbers to fractions
Adding and subtracting fractions
Multiplying and dividing fractions
Finding the percent of a number
Finding a number when a percent of the number is known

Starting the Lesson

Problem Solving Have the students make a drawing to help answer each customer's question at the top of page 346. The drawing in problem 3 is also on ■ **Visual Aid 63** (copymaster or transparency S161). You may wish to use the visual aid and work through problem 3 before assigning the other problems.

Cumulative Skill Practice Write these four answers on the chalkboard:

6.75 $1\frac{3}{10}$ $\frac{7}{24}$ $\frac{1}{9}$

Challenge the students to an answer hunt by saying, "Look at exercises 1–44 on page 347. Find the four exercises that have these answers. You have five minutes to find as many of the exercises as you can." (Exercises 12, 21, 32, and 36)

346

Problem solving
Using a drawing
YOU'RE THE CLERK!

Use a drawing to help answer each customer's question. Decide when a calculator would be useful.

MEL'S HARDWARE WEEKEND SPECIALS
Wire Fence	$1 per foot
Steel Posts	$4.50 each
Gates	$25.50 each

1. Is 200 feet of fence enough to build a pen that is 60 feet long by 30 feet wide? **Yes**

2. How many feet of fence will I need to fence in my 120-foot by 75-foot yard? **390**

Use the weekend special prices. Find the total cost for each fencing project.

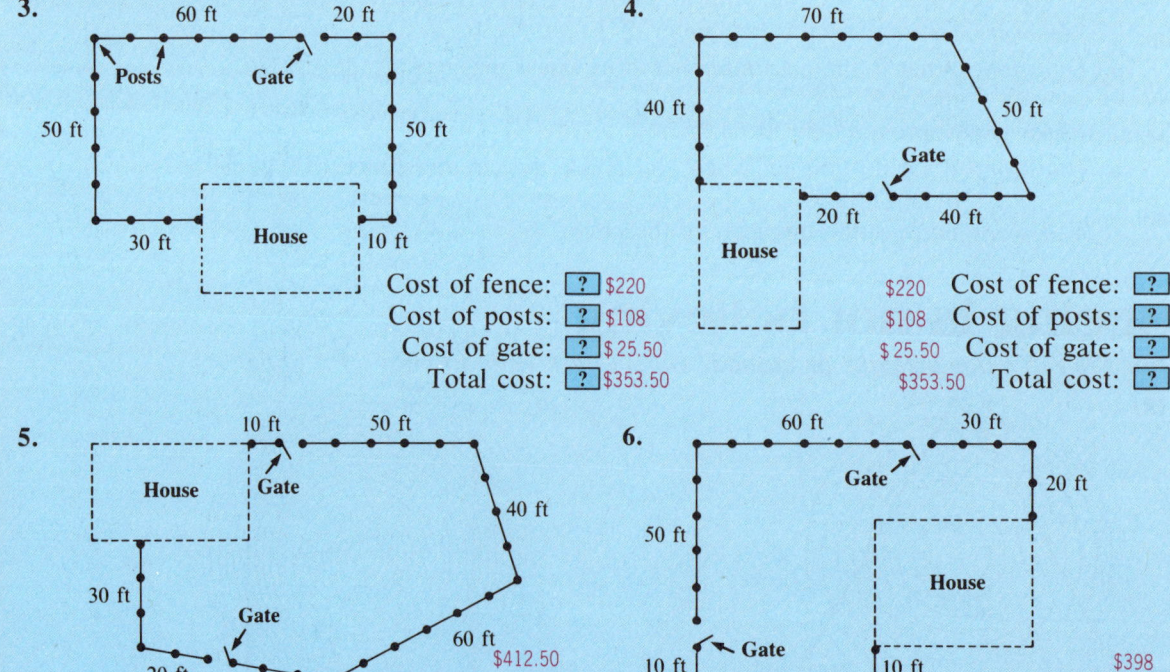

346 Chapter 13

Cumulative Skill Practice

Change to a decimal. *(page 166)*

1. $2\frac{1}{5}$ 2.2
2. $1\frac{1}{5}$ 1.2
3. $1\frac{4}{5}$ 1.8
4. $4\frac{1}{8}$ 4.125
5. $3\frac{3}{10}$ 3.3
6. $6\frac{1}{8}$ 6.125
7. $3\frac{1}{2}$ 3.5
8. $6\frac{3}{8}$ 6.375
9. $8\frac{1}{2}$ 8.5
10. $4\frac{5}{8}$ 4.625
11. $5\frac{1}{4}$ 5.25
12. $6\frac{3}{4}$ 6.75
13. $4\frac{3}{10}$ 4.3
14. $3\frac{1}{25}$ 3.04

Give the sum in simplest form. *(page 180)*

15. $\frac{3}{8} + \frac{1}{4}$ $\frac{5}{8}$
16. $\frac{3}{4} + \frac{1}{6}$ $\frac{11}{12}$
17. $\frac{1}{3} + \frac{1}{2}$ $\frac{5}{6}$
18. $\frac{3}{4} + \frac{2}{3}$ $1\frac{5}{12}$
19. $\frac{5}{8} + 4$ $4\frac{5}{8}$
20. $\frac{2}{3} + \frac{5}{9}$ $1\frac{2}{9}$
21. $\frac{1}{2} + \frac{4}{5}$ $1\frac{3}{10}$
22. $\frac{2}{3} + \frac{1}{6}$ $\frac{5}{6}$
23. $\frac{3}{5} + \frac{3}{4}$ $1\frac{7}{20}$
24. $3 + \frac{3}{10}$ $3\frac{3}{10}$

Give the difference in simplest form. *(page 188)*

25. $\frac{5}{6} - \frac{1}{3}$ $\frac{1}{2}$
26. $\frac{5}{12} - \frac{1}{6}$ $\frac{1}{4}$
27. $\frac{5}{9} - \frac{1}{3}$ $\frac{2}{9}$
28. $\frac{7}{8} - \frac{3}{4}$ $\frac{1}{8}$
29. $\frac{7}{10} - \frac{2}{5}$ $\frac{3}{10}$
30. $\frac{8}{9} - \frac{5}{6}$ $\frac{1}{18}$
31. $\frac{5}{3} - \frac{3}{4}$ $\frac{11}{12}$
32. $\frac{2}{3} - \frac{3}{8}$ $\frac{7}{24}$
33. $\frac{9}{10} - \frac{3}{4}$ $\frac{3}{20}$
34. $3 - \frac{1}{2}$ $2\frac{1}{2}$

Give the product in simplest form. *(page 202)*

35. $\frac{5}{8} \times \frac{2}{2}$ $\frac{5}{8}$
36. $\frac{1}{3} \times \frac{1}{3}$ $\frac{1}{9}$
37. $\frac{3}{8} \times \frac{4}{5}$ $\frac{3}{10}$
38. $\frac{3}{4} \times \frac{1}{2}$ $\frac{3}{8}$
39. $\frac{1}{4} \times 0$ 0
40. $\frac{1}{2} \times \frac{1}{3}$ $\frac{1}{6}$
41. $\frac{2}{5} \times \frac{5}{8}$ $\frac{1}{4}$
42. $\frac{2}{3} \times \frac{2}{3}$ $\frac{4}{9}$
43. $\frac{3}{8} \times \frac{1}{3}$ $\frac{1}{8}$
44. $\frac{3}{10} \times \frac{5}{6}$ $\frac{1}{4}$

Give the quotient in simplest form. *(page 212)*

45. $\frac{3}{4} \div \frac{1}{2}$ $1\frac{1}{2}$
46. $\frac{5}{9} \div \frac{5}{3}$ $\frac{1}{3}$
47. $\frac{2}{5} \div \frac{3}{5}$ $\frac{2}{3}$
48. $\frac{2}{3} \div \frac{1}{4}$ $2\frac{2}{3}$
49. $\frac{5}{6} \div \frac{5}{6}$ 1
50. $\frac{3}{8} \div 2$ $\frac{3}{16}$
51. $\frac{5}{8} \div \frac{3}{4}$ $\frac{5}{6}$
52. $\frac{5}{6} \div 3$ $\frac{5}{18}$
53. $\frac{4}{3} \div \frac{3}{4}$ $1\frac{7}{9}$
54. $0 \div \frac{2}{3}$ 0

MIXED PRACTICE
Solve.

55. 25% of 32 = n 8
56. 110% of n = 55 50
57. 12% of 74 = n 8.88
58. 20% of n = 12 60
59. 18% of 60 = n 10.8
60. 40% of 25 = n 10
61. 6% of n = 18 300
62. 20% of 45 = n 9
63. 125% of n = 40 32
64. $33\frac{1}{3}$% of 42 = n 14
65. 8.4% of n = 12.6 150
66. 4.6% of 90 = n 4.14
67. $37\frac{1}{2}$% of 56 = n 21
68. 0.5% of n = 2 400
69. 14.5% of 74 = n 10.73

Geometry, Perimeter, and Area

Problem-Solving Worksheet
Workbook S317, Copymaster S317, or Duplicating Master S317

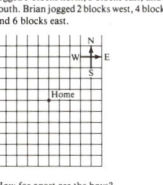

Challenge Problems
Unscramble the letters of the geometry words.

1. ETAUC ENAGL — Acute angle
2. LLRAPRALE — Parallel
3. ERQSUA — Square
4. TEREPIMRE — Perimeter
5. UIARDS — Radius
6. ERUCICRMFEECN — Circumference

Copymaster S459

Class Starter Quiz 137
on previous lesson

Solve. Use the weekend special prices on page 346.

1. What is the cost of the wire fence that is needed to build a 100-foot by 60-foot pen? $320
2. Is $70 enough money for the steel posts needed to fence in a 40-foot by 40-foot yard if the posts are spaced 10 feet apart? No

Copymaster S419

Lesson Objective
To compute the areas of squares and other rectangles

Problem-Solving Skills
Finding information in a drawing
Using a formula
Choosing the correct operation

Starting the Lesson
Write this sentence on the chalkboard:

The <u>ERAA</u> of a region is the number of <u>QARSUE</u> units that it takes to cover the region.

Before the students open their books, challenge them to unscramble the underlined words so the sentence will make sense. Then have them read the sentence at the top of page 348 to see if they correctly unscrambled the letters.

Here's How Note
Use of Concrete Materials You may wish to use the area grid and area tiles from copymasters or transparencies S532–S534 and S540 to demonstrate finding the area of squares and rectangles. See ■ **Manipulative Activity 23** on copymaster S522 in the Teacher's Resource Binder.

348

Area—squares and rectangles

The **area** of a region is the number of square units that it takes to cover the region.

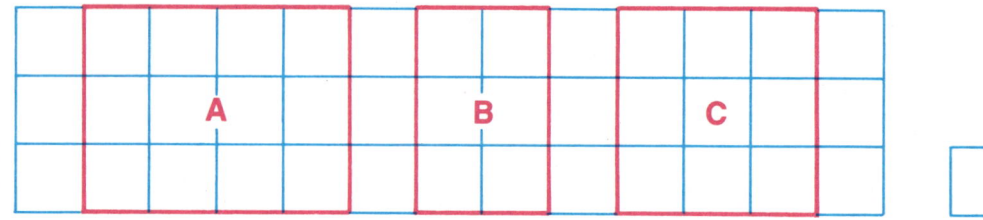

1. Count the squares. Which rectangle has an area of 12 square centimeters? A
2. Which rectangle has an area of 6 square centimeters? B
3. What is the area of square C? 9 square centimeters

Here's how to use formulas to find the area of rectangles and squares.

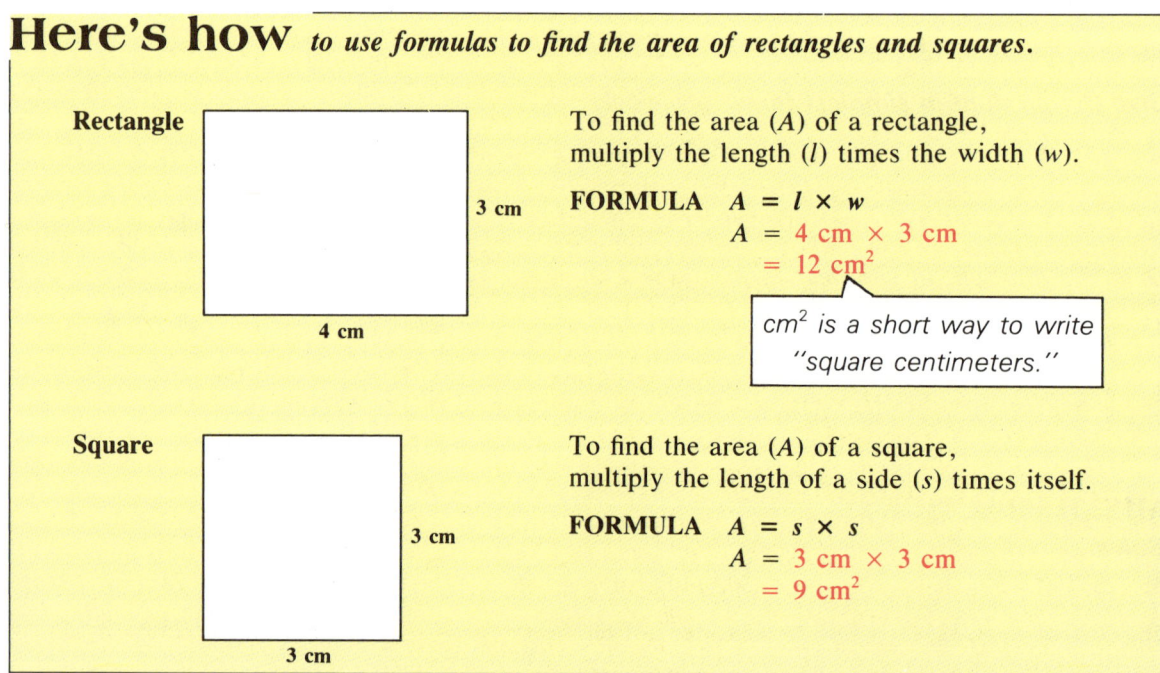

4. Look at the *Here's how*. What is the formula for the area of a rectangle? What does each letter stand for? $A = l \times w$, A = area, l = length, w = width
5. If the length and width of a rectangle are 10 centimeters and 5 centimeters, the area is 50 [?] centimeters. square
6. If the side of a square is 6 centimeters, its area is 36 square [?]. centimeters

348 *Chapter 13*

EXERCISES

Find the area.

7. 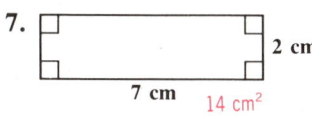 2 cm, 7 cm, 14 cm²

8. 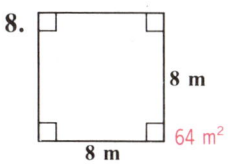 8 m, 8 m, 64 m²

9. 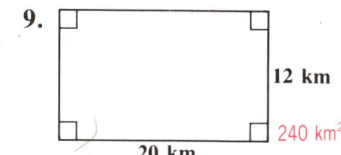 12 km, 20 km, 240 km²

10. 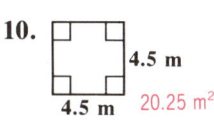 4.5 m, 4.5 m, 20.25 m²

11. 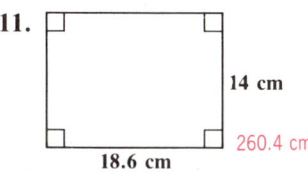 14 cm, 18.6 cm, 260.4 cm²

12. 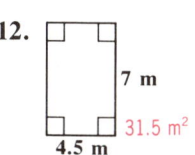 7 m, 4.5 m, 31.5 m²

13. 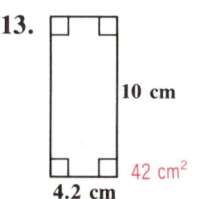 10 cm, 4.2 cm, 42 cm²

14. 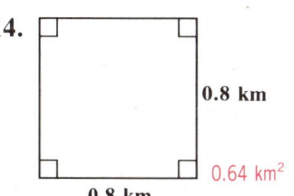 0.8 km, 0.8 km, 0.64 km²

15. 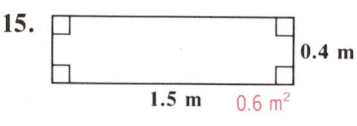 0.4 m, 1.5 m, 0.6 m²

16. One of the squares above has a perimeter of 18 meters. Which one is it? *Number 10*

17. One of the rectangles above has a perimeter of 23 meters. Which one is it? *Number 12*

You decide!

First tell whether the problem is about perimeter or area. Then solve the problem.

18. How much fencing do I need to enclose a field that is 20 yards long and 15 yards wide? *Perimeter, 70 yards*

19. How many 1-foot-square tiles are needed to tile my 15-foot by 12-foot kitchen floor? *Area, 180 tiles*

20. How much sod is needed to cover a 20-yard by 40-yard lawn? *Area, 800 yd²*

21. How much molding is needed to go around a 25-foot by 15-foot ceiling? *Perimeter, 80 feet*

22. How much paint is needed to cover a floor that is 25 feet by 14 feet? A quart of paint covers about 50 square feet. *Area, 7 quarts*

23. How much does it cost to frame a square painting that is $2\frac{1}{2}$ feet on each side? Framing costs $.89 a foot. *Perimeter, $8.90*

Geometry, Perimeter, and Area **349**

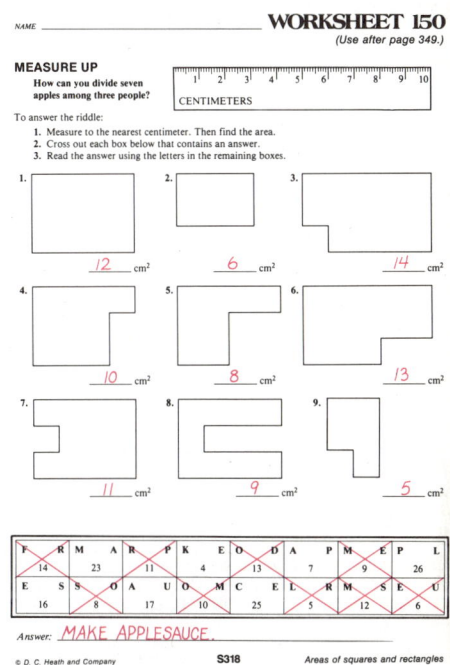

Area—parallelograms

In each of these parallelograms, the **base** (*b*) is 4 centimeters and the **height** (*h*) is 3 centimeters.

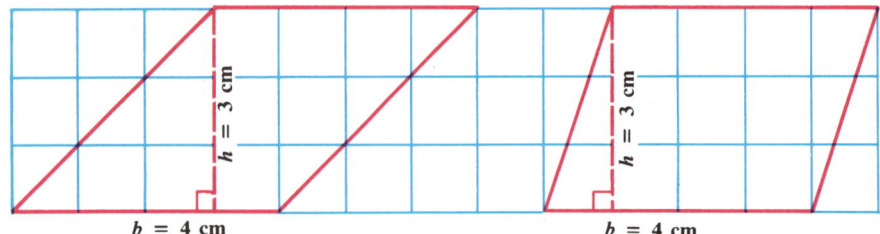

Count the squares. Each parallelogram has an area of 12 square centimeters.

Here's how to use a formula to find the area of a parallelogram.

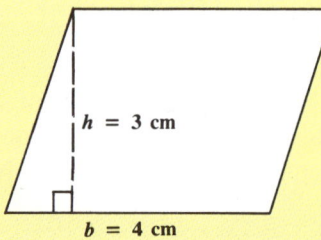

Notice that the height (*h*) is perpendicular to the base (*b*).

To find the area (*A*) of a parallelogram, multiply the base (*b*) times the height (*h*).

FORMULA $A = b \times h$
$A = 4 \text{ cm} \times 3 \text{ cm}$
$= 12 \text{ cm}^2$

The area of the parallelogram is 12 square centimeters.

1. Look at the *Here's how*. What is the formula for the area of a parallelogram? $A = b \times h$
What does each letter stand for? A = area, b = base, h = height

2. If the base and height of a parallelogram are 5 meters and 9 meters, the area is 45 square ?. meters

EXERCISES
Find the area.

3.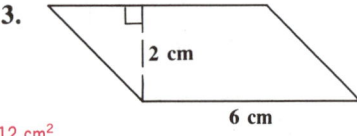
12 cm²

4.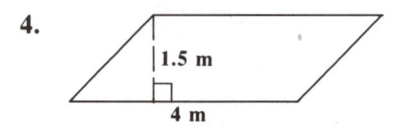
6 m²

5.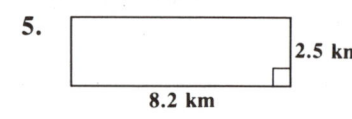
20.5 km²

6. 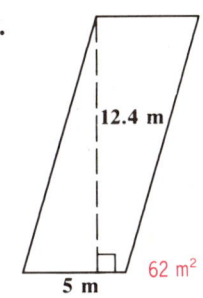 12.4 m, 5 m, 62 m²

7. 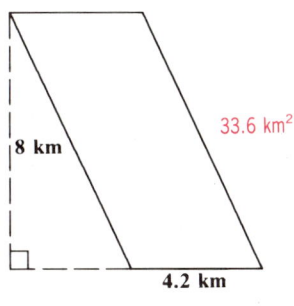 8 km, 4.2 km, 33.6 km²

8. 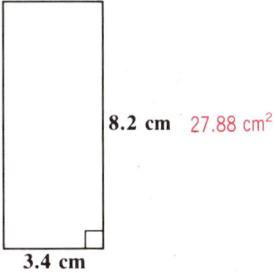 8.2 cm, 3.4 cm, 27.88 cm²

9. 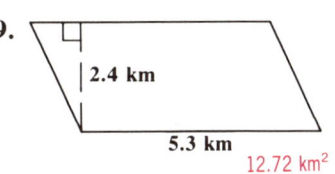 2.4 km, 5.3 km, 12.72 km²

10. 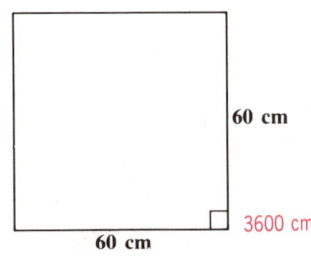 60 cm, 60 cm, 3600 cm²

11. 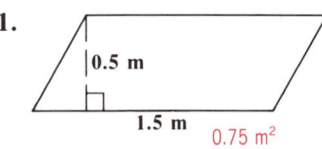 0.5 m, 1.5 m, 0.75 m²

Use the formula $A = b \times h$. Find the area of each parallelogram.

12. $b = 15$ cm
    $h = 3$ cm   45 cm²

13. $b = 20$ cm
    $h = 6.5$ cm   130 cm²

14. $b = 10$ m
    $h = 4.4$ m   44 m²

15. $b = 2.1$ km
    $h = 5$ km   10.5 km²

16. $b = 100$ m
    $h = 65$ m   6500 m²

17. $b = 8.4$ km
    $h = 2.2$ km   18.48 km²

18. $b = 6.1$ m
    $h = 12$ m   73.2 m²

19. $b = 500$ m
    $h = 35$ m   17,500 m²

20. $b = 85$ cm
    $h = 25$ cm   2125 cm²

21. $b = 16$ cm
    $h = 40$ cm   640 cm²

22. $b = 15$ km
    $h = 13$ km   195 km²

23. $b = 68.3$ m
    $h = 5.4$ m   368.82 m²

Pick up on these Visual thinking

Toothpicks were used to make this array of 9 small squares.

24. Draw a picture. Show how to remove 2 toothpicks to get 7 small squares.

25. Draw another picture. Show how to remove 4 toothpicks to get 5 small squares.

26. Show how to remove 4 toothpicks to get 6 small squares.

27. Show how to remove 8 toothpicks to get 5 small squares.

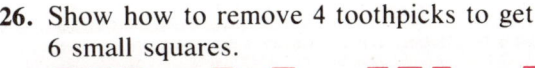
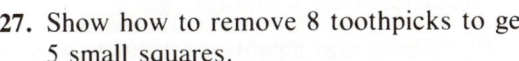
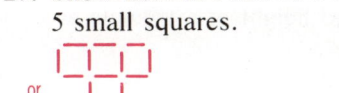
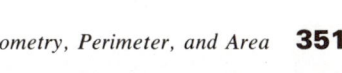

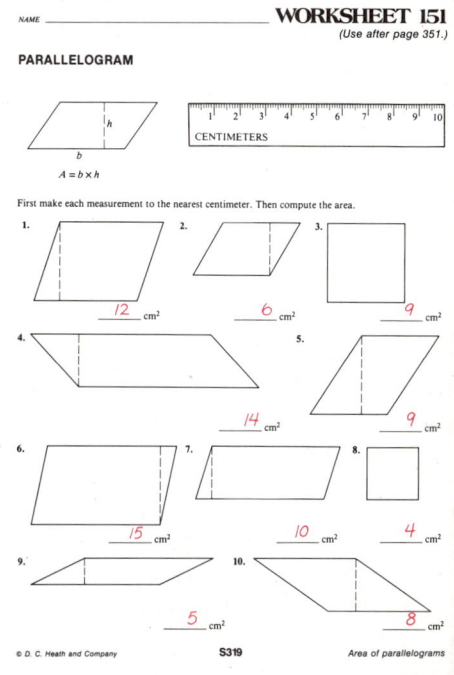

Area—triangles

CAN YOU CUT IT?

Think about cutting a parallelogram in half.

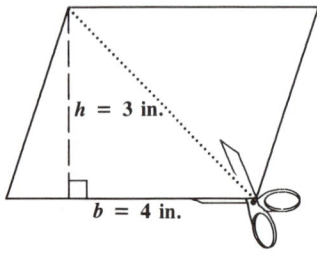 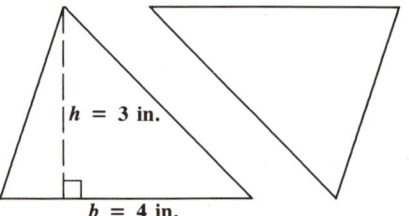

Two triangles are formed. The area of each triangle is half the area of the parallelogram.

1. The area of the parallelogram is 12 square ?. inches
2. The area of each triangle is ? square inches. 6

Here's how to use a formula to find the area of a triangle.

EXAMPLE 1.

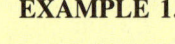

To find the area (A) of a triangle, multiply $\frac{1}{2}$ times the base (b) times the height (h).

FORMULA $A = \frac{1}{2} \times b \times h$

$A = \frac{1}{2} \times 4 \text{ in.} \times 3 \text{ in.}$

$= \frac{1}{2} \times 12 \text{ in.}^2$

$= 6 \text{ in.}^2$

EXAMPLE 2.

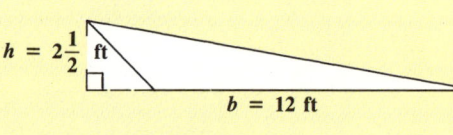

FORMULA $A = \frac{1}{2} \times b \times h$

$A = \frac{1}{2} \times 12 \text{ ft} \times 2\frac{1}{2} \text{ ft}$

$= \frac{1}{2} \times 30 \text{ ft}^2$

$= 15 \text{ ft}^2$

3. Look at the *Here's how*. What is the formula for the area of a triangle? What does each letter stand for? $A = \frac{1}{2} \times b \times h$, $A =$ area, $b =$ base, $h =$ height

4. If the base and height of a triangle are 5 yards and 6 yards, the area is ? square yards. 15

EXERCISES

Find the area. Use the formula $A = \frac{1}{2} \times b \times h$.

Here are scrambled answers for the next three exercises: 45 ft², 12 ft², 35 ft²

5. 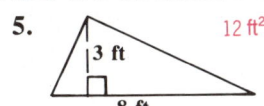 12 ft²

6. 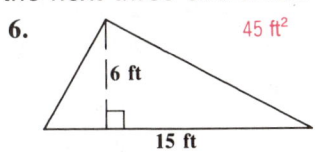 45 ft²

7. 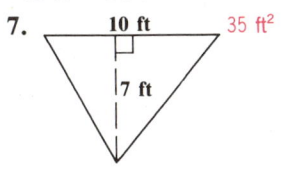 35 ft²

8. 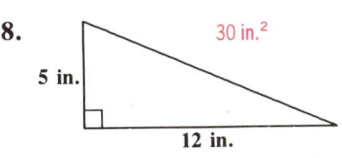 30 in.²

9. 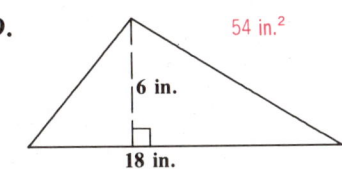 54 in.²

10. 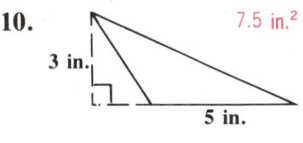 7.5 in.²

11. 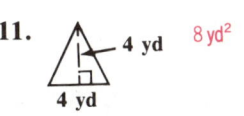 8 yd²

12. 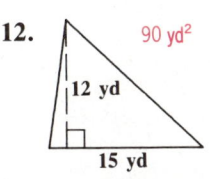 90 yd²

13. 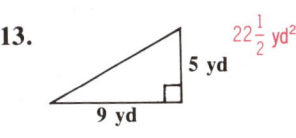 22½ yd²

Use $A = \frac{1}{2} \times b \times h$. Find the area of each triangle.

14. $b = 9$ ft
    $h = 4$ ft   18 ft²

15. $b = 21$ ft
    $h = 10$ ft   105 ft²

16. $b = 40$ yd
    $h = 7$ yd   140 yd²

17. $b = 4$ in.
    $h = 5$ in.   10 in.²

18. $b = 20$ ft
    $h = \frac{1}{2}$ ft   5 ft²

19. $b = 10$ yd
    $h = 2\frac{1}{2}$ yd   12½ yd²

Crack the code

20. Use the code to get the answer.

    RIDDLE: What is the easiest way to double your money?

    ANSWER: 6 cm²* 2 cm²* 4 cm²* 5 cm²
    12 cm²* 8 cm²

    Fold it.

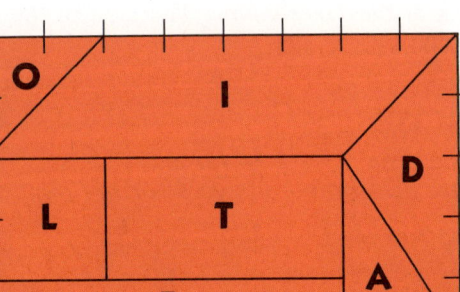

Geometry, Perimeter, and Area 353

Practice Worksheet

Workbook S320, Copymaster S320, or Duplicating Master S320

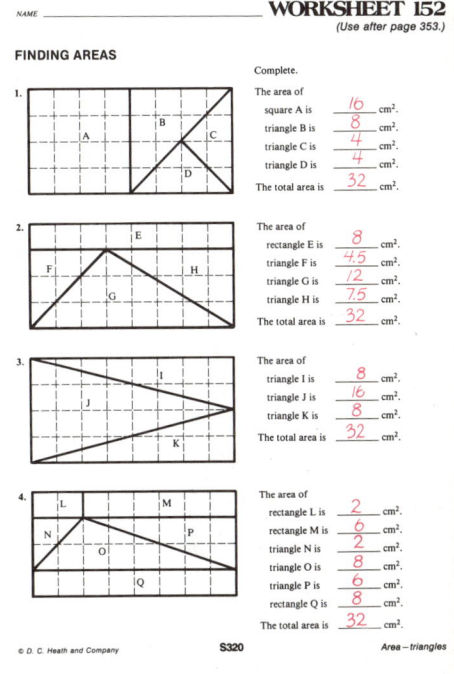

Project

Finding area

Make some cutouts of triangles and use a ruler to measure their bases and their heights. Then compute the area of each triangle.

Copymaster S485

Class Starter Quiz 140
on previous lesson

Find the area.

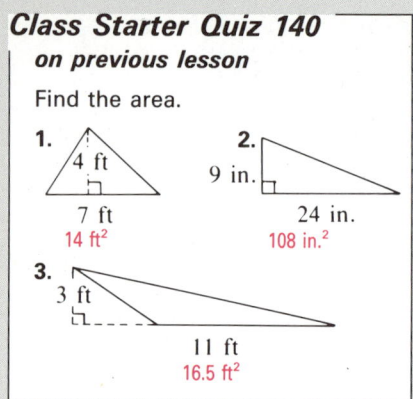

1. 4 ft, 7 ft — 14 ft²
2. 9 in., 24 in. — 108 in.²
3. 3 ft, 11 ft — 16.5 ft²

Copymaster S420

Lesson Objective
To compute the area of a circle

Problem-Solving Skills
Using a formula
Using a drawing
Using a guess-and-check strategy

Starting the Lesson
Have the students take the Circle Quiz at the top of the page. Give the answers to quiz questions 1 and 2. Tell the students to study the *Here's how* and do exercise 4 to check their answer to quiz question 3.

Here's How Note
Let the students practice using the formula by finding the area of these circles.

2 ft — 12.56 ft²
6 ft — 113.04 ft²
10 ft — 78.5 ft²

Area—circles

CIRCLE QUIZ

Choose the correct letter.

1. What caused these circles?
   a. Spaceships
   b. Sprinkler irrigation

2. What is the radius of each circle?
   a. 250 yards
   b. 500 yards

3. What is the area of each circle?
   a. Less than 1,000,000 yd²
   b. More than 1,000,000 yd²

Here's how to use a formula to find the area of a circle.

To find the area (A) of a circle, multiply π (about 3.14) times the radius (r) times the radius.

FORMULA $A = \pi \times r \times r$ or πr^2

$A \approx 3.14 \times 500 \text{ yd} \times 500 \text{ yd}$
$\approx 3.14 \times 250{,}000 \text{ yd}^2$
$\approx 785{,}000 \text{ yd}^2$

4. Look at the *Here's how*. The area of the circle is approximately 785,000 square [?]. yards

EXERCISES

Find the area. Use 3.14 for π.
Here are scrambled answers for the next row of exercises: 200.96 ft²   314 ft²   50.24 ft²

5.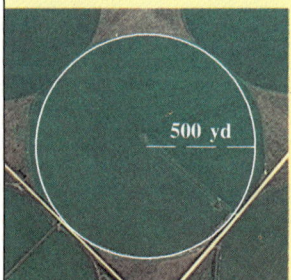
4 ft — 50.24 ft²

6.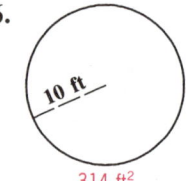
10 ft — 314 ft²

7.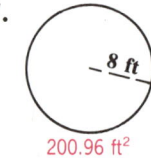
8 ft — 200.96 ft²

354   Chapter 13

8.
28.26 in.²

9.
78.5 ft²

10.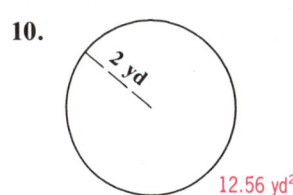
12.56 yd²

11. 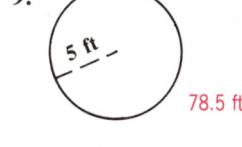 Be careful! How long is the radius?
113.04 in.²

12.
3.14 yd²

13.
19.625 ft²

Use the area clues. Find the area of each red region.
Note that you can add and subtract areas.

14.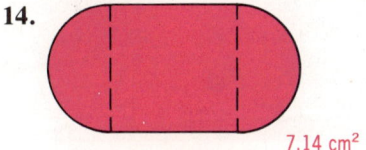
7.14 cm²

15.
4 cm²

16.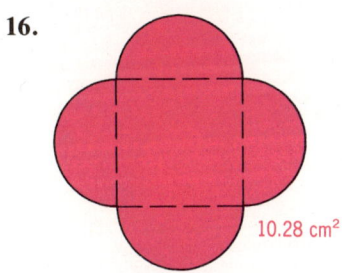
10.28 cm²

17.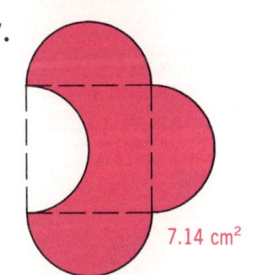
7.14 cm²

AREA CLUES
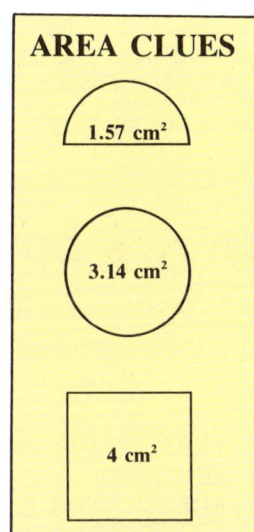
1.57 cm²
3.14 cm²
4 cm²

Area hunt Guess and check

18. Look on page 343. Find the rectangle that has an area of this many square centimeters.

Number 5

19. Look on page 345. Find the circle that has an area of this many square yards.

Number 12

20. Find the circle on page 345 that has an area of this many square feet.
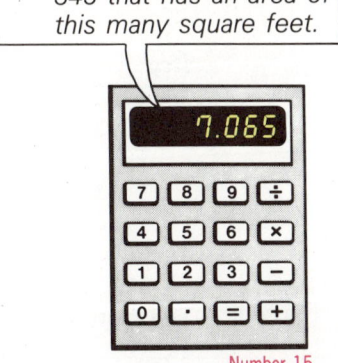
Number 15

Extra Practice
Page 498 Skill 56

Practice Worksheet
Workbook S321, Copymaster S321, or Duplicating Master S321

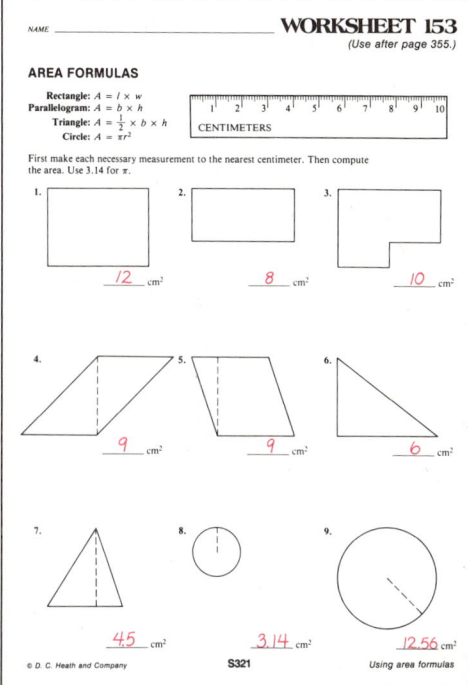

Challenge Problem
What is the area of the largest circle that can be cut out of a 10-cm by 10-cm square? 78.5 cm²

Copymaster S460

Geometry, Perimeter, and Area **355**

Class Starter Quiz 141
on previous lesson

Find the area. Use 3.14 for π.

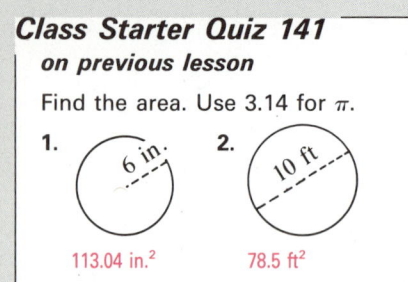

Copymaster S420

Lesson Objective
To solve problems by making a drawing

Starting the Lesson
Problem-Solving Cover-up Use the chalkboard or mask ■ **Visual Aid 64** (copymaster or transparency S162).

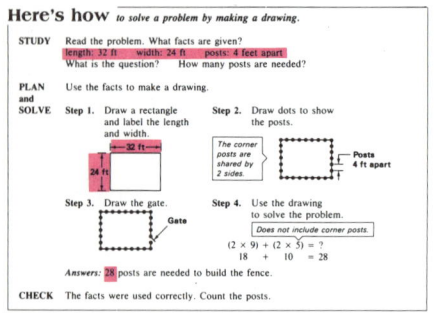

Have the students, working in small groups, study the problem on page 356 and the problem-solving steps for several minutes. Then have them close their books, look at the visual aid, and tell what has been covered up.

356

PROBLEM SOLVING *making a drawing*

The Alamo Fence Company builds fences for homes, ranches, farms, and businesses. An accurate well-labeled drawing helps the Alamo workers plan and complete each fencing project.

Problem
A rancher hired the Alamo Fence Company to build a wire fence around a rectangular calf pen. The pen is 32 feet by 24 feet. The rancher wants the fence to have steel posts every 4 feet. She also wants a gate that is 4 feet wide on one end of the pen. How many posts are needed?

Here's how *to solve a problem by making a drawing.*

STUDY   Read the problem. What facts are given?
            length: 32 ft   width: 24 ft   posts: 4 feet apart
            What is the question?   How many posts are needed?

PLAN and SOLVE

Step 1. Draw a rectangle and label the length and width.

Step 2. Draw dots to show the posts.

The corner posts are shared by 2 sides.

Posts 4 ft apart

Step 3. Draw the gate.

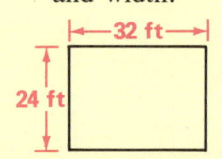

Step 4. Use the drawing to solve the problem.

Does not include corner posts.

$(2 \times 9) + (2 \times 5) = ?$
$18 + 10 = 28$

Answers: 28 posts are needed to build the fence.

CHECK   The facts were used correctly. Count the posts.

356   *Chapter 13*

PROBLEMS

Solve.

1. The fence described on page 356 will have a steel pipe around the top. The pipe comes in 4-foot lengths.

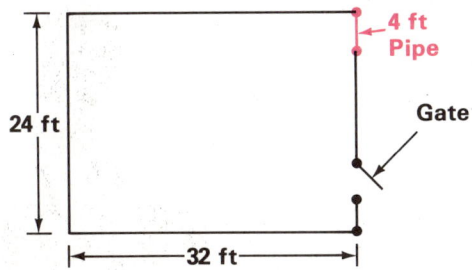

   a. Complete the drawing to show all the 4-foot lengths of pipe. How many 4-foot lengths are needed? **28**
   b. How many feet of fencing will be needed to enclose the calf pen? **112 ft**
   c. What is the area of the pen? **768 ft²**

2. The Alamo Fence Company sells sections of fence already assembled. Wire clamps are used to attach the fencing to the pipe frame.

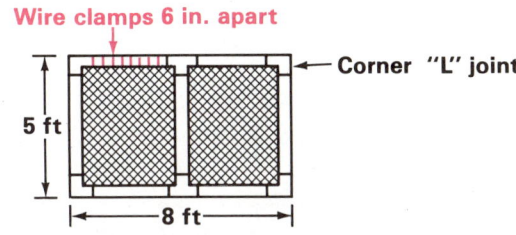

   a. Complete the drawing to show all the wire clamps. How many clamps are needed? **52**
   b. How many square feet of fencing are needed to make each section? **40**
   c. How many corner "L" joints are needed for 8 of these fence sections? **32**

3. To the right is an incomplete drawing of a fence designed for a home owner. The fence will have steel posts every 5 feet.
   a. Complete the drawing to show all the posts. How many posts are needed? **24**
   b. The fencing costs $1.25 per foot and posts cost $4.50 each. What is the total cost of the fence? **$258**
   c. How many square feet are inside the fence? *Hint: What area formula should you use?* **600, $A = \frac{1}{2}bh$**

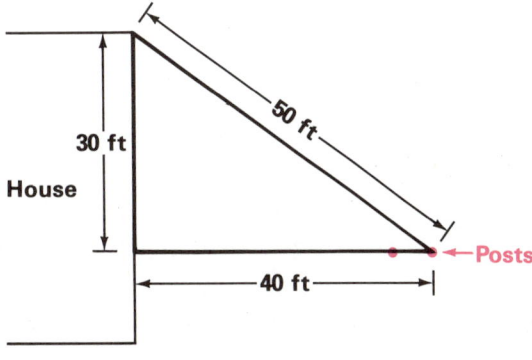

The dog pen!

4. The Alamo Fence Company sells an unassembled dog pen. Make a drawing that shows how the pieces to the right can be assembled to make a frame for the gate to this pen. The assembled gate is a little larger than 6 feet by 4 feet.

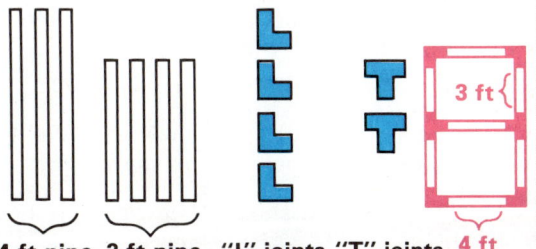

Geometry, Perimeter, and Area **357**

Practice Worksheet
Workbook S322, Copymaster S322, or Duplicating Master S322

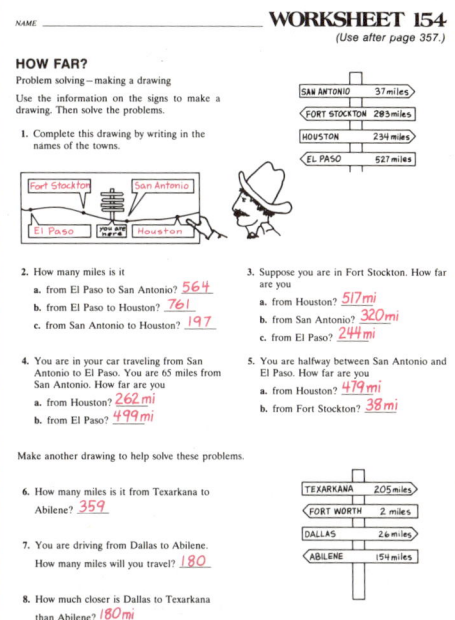

Challenge Problem

Write a question that fits the answer.

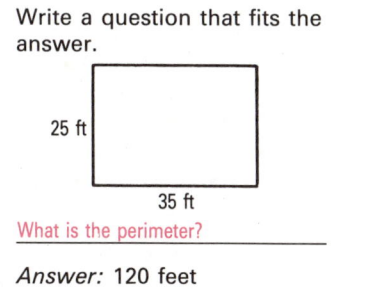

What is the perimeter?

Answer: 120 feet

Copymaster S460

Class Starter Quiz 142
on previous lesson

Complete the drawing. Then solve the problem.

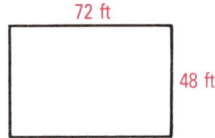

A lawn is 48 feet by 72 feet. The lawn is surrounded by a fence with a post every 8 feet.

1. How many posts are there? 30
2. How many feet of fencing are there? 240
3. What is the area of the lawn? 3456 ft²

Copymaster S420

Problem-Solving Skill
Selecting information from a drawing

Skills Reviewed
Making conversions between metric units of measure
Solving proportions
Changing percents to fractions
Changing decimals to percents
Solving percent problems

Starting the Lesson

Problem Solving Have the students read the paragraph at the top of the page. Then have the students close their books and sketch from memory a stick figure that shows the preferred angles for the wrist and elbow during a typical shot. Tell the students to compare their drawing with the stick figure at the top of the page.

Cumulative Skill Practice Write these five answers on the chalkboard:

3.825   16 $2\frac{1}{2}$ $62\frac{1}{2}\%$ 23

Challenge the students to an answer hunt by saying, "Look at exercises 1–64 on page 359. Find the five exercises that have these answers. You have five minutes to find as many of the exercises as you can." (Exercises 4, 15, 28, 53, and 64)

358

Problem solving

COMPUTERS AND SPORTS

Coaches sometimes use computer-produced stick figures to help athletes analyze and improve their performance. The stick figure superimposed on the photo at the right can be used by basketball coaches. It shows the preferred angles for the wrist and elbow during a typical shot.

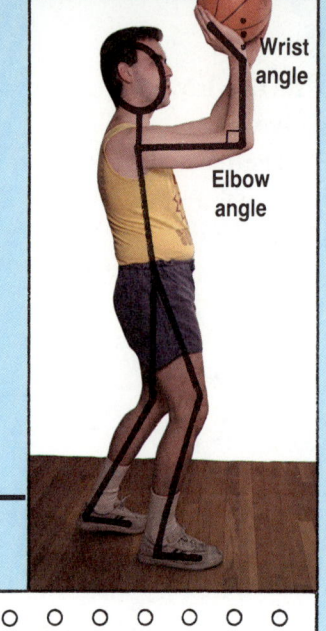

1. What is the elbow angle? 90°

2. Is the wrist angle about 45°? Yes

Study these computer-produced stick figures. Then match each of the coach's comments with the right player.

PLAYER: JOE WILL   PLAYER: ED LYNCH   PLAYER: JOE DEBOLD

3. "You need to bend your wrist less and elbow more." Joe Debold

4. "Your wrist is just right, but your elbow is bent too much!" Joe Will

5. "Bend your wrist a little more, but keep your elbow the same." Ed Lynch

Use the stick figure of this tennis player to complete the coach's statements.

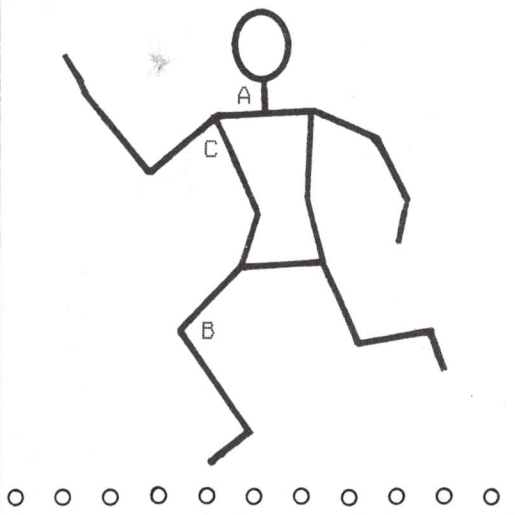

6. "Angle [?] in the figure is about 80°." C

7. "The [?] is bent at an angle of about 100°. It is labeled angle [?] in the figure." Knee, B

8. "The neck and shoulders form angle [?]. It is a(n) [?] angle, since it measures 90°." A, right

358 Chapter 13

Cumulative Skill Practice

Complete. *(page 228)*

1. 36 cm = ? m 0.36
2. 63 mm = ? cm 6.3
3. 24 km = ? m 24,000
4. 3825 m = ? km 3.825
5. 8.2 km = ? m 8200
6. 9.6 cm = ? mm 96
7. 52 cm = ? mm 520
8. 16 m = ? cm 1600
9. 2.4 m = ? cm 240

Solve. *(page 262)*

10. $\frac{8}{9} = \frac{5}{n}$ $5\frac{5}{8}$
11. $\frac{9}{8} = \frac{n}{6}$ $6\frac{3}{4}$
12. $\frac{4}{n} = \frac{5}{9}$ $7\frac{1}{5}$
13. $\frac{7}{n} = \frac{5}{9}$ $12\frac{3}{5}$
14. $\frac{6}{8} = \frac{n}{4}$ 3
15. $\frac{n}{6} = \frac{8}{3}$ 16
16. $\frac{n}{5} = \frac{3}{5}$ 3
17. $\frac{n}{4} = \frac{5}{10}$ 2
18. $\frac{8}{4} = \frac{7}{n}$ $3\frac{1}{2}$
19. $\frac{n}{6} = \frac{3}{9}$ 2

Change to a fraction in simplest form. *(page 282)*

20. 20% $\frac{1}{5}$
21. 50% $\frac{1}{2}$
22. 25% $\frac{1}{4}$
23. 75% $\frac{3}{4}$
24. 30% $\frac{3}{10}$
25. 8% $\frac{2}{25}$
26. 120% $1\frac{1}{5}$
27. 125% $1\frac{1}{4}$
28. 250% $2\frac{1}{2}$
29. 175% $1\frac{3}{4}$
30. 290% $2\frac{9}{10}$
31. 135% $1\frac{7}{20}$
32. $33\frac{1}{3}$% $\frac{1}{3}$
33. $66\frac{2}{3}$% $\frac{2}{3}$
34. $12\frac{1}{2}$% $\frac{1}{8}$
35. $83\frac{1}{3}$% $\frac{5}{6}$
36. $62\frac{1}{2}$% $\frac{5}{8}$
37. $37\frac{1}{2}$% $\frac{3}{8}$

Change to a percent. *(page 287)*

38. 0.05 5%
39. 0.01 1%
40. 0.42 42%
41. 0.75 75%
42. 0.4 40%
43. 0.1 10%
44. 2.5 250%
45. 0.25 25%
46. 3.4 340%
47. 1.6 160%
48. 2.375 237.5%
49. 0.02 2%
50. $0.16\frac{2}{3}$ $16\frac{2}{3}$%
51. $0.33\frac{1}{3}$ $33\frac{1}{3}$%
52. $0.12\frac{1}{2}$ $12\frac{1}{2}$%
53. $0.62\frac{1}{2}$ $62\frac{1}{2}$%
54. $1.87\frac{1}{2}$ $187\frac{1}{2}$%
55. $1.36\frac{1}{2}$ $136\frac{1}{2}$%

Solve. *(page 296)*

56. 10% of n = 12 120
57. 30% of n = 15 50
58. 50% of n = 16 32
59. 20% of n = 18 90
60. 60% of n = 30 50
61. 6% of n = 9 150
62. 120% of n = 75 62.5
63. 150% of n = 72 48
64. 200% of n = 46 23

MIXED PRACTICE

Complete.

65. $\frac{2}{3} + \frac{1}{4}$ = ? $\frac{11}{12}$
66. $\frac{5}{8} - \frac{1}{4}$ = ? $\frac{3}{8}$
67. $5 \times \frac{2}{3}$ = ? $3\frac{1}{3}$
68. $2\frac{5}{8} + 4$ = ? $6\frac{5}{8}$
69. $8 - 4\frac{3}{5}$ = ? $3\frac{2}{5}$
70. $\frac{4}{5} \div 6$ = ? $\frac{2}{15}$
71. $\frac{3}{8} \times \frac{4}{5}$ = ? $\frac{3}{10}$
72. $2\frac{1}{2} \times 3$ = ? $7\frac{1}{2}$
73. $6\frac{2}{3} - 1\frac{4}{5}$ = ? $4\frac{13}{15}$
74. $6 \div 2\frac{2}{3}$ = ? $2\frac{1}{4}$
75. $\frac{5}{6} - \frac{2}{5}$ = ? $\frac{13}{30}$
76. $3\frac{1}{2} + 2\frac{2}{3}$ = ? $6\frac{1}{6}$
77. $1\frac{1}{4} \times 1\frac{1}{4}$ = ? $1\frac{9}{16}$
78. $\frac{5}{9} \div \frac{2}{3}$ = ? $\frac{5}{6}$
79. $4\frac{3}{4} \div 2\frac{1}{2}$ = ? $1\frac{9}{10}$

Geometry, Perimeter, and Area

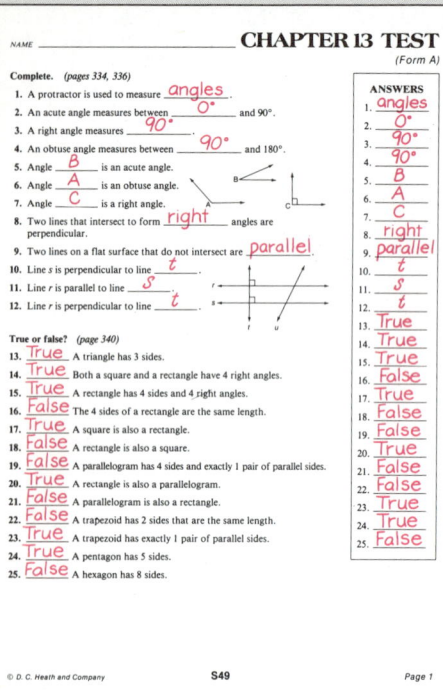

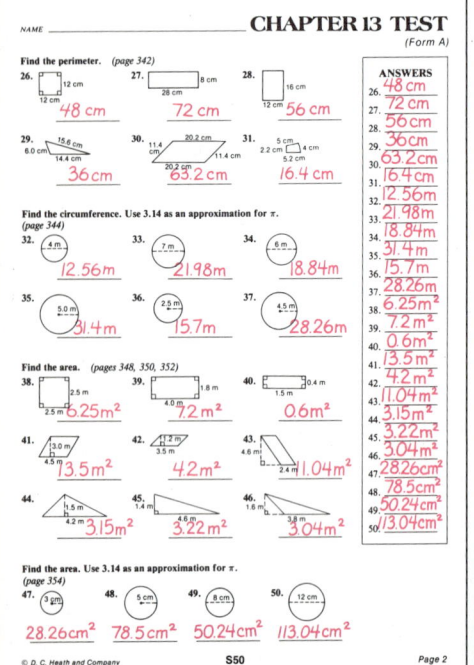

Chapter REVIEW

Here are scrambled answers for the review exercises:

18	72	π	height	obtuse	perimeter	width
36	314	acute	hexagon	parallel	perpendicular	
50	628	base	multiply	parallelogram	trapezoid	

1. An 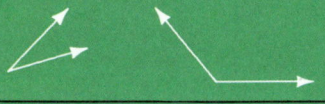 angle measures between 0° and 90°. An ? angle measures between 90° and 180°. *(page 334)*

2. Line r is ? to line t, and line s is ? to line t. *(page 336)*

3. A rectangle is also a ?. A ? has exactly 1 pair of parallel sides. A ? has 6 sides. *(page 340)*

4. The ? of a figure is the distance around the figure. The perimeter of this rectangle is ? cm. *(page 342)*

5. To find the circumference of a circle, you would ? the diameter (d) by pi (π). The circumference of this circle is ? ft. *(page 344)*

6. To find the area of a rectangle, you would multiply the length (l) times the ? (w). The area of this square is ? cm². *(page 348)*

7. To find the area of a parallelogram, you would multiply the base (b) times the ? (h). The area of this parallelogram is ? cm². *(page 350)*

8. To find the area of a triangle, you would multiply $\frac{1}{2}$ times the ? (b) times the height (h). The area of this triangle is ? ft². *(page 352)*

9. To find the area of a circle, you would multiply ? times the radius times the radius. The area of this circle is ? ft². *(page 354)*

360 Chapter 13

1. acute, obtuse
2. parallel, perpendicular
3. parallelogram, trapezoid, hexagon
4. perimeter, 72
5. multiply, 628
6. width, 36
7. height, 50
8. base, 18
9. π, 314

Chapter TEST

Complete. (pages 334, 336, 340)

1. A **?** is used to measure angles. *protractor*
2. A right angle measures **?**. *90°*
3. An acute angle measures between 0° and **?**. *90°*
4. An obtuse angle measures between **?** and 180°. *90°*
5. Two lines that intersect to form right angles are **?**. *perpendicular*
6. Two lines on a flat surface that do not intersect are **?**. *parallel*
7. A triangle has **?** sides. *3*
8. A rectangle has 4 sides and 4 **?** angles. *right*
9. A parallelogram has 2 pairs of **?** sides. *parallel*
10. A trapezoid has exactly 1 pair of **?** sides. *parallel*
11. A pentagon has **?** sides. *5*
12. A hexagon has **?** sides. *6*

Find the perimeter. (page 342)

13. 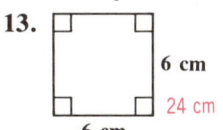 6 cm; 6 cm; *24 cm*
14. 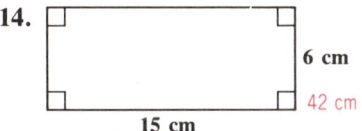 6 cm; 15 cm; *42 cm*
15. 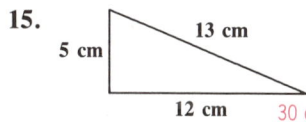 5 cm, 13 cm, 12 cm; *30 cm*

Find the circumference. Use 3.14 as an approximation for π. (page 344)

16. 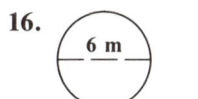 6 m; *18.84 m*
17. 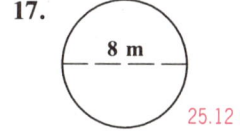 8 m; *25.12 m*
18. 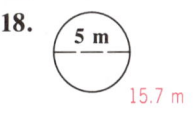 5 m; *15.7 m*

Find the area. (pages 348, 350, 352)

19. 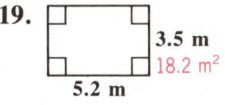 3.5 m, 5.2 m; *18.2 m²*
20. 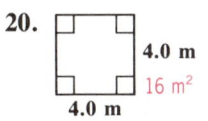 4.0 m, 4.0 m; *16 m²*
21. 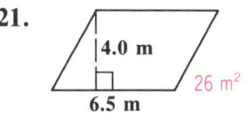 4.0 m, 6.5 m; *26 m²*
22. 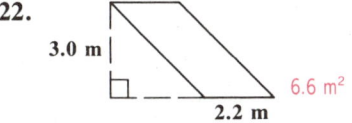 3.0 m, 2.2 m; *6.6 m²*
23. 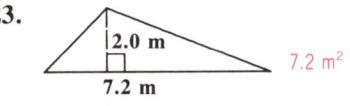 2.0 m, 7.2 m; *7.2 m²*
24. 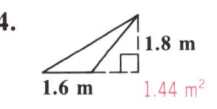 1.8 m, 1.6 m; *1.44 m²*

Find the area. Use 3.14 as an approximation for π. (page 354)

25. 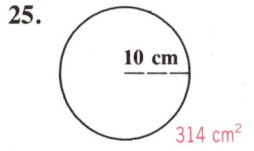 10 cm; *314 cm²*
26. 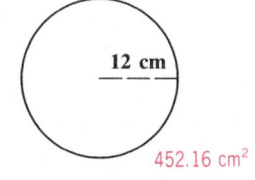 12 cm; *452.16 cm²*
27. 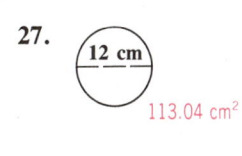 12 cm; *113.04 cm²*

Geometry, Perimeter, and Area **361**

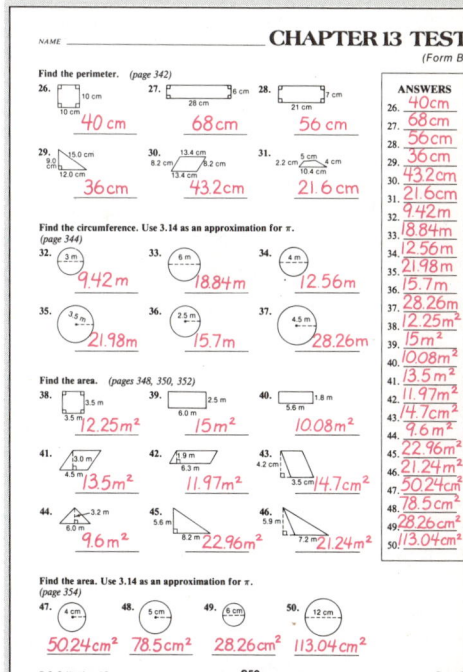

Copymasters S51 and S52 or Duplicating Masters S51 and S52

Cumulative Test
(Chapters 1–13)

Use Copymaster S109 to provide the students with an answer sheet in standardized test format.

Answers for Cumulative Test, Chapters 1–13

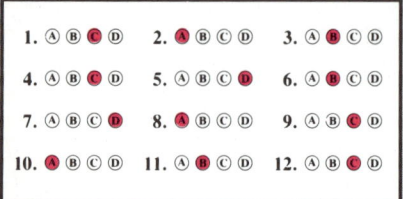

The table below correlates test items with student text pages.

Test Item	Page Taught	Skill Practice
1	166	p. 347, exercises 1–14
2	180	p. 347, exercises 15–24
3	188	p. 347, exercises 25–34
4	202	p. 347, exercises 35–44
5	212	p. 347, exercises 45–54
6	228	p. 359, exercises 1–9
7	262	p. 359, exercises 10–19
8	282	p. 359, exercises 20–37
9	287	p. 359, exercises 38–55
10	290	p. 359, exercises 56–64
11	310	
12	352	

362

Cumulative TEST — Standardized Format

Choose the correct letter.

1. Change to a decimal.

$5\frac{3}{4} = ?$

A. 0.75
B. 5.25
(C.) 5.75
D. none of these

2. Give the sum.

$\frac{5}{12} + \frac{7}{8}$

(A.) $1\frac{7}{24}$
B. $\frac{3}{5}$
C. $1\frac{1}{4}$
D. none of these

3. Give the difference.

$\frac{5}{6} - \frac{3}{4}$

A. 1
(B.) $\frac{1}{12}$
C. $\frac{1}{3}$
D. none of these

4. Give the product.

$\frac{5}{9} \times \frac{2}{3}$

A. $\frac{5}{6}$
B. $\frac{7}{27}$
(C.) $\frac{10}{27}$
D. none of these

5. Give the quotient.

$\frac{7}{8} \div \frac{1}{3}$

A. $\frac{7}{24}$
B. $2\frac{5}{16}$
C. $\frac{8}{21}$
(D.) none of these

6. 3475 m = ? km

A. 34.75
(B.) 3.475
C. 0.3475
D. none of these

7. Solve. $\frac{9}{n} = \frac{14}{5}$

A. 12
B. 11.25
C. 7.2
(D.) none of these

8. Change to a fraction.

$37\frac{1}{2}\% = ?$

(A.) $\frac{3}{8}$
B. $\frac{5}{8}$
C. $\frac{2}{5}$
D. none of these

9. Change to a percent.

0.06 = ?

A. 0.06%
B. 0.6%
(C.) 6%
D. none of these

10. Solve.

24% of n = 30

(A.) 125
B. 7.2
C. 22.8
D. none of these

11. A radio that regularly sells for $124 is on sale for 40% off. What is the sale price?

A. $49.60 (B.) $74.40
C. $173.60 D. none of these

12. Find the area.

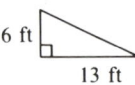

6 ft
13 ft

A. 78 ft² B. 9.5 ft²
(C.) 39 ft² D. none of these

362 Chapter 13

Surface Area and Volume

Chapter 14
Surface Area and Volume

Resources

- ***Class Starter Quizzes 143-150*** *(Copymasters S421-S422)*
- ***Visual Aids 65*** *(Copymaster or Transparency S163)*
- ***Manipulatives***
  Manipulative Activity 26 *(Copymaster S524)*
  Area tiles *(Copymasters or Transparencies S532-S534)*
- ***Worksheets 156-164*** *(Copymasters, Duplicating Masters, or Workbook pages S324-S332)*
- ***Challenge Problems*** for pages 365, 367, 371, 379, 381 *(Copymasters S460-S461)*
- ***Projects*** for pages 375, 377 *(Copymaster S485)*
- ***Tests*** *(Copymasters or Duplicating Masters S53-S56)*

Lesson Objective

To classify space figures by their faces, corners, and edges

Problem-Solving Skills

Finding information in a drawing
Using logical reasoning
Making geometric visualizations

Starting the Lesson

Have the students use the packages at the top of the page to answer questions 1–3. Then have them use the *Here's how* information to do exercise 4.

Exercise Note

Use of Concrete Materials Some students may have difficulty visualizing the cubes in exercises 15–23. Suggest that they draw each figure on graph paper, cut it out, and fold it into a cube.

Space figures
LOOK THEM OVER

1. Which package is mine? My package has 5 faces (sides), 6 corners, and 9 edges. C

2. Which package has 6 square faces? B

3. Which package has 5 faces and 5 corners? D

Here's how space figures are named.

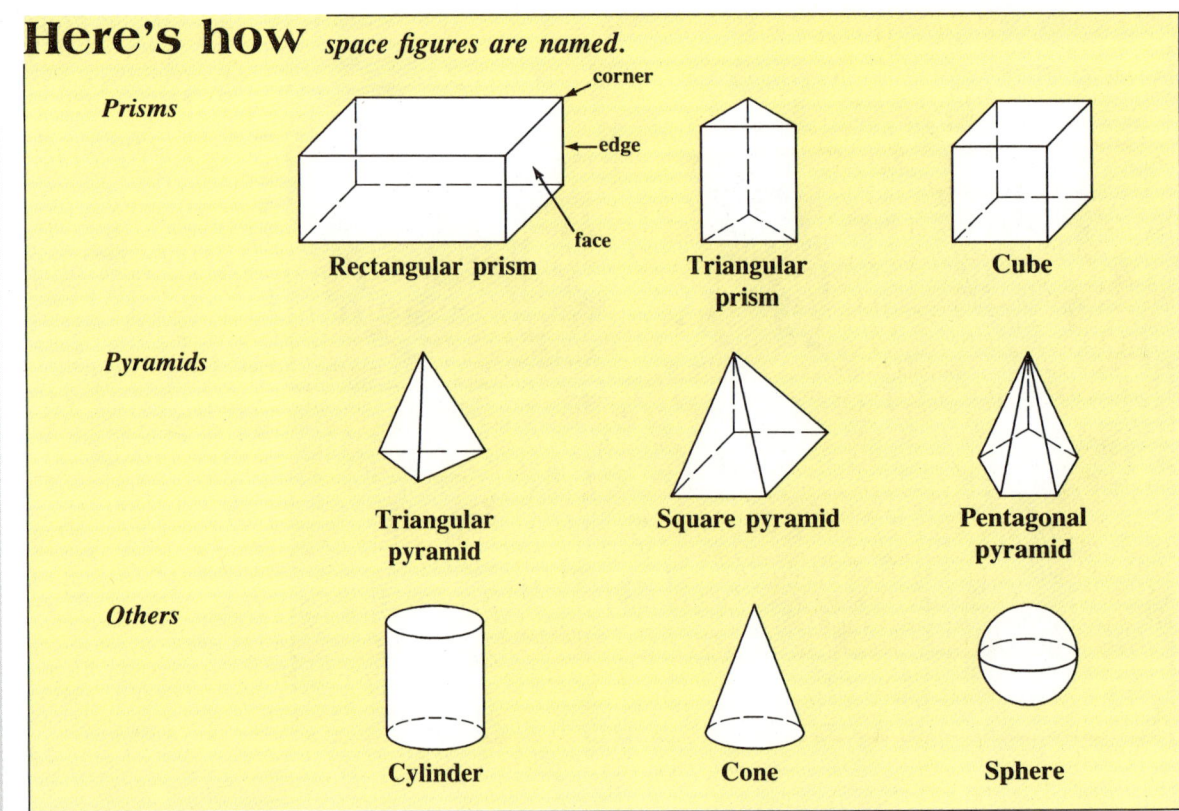

Prisms: Rectangular prism, Triangular prism, Cube

Pyramids: Triangular pyramid, Square pyramid, Pentagonal pyramid

Others: Cylinder, Cone, Sphere

4. Use the packages and the *Here's how* chart to answer the questions.
   a. Which package is a cube? B
   b. Which package is a cylinder? A
   c. Which package is a rectangular prism? E
   d. Which package is a square pyramid? D
   e. Which package is a triangular prism? C
   f. Which 3 packages are prisms? B,C,E

Chapter 14

EXERCISES

Use the clues and the drawings on page 364. Name each space figure.

5. Clues: *Cube*
- This space figure is a prism.
- All its faces are squares.

6. Clues: *Triangular pyramid*
- This space figure is a pyramid.
- It has 4 corners.

7. Clues: *Square pyramid*
- This space figure has 5 faces.
- One of its faces is square.
- It has 8 edges.

8. Clues: *Rectangular prism*
- This space figure has 6 faces.
- None of its faces are square.
- It has 8 corners.

9. Clues: *Pentagonal pyramid*
- This space figure has 6 faces.
- One of its sides is a pentagon.

10. Clues: *Triangular prism*
- This space figure has 5 faces.
- It has 9 edges.

11. Clues: *Cube*
- This space figure has 8 corners.
- All its edges are the same length.

12. Clues: *Cylinder*
- This space figure has no corners.
- Two of its faces are circles.

13. Clues: *Sphere*
- This space figure has no corners.
- If you cut this figure using one straight cut, the shape that is formed is always a circle.

14. Clues: *Cone*
- Cut this space figure one way and the shape that is formed is a circle.
- Cut it another way and the shape that is formed is a triangle.

What's on top? Visual thinking

Each of these patterns can be folded to form a cube. If the red face is the bottom of the cube, which face is the top?

15. 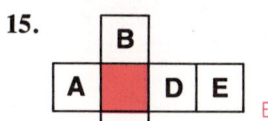 *E*

16.  *B*

17. 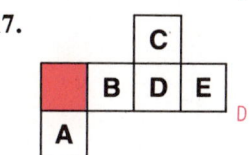 *D*

18.  *C*

19. 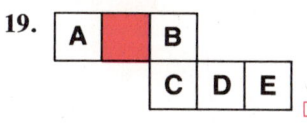 *D*

20.  *A*

21.  *E*

22.  *C*

23. 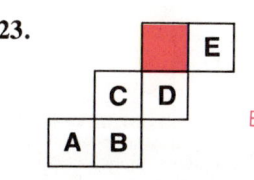 *B*

Surface Area and Volume

Practice Worksheet

Workbook S324, Copymaster S324, or Duplicating Master S324

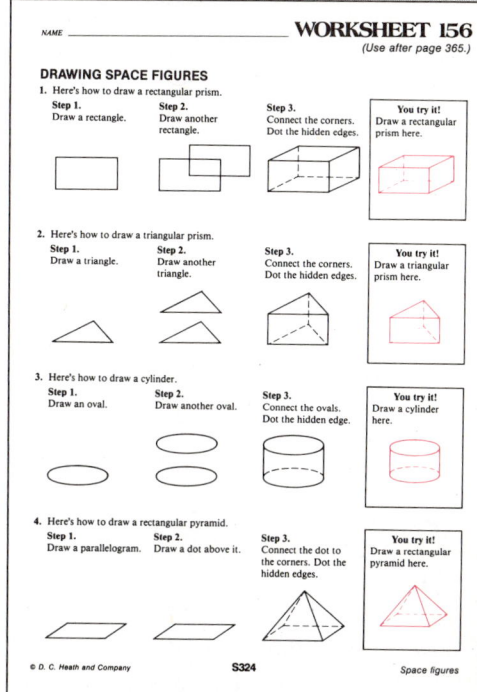

Challenge Problems

The six sides of a cube are lettered A through F. Here are three views of the cube:

 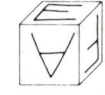

1. Which letter is opposite side A? *D*
2. Which letter is opposite side C? *F*
3. Which letter is opposite side B? *E*

Copymaster S460

Class Starter Quiz 143
on previous lesson

Name each space figure.

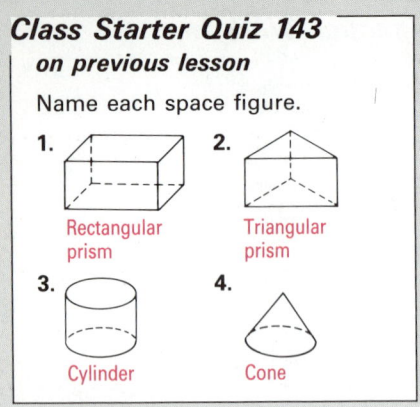

Copymaster S421

Lesson Objective
To visualize how triangular, square, and rectangular faces are used to build three-dimensional models

Problem-Solving Skills
Selecting information from a display
Making geometric visualizations

Starting the Lesson
Sketch these shapes on the chalkboard:

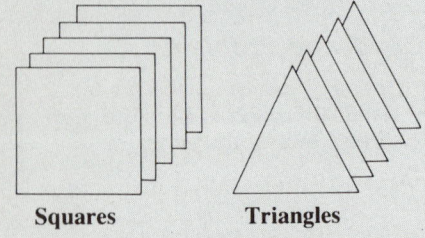

Before discussing questions 1–4, ask the students to describe what space figures can be made with

- 6 square pieces. (Cube)
- 4 triangular pieces. (Triangular pyramid)
- 3 square pieces and 2 triangular pieces. (Triangular prism)

366

More on space figures
PIECES & PRICES

   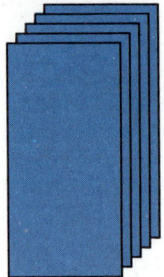

Square **Triangle A** **Triangle B** **Rectangle**
25¢   11¢   24¢   50¢

Here is a model that was made from some pieces that are shown above.

1. How many pieces were used to make the model? 5
2. a. Which piece was used for the top and bottom?
   b. What is the total cost of the top and bottom?
                          a. Triangle A   b. 22¢
3. a. Which piece was used for the other faces? Square
   b. What is their total cost? 75¢
4. What is the total cost of the model? 97¢

EXERCISES
Find the total cost of each model.

5. 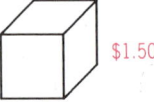 $1.50
6.  44¢
7. 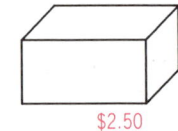 $2.50

8.  69¢
9.  $1.21
10. 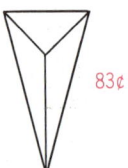 83¢

11. 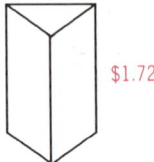 $1.72
12. 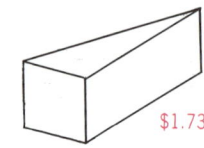 $1.73
13. 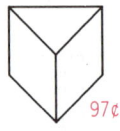 97¢

366 *Chapter 14*

14. 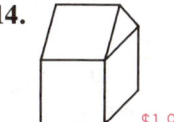 $1.97
15.  $1.69
16. 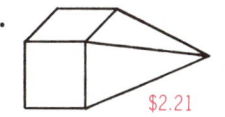 $2.21
17. 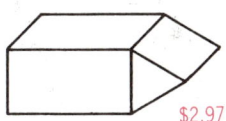 $2.97
18. 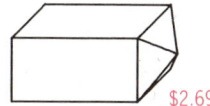 $2.69
19. 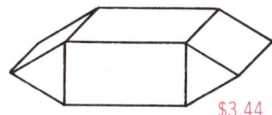 $3.44
20. 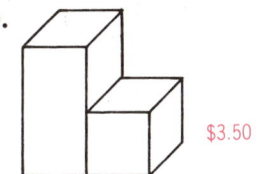 $3.50
21. 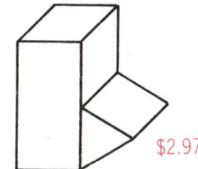 $2.97
22. 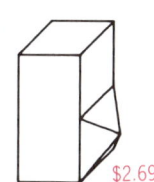 $2.69

Solve.

23. Which 5 pieces would you use to make the least expensive prism? 3 squares, 2 Triangle A's
24. Which pieces would you use to make the least expensive pyramid? 4 Triangle A's
25. Which pieces would you use to make the most expensive pyramid? 1 square, 4 Triangle B's
26. What is the price of the least expensive model that you can build that has 6 faces? *Hint: The answer is not $1.50.* 66¢

Name that product Visual thinking

Can you identify, just from the shape, what product is in each can?

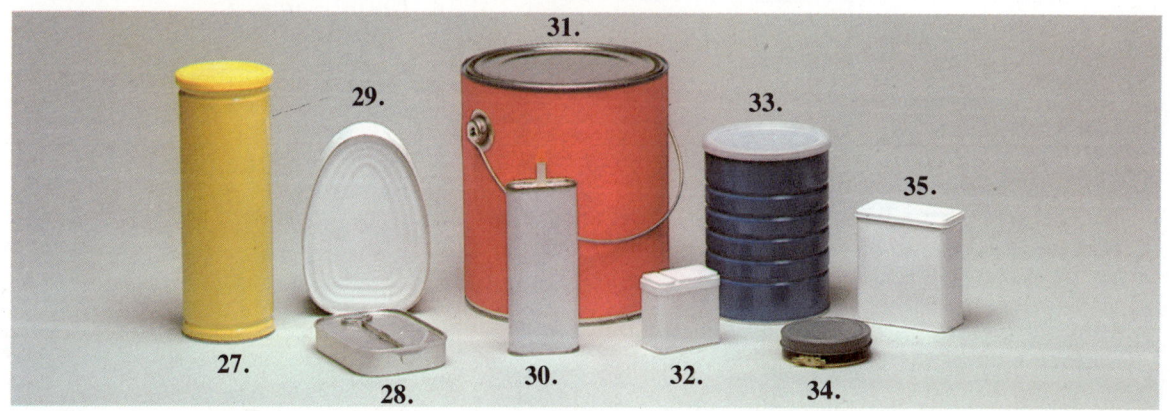

27. Tennis balls 28. Sardines 29. Ham 30. Oil
31. Paint 32. Spice 33. Coffee 34. Shoe polish 35. Adhesive bandages

Surface Area and Volume **367**

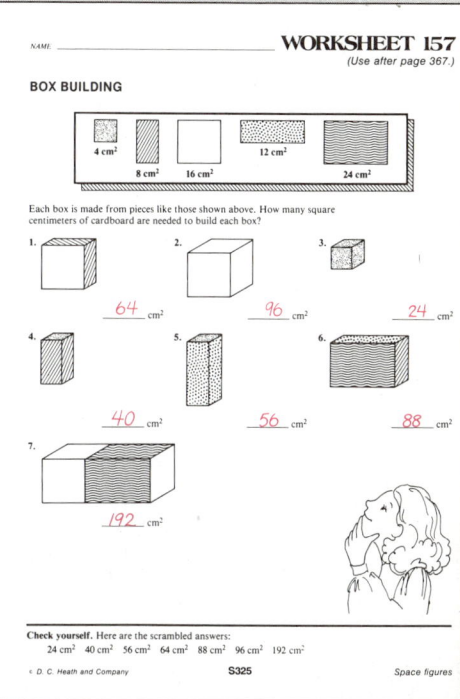

Practice Worksheet
Workbook S325, Copymaster S325, or Duplicating Master S325

Challenge Problem

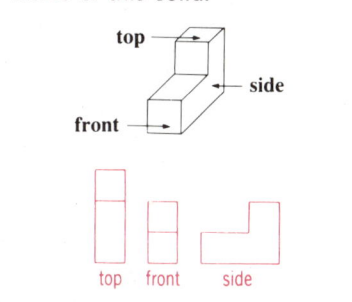

Sketch the top, front, and side views of this solid.

Copymaster S461

Class Starter Quiz 144
on previous lesson

Find the total cost of each model. Use the pieces and prices on page 366.

1. $2.50
2. $1.69
3. $3.44

Copymaster S421

Lesson Objective
To compute the surface area of rectangular prisms and cubes

Problem-Solving Skills
Finding information in a drawing
Making geometric visualizations

Starting the Lesson
Have the students study the photo boxes at the top of the page and answer question 1. Go over the *Here's how* and then discuss questions 2–4.

368

Surface area— rectangular prisms and cubes

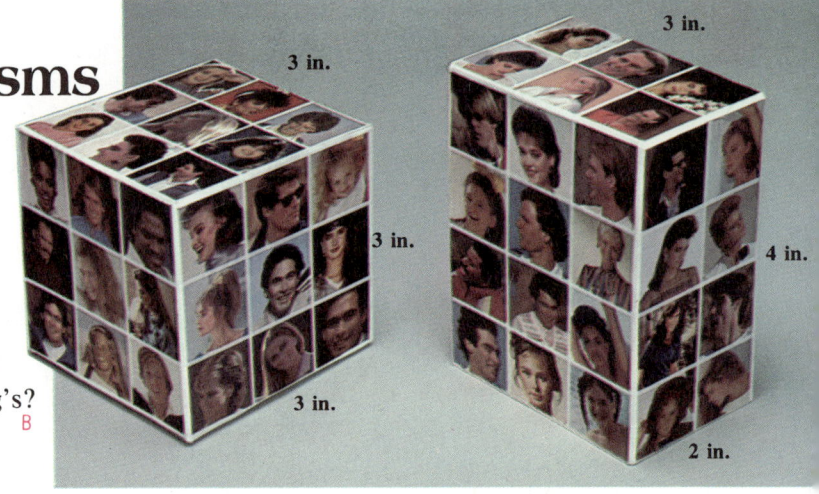

Photo box A   Photo box B

CAN YOU PICTURE IT?

1. Greg used 52 pictures to cover the 6 faces of his photo box. Which photo box is Greg's? **B**

Here's how *to find the surface area of a rectangular prism.*

Think about unfolding **photo box B**. To find the surface area, compute the area of each face by multiplying its length by its width. Then add all six areas.

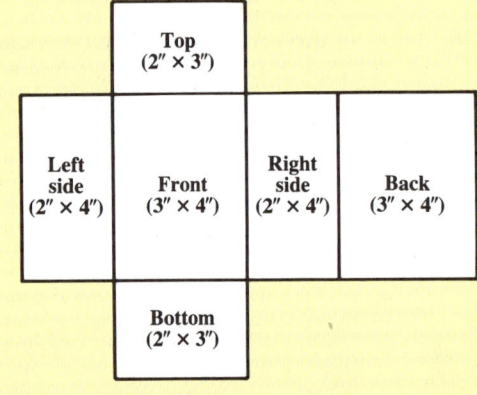

Area of front .... 12 in.²
back .... 12 in.²
top .... 6 in.²
bottom .... 6 in.²
left side ...... 8 in.²
right side ..... 8 in.²
Surface area = **52 in.²**

2. Look at the *Here's how*.
   a. The area of the front is the same as the area of the [?]. **back**
   b. The area of the top is the same as the area of the [?]. **bottom**
   c. The area of the left side is the same as the area of the [?] side. **right**

3. Look at photo box A. If the area of each face is 9 square inches, then the surface area of the cube is [?] square inches. **54**

4. Which photo box has the greater surface area? **A**

368 *Chapter 14*

EXERCISES

Find the surface area of each box.

5.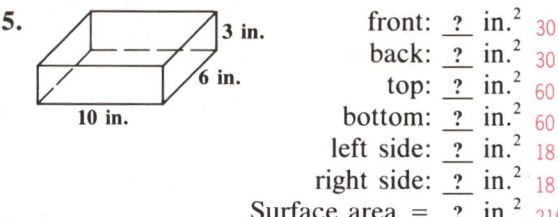
front: ? in.² 30
back: ? in.² 30
top: ? in.² 60
bottom: ? in.² 60
left side: ? in.² 18
right side: ? in.² 18
Surface area = ? in.² 216

6. 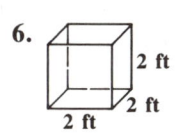 24 ft²

Hint: To find the surface area of a cube, first find the area of one face, then multiply by 6.

7. 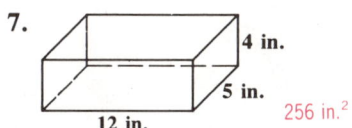 256 in.²

8. 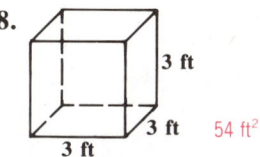 54 ft²

9. 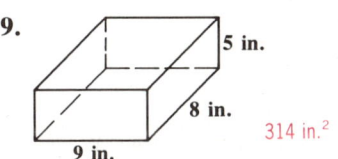 314 in.²

10. 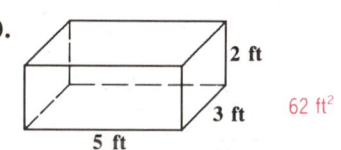 62 ft²

11. 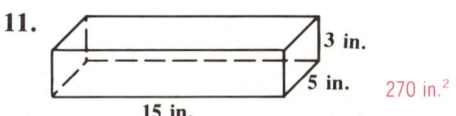 270 in.²

12. 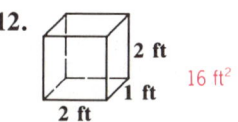 16 ft²

Solve.

13. How many 1-inch-square pictures are needed to cover a photo box 5 inches long, 4 inches wide, and 6 inches tall? 148

14. How many 1-inch-square pictures are needed to cover a photo cube that is 5 inches on an edge? 150

Blockheads Visual thinking

15. This 3-inch photo cube is cut into 27 1-inch cubes. How many of the 1-inch cubes will have
    a. photos on 3 of the faces? 8
    b. photos on 2 of the faces? 12
    c. photos on 1 of the faces? 6
    d. no photos on any of the faces? 1

Surface Area and Volume **369**

Practice Worksheet

Workbook S326, Copymaster S326, or Duplicating Master S326

WORKSHEET 158
(Use after page 369.)

SURFACE AREA
Complete the table.

		FRONT FACE	BACK FACE	TOP FACE	BOTTOM FACE	RIGHT FACE	LEFT FACE	SURFACE AREA
1.	2 cm × 2 cm × 1 cm	2 cm²	2 cm²	4 cm²	4 cm²	2 cm²	2 cm²	16 cm²
2.	3 cm × 2 cm × 2 cm	6 cm²	6 cm²	6 cm²	6 cm²	4 cm²	4 cm²	32 cm²
3.	1 cm × 1 cm × 1 cm	1 cm²	1 cm²	1 cm²	1 cm²	1 cm²	1 cm²	6 cm²
4.	2 cm × 2 cm × 1 cm	4 cm²	4 cm²	2 cm²	2 cm²	2 cm²	2 cm²	16 cm²
5.	1.5 cm × 1 cm × 1 cm	1.5 cm²	1.5 cm²	1.5 cm²	1.5 cm²	1 cm²	1 cm²	8 cm²
6.	2 cm × 1.5 cm × 1 cm	3 cm²	3 cm²	1.5 cm²	1.5 cm²	2 cm²	2 cm²	13 cm²
7.	1.5 cm × 1.5 cm × 1.5 cm	2.25 cm²	2.25 cm²	2.25 cm²	2.25 cm²	2.25 cm²	2.25 cm²	13.5 cm²

Surface area of rectangular prisms and cubes

© D. C. Heath and Company   S326

Group Project

Making measurements

Provide the students with boxes of varying sizes. Have the students measure the length, width, and height of each box to the nearest inch and then compute the surface area of each box.

Class Starter Quiz 145
on previous lesson

Find the surface area of each box.

1. 3 ft × 3 ft × 3 ft 54 ft²
2. 5 ft × 2 ft × 3 ft 62 ft²

Copymaster S421

Problem-Solving Skills
Finding information in an ad
Using a drawing to solve a problem
Solving a multi-step problem

Skills Reviewed
Multiplying and dividing decimals
Changing fractions to decimals
Subtracting mixed numbers
Dividing mixed numbers
Writing numbers with exponents as standard numerals and vice versa
Finding the prime factorization of a number
Finding the greatest common factor and least common multiple of pairs of numbers

Starting the Lesson
Problem Solving Have the students use the sale ad at the top of page 370 to answer questions like these:

- How much does a square of shake shingles cost? ($89.90)
- How many squares of asphalt shingles are needed to cover 400 square feet? (4)
- Can you buy 10 squares of asphalt shingles for $500? (No)

Cumulative Skill Practice Challenge the students to an estimation hunt by saying, "Find the two products in exercises 1–9 that are greater than 100." (Exercises 4 and 8) Then have the students find the quotient in exercises 10–18 that is greater than 100. (Exercise 12)

370

Problem solving
YOU'RE THE ROOFER!

Use the newspaper ad to answer these customers' questions.

SHINGLE SALE
Asphalt shingles .. **$59.90** per square
Shake shingles ... **$89.90** per square

Each "square" of shingles will cover 100 square feet of roof.

1. How many square feet will a square of shingles cover? 100

2. Can I buy 6 squares of asphalt shingles for $350? No

3. Are 7 squares of shingles enough to shingle my house? My roof has an area of 720 square feet. No

4. My roof has an area of 900 square feet. How much will it cost to shingle my roof with shakes? $809.10

Use the pictures to answer these questions. Decide when a calculator would be useful.

5. a. What is the area of the roof? Remember: There are two parts to the roof. 900 ft²
   b. How many squares of shingles will it take to cover the roof? 9
   c. How much will it cost to shingle the roof with asphalt shingles? $539.10

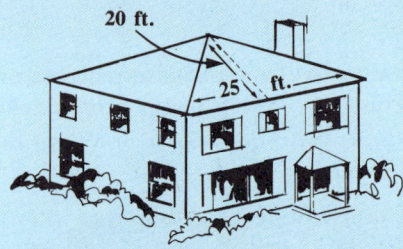

6. a. Each side of the roof is a triangle with a base of 25 feet and a height of 20 feet. What is the area of the roof? 1000 ft²
   b. How many squares of shingles will it take to cover the roof? 10
   c. How much will it cost to shingle the roof with shake shingles? $899

Cumulative Skill Practice

Give the product. *(page 68)*

1. 5.1 × 0.4 2.04
2. 6.3 × 0.42 2.646
3. 0.59 × 0.8 0.472
4. 45 × 2.7 121.5
5. 0.82 × 6.2 5.084
6. 2.18 × 4.9 10.682
7. 7.94 × 6 47.64
8. 52.6 × 9.1 478.66
9. 53.1 × 0.62 32.922

Give the quotient rounded to the nearest tenth. *(page 104)*

10. 17.4 ÷ 0.9 19.3
11. 4.65 ÷ 0.2 23.3
12. 6.47 ÷ 0.03 215.7
13. 5.27 ÷ 1.6 3.3
14. 3.059 ÷ 2.1 1.5
15. 6.07 ÷ 0.35 17.3
16. 16.083 ÷ 0.39 41.2
17. 4.003 ÷ 0.26 15.4
18. 7.03 ÷ 5.4 1.3

Change to a decimal rounded to the nearest hundredth. *(page 166)*

19. $\frac{1}{3}$ 0.33
20. $\frac{1}{6}$ 0.17
21. $\frac{1}{9}$ 0.11
22. $\frac{2}{3}$ 0.67
23. $\frac{2}{9}$ 0.22
24. $\frac{5}{6}$ 0.83
25. $\frac{1}{8}$ 0.13
26. $\frac{1}{12}$ 0.08
27. $\frac{11}{6}$ 1.83
28. $\frac{5}{12}$ 0.42
29. $\frac{5}{3}$ 1.67
30. $\frac{5}{8}$ 0.63
31. $\frac{13}{6}$ 2.17
32. $\frac{20}{6}$ 3.33

Subtract. Give the difference in simplest form. *(page 192)*

33. $3\frac{5}{6} - 1\frac{1}{6}$ $2\frac{2}{3}$
34. $5\frac{3}{4} - 3\frac{1}{4}$ $2\frac{1}{2}$
35. $7 - 2\frac{1}{2}$ $4\frac{1}{2}$
36. $8\frac{1}{3} - 3\frac{3}{8}$ $4\frac{23}{24}$
37. $6 - 4\frac{3}{4}$ $1\frac{1}{4}$
38. $9\frac{2}{5} - 1\frac{7}{10}$ $7\frac{7}{10}$

Give the quotient in simplest form. *(page 214)*

39. $1\frac{1}{3} \div 4$ $\frac{1}{3}$
40. $4\frac{1}{2} \div 2$ $2\frac{1}{4}$
41. $3 \div 1\frac{1}{2}$ 2
42. $4 \div 1\frac{1}{4}$ $3\frac{1}{5}$
43. $8 \div 1\frac{1}{2}$ $5\frac{1}{3}$
44. $2\frac{5}{6} \div 1\frac{1}{8}$ $2\frac{14}{27}$
45. $5\frac{7}{8} \div 4$ $1\frac{15}{32}$
46. $12\frac{1}{2} \div 2\frac{1}{2}$ 5
47. $6\frac{2}{3} \div 3\frac{1}{3}$ 2
48. $2\frac{1}{2} \div 3\frac{1}{2}$ $\frac{5}{7}$

MIXED PRACTICE
Complete.

49. The standard numeral for 4^3 is __?__ 64
50. 3 × 3 × 3 × 3 written using an exponent is __?__ 3^4
51. The prime factorization of 20 is __?__ 2 × 2 × 5 or 2^2 × 5
52. The greatest common factor of 16 and 24 is __?__ 8
53. The least common multiple of 12 and 20 is __?__ 60

Surface Area and Volume

Problem-Solving Worksheet
Workbook S327, Copymaster S327, or Duplicating Master S327

WORKSHEET 159
(Use after page 370.)

NAME _____

SHOW TIME

CINEMA I
Monster Kid
PG 1:00 4:00 7:50

Wendy went to the last showing of *Monster Kid*. The movie was 1 hour 36 minutes long. What time did the movie end? 9:26

LOGICAL REASONING

How much money is in the bank? $6.60
Clues:
- The bank is full of nickels and dimes.
- There are the same number of nickels and dimes.
- There are 88 coins in all.

GUESS AND CHECK

Use X's to show how you can score exactly 75 points with 4 darts.

SUM DIFFERENCE
Complete the table.

Sum	Numbers		Difference
12	10	2	8
9	8	1	7
8	5	3	2
10	9	1	8
14	7	7	0
16	9	7	2
9	5	4	1

MEASURE IT
How many $\frac{1}{2}$ cups in a gallon? 32

COUNT IT!
How many blocks? 16

© D. C. Heath and Company   S327   Problem solving

Challenge Problems

Unscramble the letters of these space figures.

1. EUCB   Cube
2. HPSREE   Sphere
3. MYPRADI   Pyramid
4. SRPIM   Prism
5. ELYCINDR   Cylinder

Copymaster S461

Class Starter Quiz 146
on previous lesson

Solve. Use the shingle sale ad on page 370.

1. How much will it cost for 5 squares of asphalt shingles? **$299.50**
2. How much will it cost to shingle a 25-foot by 40-foot section of a roof with shake shingles? **$899**

Copymaster S421

Lesson Objective
To compute the volume of rectangular prisms and cubes

Problem-Solving Skills
Finding information in a drawing
Using a formula
Choosing the correct operation
Making geometric visualizations

Starting the Lesson
Write these sentences on the chalkboard:

The amount that a space figure holds is called its <u>MOVLUE</u>. Volume is measured in <u>UICBC</u> units.

Before the students open their books, challenge them to unscramble the underlined words so the sentences make sense. Then have them read the sentences at the top of page 372 to see if they correctly unscrambled the letters.
  Next, discuss questions 1–6, using the drawings and the *Here's how*.

Volume—rectangular prisms and cubes

WHAT'S THE VOLUME?

The amount that a space figure holds is called its **volume**. Volume is measured in cubic units.

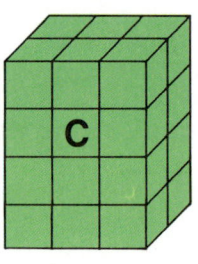
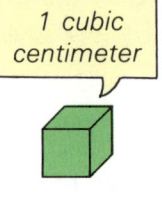

1 cubic centimeter

1. Count the cubes. Which prism has a volume of 6 cubic centimeters? **A**
2. Which prism has a volume of 8 cubic centimeters? **B**
3. What is the volume of prism C? **24 cubic centimeters**

Here's how *to use a formula to find the volume of a prism.*

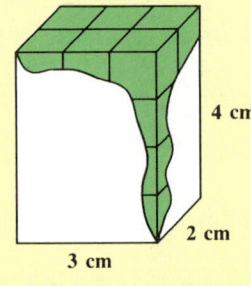

To find the volume (V) of a prism, find the area of the base (B), which is the number of cubes in one layer, and multiply by the height (h), which is the number of layers.

FORMULA $V = B \times h$ *area of the base*
$V = (l \times w) \times h$
$V = 3$ cm $\times 2$ cm $\times 4$ cm
$= 24$ cm^3

cm^3 is a short way to write "cubic centimeters"

Note: In a cube, all sides (s) are the same length, so $V = l \times w \times h = s \times s \times s = s^3$.

4. Look at the *Here's how*. What is the formula for the volume of a rectangular prism? What does each letter stand for? $V = (l \times w) \times h$

V = volume
l = length
w = width
h = height

5. If the length, width, and height of a rectangular prism are 4 centimeters, 5 centimeters, and 6 centimeters, the volume is 120 [?] centimeters.   **cubic**

6. If the edge of a cube is 5 inches long, its volume is 125 cubic [?].   **inches**

372   Chapter 14

EXERCISES

Use the formula $V = B \times h$. Find the volume.

7. 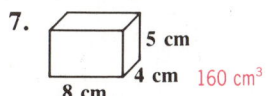 8 cm, 4 cm, 5 cm 160 cm³

8. 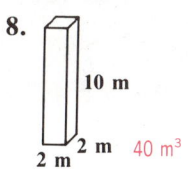 2 m, 2 m, 10 m 40 m³

9. 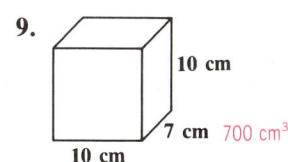 10 cm, 10 cm, 7 cm 700 cm³

10. 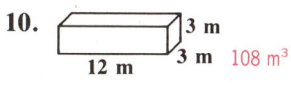 12 m, 3 m, 3 m 108 m³

11. 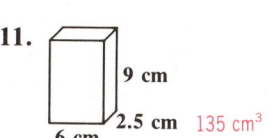 6 cm, 2.5 cm, 9 cm 135 cm³

12. 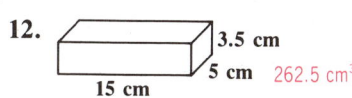 15 cm, 5 cm, 3.5 cm 262.5 cm³

13. 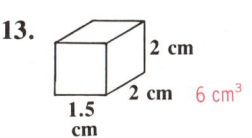 1.5 cm, 2 cm, 2 cm 6 cm³

14. 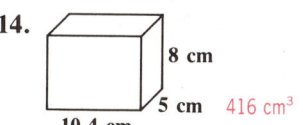 10.4 cm, 5 cm, 8 cm 416 cm³

15. 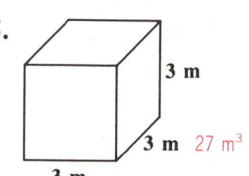 3 m, 3 m, 3 m 27 m³

Solve.

16. Which box above has a surface area of 88 square meters? **Number 8**

17. Which box above has a surface area of 54 square meters? **Number 15**

Math on the job Visual thinking

Your job is to take an inventory of all the nails. There are three stacks of boxes.

18. a. How many boxes? 13
    b. How many nails? 2600

19. a. How many boxes? 33
    b. How many nails? 4950

20. a. How many boxes? 55
    b. How many nails? 5500

 200 NAILS IN EACH BOX

 150 NAILS IN EACH BOX

 100 NAILS IN EACH BOX

Check your calculations. Your inventory should total 13,050 nails.

Practice Worksheet

Workbook S328, Copymaster S328, or Duplicating Master S328

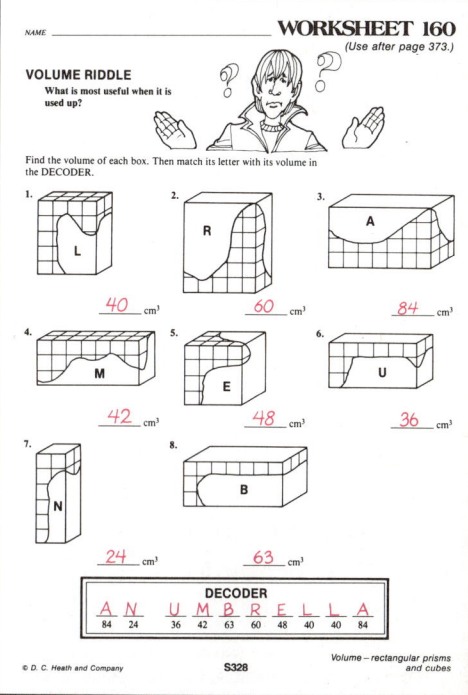

Group Project

Making measurements

Display a collection of rectangular boxes. Have the students measure the dimensions of each box to the nearest centimeter. Then have them compute the volume of each box.

Class Starter Quiz 147
on previous lesson

Find the volume.

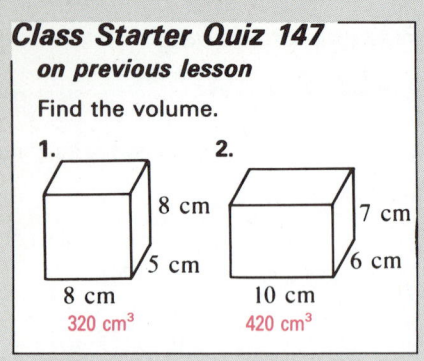

1. 8 cm / 5 cm / 8 cm — 320 cm³
2. 7 cm / 6 cm / 10 cm — 420 cm³

Copymaster S422

Lesson Objective
To compute the volume of a cylinder

Problem-Solving Skills
Finding information in a drawing
Using a formula

Starting the Lesson
Use of Concrete Materials Demonstrate the two ways to roll a sheet of paper to make cylinders like those shown at the top of the page. Ask students whether they think Cylinder A or Cylinder B has the greater volume. Then have the students go over the *Here's how* and answer questions 1–4.

Volume—cylinders

ROLL IT UP!

Here are two ways to roll a sheet of paper to make a cylinder.

Which cylinder do you think has the greater volume?

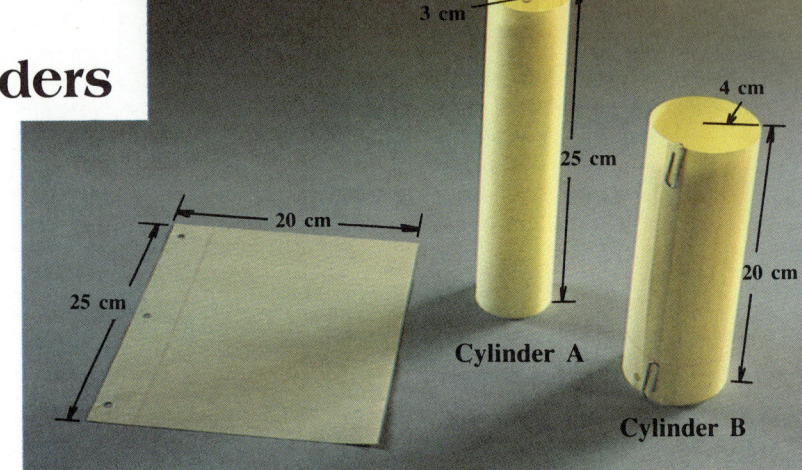

Cylinder A

Cylinder B

Here's how *to use a formula to find the volume of a cylinder.*

Cylinder A

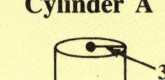

3 cm
25 cm

The area of the base gives the number of cubes needed to cover the bottom.

To find the volume (V) of a cylinder, multiply the area of the base (B) times the height (h).

FORMULA $V = B \times h$ — area of the base
$V = (\pi \times r \times r) \times h$
$V \approx (3.14 \times 3 \text{ cm} \times 3 \text{ cm}) \times 25 \text{ cm}$
$\approx (3.14 \times 9 \text{ cm}^2) \times 25 \text{ cm}$
$\approx 28.26 \text{ cm}^2 \times 25 \text{ cm}$
$\approx 706.5 \text{ cm}^3$

Cylinder B

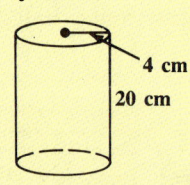

4 cm
20 cm

FORMULA $V = B \times h$
$V = (\pi \times r \times r) \times h$
$V \approx (3.14 \times 4 \text{ cm} \times 4 \text{ cm}) \times 20 \text{ cm}$
$\approx (3.14 \times 16 \text{ cm}^2) \times 20 \text{ cm}$
$\approx 1004.8 \text{ cm}^3$

1. In the formula $V = (\pi \times r \times r) \times h$, what do the letters V, π, r, and h stand for?
   V = volume, π = 3.14, r = radius, h = height
2. Look at the *Here's how*. If the radius and height of a cylinder are 3 centimeters and 25 centimeters, the volume is approximately [?] cubic centimeters. 706.5
3. If the radius and height of a cylinder are 4 centimeters and 20 centimeters, the volume is approximately [?] cubic centimeters. 1004.8
4. Which has the greater volume, Cylinder A or Cylinder B? B

EXERCISES

The area of each base is given. Use V = B × h to find the volume.

5.

6.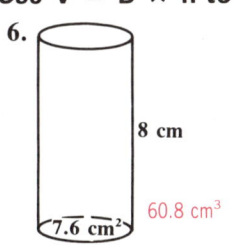

7.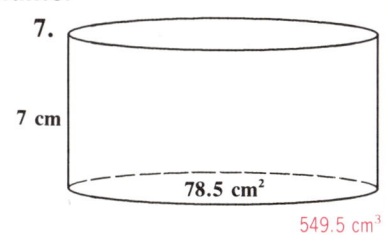

6 cm
28.56 cm²
171.36 cm³

8 cm
7.6 cm²
60.8 cm³

7 cm
78.5 cm²
549.5 cm³

The radius and height are given. Use V = π × r × r × h to find the volume. Use 3.14 as an approximation for π.

8.

9.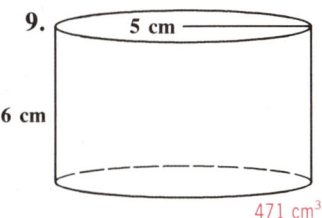

10.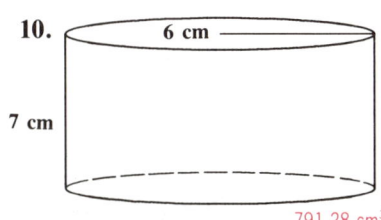

2 cm
8 cm
100.48 cm³

5 cm
6 cm
471 cm³

6 cm
7 cm
791.28 cm³

11.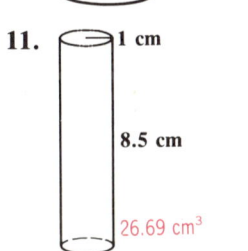

12.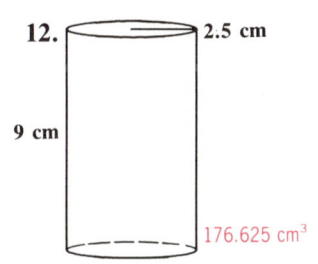

13.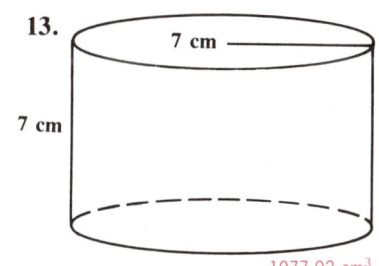

1 cm
8.5 cm
26.69 cm³

2.5 cm
9 cm
176.625 cm³

7 cm
7 cm
1077.02 cm³

You decide!

Is the question about perimeter, area, or volume?

14. How much fencing is needed to fence a patio? **Perimeter**

15. How many flowers are needed to border a garden? **Perimeter**

16. "How much paper is needed to gift-wrap a box?" **Area**

17. "How much space is needed to store a corn harvest?" **Volume**

Surface Area and Volume **375**

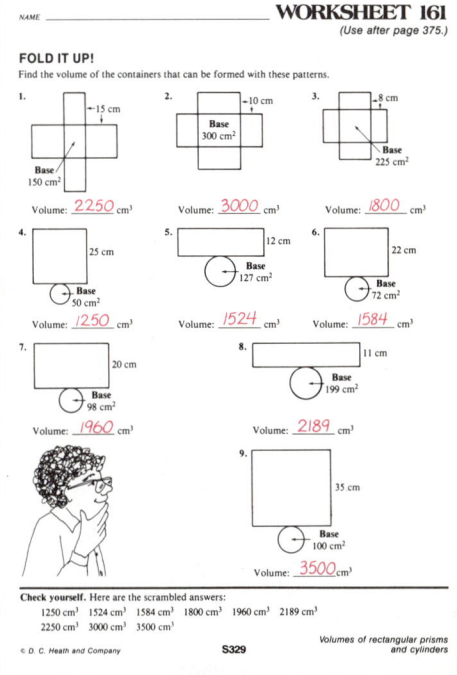

Practice Worksheet
Workbook S329, Copymaster S329, or Duplicating Master S329

Project

Making measurements

Follow these steps to find the volume of a cylinder.

First, roll a sheet of paper to make a cylinder. Second, use paper clips to hold the cylinder in place. Third, measure the radius of the base and the height. Fourth, use a calculator to compute the volume.

Use the clips to adjust the radius of the base so that the volume of the cylinder is 500 cm³.

Copymaster S485

375

Class Starter Quiz 148
on previous lesson

Use $V = \pi \times r \times r \times h$ to find the volume. Use 3.14 for π.

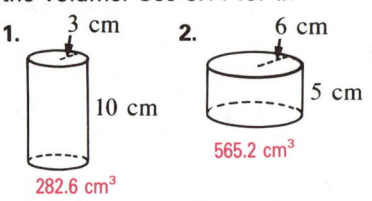

1. 3 cm, 10 cm; 282.6 cm³
2. 6 cm, 5 cm; 565.2 cm³

Copymaster S422

Lesson Objective
To find the volume of a pyramid or cone

Problem-Solving Skills
Finding information in a drawing
Using a formula
Using a guess-and-check strategy

Starting the Lesson
Use of Concrete Materials If available, use models of cylinders, cones, pyramids, and prisms that have the same base and height. Fill a pyramid (cone) with water and empty it into the prism (cylinder). Do this three times to demonstrate that the volume of the prism (cylinder) is three times that of the pyramid (cone).

Exercise Note
Problem Solving Encourage the students to use a guess-and-check approach to solve problems 23–25. The thinking for problem 23 might be as follows:

$8 \times 8 \times 8 = 512$ ← too small
$10 \times 10 \times 10 = 1000$ ← too large
$9 \times 9 \times 9 = 729$ ← perfect

Use 3.14 as an approximation for π in exercises 8–10 and exercises 17–22.

376

Volume—pyramids and cones

1. It takes 3 pyramids of sand to fill the prism. So, the volume of a pyramid is [?] the volume of a prism having the same base and height. $\frac{1}{3}$

2. It takes 3 cones of sand to fill the cylinder. So, the volume of a cone is [?] the volume of a cylinder having the same base and height. $\frac{1}{3}$

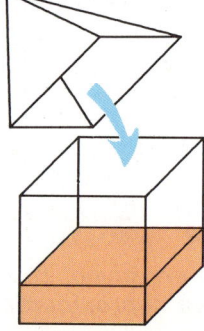
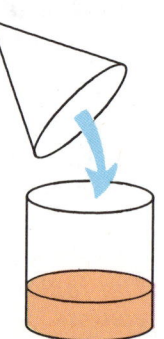

Here's how *to use a formula to find the volume of a pyramid or a cone.*

Pyramid

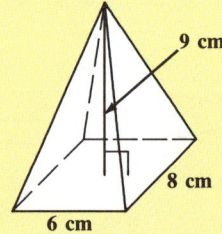

9 cm, 8 cm, 6 cm
Rectangular Base

The volume (V) of a pyramid is $\frac{1}{3}$ times the area of the base (B) times the height (h).

FORMULA $V = \frac{1}{3} \times B \times h$ (area of the base)

$V = \frac{1}{3} \times (l \times w) \times h$

$V = \frac{1}{3} \times 6$ cm $\times 8$ cm $\times 9$ cm

$= 144$ cm³

Cone

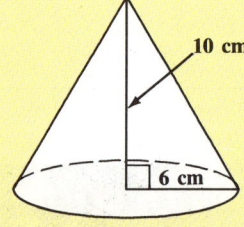

10 cm, 6 cm

The volume (V) of a cone is $\frac{1}{3}$ times the area of the base (B) times the height (h).

FORMULA $V = \frac{1}{3} \times B \times h$ (area of the base)

$V = \frac{1}{3} \times (\pi \times r \times r) \times h$

$V \approx \frac{1}{3} \times 3.14 \times 6$ cm $\times 6$ cm $\times 10$ cm

$\approx \frac{1}{3} \times 1130.4$ cm³

≈ 376.8 cm³

3. Look at the *Here's how*. If the area of the base of a pyramid is 48 square centimeters and the height is 9 centimeters, the volume is [?] cubic centimeters. 144

4. If the radius and height of a cone are 6 centimeters and 10 centimeters, the volume is approximately 376.8 cubic [?]. cm

376 Chapter 14

EXERCISES

The area of each base is given. Use the formula $V = \frac{1}{3} \times B \times h$ to find the volume. Use 3.14 for π.

5.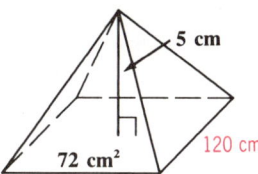
72 cm², 5 cm, 120 cm³

6.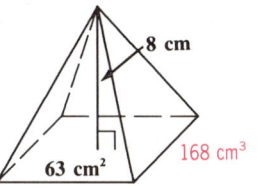
63 cm², 8 cm, 168 cm³

7.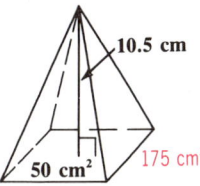
50 cm², 10.5 cm, 175 cm³

8.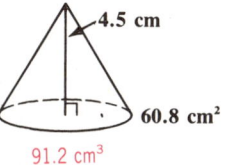
60.8 cm², 4.5 cm, 91.2 cm³

9.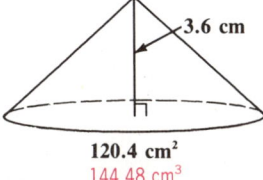
120.4 cm², 3.6 cm, 144.48 cm³

10.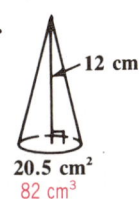
20.5 cm², 12 cm, 82 cm³

Each pyramid described below has a rectangular base. Find the volume.

11. $l = 4$ m   24 m³
    $w = 6$ m
    $h = 3$ m

12. $l = 6$ m   80 m³
    $w = 5$ m
    $h = 8$ m

13. $l = 2$ m   88 m³
    $w = 12$ m
    $h = 11$ m

14. $l = 2.5$ m   20 m³
    $w = 4$ m
    $h = 6$ m

15. $l = 7$ m   84 m³
    $w = 4.5$ m
    $h = 8$ m

16. $l = 1.5$ m   21 m³
    $w = 10$ m
    $h = 4.2$ m

Find the volume of each cone. Round the answer to the nearest tenth.

17. $r = 2$ cm   20.9 cm³
    $h = 5$ cm

18. $r = 3$ cm   37.7 cm³
    $h = 4$ cm

19. $r = 4$ cm   100.5 cm³
    $h = 6$ cm

20. $r = 6$ cm   376.8 cm³
    $h = 10$ cm

21. $r = 8$ cm   334.9 cm³
    $h = 5$ cm

22. $r = 10$ cm   1046.7 cm³
    $h = 10$ cm

Volume hunt — Guess and check

23. What size cube would have a volume of this many cubic meters?

729
9-meter cube

24. What size cube would have a volume of this many cubic meters?
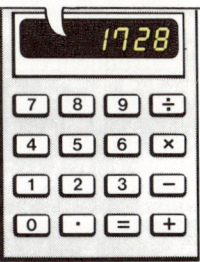
1728
12-meter cube

25. What size cube would have a volume of this many cubic meters?
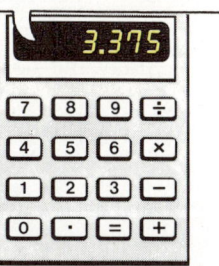
3.375
1.5-meter cube

Surface Area and Volume

Extra Practice
Page 499   Skill 57

Practice Worksheet
Workbook S330, Copymaster S330, or Duplicating Master S330

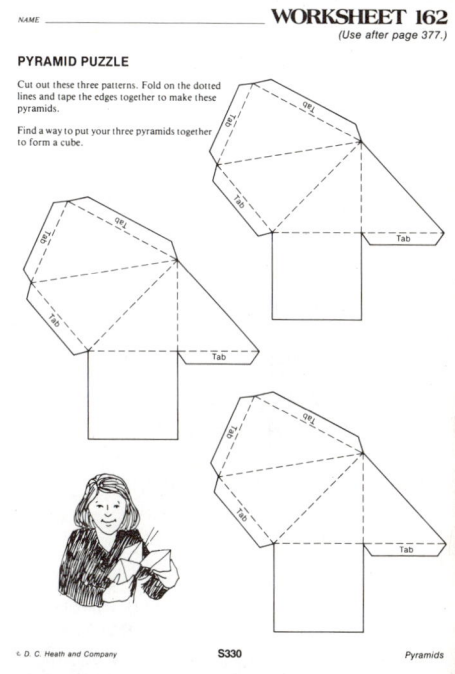

Project

Using library resources

Use a book of world records to find the dimensions and the volume of the world's largest pyramid.

The largest pyramid is the Quetzalcoatl near Mexico City. It is 177 feet tall, and its base covers an area of nearly 45 acres. Its total volume has been estimated at 4,300,000 yd³.

Copymaster S485

Class Starter Quiz 149
on previous lesson

Find the volume.

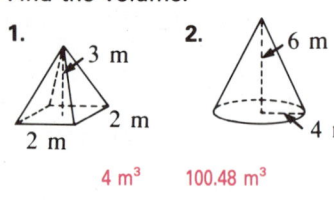

1. 4 m³
2. 100.48 m³

Copymaster S422

Lesson Objective
To solve problems using a pattern

Problem-Solving Skill
Using visual thinking

Starting the Lesson
Problems A and B on page 378 are also on ■ **Visual Aid 65** (copymaster or transparency S163). Use the visual aid when discussing the problems.

Here's How Note
Use of Concrete Materials You may wish to use the area tiles from copymasters or transparencies S532–S534 to demonstrate patterns. See ■ **Manipulative Activity 26** on copymaster S524 in the Teacher's Resource Binder.

378

PROBLEM SOLVING *finding and using patterns*

You can solve some problems by first finding a pattern then using that pattern to answer the question.

Here's how *to solve problems by finding and using a pattern.*

Problem A
What is the tenth figure in this sequence?

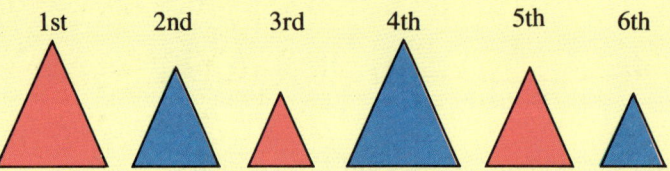

Study the figures to see what in the sequence changes.
Compare color, size, and shape.
How are the figures different? How are they the same?

The **color pattern** is red, blue, red, blue, and so on. *Hint: Odd numbered triangles are red and even numbered triangles are blue.*
The **size pattern** is large, medium, small, and so on.
The **shapes** of all the triangles are the same.

Answer: The tenth figure in the sequence is a large blue triangle.

Problem B
What is the twelfth figure in this sequence?

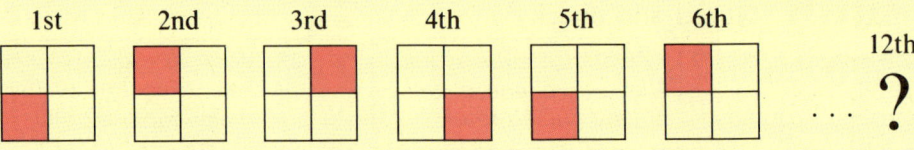

Study the position of the ▨ in the sequence.

The ▨ pattern moves up ↑, right →, down ↓, left ←, up ↑, and so on.

Answer: The twelfth figure in the sequence looks like this ▨

378 *Chapter 14*

PROBLEMS

Choose the figure that comes next in each sequence.

1.
2.
3.

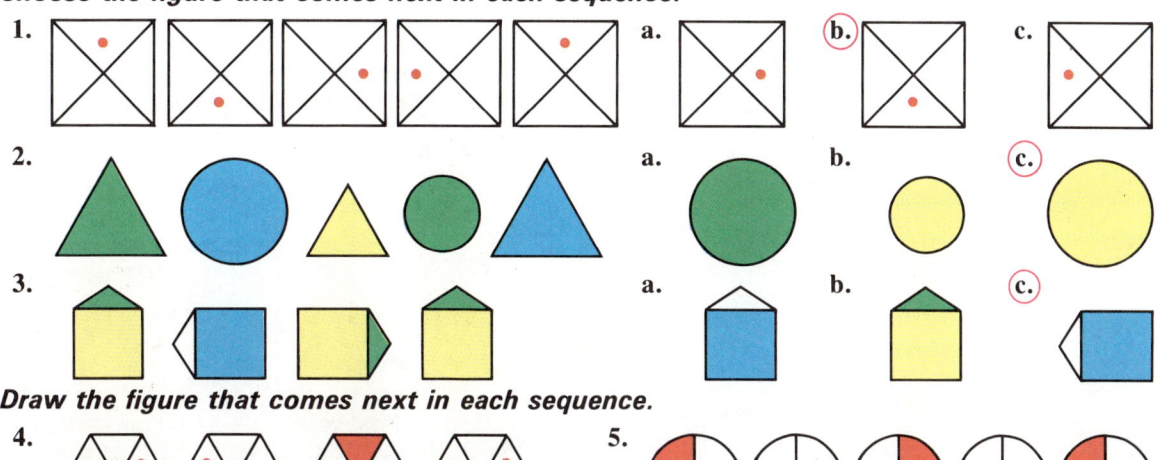

Draw the figure that comes next in each sequence.

4.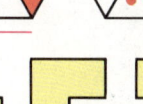

5.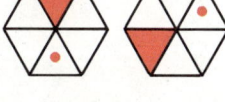

6.

7.

  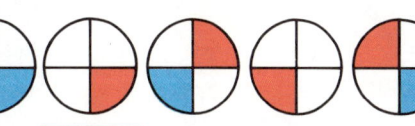

Solve.

8. Draw the 11th figure in this sequence.

9. Draw the 10th figure in this sequence.

Mix and match Visual thinking

10. Which figure does not belong in each group? Why? *Answers will vary.*

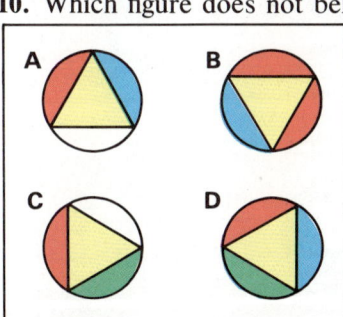
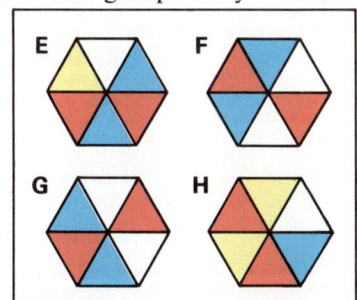

Practice Worksheet
Workbook S331, Copymaster S331, or Duplicating Master S331

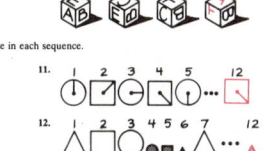

Challenge Problem

Write a question that fits the answer.

A board 8 feet long is cut into two pieces. One piece is 2 feet longer than the other.

How long is the shorter piece?

Answer: 3 feet

Copymaster S461

Surface Area and Volume **379**

Class Starter Quiz 150
on previous lesson

Draw the figure that comes next in each sequence.

1.

2.

Copymaster S422

Problem-Solving Skill
Making geometric visualizations

Skills Reviewed
Making conversions between metric units of measure
Changing percents to decimals
Solving percent problems
Computing simple interest
Adding and subtracting with fractions and mixed numbers
Multiplying and dividing with fractions and mixed numbers

Starting the Lesson

Problem Solving Have the students look at the space figures at the top of page 380. Ask, "Which space figure is a cone?" (Figure E) "Which space figure is a cylinder?" (Figure G) "Which space figure is a sphere?" (Figure A)

Cumulative Skill Practice Write these four answers on the chalkboard:

$0.275 \quad 1.62\frac{1}{2} \quad 108 \quad 160$

Challenge the students to an answer hunt by saying, "Look at exercises 1–37 on page 381. Find the four exercises that have these answers. You have five minutes to find as many of the exercises as you can." (Exercises 8, 18, 22, and 27)

Problem solving

COMPUTER GRAPHICS
A computer can be programmed to sketch a space figure. These sketches were generated by a computer:

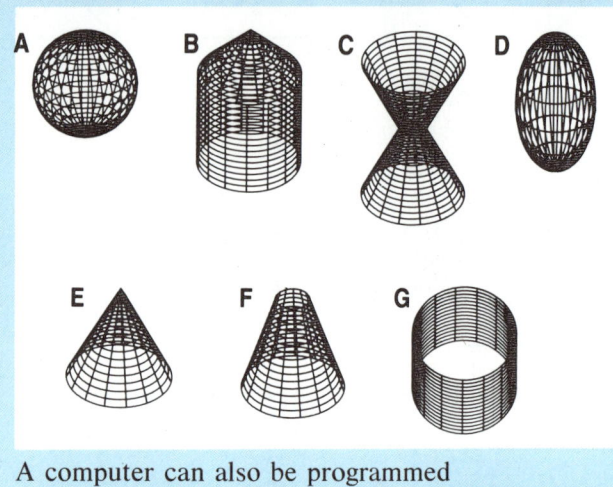

The screen shows the top view and side view of figure A.

A computer can also be programmed to show several views of the same space figure.

Match the top and side views with one of the figures shown above.

1. TOP VIEW SIDE VIEW 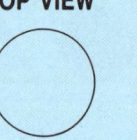 G

2. TOP VIEW SIDE VIEW  B

3. TOP VIEW SIDE VIEW  E

4. TOP VIEW SIDE VIEW 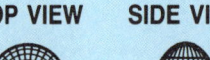 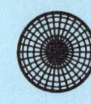 D

5. TOP VIEW SIDE VIEW 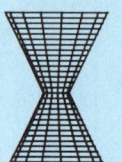 C

6. TOP VIEW SIDE VIEW  F

Chapter 14

Cumulative Skill Practice

Complete. *(page 232)*

1. 5000 mg = ? g   5
2. 3452 mg = ? g   3.452
3. 845 mg = ? g   0.845
4. 2000 g = ? kg   2
5. 6428 g = ? kg   6.428
6. 425 g = ? kg   0.425
7. 6.3 kg = ? g   6300
8. 275 g = ? kg   0.275
9. 444 mg = ? g   0.444

Change to a decimal. *(page 286)*

10. 8%   0.08
11. 37%   0.37
12. 125%   1.25
13. 212%   2.12
14. 0.5%   0.005
15. 0.06%   0.0006
16. 1.3%   0.013
17. $16\frac{2}{3}$% $0.16\frac{2}{3}$
18. $162\frac{1}{2}$% $1.62\frac{1}{2}$
19. $166\frac{2}{3}$% $1.66\frac{2}{3}$

Solve. *(page 294)*

20. 20% of 40 = n 8
21. 7% of 23.5 = n 1.645
22. 150% of 72 = n 108
23. 8.5% of 110 = n 9.35
24. 22.4% of 80 = n 17.92
25. $62\frac{1}{2}$% of 74 = n 46.25

Solve. *(page 296)*

26. 20% of n = 15   75
27. 10% of n = 16   160
28. 40% of n = 60   150
29. 6% of n = 24   400
30. 80% of n = 20   25
31. 12% of n = 3   25

Solve. Round each answer to the nearest tenth. *(page 296)*

32. 12.5% of n = 8.4   67.2
33. 8.5% of n = 11.2   131.8
34. 9.6% of n = 15   156.3
35. 1.4% of n = 2.6   185.7
36. 125% of n = 43   34.4
37. 132% of n = 7.5   5.7

Compute the interest. Round to the nearest cent. *(page 322)*

38. Principal = $1000 $450
    Rate = 15% per year
    Time = 3 years
39. Principal = $700 $42
    Rate = 12% per year
    Time = 6 months
40. Principal = $100 $4.50
    Rate = 1.5% per month
    Time = 3 months

MIXED PRACTICE
Complete.

41. $\frac{3}{4} + \frac{1}{3}$ = ? $1\frac{1}{12}$
42. $\frac{5}{6} - \frac{1}{3}$ = ? $\frac{1}{2}$
43. $4 \times \frac{3}{8}$ = ? $1\frac{1}{2}$
44. $1\frac{5}{8} + 4$ = ? $5\frac{5}{8}$
45. $9 - 3\frac{2}{5}$ = ? $5\frac{3}{5}$
46. $\frac{4}{9} \div 6$ = ? $\frac{2}{27}$
47. $\frac{2}{3} \times \frac{4}{5}$ = ? $\frac{8}{15}$
48. $2\frac{2}{3} \times 4$ = ? $10\frac{2}{3}$
49. $6\frac{2}{5} - 3\frac{5}{6}$ = ? $2\frac{17}{30}$
50. $8 \div 2\frac{1}{4}$ = ? $3\frac{5}{9}$
51. $\frac{11}{12} - \frac{5}{8}$ = ? $\frac{7}{24}$
52. $3\frac{1}{3} + 2\frac{7}{8}$ = ? $6\frac{5}{24}$
53. $1\frac{1}{3} \times 1\frac{1}{3}$ = ? $1\frac{7}{9}$
54. $\frac{5}{12} \div \frac{3}{4}$ = ? $\frac{5}{9}$
55. $5\frac{1}{2} \div 3\frac{1}{4}$ = ? $1\frac{9}{13}$

Surface Area and Volume **381**

Problem-Solving Worksheet
Workbook S332, Copymaster S332, or Duplicating Master S332

WORKSHEET 164 (Use after page 380.)

HELLO! Long-distance rate. First minute, $.40. Each additional minute, $.25.
How much would it cost you to make an 8-minute long-distance call? $2.15

GUESS AND CHECK Erika gave the clerk 3 coins and got back 2 coins. Which book did she buy?   Jokes

WHAT'S THE SALE PRICE? $\frac{1}{4}$ OFF   Reg. $280.00   Now $ 210

WHAT'S THE ANSWER? Use the code to answer the riddle.

FOLD IT! This pattern can be folded to make a box. What will be the volume of the box?   1280 in.³

Riddle: Why did the baby pig eat so much?
Answer: TO MAKE A HOG OF HIMSELF

S332   Problem solving

Challenge Problems

Unscramble these words to get a mathematical definition. Then write the math word that fits the definition.

1. DISTANCE A CIRCLE ACROSS THE CENTER THROUGH ITS
   The distance across a circle through its center; diameter

2. RECTANGULAR PRISM FACES ARE A SIX SQUARES WHOSE
   A rectangular prism whose six faces are squares; cube

3. OF SPACE A INSIDE MEASURE THE FIGURE THE SPACE
   The measure of the space inside a space figure; volume

Copymaster S461

381

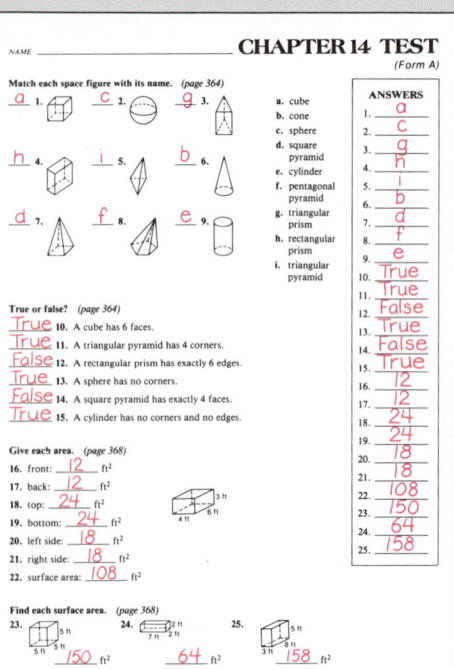

Chapter REVIEW

Here are scrambled answers for the review exercises:

12	125	add	cylinder	sphere
36	209	area	height	square pyramid
90	314	cone	rectangular prism	triangular prism

1. triangular prism   **2.** square pyramid   **3.** sphere   **4.** cylinder   **5.** rectangular prism

1. Figure A is called a ?.
2. Figure B is called a ?.
3. Figure C is called a ?.
4. Figure D is called a ?.
5. Figure E is called a ?. (page 364)

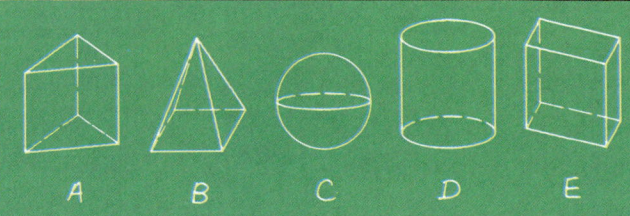

6. add, 90   **7.** height, 36   **8.** 125

6. To find the surface area of this figure, you would first compute the area of each face. Then you would ? the six areas. The surface area of this figure is ? in.² (page 368)

7. To find the volume of a rectangular prism, you multiply the length (l) times the width (w) times the ? (h). The volume of this rectangular prism is ? ft³. (page 372)

8. The volume of this cube is ? ft³. (page 372)

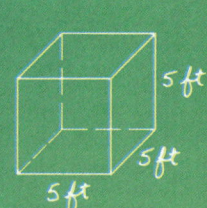

9. area, 314   **10.** cone, 12   **11.** 209

9. To find the volume of a cylinder, you multiply the ? of the base (B) times the height (h). Using 3.14 as an approximation for π, the volume of this cylinder is about ? in.³ (page 374)

10. To find the volume of a pyramid or ?, you multiply $\frac{1}{3}$ times the area of the base (B) times the height (h). The volume of this pyramid is ? in.³ (page 376)

11. Using 3.14 as an approximation for π, the volume of this cone is about ? in.³ (page 376)

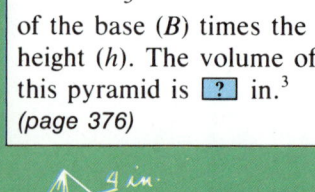
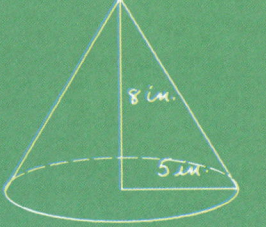

382   Chapter 14

Chapter TEST

Match each space figure with its name. *(page 364)*

1.  C
2. 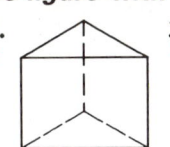 H
3. 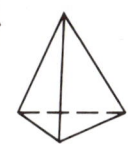 I
4. 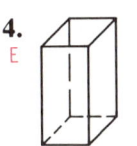 E
5.  A
6.  D
7. 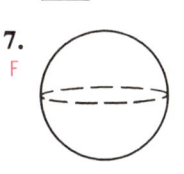 F
8. 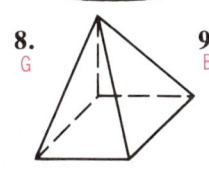 G
9. 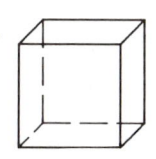 B

A. Cone
B. Cube
C. Cylinder
D. Pentagonal pyramid
E. Rectangular prism
F. Sphere
G. Square pyramid
H. Triangular prism
I. Triangular pyramid

Find the surface area. *(page 368)*

10. 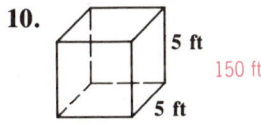 5 ft, 5 ft, 5 ft — 150 ft²
11. 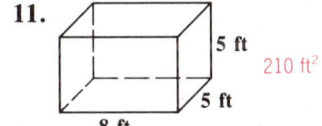 5 ft, 5 ft, 8 ft — 210 ft²
12. 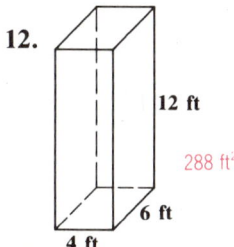 12 ft, 6 ft, 4 ft — 288 ft²

Find the volume. Use 3.14 as an approximation for π. *(pages 372, 374, 376)*

13. 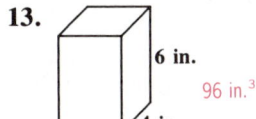 6 in., 4 in., 4 in. — 96 in.³
14. 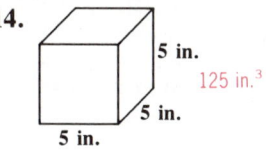 5 in., 5 in., 5 in. — 125 in.³
15. 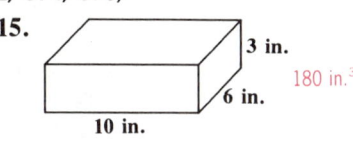 3 in., 6 in., 10 in. — 180 in.³

16. 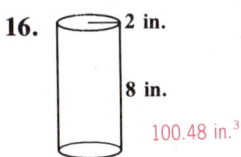 2 in., 8 in. — 100.48 in.³
17. 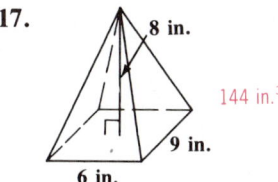 8 in., 9 in., 6 in. — 144 in.³
18. 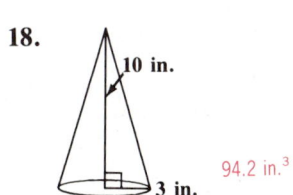 10 in., 3 in. — 94.2 in.³

Surface Area and Volume **383**

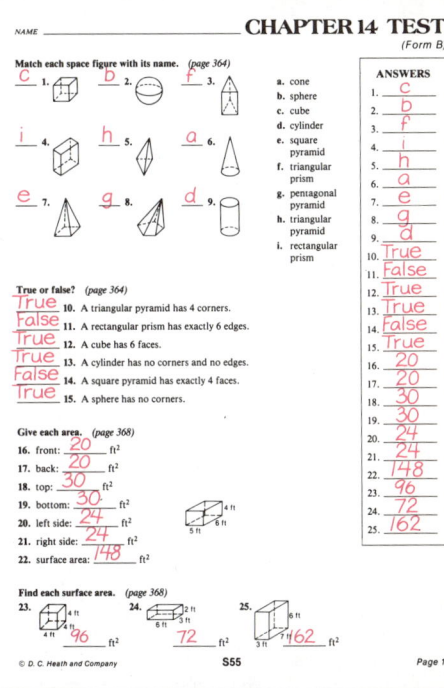

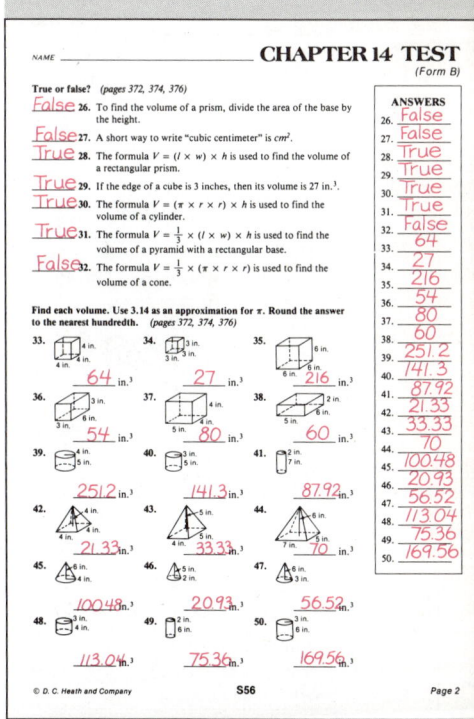

Copymasters S55 and S56
or Duplicating Masters S55 and S56

383

Cumulative TEST — Standardized Format

Choose the correct letter.

1. Give the product.

29.04 × 1.09

A. 31.6536
B. 55.1760
C. 29.2506
D. none of these

2. Give the quotient rounded to the nearest tenth.

64.08 ÷ 3.1

A. 20.67   B. 20.6
C. 20.7   D. none of these

3. Change to a decimal rounded to the nearest hundredth.

$\frac{5}{12} = ?$

A. 2.40   B. 0.42
C. 0.41   D. none of these

4. Subtract. $4\frac{1}{3}$
$-2\frac{3}{4}$

A. $2\frac{5}{12}$ B. $2\frac{7}{12}$
C. $1\frac{7}{12}$ D. none of these

5. Give the quotient.

$3\frac{1}{4} \div 1\frac{1}{2}$

A. $2\frac{1}{6}$ B. $4\frac{7}{8}$
C. $\frac{6}{13}$ D. none of these

6. 525 mg = ? g

A. 5.25
B. 52.5
C. 5250
D. none of these

7. Change to a decimal.

$37\frac{1}{2}\% = ?$

A. 37.5
B. 3.75
C. 0.375
D. none of these

8. Solve.

$33\frac{1}{3}\%$ of 45 = n

A. 15
B. 30
C. 135
D. none of these

9. Solve.

25% of n = 17

A. 4.25
B. 68
C. 12.75
D. none of these

10. Compute the interest.

Principal = $820
Rate = 14% per year
Time = 9 months

A. $114.80
B. $28.70
C. $1033.20
D. none of these

11. Find the area of this parallelogram.

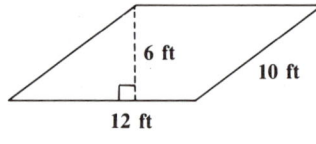

A. 72 ft²
B. 120 ft²
C. 44 ft²
D. none of these

12. Find the volume of this rectangular prism.

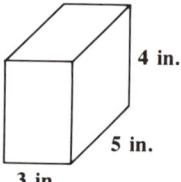

A. 12 in.³
B. 35 in.³
C. 60 in.³
D. none of these

Probability

Resources

- **Class Starter Quizzes 151-159** *(Copymasters S423-S425)*
- **Manipulatives**
  Manipulative Activities 27-29 *(Copymasters S524-S526)*
  Area tiles *(Copymasters or Transparencies S532-S534)*
- **Worksheets 165-174** *(Copymasters, Duplicating Masters, or Workbook pages S333-S342)*
- **Challenge Problems** for pages 395, 403, 405
  (Copymaster S462)
- **Projects** for pages 387, 389, 391, 401 *(Copymasters S485-S486)*
- **Tests** *(Copymasters or Duplicating Masters S57-S60)*

Lesson Objectives

To find the total number of outcomes by using a tree diagram

To use a basic counting principle to determine the number of outcomes of a compound event

Problem-Solving Skill

Drawing a tree diagram

Starting the Lesson

Use questions 1–5 and the *Here's how* to introduce the terms *tree diagram* and *basic counting principle*. Go through the diagram and counting principle carefully so that the students see how they show 6 possible outfits.

Here's How *Note*

Use of Concrete Materials You may wish to use the area tiles from copymasters or transparencies S532–S534 to demonstrate the outcomes of an event. See ■ **Manipulative Activity 27** on copymasters S524–S525 in the Teacher's Resource Binder.

You may also wish to use this additional chalkboard example before assigning exercises 6–12.

How many possible outcomes are there if you first flip a coin and then spin this spinner?

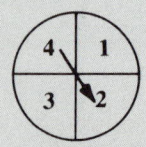

Tree diagram

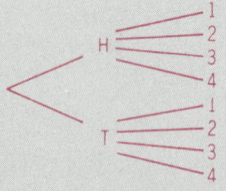

Basic counting principle

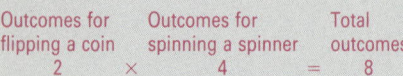

A basic counting principle

DECISIONS, DECISIONS!

1. How many different outfits (a skirt and a blouse) can be made if the blue skirt is worn? 3

2. How many different outfits can be made if the gray skirt is worn? 3

3. How many different outfits are there in all? 6

Here's how to show all the possible outfits with a *tree diagram*.

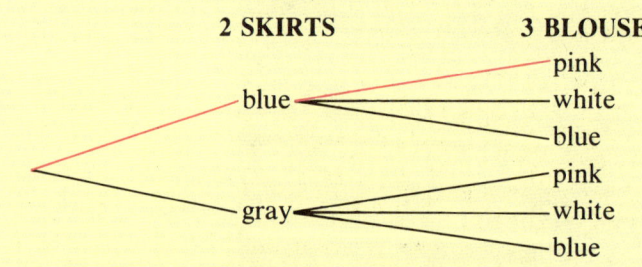

The red "branch" represents a blue skirt and a pink blouse.
You can determine the number of outfits by counting the branches of the tree diagram.

4. Look at the *Here's how*. What outfit is represented by the top branch? Blue skirt, pink blouse
   The bottom branch? Gray skirt, blue blouse

5. How many possible outfits are there? *Hint: Count the branches.* 6

Here's how to use a *basic counting principle* to compute the total number of outfits.

To compute the total number of ways that several decisions can be made, multiply the number of choices for each of the decisions.

Choices in First Decision (skirts)	Choices in Second Decision (blouses)	Total Choices (outfits)
2 ×	3 =	6

386 Chapter 15

EXERCISES

Solve.

6. a. How many choices of pants are there? How many choices of sweaters are there? 2, 4
   b. Draw a tree diagram to show all possible outfits.
   c. How many possible outfits are there? 8
   d. How many different outfits can be made if the black pants are worn? 4
   e. How many different outfits can be made if the red sweater is not worn? 6

7. You decide to wear a pair of corduroys and a sweater. How many outfits do you have if you have 4 pairs of corduroys and 5 sweaters? 20

8. How many outfits can you make from 3 pairs of pants, 4 shirts, and 2 sweaters if each outfit consists of pants, a shirt, and a sweater? 24

9. In an election of your class officers, 3 students are running for president, 3 for vice president, 2 for secretary, and 2 for treasurer. How many different combinations are possible? 36

10. You decide to buy a stereo system. You can choose from 5 amplifiers, 6 kinds of speakers, and 4 turntables. How many different systems are possible? 120

11. You are in Chicago and win a free trip to Waikiki Beach in Honolulu. The "map" shows the choices of ways to travel.
    a. How many ways can you travel from Chicago to Los Angeles? 3
    b. How many ways can you travel from Chicago to Honolulu? 6
    c. How many ways can you travel from Chicago to Waikiki Beach? 12
    d. How many choices would you have for making the round trip? 144

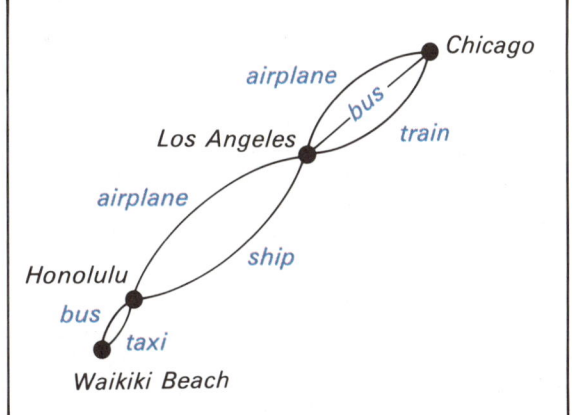

You're the detective!

12. A witness observed a speeding car leaving the scene of an accident. One digit and two letters on the license plate were covered with mud.

    What is the greatest number of license plates that you would have to check to find the owner of the car? 6760

Practice Worksheet

Workbook S333, Copymaster S333, or Duplicating Master S333

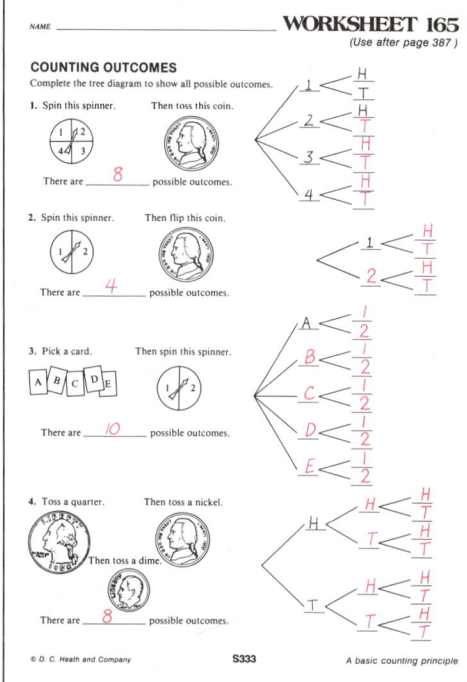

Project

Researching information

Find how many letters and numerals are combined to make license plates in the state where you live. Then figure out the number of different license plates that can be made.

Copymaster S485

Class Starter Quiz 151
on previous lesson

Solve.
1. How many possible outfits do you have if you have 3 pairs of slacks and 5 sweaters? 15
2. How many outfits can you make from 4 pairs of slacks, 5 shirts, and 3 sweaters? 60

Copymaster S423

Lesson Objective
To compute the number of permutations (possible arrangements of things in a definite order)

Problem-Solving Skill
Selecting information from a display

Starting the Lesson
Have the students guess how many ways the 3 people shown at the top of the page can be arranged for the photograph. Then go over questions 1–4 so that the students can check their guesses. Go over question 5 carefully before assigning the exercises on page 389.

Here's How Note
Use of Concrete Materials You may wish to use the area tiles from copymasters or transparencies S532–S534 to demonstrate the number of permutations of an event. See ■ **Manipulative Activity 28** on copymaster S525 in the Teacher's Resource Binder.

388

Permutations

Bob, Ann, and Mary had this photograph taken at an amusement park.

1. Who is on the left? The right? Who is in the middle? Bob, Mary, Ann
2. The letters BAM may be used to describe the order they were in when the photo was taken. Use the letters to list all possible orders. BAM, BMA, ABM, AMB, MAB, MBA
3. How many ways can the 3 people be arranged for the photograph? 6

An arrangement of things in a **definite order** is called a **permutation**.

Here's how to compute the number of permutations (possible arrangements).

Number of people to choose from for first position		Number left to choose from for second position		Number left to choose from for third position		Number of possible arrangements
↓		↓		↓		↓
3	×	2	×	1	=	6

The notation 3!, read as "3 factorial," can be used to show the above product.
$$3! = 3 \times 2 \times 1 = 6$$

4. Look at the *Here's how*. To compute the number of permutations (possible arrangements) of 3 things, you would multiply what three numbers? 3, 2, 1
5. Think about 4 people and a comic photo scene that requires 4 faces.
   a. How many people would there be to choose from for the first position? 4
   b. How many people would there be left to choose from for the second position? The third position? The fourth position? 3, 2, 1
   c. The number of permutations is: $4! = 4 \times 3 \times 2 \times 1 = ?$ 24

388 Chapter 15

EXERCISES

Give the product for each factorial.

6. 5! 120 **7.** 4! 24 **8.** 2! 2 **9.** 3! 6 **10.** 8! 40,320 **11.** 6! 180

Solve.

12. Bob, Ann, and Mary decided to buy some ride tickets at the amusement park. How many ways could they line up to purchase their tickets? 6

WHIP $1.15
INDY 500 $1.35
BOBSLED $1.25
SUBMARINE $.85
ROCKET SHIP $.90

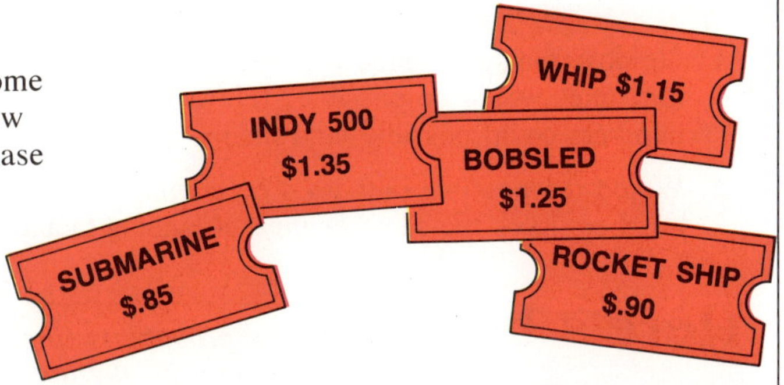

13. Bob bought the tickets shown.

a. For how many rides did he buy tickets? 5

b. What was the average price of the tickets? $1.10

c. In how many different orders could Bob take the 5 rides? 120

14. Each bobsled seated 4 people in a row. In how many ways could 4 people be seated on a bobsled? 24

15. a. Ann decided to buy a sandwich and a drink for lunch. How many different lunches could she buy? 20

b. Suppose that you wanted to buy a sandwich and a drink and that you decided not to order a fish sandwich. How many different lunches could you buy? 15

c. How many different lunches could Bob order if he decided to order a sandwich, a drink, and a dessert? 60

d. Mary ordered a cheeseburger, milk, and a dessert. She gave the clerk $5 and got $1.90 in change. What dessert did she order? Brownie

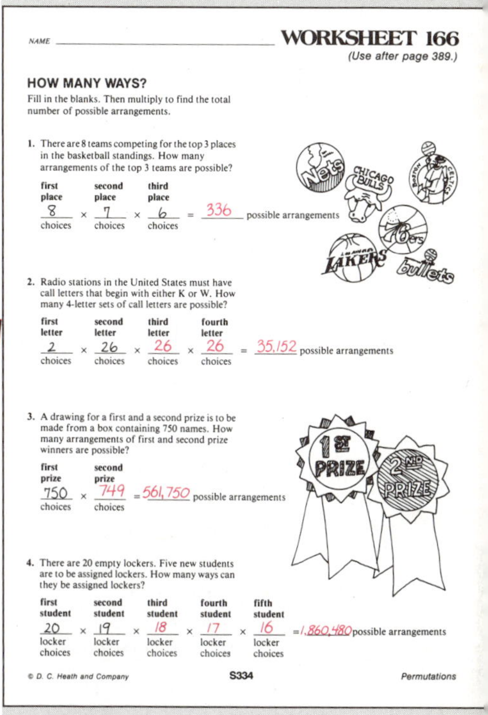

JOYLAND SNACK BAR	
SANDWICHES	
Hamburger	$1.45
Cheeseburger	1.65
Hot Dog	1.30
Fish	1.45
DRINKS	
Coffee	.60
Cola	.65
Root Beer	.65
Lemonade	.75
Milk	.70
DESSERTS	
Ice Cream	.95
Brownie	.75
Apple Pie	1.30

Please be seated!

16. Each car on the Grand Canyon Railroad holds 10 people. Suppose that you and 9 of your friends were in one of the cars.

a. How many ways could the group be seated? 3,628,800

b. Suppose that the group could change the seating order every 5 seconds. How many seconds would it take for the group to sit in all possible orders? 18,144,000

c. How many minutes would it take? How many hours? How many days? 302,400; 5040; 210

Practice Worksheet

Workbook S334, Copymaster S334, or Duplicating Master S334

NAME ______________________ **WORKSHEET 166**
(Use after page 389.)

HOW MANY WAYS?
Fill in the blanks. Then multiply to find the total number of possible arrangements.

1. There are 8 teams competing for the top 3 places in the basketball standings. How many arrangements of the top 3 teams are possible?

first place 8 choices × second place 7 choices × third place 6 choices = 336 possible arrangements

2. Radio stations in the United States must have call letters that begin with either K or W. How many 4-letter sets of call letters are possible?

first letter 2 choices × second letter 26 choices × third letter 26 choices × fourth letter 26 choices = 35,152 possible arrangements

3. A drawing for a first and a second prize is to be made from a box containing 750 names. How many arrangements of first and second prize winners are possible?

first prize 750 choices × second prize 749 choices = 561,750 possible arrangements

4. There are 20 empty lockers. Five new students are to be assigned lockers. How many ways can they be assigned lockers?

first student 20 locker choices × second student 19 locker choices × third student 18 locker choices × fourth student 17 locker choices × fifth student 16 locker choices = 1,860,480 possible arrangements

© D. C. Heath and Company S334 Permutations

Project

Researching information

Find how many telephone prefix numbers are used in the town or city where you live. Then figure out the number of different telephone numbers that can be assigned.

Copymaster S486

Class Starter Quiz 152

on previous lesson

Solve.

1. How many ways could 4 people line up to buy movie tickets? 24
2. How many ways could you buy a sandwich and a drink for lunch if there were 5 sandwich choices and 4 drink choices? 20

Copymaster S423

Lesson Objective

To determine the probabilities of outcomes of simple events

Problem-Solving Skills

Selecting information from a display
Conducting an experiment and collecting data

Starting the Lesson

Sketch this pattern on the chalkboard:

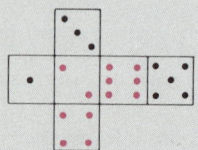

Tell the students that the pattern can be folded to make a die. Ask students to fill in the missing dots on the 3 blank faces. (Remind them that dots on opposite faces of a die add to 7.) Then go over questions 1 and 2, using the pattern on the chalkboard and the *Here's how*.

Here's How Note

Use of Concrete Materials You may wish to use the area tiles from copymasters or transparencies S532–S534 to demonstrate the outcomes of a probability experiment. See ■ **Manipulative Activity 29** on copymaster S526 in the Teacher's Resource Binder.

Probability

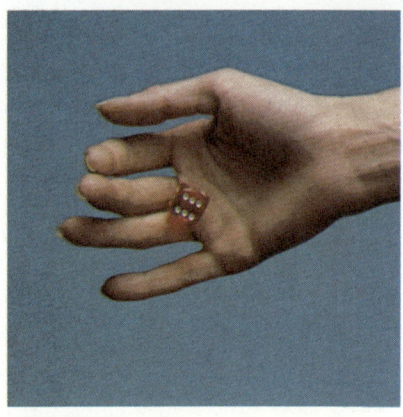

WHAT ARE YOUR CHANCES?

1. Suppose that you rolled the die. How many different outcomes would be possible? (How many different numbers could possibly land facing up?) 6

2. Would the possible outcomes be equally likely? (Would all numbers have the same chance of landing facing up?) Yes

Here's how to find the probability (the chance) of a given outcome when all the outcomes are equally likely.

Probability of rolling a 1

$P(1) = \dfrac{\text{number of ways of rolling a 1}}{\text{number of possible outcomes}}$

$P(1) = \dfrac{1}{6}$ — The probability of 1 equals $\dfrac{1}{6}$ or 1 out of 6.

Probability of rolling either a 5 or a 6

$P(5 \text{ or } 6) = \dfrac{2}{6}$ — There are two ways to get either a 5 or a 6.

$P(5 \text{ or } 6) = \dfrac{1}{3}$

3. Look at the *Here's how*. What is the probability of rolling a 1? $\frac{1}{6}$
4. What is the probability of rolling a 5 or a 6? $\frac{1}{3}$

EXERCISES

Give each probability as a fraction in simplest form.

Think about rolling a die.

5. P(3) $\frac{1}{6}$
6. P(6) $\frac{1}{6}$
7. P(1 or 2) $\frac{1}{3}$
8. P(even) $\frac{1}{2}$
9. P(odd) $\frac{1}{2}$
10. P(1, 2, or 3) $\frac{1}{2}$
11. P(3 or less) $\frac{1}{2}$
12. P(greater than 2) $\frac{2}{3}$
13. P(not 1) $\frac{5}{6}$
14. P(not 5) $\frac{5}{6}$
15. P(7) 0
16. P(6 or less) 1

17. Look at your answer to exercise 15.
What is the probability of an impossible outcome? 0

18. Look at your answer to exercise 16.
What is the probability of an outcome that is certain to occur? 1

Give each probability as a fraction in simplest form.

Think about spinning this spinner.

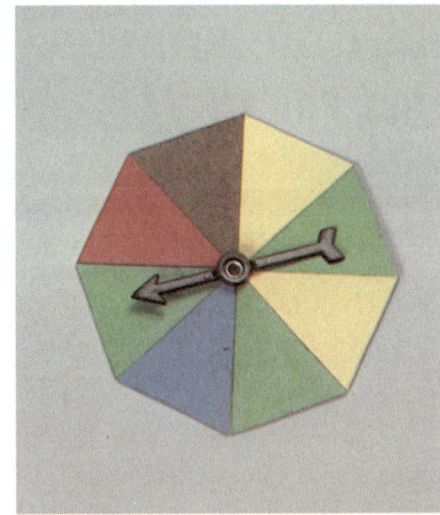

19. P(red) $\frac{1}{8}$
20. P(not red) $\frac{7}{8}$
21. P(blue) $\frac{1}{8}$
22. P(not blue) $\frac{7}{8}$
23. P(yellow) $\frac{1}{4}$
24. P(not yellow) $\frac{3}{4}$
25. P(green) $\frac{3}{8}$
26. P(not green) $\frac{5}{8}$
27. P(black) 0
28. P(not black) 1
29. P(brown or red) $\frac{1}{4}$
30. P(yellow or blue) $\frac{3}{8}$
31. P(yellow or green) $\frac{5}{8}$
32. P(red or green) $\frac{1}{2}$
33. P(blue or green) $\frac{1}{2}$
34. P(black or yellow) $\frac{1}{4}$
35. P(yellow, blue, or green) $\frac{3}{4}$
36. P(brown, blue, or red) $\frac{3}{8}$

Think about shuffling these cards and then turning one of the cards face up.

37. P(M) $\frac{1}{16}$
38. P(C) $\frac{1}{16}$
39. P(E) $\frac{1}{8}$
40. P(S) $\frac{3}{16}$
41. P(O) $\frac{3}{16}$
42. P(not E) $\frac{7}{8}$
43. P(B or L) $\frac{1}{8}$
44. P(O or U) $\frac{1}{4}$
45. P(O or S) $\frac{3}{8}$
46. P(vowel) $\frac{7}{16}$
47. P(consonant) $\frac{9}{16}$
48. P(vowel or consonant) 1

Making a prediction

When you toss a thumbtack, there are two possible outcomes: landing point up or landing point down.

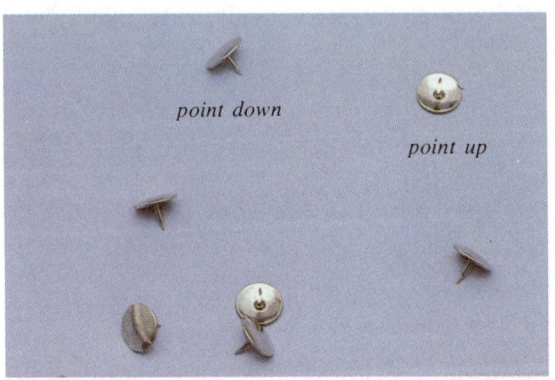

49. Predict which outcome is more likely. *point down or point up*
50. Toss a thumbtack 50 times and keep a record of the outcomes.
51. Did your prediction agree or disagree with the results of your experiment? *Answers will vary.*

Practice Worksheet
Workbook S335, Copymaster S335, or Duplicating Master S335

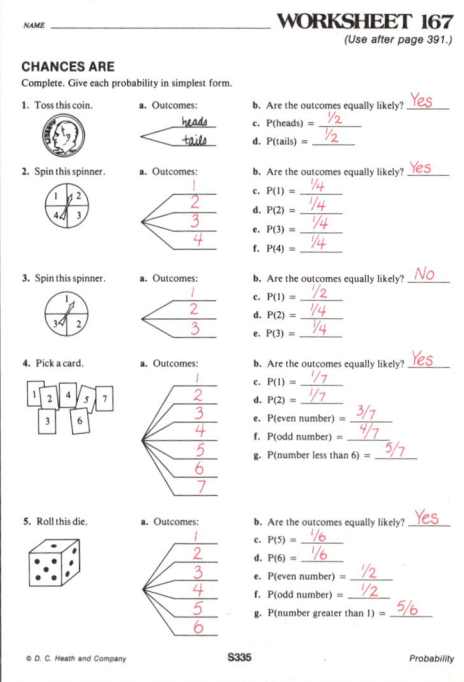

Project

Researching information

Look at a sequence of 200 names in a telephone directory to find how many first names have the same initial letter as your first name. Then figure out how likely it is that you will meet a stranger with a first name that has the same initial letter as your first name.

Copymaster S486

Class Starter Quiz 153
on previous lesson

Give the probability as a fraction in simplest form.
Think about spinning this spinner.

1. P(2) $\frac{1}{4}$
2. P(1) $\frac{1}{4}$
3. P(1 or 3) $\frac{1}{2}$
4. P(not 1) $\frac{3}{4}$
5. P(even number) $\frac{1}{2}$

Copymaster S423

Lesson Objective
To find the probability of an outcome

Problem-Solving Skills
Selecting information from a display
Using a tree diagram

Starting the Lesson
Have the students read the introductory paragraph and answer questions 1 and 2. Then go over the *Here's how* and use the sample space to answer questions 3–5.

Sample spaces
FREE FRIES!

Each time you buy a hamburger or hot dog at BOB'S BURGER HOUSE, you get a card like the one shown. When you rub each square on your card, a picture of a hamburger is as likely to appear as a picture of a hot dog.

1. What is the prize for a winning card? A free order of fries
2. Is the card shown above a winning card? No

To find how many different kinds of cards are possible, you can list the **sample space** (all possible outcomes).

Here's how to use a tree diagram to show the sample space.

Note: B stands for hamburger.
D stands for hot dog.

Sample Space: BBB, BBD, BDB, BDD, DBB, DBD, DDB, DDD

3. Look at the *Here's how*. How many outcomes are in the sample space? 8
4. How many of the outcomes are winners (all of the pictures match)? 2
5. What is the probability (in simplest form) of getting a winning card? $\frac{1}{4}$

EXERCISES
Use the sample space above. Give each probability in simplest form.
What is the probability of getting a card having

6. exactly 1 hamburger? $\frac{3}{8}$
7. exactly 2 hamburgers? $\frac{3}{8}$
8. a hot dog in the first square? $\frac{1}{2}$
9. a hamburger in the middle square? $\frac{1}{2}$
10. What is the probability of getting a losing card? $\frac{3}{4}$

Solve.

11. a. One day BOB'S BURGER HOUSE gave away 296 cards. Suppose that one fourth of the cards were winning cards. How many orders of fries were given away? 74

    b. It costs BOB'S BURGER HOUSE 23¢ to buy, prepare, and serve an order of fries. How much did the winning cards cost BOB'S? $17.02

12. a. Of the 296 hamburgers and hot dogs sold that day, 183 were hamburgers. If each hamburger sold for $1.35, how much money did BOB'S get from the sale of hamburgers? $247.05

    b. How many hot dogs were sold? 113

    c. If $107.35 was received from the sale of hot dogs, what was the price of each hot dog? $.95

If you buy an ice cream cone at BOB'S BURGER HOUSE you get one of the cards shown below. When you rub each square on your card, a vanilla, chocolate, or strawberry cone will appear. Each flavor is equally likely to appear.

13. Copy and complete the tree diagram to show the sample space.

Sample Space

```
        V----→VV
    V<— C       VC
        S   ?   VS
        V   ?   CV
    C<— C   ?   CC
        S   ?   CS
        V   ?   SV
    S<— C   ?   SC
        S   ?   SS
```

14. Use the sample space from exercise 13. Give each probability in simplest form.
    a. P(exactly 1 vanilla) $\frac{4}{9}$
    b. P(2 strawberry) $\frac{1}{9}$
    c. P(no chocolate) $\frac{4}{9}$
    d. P(winning) $\frac{1}{3}$
    e. P(losing) $\frac{2}{3}$

15. a. During one day, 141 cones were sold. If one third of the cards were winners, how many cones were sold for 25¢ off the regular price? 47

    b. The regular price for a cone is 75¢. How much money was received from cone sales that day? $94

Cone count

16. A certain ice-cream store sells 31 flavors. A customer decides to order a 3-dip cone. He first orders the flavor of the bottom dip, next orders the flavor of the middle dip, and last orders the flavor of the top dip. How many different such orders could he give? *Hint: He may order 3 scoops of the same flavor.* 29,791

Probability

Practice Worksheet
Workbook S336, Copymaster S336, or Duplicating Master S336

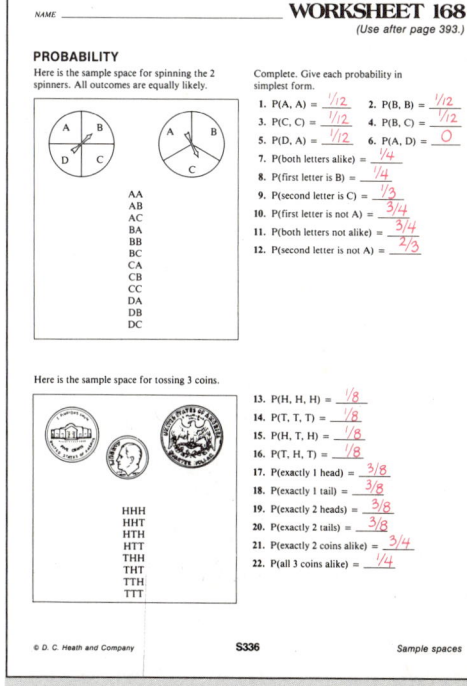

Group Project
Collecting data

Have the students, working in small groups, toss two coins 80 times and keep a tally of the number of times both coins turned up heads. Have them compare the results with the probability predicted by this tree diagram $\left(\frac{1}{4}\right)$.

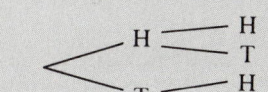

Class Starter Quiz 154
on previous lesson

Use the sample space in the *Here's how* section on page 392. What is the probability of getting a:

1. card having exactly 1 hot dog? $\frac{3}{8}$
2. card having exactly 2 hot dogs? $\frac{3}{8}$
3. card having a hamburger in the first square? $\frac{1}{2}$
4. winning card? $\frac{1}{4}$

Copymaster S423

Problem-Solving Skills
Reading an ad
Choosing the correct operation
Solving a multistep problem

Skills Reviewed
Rounding whole numbers
Multiplying and dividing decimals
Adding and subtracting fractions
Solving percent problems

Starting the Lesson
Problem Solving Use the sale ad and ask questions like these:

- What is the regular price of a red oak? ($30)
- Which trees are on sale at a 25% discount? (Pin oak, sweet gum, and white birch)
- What is the cost of 2 pounds of grass seed? ($3.30)
- How much is the lawn tractor if you pay cash? ($900)

Cumulative Skill Practice Write these five answers on the chalkboard:

390,000   392.49   82 $1\frac{1}{6}$ $\frac{1}{4}$

Challenge the students to an answer hunt by saying, "Look at exercises 1–59 on page 395. Find the five exercises that have these answers. You have five minutes to find as many of the exercises as you can." (Exercises 11, 27, 31, 46, and 50)

Problem solving   Reading an ad

SPRING SPECIAL   **ONLY $900**

INSTALLMENT PLAN
$228 down payment
$65 per month for 12 months

SALE

SHADE TREES	Regular Price	THIS WEEK ONLY
Pin Oak	$24	25% OFF
Red Oak	$30	30% OFF
Sugar Maple	$40	20% OFF
Sweet Gum	$28	25% OFF
White Birch	$36	25% OFF

Grass Seed .... $1.65 per lb
Fertilizer .... 10 lb for $9.60

Answer each customer's question. Decide when a calculator would be useful.

1. What is the sale price of a white birch tree? $27

2. How much do I save by buying a pin oak on sale? $6

3. "How much will 8 pounds of grass seed and 20 pounds of fertilizer cost?" $32.40

4. "How much do I save if I buy a sugar maple and a sweet gum on sale?" $15

5. "If I use 2 pounds of grass seed for every 600 square feet, how many pounds will I need to seed a square lawn that measures 60 feet on a side?"   12

6. "If I use 4 pounds of grass seed for every 1000 square feet, how much will it cost to seed a rectangular yard that is 50 feet by 80 feet?" $26.40

7. "What is the total cost of a red oak, a sweet gum, and a sugar maple on sale?" $74

8. "How much is the total installment cost for the lawn tractor?" $1008

9. "If I could borrow $900 at 12% per year for 1 year, would that be less expensive than buying the tractor on your installment plan?"   No, the same

10. "If 10 pounds of fertilizer will cover 5000 square feet of lawn, how many pounds will I use to fertilize a lawn that is 50 feet by 70 feet?"   7

Chapter 15

Cumulative Skill Practice

Round to the nearest ten thousand. (page 10)

1. 12,750 10,000
2. 46,394 50,000
3. 75,008 80,000
4. 45,000 50,000
5. 59,981 60,000
6. 56,000 60,000
7. 94,999 90,000
8. 95,000 100,000
9. 97,074 100,000
10. 98,241 100,000
11. 386,381 390,000
12. 283,077 280,000
13. 439,500 440,000
14. 705,000 710,000
15. 164,138 160,000

Give the product. (page 68)

16. 4.6 × 0.7 3.22
17. 3.9 × 0.31 1.209
18. 0.65 × 0.6 0.390
19. 4.5 × 3.1 13.95
20. 52 × 2.6 135.2
21. 0.74 × 8.1 5.994
22. 5.13 × 1.4 7.182
23. 2.04 × 0.1 0.204
24. 8.72 × 0.16 1.395
25. 3.85 × 7.4 28.49
26. 5.63 × 0.75 4.223
27. 62.3 × 6.3 392.49

Give the quotient rounded to the nearest tenth. (page 104)

28. 18.2 ÷ 0.6 30.3
29. 2.95 ÷ 0.2 14.8
30. 8.3 ÷ 0.03 276.7
31. 16.4 ÷ 0.2 82
32. 34.1 ÷ 0.5 68.2
33. 8.64 ÷ 0.07 123.4
34. 9.2 ÷ 0.08 115
35. 3.95 ÷ 0.03 131.7
36. 6.23 ÷ 1.4 4.5
37. 2.036 ÷ 3.1 0.7
38. 5.02 ÷ 0.35 14.3
39. 4.61 ÷ 0.21 22.0

Give the sum in simplest form. (page 180)

40. $\frac{2}{9} + \frac{4}{9}$ $\frac{2}{3}$
41. $\frac{1}{8} + \frac{3}{8}$ $\frac{1}{2}$
42. $\frac{1}{2} + \frac{1}{4}$ $\frac{3}{4}$
43. $\frac{1}{2} + \frac{1}{3}$ $\frac{5}{6}$
44. $\frac{3}{8} + \frac{1}{4}$ $\frac{5}{8}$
45. $\frac{3}{10} + \frac{3}{10}$ $\frac{3}{5}$
46. $\frac{5}{12} + \frac{3}{4}$ $1\frac{1}{6}$
47. $\frac{9}{16} + \frac{7}{8}$ $1\frac{7}{16}$
48. $\frac{4}{5} + \frac{3}{4}$ $1\frac{11}{20}$
49. $\frac{8}{15} + \frac{9}{10}$ $1\frac{13}{30}$

Give the difference in simplest form. (page 188)

50. $\frac{1}{2} - \frac{1}{4}$ $\frac{1}{4}$
51. $\frac{2}{3} - \frac{1}{4}$ $\frac{5}{12}$
52. $\frac{1}{2} - \frac{1}{3}$ $\frac{1}{6}$
53. $\frac{2}{5} - \frac{1}{4}$ $\frac{3}{20}$
54. $\frac{3}{4} - \frac{2}{3}$ $\frac{1}{12}$
55. $3 - \frac{9}{10}$ $2\frac{1}{10}$
56. $\frac{11}{16} - \frac{1}{2}$ $\frac{3}{16}$
57. $\frac{11}{12} - \frac{5}{8}$ $\frac{7}{24}$
58. $\frac{7}{10} - \frac{1}{4}$ $\frac{9}{20}$
59. $\frac{9}{16} - \frac{3}{8}$ $\frac{3}{16}$

MIXED PRACTICE

Solve.

60. 20% of 55 = n 11
61. 125% of n = 50 40
62. 15% of 46 = n 6.9
63. 40% of n = 20 50
64. 17% of 63 = n 10.71
65. 75% of 36 = n 27
66. 5% of n = 10 200
67. 50% of 78 = n 39
68. 150% of n = 72 48
69. $33\frac{1}{3}$% of 57 = n 19
70. 7.6% of n = 11.4 150
71. 5.9% of 82 = n 4.838
72. $62\frac{1}{2}$% of 40 = n 25
73. 0.3% of n = 0.45 150
74. 22.5% of 95 = n 21.375

Probability 395

Problem-Solving Worksheet
Workbook S337, Copymaster S337, or Duplicating Master S337

NAME _____ **WORKSHEET 169**
(Use after page 394.)

NAIL IT DOWN
If you bought 10 of these boxes of nails, how many boxes would be left? 8

PAPER ROUTE
Andy delivered 36 papers daily 6 days a week. He also delivers 48 Sunday papers. How many papers does he deliver each week? 264

YOU'RE THE COOK
A recipe calls for $\frac{3}{4}$ cup of milk. If you triple the recipe, how many cups of milk will you need? $2\frac{1}{4}$

WORK BACKWARD
Three dimes, some nickels, 2 quarters, and 4 pennies total $1.19. How many nickels are there? 7

PEARS FOR SALE
How much do you save if you buy 3 cans of pears at the sale price? 19¢

WHAT'S THE RULE?
Find the pattern and fill in the blanks.
Hint: Try × and +.
1 → 5
2 → 8
6 → 20
4 → 14
10 → 32
8 → 26
×3+2

© D. C. Heath and Company S337 Problem solving

Challenge Problem

Look at page 395. Find the product in exercises 16–27 that fits the clues.

Clues:
- The product is less than 7.
- Rounded to the nearest ten, the product is 10.

5.994 (Exercise 21)

Copymaster S462

Class Starter Quiz 155
on previous lesson

Solve. Use the sale ad on page 394.

1. How much do you save by buying a white birch on sale? **$9**
2. What is the total cost of a pin oak and a sugar maple on sale? **$50**

Copymaster S424

Lesson Objective
To compute the probability of a compound outcome

Problem-Solving Skills
Using a tree diagram
Selecting information from a display

Starting the Lesson
Before the students open their books, have them draw a tree diagram showing all possible outcomes of first rolling a die and then tossing a coin. Then have them compare their tree diagrams with the diagram shown at the top of page 396.

Discuss questions 1–5, using the sample space and the *Here's how*.

Probability— more than 1 event

The tree diagram shows all possible outcomes of a first event (rolling a die) followed by a second event (tossing a coin).

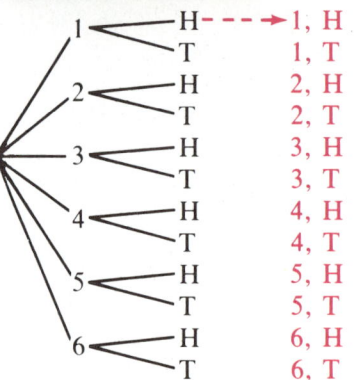

Sample Space

1, H
1, T
2, H
2, T
3, H
3, T
4, H
4, T
5, H
5, T
6, H
6, T

1. Think about the first event, rolling a die. What is the probability of rolling a 1? $\frac{1}{6}$

2. Think about the second event, tossing a coin. What is the probability of tossing heads (H)? $\frac{1}{2}$

3. Now look at the tree diagram. What is the probability of first rolling a 1 and then tossing heads (H)? $\frac{1}{12}$

Here's how to compute the probability of rolling a 1 and then tossing heads (H).

Multiply the probability of the first outcome's occurring by the probability of the second outcome's occurring.

the probability of 1 and then heads

$$P(1, H) = \frac{1}{6} \times \frac{1}{2}$$
$$= \frac{1}{12}$$

4. Look at the *Here's how*. What is P(1)? What is P(H)? $\frac{1}{6}, \frac{1}{2}$

5. What is P(1, H)? $\frac{1}{12}$

EXERCISES
Give each probability as a fraction in simplest form.

Think about first rolling a die and then tossing a coin.

6. a. P(2) $\frac{1}{6}$
   b. P(T) $\frac{1}{2}$
   c. P(2, T) $\frac{1}{12}$

7. a. P(even number) $\frac{1}{2}$
   b. P(H) $\frac{1}{2}$
   c. P(even number, H) $\frac{1}{4}$

8. P(number less than 3, T) $\frac{1}{6}$

9. P(not 5, T) $\frac{5}{12}$

Think about first rolling the die and then spinning the spinner.

10. P (4, red) $\frac{1}{48}$
11. P (6, yellow) $\frac{1}{24}$
12. P (even number, yellow) $\frac{1}{8}$
13. P (not 6, blue) $\frac{5}{48}$
14. P (2, not yellow) $\frac{1}{8}$
15. P (not 1, not brown) $\frac{35}{48}$

Think about spinning the above spinner once and then spinning it again.

16. P (red, red) $\frac{1}{64}$
17. P (yellow, yellow) $\frac{1}{16}$
18. P (green, green) $\frac{9}{64}$
19. P (brown, blue) $\frac{1}{64}$
20. P (red, yellow) $\frac{1}{32}$
21. P (yellow, green) $\frac{3}{32}$

Think about placing the 6 marbles in a bag and thoroughly mixing them up. Suppose that, without looking, you picked out a first marble, **put it back into the bag,** and then picked out a second marble.

22. P (green, blue) $\frac{1}{18}$
23. P (orange, green) $\frac{1}{12}$
24. P (blue, red) 0
25. P (green, not orange) $\frac{1}{12}$
26. P (not orange, green) $\frac{1}{12}$
27. P (blue, not blue) $\frac{2}{9}$

Think about placing the 6 marbles in a bag and thoroughly mixing them up. Suppose that without looking, you picked out a first marble, **left it out of the bag,** and then picked out a second marble.

28. P (green, blue) = $\frac{1}{6} \times \frac{2}{5}$ = ? $\frac{1}{15}$

> There are 2 blue marbles among the 5 remaining marbles.

29. P (orange, green) $\frac{1}{10}$
30. P (blue, orange) $\frac{1}{5}$
31. P (orange, orange) $\frac{1}{5}$
32. P (green, blue) $\frac{1}{15}$
33. P (blue, blue) $\frac{1}{15}$
34. P (green, green) 0

Making a prediction

When you toss 2 dice, you get one of these sums: 2, 3, 4, 5, 6, 7, 8, 9, 10, 11, or 12.

35. Predict which is more likely, a sum of 7 or a sum of 10. 7
36. Toss 2 dice 50 times and keep a record of the sums.
37. Did your prediction agree or disagree with the result of your experiment? Answers will vary.

Probability **397**

Practice Worksheet
Workbook S338, Copymaster S338, or Duplicating Master S338

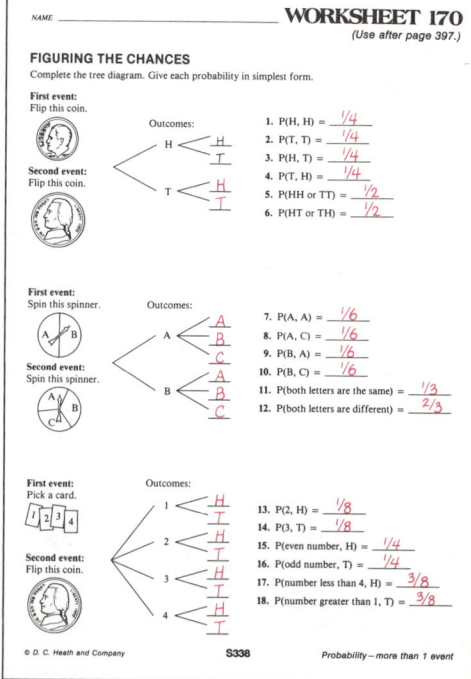

Group Project
Collecting data
Have the students, working in small groups, toss two dice 72 times and keep a tally of the number of times they get a sum of 7. Have them compare the results with the probability predicted by the 36 outcomes in this sample space $\left(\frac{1}{6}\right)$.

(1, 1) (2, 1) (3, 1) (4, 1) (5, 1) (6, 1)
(1, 2) (2, 2) (3, 2) (4, 2) (5, 2) (6, 2)
(1, 3) (2, 3) (3, 3) (4, 3) (5, 3) (6, 3)
(1, 4) (2, 4) (3, 4) (4, 4) (5, 4) (6, 4)
(1, 5) (2, 5) (3, 5) (4, 5) (5, 5) (6, 5)
(1, 6) (2, 6) (3, 6) (4, 6) (5, 6) (6, 6)

397

Class Starter Quiz 156
on previous lesson

Give the probability as a fraction in simplest form.
Think about first rolling a die and then tossing a coin.

1. P(2, tails) $\frac{1}{12}$
2. P(odd number, heads) $\frac{1}{4}$
3. P(number less than 5, tails) $\frac{1}{3}$
4. P(number greater than 1, heads) $\frac{5}{12}$
5. P(not 3, tails) $\frac{5}{12}$

Copymaster S424

Lesson Objective
To compute the odds of an outcome

Problem-Solving Skill
Reading a chart

Starting the Lesson
Discuss questions 1–5, using the chart and the *Here's how* example on page 398. Before assigning the exercises, let the students practice computing the odds in favor of getting:

- an orange gumball $\left(\frac{120}{600} \text{ or } \frac{1}{5}\right)$
- either a green or a yellow gumball $\left(\frac{220}{600} \text{ or } \frac{11}{30}\right)$

Exercise Note
Tell the students to study the *Here's how* example on page 399 before attempting exercises 12–23.

Odds

The table shows the contents of the gum-ball machine.

COLOR	NUMBER OF GUM BALLS
Green	80
Blue	90
White	70
Yellow	140
Red	100
Orange	120

1. Suppose that the gum balls have been thoroughly mixed. Which color do you have the best chance of getting? The worst chance of getting? Yellow, White
2. How many gum balls are red? 100
3. How many gum balls are not red? 500

Here's how to find the odds in favor of getting a red gum ball with your first penny.

To find the **odds in favor of an outcome,** write the ratio of the number of ways the outcome can occur to the number of ways that the outcome cannot occur.

number of ways outcome **can occur**
number of ways outcome **cannot occur**

$\frac{100}{500} = \frac{1}{5}$

The odds in favor of getting a red gum ball are 1 to 5.

4. Look at the *Here's how*. What is the number of ways the outcome can occur? 100 Cannot occur? 500
5. What are the odds in favor of getting a red gum ball? $\frac{1}{5}$

EXERCISES
Give the odds as a fraction in lowest terms.

Think about putting your first penny into the gum-ball machine.

6. Odds in favor of getting a green $\frac{2}{13}$
7. Odds in favor of getting a blue $\frac{3}{17}$
8. Odds in favor of getting a white $\frac{7}{53}$
9. Odds in favor of getting a yellow $\frac{7}{23}$
10. Odds in favor of getting either a red or an orange $\frac{11}{19}$
11. Odds in favor of getting either a green or a white $\frac{1}{3}$

Here's how *to find the odds against getting a red gum ball with your first penny.*

To find the **odds against an outcome,** write the ratio of the number of ways that the outcome cannot occur to the number of ways that the outcome can occur.

number of ways outcome **cannot occur**
number of ways outcome **can occur**

$$\frac{500}{100} = \frac{5}{1}$$

The odds against getting a red gum ball are 5 to 1

Give the odds in lowest terms.
Think about putting your first penny into the gum-ball machine.

12. Odds against getting red $\frac{5}{1}$
13. Odds against getting blue $\frac{17}{3}$
14. Odds against getting white $\frac{53}{7}$
15. Odds against getting yellow $\frac{23}{7}$
16. Odds against getting either a blue or a white $\frac{11}{4}$
17. Odds against getting either a yellow or a red $\frac{3}{2}$
18. Suppose that your favorite color is yellow. What are the odds in favor of your getting your favorite color? The odds against? $\frac{7}{23}, \frac{23}{7}$
19. A friend likes either white or orange. What are the odds in favor of your friend's getting one of his favorite colors? The odds against? $\frac{19}{41}, \frac{41}{19}$

Solve.

20. Suppose that in another gum-ball machine the odds in favor of your getting your favorite color are 1 to 7. What would be the odds against your getting your favorite color? $\frac{7}{1}$
21. If the odds against getting a black gum ball are 11 to 3, what are the odds in favor of getting a black gum ball? $\frac{3}{11}$

Something to chew on!

22. The odds in favor of getting a red gum ball are 5 to 23. What is the probability of getting a red gum ball? $\frac{5}{28}$
23. The odds against getting a green gum ball are 25 to 3. What is the probability of getting a green gum ball? $\frac{3}{28}$

Probability

Class Starter Quiz 157
on previous lesson

Give the odds in simplest form. Think about spinning this spinner.

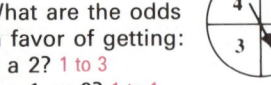

1. What are the odds in favor of getting:
   - a 2? 1 to 3
   - a 1 or 2? 1 to 1
   - an odd number? 1 to 1
2. What are the odds against getting:
   - a 2? 3 to 1
   - 1 or 2? 1 to 1
   - an odd number? 1 to 1

Copymaster S424

Lesson Objective
To compute the expectation of an outcome

Problem-Solving Skills
Selecting information from a display
Using a guess-and-check strategy

Starting the Lesson
Have the students read the introductory paragraph and answer questions 1–4. Then go over the *Here's how* and discuss questions 5–8.

Applying probability— expectation

PLAY MATCH-UP! BETH NANCY SUE

A set of triplets gave these childhood pictures to their school carnival committee. Each contestant is charged $2 to match the pictures with the names.

Each contestant who correctly matchs all three pictures wins an $8 calculator.

1. How much does it cost to play MATCH-UP? $2
2. How many ways can the pictures be matched with the names? 6
3. What is the probability of winning (matching all the pictures with the right names)? Assume that all matchings are equally likely. $\frac{1}{6}$
4. What is the value of the prize? $8

Here's how *to compute the* **expectation** *for a game of MATCH-UP.*

To find the expectation, multiply the probability of winning the prize times the value of the prize.

$$\text{Expectation} = \underset{P(\text{winning})}{\frac{1}{6}} \times \underset{\text{Value of Prize}}{\$8} = \$1.33$$

The expectation for a game of MATCH-UP is $1.33

5. Look at the *Here's how*. What is the expectation for a game of MATCH-UP? $1.33
6. To decide whether such a game is a good deal for the player, you compare the cost of playing with the expectation. Is the expectation less than or greater than the cost of playing? Less than
7. If the expectation is less than the cost of playing, then such a game is a "bad deal" for the player. Is the game MATCH-UP a "bad deal" for the player? Yes
8. How much would the expectation have to be for the game to be considered a "good deal" for the player? Over $2

EXERCISES

Here is another game that was played at the school carnival. A player pays $1.50 to spin this wheel. If the wheel stops on yellow, the player wins a $4.00 movie ticket.

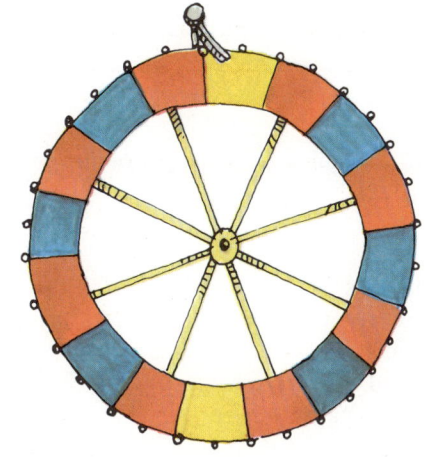

9. What is the probability (in simplest form) of winning? $\frac{1}{8}$

10. What is the value of the movie ticket? $4.00

11. How much is the expectation? $.50

12. How much does it cost to play? $1.50

13. Is the expectation less than or greater than the cost of playing the game? Less than

14. Is the game a good deal or a bad deal? Bad deal

BUY A CHANCE ON A UNICYCLE WORTH $220!

A local merchant donated this unicycle to the school carnival. The carnival committee decided to sell 135 chances on it, at $2 each.

15. What is the cost of one chance? $2

16. If a person bought one chance, what would be that person's probability of winning? $\frac{1}{135}$

17. What is the value of the unicycle? $220

18. What is the expectation rounded to the nearest cent? $1.63

19. Is the expectation less than or greater than the cost of a chance? Less than

20. Is buying a chance a good deal or a bad deal? Bad deal

Which for what? Guess and check

21. John bought 2 chances on the watch and 1 chance on the radio. He spent $4.00. Sue bought 1 chance on the watch and 2 chances on the radio. She spent $3.50. How much would one chance on each have cost? $2.50 (watch $1.50, radio $1.00)

Probability **401**

Practice Worksheet
Workbook S340, Copymaster S340, or Duplicating Master S340

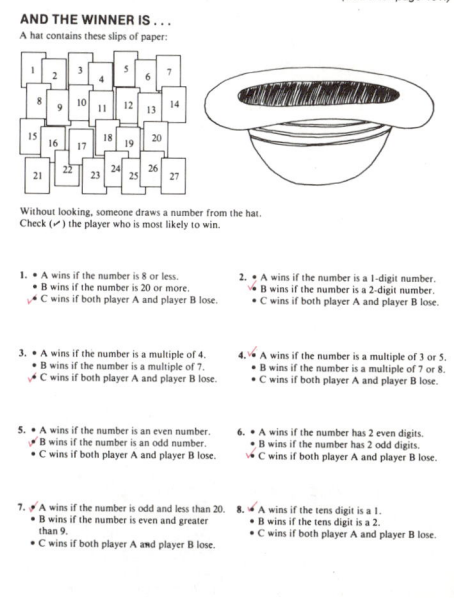

Project

Making predictions

Make a frequency table of the number of times each letter appears in a newspaper article. Then decide which letter key on a typewriter will need to be replaced most often.

Since the letter *e* is the most commonly used letter, the *e* key will most likely need to be replaced first.

Copymaster S486

Class Starter Quiz 158
on previous lesson

A player pays $2 to spin this spinner. If the pointer stops on an even number, the player wins $3.

1. What is the probability (in simplest form) of winning? $\frac{1}{2}$
2. How much is the expectation? $1.50
3. Is the game a good deal or a bad deal? Bad

Copymaster S424

Lesson Objective
To solve problems using income tax tables

Starting the Lesson
The income tax table on page 402 is also on ■ **Visual Aid 66** (copymaster or transparency S163). Use the table when discussing the problem on page 402.

PROBLEM SOLVING *using income tax tables*

You can use tax tables to solve problems about income taxes.

Problem
When Derek Foss completed his income tax forms for a recent year, he determined that his taxable income was $20,433. Derek was not married and lived in an apartment. How much tax did he pay that year?

Here's how *to use an income tax table.*

STUDY Read the problem carefully. What are the facts? What is the question?

Taxable income: $20,433
Filing status: Single
Question: How much tax was paid?

PLAN and SOLVE Use an income tax table.

Read down the appropriate **filing status** column. Then read across the appropriate **taxable income** bracket.

TAXABLE INCOME
Derek's taxable income was between these two amounts.

Answer: Derek's tax for the year was $3,158.

CHECK The correct filing status was used. The correct taxable income bracket was used.

FILING STATUS
Derek's filing status was single.

If line 37 (taxable income) is—		And you are—			
At least	But less than	Single	Married filing jointly *	Married filing separately	Head of a household
		Your tax is—			
20,000					
20,000	20,050	3,054	2,375	3,726	2,827
20,050	20,100	3,067	2,384	3,742	2,839
20,100	20,150	3,080	2,393	3,759	2,851
20,150	20,200	3,093	2,402	3,775	2,863
20,200	20,250	3,106	2,411	3,792	2,875
20,250	20,300	3,119	2,420	3,808	2,887
20,300	20,350	3,132	2,429	3,825	2,899
20,350	20,400	3,145	2,438	3,841	2,911
20,400	20,450	3,158	2,447	3,858	2,923
20,450	20,500	3,171	2,456	3,874	2,935
20,500	20,550	3,184	2,465	3,891	2,947
20,550	20,600	3,197	2,474	3,907	2,959
20,600	20,650	3,210	2,483	3,924	2,971
20,650	20,700	3,223	2,492	3,940	2,983
20,700	20,750	3,236	2,501	3,957	2,995
20,750	20,800	3,249	2,510	3,973	3,007
20,800	20,850	3,262	2,519	3,990	3,019
20,850	20,900	3,275	2,528	4,006	3,031
20,900	20,950	3,288	2,537	4,023	3,043
20,950	21,000	3,301	2,546	4,039	3,055

PROBLEMS

Solve. Use the tax table on page 402.

1. A single taxpayer has a taxable income of $20,350. Which of these amounts is her tax?
   a. $3,145   b. $2,438   c. $3,841

2. A married couple has a taxable income of 20,750. They file jointly. Which of these amounts is their tax?
   a. $3,249   **b.** $2,510   c. $3,973

3. Kao Chung-Chou is married, but he and his wife file separately. His taxable income is $20,675. Which of these amounts is his tax?
   a. $3,223   b. $2,492   **c.** $3,940

4. Emilia is the head of a household. Her taxable income is $20,178. Which of these amounts is her tax?
   a. $2,402   b. $3,775   **c.** $2,863

5. Stacy Larson and her husband file a joint return. Their taxable income is $20,993. How much is their tax? $2,546

6. Bob Dorr is single and has a taxable income of $20,596. How much is his tax? $3,197

7. Mateo and his wife, Tara, file separately. His taxable income is $20,173. What is his tax? $3,775

8. Ann Westlake is the head of a household. Her taxable income is $20,663. What is her tax? $2,983

9. A single taxpayer and a married couple have a taxable income of $20,800 each. The married couple files jointly. How much more tax did the single taxpayer pay than the married couple? $743

10. Jean and her husband file a joint return. Their tax is $2,528. Which of these amounts is their taxable income?
    a. $20,200   b. $20,375   **c.** $20,876

A taxing situation

Using a table

11. Employers use tables to determine the amount of Social Security Tax that must be withheld from an employee's paycheck. Rick earns $5.85 an hour. Last week he worked 17 hours and his employer withheld $7.12. Was that the right amount?   No, he should have withheld $7.11.

Wages at least	But less than	Tax to be withheld
98.82	98.96	7.07
98.96	99.10	7.08
99.10	99.24	7.09
99.24	99.38	7.10
99.38	99.52	7.11
99.52	99.66	7.12
99.66	99.80	7.13
99.80	99.94	7.14
99.94	100.00	7.15

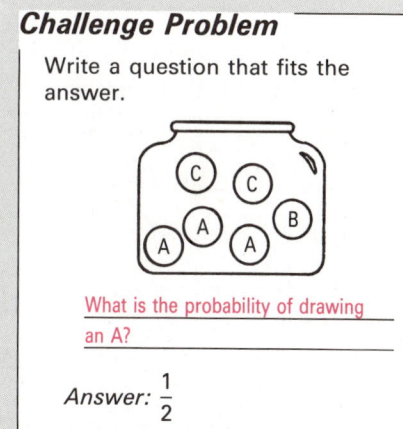

Challenge Problem

Write a question that fits the answer.

What is the probability of drawing an A?

Answer: $\frac{1}{2}$

Class Starter Quiz 159
on previous lesson

Solve. Use the tax table on page 402.

1. Kelly is head of a household. Her taxable income is $20,860. How much is her tax? **$3,031**
2. Gene and his wife file a joint return. Their tax is $2,546. Between which two amounts is their taxable income? **$20,950 and $21,000**

Copymaster S425

Problem-Solving Skills
Reading a blueprint
Choosing the correct operation
Solving a multistep problem

Skills Reviewed
Multiplying and dividing fractions
Solving proportions
Changing fractions to percents
Solving percent problems
Adding and subtracting decimals
Multiplying and dividing decimals
Simplifying expressions

Starting the Lesson
Problem Solving Help the students read the blueprint by asking questions such as these:

- Which room is 22 feet by 15 feet? (Living room)
- What are the dimensions of the garage? (10 feet by 15 feet)
- Which room has an area of 120 square feet? (Bedroom C)

Cumulative Skill Practice Challenge the students to an estimation hunt by saying, "Find the product in exercises 1–10 that is greater than 1." (Exercise 10) Then have the students find the three quotients in exercises 11–20 that are less than 1. (Exercises 13, 16, and 19)

Problem solving
COMPUTERS AND ARCHITECTS

Computers are used to draw "blueprints" of an architect's design. The computer can be programmed to display a floor plan on a screen or printout.

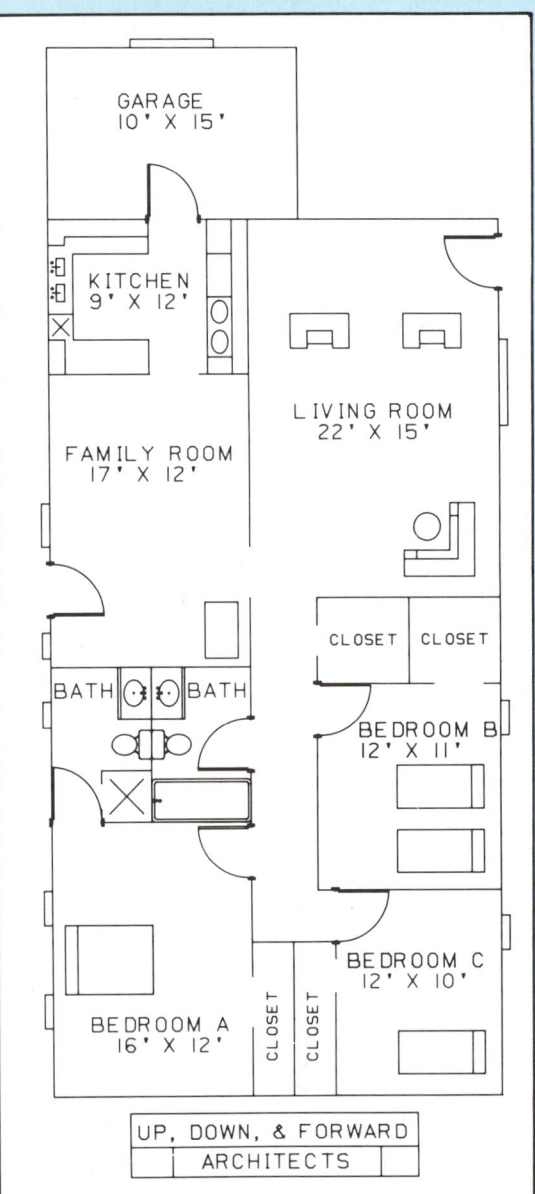

Use the floor plan to answer the questions. Decide when a calculator would be useful.

1. Which room is 17 feet by 12 feet? **Family Room**
2. Which room is 16 feet by 12 feet? **Bedroom A**
3. What are the length and width of the living room? **22 feet by 15 feet**
4. What are the dimensions of the kitchen? **9 feet by 12 feet**
5. Which room has an area of 132 square feet? **Bedroom B**
6. What is the area of the smallest bedroom? **120 ft²**
7. What is the area of the largest bedroom? **192 ft²**
8. Carpet costs $1.50 per square foot. How much will it cost to carpet bedroom B? *Hint: First find the area; then find the cost.* **$198**
9. At $1.75 per square foot, how much will it cost to carpet the living room? **$577.50**
10. A floor tile 1 foot by 1 foot costs $.49. How much will it cost to tile the family-room floor? **$99.96**
11. A contractor said this house could be built for $50 per square foot. Would you expect the cost to be more or less than $60,000? **More**

Cumulative Skill Practice

Give the product in simplest form. (page 202)

1. $\frac{1}{2} \times \frac{1}{4}$ $\frac{1}{8}$
2. $\frac{1}{3} \times \frac{1}{3}$ $\frac{1}{9}$
3. $\frac{7}{8} \times \frac{4}{5}$ $\frac{7}{10}$
4. $\frac{3}{8} \times \frac{2}{3}$ $\frac{1}{4}$
5. $\frac{2}{5} \times \frac{2}{5}$ $\frac{4}{25}$
6. $\frac{9}{10} \times \frac{1}{3}$ $\frac{3}{10}$
7. $\frac{5}{6} \times \frac{3}{3}$ $\frac{5}{6}$
8. $\frac{4}{5} \times \frac{1}{2}$ $\frac{2}{5}$
9. $\frac{5}{6} \times \frac{0}{2}$ 0
10. $\frac{3}{8} \times 4$ $1\frac{1}{2}$

Give the quotient in simplest form. (page 212)

11. $3 \div \frac{1}{2}$ 6
12. $4 \div \frac{1}{3}$ 12
13. $\frac{1}{2} \div \frac{2}{3}$ $\frac{3}{4}$
14. $\frac{2}{3} \div \frac{1}{2}$ $1\frac{1}{3}$
15. $3 \div \frac{2}{3}$ $4\frac{1}{2}$
16. $\frac{4}{5} \div 3$ $\frac{4}{15}$
17. $\frac{3}{4} \div \frac{3}{8}$ 2
18. $\frac{3}{5} \div \frac{3}{8}$ $1\frac{3}{5}$
19. $\frac{3}{8} \div \frac{3}{4}$ $\frac{1}{2}$
20. $\frac{5}{8} \div \frac{1}{4}$ $2\frac{1}{2}$

Solve. (page 262)

21. $\frac{5}{n} = \frac{2}{7}$ $17\frac{1}{2}$
22. $\frac{n}{3} = \frac{5}{2}$ $7\frac{1}{2}$
23. $\frac{1}{4} = \frac{n}{6}$ $1\frac{1}{2}$
24. $\frac{10}{n} = \frac{3}{5}$ $16\frac{2}{3}$
25. $\frac{3}{8} = \frac{n}{12}$ $4\frac{1}{2}$
26. $\frac{11}{n} = \frac{8}{21}$ $28\frac{7}{8}$
27. $\frac{3}{15} = \frac{5}{n}$ 25
28. $\frac{2}{3} = \frac{9}{n}$ $13\frac{1}{2}$
29. $\frac{n}{9} = \frac{5}{12}$ $3\frac{3}{4}$
30. $\frac{6}{15} = \frac{4}{n}$ 10

Change to a percent. (page 284)

31. $\frac{1}{10}$ 10%
32. $\frac{2}{5}$ 40%
33. $\frac{1}{4}$ 25%
34. $\frac{1}{5}$ 20%
35. $\frac{1}{2}$ 50%
36. $\frac{1}{7}$ $14\frac{2}{7}$%
37. $\frac{3}{4}$ 75%
38. $\frac{1}{3}$ $33\frac{1}{3}$%
39. 2 200%
40. $\frac{1}{6}$ $16\frac{2}{3}$%
41. $\frac{7}{8}$ $87\frac{1}{2}$%
42. $\frac{2}{9}$ $22\frac{2}{9}$%

Solve. (page 294)

43. 15% of 46 = n 6.9
44. 9% of 36.5 = n 3.285
45. 125% of 64 = n 80
46. 7.5% of 112 = n 8.4
47. 12.3% of 60 = n 7.38
48. 22.5% of 72.5 = n 16.3125
49. $62\frac{1}{2}$% of 62 = n 38.75
50. $33\frac{1}{3}$% of 126 = n 42
51. $66\frac{2}{3}$% of 159 = n 106

MIXED PRACTICE
Complete.

52. 3.87 + 15.9 = _?_ 19.77
53. 38 × 1000 = _?_ 38,000
54. 45.23 − 9.4 = _?_ 35.83
55. 8.1 × 0.02 = _?_ 0.162
56. 20 − 13.65 = _?_ 6.35
57. 21.06 ÷ 0.03 = _?_ 702
58. 36.7 × 10 = _?_ 367
59. 3.1 + 200 + 4.85 = _?_ 207.95
60. 76.32 ÷ 2.4 = _?_ 31.8
61. 0.77 × 100 = _?_ 77
62. 61.5 ÷ 10 = _?_ 6.15
63. 0.52 ÷ 1000 = _?_ 0.00052
64. (14.2 − 3) − 1.4 = _?_ 9.8
65. 14.2 − (3 − 1.4) = _?_ 12.6
66. 14.2 − (3 + 1.4) = _?_ 9.8

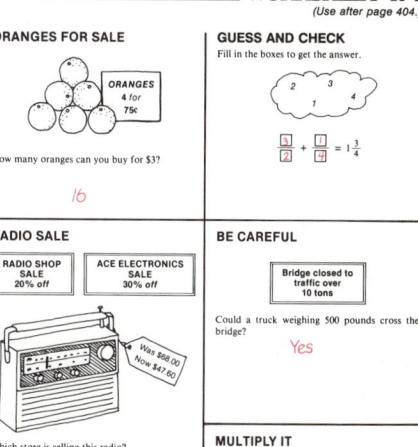

Chapter REVIEW

Here are scrambled answers for the review exercises:

| $\frac{1}{12}$ | $\frac{1}{3}$ | 5 | 18 | permutation | sample |
| $\frac{3}{10}$ | 1 | 8 | 24 | possible | value |

1. A certain automobile is available in 3 different models, and each model comes in 6 different colors. If you decided to order such an automobile, you would have ? choices. *(page 386)*

2. An arrangement of things in a definite order is called a ? . Four students decide to go to lunch. They can line up in ? different ways. *(page 388)*

3. If you roll a die, the probability of rolling a 2 is equal to
$$\frac{\text{number of ways of rolling a 2}}{\text{number of ? outcomes}}$$
(page 390)

4. If you roll a die, the probability of rolling a number greater than 4 is ? . *(page 390)*

5. A ? space is the set of all possible outcomes. If you listed the sample space for tossing a coin 3 times, the sample space would have ? outcomes. *(page 392)*

6. Think about first rolling a die and then tossing a coin. P(6, H) = ? . *(page 396)*

7. Think about placing these marbles in a bag, thoroughly mixing them up, picking a marble, not replacing it, and then picking a second marble. P(red, blue) = ? . *(page 396)*

8. Think about rolling a die. The odds in favor of rolling a 1 are ? to ? . *(page 398)*

9. You can find the expectation by multiplying the probability of winning a prize by the ? of the prize. *(page 400)*

406 Chapter 15   **1.** 18   **2.** permutation, 24   **3.** possible   **4.** $\frac{1}{3}$ **5.** sample, 8   **6.** $\frac{1}{12}$ **7.** $\frac{3}{10}$ **8.** 1, 5   **9.** value

Chapter TEST

Solve. *(pages 386, 388)*

1. You have 4 different colors of slacks and 6 different colors of shirts. How many different outfits can you make? 24

2. You and 5 friends decide to go to a movie. How many ways can you line up at the ticket office? 720

Give each probability in simplest form. *(page 390)*
Think about spinning this spinner.

3. P(red) $\frac{1}{8}$
4. P(yellow) $\frac{1}{4}$
5. P(green) $\frac{3}{8}$
6. P(blue) $\frac{1}{8}$
7. P(red or brown) $\frac{1}{4}$
8. P(green or blue) $\frac{1}{2}$
9. P(not brown) $\frac{7}{8}$
10. P(not yellow) $\frac{3}{4}$

Solve. *(page 392)*
Think about a family with 3 children.

11. List the sample space for the children in the family. *Hint: Use GBB to represent a girl as the oldest, a boy as the "middle" child, and a boy as the youngest child.*
BBB, BBG, BGB, GB, B, B, G, G, GBG, G, G, B, GGG

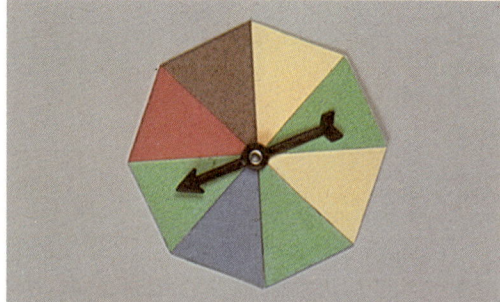

12. a. What is the probability that in a family with 3 children, exactly 2 will be girls? $\frac{3}{8}$

    b. What is the probability that in a family with 3 children, exactly 1 will be a girl? $\frac{3}{8}$

Give each probability in simplest form. *(page 396)*
Think about first flipping the coin and then tossing the die.

13. P(H, 3) $\frac{1}{12}$
14. P(T, 4) $\frac{1}{12}$
15. P(H, even number) $\frac{1}{4}$
16. P(T, number greater than 3) $\frac{1}{4}$

Give the odds as a fraction in lowest terms. *(pages 398, 399)*
Think about spinning the spinner shown above.

17. Odds in favor of spinning red $\frac{1}{7}$
18. Odds in favor of spinning green $\frac{3}{5}$
19. Odds against spinning blue $\frac{7}{1}$
20. Odds against spinning yellow $\frac{3}{1}$

Solve. *(page 400)*

21. a. Eighty chances were sold for $2 each. Suppose you bought a chance. What would be your expectation? $1.50

    b. Is buying a chance on the binoculars a good deal? No

BUY A CHANCE!
BINOCULARS
WORTH $120

Probability

Cumulative Test
(Chapters 1–15)

Use Copymaster S109 to provide the students with an answer sheet in standardized test format.

Answers for Cumulative Test, Chapters 1–15

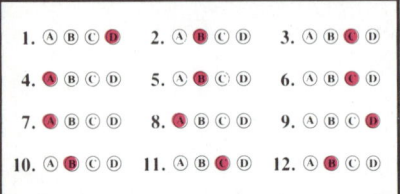

The table below correlates test items with student text pages.

Test Item	Page Taught	Skill Practice
1	10	p. 395, exercises 1–15
2	68	p. 395, exercises 16–27
3	104	p. 395, exercises 28–39
4	180	p. 395, exercises 40–49
5	188	p. 395, exercises 50–59
6	202	p. 405, exercises 1–10
7	212	p. 405, exercises 11–20
8	262	p. 405, exercises 21–30
9	284	p. 405, exercises 31–42
10	294	p. 405, exercises 43–51
11	344	
12	396	

408

Cumulative TEST — Standardized Format

Choose the correct letter.

1. 398,520 rounded to the nearest ten thousand is
  A. 399,000
  B. 390,000
  C. 398,500
  D. 400,000

2. Give the product.
  52.4×2.07
  A. 141.48
  B. 108.468
  C. 14.288
  D. none of these

3. Give the quotient rounded to the nearest tenth.
  $30.89 \div 3.7$
  A. 8.35
  B. 8.4
  C. 8.3
  D. none of these

4. Give the sum.
  $\frac{7}{16} + \frac{5}{8}$
  A. $1\frac{1}{16}$
  B. $\frac{1}{2}$
  C. $\frac{3}{4}$
  D. none of these

5. Give the difference.
  $\frac{5}{6} - \frac{2}{3}$
  A. 1
  B. $\frac{1}{6}$
  C. $\frac{1}{2}$
  D. none of these

6. Give the product.
  $\frac{5}{12} \times \frac{3}{4}$
  A. $\frac{5}{9}$
  B. $1\frac{1}{4}$
  C. $\frac{5}{16}$
  D. none of these

7. Give the quotient.
  $\frac{9}{10} \div \frac{3}{2}$
  A. $\frac{3}{5}$
  B. $1\frac{2}{3}$
  C. $1\frac{7}{20}$
  D. none of these

8. Solve.
  $\frac{5}{8} = \frac{n}{6}$
  A. 3.75
  B. 9.6
  C. 240
  D. none of these

9. Change to a percent.
  $\frac{3}{8} = ?$
  A. $33\frac{1}{3}\%$
  B. $38\frac{1}{2}\%$
  C. $62\frac{1}{2}\%$
  D. none of these

10. Solve.
  32% of 18 = n
  A. 56.25
  B. 5.76
  C. 57.6
  D. none of these

11. How many feet of fencing will it take to go around a circular garden that has a radius of 7 feet? Use 3.14 for π.
  A. 21.98
  B. 153.86
  C. 43.96
  D. none of these

12. If you toss a penny and a dime, what is the probability that both will land heads?
  A. $\frac{1}{2}$
  B. $\frac{1}{4}$
  C. $\frac{1}{3}$
  D. none of these

Integers

Chapter 16
Integers

Resources

- **Class Starter Quizzes 160-168** *(Copymasters S425-S427)*
- **Visual Aids 67-68** *(Copymasters or Transparencies S164-S165)*
- **Manipulatives**
  Manipulative Activities 30-32 *(Copymasters S526-S528)*
  Positive and negative charge models *(Copymaster or Transparency S542)*
- **Worksheets 175-184** *(Copymasters, Duplicating Masters, or Workbook pages S343-S352)*
- **Challenge Problems** for pages 411, 417, 419, 421, 423, 425, 427, 429 *(Copymasters S462-S464)*
- **Projects** for pages 413, 415 *(Copymaster S487)*
- **Mental Math Extensions** for Skills 59-63 *(Copymasters S511-S513)*
- **Tests** *(Copymasters or Duplicating Masters S61-S64)*

Lesson Objectives
To compare integers
To find the absolute value of integers

Problem-Solving Skill
Reading a map

Starting the Lesson
Use the map at the top of the page and ask questions like these:
- Which city has a high temperature of positive 113 degrees? (Dallas)
- Which city has a low temperature of negative 24 degrees? (Duluth)
- How many cities have low temperatures below 20 degrees? (12)
- How many cities have high temperatures above Denver's high? (8)

Go over exercises 1–4, using the map and the *Here's how*. The number line in the *Here's how* is on ■ **Visual Aid 67** (copymaster or transparency S164).

Comparing integers/absolute value

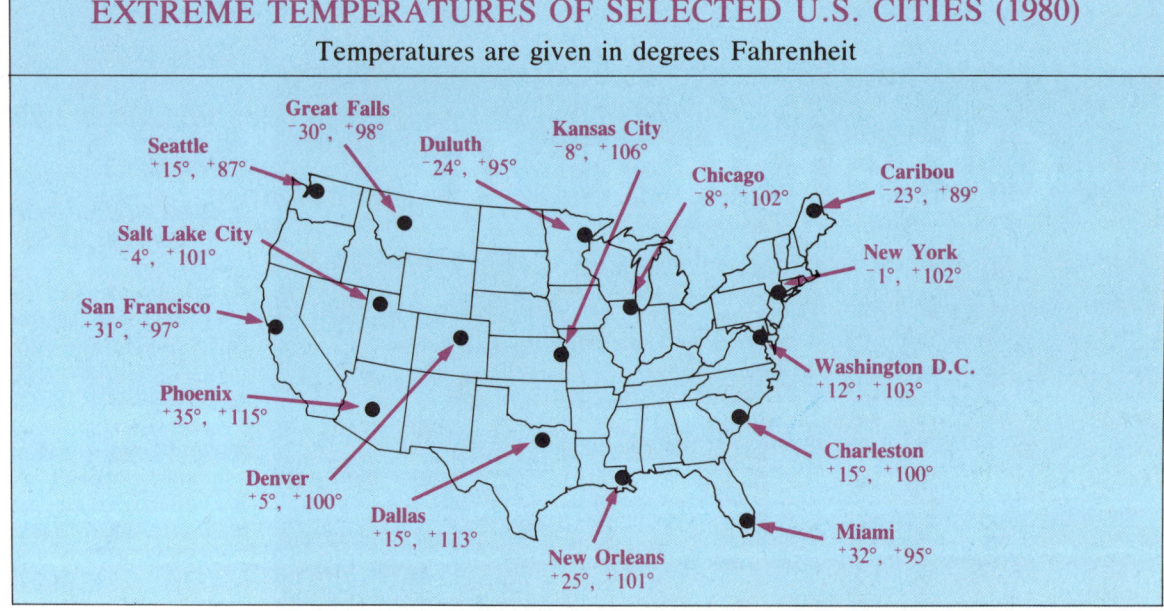

1. The low temperature for Seattle is 15 degrees above 0, which can be written as $^+15°$ ("positive fifteen degrees"). What is the low temperature for San Francisco? $^+31°$

2. The low temperature for Chicago is 8 degrees below 0, which can be written as $^-8°$ ("negative eight degrees"). What is the low temperature for Great Falls? $^-30°$

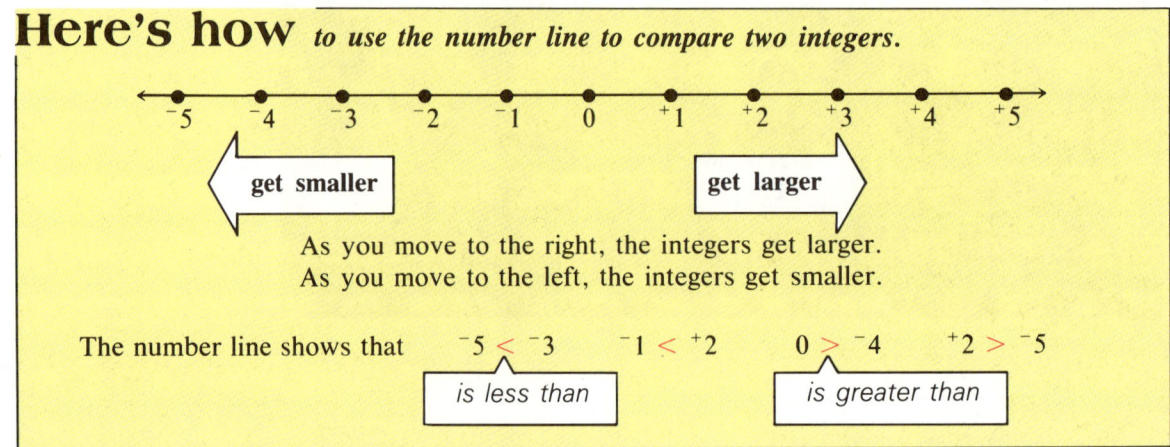

3. Look at the *Here's how*. Is $^-5$ less than or greater than $^-3$? Less than

4. Is 0 less than or greater than $^-4$? Greater than

410 *Chapter 16*

The number of units that an integer is from 0 is called the **absolute value** of the integer.

Here's how to use the number line to find the absolute value of an integer.

```
←————4 units————→ ←————4 units————→
─┼──┼──┼──┼──┼──┼──┼──┼──┼──┼─
⁻5 ⁻4 ⁻3 ⁻2 ⁻1  0 ⁺1 ⁺2 ⁺3 ⁺4 ⁺5
```

The absolute value of ⁻4 is 4.   The absolute value of ⁺4 is 4.
We write: |⁻4| = 4   We write: |⁺4| = 4

5. Look at the *Here's how*. What is the absolute value of ⁻4? 4 of ⁺4? 4

EXERCISES
< or >?

6. ⁺3 ● ⁺4 <   7. ⁺3 ● ⁻4 >   8. ⁻3 ● ⁺4 <   9. ⁻3 ● ⁻4 >   10. ⁻8 ● ⁺9 <

11. ⁺5 ● ⁻2 >   12. ⁻5 ● ⁺2 <   13. ⁻5 ● ⁻2 <   14. ⁺5 ● ⁺2 >   15. ⁻2 ● 0 <

16. ⁺8 ● ⁺11 <   17. ⁻12 ● ⁺10 <   18. ⁺15 ● ⁻19 >   19. ⁻20 ● 0 <   20. 0 ● ⁺15 <

Give each absolute value.

21. |⁺2| 2   22. |⁻3| 3   23. |⁺5| 5   24. |⁻5| 5   25. |0| 0

26. |⁻8| 8   27. |⁺7| 7   28. |⁻15| 15   29. |⁻20| 20   30. |⁺16| 16

Solve. Use the information on page 410.

31. Which city has the highest high temperature? Phoenix

32. Which city has the lowest high temperature? Seattle

33. List the low temperatures in order from least to greatest. (If two or more cities have the same low temperature, tell the number of cities with that low temperature.) ⁻30, ⁻24, ⁻23, ⁻8 (2 cities), ⁻4, ⁻1, ⁺5, ⁺12, ⁺15 (3 cities), ⁺25, ⁺31, ⁺32, ⁺35

34. Which cities have high temperatures that are higher than the high for Washington, D.C.? Kansas City, Dallas, Phoenix

It was a record!

35. A record for extreme temperatures in a 24-hour period was set at Browning, Montana, in 1916. The high was 44°F. During the next 24 hours the temperature dropped 100 degrees. What was the low? −56°F

Records may vary from year to year.

Class Starter Quiz 160
on previous lesson

< or >?

1. ⁺4 ● ⁻4   >
2. ⁻3 ● ⁻2   <
3. ⁻1 ● 0   <
4. 0 ● ⁻5   >
5. ⁻7 ● ⁺9   <
6. ⁻21 ● ⁺12 <
7. ⁻32 ● ⁻23 <
8. ⁺24 ● ⁺23 >
9. ⁻7 ● ⁻5   <
10. ⁺4 ● ⁻3 >

Copymaster S425

Lesson Objective
To add integers

Problem-Solving Skills
Finding information in a display
Applying an addition model
Working backward

Starting the Lesson

Use of Concrete Materials You may wish to use the positive and the negative charge models from copymaster or transparency S542 to demonstrate adding integers. See ■ **Manipulative Activity 30** on copymaster S526 in the Teacher's Resource Binder.

Calculator You may want to show the students how to use the +/− (change sign) key to add integers.

Example. Find ⁻15 + ⁺4.

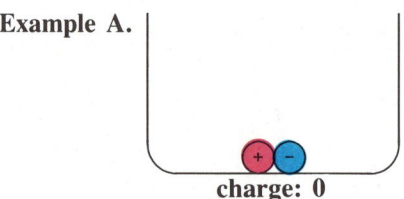

Example. Find ⁻10 + ⁻21.

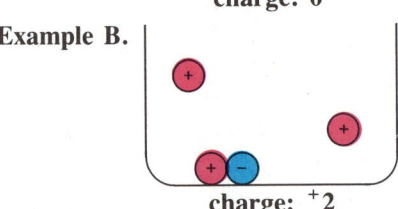

Adding integers

Imagine some small particles that have either a positive electrical charge or a negative electrical charge. The positive charges and negative charges are opposites. This means that when one positive charge and one negative charge are put together, the result is no charge, or a charge of 0.

1. Look at Example A. What is the charge when one positive charge and one negative charge are combined? 0

2. Look at Example B. What is the charge when three positive charges and one negative charge are combined? ⁺2

3. Look at Example C. What is the charge when two positive charges and five negative charges are combined? ⁻3

Example A. charge: 0

Example B. charge: ⁺2

Example C.

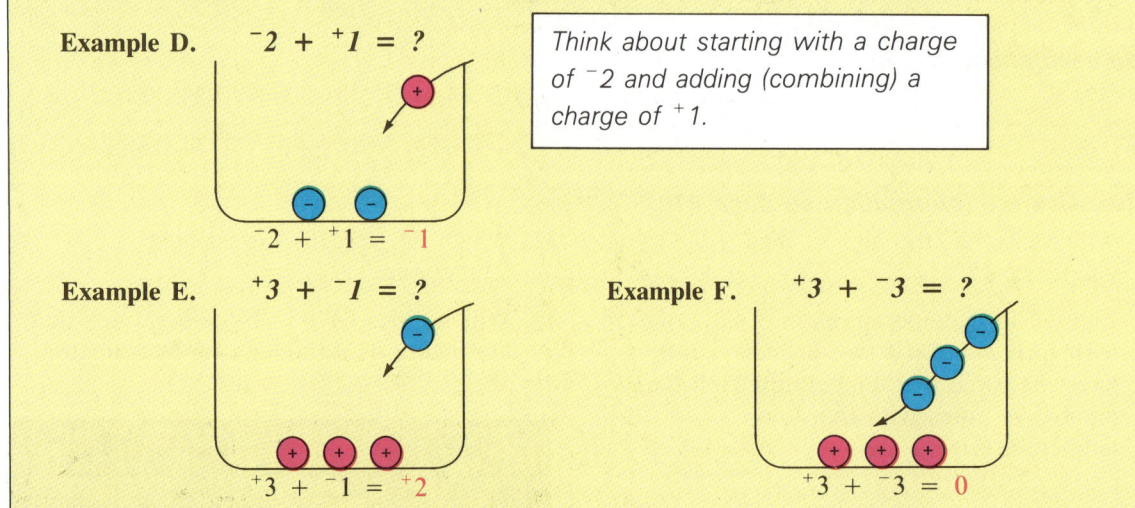

Here's how *to add integers by thinking about combining charges.*

Example D. ⁻2 + ⁺1 = ?

⁻2 + ⁺1 = ⁻1

Think about starting with a charge of ⁻2 and adding (combining) a charge of ⁺1.

Example E. ⁺3 + ⁻1 = ?

⁺3 + ⁻1 = ⁺2

Example F. ⁺3 + ⁻3 = ?

⁺3 + ⁻3 = 0

4. Look at the *Here's how.* What is the sum of ⁻2 and ⁺1? ⁻1
   What is the sum of ⁺3 and ⁻1? ⁺2

5. If the sum of the two numbers is 0, then one number is the **opposite** of the other. Look at Example F. Since the sum of the two numbers is 0, you know that the opposite of ⁺3 is what number? What is the opposite of ⁻3? ⁻3, ⁺3

Chapter 16

EXERCISES

Give each sum.
Here are scrambled answers for the next row of exercises: 0 ⁺5 ⁺4 ⁻2 ⁻6

6. ⁺3 + ⁺2 **⁺5**   7. ⁺4 + ⁻2 **⁺2**   8. ⁻3 + ⁺3 **0**   9. ⁻1 + ⁻5 **⁻6**   10. ⁺5 + ⁻1 **⁺4**
11. ⁺6 + ⁻1 **⁺5**   12. ⁻6 + ⁺1 **⁻5**   13. ⁻6 + ⁻1 **⁻7**   14. ⁺6 + ⁺1 **⁺7**   15. ⁺6 + ⁻6 **0**
16. ⁻5 + ⁺4 **⁻1**   17. ⁺5 + ⁺4 **⁺9**   18. ⁻5 + ⁻4 **⁻9**   19. ⁺5 + ⁻4 **⁺1**   20. ⁺4 + ⁻9 **⁻5**
21. ⁺7 + ⁻3 **⁺4**   22. ⁻7 + ⁻3 **⁻10**   23. ⁺7 + ⁺3 **⁺10**   24. ⁻7 + ⁺3 **⁻4**   25. ⁺5 + ⁻8 **⁻3**
26. ⁺4 + ⁻4 **0**   27. ⁻5 + ⁺5 **0**   28. ⁺8 + ⁻8 **0**   29. ⁻9 + ⁺9 **0**   30. ⁻2 + ⁻2 **⁻4**
31. ⁺7 + 0 **⁺7**   32. ⁻6 + 0 **⁻6**   33. 0 + ⁺5 **⁺5**   34. 0 + 0 **0**   35. ⁻7 + ⁺7 **0**
36. ⁻12 + ⁻10 **⁻22**   37. ⁺15 + ⁺18 **⁺33**   38. ⁺17 + ⁻11 **⁺6**   39. ⁻16 + ⁺19 **⁺3**   40. ⁺13 + ⁻19 **⁻6**
41. ⁻20 + ⁺12 **⁻8**   42. ⁻24 + ⁻24 **⁻48**   43. ⁺28 + ⁺21 **⁺49**   44. ⁺26 + ⁻28 **⁻2**   45. ⁻34 + ⁻11 **⁻45**
46. ⁺30 + ⁻24 **⁺6**   47. ⁻24 + ⁺30 **⁺54**   48. ⁻36 + ⁺32 **⁻4**   49. ⁻36 + ⁻32 **⁻68**   50. ⁺46 + ⁻18 **⁺28**
51. ⁻35 + 0 **⁻35**   52. ⁻37 + ⁺37 **0**   53. ⁺39 + ⁻30 **⁺9**   54. ⁻38 + ⁺34 **⁻4**   55. ⁺25 + ⁻25 **0**
56. ⁻40 + ⁻48 **⁻88**   57. ⁺40 + ⁻48 **⁻8**   58. ⁻40 + ⁺48 **⁺8**   59. ⁺40 + ⁺48 **⁺88**   60. 0 + ⁻88 **⁻88**

True or false?

61. The sum of two positive numbers is always positive. **True**

62. The sum of two negative numbers is always positive. **False**

63. The sum of a positive number and a negative number is always positive. **False**

64. The sum of a positive number and a negative number is always negative. **False**

65. The sum of a positive number and a negative number may be positive, may be negative, or may be 0. **True**

66. The sum of two opposites is 0. **True**

Magic squares Working backward

67. Add the numbers in each row, column, and diagonal. If the sums are the same, the square is a Magic Square. Is this a Magic Square? **Yes. (The sum is ⁺6.)**

68. Copy and complete this Magic Square.

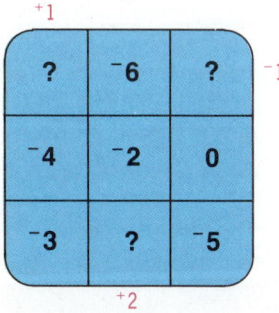

Extra Practice
Page 500 Skill 60

Practice Worksheet
Workbook S344, Copymaster S344, or Duplicating Master S344

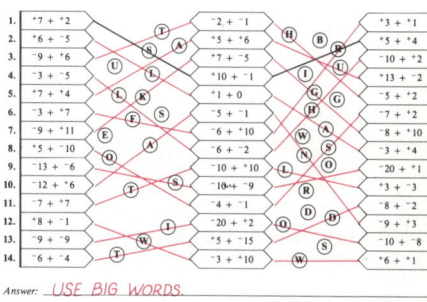

Project

Adding integers

Mark two identical integer number lines from ⁻10 to ⁺10. To find the sum of ⁻4 and ⁻5, place 0 on Scale A over ⁺4 on Scale B. The sum appears on Scale B under ⁻5 on Scale A.
⁺4 + ⁻5 = ?

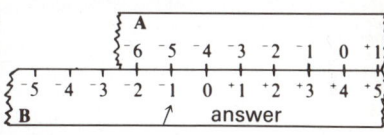

⁺4 + ⁻5 = ⁻1
Use your number lines to find these sums.

1. ⁺6 + ⁻9 **⁻3**   2. ⁻6 + ⁺5 **⁻1**
3. ⁻3 + ⁻2 **⁻5**   4. ⁺8 + ⁻2 **⁺6**
5. ⁺4 + ⁻3 **⁺1**   6. ⁻1 + ⁻6 **⁻7**

Copymaster S487

Integers

Class Starter Quiz 161
on previous lesson

Give each sum.

1. $^+6 + {}^-2$ $^+4$
2. $^-7 + {}^-3$ $^-10$
3. $^-7 + {}^+5$ $^-2$
4. $^+5 + {}^-5$ 0
5. $^-6 + 0$ $^-6$
6. $^-9 + {}^+10$ $^+1$
7. $^+14 + {}^-19$ $^-5$
8. $^-35 + {}^+12$ $^-23$
9. $^+20 + {}^-16$ $^+4$
10. $^-20 + {}^-16$ $^-36$

Copymaster S425

Lesson Objective
To subtract integers

Starting the Lesson
Use of Concrete Materials You may wish to use the positive and negative charge models from copymaster or transparency S542 to demonstrate subtracting integers. See ■ **Manipulative Activity 31** on copymasters S527–S528 in the Teacher's Resource Binder.

Here's How Note
You may want to follow these steps in going over the examples.

Example A. $^-3 - {}^-2 = ?$

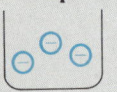 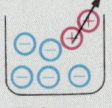

Do not have 2 ⊕ to take out. | Put in 2 ⊕⊖ pairs. | Take out 2 ⊕.

Example B. $^+2 - {}^-1 = ?$

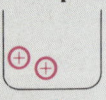 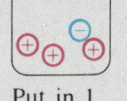 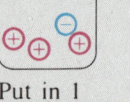 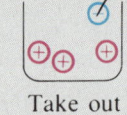

No ⊖ to take out. | Put in 1 ⊕⊖ pair. | Take out 1 ⊖.

Example C. $^-2 - {}^-3 = ?$

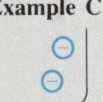 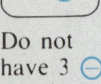 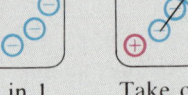

Do not have 3 ⊖ to take out. | Put in 1 ⊕⊖ pair. | Take out 3 ⊖.

Subtracting integers

Look at the picture at the right to answer the following questions.

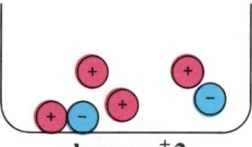

charge: $^+2$

1. What is the charge of the particles in the container? $^+2$
2. Suppose that you removed a charge of $^+1$. What would the charge be then? $^+1$
3. Suppose instead that you removed a charge of $^-2$. What would the charge be then? $^+4$

Here's how *to subtract integers by thinking about removing charges.*

Example A. $^-3 - {}^+2 = ?$

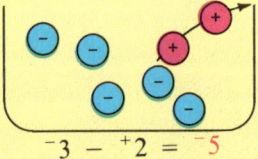

$^-3 - {}^+2 = {}^-5$

Think about starting with a charge of $^-3$ and subtracting (removing) a charge of $^+2$.

Example B. $^+2 - {}^-1 = ?$

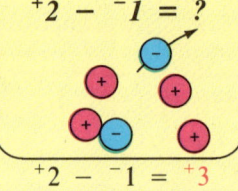

$^+2 - {}^-1 = {}^+3$

Example C. $^-2 - {}^-3 = ?$

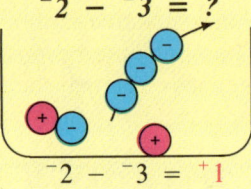

$^-2 - {}^-3 = {}^+1$

---

4. Look at Example A in the *Here's how*. Adding a charge of $^-2$ would give the same result as subtracting a charge of $^+2$.

   $^-3 - {}^+2 = {}^-5$
   $^-3 + \boxed{?} = {}^-5$ $^-2$

5. Look at Example B. Adding a charge of $^+1$ would give the same result as subtracting a charge of $^-1$.

   $^+2 - {}^-1 = {}^+3$
   $^+2 + \boxed{?} = {}^+3$ $^+1$

6. Look at Example C.

   $^-2 - {}^-3 = {}^+1$
   $^-2 + \boxed{?} = {}^+1$ $^+3$

Here's how to subtract integers.

To subtract an integer, add the opposite of the integer.

$^+5 - {}^-4 = {}^+5 + {}^+4$ $^-6 - {}^+2 = {}^-6 + {}^-2$ $^+3 - {}^+7 = {}^+3 + {}^-7$
$\quad\quad\quad = {}^+9$ $= {}^-8$ $= {}^-4$

EXERCISES

7. The result of subtracting $^-4$ is the same as adding [?]. $^+4$

8. The result of subtracting $^+2$ is the same as adding [?]. $^-2$

9. To subtract $^+7$, you would add [?]. $^-7$

10. To subtract $^-9$, you would add [?]. $^+9$

Give each difference.
Here are scrambled answers for the next row of exercises: $^-3$ $^+10$ $^-8$ $^-15$ $^-4$

11. $^+7 - {}^-3$ $^+10$ 12. $^-9 - {}^-5$ $^-4$ 13. $^-2 - {}^+6$ $^-8$ 14. $^+4 - {}^+7$ $^-3$ 15. $^-10 - {}^+5$ $^-15$

16. $^-4 - {}^-9$ $^+5$ 17. $^+6 - {}^+2$ $^+4$ 18. $^-6 - {}^+6$ $^-12$ 19. $^+7 - 0$ $^+7$ 20. $0 - {}^+8$ $^-8$

21. $^+7 - {}^-5$ $^+12$ 22. $0 - {}^-6$ $^+6$ 23. $^+1 - {}^+9$ $^-8$ 24. $^-4 - {}^+4$ $^-8$ 25. $^-3 - {}^+4$ $^-7$

26. $^-6 - {}^+3$ $^-9$ 27. $^+6 - {}^+3$ $^+3$ 28. $^+6 - {}^-3$ $^+9$ 29. $^-6 - {}^-3$ $^-3$ 30. $^-9 - {}^-5$ $^-14$

31. $^+6 - 0$ $^+6$ 32. $0 - {}^+6$ $^-6$ 33. $^-7 - {}^+4$ $^-11$ 34. $^+4 - {}^+9$ $^-5$ 35. $^-11 - {}^-4$ $^-7$

36. $^-4 - {}^+9$ $^-13$ 37. $^-9 - {}^+4$ $^-13$ 38. $^+8 - {}^-3$ $^+11$ 39. $^+3 - {}^+8$ $^-5$ 40. $^-14 - {}^+8$ $^-22$

41. $^-10 - {}^+13$ $^-23$ 42. $^+11 - {}^-12$ $^+23$ 43. $^-14 - {}^-16$ $^+2$ 44. $^+12 - {}^+12$ 0 45. $^-21 - {}^-3$ $^-18$

46. $^+16 - {}^+18$ $^-2$ 47. $^+18 - {}^+16$ $^+2$ 48. $^-17 - {}^+11$ $^-28$ 49. $^+17 - {}^-11$ $^+28$ 50. $^-14 - {}^+18$ $^-32$

Give each sum or difference.

51. $^+10 + {}^+11$ $^+21$ 52. $^-12 - {}^+15$ $^-27$ 53. $^-16 - {}^-14$ $^-2$ 54. $^+18 + {}^-12$ $^+6$ 55. $^+13 - {}^-12$ $^+25$

56. $^-14 + 0$ $^-14$ 57. $0 + {}^+17$ $^+17$ 58. $^-11 + {}^-11$ $^-22$ 59. $^-11 - {}^-11$ 0 60. $^-30 + {}^-15$ $^-45$

61. $^+20 - {}^-16$ $^+36$ 62. $^-23 - {}^+23$ $^-46$ 63. $^+24 + {}^+16$ $^+40$ 64. $^-25 + {}^+25$ 0 65. $^-24 - {}^+18$ $^-42$

Add across. Subtract down. Working backward

Copy and complete these addition-subtraction boxes.

66.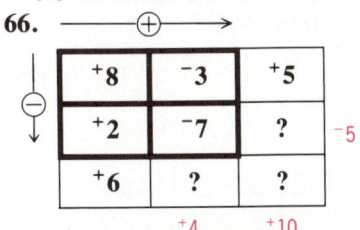

67.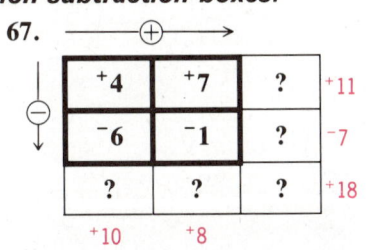

68.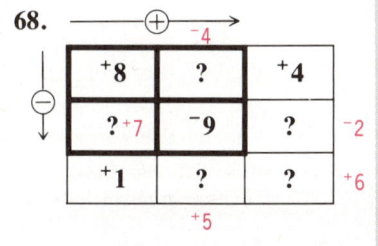

Extra Practice
Page 501   Skill 61

Practice Worksheet
Workbook S345, Copymaster S345, or Duplicating Master S345

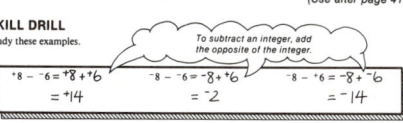

Project

Subtracting integers

Mark two identical integer number lines from $^-10$ to $^+10$. To find $^+2 - {}^-1$, place $^+2$ on Scale A over $^-1$ on Scale B. Find 0 on Scale B and read the answer on Scale A.

$^+2 - {}^-1 = ?$

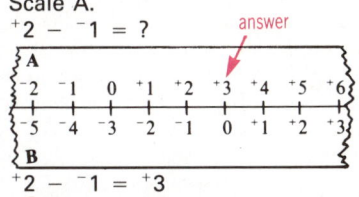

$^+2 - {}^-1 = {}^+3$

Use your number lines to find these differences.

1. $^+3 - {}^-2$ $^+5$ 2. $^+3 - {}^+4$ $^-1$
3. $^-2 - {}^+3$ $^-5$ 4. $^-4 - {}^-2$ $^-2$
5. $^+5 - {}^-2$ $^+7$ 6. $^-3 - {}^-3$ 0

Copymaster S487

Class Starter Quiz 162
on previous lesson

Give the difference.

1. $^+6 - {}^-2$ $^+8$
2. $^-8 - {}^-5$ $^-3$
3. $^-3 - {}^+7$ $^-10$
4. $^+5 - {}^+7$ $^-2$
5. $^-9 - {}^+4$ $^-13$
6. $^-3 - {}^-8$ $^+5$
7. $^+6 - {}^-2$ $^+8$
8. $^-6 - {}^+6$ $^-12$
9. $^+7 - 0$ $^+7$
10. $0 - {}^+8$ $^-8$

Copymaster S425

Problem-Solving Skills

Finding information in a price list
Choosing the correct operation
Solving a multistep problem

Skills Reviewed

Multiplying and dividing by 10, 100, or 1000
Changing fractions to decimals
Adding mixed numbers
Multiplying mixed numbers
Solving percent problems

Starting the Lesson

Problem Solving Use the pizza prices and ask questions like these:

- What is the cost of a medium pepperoni pizza? ($5.25)
- How much is a small sausage pizza with chili peppers? ($4.25)
- What kind of large pizza with green peppers can you buy for exactly $7.25? (Pepperoni)

Cumulative Skill Practice Write these five answers on the chalkboard:

0.04   0.71   0.33 $18\frac{3}{20}$ 4

Challenge the students to an answer hunt by saying, "Look at exercises 1–46 on page 417. Find the five exercises that have these answers. You have five minutes to find as many of the exercises as you can." (Exercises 8, 17, 19, 37, and 39)

416

Problem solving

YOU'RE THE PIZZA MAKER!

THE PIZZA WITH PIZZAZZ!

	small 10-inch diameter	medium 12-inch diameter	large 14-inch diameter
CHEESE	$3.00	$4.50	$5.50
BACON	$3.60	$4.80	$6.25
PEPPERONI	$4.00	$5.25	$6.75
SAUSAGE	$3.75	$5.00	$6.50

Add **50¢** for each topping:
chili peppers
green peppers
mushrooms
onions

Use the information on the sign to answer these customers' questions.

1. How much will a large pepperoni pizza with mushrooms cost? $7.25

2. I have $10. Do I have enough to buy 2 small cheese pizzas and 1 small sausage pizza with onions? No

3. "What will 2 medium bacon and 3 large sausage pizzas cost?" $29.10

4. "We have $20. How much more will we need to buy 3 large pepperoni pizzas with green peppers and mushrooms?" $3.25

Solve. Decide when a calculator would be useful.

5. On Wednesday, all large pizzas are 20% off. What would you charge a customer who orders a large bacon, a large cheese, and a small pepperoni pizza? $13.40

6. You get a special order for 18 medium pepperoni pizzas. It costs you $43.20 to make the pizzas. How much profit do you make? $51.30

7. You hire 2 part-time employees to work from 4:00 to 8:00 each day. If you pay each $4.50 an hour, how much does your part-time help cost per day? $36

8. You borrow $1150 for a pizza oven. How much interest will you have to pay if you borrow the money for 9 months at the yearly rate of 16%? $138

9. a. What is the area of a 10-inch pizza? Use 3.14 for π. 78.5 in.^{2}
   b. What is the price per square inch of a small cheese pizza? Give the answer to the nearest tenth of a cent. 3.8¢ or $.038

10. Which pizza is the better deal (costs less per square inch), a medium sausage or a large sausage? Large sausage

416   Chapter 16

Cumulative Skill Practice

Give the product. *(page 72)*

1. 8.23 × 10   82.3
2. 8.23 × 100   823
3. 8.23 × 1000   8230
4. 4.5 × 1000   4500
5. 4.5 × 10   45
6. 4.5 × 100   450
7. 0.004 × 100   0.4
8. 0.004 × 10   0.04
9. 0.004 × 1000   4

Give the quotient. *(page 100)*

10. 789.5 ÷ 10   78.95
11. 789.5 ÷ 100   7.895
12. 789.5 ÷ 1000   0.7895
13. 297 ÷ 100   2.97
14. 297 ÷ 10   29.7
15. 297 ÷ 1000   0.297
16. 7.1 ÷ 100   0.071
17. 7.1 ÷ 10   0.71
18. 7.1 ÷ 1000   0.0071

Change to a decimal rounded to the nearest hundredth. *(page 166)*

19. $\frac{1}{3}$ 0.33
20. $\frac{1}{6}$ 0.17
21. $\frac{1}{8}$ 0.13
22. $\frac{1}{12}$ 0.08
23. $\frac{5}{6}$ 0.83
24. $\frac{3}{8}$ 0.38
25. $\frac{7}{9}$ 0.78
26. $\frac{5}{3}$ 1.67
27. $\frac{7}{6}$ 1.17
28. $\frac{5}{8}$ 0.63
29. $\frac{16}{9}$ 1.78
30. $\frac{4}{3}$ 1.33
31. $\frac{11}{8}$ 1.38
32. $\frac{13}{3}$ 4.33

Give the sum in simplest form. *(page 182)*

33. $5\frac{1}{2} + 4\frac{1}{3} = 9\frac{5}{6}$
34. $6\frac{2}{3} + 3\frac{1}{2} = 10\frac{1}{6}$
35. $5\frac{5}{8} + 1\frac{3}{4} = 7\frac{3}{8}$
36. $7\frac{5}{6} + 7\frac{2}{3} = 15\frac{1}{2}$
37. $9\frac{3}{4} + 8\frac{2}{5} = 18\frac{3}{20}$
38. $6\frac{1}{4} + 3\frac{1}{2} = 9\frac{3}{4}$

Give the product in simplest form. *(page 206)*

39. $3 \times 1\frac{1}{3}$ 4
40. $2 \times 2\frac{1}{2}$ 5
41. $1\frac{3}{4} \times 4$ 7
42. $3\frac{1}{3} \times 3$ 10
43. $1\frac{1}{3} \times 1\frac{1}{2}$ 2
44. $2\frac{1}{2} \times 2\frac{1}{3}$ $5\frac{5}{6}$
45. $3\frac{1}{2} \times 2\frac{1}{4}$ $7\frac{7}{8}$
46. $3\frac{1}{2} \times 2\frac{1}{2}$ $8\frac{3}{4}$

MIXED PRACTICE

Complete.

47. 25% of 48 = n   12
48. 150% of n = 27   18
49. 17% of 51 = n   8.67
50. 60% of n = 90   150
51. 42% of 83 = n   34.86
52. 60% of 60 = n   36
53. 8% of n = 10   125
54. 80% of 75 = n   60
55. 125% of n = 25   20
56. $66\frac{2}{3}$% of 51 = n   34
57. 8.4% of n = 2.1   25
58. 3.7% of 112 = n   4.144
59. $37\frac{1}{2}$% of 112 = n   42
60. 0.9% of n = 0.9   100
61. 31.6% of 93 = n   29.388

Integers   **417**

Class Starter Quiz 163
on previous lesson

Solve. Use the pizza price list on page 416.

1. How much will a medium bacon pizza with onions cost? $5.30
2. You have $10. How much more money will you need to buy 2 large cheese pizzas with green peppers and mushrooms? $3

Copymaster S426

Lesson Objective
To multiply integers

Problem-Solving Skills
Finding information in a display
Applying a multiplication model
Using a guess-and-check strategy

Starting the Lesson
Use of Concrete Materials You may wish to use the positive and negative charge models from copymaster or transparency S542 to demonstrate multiplying integers. See ■ **Manipulative Activity 32** on copymaster S528 in the Teacher's Resource Binder.

Multiplying integers

Look at the picture at the right to answer the following questions

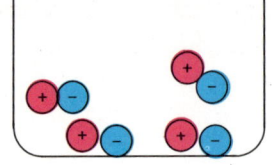

1. What is the charge of the particles in the container? 0
2. Suppose that you put in 2 sets of $^+2$ charges. What would the charge be then? $^+4$

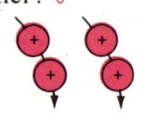

Look again at the container with a charge of 0.

3. Suppose that you took out 2 sets of $^+2$ charges. What would the charge be then? $^-4$
4. What would the charge be if instead you put in 2 sets of $^-2$? $^-4$
5. What would the charge be if instead you took out 2 sets of $^-2$? $^+4$

Here's how to multiply integers by thinking about putting in or taking out sets of charges.

To multiply, we will think of "putting charges in" as positive and "taking charges out" as negative.

Example A. $^+3 \times {}^+2 = ?$

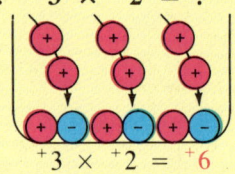

Start with a charge of 0 and put in 3 sets of $^+2$.

$^+3 \times {}^+2 = {}^+6$

Example B. $^+3 \times {}^-2 = ?$

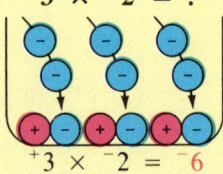

$^+3 \times {}^-2 = {}^-6$

Example C. $^-3 \times {}^+2 = ?$

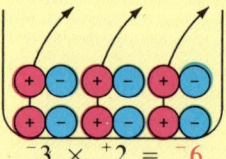

$^-3 \times {}^+2 = {}^-6$

Example D. $^-3 \times {}^-2 = ?$

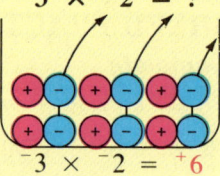

$^-3 \times {}^-2 = {}^+6$

6. Look at the *Here's how*. In Examples A and D, we multiplied two integers with the same signs (both positive or both negative). Was the product positive or negative? Positive
7. In Examples B and C, we multiplied two integers with different signs. Was the product positive or negative? Negative

Here's how to multiply integers.

The product of two integers with the **same** signs is **positive**.
The product of two integers with **different** signs is **negative**.
The product of any integer and 0 is 0.

EXERCISES

Positive, negative, or zero?

8. The product of a positive integer and a positive integer is a ? integer. **positive**
9. The product of a positive integer and a negative integer is a ? integer. **negative**
10. The product of a negative integer and a negative integer is a ? integer. **positive**
11. The product of an integer and 0 is ? . **0**

Give each product.
Here are scrambled answers for the next row of exercises: ⁻12 ⁺8 ⁺9 ⁻12 ⁻10

12. ⁻2 × ⁺5 **⁻10**
13. ⁺4 × ⁺2 **⁺8**
14. ⁻3 × ⁻3 **⁺9**
15. ⁺2 × ⁻6 **⁻12**
16. ⁻3 × ⁺4 **⁻12**
17. ⁺3 × ⁺5 **⁺15**
18. ⁺3 × ⁻5 **⁻15**
19. ⁻3 × ⁺5 **⁻15**
20. ⁻3 × ⁻5 **⁺15**
21. ⁺1 × ⁻8 **⁻8**
22. ⁻6 × ⁺4 **⁻24**
23. ⁺6 × ⁺4 **⁺24**
24. ⁻6 × ⁻4 **⁺24**
25. ⁺6 × ⁻4 **⁻24**
26. ⁻4 × ⁻7 **⁺28**
27. ⁻6 × 0 **0**
28. 0 × ⁺4 **0**
29. ⁺8 × 0 **0**
30. 0 × 0 **0**
31. 0 × ⁻3 **0**
32. ⁻5 × ⁺5 **⁻25**
33. ⁺7 × ⁺6 **⁺42**
34. ⁻6 × ⁻6 **⁺36**
35. ⁺5 × ⁻9 **⁻45**
36. ⁻3 × ⁺8 **⁻24**
37. ⁻8 × ⁻6 **⁺48**
38. ⁺9 × ⁻6 **⁻54**
39. 0 × ⁺3 **0**
40. ⁺9 × ⁺5 **⁺45**
41. ⁺4 × ⁺8 **⁺32**
42. ⁺12 × ⁺12 **⁺144**
43. ⁺12 × ⁻12 **⁻144**
44. ⁻12 × ⁺12 **⁻144**
45. ⁻12 × ⁻12 **⁺144**
46. ⁻11 × ⁻11 **⁺121**
47. ⁻11 × ⁺11 **⁻121**
48. ⁺15 × 0 **0**
49. ⁻16 × ⁻14 **⁺224**
50. ⁺20 × ⁻12 **⁻240**
51. ⁻4 × ⁺8 **⁻32**

Build an expression Guess and check

Use all the cards to build an expression for each of the following numbers.

52. ⁻9   ⁺3 × (⁻2 + ⁻1)
53. ⁺5   ⁺3 + (⁻1 × ⁻2)
54. ⁻4   ⁻2 × (⁺3 + ⁻1)
55. ⁻7   ⁻1 + (⁺3 × ⁻2)

Cards: ⁺3 ⁻1 ⁻2 + × ( )

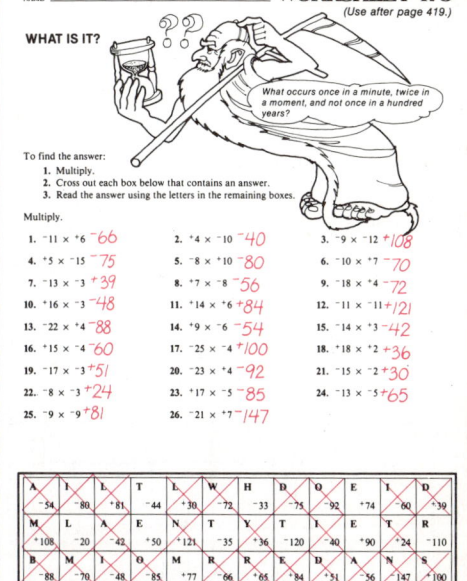

Extra Practice
Page 501 Skill 62

Practice Worksheet
Workbook S347, Copymaster S347, or Duplicating Master S347

WORKSHEET 179 (Use after page 419.)

WHAT IS IT?
What occurs once in a minute, twice in a moment, and not once in a hundred years?

To find the answer:
1. Multiply.
2. Cross out each box below that contains an answer.
3. Read the answer using the letters in the remaining boxes.

Multiply.
1. ⁻11 × ⁺6 **⁻66**
2. ⁺4 × ⁻10 **⁻40**
3. ⁻9 × ⁻12 **⁺108**
4. ⁺5 × ⁻15 **⁻75**
5. ⁻8 × ⁺10 **⁻80**
6. ⁻10 × ⁺7 **⁻70**
7. ⁻13 × ⁻3 **⁺39**
8. ⁺7 × ⁻8 **⁻56**
9. ⁻18 × ⁻4 **⁺72**
10. ⁺16 × ⁻3 **⁻48**
11. ⁻14 × ⁻6 **⁺84**
12. ⁻11 × ⁻11 **⁺121**
13. ⁻22 × ⁺4 **⁻88**
14. ⁺9 × ⁻6 **⁻54**
15. ⁻14 × ⁺3 **⁻42**
16. ⁺15 × ⁻4 **⁻60**
17. ⁻25 × ⁻4 **⁺100**
18. ⁺18 × ⁺2 **⁺36**
19. ⁻17 × ⁻3 **⁺51**
20. ⁻23 × ⁺4 **⁻92**
21. ⁻15 × ⁻2 **⁺30**
22. ⁻8 × ⁻3 **⁺24**
23. ⁺17 × ⁻5 **⁻85**
24. ⁻13 × ⁻5 **⁺65**
25. ⁻9 × ⁻9 **⁺81**
26. ⁻21 × ⁺7 **⁻147**

Answer: **THE LETTER M**

Challenge Problem

Multiply each number in this Magic Square by ⁻2 to get a new square.

⁺3	⁻4	⁺1
⁻2	0	⁺2
⁻1	⁺4	⁻3

⁻6	⁺8	⁻2
⁺4	0	⁻4
⁺2	⁻8	⁺6

Is your new square a Magic Square? **Yes**

Copymaster S463

Class Starter Quiz 164
on previous lesson

Give each product.

1. ⁻3 × ⁺5 ⁻15
2. ⁺6 × ⁺2 ⁺12
3. ⁻4 × ⁻4 ⁺16
4. ⁺4 × ⁻7 ⁻28
5. ⁻4 × ⁺5 ⁻20
6. ⁺7 × ⁺9 ⁺63
7. ⁺3 × ⁻8 ⁻24
8. ⁻7 × 0 0
9. ⁺9 × ⁻6 ⁻54
10. ⁻6 × ⁻9 ⁺54

Copymaster S426

Lesson Objective
To divide integers

Problem-Solving Skills
Discovering numerical relationships
Working backward

Starting the Lesson
Write this table on the chalkboard:

×	⁺2	⁺1	0	⁻1	⁻2
⁺3	⁺6	⁺3	0	⁻3	⁻6
⁻3	⁻6	⁻3	0	⁺3	⁺6

Ask the students how they can use the integer multiplication table to find ⁺6 ÷ ⁻2. (To find ⁺6 ÷ ⁻2, find ⁻2 in the top row and go down to ⁺6. The number at the left of this row is the answer. ⁺6 ÷ ⁻2 = ⁻3.) Have the students use the table to find these quotients.

⁻6 ÷ ⁻2 = ?
⁺6 ÷ ⁺2 = ?
⁻6 ÷ ⁺2 = ?

420

Dividing integers

1. What would you multiply by ⁺3 to get ⁺18? ⁺6
2. What would you multiply by ⁻4 to get ⁺28? ⁻7
3. What would you multiply by ⁻9 to get 0? 0

Here's how *to divide integers by finding a missing factor.*

Example A. ⁺18 ÷ ⁺3 = ?
⁺18 ÷ ⁺3 = ⁺6
because
⁺3 × ⁺6 = ⁺18

Example B. ⁺18 ÷ ⁻3 = ?
⁺18 ÷ ⁻3 = ⁻6
because
⁻3 × ⁻6 = ⁺18

Example C. ⁻18 ÷ ⁺3 = ?
⁻18 ÷ ⁺3 = ⁻6
because
⁺3 × ⁻6 = ⁻18

Example D. ⁻18 ÷ ⁻3 = ?
⁻18 ÷ ⁻3 = ⁺6
because
⁻3 × ⁺6 = ⁻18

4. Look at the *Here's how*. In Example A, we divided a positive integer by a positive integer. Was the quotient positive or negative? Positive

5. If you divide a positive integer by a negative integer, will the quotient be positive or negative? Negative

6. If you divide a negative integer by a positive integer, will the quotient be positive or negative? Negative

7. If you divide a negative integer by a negative integer, will the quotient be positive or negative? Positive

Here's how *to divide integers.*

The quotient of two integers with the **same** signs is **positive.**

The quotient of two integers with **different** signs is **negative.**

The quotient of 0 divided by any nonzero integer is 0.

EXERCISES

Give each quotient.
Here are scrambled answers for the next row of exercises: $^-8$ $^+4$ $^-10$ $^-6$ $^-8$

8. $^+20 \div ^-2$ $_{-10}$
9. $^+24 \div ^+3$ $_{+8}$
10. $^-30 \div ^+5$ $_{-6}$
11. $^-36 \div ^-9$ $_{+4}$
12. $^-56 \div ^+7$ $_{-8}$
13. $^+14 \div ^+2$ $_{+7}$
14. $^+14 \div ^-2$ $_{-7}$
15. $^-14 \div ^+2$ $_{-7}$
16. $^-14 \div ^-2$ $_{+7}$
17. $^+15 \div ^-5$ $_{-3}$
18. $^+15 \div ^-3$ $_{-5}$
19. $^-15 \div ^+3$ $_{-5}$
20. $^-18 \div ^-6$ $_{+3}$
21. $^+15 \div ^+3$ $_{+5}$
22. $^-15 \div ^-3$ $_{+5}$
23. $^+24 \div ^+6$ $_{+4}$
24. $^+24 \div ^-6$ $_{-4}$
25. $^+12 \div ^+6$ $_{+2}$
26. $^-24 \div ^+6$ $_{-4}$
27. $^-24 \div ^-6$ $_{+4}$
28. $^-30 \div ^+6$ $_{-5}$
29. $0 \div ^-6$ $_{0}$
30. $^+18 \div ^-6$ $_{-3}$
31. $^-30 \div ^-5$ $_{+6}$
32. $^+30 \div ^+6$ $_{+5}$
33. $^+30 \div ^-5$ $_{-6}$
34. $0 \div ^+8$ $_{0}$
35. $^-12 \div ^-4$ $_{+3}$
36. $^+10 \div ^-2$ $_{-5}$
37. $^-16 \div ^-4$ $_{+4}$
38. $^-24 \div ^+8$ $_{-3}$
39. $^+28 \div ^-4$ $_{-7}$
40. $^+32 \div ^-8$ $_{-4}$
41. $^+25 \div ^+5$ $_{+5}$
42. $^-27 \div ^-3$ $_{+9}$
43. $^-45 \div ^-5$ $_{+9}$
44. $^+54 \div ^-6$ $_{-9}$
45. $^+36 \div ^-6$ $_{-6}$
46. $^-48 \div ^+8$ $_{-6}$
47. $^+56 \div ^+7$ $_{+8}$
48. $^+81 \div ^+9$ $_{+9}$
49. $0 \div ^-12$ $_{0}$
50. $^+81 \div ^-9$ $_{-9}$
51. $^-64 \div ^+8$ $_{-8}$
52. $^+54 \div ^+9$ $_{+6}$
53. $^-72 \div ^-9$ $_{+8}$
54. $^+121 \div ^+11$ $_{+11}$
55. $^+144 \div ^+12$ $_{+12}$
56. $^+150 \div ^-10$ $_{-15}$
57. $^-132 \div ^-12$ $_{+11}$
58. $^+124 \div ^-4$ $_{-31}$
59. $^-176 \div ^-16$ $_{+11}$
60. $^-120 \div ^+20$ $_{-6}$
61. $^+162 \div ^+18$ $_{+9}$
62. $^-147 \div ^+21$ $_{-7}$

Simplify.
Here are scrambled answers for the next row of exercises: $^-3$ $^+4$ $^+3$ $^+12$ $^+8$

63. $^+6 + ^+2$ $_{+8}$
64. $^+6 - ^+2$ $_{+4}$
65. $^-6 \div ^+2$ $_{-3}$
66. $^+6 \times ^+2$ $_{+12}$
67. $^+6 \div ^+2$ $_{+3}$
68. $^+8 + ^-4$ $_{+4}$
69. $^+8 - ^-4$ $_{+12}$
70. $^-8 - ^+4$ $_{-12}$
71. $^+8 \times ^-4$ $_{-32}$
72. $^+8 \div ^-4$ $_{-2}$
73. $^-15 + ^+3$ $_{-12}$
74. $^-15 - ^+3$ $_{-18}$
75. $^+15 \div ^-3$ $_{-5}$
76. $^-15 \times ^+3$ $_{-45}$
77. $^-15 \div ^+3$ $_{-5}$
78. $^+8 + 0$ $_{+8}$
79. $^+8 - 0$ $_{+8}$
80. $0 \times ^-8$ $_{0}$
81. $^+8 \times 0$ $_{0}$
82. $0 \div ^+8$ $_{0}$
83. $^-24 + ^-6$ $_{-30}$
84. $^-24 - ^-6$ $_{-18}$
85. $^+24 + ^-6$ $_{+18}$
86. $^-24 \times ^-6$ $_{+144}$
87. $^-24 \div ^-6$ $_{+4}$
88. $^+18 + ^-9$ $_{+9}$
89. $^+18 - ^-9$ $_{+27}$
90. $^+18 \div ^+9$ $_{+2}$
91. $^+18 \times ^-9$ $_{-162}$
92. $^+18 \div ^-9$ $_{-2}$
93. $^-21 + ^+3$ $_{-18}$
94. $^-21 - ^+3$ $_{-24}$
95. $^+21 \div ^-3$ $_{-7}$
96. $^-21 \times ^+3$ $_{-63}$
97. $^-21 \div ^+3$ $_{-7}$
98. $^+20 + ^+5$ $_{+25}$
99. $^+20 - ^+5$ $_{+15}$
100. $^+20 \times ^-5$ $_{-100}$
101. $^+20 \times ^+5$ $_{+100}$
102. $^+20 \div ^+5$ $_{+4}$

Multiply across. Divide down. Working backward

Copy and complete these multiplication-division boxes.

103.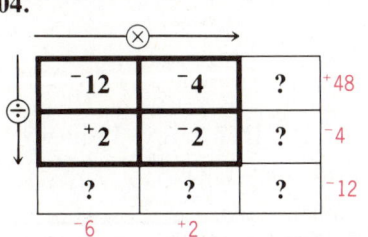

104.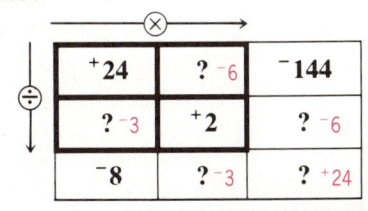

105.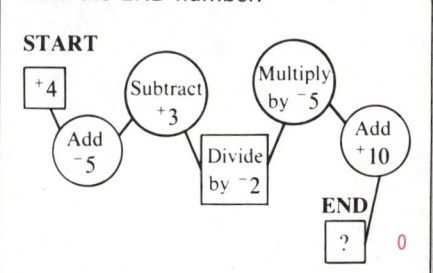

Integers **421**

Extra Practice
Page 502 Skill 63

Practice Worksheet
Workbook S348, Copymaster S348, or Duplicating Master S348

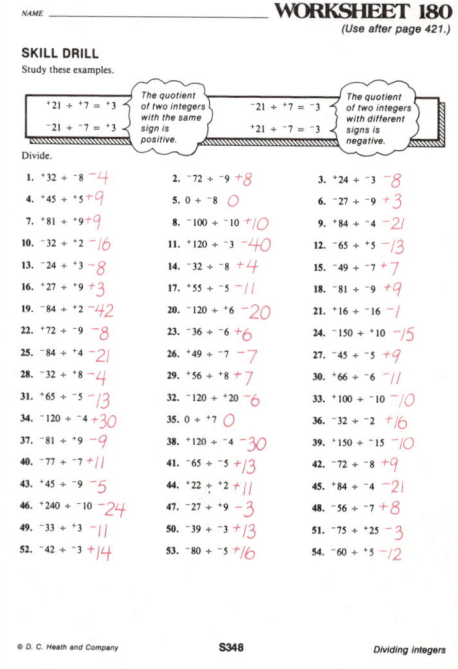

Challenge Problem
Find the END number.

START $^+4$ → Add $^-5$ → Subtract $^+3$ → Divide by $^-2$ → Multiply by $^-5$ → Add $^+10$ → END ? $_0$

Copymaster S463

421

Class Starter Quiz 165
on previous lesson

Give the quotient.

1. $^{+}18 \div {}^{-}3$ 2. $^{-}18 \div {}^{-}3$
   $^{-}6$ $^{+}6$
3. $^{+}18 \div {}^{+}3$ 4. $^{-}18 \div {}^{+}3$
   $^{+}6$ $^{-}6$

Simplify.

5. $^{+}6 + {}^{-}7$ 6. $^{-}8 + {}^{-}3$
   $^{-}1$ $^{-}11$
7. $^{+}8 - {}^{+}10$ 8. $^{-}4 - {}^{-}5$
   $^{-}2$ $^{+}1$
9. $^{-}6 \times {}^{-}7$ 10. $^{-}5 \times {}^{+}8$
   $^{+}42$ $^{-}40$

Copymaster S426

Lesson Objective
To recognize and use the basic properties of addition and multiplication

Problem-Solving Skills
Collecting, organizing, and analyzing data

Starting the Lesson
Discuss exercises 1 and 2 using the equations at the top of page 422. Go over the properties of addition and multiplication. You may want to have the students make up other numerical examples to illustrate each property.

422

Properties of addition and multiplication

The same number was written on all of the red cards before they were turned facedown. All the blue cards had the same number written on them before they were turned facedown.

1. If you turned these cards faceup, would you see true equations? yes

2. If you turned these cards faceup, would you see true equations? yes

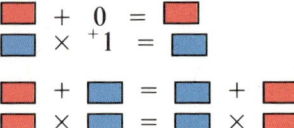

Here are some properties of addition and multiplication.

The Adding 0 Property

The sum of any number and 0 is that number.

$$^{+}12 + 0 = {}^{+}12$$

The Multiplying by 1 Property

The product of any number and 1 is that number.

$$^{+}12 \times {}^{+}1 = {}^{+}12$$

The Commutative Property of Addition

Changing the order of the addends does not change the sum.

$$^{+}8 + {}^{-}3 = {}^{-}3 + {}^{+}8$$

The Commutative Property of Multiplication

Changing the order of the factors does not change the product.

$$^{+}8 \times {}^{-}3 = {}^{-}3 \times {}^{+}8$$

The Associative Property of Addition

Changing the grouping of the addends does not change the sum.

$$(^{+}5 + {}^{+}2) + {}^{-}6 = {}^{+}5 + ({}^{+}2 + {}^{-}6)$$

The Associative Property of Multiplication

Changing the grouping of the factors does not change the product.

$$(^{+}5 \times {}^{+}2) \times {}^{-}6 = {}^{+}5 \times ({}^{+}2 \times {}^{-}6)$$

The Distributive Property

$$^{+}3 \times ({}^{+}8 + {}^{-}2) = ({}^{+}3 \times {}^{+}8) + ({}^{+}3 \times {}^{-}2)$$

422   Chapter 16

EXERCISES

Match each property with its example.

3. The commutative property of addition b
4. The associative property of addition f
5. The adding 0 property a
6. The commutative property of multiplication d
7. The associative property of multiplication g
8. The multiplying by 1 property e
9. The distributive property c

a. $^+9 + 0 = {}^+9$
b. $^-12 + {}^-10 = {}^-10 + {}^-12$
c. $^+4 \times ({}^-5 + {}^-3) = ({}^+4 \times {}^-5) + ({}^+4 \times {}^-3)$
d. $^+6 \times {}^+12 = {}^+12 \times {}^+6$
e. $^-15 \times {}^+1 = {}^-15$
f. $({}^+8 + {}^-10) + {}^+6 = {}^+8 + ({}^-10 + {}^+6)$
g. $({}^-7 \times {}^+5) \times {}^+1 = {}^-7 \times ({}^+5 \times {}^+1)$

Copy and complete these examples of the properties.

10. $({}^+9 + {}^+18) + {}^-27 = {}^+9 + ({}^+18 + \underline{?})$ 27
11. $^-10 \times {}^-4 = {}^-4 \times \underline{?}$ $^-10$
12. $^+24 \times {}^+1 = \underline{?}$ $^+24$
13. $({}^-5 \times {}^-6) \times {}^-2 = {}^-5 \times ({}^-6 + \underline{?})$ $^-2$
14. $^-17 + 0 = \underline{?}$ $^-17$
15. $^+20 + {}^-16 = {}^-16 + \underline{?}$ $^+20$
16. $^+5 \times ({}^-4 + {}^-6) = ({}^+5 \times {}^-4) + ({}^+5 \times \underline{?})$ $^-6$
17. $(\underline{?} + {}^-11) + {}^+10 = {}^-9 + ({}^-11 + {}^+10)$ $^-9$
18. $^+18 \times \underline{?} = {}^-6 \times {}^+18$ $^-6$
19. $\underline{?} + {}^+16 = {}^+16 + {}^+29$ $^+29$
20. $\underline{?} + 0 = {}^-30$ $^-30$
21. $^-6 \times ({}^+10 + {}^-4) = ({}^-6 \times {}^+10) + ({}^-6 \times \underline{?})$ $^-4$
22. $({}^-15 \times \underline{?}) \times {}^+10 = {}^-15 \times ({}^-2 \times {}^+10)$ $^-2$
23. $\underline{?} \times {}^+1 = {}^+45$ $^+45$

Project — Collecting, organizing, and analyzing data

24. Ask your classmates how much time they average sleeping each day. Have them round their times to the nearest half hour. Answers will vary.
25. Show the results on a graph.
26. Write some facts shown by your graph.

Integers **423**

Practice Worksheet

Workbook S349, Copymaster S349, or Duplicating Master S349

WORKSHEET 181
(Use after page 423.)

NAME _____

SKILL DRILL
Match each property with its example.

d 1. Adding 0 Property
f 2. Associative Property of Multiplication
a 3. Commutative Property of Addition
c 4. Multiplying by 1 Property
g 5. Associative Property of Addition
b 6. Distributive Property
e 7. Commutative Property of Multiplication

a. $^-4 + {}^+6 = {}^+6 + {}^-4$
b. $^+5 \times ({}^+7 + {}^+3) = ({}^+5 \times {}^+7) + ({}^+5 \times {}^+3)$
c. $^-8 \times {}^+1 = {}^-8$
d. $^+3 + 0 = {}^+3$
e. $^+9 \times {}^-2 = {}^-2 \times {}^+9$
f. $({}^+6 \times {}^-2) \times {}^-3 = {}^+6 \times ({}^-2 \times {}^-3)$
g. $({}^-7 + {}^+4) + {}^+3 = {}^-7 + ({}^+4 + {}^+3)$

j 8. Commutative Property of Addition
m 9. Associative Property of Addition
h 10. Commutative Property of Multiplication
k 11. Addition 0 Property
n 12. Distributive Property
i 13. Associative Property of Multiplication
l 14. Multiplying by 1 Property

h. $^-22 \times {}^-18 = {}^-18 \times {}^-22$
i. $({}^+5 \times {}^-9) \times {}^+6 = {}^+5 \times ({}^-9 \times {}^+6)$
j. $^+12 + {}^-31 = {}^-31 + {}^+12$
k. $^-41 + 0 = {}^-41$
l. $^+35 \times {}^+1 = {}^+35$
m. $({}^-8 + {}^+6) + {}^-9 = {}^-8 + ({}^+6 + {}^-9)$
n. $^+7 \times ({}^+3 + {}^-4) = ({}^+7 \times {}^+3) + ({}^+7 \times {}^-4)$

u 15. Associative Property of Multiplication
q 16. Distributive Property
s 17. Multiplying by 1 Property
p 18. Commutative Property of Addition
t 19. Commutative Property of Multiplication
o 20. Adding 0 Property
r 21. Associative Property of Addition

o. $^+112 + 0 = {}^+112$
p. $^-132 + {}^+110 = {}^+110 + {}^-132$
q. $^+5 \times ({}^-15 + {}^+9) = ({}^+5 \times {}^-15) + ({}^+5 \times {}^+9)$
r. $({}^+8 + {}^-8) + {}^+12 = {}^+8 + ({}^-8 + {}^+12)$
s. $^-115 \times {}^+1 = {}^-115$
t. $^-110 \times {}^+48 = {}^+48 \times {}^-110$
u. $({}^-6 \times {}^+11) \times {}^-2 = {}^-6 \times ({}^+11 \times {}^-2)$

© D.C. Heath and Company   S349   Properties of addition and multiplication

Challenge Problem

Draw the path from START to END.

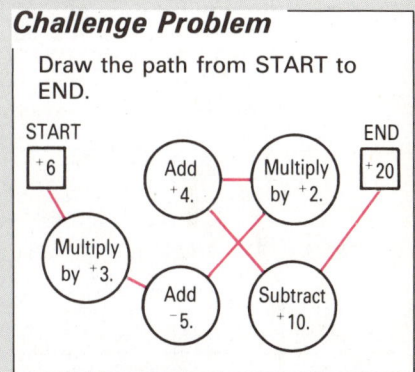

Copymaster S463

423

Class Starter Quiz 166
on previous lesson

Complete these examples of the properties.

1. $(^-6 + {}^+7) + {}^-3 =$
   $^-6 + ({}^+7 + \underline{})$ $^-3$
2. $^+8 \times {}^+1 = \underline{}$ $^+8$
3. $\underline{} + {}^-7 = {}^-7 + {}^-6$ $^-6$
4. $^+3 \times ({}^-2 + {}^-1) =$
   $({}^+3 \times {}^-2) + ({}^+3 \times \underline{})$ $^-1$
5. $\underline{} + 0 = {}^-15$ $^-15$
6. $^+3 \times \underline{} = {}^-5 \times {}^+3$ $^-5$

Copymaster S426

Lesson Objectives
To give the ordered pair for the graph of a point
To draw the graph of an ordered pair of integers

Problem-Solving Skills
Reading a graph
Interpreting directions

Starting the Lesson
Discuss questions 1–9, using the coordinate grid and the *Here's how*. The coordinate grid is on ■ *Visual Aid 68* (copymaster or transparency S165).

Answers for page 425. (34–57)

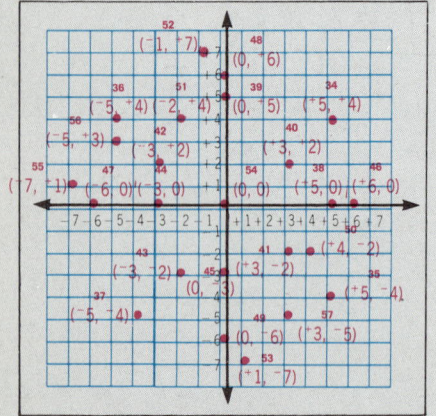

Graphing ordered pairs

Look at the picture on the right to answer the following questions.

1. What is the red number line called? Horizontal axis
2. What is the blue number line called? Vertical axis
3. What is the point called where the two axes intersect? Origin

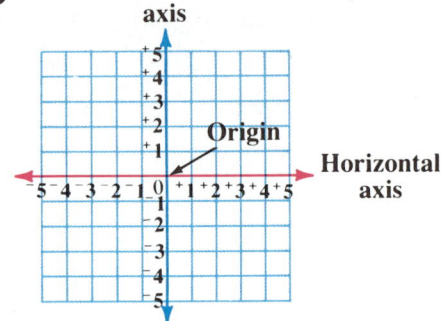

Here's how *to graph ordered pairs of integers.*

Example A. Graph the ordered pair $({}^+4, {}^-3)$

   Step 1. Start at the origin.
            Go 4 units to the right $({}^+4)$.

   Step 2. Now go 3 units down $({}^-3)$.

Example B. Graph the ordered pair $({}^-5, {}^+2)$

   Step 1. Start at the origin.
            Go 5 units to the left $({}^-5)$.

   Step 2. Now go 2 units up $({}^+2)$.

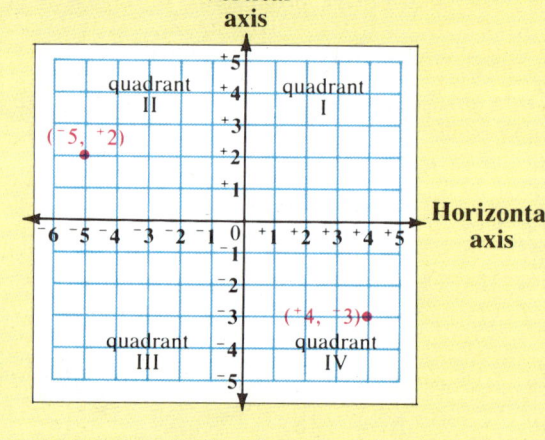

4. Look at the *Here's how*. Which ordered pair is graphed in quadrant II? $({}^-5, {}^+2)$

5. Which ordered pair is graphed in quadrant IV? $({}^+4, {}^-3)$

6. The ordered pair $({}^+4, {}^+5)$ would be graphed in quadrant [?]. I

7. The ordered pair $({}^-3, {}^-4)$ would be graphed in quadrant [?]. III

8. $(0, {}^+4)$ would be graphed on the [?] axis. Vertical

9. Where would the ordered pair $(0, 0)$ be graphed? At the origin

EXERCISES

Give the ordered pair for each point.

10. A (⁺3, ⁺3)
11. B (⁻3, ⁺4)
12. C (⁻6, ⁻4)
13. D (⁺5, ⁻7)
14. E (⁻4, ⁻7)
15. F (⁻5, ⁺7)
16. G (⁺6, ⁺7)
17. H (⁺3, ⁻4)
18. I (⁺5, 0)
19. J (0, ⁻7)
20. K (⁻6, 0)
21. L (0, ⁺6)
22. M (⁺6, ⁻3)
23. N (⁺5, ⁺5)
24. P (⁺1, 0)
25. Q (⁻4, ⁻4)
26. R (⁺2, ⁻2)
27. S (⁻6, ⁺4)
28. T (⁻7, ⁻7)
29. U (⁻7, ⁺6)
30. V (⁺3, ⁺6)
31. W (⁻7, ⁻2)
32. X (⁺1, ⁻5)
33. Y (⁻4, ⁺2)

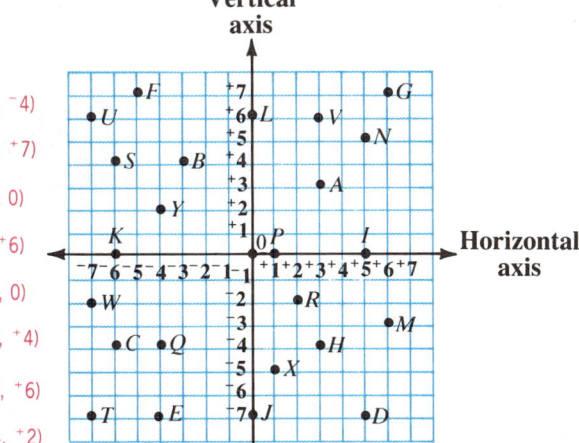

Graph these ordered pairs. Label each point with its ordered pair.

34. (⁺5, ⁺4)
35. (⁺5, ⁻4)
36. (⁻5, ⁺4)
37. (⁻5, ⁻4)
38. (⁺5, 0)
39. (0, ⁺5)
40. (⁺3, ⁺2)
41. (⁺3, ⁻2)
42. (⁻3, ⁺2)
43. (⁻3, ⁻2)
44. (⁻3, 0)
45. (0, ⁻3)
46. (⁺6, 0)
47. (⁻6, 0)
48. (0, ⁺6)
49. (0, ⁻6)
50. (⁺4, ⁻2)
51. (⁻2, ⁺4)
52. (⁻1, ⁺7)
53. (⁺1, ⁻7)
54. (0, 0)
55. (⁻7, ⁺1)
56. (⁻5, ⁺3)
57. (⁺3, ⁻5)

Triangle tricks

58. Copy the triangle on graph paper and give the ordered pairs for points A, B, and C. (⁺2, ⁺2) (⁺4, ⁺6) (⁺6, ⁺2)

59. Add ⁺3 to the second number of each ordered pair. Graph the new ordered pairs. Connect the points to make a new triangle.

60. Multiply the second number of each ordered pair (for points A, B, and C) by ⁻1. Graph the ordered pairs and draw the triangle.

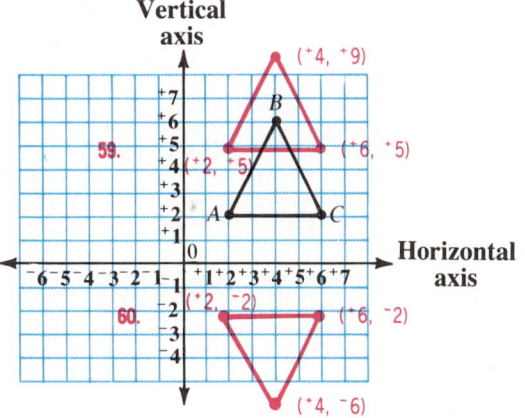

Integers **425**

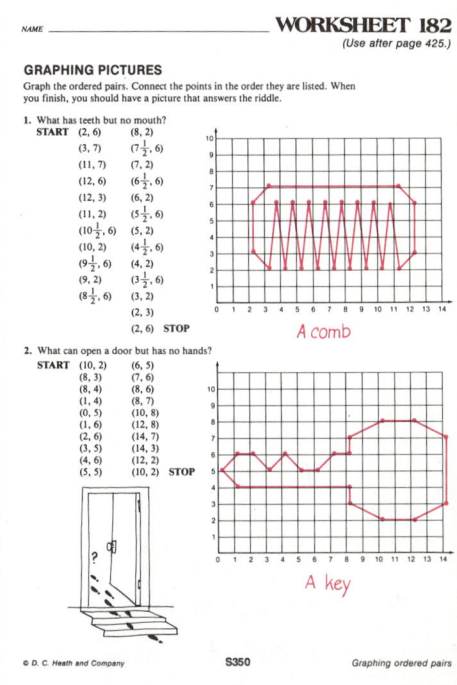

Practice Worksheet
Workbook S350, Copymaster S350, or Duplicating Master S350

Challenge Problems

Graph each geometric figure by graphing the ordered pairs and connecting the points in the order given.

1. Hexagon
   (0, 0) (1, 3) (4, 3) (5, 0) (4, ⁻3) (1, ⁻3)

2. Rectangle
   (⁻1, 3) (3, ⁻1) (1, ⁻3) (⁻3, 1)

3. Octagon
   (1, 4) (3, 2) (3, 0) (1, ⁻2) (⁻1, ⁻2) (⁻3, 0) (⁻3, 2) (⁻1, 4)

4. Trapezoid
   (2, 4) (2, 0) (⁻2, ⁻2) (⁻4, 1)

Copymaster S464

425

Class Starter Quiz 167
on previous lesson

Graph these ordered pairs. Label each point with the ordered pair.

1. (⁺3, ⁺2) 2. (⁻4, 0)
3. (⁻3, ⁻4) 4. (⁺2, ⁻3)
5. (0, ⁻4)

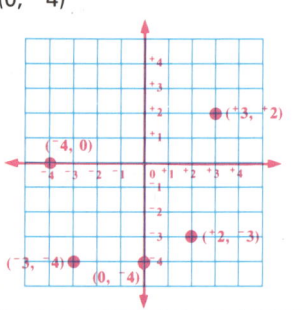

Copymaster S427

Lesson Objective
To solve problems using scoring rules

Problem-Solving Skill
Finding information in a display

Starting the Lesson
Before the students open their books, ask them these scoring-rule questions. In golf, how many strokes under or over par is:
- an eagle?
- a bogey?
- a birdie?

Then have them look at the scoring rules on page 426 to check their answers.

426

PROBLEM SOLVING *applications*

You often use scoring rules to solve problems about sporting events.

GOLF SCORING TERMS

Par: The number of strokes it should take a player to hit the ball from the tee into the hole.
Ace or hole-in-one: Hitting the ball into the hole with one stroke.
Birdie: One stroke under par (−1).
Eagle: Two strokes under par (−2).
Bogey: One stroke over par (+1). Double bogey: two strokes over par (+2). Triple bogey: three strokes over par (+3), etc.

Problem
Nancy Lopez-Knight made these scores on the first nine holes of a recent golf tournament: par, birdie, birdie, bogey, par, birdie, par, birdie, and eagle. How would her score be reported?

Here's how *to solve a golf scoring problem.*

First write the scores using (+), (−), and 0.

par	birdie	birdie	bogey	par	birdie	par	birdie	eagle
0	−1	−1	+1	0	−1	0	−1	−2

Find the total.

0 + (−1) + (−1) + (+1) + 0 + (−1) + 0 + (−1) + (−2) = −5

Answer: Nancy was 5 strokes under par for the nine holes.

426 Chapter 16

PROBLEMS

Solve. Use the scoring rules on page 426.

1. Last week Pat Bradley had these scores in a golf tournament: par, par, birdie, birdie, par, birdie, eagle, par, birdie, birdie, par, birdie, par, par, bogey, par, par, birdie. How would her score be reported?
   a. -3 (b.) -8 c. $+2$

2. During a recent tournament on the Senior's Golf Tour, Lee Trevino had a -3 total for the first 9 holes. His scores on the next 9 holes were double bogey, par, par, birdie, birdie, par, bogey, par, birdie. What was Lee's total?
   a. -2 (b.) -3 c. $+2$

Solve.

3. In football, a gain in yardage is positive ($+$) and loss of yardage is negative ($-$). During the Cowboy's last game, Tony Dorsett carried the ball 9 times. His runs were $+4$, $+7$, -2, $+17$, $+8$, -3, $+2$, 0, and -4. What was his total running yardage for the game? $+29$

4. In one of last year's games, Houston's quarterback completed passes for the following yardage: $+9$, $+12$, $+4$, -7, $+5$, -1, $+25$, and $+4$. What was his total passing yardage for that game? $+51$

5. | Football Scoring Rules | |
   | --- | --- |
   | Touchdown (TD) | 6 points |
   | Field Goal (FG) | 3 points |
   | Extra Point | 1 point |
   | Safety | 2 points |

   The Houston Oilers scored 4 touchdowns, 3 extra points, and 2 field goals in their last game. What was their total score?
   33 points

6. | Basketball Scoring Rules | |
   | --- | --- |
   | Field Goal (FG) | 2 points |
   | 3-point FG | 3 points |
   | Free Throw (FT) | 1 point |

   Larry Bird of the Boston Celtics scored 83 3-point FGs, 441 FTs, and 714 FGs during a recent season. What was his total number of points?   2,118

Get the point

7. After three quarters, the Cowboys were trailing the Redskins 20 to 17. Both teams scored 3 times in the fourth quarter, but the Cowboys rallied to win the game 30 to 28. How did the two teams score their points in the fourth quarter?
   Cowboys: 2 TDs, 1 Extra Point
   Redskins: 2 FGs, 1 Safety

Integers **427**

Class Starter Quiz 168
on previous lesson

Solve. Use the chart on page 426.

1. In golf, how would the score be reported for these scores on the first 9 holes: par, birdie, par, bogie, birdie, eagle, par, double bogie, par? ⁻1

2. The Bears had 9 running plays in the first quarter. The yardage was ⁺7, ⁻3, ⁺4, ⁺5, ⁻2, ⁺1, ⁻4, ⁺6, ⁺12. What was the total running yardage? ⁺26

Copymaster S427

Problem-Solving Skill
Identifying differences in two photographs

Skills Reviewed
Making conversions between metric units of length
Solving proportions
Changing percents to fractions
Solving a percent problem

Starting the Lesson

Problem Solving Have the students look at the photos on page 428. Ask them to identify the city in the photos. (New York) Then ask "Which photo do you think shows the real New York skyline?" (The top photo) Have students read the introductory paragraphs and then work on question 1.

Cumulative Skill Practice Write these five answers on the chalkboard:

9600   50 $\frac{1}{10}$ $\frac{1}{3}$ 4

Challenge the students to an answer hunt by saying, "Look at exercises 1–49 on page 429. Find the five exercises that have these answers. You have five minutes to find as many of the exercises as you can." (Exercises 5, 19, 20, 27, and 32)

428

Problem solving

COMPUTERS AND PHOTOGRAPHY

Look at the two photographs. The top photograph was altered to produce the bottom photograph. This "trick" photography was performed by an image processor, a special machine that uses a large computer.

The computer divides the photograph into tiny squares called pixels. (If you take a close look at a television screen, you can see large pixels that make up a TV image.) Each of the pixels in the photograph is given a number. The numbers can then be rearranged to create a new photograph.

Use the two photographs to answer these questions.

Decide when a calculator would be useful.

1. Can you find seven ways that the bottom photograph differs from the top photograph? (*Hint: Look at each building identified by a letter.*)

2. What is the area of the original photograph? Give the answer to the nearest square inch. 9 in.²

3. There are more than 100,000 pixels in each square inch of the photograph. Does the photograph have more or less than 1 million pixels? Less

4. To change each square inch of the photograph, the computer used 360,000 bytes (parts) of its memory.
   a. How many bytes of the computer's memory were used to change the entire photograph? 3,240,000
   b. How many bytes of the computer's memory would be used to change an 8-inch by 10-inch photograph? 28,800,000

428 Chapter 16

Cumulative Skill Practice

Complete. *(page 228)*

1. 49 cm = ? m   0.49
2. 58 mm = ? cm   5.8
3. 36 km = ? m   36,000
4. 2250 m = ? km   2.250
5. 9.6 km = ? m   9600
6. 7.4 cm = ? mm   74
7. 0.8 m = ? mm   800
8. 700 mm = ? m   0.7
9. 5.3 m = ? cm   530

Solve. *(page 262)*

10. $\frac{n}{8} = \frac{3}{4}$ 6
11. $\frac{9}{n} = \frac{8}{5}$ $5\frac{5}{8}$
12. $\frac{16}{5} = \frac{n}{3}$ $9\frac{3}{5}$
13. $\frac{4}{12} = \frac{3}{n}$ 9
14. $\frac{5}{n} = \frac{10}{6}$ 3
15. $\frac{7}{n} = \frac{3}{5}$ $11\frac{2}{3}$
16. $\frac{n}{4} = \frac{13}{10}$ $5\frac{1}{5}$
17. $\frac{15}{8} = \frac{n}{6}$ $11\frac{1}{4}$
18. $\frac{2}{3} = \frac{11}{n}$ $16\frac{1}{2}$
19. $\frac{3}{5} = \frac{30}{n}$ 50

Change to a fraction, whole number, or mixed number in simplest form. *(page 282)*

20. 10% $\frac{1}{10}$
21. 20% $\frac{1}{5}$
22. 25% $\frac{1}{4}$
23. 40% $\frac{2}{5}$
24. 150% $1\frac{1}{2}$
25. 275% $2\frac{3}{4}$
26. $12\frac{1}{2}$% $\frac{1}{8}$
27. $33\frac{1}{3}$% $\frac{1}{3}$
28. $37\frac{1}{2}$% $\frac{3}{8}$
29. $87\frac{1}{2}$% $\frac{7}{8}$
30. $166\frac{2}{3}$% $1\frac{2}{3}$
31. $66\frac{2}{3}$% $\frac{2}{3}$

Solve. *(page 294)*

32. 10% of 40 = n   4
33. 25% of 36 = n   9
34. 12% of 44 = n   55
35. 9% of 79 = n   7.11
36. 8.5% of 23 = n   1.955
37. $66\frac{2}{3}$% of 144 = n   96

Solve. *(page 296)*

38. 20% of n = 19   95
39. 25% of n = 23   92
40. 50% of n = 37   74
41. 75% of n = 51   68
42. 125% of n = 75   60
43. 150% of n = 81   54

Solve. Round each answer to the nearest tenth. *(page 296)*

44. 18% of n = 37   205.6
45. 21% of n = 45   214.3
46. 35% of n = 29   82.9
47. 42% of n = 53   126.2
48. 12% of n = 25   208.3
49. 15% of n = 41   273.3

MIXED PRACTICE

Complete.

50. $\frac{1}{2} + \frac{5}{8} = $? $1\frac{1}{8}$
51. $\frac{2}{3} - \frac{1}{4} = $? $\frac{5}{12}$
52. $3 \times \frac{5}{6} = $? $2\frac{1}{2}$
53. $5 + 1\frac{2}{3} = $? $6\frac{2}{3}$
54. $7 - 4\frac{1}{3} = $? $2\frac{2}{3}$
55. $\frac{5}{8} \div 4 = $? $\frac{5}{32}$
56. $\frac{3}{8} \times \frac{4}{5} = $? $\frac{3}{10}$
57. $2\frac{3}{8} \times 2 = $? $4\frac{3}{4}$
58. $5\frac{1}{3} - 2\frac{3}{4} = $? $2\frac{7}{12}$
59. $6 \div 1\frac{3}{4} = $? $3\frac{3}{7}$
60. $\frac{9}{10} - \frac{3}{8} = $? $\frac{21}{40}$
61. $2\frac{5}{6} + 4\frac{1}{2} = $? $7\frac{1}{3}$
62. $2\frac{1}{4} \times 2\frac{1}{4} = $? $5\frac{1}{16}$
63. $\frac{11}{12} \div \frac{2}{3} = $? $1\frac{3}{8}$
64. $4\frac{5}{6} \div 2\frac{1}{2} = $? $1\frac{14}{15}$

Integers **429**

Chapter REVIEW

Here are scrambled answers for the review exercises:

−3	2	absolute	down	nonzero	positive
−1	+3	add	less	opposite	right
0	+4	different	negative	origin	same

1. As you move to the [?] on this number line, the numbers get larger. (page 410)

2. The number line shows that negative 3 is [?] than positive one. (page 410)

3. The number of units that an integer is from 0 is called the [?] value of the integer. $|{-2}| = $ [?] (page 411)

4. The picture shows that $^-2 + {^+1} = $ [?]. If the sum of two numbers is 0, then one number is the [?] of the other. The opposite of $^+3$ is [?]. (page 412)

5. The picture shows that $^+2 - {^-1} = $ [?]. To subtract an integer, you [?] the opposite of the integer. (pages 414, 415)

6. The picture shows that $^-2 \times {^-2} = $ [?]. (page 418)

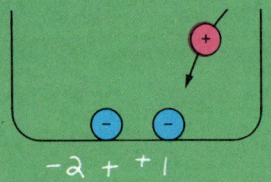 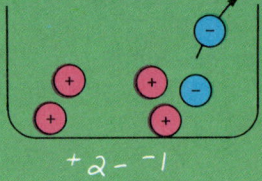 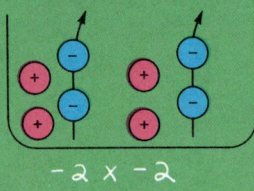

7. The product of two integers with the same signs is [?]. The product of two integers with different signs is [?]. The product of any integer and 0 is [?]. (page 419)

8. The quotient of two integers with the [?] signs is positive. The quotient of two integers with [?] signs is negative. The quotient of 0 divided by any [?] integer is 0. (page 420)

9. To graph the ordered pair ($^+3$, $^-2$) you would start at the [?] and go 3 units to the right ($^+3$). Then you would go 2 units [?] ($^-2$). (page 424)

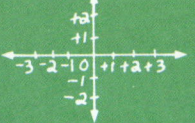

1. right   2. less   3. absolute, 2   4. $^-1$, opposite, $^-3$
5. $^+3$, add   6. $^+4$ 7. positive, negative, 0
8. same, different, nonzero   9. origin, down

Chapter TEST

< or >? *(page 410)*

1. $^+4 < ^+5$
2. $^+3 > ^-6$
3. $^-6 < ^+2$
4. $^-8 < ^-1$
5. $^-6 < ^+3$
6. $^+20 > ^-23$
7. $^-25 < ^-22$
8. $^-26 < ^+21$
9. $^+24 > ^-24$
10. $^-2 < 0$

Give each absolute value. *(page 411)*

11. $|^+6| \; 6$
12. $|^-3| \; 3$
13. $|0| \; 0$
14. $|^+8| \; 8$
15. $|^-11| \; 11$

Give each sum. *(page 412)*

16. $^+6 + ^+2 \; ^+8$
17. $^+6 + ^-2 \; ^+4$
18. $0 + ^-2 \; ^-2$
19. $^-4 + ^-8 \; ^-12$
20. $^-4 + ^+8 \; ^+4$
21. $^-7 + ^-10 \; ^-17$
22. $^+12 + ^-12 \; 0$
23. $^+15 + ^+11 \; ^+26$
24. $^-18 + ^+16 \; ^-2$
25. $^-18 + ^-16 \; ^-34$

Give each difference. *(pages 414, 415)*

26. $^+4 - ^+2 \; ^+2$
27. $^+4 - ^-2 \; ^+6$
28. $^-5 - ^+8 \; ^-13$
29. $^+5 - ^+8 \; ^-3$
30. $^-8 - 0 \; ^-8$
31. $^-12 - ^+10 \; ^-22$
32. $^-13 - ^-13 \; 0$
33. $^+18 - ^-11 \; ^+29$
34. $^+15 - ^+19 \; ^-4$
35. $^+21 - ^-3 \; ^+24$

Give each product. *(pages 418, 419)*

36. $^+3 \times ^+5 \; ^+15$
37. $^+3 \times ^-5 \; ^-15$
38. $^-7 \times ^-4 \; ^+28$
39. $^+7 \times ^+4 \; ^+28$
40. $^+7 \times ^-1 \; ^-7$
41. $^-11 \times ^+8 \; ^-88$
42. $^-12 \times ^-12 \; ^+144$
43. $^+15 \times ^-6 \; ^-90$
44. $^+16 \times ^+10 \; ^+160$
45. $^-21 \times ^+10 \; ^-210$

Give each quotient. *(page 420)*

46. $^+12 \div ^+3 \; ^+4$
47. $^+12 \div ^-3 \; ^-4$
48. $^-12 \div ^+3 \; ^-4$
49. $^+8 \div ^+1 \; ^+8$
50. $^+16 \div ^+2 \; ^+8$
51. $^-32 \div ^-4 \; ^+8$
52. $^-36 \div ^+9 \; ^-4$
53. $^+45 \div ^+5 \; ^+9$
54. $^+54 \div ^-9 \; ^-6$
55. $^-63 \div ^-9 \; ^+7$

Give the ordered pair for each point. *(page 424)*

56. $A \; (^-3, ^+1)$
57. $B \; (^+2, ^+2)$
58. $C \; (^+7, ^+1)$
59. $D \; (0, ^-4)$
60. $E \; (^+5, 0)$
61. $F \; (^-6, ^-4)$
62. $G \; (^+5, ^-4)$
63. $H \; (^-8, 0)$

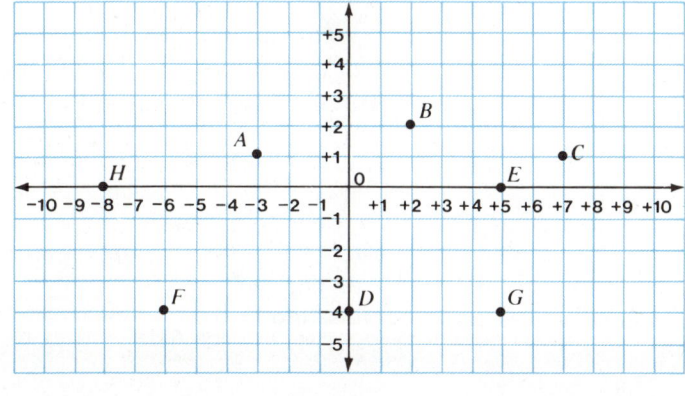

Cumulative Test
(Chapters 1–16)

Use Copymaster S109 to provide the students with an answer sheet in standardized test format.

Answers for Cumulative Test, Chapters 1–16

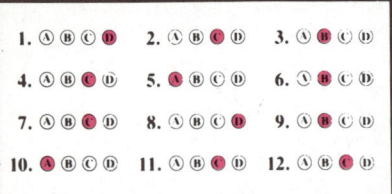

The table below correlates test items with student text pages.

Test Item	Page Taught	Skill Practice
1	72	p. 417, exercises 1–9
2	100	p. 417, exercises 10–18
3	166	p. 417, exercises 19–32
4	182	p. 417, exercises 33–38
5	206	p. 417, exercises 39–46
6	228	p. 429, exercises 1–9
7	262	p. 429, exercises 10–19
8	282	p. 429, exercises 20–31
9	294	p. 429, exercises 32–37
10	296	p. 429, exercises 38–43
11	368	
12	414	

432

Cumulative TEST — Standardized Format

Choose the correct letter.

1. Give the product.

2.408 × 100

A. 24.08
B. 0.02408
C. 2408
D. none of these

2. Give the quotient.

78.95 ÷ 1000

A. 78950
B. 0.7895
C. 0.07895
D. none of these

3. Change to a decimal rounded to the nearest hundredth.

$\frac{5}{6} = ?$

A. 0.17
B. 0.83
C. 0.67
D. none of these

4. Add. $3\frac{1}{2}$
         $+2\frac{2}{3}$

A. $5\frac{3}{5}$ B. $5\frac{1}{6}$
C. $6\frac{1}{6}$ D. none of these

5. Give the product.

$2\frac{2}{3} \times 1\frac{3}{4}$

A. $4\frac{2}{3}$ B. $\frac{21}{32}$
C. $1\frac{11}{32}$ D. none of these

6. 345 cm = ? m

A. 34.5
B. 3.45
C. 345
D. none of these

7. Solve.

$\frac{n}{3} = \frac{11}{4}$

A. 3.5
B. 14.7
C. 8.25
D. none of these

8. Change to a fraction.

$37\frac{1}{2}\% = ?$

A. $\frac{1}{3}$
B. $\frac{5}{8}$
C. $\frac{2}{5}$
D. none of these

9. Solve.

25% of 24.5 = n

A. 98
B. 6.125
C. 61.25
D. none of these

10. Solve.

40% of n = 36

A. 90
B. 14.4
C. 9.0
D. none of these

11. Find the surface area.

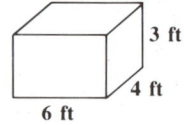

A. 54 ft²   B. 72 ft²
C. 108 ft²   D. none of these

12. Give the difference.

⁺12 − ⁻10

A. ⁺2
B. ⁻2
C. ⁺22
D. none of these

432   Chapter 16

Algebra

Chapter 17
Algebra

Resources

- **Class Starter Quizzes 169-179** (Copymasters S427-S430)
- **Visual Aids 69-72** (Copymasters or Transparencies S166-S168)
- **Worksheets 185-196** (Copymasters, Duplicating Masters, or Workbook pages S353-S364)
- **Challenge Problems** for pages 435, 437, 443, 445, 451, 453, 455, 457 (Copymasters S464-S466)
- **Projects** for pages 439, 447 (Copymaster S487)
- **Tests** (Copymasters or Duplicating Masters S65-S68, S81-S91)

Lesson Objective

To write algebraic expressions for word phrases

Problem-Solving Skills

Interpreting and checking information
Using logical reasoning

Starting the Lesson

Have the students read the age clues at the top of the page. Ask them who is the oldest (Craig) and who is the youngest (Chris). Then go over the phrases for the expressions on page 434.

Writing expressions

Ann   David   Chris   Craig   Beth

Who's the oldest?   Who's the youngest?

The letter *n* is a variable. It represents Ann's age in years.

Variables, numbers, and operation signs can be combined to form **mathematical expressions.** Look at the cards below. First a mathematical expression is given in red. Then several different word expressions are given for the mathematical expression.

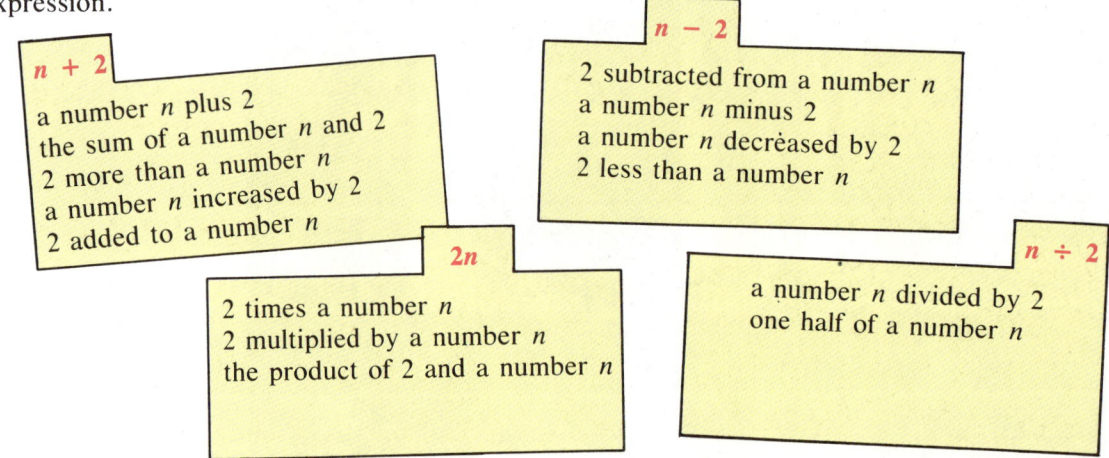

Other examples:

Word Expression	Mathematical Expression
a number s increased by 6	$s + 6$
9 less than a number h	$h - 9$
4 times a number d	$4d$
one third of a number r	$r \div 3$ or $\frac{r}{3}$ or $\frac{1}{3}r$
5 times a number x, plus 4	$5x + 4$

Chapter 17

EXERCISES

Write a mathematical expression for each word expression.

1. 5 more than a number n $n + 5$
2. 10 more than a number r $r + 10$
3. 11 times a number t $11t$
4. a number x minus 6 $x - 6$
5. the sum of a number y and 4 $y + 4$
6. a number b divided by 4 $b \div 4$ ($\frac{b}{4}, \frac{1}{4}b$)
7. 15 times a number c $15c$
8. a number t decreased by 4 $t - 4$
9. a number d increased by 6 $d + 6$
10. 15 less than a number x $x - 15$
11. 40 divided by a number s $40 \div s$ ($\frac{40}{s}, \frac{1}{40}s$)
12. the product of 2 and a number n $2n$
13. 3 times a number m, plus 2 $3m + 2$
14. 4 times a number c, minus 6 $4c - 6$
15. 6 multiplied by a number r, plus 8 $6r + 8$
16. a number e divided by 5, minus 3 $(e \div 5) - 3$ ($\frac{e}{5} - 3, \frac{1}{5}e - 3$)

Let n be the number of letters you wrote last year. Write a mathematical expression for the number of letters that is

17. 2 more than your number of letters. $n + 2$
18. 5 letters less than your number of letters. $n - 5$
19. one third the number of your letters. $\frac{1}{3}n$
20. your number of letters decreased by 8. $n - 8$
21. one fourth the number of your letters. $\frac{1}{4}n$
22. 4 less than your number of letters. $n - 4$
23. your number of letters increased by 3 letters. $n + 3$
24. your number of letters divided by 9. $n \div 9$ ($\frac{1}{9}n, \frac{n}{9}$)
25. 4 times the number of your letters, plus 5 more letters. $4n + 5$
26. 3 times the number of your letters, plus 2 more letters. $3n + 2$

Who is it? Logical reasoning

27. One of these people is not telling his/her true age. Who is it?   Beth
    Hint: Use the clues at the top of page 434.

I'm 22 years old. — Ann
I'm 24 years old. — David
I'm 11 years old. — Chris
I'm 44 years old. — Craig
I'm 18 years old. — Beth

Practice Worksheet

Workbook S353, Copymaster S353, or Duplicating Master S353

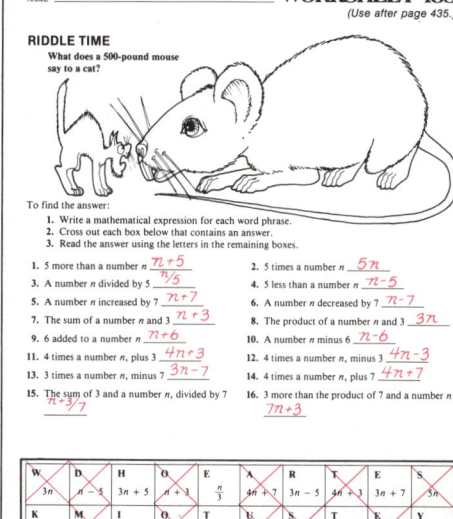

Challenge Problem

Follow the instructions for this number trick.

- Pick any number.
- Add 5.
- Multiply by 2.
- Subtract 4.
- Divide by 2.
- Subtract the number you picked.

Your answer is always 3.

Copymaster S464

Class Starter Quiz 169
on previous lesson

Write a mathematical expression for each word expression.

1. 3 more than a number *n*
   $n + 3$
2. a number *n* decreased by 10
   $n - 10$
3. the product of a number *n* and 7 $7n$
4. the sum of a number *n* and 5
   $n + 5$
5. a number *n* divided by 3, minus 2 $n \div 3 - 2$

Copymaster S427

Lesson Objective
To substitute numbers for variables and then evaluate the resulting expression

Problem-Solving Skills
Finding information in a display
Comparing prices
Choosing the correct operation
Using a guess-and-check strategy

Starting the Lesson
Group Activity Before going over exercises 1–4 and the *Here's how*, ask the students, working in small groups, which rental plan they would choose if they planned to rent a car for 3 days and drive it 500 miles.

Exercise Note
Point out to the students that from now on in the textbook, the raised plus sign for positive numbers will be omitted.

436

Evaluating expressions

WHAT DO YOU THINK?

Which is the cheapest plan?

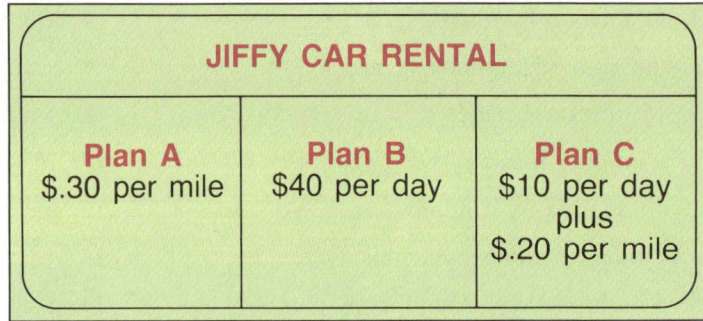

JIFFY CAR RENTAL

Plan A	Plan B	Plan C
$.30 per mile	$40 per day	$10 per day plus $.20 per mile

1. Let *d* be the number of days and *m* be the number of miles. Which rental plan would cost $10d + .20m$ dollars? C
2. Which plan would cost $.30m$ dollars? A
3. What expression would you use for the cost of Plan B? $40d$

Here's how *to evaluate mathematical expressions.*

What will it cost to rent a car for 3 days? I plan to drive it 500 miles.

To find the cost (in dollars) for each plan, substitute 3 for *d* and/or 500 for *m* in the expression and simplify.

Plan A. $.30m$
$.30 \times 500 = 150$ So Plan A would cost $150

Plan B. $40d$
$40 \times 3 = 120$ So Plan B would cost $120

Plan C. $10d + .20m$
$10 \times 3 + .20 \times 500$
$30 + 100 = 130$ So Plan C would cost $130

4. Look at the *Here's how*. Which is the cheapest plan for renting a car for 3 days and 500 miles? B

Chapter 17

EXERCISES

Evaluate each expression for n = 12.
Here are scrambled answers for the next row of exercises: 47 31 38 22 6

5. $n + 10$ 22
6. $3n - 5$ 31
7. $4n - 1$ 47
8. $n \div 2$ 6
9. $3n + 2$ 38

> From now on, let's agree not to write the raised plus sign when writing a positive number.

10. $2n + 1$ 25
11. $4 + n$ 16
12. $n - 11$ 1
13. $n + 12$ 24
14. $8n$ 96
15. $2n - 20$ 4
16. $n \div 12$ 1
17. $12n + 6$ 150

Evaluate each expression for r = 6 and s = 5.
Here are scrambled answers for the next row of exercises: 28 27 30 11 1

18. $r + s$ 11
19. $r - s$ 1
20. rs $\boxed{r \times s}$ 30
21. $2r + 3s$ 27
22. $rs - 2$ 28
23. $3r$ 18
24. $4s$ 20
25. $r \div 3$ 2
26. $\dfrac{10}{s}$ 2
27. $\dfrac{15}{s}$ 3
28. $r + s + 5$ 16
29. $2rs$ 60
30. $3r - 2s$ 8
31. $12s - r$ 54
32. $14r - 3$ 81

Evaluate each expression for a = 10 and b = ⁻2.
Here are scrambled answers for the next row of exercises: ⁻6 8 ⁻20 28 12

33. $a + b$ 8
34. ab ⁻20
35. $3a + b$ 28
36. $3b$ ⁻6
37. $a - b$ 12
38. $4a$ 40
39. $5a + 10$ 60
40. $b + a$ 8
41. $\dfrac{20}{a}$ 2
42. $ab - 5$ ⁻25
43. $10 \div b$ ⁻5
44. $ab + 20$ 0
45. $4a + 2b$ 36
46. $3ab$ ⁻60
47. $4ab - 8$ ⁻88

Solve. Use the car rental plans on page 436.

48. Using Plan A, how much would it cost to rent a car for 600 miles? $180
49. Using Plan B, how much would it cost to rent a car for 6 days? $240
50. Using Plan C, how much would it cost to rent a car for 4 days and drive it 1000 miles? $240
51. You want to rent a car for 5 days and drive it 2000 miles. Which is the cheapest plan for you? Plan B

Name the rental plan Guess and check

52. Which rental plan did I use? It cost me $80 to rent a car for 2 days and 100 miles. Plan B
53. Which rental plan did I use? I paid $100 to rent a car for 4 days. I drove it 300 miles. Plan C

Practice Worksheet
Workbook S354, Copymaster S354, or Duplicating Master S354

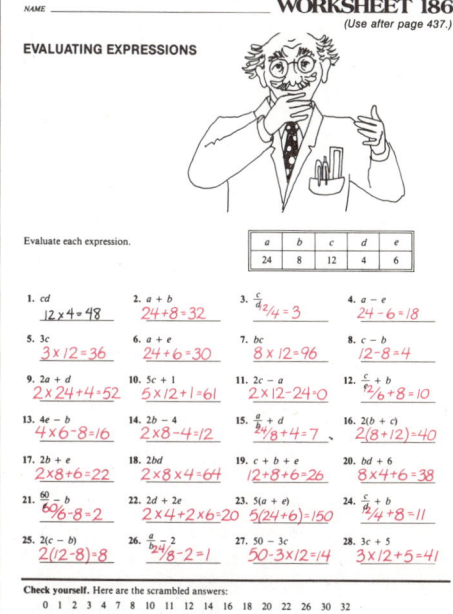

Challenge Problems

The formula below gives the distance, d, that a car traveling 50 miles per hour would go in a certain time, t.

$$d = 50t$$

Have the students use the formula to find how far a car would go in these times:

1. 3 hours   150 miles
2. 2.5 hours   125 miles
3. 0.4 hour   20 miles
4. 2 days driving 7 hours each day   700 miles

Copymaster S465

Class Starter Quiz 170
on previous lesson

Evaluate each expression for $r = 4$ and $s = 10$.

1. $r + s$	14	2. $rs - 5$	35
3. $7s$	70	4. $r \div 2$	2
5. $3r + s$	22	6. $4rs - 150$	10
7. $\dfrac{s + 2}{r}$	3	8. $5r \div s$	2

Copymaster S428

Lesson Objective
To solve addition equations

Problem-Solving Skill
Using equations to solve problems

Starting the Lesson
Use of Concrete Materials Use questions 1 and 2 and the *Here's how* to introduce solving addition equations by subtracting the same number from both sides. The balance-scale model is on ■ **Visual Aid 69** (copymaster or transparency S166).

Go over the examples in exercise 3 on the chalkboard to show the students the steps that they are expected to use. Be sure the students understand that they are learning a method that they will later use to solve more difficult equations. It is important for them to show their work.

Solving addition equations
A BALANCING ACT!

Each of the marbles weighs the same.
1. The marbles in the red box plus the 4 extra marbles weigh the same as how many marbles? 150

If we let m be the number of marbles in the red box, we can write the equation
$$m + 4 = 150$$
To **solve the equation** is to find the number that we can substitute for m to make the equation true.

Here's how *to solve an addition equation.*

Equation:	$m + 4 = 150$
Subtract 4 from both sides of the equation:	$m + 4 - 4 = 150 - 4$
Simplify both sides:	$m = 146$

Check the solution by substituting 146 for m in the equation $m + 4 = 150$: $146 + 4 = 150$ It checks!

2. Look at the *Here's how*. To find m, what number was subtracted from both sides of the equation? How many marbles are in the red box? 4, 146

3. Check these examples. Has each equation been solved correctly? Yes

  a. $a + 7 = 20$
  $a + 7 - 7 = 20 - 7$
  $a = 13$

  b. $b + 8 = 2$
  $b + 8 - 8 = 2 - 8$
  $b = {}^-6$

  c. $c + 10 = {}^-6$
  $c + 10 - 10 = {}^-6 - 10$
  $c = {}^-16$

EXERCISES

Solve and check.
Here are scrambled answers for the next 2 rows of exercises:
2   13   ⁻14   25   ⁻1   22   ⁻10   ⁻21

4. $n + 16 = 38$ 22
5. $r + 10 = {}^-4$ ⁻14
6. $x + {}^-8 = 5$ 13
7. $t + {}^-2 = {}^-3$ ⁻1

Hint: Subtract 16 from both sides. | Hint: Subtract 10 from both sides. | Hint: Subtract ⁻8 from both sides. | Hint: Subtract ⁻2 from both sides.

8. $n + 28 = 30$ 2
9. $y + {}^-20 = 5$ 25
10. $x + {}^-15 = {}^-25$ ⁻10
11. $z + 13 = {}^-8$ ⁻21
12. $m + 9 = 19$ 10
13. $r + {}^-6 = {}^-4$ 2
14. $w + 7 = {}^-9$ ⁻16
15. $m + 4 = 26$ 22
16. $y + {}^-5 = {}^-1$ 4
17. $d + {}^-15 = {}^-10$ 5
18. $x + 18 = 25$ 7
19. $y + {}^-5 = {}^-8$ ⁻3
20. $27 + x = 68$ 41
21. $42 + p = 40$ ⁻2
22. $19 + s = 11$ ⁻8
23. $g + 8 = {}^-7$ ⁻15
24. $x + 4 = {}^-6$ ⁻10
25. $m + {}^-11 = 40$ 51
26. $t + 5 = {}^-7$ ⁻12
27. $h + {}^-3 = {}^-3$ 0
28. $g + {}^-8 = {}^-10$ ⁻2
29. $s + {}^-3 = {}^-4$ ⁻1
30. $r + 15 = {}^-50$ ⁻65
31. $b + 4 = 10$ 6
32. $m + {}^-8 = 50$ 58
33. $x + 16 = 14$ ⁻2
34. $g + {}^-20 = {}^-21$ ⁻1
35. $n + 6 = 6$ 0
36. $t + 97 = 80$ ⁻17
37. $c + 46 = 58$ 12
38. $x + 24 = 24$ 0
39. $f + 21 = {}^-2$ ⁻23
40. $a + 62 = 75$ 13
41. $y + {}^-10 = {}^-10$ 0
42. $k + 3 = {}^-32$ ⁻35
43. $c + {}^-4 = 5$ 9
44. $r + 15 = {}^-25$ ⁻40
45. $k + 20 = 0$ ⁻20
46. $c + {}^-46 = 0$ 46
47. $p + 5 = {}^-8$ ⁻13
48. $x + 16 = 9$ ⁻7
49. $f + 17 = 17$ 0
50. $d + {}^-18 = 4$ 22
51. $k + 4 = 4$ 0
52. $t + 12 = 15$ 3
53. $r + {}^-9 = {}^-10$ ⁻1
54. $6 + k = {}^-5$ ⁻11
55. $a + 18 = {}^-15$ ⁻33
56. $d + 0 = 9$ 9
57. $w + 7 = 23$ 16
58. $n + {}^-4 = {}^-9$ ⁻5
59. $t + {}^-14 = 7$ 21
60. $m + 6 = {}^-8$ ⁻14
61. $q + {}^-9 = {}^-1$ 8
62. $b + {}^-32 = 0$ 32
63. $e + 10 = 3$ ⁻7
64. $v + {}^-15 = 7$ 22
65. $n + 18 = 18$ 0
66. $s + 8 = {}^-2$ ⁻10
67. $r + {}^-2 = {}^-21$ ⁻19
68. $a + 24 = 0$ ⁻24
69. $s + 9 = {}^-17$ ⁻26
70. $p + 14 = {}^-5$ ⁻19
71. $m + {}^-8 = 2$ 10
72. $w + {}^-3 = {}^-9$ ⁻6
73. $6 + y = {}^-2$ ⁻8
74. $y + {}^-10 = 10$ 20
75. $s + 14 = {}^-13$ ⁻27
76. $t + 5 = 1$ ⁻4
77. $c + 0 = {}^-7$ ⁻7
78. $k + {}^-1 = 6$ 7
79. ${}^-20 + d = {}^-1$ 19

Balance it!

Find the weight n that is needed to make each scale balance.

80.
  19

81.
  21

Practice Worksheet

Workbook S355, Copymaster S355, or Duplicating Master S355

Project

Writing equations

How many addition equations can you write that have a solution of 10? Here are some examples:

$n + 9 = 19$

$n + 3\frac{1}{2} = 13\frac{1}{2}$

Copymaster S487

Class Starter Quiz 171
on previous lesson

Solve and check.

1. $n + 15 = 22$ 7
2. $r + 10 = 12$ 2
3. $t + 10 = 44$ 34
4. $x + 4 = 0$ ⁻4
5. $m + 9 = 2$ ⁻7
6. $m + 25 = 10$ ⁻15
7. $s + 3 = 2$ ⁻1
8. $n + 15 = {}^-10$ ⁻25

Copymaster S428

Lesson Objective
To solve subtraction equations

Problem-Solving Skill
Choosing the correct equation

Starting the Lesson
Discuss questions 1–3, using the sale ad and the *Here's how* example. Then go over the examples in exercise 4. Be sure the students understand the technique of adding the same number to both sides of the equation.

Solving subtraction equations

1. What is the sale price of the cross-country ski package? $109

2. Is the regular price of the ski package more or less than $109? More

If we let r be the regular price (in dollars), we can write the equation

$$r - 55 = 109$$

and solve the equation to find the regular price.

X-COUNTRY SKI PACKAGE
$55 OFF THE REGULAR PRICE
NOW $109

Here's how *to solve a subtraction equation.*

Equation:	$r - 55 = 109$
Add 55 to both sides of the equation:	$r - 55 + 55 = 109 + 55$
Simplify both sides:	$r = 164$
Check:	$164 - 55 = 109$ It checks!

3. Look at the *Here's how*. To find r, what number was added to both sides of the equation? What is the regular price of the ski package? 55, $164

4. Check these examples. Has each equation been solved correctly? Yes

   a. $\quad a - 5 = 12$
        $a - 5 + 5 = 12 + 5$
             $a = 17$

   b. $\quad b - 6 = {}^-2$
        $b - 6 + 6 = {}^-2 + 6$
            $b = 4$

   c. $\quad c - {}^-4 = {}^-1$
        $c - {}^-4 + {}^-4 = {}^-1 + {}^-4$
             $c = {}^-5$

Chapter 17

EXERCISES

Solve and check. Here are scrambled answers for the next two rows of exercises: 3 ⁻1 ⁻5 40 37 70 41 63

5. $n - 15 = 25$ 40
6. $r - 46 = {}^-5$ 41
7. $n - 10 = 60$ 70
8. $p - {}^-3 = 6$ 3

Hint: Add 15 to both sides.
Hint: Add 46 to both sides.
Hint: Add 10 to both sides.
Hint: Add ⁻3 to both sides.

9. $s - 19 = 18$ 37
10. $g - 5 = {}^-6$ ⁻1
11. $b - 35 = {}^-40$ ⁻5
12. $n - 25 = 38$ 63
13. $m - 23 = 49$ 72
14. $n - 42 = 76$ 118
15. $a - 65 = 61$ 126
16. $r - 24 = {}^-8$ 16
17. $x - 6 = 2$ 8
18. $d - 9 = {}^-1$ 8
19. $y - 20 = 7$ 27
20. $k - 125 = 0$ 125
21. $x - 18 = {}^-2$ 16
22. $b - 33 = 33$ 66
23. $s - 5 = {}^-5$ 0
24. $n - 2 = 5$ 7
25. $c - 61 = 62$ 123
26. $x - 2 = {}^-8$ ⁻6
27. $y - 17 = 20$ 37
28. $t - 14 = 2$ 16
29. $y - 17 = 14$ 31
30. $b - 3 = {}^-4$ ⁻1
31. $y - 20 = 1$ 21
32. $n - 4 = {}^-1$ 3
33. $p - 8 = {}^-3$ 5
34. $m - 8 = {}^-2$ 6
35. $t - 1 = 5$ 6
36. $x - 45 = 54$ 99
37. $c - {}^-7 = {}^-3$ ⁻10
38. $m - 100 = 0$ 100
39. $a - {}^-2 = 9$ 7
40. $y - 6 = {}^-11$ ⁻5
41. $c - {}^-4 = 10$ 6
42. $x - 12 = 6$ 18
43. $e - 9 = 0$ 9
44. $t - 16 = {}^-8$ 8
45. $n - {}^-7 = 16$ 9
46. $z - 5 = 4$ 9
47. $d - {}^-8 = 0$ ⁻8
48. $w - 6 = 20$ 26

You decide! Choosing the equation

Decide whether Equation A or B would be used to solve each problem. Then solve the problem.

Equation A: $n + 15 = 66$
Equation B: $n - 15 = 66$

49. $15 off the regular price is $66. What is the regular price? B, $81
50. $15 more than the sale price is $66. What is the sale price? A, $51
51. The regular price decreased by $15 is $66. What is the regular price? B, $81
52. The sum of the sale price and a $15 discount is $66. What is the sale price? A, $51
53. Sarah gave her brother $15. She then had $66. How much money did Sarah have before she gave her brother the money? B, $81
54. James got paid $15. He then had $66. How much money did he have before he got paid? A, $51

Algebra 441

Practice Worksheet

Workbook S356, Copymaster S356, or Duplicating Master S356

WORKSHEET 188 (Use after page 441.)

NAME _____

RIDDLE TIME — Why is noon like the letter A?

To find the answer:
1. Solve each question.
2. Write its letter under its matching number in the DECODER.

1. $n - 15 = 38$ $n-15+15=38+15$ $n=53$ R	2. $n - 23 = 49$ $n=72$ L	3. $n - 71 = 90$ $n=161$ T	4. $n - 65 = 61$ $n=126$ E
5. $n - 36 = 28$ $n=64$ I	6. $n - 62 = 55$ $n=117$ D	7. $n - 62 = 41$ $n=103$ F	8. $n - 33 = 26$ $n=59$ H
9. $n - 28 = 45$ $n=73$ E	10. $n - 15 = 69$ $n=84$ Y	11. $n - 58 = 19$ $n=77$ T	12. $n - 30 = 53$ $n=83$ O
13. $n - 39 = 61$ $n=100$ N	14. $n - 120 = 44$ $n=164$ E	15. $n - 27 = 23$ $n=50$ H	16. $n - 15 = 25$ $n=40$ I
17. $n - 42 = 112$ $n=154$ O	18. $n - 39 = 67$ $n=106$ D	19. $n - 53 = 42$ $n=95$ A	20. $n - 72 = 19$ $n=91$ B
21. $n - 31 = 27$ $n=58$ M	22. $n - 44 = 3$ $n=47$ D	23. $n - 100 = 1$ $n=101$ A	

DECODER

91	83	161	50	95	53	73	40	100	77	59	126
B	O	T	H	A	R	E	I	N	T	H	E

58	64	106	117	72	164		154	103	47	101	84
M	I	D	D	L	E		O	F		D	A	Y

© D. C. Heath and Company S356 Solving subtraction equations

Group Project

Writing and solving equations

Give the students a riddle such as this: "I am thinking of a number. If I subtract 7 from it, I get 2. What is the number?" Have the students write an equation and then solve it. Next, let groups of students make up equation riddles for others to solve.

Class Starter Quiz 172
on previous lesson

Solve and check.

1. $n - 12 = 23$ 2. $r - 6 = 10$
   35   16
3. $m - 10 = 2$ 4. $t - 3 = 2$
   12   5
5. $s - 5 = 7$ 6. $n - 5 = 1$
   12   6
7. $n - 4 = {}^-1$ 8. $x - 6 = {}^-11$
   3                  ${}^-5$

Copymaster S428

Problem-Solving Skills

Choosing the correct operation
Solving a multistep problem
Writing equations

Skills Reviewed

Multiplying and dividing decimals
Adding and subtracting fractions
Multiplying fractions
Adding and subtracting integers
Multiplying and dividing integers

Starting the Lesson

Problem Solving Do exercises 1–4 on page 442 with the class. Assign the other exercises for independent work.

Cumulative Skill Practice Challenge the students to an estimation hunt by saying, "Find the two products in exercises 1–12 that are greater than 100." (Exercises 5 and 10) Then have the students find the two quotients in exercises 13–24 that are greater than 100. (Exercises 15 and 18)

Problem solving Writing equations

Decide which operation sign $(+, -, \times, \div)$ should replace the question mark to give the correct equation. Then solve the equation.

1. One long-stemmed rose costs $2. What is the price (p) of 12 long-stemmed roses?

   $2 \;\boxed{?}\; 12 = p \quad \times, 24$

2. Six carnations cost $4.50. What is the price (p) of one carnation?

   $4.50 \;\boxed{?}\; 6 = p \quad \div, \$.75$

3. A dozen strawflowers cost $2.40. What is the price (p) of each flower?

   $2.40 \;\boxed{?}\; 12 = p \quad \div, \$.20$

4. Five tulips cost $1.50. What is the price (p) of 3 tulips?

   $(1.50 \;\boxed{?}\; 5) \;\boxed{?}\; 3 = p \quad \div, \times, \$.90$

5. A customer bought 2 bunches of baby's breath at $.89 per bunch. How much change (c) should he get from a $10 bill?

   $10 \;\boxed{?}\; (2 \;\boxed{?}\; .89) = c \quad -, \times, \$8.22$

Write an equation and solve the problem. Decide when a calculator would be useful.

6. Mini carnations are priced at $3.89 per bunch. How much would you pay for 3 bunches of mini carnations? $3.89 \times 3 = p, \$11.67$

7. A bunch of 15 daisies costs $3.45. What is the cost per daisy? $345 \div 15 = c, \$.23$

8. A mixed bouquet costs $4.95. What is the cost of 2 mixed bouquets? $4.95 \times 2 = c, \$9.90$

9. A customer bought 4 mixed bouquets at $4.95 each. How much change should she get from a $20 bill? $20 - (4.95 \times 4) = c, \$.20$

10. Three dozen daffodils cost $9.60. What is the cost of 4 dozen daffodils? $9.60 \div 3 \times 4 = c,$
    $\$12.80$

11. Gladioli are selling at 3 for $2.37. How much would you expect to pay for 5 gladioli? $2.37 \div 3 \times 5 = c, \$3.95$

12. Marcia bought 3 dozen chrysanthemums at $10.50 per dozen and some fern for $1.25. How much did she spend in all?
    $10.50 \times 3 + 1.25 = p, \$32.75$

13. Marty paid $3.30 for some sweetheart roses. If 3 sweetheart roses cost $1.65, how many did Marty buy?
    $3.30 \div 1.65 \times 3 = c, 6$

Cumulative Skill Practice

Give the product. *(page 68)*

1. 6.2×0.5 3.10
2. 8.4×0.61 5.124
3. 0.72×0.4 0.288
4. 6.2×0.83 5.146
5. 32×5.3 169.6
6. 0.79×5.3 4.187
7. 2.05×1.7 3.485
8. 3.01×1.5 4.515
9. 9.46×8 75.68
10. 47.1×6.2 292.02
11. 47.9×0.38 18.202
12. 48.6×0.11 5.346

Give the quotient rounded to the nearest tenth. *(page 104)*

13. $19.3 \div 0.9$ 21.4
14. $5.47 \div 0.2$ 27.4
15. $8.09 \div 0.03$ 269.7
16. $8.65 \div 0.2$ 43.3
17. $15.7 \div 0.5$ 31.4
18. $8.43 \div 0.07$ 120.4
19. $5.405 \div 0.08$ 67.6
20. $3.09 \div 0.04$ 77.3
21. $8.41 \div 1.7$ 4.9
22. $3.047 \div 2.9$ 1.1
23. $5.09 \div 0.35$ 14.5
24. $4.65 \div 0.21$ 22.1

Give the sum in simplest form. *(page 180)*

25. $\frac{1}{2} + \frac{1}{4}$ $\frac{3}{4}$
26. $\frac{1}{4} + \frac{5}{8}$ $\frac{7}{8}$
27. $\frac{1}{5} + \frac{3}{10}$ $\frac{1}{2}$
28. $\frac{2}{3} + \frac{1}{3}$ 1
29. $\frac{1}{3} + \frac{2}{5}$ $\frac{11}{15}$
30. $\frac{3}{7} + \frac{4}{7}$ 1
31. $\frac{1}{2} + \frac{5}{8}$ $1\frac{1}{8}$
32. $\frac{7}{8} + \frac{3}{5}$ $1\frac{19}{40}$
33. $\frac{1}{2} + \frac{3}{8}$ $\frac{7}{8}$
34. $\frac{7}{8} + 0$ $\frac{7}{8}$
35. $\frac{6}{11} + \frac{3}{11}$ $\frac{9}{11}$
36. $\frac{3}{5} + \frac{1}{2}$ $1\frac{1}{10}$

Give the difference in simplest form. *(page 188)*

37. $\frac{3}{4} - \frac{1}{4}$ $\frac{1}{2}$
38. $\frac{3}{4} - \frac{1}{8}$ $\frac{5}{8}$
39. $\frac{1}{2} - \frac{1}{3}$ $\frac{1}{6}$
40. $\frac{5}{8} - \frac{1}{2}$ $\frac{1}{8}$
41. $\frac{6}{7} - \frac{1}{2}$ $\frac{5}{14}$
42. $\frac{1}{2} - \frac{1}{4}$ $\frac{1}{4}$
43. $\frac{3}{5} - \frac{1}{4}$ $\frac{7}{20}$
44. $\frac{3}{2} - \frac{2}{3}$ $\frac{5}{6}$
45. $\frac{3}{4} - \frac{1}{2}$ $\frac{1}{4}$
46. $\frac{9}{10} - \frac{1}{5}$ $\frac{7}{10}$
47. $\frac{4}{5} - \frac{1}{10}$ $\frac{7}{10}$
48. $\frac{2}{3} - \frac{1}{9}$ $\frac{5}{9}$

Give the product in simplest form. *(page 202)*

49. $\frac{1}{2} \times \frac{1}{3}$ $\frac{1}{6}$
50. $\frac{1}{2} \times \frac{1}{4}$ $\frac{1}{8}$
51. $\frac{1}{3} \times \frac{3}{5}$ $\frac{1}{5}$
52. $\frac{3}{4} \times \frac{2}{3}$ $\frac{1}{2}$
53. $\frac{2}{3} \times \frac{1}{5}$ $\frac{2}{15}$
54. $\frac{2}{7} \times \frac{1}{2}$ $\frac{1}{7}$
55. $\frac{3}{2} \times \frac{3}{4}$ $1\frac{1}{8}$
56. $\frac{5}{9} \times \frac{3}{8}$ $\frac{5}{24}$
57. $\frac{1}{2} \times \frac{2}{5}$ $\frac{1}{5}$
58. $2 \times \frac{7}{8}$ $1\frac{3}{4}$
59. $3 \times \frac{5}{9}$ $1\frac{2}{3}$
60. $\frac{3}{5} \times \frac{1}{4}$ $\frac{3}{20}$

MIXED PRACTICE
Complete.

61. $^+9 + {}^-3 = \underline{?}$ $^+6$
62. $^+8 - {}^+3 = \underline{?}$ $^+5$
63. $^+9 \times 0 = \underline{?}$ 0
64. $^+24 \div {}^+3 = \underline{?}$ $^+8$
65. $^-11 + {}^-4 = \underline{?}$ $^-15$
66. $^-6 - {}^+10 = \underline{?}$ $^-16$
67. $^-4 \times {}^-11 = \underline{?}$ $^+44$
68. $^-48 \div {}^-12 = \underline{?}$ $^+4$
69. $^+17 + {}^+17 = \underline{?}$ $^+34$
70. $^-5 - {}^-9 = \underline{?}$ $^+4$
71. $^+10 \times {}^-11 = \underline{?}$ $^-110$
72. $^-60 \div {}^+5 = \underline{?}$ $^-12$
73. $^-20 + {}^+12 = \underline{?}$ $^-8$
74. $^-8 - {}^+15 = \underline{?}$ $^-23$
75. $^+10 \times {}^-15 = \underline{?}$ $^-150$

Algebra **443**

Problem-Solving Worksheet
Workbook S357, Copymaster S357, or Duplicating Master S357

WORKSHEET 189
(Use after page 442.)

WHAT'S THE COST?
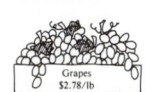
Grapes $2.78/lb
How much do 2.5 pounds of grapes cost?
$6.95

MAKE A LIST
When Sara counted some chickens and cats, there were 13. When she counted legs, there were 40. How many chickens were there?
6

SQUARE COUNT
How many squares can you count? *Hint: There are more than 10.*

12

WORK BACKWARD
Divide my age by 2. Then add 8 and multiply by 3. You get 39. How old am I?
10

A PIECE OF CAKE
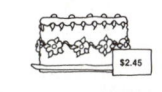
$2.45
Shawn paid for the cake with dimes and quarters. He gave the clerk the same number of dimes as quarters. How many quarters did Shawn give the clerk?
7

CRACK THE CODE

	CODE
9 10	$4\uparrow = 20$
8 7	$6\uparrow = 42$
5 6	$2\uparrow = 6$
4 3	$3\uparrow = 18$
1 2	$5\uparrow = 40$
	$7\uparrow = 70$

© D. C. Heath and Company S357 Problem solving

Challenge Problems
Unscramble the letters of these algebra words.

1. OSEXEPRSIN — Expression
2. TAVELUAE — Evaluate
3. IUQEATNO — Equation
4. TUSBTSIUTE — Substitute

Copymaster S465

Class Starter Quiz 173
on previous lesson

Solve.

1. A customer bought 3 bunches of daisies at $2.29 per bunch. How much change should she get back from a $10 bill? $3.13
2. Roses are selling at 6 for $9. How much would you expect to pay for 4 roses? $6

Copymaster S428

Lesson Objective
To solve multiplication equations

Problem-Solving Skill
Choosing the correct equation

Starting the Lesson
Use of Concrete Materials Discuss questions 1 and 2, using the balance-scale model and the *Here's how*. The balance-scale model is on ■ **Visual Aid 69** (copymaster or transparency S166). Go over the examples in exercise 3 before assigning the exercises on page 445.

Solving multiplication equations
ANOTHER BALANCING ACT!

Each box contains the same number of marbles.
1. Three times the number of marbles in 1 box weighs the same as how many marbles? 12

If we let m be the number of marbles in each box, we can write the equation

$$3m = 12$$

Here's how *to solve a multiplication equation.*

Equation:	$3m = 12$
Divide both sides of the equation by 3:	$\dfrac{3m}{3} = \dfrac{12}{3}$
Simplify both sides:	$m = 4$

Remember that a fraction bar can be used to show division.

Check the solution by substituting 4 for m in the equation $3m = 12$: $3 \times 4 = 12$ It checks!

2. Look at the *Here's how*. To find m, both sides of the equation were divided by what number? How many marbles are in each box? 3, 4

3. Check these examples. Has each equation been solved correctly? Yes

   a. $5a = 60$
   $\dfrac{5a}{5} = \dfrac{60}{5}$
   $a = 12$

   b. $3b = {}^-30$
   $\dfrac{3b}{3} = \dfrac{{}^-30}{3}$
   $b = {}^-10$

   c. ${}^-2c = 14$
   $\dfrac{{}^-2c}{{}^-2} = \dfrac{14}{{}^-2}$
   $c = {}^-7$

444   Chapter 17

EXERCISES

Solve and check.
Here are scrambled answers for the next row of exercises: $^-7$ 5 6 $^-3$

4. $4n = 24$ **6**
5. $^-5t = 35$ **$^-7$**
6. $2r = ^-6$ **$^-3$**
7. $^-3x = ^-15$ **5**

Hint: Divide both sides by 4.
Hint: Divide both sides by $^-5$.
Hint: Divide both sides by 2.
Hint: Divide both sides by $^-3$.

8. $9n = 27$ **3**
9. $^-7n = 14$ **$^-2$**
10. $5s = 55$ **11**
11. $3y = ^-21$ **$^-7$**

12. $4t = 48$ **12**
13. $8w = 120$ **15**
14. $^-6n = 42$ **$^-7$**
15. $^-2r = ^-22$ **11**

16. $9a = 108$ **12**
17. $12n = 240$ **20**
18. $^-3x = 15$ **$^-5$**
19. $9y = ^-27$ **$^-3$**

20. $^-2x = 6$ **$^-3$**
21. $30t = 30$ **1**
22. $7x = ^-21$ **$^-3$**
23. $^-15a = 30$ **$^-2$**

24. $^-4s = 40$ **$^-10$**
25. $3s = 57$ **19**
26. $6c = 66$ **11**
27. $11n = 88$ **8**

28. $^-4c = ^-40$ **10**
29. $8n = ^-16$ **$^-2$**
30. $^-2x = ^-2$ **1**
31. $7n = 7$ **1**

32. $^-3b = ^-33$ **11**
33. $9y = 189$ **21**
34. $25t = 100$ **4**
35. $20n = ^-60$ **$^-3$**

36. $7h = ^-77$ **$^-11$**
37. $^-5s = ^-5$ **1**
38. $15t = 30$ **2**
39. $25a = 0$ **0**

40. $^-15x = ^-15$ **1**
41. $12y = 132$ **11**
42. $^-20a = ^-100$ **5**
43. $25c = 175$ **7**

44. $10w = 130$ **13**
45. $^-16r = 0$ **0**
46. $14x = ^-154$ **$^-11$**
47. $^-9w = 270$ **$^-30$**

48. $^-8b = 168$ **$^-21$**
49. $11v = 165$ **15**
50. $7r = ^-84$ **$^-12$**
51. $^-16n = 224$ **$^-14$**

52. $30t = ^-210$ **$^-7$**
53. $^-15y = ^-330$ **22**
54. $32k = 0$ **0**
55. $^-24w = ^-192$ **8**

56. $^-17x = 153$ **$^-9$**
57. $16c = ^-272$ **$^-17$**
58. $15y = 15$ **1**
59. $^-26s = 0$ **0**

You decide! Choosing the equation

Decide whether Equation A, B, or C could be used to solve each problem. Then solve the problem.

Equation A: $s + .99 = 1.65$
Equation B: $3s = 5.28$
Equation C: $2s = 1.14$

60. What is the price of a box of cereal? Clue: 3 boxes of cereal would cost $5.28. **$1.76**

61. What is the price of a can of soup? Clue: A can of soup and a jar of pickles cost a total of $1.65. **A, $.66**

62. What is the price of a can of juice? Clue: 2 cans of juice cost $.15 more than a jar of pickles. **C, $.57**

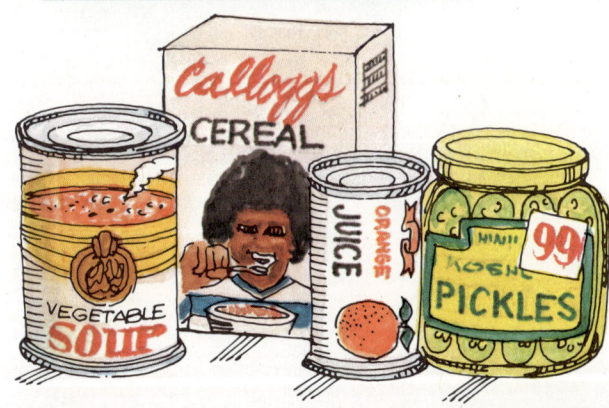

Class Starter Quiz 174
on previous lesson

Solve and check.

1. $9n = 36$ 4
2. $5t = 65$ 13
3. $7h = 140$ 20
4. $2n = 130$ 65
5. $4r = {}^-20$ $^-5$
6. $15a = {}^-45$ $^-3$
7. $^-5a = 35$ $^-7$
8. $^-10r = {}^-60$ 6

Copymaster S429

Lesson Objective
To solve division equations

Problem-Solving Skill
Choosing the correct equation

Starting the Lesson
Discuss questions 1–3, using the travel ad and the *Here's how*. Go over the examples in exercise 4 before assigning the exercises on page 447.

Solving division equations

See your Holiday Travel Agent for your **Florida vacation.**

HAVE YOUR DAY IN THE SUN

Pay no money now. Use our easy 12-month payment plan. Pay only $80 a month.

1. How many monthly payments are there? 12
2. How much is each monthly payment? $80

If we let t be the total cost (in dollars) of the Florida vacation,
we can write the equation $\frac{t}{12} = 80$

Here's how *to solve a division equation.*

Equation: $\quad \frac{t}{12} = 80$

Multiply both sides by 12: $\quad 12 \times \frac{t}{12} = 12 \times 80$

Simplify both sides: $\quad t = 960$

Check: $\quad \frac{960}{12} = 80 \quad$ It checks!

3. Look at the *Here's how*. To find t, both sides of the equation were multiplied by what number? What is the total cost of the Florida vacation? 12, $960

4. Check these examples. Has each equation been solved correctly? Yes

    a. $\quad \frac{a}{4} = 20$ b. $\quad \frac{b}{^-2} = 16$ c. $\quad \frac{c}{^-5} = {}^-4$

    $4 \times \frac{a}{4} = 4 \times 20 \quad\quad {}^-2 \times \frac{b}{^-2} = {}^-2 \times 16 \quad\quad {}^-5 \times \frac{c}{^-5} = {}^-5 \times {}^-4$

        $a = 80 \quad\quad\quad\quad\quad\quad b = {}^-32 \quad\quad\quad\quad\quad\quad c = 20$

446   Chapter 17

EXERCISES

Solve and check.
Here are scrambled answers for the next row of exercises: 10   ⁻18   ⁻12   64   35

5. $\frac{x}{2} = 32$ 64
6. $\frac{b}{-3} = 4$ ⁻12
7. $\frac{n}{6} = -3$ ⁻18
8. $\frac{n}{-2} = -5$ 10
9. $\frac{d}{7} = 5$ 35

10. $\frac{y}{5} = 3$ 15
11. $\frac{m}{9} = -3$ ⁻27
12. $\frac{k}{10} = 15$ 150
13. $\frac{n}{5} = 12$ 60
14. $\frac{a}{20} = 2$ 40

15. $\frac{h}{15} = 3$ 45
16. $\frac{c}{-3} = -4$ 12
17. $\frac{t}{12} = -6$ ⁻72
18. $\frac{x}{10} = -8$ ⁻80
19. $\frac{a}{2} = 15$ 30

20. $\frac{x}{4} = -9$ ⁻36
21. $\frac{n}{7} = 9$ 63
22. $\frac{y}{100} = 5$ 500
23. $\frac{t}{-8} = -1$ 8
24. $\frac{n}{50} = 6$ 300

25. $\frac{c}{-3} = -40$ 120
26. $\frac{n}{30} = 4$ 120
27. $\frac{x}{4} = -7$ ⁻28
28. $\frac{n}{-8} = -2$ 16
29. $\frac{x}{4} = -10$ ⁻40

30. $\frac{f}{15} = 8$ 120
31. $\frac{c}{-7} = 1$ ⁻7
32. $\frac{r}{5} = -25$ ⁻125
33. $\frac{y}{-5} = -2$ 10
34. $\frac{a}{4} = 300$ 1200

35. $\frac{x}{-9} = -2$ 18
36. $\frac{b}{50} = 1$ 50
37. $\frac{a}{-8} = -9$ 72
38. $\frac{n}{6} = 0$ 0
39. $\frac{x}{-9} = -11$ 99

40. $\frac{t}{4} = -5$ ⁻20
41. $\frac{r}{3} = 12$ 36
42. $\frac{b}{-10} = 10$ ⁻100
43. $\frac{n}{12} = -3$ ⁻36
44. $\frac{c}{-7} = -15$ 105

45. $\frac{k}{9} = 0$ 0
46. $\frac{y}{-6} = -11$ 66
47. $\frac{w}{-5} = 20$ ⁻100
48. $\frac{s}{8} = 21$ 168
49. $\frac{m}{18} = -13$ ⁻234

You decide!   Choosing the equation

Decide whether Equation A, B, C, or D could be used to solve each problem. Then solve the problem.

50. I spent $450 for 5 nights' lodging at the resort. How much was it for 1 night's lodging?   A, $90

Equation A: $5n = 450$
Equation B: $\frac{n}{5} = 450$
Equation C: $n + 50 = 450$
Equation D: $n - 50 = 450$

51. Karen spent $50 more for her trip than Alan did. She spent $450. How much did Alan spend?   C, $400

52. After spending $50 at the ski resort, Brian had $450 left. How much money did he have to start with?   D, $500

53. A vacation cottage at the ocean rents for $450 a week. How much rent is paid in 5 weeks?   B, $2250

Algebra **447**

Practice Worksheet
Workbook S359, Copymaster S359, or Duplicating Master S359

WORKSHEET 191   (Use after page 447.)

RIDDLE
If all the cars in the United States were pink, what would we have?

To find the answer:
1. Solve each equation.
2. Cross out each box below that contains an answer.
3. Read the answer using the letters in the remaining boxes.

1. $\frac{n}{5} = 13$ $5 \times \frac{n}{5} = 13 \times 5$ $n = 65$	2. $\frac{n}{8} = 6$ $n = 48$	3. $\frac{n}{15} = 3$ $n = 45$	4. $\frac{n}{20} = 7$ $n = 140$
5. $\frac{n}{4} = 13$ $n = 52$	6. $\frac{n}{7} = 12$ $n = 84$	7. $\frac{n}{9} = 11$ $n = 99$	8. $\frac{n}{5} = 15$ $n = 75$
9. $\frac{n}{2} = -6$ $n = -12$	10. $\frac{n}{7} = 10$ $n = 70$	11. $\frac{n}{12} = 12$ $n = 144$	12. $\frac{n}{7} = -3$ $n = -21$
13. $\frac{n}{4} = 25$ $n = 100$	14. $\frac{n}{5} = 32$ $n = 160$	15. $\frac{n}{3} = -15$ $n = -45$	16. $\frac{n}{7} = 0$ $n = 0$
17. $\frac{n}{6} = 30$ $n = 180$	18. $\frac{n}{4} = 19$ $n = 76$	19. $\frac{n}{14} = 1$ $n = 14$	20. $\frac{n}{-3} = -3$ $n = 9$
21. $\frac{n}{25} = 8$ $n = 200$	22. $\frac{n}{8} = 12$ $n = 96$		

Answer: A PINK CARNATION

© D. C. Heath and Company   S359   *Solving division equations*

Project

Writing and solving equations

Use a calculator to solve these equations:

1. $465n = 2418$ 5.2
2. $\frac{n}{7.8} = 23.5$ 183.3
3. $n + 9.12 = 275.6$ 266.48
4. $4.6n = 33.35$ 7.25
5. $n - 0.265 = 31.28$ 31.545

Write equations for others to solve with a calculator.

Copymaster S487

Solving two-step equations

To find the number of reprints he can get for $20, solve this two-step equation:

$$\underset{\substack{\uparrow \\ \text{number} \\ \text{of} \\ \text{reprints}}}{3r} + \underset{\substack{\uparrow \\ \text{handling} \\ \text{charge}}}{2} = \underset{\substack{\uparrow \\ \text{total} \\ \text{cost}}}{20}$$

cost of 1 reprint

Here's how to solve a two-step equation.

To solve the equation, get an equation with only the variable r on one side of the equal sign.

Equation: $3r + 2 = 20$

First subtract 2 from both sides: $3r + 2 - 2 = 20 - 2$

Then divide both sides by 3: $\frac{3r}{3} = \frac{18}{3}$

$r = 6$

Check: $3 \times 6 + 2 = 20$ It checks!

1. Look at the *Here's how*. To find r, first subtract [?] from both sides of the equation and then divide both sides by [?].

2. Check the solution. How many reprints can be bought for $20?

3. Check these examples. Has each two-step equation been solved correctly?

a. $2a + 4 = 26$
$2a + 4 - 4 = 26 - 4$
$2a = 22$
$\frac{2a}{2} = \frac{22}{2}$
$a = 11$

b. $4b - 3 = 21$
$4b - 3 + 3 = 21 + 3$
$4b = 24$
$\frac{4b}{4} = \frac{24}{4}$
$b = 6$

c. $\frac{c}{2} + 7 = 18$
$\frac{c}{2} + 7 - 7 = 18 - 7$
$\frac{c}{2} = 11$
$2 \times \frac{c}{2} = 2 \times 11$
$c = 22$

448   Chapter 17

EXERCISES

Copy and finish solving each equation. Check your solution.

4. $2y + 8 = 14$
$2y + 8 - 8 = 14 - ?\ 8$
$2y = ?\ 6$
$y = ?\ 3$

5. $5t - 7 = 28$
$5t - 7 + 7 = 28 + ?\ 7$
$5t = ?\ 35$
$t = ?\ 7$

6. $\frac{n}{3} + 6 = 12$
$\frac{n}{3} + 6 - 6 = 12 - ?\ 6$
$\frac{n}{3} = ?\ 6$
$n = ?\ 18$

Solve and check.
Here are scrambled answers for the next two rows of exercises: 17 9 4 11 5 2 6 7

7. $4x + 3 = 27$ 6
8. $2n + 5 = 39$ 17
9. $4t + 8 = 28$ 5
10. $3m - 2 = 25$ 9
11. $6y - 4 = 20$ 4
12. $6k - 3 = 39$ 7
13. $2y + 10 = 32$ 11
14. $8y - 2 = 14$ 2
15. $6a + 4 = 28$ 4
16. $3y - 5 = 7$ 4
17. $5t + 7 = 52$ 9
18. $5c - 2 = 33$ 7
19. $\frac{m}{3} + 7 = 9$ 6
20. $\frac{n}{2} + 2 = 5$ 6
21. $\frac{t}{7} + 4 = 6$ 14
22. $\frac{d}{3} - 4 = 5$ 27
23. $\frac{n}{3} + 20 = 21$ 3
24. $5n + {}^-3 = 7$ 2
25. $\frac{t}{6} - 9 = {}^-3$ 36
26. $6a - 12 = 30$ 7
27. $\frac{c}{{}^-3} + 7 = {}^-8$ 45
28. ${}^-7x + 4 = 32$ ${}^-4$
29. $\frac{y}{{}^-4} - {}^-2 = 10$ ${}^-32$
30. $9a - 4 = 23$ 3
31. $\frac{d}{9} + 9 = 0$ ${}^-81$
32. $8r + {}^-5 = 11$ 2
33. ${}^-10y + 6 = {}^-14$ 2
34. $\frac{s}{{}^-7} - 3 = 8$ ${}^-77$
35. ${}^-4x + 8 = 0$ 2
36. $\frac{n}{{}^-4} + {}^-6 = 1$ ${}^-28$
37. $12s - {}^-3 = {}^-9$ ${}^-1$
38. $\frac{n}{7} + {}^-5 = 0$ 35

Mystery numbers Writing equations

Write an equation and solve the problem.

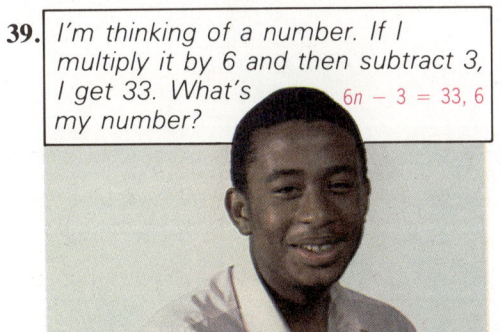

39. I'm thinking of a number. If I multiply it by 6 and then subtract 3, I get 33. What's my number?
$6n - 3 = 33,\ 6$

40. If you multiply a number by 4 and then add 2, you get 22. What is the number?
$4n + 2 = 22,\ 5$

41. A number divided by 3 and then decreased by 10 is 2. What is the number?
$\frac{n}{3} - 10 = 2,\ 36$

42. If you divide a number by 8 and then add 4, you get 5. What is the number?
$\frac{n}{8} + 4 = 5,\ 8$

43. A number multiplied by 7 and then increased by 4 is 25. What is the number?
$7n + 4 = 25,\ 3$

Algebra **449**

Extra Practice
Page 502 Skill 64

Practice Worksheet
Workbook S360, Copymaster S360, or Duplicating Master S360

WORKSHEET 192
(Use after page 449.)

NAME _____

SKILL DRILL
Solve.

1. $6n + 8 = 50$
$6n + 8 - 8 = 50 - 8$
$6n = 42$
$\frac{6n}{6} = \frac{42}{6}$
$n = 7$

2. $12n - 4 = 68$
$n = 6$

3. $5n + 15 = 60$
$n = 9$

4. $9n - 16 = 92$
$n = 12$

5. $3n + 13 = 46$
$n = 11$

6. $7n + 5 = 110$
$n = 15$

7. $13n + 14 = 40$
$n = 2$

8. $4n - 30 = 46$
$n = 19$

9. $15n + 15 = 30$
$n = 1$

10. $\frac{n}{4} + 11 = 31$
$\frac{n}{4} + 11 - 11 = 31 - 11$
$\frac{n}{4} = 20$
$4 \times \frac{n}{4} = 20 \times 4$
$n = 80$

11. $\frac{n}{3} - 5 = 2$
$n = 21$

12. $\frac{n}{2} - 8 = 7$
$n = 30$

13. $\frac{n}{3} + 16 = 33$
$n = 51$

14. $\frac{n}{5} + 6 = 13$
$n = 35$

15. $\frac{n}{2} - 8 = 6$
$n = 28$

16. $\frac{n}{5} + 2 = 10$
$n = 40$

17. $\frac{n}{7} - 3 = 8$
$n = 77$

18. $\frac{n}{4} - 3 = 2$
$n = 20$

Check yourself. Here are the scrambled answers:
1 2 6 7 9 11 12 15 19 20 21 28 30 35 40 51 77 80

© D. C. Heath and Company   S360   Solving two-step equations

Group Project

Writing and solving equations
Have the students write true equations—for example,

$$4 \times 9 - 7 = 29$$

Then have them erase one of the numbers and replace it with a letter. For example, from the equation above they might write one of these equations:

$4n - 7 = 29$ $4 \times 9 - n = 29$

Ask them to solve each other's equations.

Class Starter Quiz 176
on previous lesson

Solve and check.

1. $4n + 2 = 22$ 5
2. $2n + 13 = 33$ 10
3. $3t - 2 = 10$ 4
4. $\frac{m}{3} + 7 = 11$ 12

Copymaster S429

Lesson Objectives

To find the square root of a perfect square

To use the divide-and-average method to find a decimal approximation of a square root

Problem-Solving Skill

Using a guess-and-check strategy

Starting the Lesson

Sketch these squares on the chalkboard:

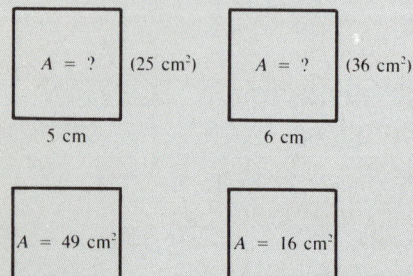

Before the students open their books, ask them to give the missing area or length for each square. Then have the students read the introductory paragraph at the top of the page. Discuss exercises 1 and 2, using the drawings and the examples in the *Here's how*.

450

Square roots

Ancient mathematicians thought about numbers in geometric terms. So when they multiplied a number by itself, they thought about finding the area of a square. That is the reason we still talk about **squaring** numbers.

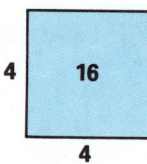

Since $4^2 = 16$, we say that **the square root** of 16 is 4. We write: $\sqrt{16} = 4$.

1. Think about a square that has an area of 22.
    a. Would each side be between 4 and 5? yes
    b. Is $\sqrt{22}$ between 4 and 5? yes

Some square roots such as $\sqrt{22}$ have only decimal approximations.

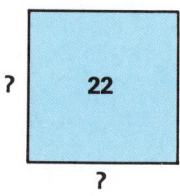

Here's how to use the divide-and-average method to find a decimal approximation of a square root.

Step 1. Estimate $\sqrt{22}$.

$\sqrt{16} = 4$
$\sqrt{22} = ?$
$\sqrt{25} = 5$

$\sqrt{22}$ is between 4 and 5.
We will try 4.5.

Step 2. Divide 22 by 4.5.

```
          4.8
    4.5 ) 22.0,0
         -180
          400
         -360
           40
```

So, $\sqrt{22}$ is between 4.5 and 4.8.

Step 3. Average 4.5 and 4.8.

$\frac{4.5 + 4.8}{2} = 4.65$

Step 4. Divide 22 by 4.65.

```
           4.73
    4.65 ) 22.00,000
          -1860
           3400
          -3255
           1450
          -1395
             55
```

So, $\sqrt{22}$ is between 4.65 and 4.73.
$\sqrt{22} = 4.7$ to the nearest tenth.

2. Look at the *Here's how*. What is $\sqrt{22}$ rounded to the nearest tenth? 4.7

EXERCISES

Find the following square roots.
Here are scrambled answers for the next row of exercises: 1 4 8 9 10

3. $\sqrt{64}$ 8
4. $\sqrt{1}$ 1
5. $\sqrt{16}$ 4
6. $\sqrt{100}$ 10
7. $\sqrt{81}$ 9
8. $\sqrt{36}$ 6
9. $\sqrt{9}$ 3
10. $\sqrt{4}$ 2
11. $\sqrt{49}$ 7
12. $\sqrt{25}$ 5
13. $\sqrt{169}$ 13
14. $\sqrt{121}$ 11
15. $\sqrt{225}$ 15
16. $\sqrt{144}$ 12
17. $\sqrt{400}$ 20

True or false?

18. $\sqrt{6}$ is between 2 and 3.   true
19. $\sqrt{30}$ is between 5 and 6.   true
20. $\sqrt{20}$ is between 5 and 6.   false
21. $\sqrt{38}$ is between 6 and 7.   true
22. $\sqrt{90}$ is between 8 and 9.   false
23. $\sqrt{65}$ is between 7 and 8.   false
24. $\sqrt{53}$ is between 7 and 8.   true
25. $\sqrt{108}$ is between 10 and 11.   true

Use the divide-and-average method. Give a decimal approximation to the nearest tenth.

26. $\sqrt{6}$ 2.4
27. $\sqrt{30}$ 5.5
28. $\sqrt{20}$ 4.5
29. $\sqrt{38}$ 6.2
30. $\sqrt{90}$ 9.5

Key it in Guess and check

Some calculators have a $\sqrt{}$ key for finding a decimal approximation of a square root. If not, you can use a guess-and-check method to find a decimal approximation of a square root.

Example:

$\sqrt{40}$ is between 6 and 7.

Try 6.3.

6.3 [X] 6.3 [=] 39.69

0.31 from 40.

Try 6.4.

6.4 X 6.4 [=] 40.96.

0.96 from 40.

So, 6.3 is closer than 6.4.

Use the guess-and-check method to find the decimal approximation to the nearest tenth.

31. $\sqrt{56}$ 7.5
32. $\sqrt{34}$ 5.8
33. $\sqrt{85}$ 9.2
34. $\sqrt{72}$ 8.5
35. $\sqrt{112}$ 10.6

Algebra 451

Practice Worksheet
Workbook S361, Copymaster S361, or Duplicating Master S361

WORKSHEET 193
(Use after page 451.)

NAME _____

TRIVIA TIME
What city is called the Chocolate Capital of the World?
Follow these steps to find the answer.
1. Give each square root as a whole number or as a decimal approximation to the nearest tenth.
2. Cross out each box below that contains a square root.
3. Read the answer using the letters in the remaining boxes.

1. $\sqrt{4}$ = 2
2. $\sqrt{16}$ = 4
3. $\sqrt{36}$ = 6
4. $\sqrt{9}$ = 3
5. $\sqrt{1}$ = 1
6. $\sqrt{25}$ = 5
7. $\sqrt{49}$ = 7
8. $\sqrt{100}$ = 10
9. $\sqrt{81}$ = 9
10. $\sqrt{64}$ = 8
11. $\sqrt{121}$ = 11
12. $\sqrt{169}$ = 13
13. $\sqrt{144}$ = 12
14. $\sqrt{225}$ = 15
15. $\sqrt{196}$ = 14
16. $\sqrt{34}$ = 5.8
17. $\sqrt{42}$ = 6.5
18. $\sqrt{30}$ = 5.5
19. $\sqrt{46}$ = 6.8
20. $\sqrt{60}$ = 7.7
21. $\sqrt{53}$ = 7.3

Answer: HERSHEY, PENNSYLVANIA

© D.C. Heath and Company S361 Square roots

Challenge Problem

What is the area of rectangle C?   35 cm²

A	B
21 cm²	9 cm²
C	D
?	15 cm²

Copymaster S466

The Pythagorean Rule

A triangle having a right angle is called a **right triangle.** The red triangle is a right triangle. Sides a and b are called the **legs,** and the side opposite the right angle (side c) is called the **hypotenuse.**

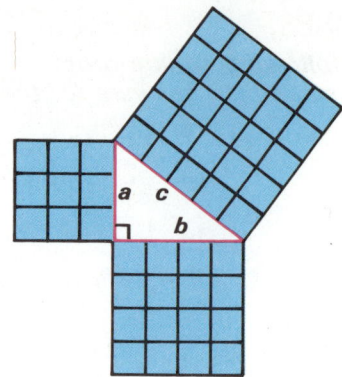

1. The area of the blue square on leg a is 9 square units. What is the area of the blue square on leg b? What is the sum of the two areas? 16, 25

2. What is the area of the blue square on the hypotenuse? 25

3. Is the sum of the areas of the squares on the legs the same as the area of the square on the hypotenuse? yes

The Pythagorean Rule

In a right triangle, the sum of the squares of the length of the legs is equal to the square of the length of the hypotenuse.

$$a^2 + b^2 = c^2$$

Here's how to use the Pythagorean Rule to find the length of a side of a right triangle when the lengths of the other two sides are known.

Example 1.

$$a^2 + b^2 = c^2$$
Substitute: $5^2 + 6^2 = c^2$
Simplify: $25 + 36 = c^2$
$$61 = c^2$$
$$\sqrt{61} = c$$

5 ft, 6 ft, c

Find the decimal approximation in the table on page 453.

$$\sqrt{61} \approx 7.810$$

Read "$\sqrt{61}$ is approximately equal to 7.810."

The approximate length of side c is 7.810 ft.

Example 2.

$$a^2 + b^2 = c^2$$
Substitute: $a^2 + 6^2 = 8^2$
Simplify: $a^2 + 36 = 64$
$$a^2 + 36 - 36 = 64 - 36$$
$$a^2 = 28$$
$$a = \sqrt{28}$$

8 m, 6 m, a

Find the decimal approximation in the table on page 453.

$$\sqrt{28} \approx 5.292$$

The approximate length of side a is 5.292 m.

---

4. Look at the examples. In which example was the length of the hypotenuse found? 1   The length of a leg found? 2

5. What was the approximate length of side c in Example 1? 7.810 ft

6. What was the approximate length of side a in Example 2? 5.292 m

452   Chapter 17

EXERCISES

Use the Pythagorean Rule to find the missing length. Use the square-root table to give a decimal approximation of the length.

Here are scrambled answers for the next two rows of exercises:

5.567 ft   5.831 cm   6.325 cm   6.708 ft

7. 5.831 cm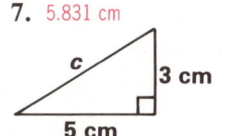

8. 6.708 ft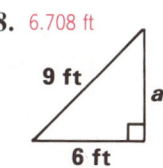

9. 6.325 cm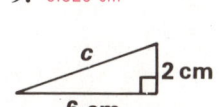

10. 5.657 ft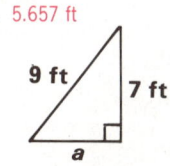

11. 8.062 yd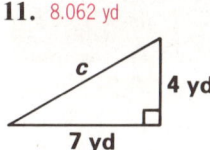

12. 6.633 m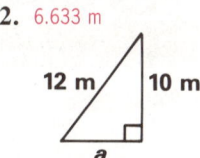

13. 8.246 ft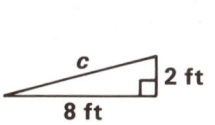

14. 4.899 in.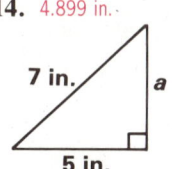

A missing length

Use the Pythagorean Rule to find the missing lengths. Use the divide-and-average method to give a decimal approximation to the nearest tenth.

15.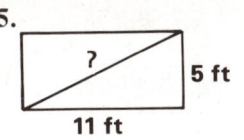
12.08 ft

16.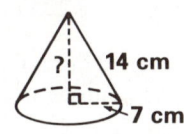
12.12 cm

Table of Square Roots

Number	Positive Square Root	Number	Positive Square Root
N	$\sqrt{N}$	N	$\sqrt{N}$
1	1	41	6.403
2	1.414	42	6.481
3	1.732	43	6.557
4	2	44	6.633
5	2.236	45	6.708
6	2.449	46	6.782
7	2.646	47	6.856
8	2.828	48	6.928
9	3	49	7
10	3.162	50	7.071
11	3.317	51	7.141
12	3.464	52	7.211
13	3.606	53	7.280
14	3.742	54	7.348
15	3.873	55	7.416
16	4	56	7.483
17	4.123	57	7.550
18	4.243	58	7.616
19	4.359	59	7.681
20	4.472	60	7.746
21	4.583	61	7.810
22	4.690	62	7.874
23	4.796	63	7.937
24	4.899	64	8
25	5	65	8.062
26	5.099	66	8.124
27	5.196	67	8.185
28	5.292	68	8.246
29	5.385	69	8.307
30	5.477	70	8.367
31	5.568	71	8.426
32	5.657	72	8.485
33	5.745	73	8.544
34	5.831	74	8.602
35	5.916	75	8.660
36	6	76	8.718
37	6.083	77	8.775
38	6.164	78	8.832
39	6.245	79	8.888
40	6.325	80	8.944

Algebra 453

Practice Worksheet

Workbook S362, Copymaster S362, or Duplicating Master S362

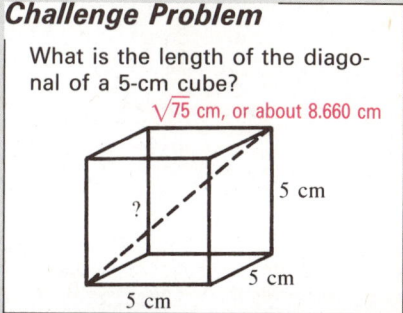

Challenge Problem

What is the length of the diagonal of a 5-cm cube?

$\sqrt{75}$ cm, or about 8.660 cm

Copymaster S466

453

Class Starter Quiz 178
on previous lesson

Use the Pythagorean rule to find the missing length.

1.   **2.**

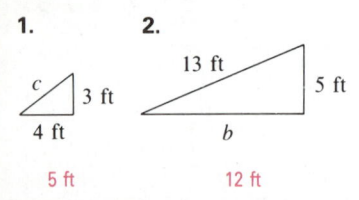

Copymaster S430

Lesson Objective
To solve problems using formulas

Starting the Lesson
Problem-Solving Cover-up Use the chalkboard or mask ■ *Visual Aid 72* (copymaster or transparency S168).

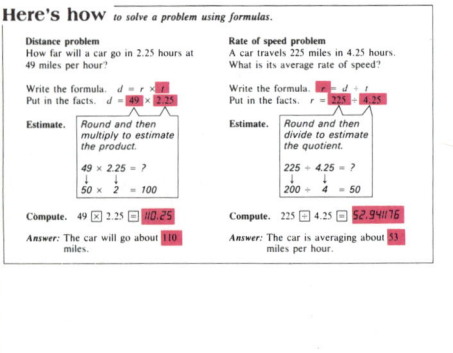

Have the students, working in small groups, study the problems on page 454 for several minutes. Then have them close their books, look at the visual aid, and tell what has been covered up.

454

PROBLEM SOLVING *using formulas*

There are three formulas involving distance, rate, and time.

The distance traveled (d) is equal to the rate of speed (r) multiplied by the time of travel (t).

FORMULA $d = r \times t$

The rate of speed (r) is equal to the distance traveled (d) divided by the time of travel (t).

FORMULA $r = d \div t$

The time of travel (t) is equal to the distance traveled (d) divided by the rate of speed (r).

FORMULA $t = d \div r$

Here's how *to solve a problem using formulas.*

Distance problem
How far will a car go in 2.25 hours at 49 miles per hour?

Write the formula. $d = r \times t$
Put in the facts. $d = 49 \times 2.25$

Estimate.

> Round and then multiply to estimate the product.
>
> $49 \times 2.25 = ?$
> ↓   ↓
> $50 \times 2 = 100$

Compute. 49 ⊠ 2.25 ⊟ *110.25*

Answer: The car will go about 110 miles.

Rate of speed problem
A car travels 225 miles in 4.25 hours. What is its average rate of speed?

Write the formula. $r = d \div t$
Put in the facts. $r = 225 \div 4.25$

Estimate.

> Round and then divide to estimate the quotient.
>
> $225 \div 4.25 = ?$
> ↓   ↓
> $200 \div 4 = 50$

Compute. 225 ⊟ 4.25 ⊟ *52.941176*

Answer: The car is averaging about 53 miles per hour.

454   Chapter 17

PROBLEMS

Put in the missing facts and solve each problem.

1. How far will a car go in 2.5 hours at 48 miles per hour?

   FORMULA $d = r \times t$
   $d = 48 \times ?$ 2.5
   $d = ?$ 120 miles

2. How long will it take to walk 10 miles at 4 miles per hour?

   FORMULA $t = d \div r$
   $t = 10 \div ?$ 4
   $t = ?$ 2.5 hours

3. How fast are you traveling if you travel 75 miles in 3 hours?

   FORMULA $r = d \div t$
   $r = ? \div ?$ 75, 3
   $r = ?$ 25 mph

4. At 52 miles per hour, how far will you go in 6 hours?

   FORMULA $d = r \times t$
   $d = ? \times ?$ 52, 6
   $d = ?$ 312 miles

Solve by using a formula.

5. If a car goes 164 miles in 4 hours, what is its average rate of speed? 41 mph

6. How many miles will an airplane go in 1.5 hours at 520 miles per hour? 780 miles

7. At 6 miles per hour, how many hours will it take to jog 9 miles? 1.5 hours

8. If a truck averages 48 miles per hour, how far will it go in 5.5 hours? 264 miles

9. A swimmer swims a distance of 880 yards in 20 minutes. How many yards per second is that? 0.73 yd/s

10. A boat is rowed at a rate of 2.4 miles per hour across a lake 6 miles wide. How many hours will it take? 2.5 hours

11. At an average rate of 2 miles per hour, how long will it take to hike 10 miles? 5 hours

12. How long will it take to bicycle 30 miles at an average rate of 12 miles per hour? 2.5 hours

You decide Reading a map

13. You leave Meade at 9:40 A.M. and drive to Winfield on highways 26 and 160. Your friend leaves Meade at 10:00 A.M. and drives to Winfield on highway 56. If you both average 50 miles per hour, who gets to Winfield first? The person who took highway 56.

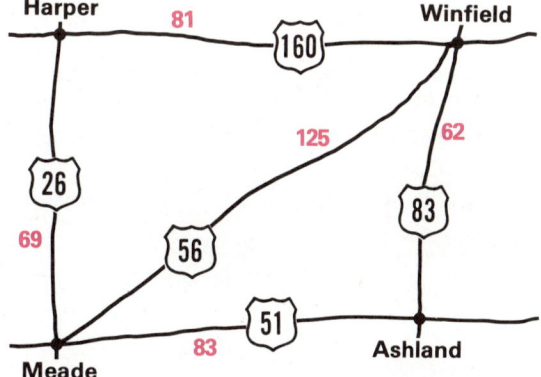

Challenge Problem

Write a question that fits the answer.

If you divide a number by 7 and then subtract 6, you get 2.

What is the number?

Answer: 56

Class Starter Quiz 179
on previous lesson

Solve by using a formula.

1. How many miles will a car go in 2.5 hours at 45 miles per hour? 112.5
2. At 50 miles per hour, how many hours will it take to go 225 miles? 4.5

Copymaster S430

Problem-Solving Skill
Making a geometric visualization

Skills Reviewed
Solving percent and interest problems
Adding integers
Operating with fractions and mixed numbers

Starting the Lesson
Problem Solving Give each of the students a piece of graph paper and have them follow these directions: "Put your pencil on a point in the middle of your paper. Move your pencil to the left 1 unit. Make a 90° turn to the right. Go forward 4 units. Make a 90° turn to the right. Go forward 4 units. Make a 90° turn to the right. Go forward 1 unit. Compare your drawing with the drawing in Screen E on page 456."

Have the students read the information at the top of the page and answer these questions:

- What computer language is being described? (Logo)
- What is the triangular printer called? (The "turtle")
- Through what angle did the turtle rotate in Screen 6? (45°)

Cumulative Skill Practice Write these four answers on the chalkboard:

    0   54   180   60

Challenge the students to an answer hunt by saying, "Look at exercises 1–36 on page 457. Find the four exercises that have these answers. You have five minutes to find as many of the exercises as you can." (Exercises 8, 20, 22, and 27)

Problem solving
LOGO—A COMPUTER LANGUAGE

Logo is a popular **computer language** that allows students to use a computer as a tool in learning, playing, and exploring. Many students learn Logo by starting with "turtle graphics." The "turtle" is a small triangular printer on the computer screen that responds to commands. Study these examples.

A
Here is the turtle pointing up.

B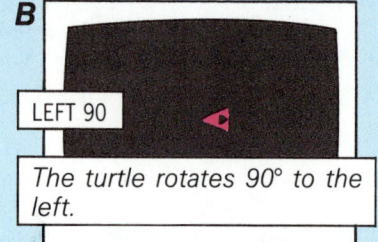
LEFT 90
The turtle rotates 90° to the left.

C
FORWARD 25
The turtle moves 25 units forward.

D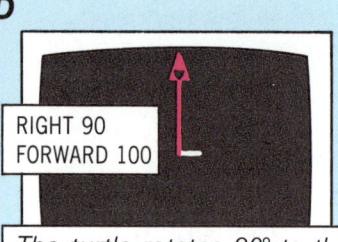
RIGHT 90
FORWARD 100
The turtle rotates 90° to the right then moves forward 100 units.

E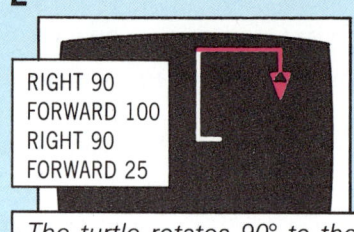
RIGHT 90
FORWARD 100
RIGHT 90
FORWARD 25
The turtle rotates 90° to the right, moves forward 100 units, rotates 90° to the right, and then moves forward 25 units.

F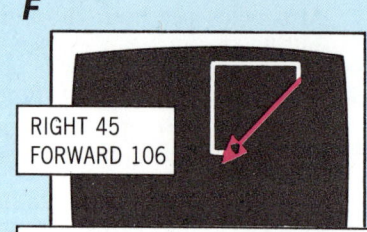
RIGHT 45
FORWARD 106
The turtle rotates 45° to the right and moves forward 106 units.

Match each of the turtle graphics with its set of Logo commands. Assume that the turtle starts pointing up.

1. 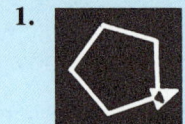 a

2.  b

3.  c

a. FORWARD 60
LEFT 72
FORWARD 60
LEFT 72
FORWARD 60
LEFT 72
FORWARD 60
LEFT 72
FORWARD 60
LEFT 72
END

b. FORWARD 100
LEFT 144
FORWARD 100
LEFT 144
FORWARD 100
LEFT 144
FORWARD 100
LEFT 144
FORWARD 100
LEFT 144
END

c. FORWARD 60
LEFT 120
FORWARD 60
LEFT 120
FORWARD 60
END

456 *Chapter 17*

Cumulative Skill Practice

Give the quotient in simplest form. *(page 212)*

1. $\frac{1}{2} \div \frac{1}{4}$ 2
2. $\frac{1}{6} \div \frac{1}{2}$ $\frac{1}{3}$
3. $\frac{3}{5} \div \frac{3}{2}$ $\frac{2}{5}$
4. $\frac{3}{4} \div \frac{2}{3}$ $1\frac{1}{8}$
5. $\frac{2}{3} \div 2$ $\frac{1}{3}$
6. $\frac{4}{5} \div \frac{1}{2}$ $1\frac{3}{5}$
7. $\frac{3}{5} \div \frac{3}{8}$ $1\frac{3}{5}$
8. $0 \div \frac{2}{3}$ 0
9. $\frac{9}{10} \div \frac{4}{5}$ $1\frac{1}{8}$
10. $\frac{5}{8} \div 2$ $\frac{5}{16}$
11. $\frac{2}{5} \div 5$ $\frac{2}{25}$
12. $4 \div \frac{1}{8}$ 32

Solve. *(page 294)*

13. 14% of 58 = n 8.12
14. 8% of 36.5 = n 2.92
15. 125% of 52 = n 65
16. 6.5% of 147 = n 9.555
17. 13.2% of 60 = n 7.92
18. 22.6% of 34.7 = n 7.8422
19. $16\frac{2}{3}$% of 78 = n 13
20. $66\frac{2}{3}$ of 81 = n 54
21. $62\frac{1}{2}$% of 120 = n 75

Solve. *(page 296)*

22. 10% of n = 18 180
23. 20% of n = 20 100
24. 25% of n = 16 64
25. 50% of n = 26 52
26. 75% of n = 60 80
27. 150% of n = 90 60
28. 6% of n = 21 350
29. 60% of n = 51 85
30. 15% of n = 30 200

Solve. Round each number to the nearest tenth. *(page 296)*

31. 3.5% of n = 8 228.6
32. 8.4% of n = 10 119.0
33. 6.2% of n = 11 177.4
34. 9.5% of n = 12 126.3
35. 32.4% of n = 12.4 38.3
36. 16.8% of n = 9.7 57.7

Compute the interest. *(page 322)*

37. Principal = $1000 $480
    Rate = 12% per year
    Time = 4 years
38. Principal = $900 $45
    Rate = 15% per year
    Time = 4 months
39. Principal = $150 $6.75
    Rate = 1.5% per month
    Time = 3 months

Give the sum. *(page 412)*

40. $^+4 + {^+6}$ $^+10$
41. $^+4 + {^-6}$ $^-2$
42. $^-4 + {^+6}$ $^+2$
43. $^-4 + {^-6}$ $^-10$
44. $^-6 + 0$ $^-6$
45. $0 + {^+9}$ $^+9$
46. $0 + {^-9}$ $^-9$
47. $^-6 + {^-7}$ $^-13$
48. $0 + 0$ 0
49. $^-7 + {^+7}$ 0

MIXED PRACTICE
Complete.

50. $3 + 2\frac{1}{2}$ = ? $5\frac{1}{2}$
51. $6\frac{3}{8} - 2\frac{1}{8}$ = ? $4\frac{1}{4}$
52. $4 \times 2\frac{1}{2}$ = ? 10
53. $6\frac{2}{3} \div 2$ = ? $3\frac{1}{3}$
54. $4\frac{2}{3} + 6$ = ? $10\frac{2}{3}$
55. $5 - 4\frac{2}{5}$ = ? $\frac{3}{5}$
56. $3\frac{2}{3} \times 3$ = ? 11
57. $9\frac{2}{3} - \frac{3}{4}$ = ? $8\frac{11}{12}$
58. $8 \div 2\frac{1}{2}$ = ? $3\frac{1}{5}$
59. $8\frac{1}{3} + 4\frac{1}{2}$ = ? $12\frac{5}{6}$
60. $1\frac{1}{3} \times 1\frac{1}{3}$ = ? $1\frac{7}{9}$
61. $6\frac{4}{5} - 1\frac{5}{8}$ = ? $5\frac{7}{40}$
62. $3\frac{5}{8} \div 1\frac{3}{4}$ = ? $2\frac{1}{14}$
63. $7\frac{2}{3} + 5\frac{7}{8}$ = ? $13\frac{13}{24}$
64. $4\frac{1}{2} \times 2\frac{3}{4}$ = ? $12\frac{3}{8}$

Algebra 457

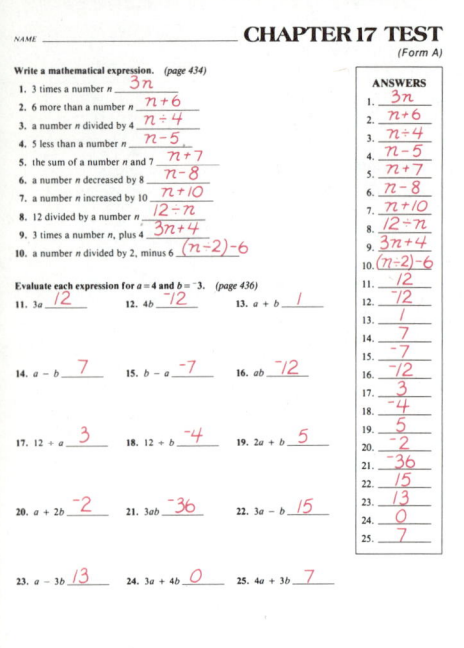
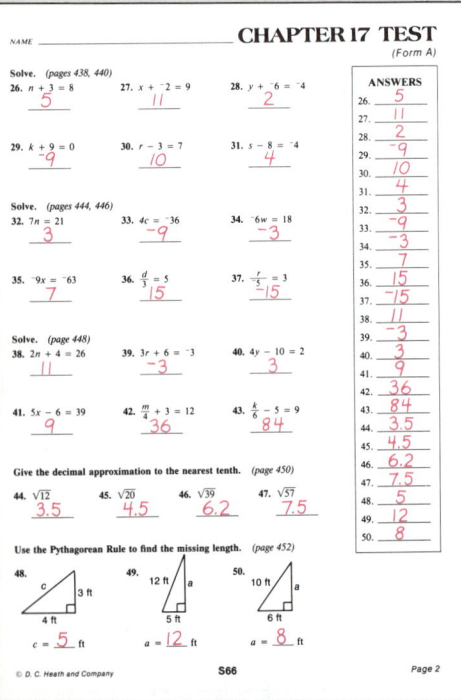

Chapter REVIEW

Here are scrambled answers for the review exercises:

⁻2	5	8	add	hypotenuse	root
⁻7	6	35	divide	more	simplified
3	7	48	expressions	multiply	variable

1. expressions, variable, more   **2.** 48, 35   **3.** 5, simplified, ⁻7

1. Variables, numbers, and operation signs can be combined to form mathematical [?]. In this expression, the letter n is called a [?]. A word expression for this mathematical expression is 6 [?] than a number n. (page 434)

$$n + 6$$

2. If you evaluate this expression for $a = 6$ and $b = 9$, you get [?]. If you evaluate this expression for $a = 3$ and $b = 10$, you get [?]. (page 436)

$$5a + 2b$$

3. To solve the equation, [?] was subtracted from both sides of the equation. Then both sides were [?]. To check the solution, [?] was substituted for x in the equation $x + 5 = {}^-2$. (page 438)

$$x + 5 = {}^-2$$
$$x + 5 - 5 = {}^-2 - 5$$
$$x = {}^-7$$
$$\text{Check: } {}^-7 + 5 = {}^-2$$

4. ⁻2   **5.** divide   **6.** multiply

4. To solve this equation, you would first add [?] to both sides of the equation and then simplify both sides. (page 440)

$$c - {}^-2 = 7$$

5. To find the solution of this equation, you would [?] both sides by 9 and then simplify both sides. (page 444)

$$9d = {}^-45$$

6. To find the solution of this equation, you would [?] both sides by ⁻4 and then simplify both sides. (page 446)

$$\frac{t}{-4} = 7$$

7. add, 3   **8.** root, 7, 8   **9.** hypotenuse, 6

7. To solve this equation, you would first [?] 4 to both sides and then divide by [?]. (page 448)

$$3y - 4 = 14$$

8. Since $7^2 = 49$, we say the square [?] of 49 is [?]. The $\sqrt{34}$ rounded to the nearest tenth is 5. [?]. (page 450)

9. The Pythagorean Rule: In a right triangle, the sum of the squares of the lengths of the legs is equal to the square of the length of the [?]. The length of side a is [?] ft. (page 452)

458   Chapter 17

Chapter TEST

Write a mathematical expression. (page 434)

1. 7 more than a number n $n+7$
2. 11 times a number n $11n$
3. the sum of a number n and 3 $n+3$
4. a number n divided by 6 $n \div 6$ ($\frac{n}{6}, \frac{1}{6}n$)
5. 12 times a number n $12n$
6. 8 less than a number n $n-8$
7. 2 more than 4 times a number n $4n+2$
8. a number n divided by 5, minus 6 $\frac{n}{5}-6$

Evaluate each expression for $a = 2$ and $b = {}^-3$. (page 436)

9. $a+b$ $^-1$
10. $a-b$ 5
11. $b-a$ $^-5$
12. ab $^-6$
13. ^-5ab 30
14. $4a$ 8
15. $ab+10$ 4
16. $16 \div a$ 8
17. $3ab$ $^-18$
18. ^-8ab 48
19. $2a+b$ 1
20. $a+4b$ $^-10$
21. $4a+3b$ $^-1$
22. $3a+4b$ $^-6$
23. $4a+5b$ $^-1$

Solve and check. (pages 438, 440)

24. $n+7=4$ $^-3$
25. $x+9={}^-3$ $^-12$
26. $a+6=6$ 0
27. $y+8=8$ 0
28. $r+9=5$ $^-4$
29. $l-12=7$ 19
30. $12+v={}^-1$ $^-13$
31. $m-9={}^-8$ 1

Solve and check. (pages 444, 446)

32. $6n=42$ 7
33. $5c={}^-35$ $^-7$
34. $8d=24$ 3
35. $^-9x=81$ $^-9$
36. $\frac{s}{3}=12$ 36
37. $\frac{v}{^-6}=8$ $^-48$
38. $\frac{y}{^-10}={}^-20$ 200
39. $\frac{x}{30}={}^-90$ $^-2700$

Solve and check. (page 448)

40. $3n+6=30$ 8
41. $4k+2=34$ 8
42. $7j+3=52$ 7
43. $7m+2=51$ 7
44. $5x-6=34$ 8
45. $6r-8=46$ 9
46. $9t-10=44$ 6
47. $12a-3=45$ 4
48. $\frac{x}{3}+2=6$ 12
49. $\frac{n}{4}+3=9$ 24
50. $\frac{c}{7}-6=0$ 42
51. $\frac{r}{4}-8=12$ 80

Give the decimal approximation to the nearest tenth. (page 450)

52. $\sqrt{10}$ 3.2
53. $\sqrt{31}$ 5.6
54. $\sqrt{52}$ 7.2
55. $\sqrt{89}$ 9.4

Use the Pythagorean Rule to find the missing length. (page 452)

56. 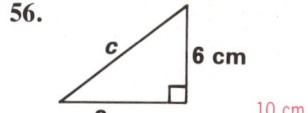 10 cm

57. 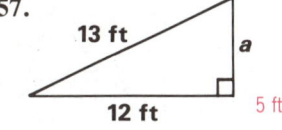 5 ft

Algebra **459**

Cumulative Test
(Chapters 1–17)

Use Copymaster S109 to provide the students with an answer sheet in standardized test format.

Answers for Cumulative Test, Chapters 1–17

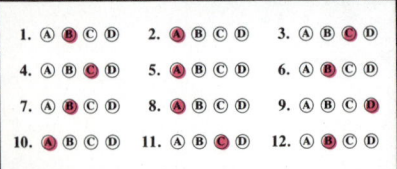

The table below correlates test items with student text pages.

Test Item	Page Taught	Skill Practice
1	68	p. 443, exercises 1–12
2	104	p. 443, exercises 13–24
3	180	p. 443, exercises 25–36
4	188	p. 443, exercises 37–48
5	202	p. 443, exercises 49–60
6	212	p. 457, exercises 1–12
7	294	p. 457, exercises 13–21
8	296	p. 457, exercises 22–30
9	322	p. 457, exercises 37–39
10	412	p. 457, exercises 40–49
11	376	
12	448	

460

Cumulative TEST — Standardized Format

Choose the correct letter.

1. Give the product.

0.26×0.031

A. 0.0806
B. 0.00806
C. 0.00104
D. none of these

2. Give the quotient rounded to the nearest tenth.

$17.43 \div 2.9$

A. 6.0
B. 6.1
C. 0.61
D. none of these

3. Give the sum.

$\frac{4}{9} + \frac{5}{6}$

A. $\frac{3}{5}$
B. 1
C. $1\frac{5}{18}$
D. none of these

4. Give the difference.

$\frac{7}{8} - \frac{2}{3}$

A. 1   B. $\frac{5}{8}$
C. $\frac{5}{24}$ D. none of these

5. Give the product.

$\frac{4}{5} \times \frac{3}{2}$

A. $1\frac{1}{5}$ B. $\frac{8}{15}$
C. $1\frac{7}{8}$ D. none of these

6. Give the quotient.

$\frac{5}{6} \div \frac{2}{3}$

A. $\frac{5}{9}$ B. $1\frac{1}{4}$
C. $\frac{4}{5}$ D. none of these

7. Solve.

$66\frac{2}{3}\%$ of $42 = n$

A. 63
B. 28
C. 14
D. none of these

8. Solve.

20% of $n = 15$

A. 75
B. 3
C. 150
D. none of these

9. Compute the interest.

Principal = $720
Rate = 12% per year
Time = 4 months

A. $86.40
B. $43.20
C. $1036.80
D. none of these

10. Give the sum.

$^-9 + {}^+7$

A. $^-2$
B. $^+2$
C. $^-16$
D. $^+16$

11. Find the volume of this cone. Use 3.14 for π.

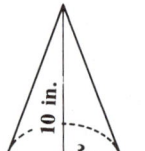

A. 62.8 in.^{3}
B. 282.6 in.^{3}
C. 94.2 in.^{3}
D. none of these

12. Solve.

$2b + 8 = 32$

A. 20
B. 12
C. 24
D. none of these

460 *Chapter 17*

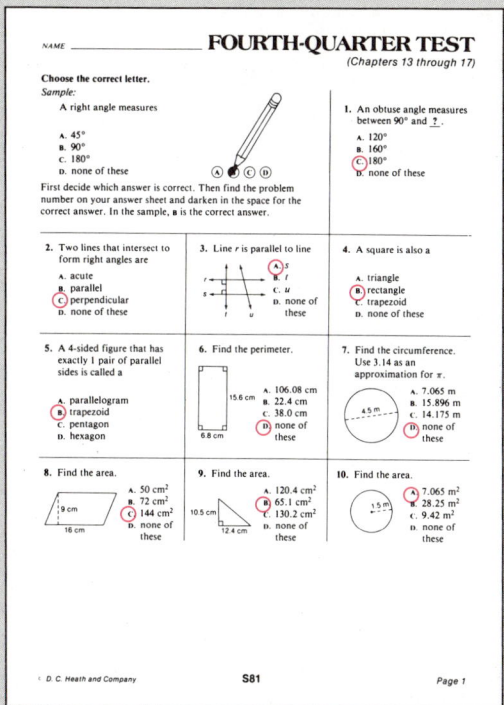
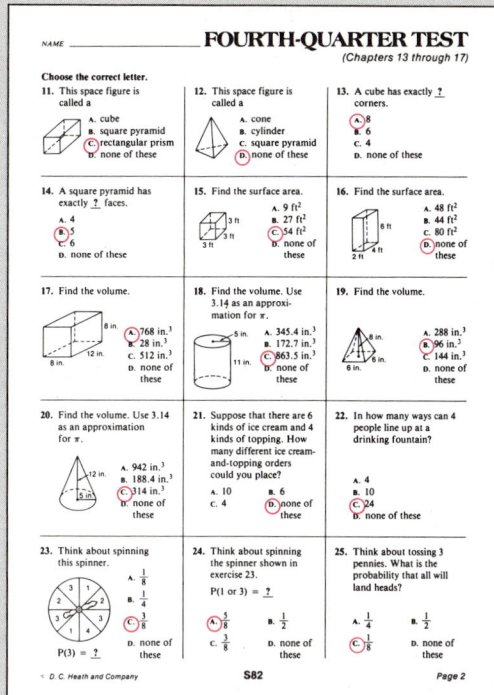

Fourth-Quarter Test

The fourth-quarter test shown on these two pages is in standardized format so that the students can become accustomed to taking standardized tests.

460A

Use Copymaster S92 or Duplicating Master S92 to provide the students with an answer sheet in standardized test format.

Copymaster S107 has a quick-score answer key for the fourth-quarter test.

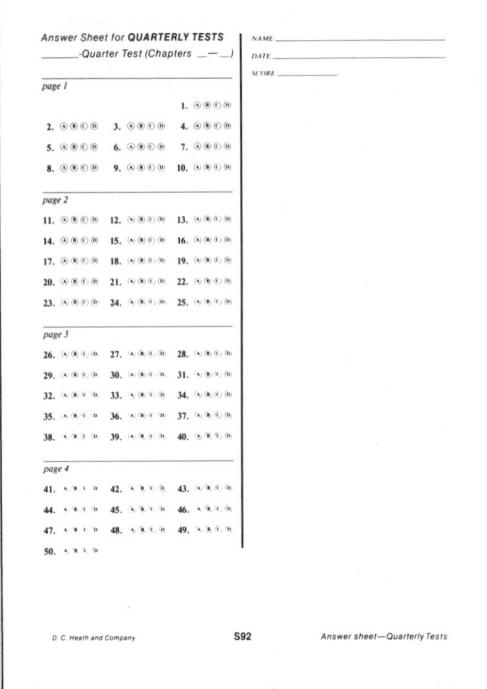

The table below correlates test items with student text pages.

Test Item	Text Page	Test Item	Text Page	Test Item	Text Page	Test Item	Text Page
1	p. 334	14	p. 364	27	p. 396	40	p. 434
2	p. 336	15	p. 368	28	p. 398	41	p. 436
3	p. 336	16	p. 368	29	p. 399	42	p. 436
4	p. 340	17	p. 372	30	p. 400	43	p. 438
5	p. 340	18	p. 374	31	p. 410	44	p. 440
6	p. 342	19	p. 376	32	p. 410	45	p. 444
7	p. 344	20	p. 376	33	p. 411	46	p. 446
8	p. 350	21	p. 386	34	p. 412	47	p. 448
9	p. 352	22	p. 388	35	p. 415	48	p. 448
10	p. 354	23	p. 390	36	p. 419	49	p. 450
11	p. 364	24	p. 390	37	p. 420	50	p. 452
12	p. 364	25	p. 396	38	p. 424		
13	p. 364	26	p. 396	39	p. 434		

Copymaster S85 or Duplicating Master S85

Copymaster S86 or Duplicating Master S86

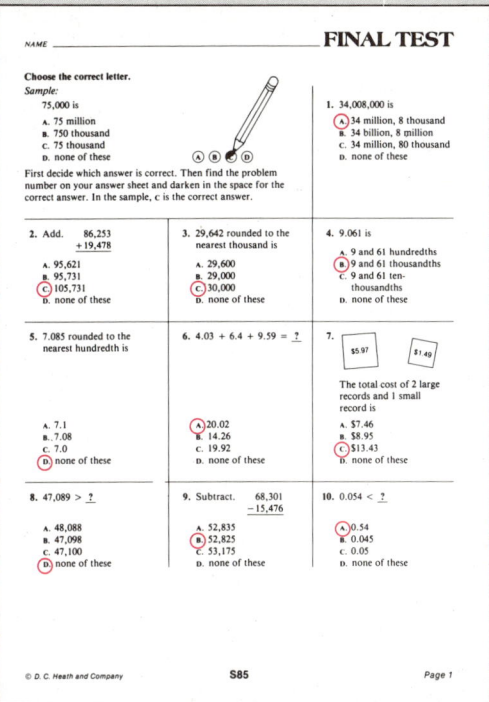

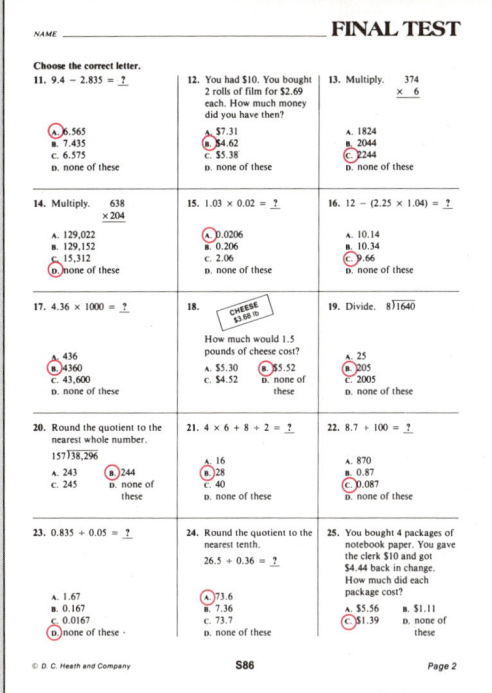

Final Test

The final test shown on these two pages is in standardized format so that the students can become accustomed to taking standardized tests.

460C

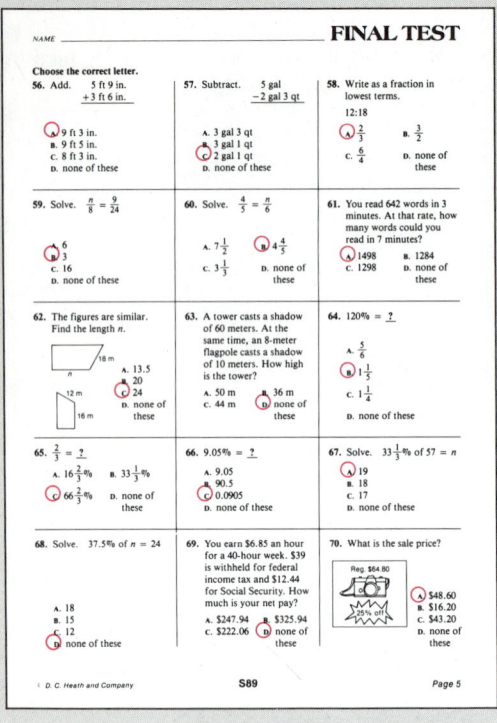

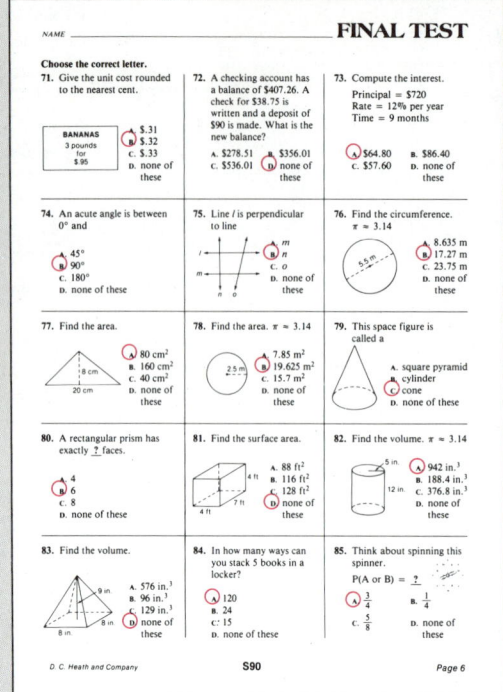

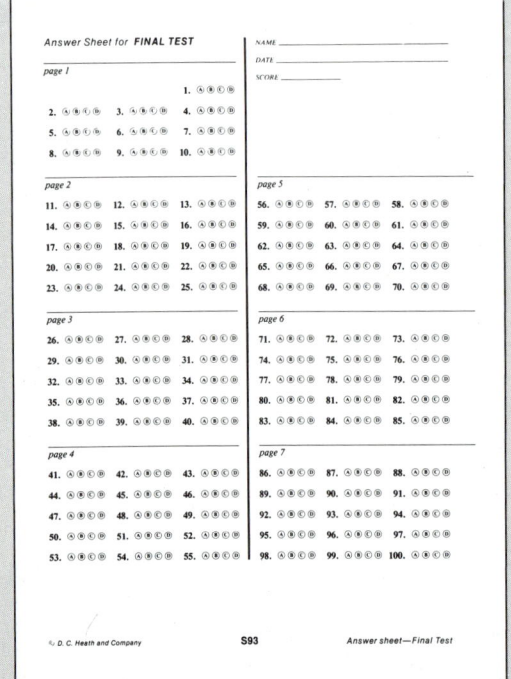

Use Copymaster S93 or Duplicating Master S93 to provide the students with an answer sheet in standardized format.

Copymaster S108 has a quick-score answer key for the final test.

460D

SKILL TEST
EXTRA PRACTICE

SKILL TEST

Pages 462–470

This test will help you find out which skills you know well and which skills you need to practice more.

EXTRA PRACTICE

Pages 471–502

These practice sets cover the skills tested on the SKILL TEST. Each set practices one skill. The skills are presented in the same order as they are in the book. Page references will help you and your teacher decide when to use them.

Minimum Competency

This Skill Test may be used to help students prepare for **minimum competency tests.** If a student responds incorrectly to two or more items on a particular skill, assign the corresponding Extra Practice set.

SKILL TEST

SKILL	TEST ITEMS			EXTRA PRACTICE
1 Adding whole numbers *page 6*	7549 + 4261 11,810 496 + 3081 + 2566 6143		80,665 + 24,364 105,029 26,245 + 6518 + 276 33,039	page 471
2 Rounding whole numbers *page 10*	Round to the nearest ten. 63 60 125 130 682 680 598 600 Round to the nearest hundred. 446 400 967 1000 2809 2800 650 700			page 471
3 Rounding decimals *page 18*	Round to the nearest tenth. 1.38 1.4 2.50 2.5 63.05 63.1 36.95 37.0 Round to the nearest hundredth. 18.342 18.34 0.375 0.38 0.496 0.50 8.640 8.64			page 472
4 Adding decimals *page 22*	2.34 + 1.7 4.04 8.04 + 7 + 9.6 24.64		5.62 + 2.94 8.56 0.483 + 1.56 + 4.4 6.443	page 472
5 Comparing whole numbers *page 32*	< or >? 743 ● 734 > 42,382 ● 42,328 >		3321 ● 3400 < 599,999 ● 600,000 <	page 473
6 Subtracting whole numbers *page 38*	78 − 37 41 802 − 378 424 500 − 142 358	80 − 19 61 600 − 374 226 300 − 62 238	972 − 514 458 4534 − 3492 1042 3800 − 1452 2348	page 473
7 Comparing decimals *page 42*	< or >? 3.57 ● 3.75 < 0.345 ● 0.3366 >	4.2 ● 3.21 > 0.4 ● 4.0 <	0.2 ● 0.19 > 0.031 ● 0.13 <	page 474
8 Subtracting decimals *page 46*	5 − 2.7 2.3 4.23 − 2.849 1.381	7 − 2.4 4.6 13 − 6.7 6.3	25.3 − 6 19.3 16.2 − 3.571 12.629	page 474

SKILL TEST

SKILL	TEST ITEMS			EXTRA PRACTICE
9 Multiplying by multiples of 10, 100, 1000 *page 56*	35 × 100 3500 78 × 1000 78,000 92 × 10 920	40 × 60 2400 70 × 500 35,000 90 × 60 5400	324 × 10 3240 500 × 600 300,000 300 × 40 12,000	*page 475*
10 Multiplying whole numbers *page 62*	58 × 5 290	304 × 62 18,848	879 × 28 24,612	2836 × 81 229,716	*page 475*
11 Multiplying decimals *page 68*	5.4 × 0.36 1.944 6.05 × 0.39 2.3595	6.3 × 1.2 7.56 2.04 × 1.6 3.264	34 × 0.88 29.92 8.25 × 2.06 16.995	*page 476*
12 Simplifying expressions *page 70*	12 − (2.8 + 7.4) 1.8 (23.97 × 1.6) + 17.4 55.752		5.6 × (9.9 − 3.8) 34.16 5 − (3.07 × 0.44) 3.6492	*page 476*
13 Multiplying decimals by 10, 100, 1000 *page 72*	0.93 × 100 93 5.28 × 1000 5280 3.8 × 10 38	0.3 × 10 3 4.1 × 1000 4100 5.22 × 100 522	4.7 × 10 47 0.004 × 100 0.4 0.03 × 1000 30	*page 477*
14 Dividing whole numbers *page 88*	6) 582 97 52) 45,388 872 R44	9) 387 43 50) 7500 150	26) 5280 203 R2 78) 60,411 774 R39	*page 477*
15 Applying rules of order of operations *page 94*	12 − 8 + 4 8 24 × (12 − 2) 240		18 ÷ 6 × 2 6 16 + 3 × 4 ÷ 2 22	*page 478*
16 Dividing decimals by 10, 100, 1000 *page 100*	9.45 ÷ 10 0.945 53.5 ÷ 1000 0.0535	8.3 ÷ 10 0.83 6.94 ÷ 1000 0.00694	450.5 ÷ 100 4.505 0.84 ÷ 100 0.0084	*page 478*
17 Dividing decimals *page 104*	0.3) 1.7 5.7	0.2) 3.7 18.5	0.06) 0.574 9.6	*page 479*

17 Round each quotient to the nearest tenth.

2.6) 3.79 1.5   1.3) 6.21 4.8   0.82) 0.097 0.1

SKILL TEST

SKILL	TEST ITEMS		EXTRA PRACTICE
18 Solving two- and three-step problems *page 106*	Beth had $10. She wanted to buy 2 records that cost $5.79 each. How much more money did she need? $1.58	Ron bought 2 records for $3.45 each and 4 records for $3.79 each. What was the total cost? $22.06	*page 479*
19 Solving problems from a bar graph *page 116*	How many more hours did Ned watch TV than Ann? 3 What was the average number of hours spent watching TV? 9	Hours Spend Watching TV in One Week (bar graph: Ann, Bill, Jon, Ned)	*page 480*
20 Solving problems from a line graph *page 118*	How much money was saved during the first three weeks? $35.50 How much more did Sam save the fifth week than the fourth week? $6.50	Sam's Weekly Savings (line graph)	*page 480*
21 Solving problems from a picture graph *page 120*	How many more tickets did Fran sell than Carl? 8 What is the total number of tickets sold by Lou and Pat? 56	Tickets Sold for the School Play — Carl, Fran, Lou, Pat. Each ticket stands for 4 tickets	*page 481*

464   *Skill Test*

SKILL TEST

SKILL	TEST ITEMS	EXTRA PRACTICE

22 Solving problems from a chart
page 128

What was Robin's three-game total? **404**
In game 3, Robin and Jenny outscored Amy and Sarah by how many points? **3**

BOWLER	Scores		
	GAME 1	GAME 2	GAME 3
Robin	126	146	132
Jenny	125	163	140
Amy	113	123	145
Sarah	138	151	124

page 481

---

23 Finding equivalent fractions
page 146

$\frac{1}{4} = \frac{?}{8}$ **2** $\quad \frac{3}{9} = \frac{?}{3}$ **1** $\quad \frac{8}{12} = \frac{?}{6}$ **4** $\quad \frac{5}{8} = \frac{?}{24}$ **15**

$\frac{1}{3} = \frac{?}{9}$ **3** $\quad \frac{4}{8} = \frac{?}{2}$ **1** $\quad \frac{4}{5} = \frac{?}{10}$ **8** $\quad \frac{5}{12} = \frac{?}{24}$ **10**

page 482

---

24 Writing fractions in lowest terms
page 148

$\frac{3}{12} = ?$ $\frac{1}{4}$ $\quad \frac{15}{20} = ?$ $\frac{3}{4}$ $\quad \frac{9}{6} = ?$ $\frac{3}{2}$ $\quad \frac{18}{10} = ?$ $\frac{9}{5}$

$\frac{2}{6} = ?$ $\frac{1}{3}$ $\quad \frac{3}{9} = ?$ $\frac{1}{3}$ $\quad \frac{8}{7} = ?$ $\frac{8}{7}$ $\quad \frac{15}{10} = ?$ $\frac{3}{2}$

page 482

---

25 Finding the least common denominator
page 150

Find the least common denominator.

$\frac{1}{3} \quad \frac{1}{2}$ **6** $\qquad \frac{1}{4} \quad \frac{3}{4}$ **4** $\qquad \frac{3}{4} \quad \frac{5}{6}$ **12** $\qquad \frac{5}{6} \quad \frac{5}{8}$ **24**

page 483

---

26 Comparing fractions
page 152

< or >?

$\frac{4}{5} \bullet \frac{3}{5}$ **>** $\qquad \frac{1}{4} \bullet \frac{3}{8}$ **<** $\qquad \frac{3}{4} \bullet \frac{2}{3}$ **>** $\qquad \frac{5}{6} \bullet \frac{7}{8}$ **<**

page 483

---

27 Writing whole and mixed numbers as fractions
page 156

Change to fourths.

$2 = ?$ $\frac{8}{4}$ $\qquad 4 = ?$ $\frac{16}{4}$ $\qquad 6 = ?$ $\frac{24}{4}$ $\qquad 3 = ?$ $\frac{12}{4}$

Change to a fraction.

$1\frac{1}{4} = ?$ $\frac{5}{4}$ $\qquad 1\frac{2}{3} = ?$ $\frac{5}{3}$ $\qquad 2\frac{3}{4} = ?$ $\frac{11}{4}$ $\qquad 3\frac{5}{6} = ?$ $\frac{23}{6}$

page 484

---

28 Writing fractions as whole or mixed numbers
page 158

Change to a whole number.

$\frac{8}{2} = ?$ **4** $\qquad \frac{9}{3} = ?$ **3** $\qquad \frac{16}{4} = ?$ **4** $\qquad \frac{18}{6} = ?$ **3**

Change to a mixed number.

$\frac{4}{3} = ?$ $1\frac{1}{3}$ $\qquad \frac{5}{2} = ?$ $2\frac{1}{2}$ $\qquad \frac{13}{5} = ?$ $2\frac{3}{5}$ $\qquad \frac{11}{4} = ?$ $2\frac{3}{4}$

page 484

Skill Test

SKILL TEST

SKILL	TEST ITEMS	EXTRA PRACTICE
29 Writing fractions and mixed numbers in simplest form *page 160*	Write in simplest form. $\frac{6}{9} = ?\ \frac{2}{3}$ $3\frac{2}{4} = ?\ 3\frac{1}{2}$ $\frac{15}{3} = ?\ 5$ $\frac{14}{6} = ?\ 2\frac{1}{3}$	*page 485*
30 Writing fractions and mixed numbers as decimals *page 166*	Change to a decimal. $\frac{2}{5} = ?\ 0.4$ $\frac{3}{2} = ?\ 1.5$ $1\frac{3}{4} = ?\ 1.75$ $2\frac{3}{8} = ?\ 2.375$ Change to a decimal rounded to the nearest hundredth. $\frac{1}{3} = ?\ 0.33$ $\frac{5}{6} = ?\ 0.83$ $\frac{10}{9} = ?\ 1.11$ $\frac{20}{3} = ?\ 6.67$	*page 485*
31 Writing decimals as fractions or mixed numbers *page 168*	Change to a fraction in simplest form. $0.4 = ?\ \frac{2}{5}$ $0.75 = ?\ \frac{3}{4}$ $0.375 = ?\ \frac{3}{8}$ $0.36 = ?\ \frac{9}{25}$ Change to a mixed number in simplest form. $1.25 = ?\ 1\frac{1}{4}$ $3.5 = ?\ 3\frac{1}{2}$ $1.625 = ?\ 1\frac{5}{8}$ $2.08 = ?\ 2\frac{2}{25}$	*page 486*
32 Adding fractions *page 180*	Give the sum in simplest form. $\frac{1}{8} + \frac{1}{8}\ \frac{1}{4}$ $\frac{1}{3} + \frac{1}{2}\ \frac{5}{6}$ $\frac{5}{6} + \frac{3}{4}\ 1\frac{7}{12}$ $\frac{9}{16} + \frac{5}{8}\ 1\frac{3}{16}$	*page 486*
33 Adding mixed numbers *page 182*	Give the sum in simplest form. $3\frac{1}{4} + 2\frac{1}{2} = 5\frac{3}{4}$ $4\frac{3}{4} + 3\frac{1}{8} = 7\frac{7}{8}$ $2\frac{2}{3} + 2\frac{1}{2} = 5\frac{1}{6}$ $4\frac{1}{4} + 3\frac{7}{8} = 8\frac{1}{8}$	*page 487*
34 Subtracting fractions *page 188*	Give the difference in simplest form. $\frac{3}{4} - \frac{1}{2}\ \frac{1}{4}$ $\frac{5}{8} - \frac{1}{3}\ \frac{7}{24}$ $\frac{5}{6} - \frac{1}{4}\ \frac{7}{12}$ $\frac{7}{6} - \frac{5}{9}\ \frac{11}{18}$	*page 487*
35 Subtracting mixed numbers *page 192*	Give the difference in simplest form. $4\frac{1}{2} - 1\frac{3}{8} = 3\frac{1}{8}$ $6 - 2\frac{1}{2} = 3\frac{1}{2}$ $8\frac{1}{4} - 3\frac{2}{3} = 4\frac{7}{12}$ $5\frac{1}{4} - 4\frac{2}{5} = \frac{17}{20}$	*page 488*

SKILL TEST

	SKILL	TEST ITEMS	EXTRA PRACTICE
36	Multiplying fractions *page 202*	Give the product in simplest form. $\frac{1}{2} \times \frac{1}{3}$ $\frac{1}{6}$ $\frac{2}{3} \times \frac{3}{4}$ $\frac{1}{2}$ $3 \times \frac{3}{4}$ $2\frac{1}{4}$ $\frac{5}{6} \times \frac{6}{5}$ 1	*page 488*
37	Finding a fraction of a number *page 204*	$\frac{1}{2}$ of $24 = ?$ 12 $\frac{1}{3}$ of $18 = ?$ 6 $\frac{2}{3}$ of $42 = ?$ 28 $\frac{3}{5}$ of $60 = ?$ 36 $\frac{2}{7}$ of $21 = ?$ 6 $\frac{3}{4}$ of $72 = ?$ 54	*page 489*
38	Multiplying mixed numbers *page 206*	Give the product in simplest form. $1\frac{1}{4} \times 1\frac{1}{2}$ $1\frac{7}{8}$ $2 \times 2\frac{1}{2}$ 5 $1\frac{2}{3} \times 1\frac{3}{4}$ $2\frac{11}{12}$ $3\frac{1}{5} \times 2\frac{3}{8}$ $7\frac{3}{5}$	*page 489*
39	Finding a fraction of a unit of measure *pages 208, 209*	$1\frac{2}{3}$ h $= \underline{?}$ min 100 $2\frac{1}{2}$ ft $= \underline{?}$ in. 30 $2\frac{2}{3}$ yd $= \underline{?}$ ft 8 $1\frac{3}{4}$ gal $= \underline{?}$ qt 7	*page 490*
40	Dividing fractions *page 212*	Give the quotient in simplest form. $\frac{2}{3} \div 4$ $\frac{1}{6}$ $\frac{3}{4} \div \frac{3}{2}$ $\frac{1}{2}$ $\frac{4}{9} \div \frac{1}{3}$ $1\frac{1}{3}$ $\frac{5}{2} \div \frac{3}{4}$ $3\frac{1}{3}$	*page 490*
41	Dividing mixed numbers *page 214*	Give the quotient in simplest form. $3 \div 2\frac{1}{3}$ $1\frac{2}{7}$ $3\frac{2}{3} \div 1\frac{1}{2}$ $2\frac{4}{9}$ $4\frac{3}{4} \div 2$ $2\frac{3}{8}$ $4\frac{1}{5} \div 2\frac{5}{8}$ $1\frac{3}{5}$	*page 491*
42	Solving measurement problems in the metric system *pages 226, 230*	Cory's arm span is 1 meter 48 centimeters. Al's arm span is 159 centimeters. How much longer is Al's arm span? 11 centimeters	How many liters of fruit punch will it take to fill 60 paper cups that hold 150 milliliters each? 9	*page 491*
43	Solving measurement problems in the customary system *pages 242, 246*	Mel bought 4 yards 2 feet of chain. If chain costs $1.10 per foot, what was the total price? $15.40	Dawn bought 8 pounds 4 ounces of hamburger. How many 6-ounce servings of hamburger could she serve? 22	*page 492*

SKILL TEST

SKILL	TEST ITEMS		EXTRA PRACTICE
44 Solving proportions *page 262*	$\frac{12}{n} = \frac{1}{4}$ 48 $\frac{n}{8} = \frac{4}{5}$ $6\frac{2}{5}$ $\frac{2}{3} = \frac{3}{n}$ $4\frac{1}{2}$ $\frac{7}{10} = \frac{n}{8}$ $5\frac{3}{5}$		page 492
45 Solving ratio and proportion problems *pages 265, 268*	A cookie recipe calls for 3 eggs to make 48 cookies. How many eggs are needed to make 112 cookies? 7	Bill can bake 24 cookies every 18 minutes. At that rate, how long will it take him to bake 120 cookies? 90 min or $1\frac{1}{2}$ h	page 493
46 Writing percents as fractions or mixed numbers *page 282*	Change to a fraction or mixed number in simplest form. 25% = ? $\frac{1}{4}$ 150% = ? $1\frac{1}{2}$ $33\frac{1}{3}$% = ? $\frac{1}{3}$ $62\frac{1}{2}$% = ? $\frac{5}{8}$		page 493
47 Writing fractions as percents *page 284*	Change to a percent. $\frac{3}{5}$ = ? 60% $\frac{3}{2}$ = ? 150% $\frac{1}{6}$ = ? $16\frac{2}{3}$% $\frac{2}{3}$ = ? $66\frac{2}{3}$%		page 494
48 Writing percents as decimals *page 286*	Change to a decimal. 8% = ? 0.08 32% = ? 0.32 37.5% = ? 0.375 5.6% = ? 0.056		page 494
49 Writing decimals as percents *page 287*	Change to a percent. 0.05 = ? 5% 0.9 = ? 90% 1.375 = ? 137.5% $0.12\frac{1}{2}$ = ? $12\frac{1}{2}$%		page 495
50 Finding a percent of a number *page 290*	25% of 44 = n 11 6.5% of 38 = n 2.47	9% of 37 = n 3.33 0.5% of 60 = n 0.3	page 495
51 Finding a percent of a number by proportion *page 294*	$33\frac{1}{3}$% of 81 = n 27 $66\frac{2}{3}$% of 96 = n 64	$12\frac{1}{2}$% of 48 = n 6 $16\frac{2}{3}$% of 72 = n 12	page 496
52 Finding the number when a percent is given *page 296*	20% of n = 16 80 Round each answer to the nearest tenth. 8.5% of n = 12.5 147.1	75% of n = 48 64 32.2% of n = 34.6 107.5	page 496

SKILL TEST

SKILL	TEST ITEMS		EXTRA PRACTICE
53 Solving percent problems *page 298*	Dava made 16 of 25 free throws. What percent of her free throws did she make? 64%	Of the 20 players on the basketball team, 12 are seniors. What percent of the players are seniors? 60%	*page 497*
54 Solving percent problems *page 300*	How many in the survey chose rock? 120 How many in the survey chose country? 60	**Favorite Kind of Music** — 15% Jazz, 10% Classical, 50% Rock, 25% Country. 240 Surveyed	*page 497*
55 Solving personal finance problems *pages 322, 326*	Sue borrowed $900 for 1 year at the yearly rate of 12%. How much will she have to pay back at the end of the year? $1008	Craig bought a blank video tape for $3.85 and some cleaner solution for $1.98. The sales tax was 5%. What was his total bill? $6.13	*page 498*
56 Finding an area *pages 348, 352, 354*	Give the area. Use 3.14 for π. Rectangle 5 m × 2.4 m — 12 m² ; Triangle base 14 cm, height 6 cm — 42 cm² ; Circle radius 3 m — 28.26 m²		*page 498*

Skill Test **469**

SKILL TEST

SKILL	TEST ITEMS			EXTRA PRACTICE
57 Finding a volume *pages 372, 374, 376*	Give the volume. Use 3.14 for π. 24 m³	452.16 cm³	200.96 m³	*page 499*
58 Solving probability problems *pages 396, 398*	A bag contains 5 red crayons and 4 blue crayons. If you pick a crayon without looking, what is the probability that it will be red? $\frac{5}{9}$		Carol had 26 cards. On each card she wrote a different letter of the alphabet. If she picks a card without looking, what is the probability that she will pick a letter in her first name? $\frac{5}{26}$	*page 499*
59 Comparing integers *page 410*	$<$ or $>$? $^-3 \bullet \ ^-2$ $<$	$^+9 \bullet 0$ $>$	$^+12 \bullet \ ^+11$ $>$ $^+10 \bullet \ ^-13$ $>$	*page 500*
60 Adding integers *page 412*	$^+4 + \ ^+7$ $^+11$	$^+9 + \ ^-3$ $^+6$	$^-8 + \ ^+8$ 0 $^-12 + \ ^-15$ $^-27$	*page 500*
61 Subtracting integers *pages 414, 415*	$^-7 - \ ^+3$ $^-10$	$^+8 - \ ^-9$ $^+17$	$^+14 - \ ^+14$ 0 $^-16 - \ ^-20$ $^+4$	*page 501*
62 Multiplying integers *pages 418, 419*	$^+6 \times \ ^-7$ $^-42$	$^+8 \times \ ^+3$ $^+24$	$^-5 \times \ ^+9$ $^-45$ $^-7 \times 0$ 0	*page 501*
63 Dividing integers *page 420*	$^-24 \div \ ^+8$ $^-3$	$^+54 \div \ ^-9$ $^-6$	$^-45 \div \ ^-5$ $^+9$ $^+72 \div \ ^+8$ $^+9$	*page 502*
64 Solving equations *page 448*	$2y + 7 = 21$ 7 $\frac{c}{2} + 8 = \ ^-3$ $^-22$		$6x - 10 = \ ^-22$ $^-2$ $\frac{w}{3} - 5 = \ ^-7$ $^-6$	*page 502*

EXTRA PRACTICE

The **Mental Math Extensions** require the student to mentally compute with rounded numbers in order to estimate the answers to exercises. Each extension is matched with a particular skill. The extension for the particular skill can be done anytime from a few days to many weeks after the student has completed the exercises for that skill. It is important that the answers to the exercises not be available to the student while an extension is being done.

SKILL 1 (Use after page 6.)

Give the sum.

1. 6438 + 8310   14,748
2. 5832 + 694   6526
3. 966 + 2947   3913
4. 3370 + 1938   5308
5. 34,006 + 8825   42,831
6. 4721 + 76,082   80,803
7. 12,500 + 38,926   51,426
8. 38,842 + 27,111   65,953
9. 493 + 3493 + 977   4963
10. 8218 + 739 + 1005   9962
11. 182 + 4200 + 3628   8010
12. 7467 + 941 + 604   9012
13. 593 + 444 + 1660   2697
14. 2741 + 8009 + 476   11,226
15. 4850 + 1188 + 2055   8093
16. 1748 + 2966 + 1826   6540
17. 54,388 + 2112 + 599   57,099
18. 4368 + 829 + 12,477   17,674
19. 29,006 + 2704 + 1822   33,532
20. 2864 + 31,000 + 8002   41,866

Here's how

245 + 92 + 3916 = ?

Line up the digits that are in the same place.

```
  245
   92
+3916
```

Add.

```
  1 1
  245
  1 92
+3916
 4253
```

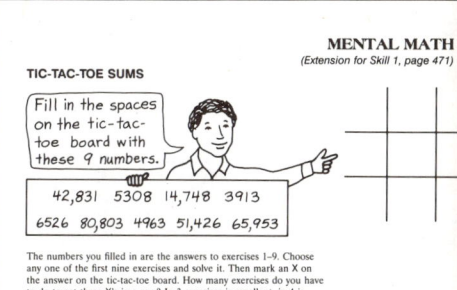

SKILL 2 (Use after page 10.)

Round to the nearest ten.

1. 74   70
2. 37   40
3. 42   40
4. 75   80
5. 183   180
6. 366   370
7. 805   810
8. 411   410
9. 4336   4340
10. 3721   3720
11. 3605   3610
12. 2398   2400

Round to the nearest hundred.

13. 276   300
14. 550   600
15. 743   700
16. 849   800
17. 3408   3400
18. 3423   3400
19. 6660   6700
20. 8050   8100
21. 20,305   20,300
22. 32,780   32,800
23. 42,912   42,900
24. 62,950   63,000

Round to the nearest thousand.

25. 4841   5000
26. 6851   7000
27. 9310   9000
28. 6500   7000
29. 35,431   35,000
30. 42,573   43,000
31. 719,527   720,000
32. 273,500   274,000

Round to the nearest ten thousand.

33. 24,146   20,000
34. 52,700   50,000
35. 56,913   60,000
36. 49,430   50,000
37. 92,604   90,000
38. 28,911   30,000
39. 249,300   250,000
40. 613,812   610,000

Here's how

Round 45,359 to the nearest hundred.

Rounding to this place
↓
45,359
↑
When the next digit to the right is 5 or greater, round up.

45,359 rounds to 45,400.

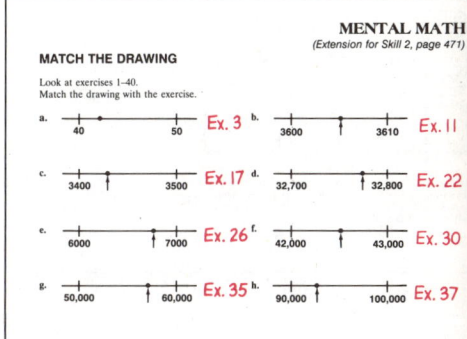

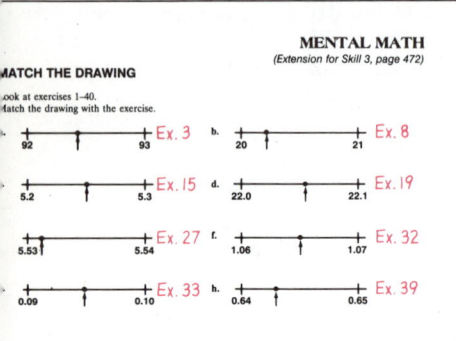

SKILL 3 (Use after page 18.)

Here's how

Round 36.417 to the nearest tenth.

Rounding to this place
↓
36.417
↑
When the next digit to the right is 5 or greater, round up.

36.417 rounds to 36.4.

Round to the nearest whole number.

1. 16.6   17
2. 38.3   38
3. 92.4   92
4. 35.5   36
5. 51.27   51
6. 38.93   39
7. 0.025   0
8. 20.19   20
9. 327.04   327
10. 118.40   118
11. 0.500   1
12. 12.099   12

Round to the nearest tenth.

13. 403.38   403.4
14. 26.10   26.1
15. 5.25   5.3
16. 3.95   4.0
17. 21.39   21.4
18. 24.188   24.2
19. 22.06   22.1
20. 7.472   7.5
21. 204.29   204.3
22. 444.484   444.5
23. 0.0592   0.1
24. 0.95   1.0

Round to the nearest hundredth.

25. 22.317   22.32
26. 56.208   56.21
27. 5.531   5.53
28. 54.325   54.33
29. 71.594   71.59
30. 6.30196   6.30
31. 0.0518   0.05
32. 1.065   1.07
33. 0.0946   0.09
34. 11.269   11.27
35. 3.9421   3.94
36. 0.097   0.10
37. 42.3381   42.34
38. 28.095   28.10
39. 0.6422   0.64
40. 1.4920   1.49

SKILL 4 (Use after page 22.)

Here's how

3 + 2.51 + 8.6 = ?

Line up the decimal points.

```
   3
 2.51
+8.6
```

Add.

```
  1
   3
 2.51
+8.6
14.11
```

Give the sum.

1. 4.64 + 3.08   7.72
2. 7.564 + 3.806   11.37
3. 6.3521 + 0.5821   6.9342
4. 721.6 + 38.4   760
5. 2.35 + 4.829   7.179
6. 5.008 + 3.62   8.628
7. 43.6 + 27.48   71.08
8. 10.88 + 9.3   20.18
9. 5.6 + 3.04 + 2.7   11.34
10. 2.64 + 5.7 + 8.8   17.14
11. 4.20 + 9.2 + 3.65   17.05
12. 6.1 + 2.22 + 6.83   15.15
13. 2.641 + 0.75 + 3.58   6.971
14. 5.34 + 0.756 + 2.84   8.936
15. 9.3645 + 2.055 + 0.221   11.6405
16. 8.471 + 0.4911 + 3.300   12.2621
17. 7.4 + 4.611 + 8.5   20.511
18. 15.966 + 8.4 + 4.8   29.166
19. 32 + 3.4 + 2.08   37.48
20. 5.7 + 41 + 6.63   53.33

Extra Practice

SKILL 5 (Use after page 32.)

Here's how
54,375 ● 54,491
Start at the left and compare digits that are in the same place.

┌─ is less than ─┐
↓   ↓
54,375   54,491

So

54,375 < 54,491

< or >?

1. 93 ● 90 >
2. 74 ● 79 <
3. 376 ● 388 <
4. 200 ● 186 >
5. 565 ● 566 <
6. 496 ● 490 >
7. 567 ● 576 <
8. 699 ● 700 <
9. 950 ● 1000 <
10. 1503 ● 999 >
11. 875 ● 1100 <
12. 1224 ● 899 >
13. 3721 ● 3615 >
14. 6732 ● 6655 >
15. 8472 ● 8427 >
16. 3818 ● 3811 >
17. 57,352 ● 58,410 <
18. 29,435 ● 29,400 >
19. 88,642 ● 89,000 <
20. 49,462 ● 50,362 <
21. 74,000 ● 73,999 >
22. 53,078 ● 53,780 <
23. 480,000 ● 479,000 >
24. 799,999 ● 800,000 <

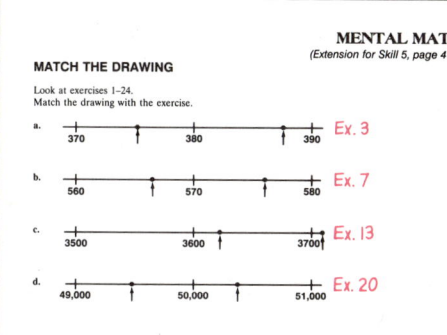

MENTAL MATH (Extension for Skill 5, page 473)

MATCH THE DRAWING
Look at exercises 1–24.
Match the drawing with the exercise.

a. 370 — 380 — 390   Ex. 3
b. 560 — 570 — 580   Ex. 7
c. 3500 — 3600 — 3700   Ex. 13
d. 49,000 — 50,000 — 51,000   Ex. 20

Copymaster S491

SKILL 6 (Use after page 38.)

Here's how
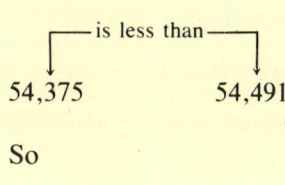

No tens! Regroup 1 hundred for 10 tens.

```
    2
  8 3̶ 0 5
 -4 0 7 8
```

Regroup 1 ten for 10 ones.

```
    2 9
  8 3̶ 0̶ 15
 -4 0 7 8
```

Subtract.

```
    2 9
  8 3̶ 0̶ 15
 -4 0 7 8
  4 2 2 7
```

Give the difference.

1. 828 − 411   417
2. 594 − 221   373
3. 710 − 463   247
4. 824 − 258   566
5. 504 − 356   148
6. 701 − 588   113
7. 806 − 529   277
8. 903 − 165   738
9. 800 − 361   439
10. 400 − 249   151
11. 700 − 318   382
12. 600 − 233   367
13. 800 − 444   356
14. 500 − 381   119
15. 4916 − 2854   2062
16. 5874 − 2222   3652
17. 3406 − 2153   1253
18. 7112 − 4338   2774
19. 2502 − 458   2044
20. 3701 − 229   3472
21. 5205 − 1286   3919
22. 6101 − 2255   3846
23. 3111 − 2478   633
24. 9055 − 3861   5194

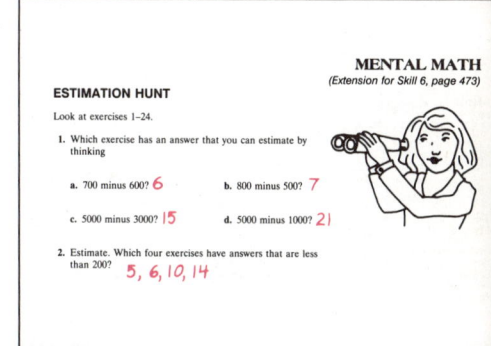

MENTAL MATH (Extension for Skill 6, page 473)

ESTIMATION HUNT
Look at exercises 1–24.

1. Which exercise has an answer that you can estimate by thinking
   a. 700 minus 600? 6
   b. 800 minus 500? 7
   c. 5000 minus 3000? 15
   d. 5000 minus 1000? 21

2. Estimate. Which four exercises have answers that are less than 200? 5, 6, 10, 14

Copymaster S491

Extra Practice **473**

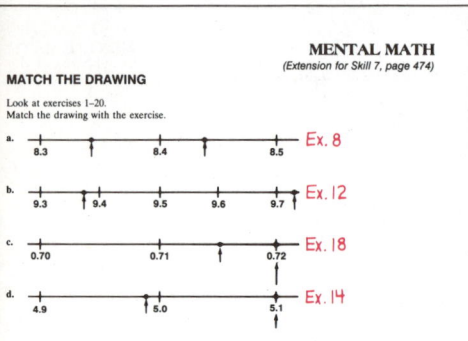

SKILL 7 (Use after page 42.)

Here's how

82.64 ● 82.39
Start at the left and compare digits that are in the same place.

┌─ is greater than ─┐
↓   ↓
82.64   82.39

So

82.64 > 82.39

< or >?

1. 0.3 ● 0.8   <
2. 0.6 ● 0.1   >
3. 0.04 ● 0.03   >
4. 0.06 ● 0.07   <
5. 0.004 ● 0.002   >
6. 0.008 ● 0.009   <
7. 15.5 ● 15.0   >
8. 8.43 ● 8.34   >
9. 0.57 ● 0.5   >
10. 0.007 ● 0.06   <
11. 0.6 ● 0.07   >
12. 9.73 ● 9.37   >
13. 42.89 ● 42.9   <
14. 5.1 ● 4.99   >
15. 4.352 ● 43.52   <
16. 0.625 ● 1.2   <
17. 3.08 ● 3.008   >
18. 0.715 ● 0.72   <
19. 51.86 ● 51.87   <
20. 33.78 ● 31.88   >

SKILL 8 (Use after page 46.)

Here's how

20 − 14.38 = ?

Line up the decimal points. Write the zeros.

2 0 . 0 0
−1 4 . 3 8

Regroup.

   1 9 9
2̸ 0̸. 0̸ 10
−1 4 . 3 8

Subtract.

   1 9 9
2̸ 0̸. 0̸ 10
−1 4 . 3 8
   5 . 6 2

Give the difference.

1. 9 − 3.2   5.8
2. 8 − 4.6   3.4
3. 15 − 7.2   7.8
4. 23 − 8.6   14.4
5. 18.01 − 9.45   8.56
6. 14.05 − 7.75   6.30
7. 9.4 − 6.73   2.67
8. 8.5 − 4.55   3.95
9. 8.3 − 6   2.3
10. 7.4 − 2   5.4
11. 10.3 − 8.4   1.9
12. 30.1 − 9.7   20.4
13. 7 − 3.44   3.56
14. 8 − 6.45   1.55
15. 8.234 − 0.749   7.485
16. 6.729 − 0.882   5.847
17. 8.5 − 3.692   4.808
18. 5.1 − 0.651   4.449
19. 42 − 8.2   33.8
20. 34 − 9.5   24.5
21. 81.64 − 33   48.64
22. 63.89 − 18   45.89
23. 100 − 44.63   55.37
24. 200 − 53.87   146.13
25. 102 − 9.4   92.6
26. 105 − 49.7   55.3

SKILL 9 (Use after page 56.)

Here's how
30 × 200 = ?
Multiply 3 × 2.
30 × 200 = 6
Copy all the zeros.
30 × 200 = 6000

Give the product.

1. 9 × 10   90
2. 6 × 100   600
3. 5 × 1000   5000
4. 3 × 100   300
5. 8 × 20   160
6. 8 × 200   1600
7. 12 × 10   120
8. 15 × 1000   15,000
9. 20 × 30   600
10. 80 × 300   24,000
11. 3 × 800   2400
12. 20 × 40   800
13. 30 × 2000   60,000
14. 40 × 400   16,000
15. 50 × 100   5000
16. 50 × 1000   50,000
17. 40 × 200   8000
18. 20 × 3000   60,000
19. 30 × 20   600
20. 40 × 300   12,000
21. 400 × 30   12,000
22. 145 × 100   14,500
23. 256 × 1000   256,000
24. 100 × 300   30,000

BEAT THE CALCULATOR — MENTAL MATH
(Extension for Skill 9, page 475)

Use a calculator and have a classmate use mental math to solve exercises 1–10. The first one to get a correct answer to each exercise scores a point. Which was faster, the calculator or mental math? Solve the exercises again, but this time you use mental math and have your classmate use the calculator.

Copymaster S493

SKILL 10 (Use after page 62.)

Here's how
Multiply by 3.
145
× 23

435

Multiply by 20.
145
× 23

435
2900

Add.
145
× 23

435
2900

3335

Multiply.

1. 34 × 2 = 68
2. 26 × 4 = 104
3. 40 × 7 = 280
4. 51 × 6 = 306
5. 75 × 8 = 600
6. 84 × 5 = 420
7. 59 × 6 = 354
8. 47 × 9 = 423
9. 43 × 55 = 2365
10. 50 × 62 = 3100
11. 78 × 18 = 1404
12. 95 × 77 = 7315
13. 125 × 31 = 3875
14. 236 × 22 = 5192
15. 304 × 58 = 17,632
16. 411 × 70 = 28,770
17. 638 × 63 = 40,194
18. 905 × 85 = 76,925
19. 731 × 17 = 12,427
20. 592 × 46 = 27,232
21. 3015 × 133 = 400,995
22. 4628 × 160 = 740,480
23. 5911 × 245 = 1,448,195
24. 6408 × 328 = 2,101,824

ESTIMATION HUNT — MENTAL MATH
(Extension for Skill 10, page 475)

Look at exercises 1–24.

1. Which exercise has an answer that you can estimate by thinking
   a. 60 times 6?   7
   b. 80 times 20?   11
   c. 600 times 60?   17
   d. 700 times 20?   19

2. Estimate. Which two exercises have answers that are closest to 30,000?   16, 20

Copymaster S493

Extra Practice **475**

ESTIMATION HUNT

Look at exercises 1–30.

1. Which exercise has an answer that you can estimate by thinking
   a. 6 times 9? 4
   b. 200 times 5? 11
   c. 6 times 8? 15
   d. 60 times 9? 24

2. Estimate. Which two exercises have answers that are closest to 10? 6, 28

Copymaster S494

MENTAL MATH *(Extension for Skill 11, page 476)*

Here's how

3.08 × 4.2 = ?

Multiply as whole numbers.

```
   3.0 8
 ×   4.2
   6 1 6
 1 2 3 2
 1 2 9 3 6
```

Count the digits to the right of the decimal points.

```
   3.0 8
 × 4.2     ⎫ 3
   6 1 6
 1 2 3 2
 1 2 9 3 6
```

Count off the same number of digits in the product.

SKILL 11 (Use after page 68.)

Give the product.

1. 4.2 × 12   50.4
2. 3.8 × 10   38
3. 2.6 × 2.6   6.76
4. 5.9 × 8.7   51.33
5. 4.06 × 0.8   3.248
6. 2.05 × 5.5   11.275
7. 0.94 × 0.34   0.3196
8. 0.95 × 0.55   0.5225
9. 58 × 0.25   14.5
10. 74 × 0.78   57.72
11. 221 × 4.6   1016.6
12. 360 × 8.2   2952
13. 3.62 × 0.95   3.439
14. 2.88 × 0.47   1.3536
15. 6.16 × 7.5   46.2
16. 2.09 × 0.8   1.672
17. 5.4 × 0.06   0.324
18. 8.8 × 0.07   0.616
19. 6.25 × 0.56   3.5
20. 8.65 × 0.44   3.806
21. 30.5 × 20.2   616.1
22. 56.7 × 18.4   1043.28
23. 55.5 × 21.6   1198.8
24. 63.2 × 8.94   565.008
25. 300 × 4.8   1440
26. 600 × 0.52   312
27. 2.54 × 2.54   6.4516
28. 3.08 × 3.08   9.4864
29. 298 × 16.1   4797.8
30. 315 × 22.5   7087.5

TIC-TAC-TOE COMPUTATIONS

Fill in the spaces on the tic-tac-toe board with these 9 numbers.

2.51 15.75 4.9 11.5 0.312
51.48 7.982 9.85 4.16

The numbers you filled in are the answers to exercises 1–9. Choose any one of the first nine exercises and solve it. Then mark an X on the answer on the tic-tac-toe board. How many exercises do you have to do to get three X's in a row? In 3 exercises is excellent, in 4 is very good, in 5 is good.

Copymaster S494

MENTAL MATH *(Extension for Skill 12, page 476)*

Here's how

2.8 + (0.2 × 0.3) = ?

Work inside the grouping symbols first.

First multiply.

```
  0.2
× 0.3
 0.06
```

Then add.

```
  0.2
× 0.3
 0.06
+2.8
 2.86
```

SKILL 12 (Use after page 70.)

Simplify.

1. 4.1 + (0.2 × 0.3)   4.16
2. 12 − (4.3 + 2.8)   4.9
3. 5.2 × (6.1 + 3.8)   51.48
4. (8.2 × 0.7) + 4.11   9.85
5. (0.5 × 0.04) + 2.49   2.51
6. 8 − (0.3 × 0.06)   7.982
7. (5.3 − 1.4) × 0.08   0.312
8. (9.22 + 4.1) − 1.82   11.5
9. (23 − 15.5) × 2.1   15.75
10. (17.9 + 8.4) + 0.9   27.2
11. 18.3 + (8.6 + 7.7)   34.6
12. (11.34 − 6.91) − 2.08   2.35
13. 14.37 − (6.02 − 3.11)   11.46
14. (7.4 × 2.3) × 1.7   28.934
15. 3.5 × (9.1 − 2.03)   24.745
16. (3.5 × 9.1) − 2.03   29.82
17. 4.2 × (3.1 × 6)   78.12
18. (4.2 × 3.1) × 6   78.12
19. 6.01 + (2.5 + 3.7)   12.21
20. (6.01 + 2.5) + 3.7   12.21
21. 22 − (6.6 + 1.9)   13.5
22. (22 − 6.6) + 1.9   17.3

Extra Practice

SKILL 13 (Use after page 72.)

Here's how

Multiplying by 10 moves the decimal point 1 place to the right.

2.47 × 10 = 24.7

Multiplying by 100 moves the decimal point 2 places to the right.

2.47 × 100 = 247

Give the product.

1. 42 × 10   420
2. 38 × 100   3800
3. 65 × 100   6500
4. 125 × 100   12,500
5. 8.2 × 10   82
6. 0.05 × 1000   50
7. 4.7 × 100   470
8. 2.95 × 10   29.5
9. 220 × 1000   220,000
10. 300 × 10   3000
11. 9.55 × 100   955
12. 8.74 × 1000   8740
13. 0.005 × 10   0.05
14. 0.002 × 100   0.2
15. 8.4 × 1000   8400
16. 7.2 × 100   720
17. 6.9 × 10   69
18. 3.74 × 100   374
19. 3.96 × 1000   3960
20. 6.66 × 10   66.6
21. 4.798 × 100   479.8
22. 4.798 × 10   47.98
23. 2.655 × 1000   2655
24. 148 × 100   14,800

MENTAL MATH (Extension for Skill 13, page 477)

BEAT THE CALCULATOR

Use a calculator and have a classmate use mental math to solve exercises 1–10. The first one to get a correct answer to each exercise scores a point. Which was faster, the calculator or mental math? Solve the exercises again, but this time you use mental math and have your classmate use the calculator.

Copymaster S495

SKILL 14 (Use after page 88.)

Here's how

Think about dividing 39 by 4. So try 9.

```
        48
48 ) 3999
 × 9
 432    432 is too big!
```

Try 8.
```
    48           8
   × 8       48 ) 3999
   384         −384
                 15
```

Think about dividing 15 by 4. So try 3.

```
             83 R15
   48     48 ) 3999
  × 3       −384
   144        159
             −144
               15
```

Divide.

1. 2)46   23
2. 3)75   25
3. 5)85   17
4. 3)789   263
5. 4)716   179
6. 8)696   87
7. 4)428   107
8. 7)735   105
9. 8)840   105
10. 7)8330   1190
11. 4)2016   504
12. 9)9396   1044
13. 64)5773   90 R13
14. 38)6310   166 R2
15. 31)6000   193 R17
16. 60)7008   116 R48
17. 75)9362   124 R62
18. 53)4867   91 R44
19. 46)7351   159 R37
20. 68)8022   117 R66
21. 99)3366   34
22. 25)4875   195
23. 35)8610   246
24. 60)8614   143 R34
25. 91)6235   68 R47
26. 35)6842   195 R17
27. 43)9286   215 R41
28. 61)8642   141 R41
29. 32)6192   193 R16
30. 60)7815   130 R15
31. 125)38,500   308
32. 150)67,800   452
33. 212)91,372   431
34. 318)57,558   181
35. 222)49,506   223
36. 300)56,100   187

MENTAL MATH (Extension for Skill 14, page 477)

ESTIMATION HUNT

Look at exercises 1–12.

1. Estimate. Which four exercises have answers that are less than 100?   1, 2, 3, 6
2. Estimate. Which two exercises have answers that are greater than 1000?   10, 12
3. Estimate. Which exercise has an answer of about 500?   11

Copymaster S495

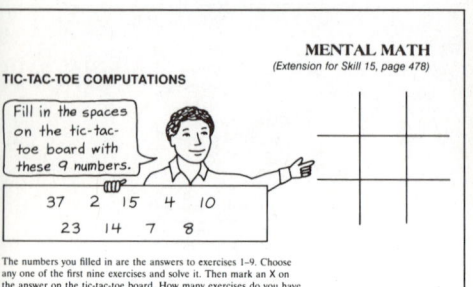

SKILL 15 (Use after page 94.)

Here's how

6 + 8 × (4 − 2) = ?

First, work within the grouping symbols.

6 + 8 × 2

Next, do the multiplication and division.

6 + 16

Last, do the addition and subtraction.

22

Simplify.

1. 6 ÷ 3 × 2   4
2. 12 − 8 + 4   8
3. 5 + 2 × 5 − 1   14
4. 5 × 2 + 10 ÷ 2   15
5. 5 + (3 + 9) ÷ 6   7
6. (4 + 5) × 2 − 8   10
7. 12 ÷ 4 − 1   2
8. 8 × 5 − 3   37
9. 24 − 4 ÷ 4   23
10. 30 − 12 − 6   12
11. 10 + 16 ÷ 4   14
12. 18 + 6 ÷ 3   20
13. 48 ÷ 8 × 2   12
14. 35 + 12 − 10   37
15. 18 − 6 + 6   18
16. 20 − 9 + 5   16
17. (12 + 18) ÷ 6   5
18. 34 × (8 − 3)   170
19. 16 + 8 ÷ 4 + 4   22
20. (16 + 8) ÷ 4 + 4   10
21. 16 + 8 ÷ (4 + 4)   17
22. 20 + 12 × 4 − 1   67

SKILL 16 (Use after page 100.)

Here's how

Dividing by 10 moves the decimal point 1 place to the left.

5.2 ÷ 10 = 0.52

Dividing by 100 moves the decimal point 2 places to the left.

5.2 ÷ 100 = 0.052

Give the quotient.

1. 34.2 ÷ 10   3.42
2. 34.2 ÷ 100   0.342
3. 458 ÷ 100   4.58
4. 458 ÷ 1000   0.458
5. 252.5 ÷ 100   2.525
6. 252.5 ÷ 10   25.25
7. 80 ÷ 10   8
8. 80 ÷ 100   0.80
9. 23.94 ÷ 10   2.394
10. 23.94 ÷ 100   0.2394
11. 2.8 ÷ 10   0.28
12. 2.8 ÷ 100   0.028
13. 2.8 ÷ 1000   0.0028
14. 9.05 ÷ 10   0.905
15. 9.05 ÷ 100   0.0905
16. 9.05 ÷ 1000   0.00905
17. 325 ÷ 100   3.25
18. 325 ÷ 10   32.5
19. 325 ÷ 1000   0.325
20. 9 ÷ 10   0.9
21. 9 ÷ 1000   0.009
22. 9 ÷ 100   0.09
23. 26.7 ÷ 100   0.267
24. 26.7 ÷ 1000   0.0267
25. 26.7 ÷ 10   2.67
26. 42.059 ÷ 100   0.42059
27. 42.059 ÷ 10   4.2059
28. 42.059 ÷ 1000   0.042059

Extra Practice

Here's how

$0.42 \overline{)0.5670}$

Move both decimal points two places to the right.

$0.42. \overline{)0.56.70}$

Divide.

```
           1.35
0.42. ) 0.56.70
       − 42
         147
       − 126
          210
        − 210
            0
```

SKILL 17 (Use after page 104.)

Divide.

1. $7 \overline{)38.36}$ 5.48
2. $6 \overline{)2.634}$ 0.439
3. $9 \overline{)5.067}$ 0.563
4. $0.08 \overline{)4.584}$ 57.3
5. $0.03 \overline{)1.473}$ 49.1
6. $0.005 \overline{)3.605}$ 721
7. $0.004 \overline{)0.2656}$ 66.4
8. $0.3 \overline{)96.30}$ 321
9. $0.04 \overline{)6.400}$ 160
10. $0.6 \overline{)350.4}$ 584
11. $0.07 \overline{)0.0644}$ 0.92
12. $0.9 \overline{)0.963}$ 1.07
13. $12 \overline{)7.8}$ 0.65
14. $3.1 \overline{)124}$ 40
15. $25 \overline{)17.5}$ 0.7
16. $0.42 \overline{)159.6}$ 380
17. $0.049 \overline{)401.8}$ 8200
18. $0.066 \overline{)48.84}$ 740

Divide. Round each quotient to the nearest tenth.

19. $0.3 \overline{)1.7}$ 5.7
20. $0.6 \overline{)1.02}$ 1.7
21. $1.3 \overline{)4.0}$ 3.1
22. $2.5 \overline{)9.1}$ 3.6
23. $3.7 \overline{)0.99}$ 0.3
24. $7.1 \overline{)52.8}$ 7.4
25. $4.3 \overline{)0.271}$ 0.1
26. $0.55 \overline{)0.382}$ 0.7
27. $0.67 \overline{)1.092}$ 1.6

MENTAL MATH (Extension for Skill 17, page 479)

TIC-TAC-TOE QUOTIENTS

Fill in the spaces on the tic-tac-toe board with these 9 numbers.

0.563	57.3	160	0.439
5.48	49.1	66.4	321	721

The numbers you filled in are the answers to exercises 1–9. Choose any one of the first nine exercises and solve it. Then mark an X on the answer on the tic-tac-toe board. How many exercises do you have to do to get three X's in a row? In 3 exercises is excellent, in 4 is very good, in 5 is good.

Copymaster S497

Here's how

David had $10. He bought 2 records for $4.69 each. How much money did he have left?

Step 1. Find the cost of the records.

```
    1 1
  $4.69
 ×    2
  $9.38
```

Step 2. Find how much money was left.

```
     9 9
   1 1
  $10.00
 −  9.38
     .62
```

He had $.62 left.

SKILL 18 (Use after page 106.)

Solve.

1. Susan works 3 hours each school night and 6 hours on Saturday. How many hours does she work in 2 weeks? **42 hours**

2. Bill lives 18 blocks from where he works. He works 6 nights each week. How many blocks does he travel to and from work during a week? **216 blocks**

3. Sandy works 3 hours each school night and 6 hours on Saturday. If she is paid $5.00 an hour, how much does she earn in a week? **$105**

4. Paul wants to buy a bicycle that costs $219. He saves $23 each week. How much more money will he need after saving for 8 weeks? **$35**

5. Jerry earned $13.05 for working 3 hours. If he had worked 5 hours and had earned the same per hour, how much would he have earned? **$21.75**

6. Beth and Jill want to share equally the cost of three magazines that cost $1.50, $2.25, and $1.75. How much should each pay? **$2.75**

Extra Practice **479**

SKILL 19 (Use after page 116.)

Here's how

How much more was saved during the first two weeks than the third week?

Step 1. Find the amount saved during the first two weeks.

$19 + $12 = **$31**

Step 2. Find how much more that amount is than the amount saved the third week.

$31 − $21 = **$10**

$10 more was saved.

Solve.

1. How much was saved during the third and fourth weeks? *$37*

2. How much more was saved during the third week than the second week? *$9*

3. What was the average weekly savings during the 4 weeks? *$17*

4. How much more was saved during the last two weeks than the first two weeks? *$6*

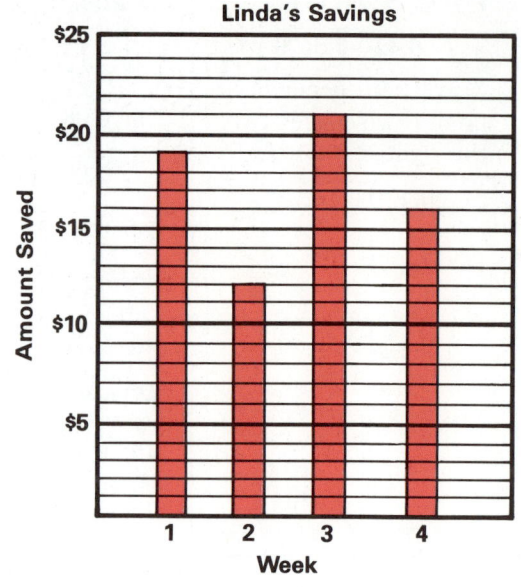

Linda's Savings

SKILL 20 (Use after page 118.)

Here's how

During the first three Saturdays they spent $60 for supplies. How much money did they have left?

Step 1. Find the total for the first three Saturdays.

$240 + $320 + $290 = **$850**

Step 2. Find how much money was left.

$850 − $60 = **$790**

$790 was left.

Solve.

1. How much money was raised during the first two Saturdays? *$560*

2. How much more money was raised the fourth Saturday than the third Saturday? *$180*

3. If they charged $5 a car, how many cars did they wash during the last two weeks? *166 cars*

4. How much less money did they make during the first two Saturdays than the last two Saturdays? *$270*

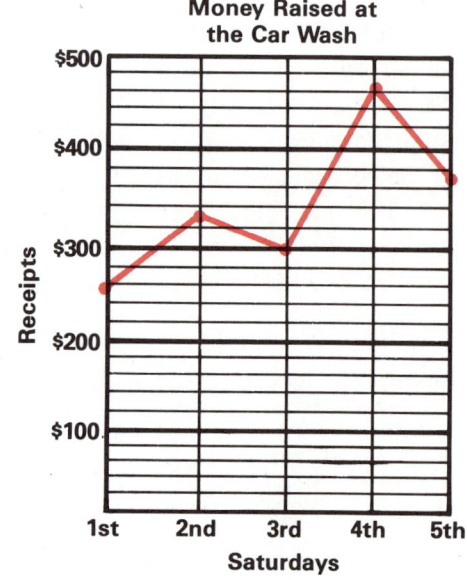

Money Raised at the Car Wash

SKILL 21 (Use after page 120.)

Here's how

How many records do Juan and Larry have together?

Step 1. Find how many records each boy has.

Juan:
$9 \times 4 = 36$

Larry:
$(7 \times 4) + 2 = 30$

Step 2. Add.

$36 + 30 = 66$

They have 66 records together.

Number of Records in Collection

Juan	●●●●●●●●
Larry	●●●●●●●◐
Nan	●●●●●●●●●
Sue	●●●●●◐

Each ● stands for 4 records.

Solve.

1. How many records does Sue have in her collection? 22

2. How many more records does Nan have than Juan? 8

3. What is the total number of records Juan and Sue have? 58

4. What is the average number of records in each collection? 33

SKILL 22 (Use after page 128.)

Here's how

How many more yards did Jim Brown have rushing than pass receiving?

$12,312 - 2,499 = ?$

```
  1 12 101
  12,312
 - 2,499
   9,813
```

9813 yards.

TOTAL YARDS GAINED

NFL HALL OF FAME PLAYER	YARDS RUSHING	YARDS PASS-RECEIVING
Jim Brown	12,312	2,499
Frank Gifford	3,069	5,434
Lenny Moore	5,174	6,039
Gale Sayers	4,957	1,309
Jim Taylor	8,597	225

Solve.

1. What was Jim Brown's total rushing and pass-receiving yardage? 14,811 yards

2. How many more yards rushing did Jim Taylor have than Gale Sayers? 3640 yards

3. How many more yards would Lenny Moore have had to gain to have had a total of 12,000 yards? 787 yards

4. How much greater was Lenny Moore's total yardage than Frank Gifford's? 2710 yards

Extra Practice 481

SKILL 23 (Use after page 146.)

Here's how

$\frac{2}{5} = \frac{?}{15}$

Multiply numerator and denominator by 3.

$\frac{2}{5} \begin{array}{c}\boxed{\times 3}\\ = \frac{6}{15}\\ \boxed{\times 3}\end{array}$

$\frac{12}{20} = \frac{?}{5}$

Divide numerator and denominator by 4.

$\frac{12}{20} \begin{array}{c}\boxed{\div 4}\\ = \frac{3}{5}\\ \boxed{\div 4}\end{array}$

Complete.

1. $\frac{1}{2} = \frac{?}{4}$ 2
2. $\frac{1}{3} = \frac{?}{6}$ 2
3. $\frac{1}{4} = \frac{?}{12}$ 3
4. $\frac{2}{3} = \frac{?}{6}$ 4
5. $\frac{6}{8} = \frac{?}{4}$ 3
6. $\frac{9}{24} = \frac{?}{8}$ 3
7. $\frac{9}{6} = \frac{?}{2}$ 3
8. $\frac{3}{9} = \frac{?}{3}$ 1
9. $\frac{1}{4} = \frac{?}{20}$ 5
10. $\frac{1}{2} = \frac{?}{6}$ 3
11. $\frac{1}{5} = \frac{?}{10}$ 2
12. $\frac{4}{5} = \frac{?}{15}$ 12
13. $\frac{8}{12} = \frac{?}{3}$ 2
14. $\frac{25}{10} = \frac{?}{2}$ 5
15. $\frac{4}{16} = \frac{?}{4}$ 1
16. $\frac{6}{9} = \frac{?}{3}$ 2
17. $\frac{1}{3} = \frac{?}{15}$ 5
18. $\frac{1}{6} = \frac{?}{18}$ 3
19. $\frac{5}{8} = \frac{?}{16}$ 10
20. $\frac{3}{4} = \frac{?}{16}$ 12
21. $\frac{4}{24} = \frac{?}{6}$ 1
22. $\frac{12}{10} = \frac{?}{5}$ 6
23. $\frac{5}{10} = \frac{?}{2}$ 1
24. $\frac{4}{20} = \frac{?}{5}$ 1
25. $\frac{5}{2} = \frac{?}{10}$ 25
26. $\frac{1}{4} = \frac{?}{8}$ 2
27. $\frac{3}{2} = \frac{?}{10}$ 15
28. $\frac{3}{5} = \frac{?}{20}$ 12
29. $\frac{6}{12} = \frac{?}{2}$ 1
30. $\frac{30}{18} = \frac{?}{3}$ 5
31. $\frac{10}{15} = \frac{?}{3}$ 2
32. $\frac{4}{12} = \frac{?}{3}$ 1

SKILL 24 (Use after page 148.)

Here's how

To write a fraction in lowest terms, divide both terms by their greatest common divisor.

$\frac{12}{18} \begin{array}{c}\boxed{\div 6}\\ = \frac{2}{3}\\ \boxed{\div 6}\end{array}$

Write in lowest terms.

1. $\frac{3}{6}$ $\frac{1}{2}$
2. $\frac{3}{9}$ $\frac{1}{3}$
3. $\frac{2}{8}$ $\frac{1}{4}$
4. $\frac{6}{9}$ $\frac{2}{3}$
5. $\frac{9}{6}$ $\frac{3}{2}$
6. $\frac{14}{16}$ $\frac{7}{8}$
7. $\frac{9}{12}$ $\frac{3}{4}$
8. $\frac{10}{12}$ $\frac{5}{6}$
9. $\frac{8}{10}$ $\frac{4}{5}$
10. $\frac{5}{15}$ $\frac{1}{3}$
11. $\frac{6}{4}$ $\frac{3}{2}$
12. $\frac{7}{14}$ $\frac{1}{2}$
13. $\frac{8}{16}$ $\frac{1}{2}$
14. $\frac{15}{6}$ $\frac{5}{2}$
15. $\frac{2}{10}$ $\frac{1}{5}$
16. $\frac{16}{12}$ $\frac{4}{3}$
17. $\frac{21}{6}$ $\frac{7}{2}$
18. $\frac{4}{8}$ $\frac{1}{2}$
19. $\frac{6}{16}$ $\frac{3}{8}$
20. $\frac{4}{12}$ $\frac{1}{3}$
21. $\frac{3}{12}$ $\frac{1}{4}$
22. $\frac{4}{6}$ $\frac{2}{3}$
23. $\frac{3}{18}$ $\frac{1}{6}$
24. $\frac{15}{18}$ $\frac{5}{6}$
25. $\frac{10}{6}$ $\frac{5}{3}$
26. $\frac{15}{20}$ $\frac{3}{4}$
27. $\frac{5}{10}$ $\frac{1}{2}$
28. $\frac{10}{15}$ $\frac{2}{3}$
29. $\frac{6}{12}$ $\frac{1}{2}$
30. $\frac{4}{20}$ $\frac{1}{5}$
31. $\frac{18}{24}$ $\frac{3}{4}$
32. $\frac{6}{18}$ $\frac{1}{3}$

MENTAL MATH (Extension for Skill 23, page 482)

SOLUTION SEARCH

Look at exercises 1–20.

1. Which exercise can you solve mentally by
   a. multiplying the numerator and the denominator of a fraction by 4? 20
   b. dividing the numerator and the denominator of a fraction by 5? 14

2. How would you mentally complete exercise 18?
   Multiply the numerator and the denominator of the fraction by 3.

Copymaster S497

MENTAL MATH (Extension for Skill 24, page 482)

TIC-TAC-TOE FRACTIONS

Fill in the spaces on the tic-tac-toe board with these 9 numbers.

$\frac{7}{8}$ $\frac{2}{3}$ $\frac{1}{3}$ $\frac{3}{4}$ $\frac{5}{6}$ $\frac{3}{2}$ $\frac{4}{5}$ $\frac{1}{4}$ $\frac{1}{2}$

The numbers you filled in are the answers to exercises 1–9. Choose any one of the first nine exercises and solve it. Then mark an X on the answer on the tic-tac-toe board. How many exercises do you have to do to get three X's in a row? In 3 exercises is excellent, in 4 is very good, in 5 is good.

Copymaster S498

SKILL 25 (Use after page 150.)

Here's how

To find the least common denominator of two fractions, find the least common multiple of the denominators.

$$\frac{2}{3} \quad \frac{3}{4}$$

3, 6, 9, **(12)**, 15
4, 8, **(12)**, 16, 20

The least common denominator is **12**.

Find the least common denominator.

1. $\frac{1}{6}$ $\frac{1}{5}$ **30**
2. $\frac{3}{4}$ $\frac{1}{2}$ **4**
3. $\frac{1}{5}$ $\frac{2}{9}$ **45**
4. $\frac{3}{7}$ $\frac{1}{6}$ **42**
5. $\frac{1}{10}$ $\frac{2}{5}$ **10**
6. $\frac{3}{20}$ $\frac{1}{10}$ **20**
7. $\frac{1}{4}$ $\frac{1}{6}$ **12**
8. $\frac{1}{2}$ $\frac{3}{7}$ **14**
9. $\frac{5}{6}$ $\frac{3}{8}$ **24**
10. $\frac{1}{6}$ $\frac{1}{8}$ **24**
11. $\frac{1}{8}$ $\frac{4}{3}$ **24**
12. $\frac{1}{10}$ $\frac{3}{4}$ **20**
13. $\frac{7}{8}$ $\frac{1}{12}$ **24**
14. $\frac{1}{6}$ $\frac{3}{4}$ **12**
15. $\frac{1}{5}$ $\frac{3}{7}$ **35**
16. $\frac{1}{3}$ $\frac{5}{6}$ **6**
17. $\frac{2}{5}$ $\frac{1}{4}$ **20**
18. $\frac{1}{10}$ $\frac{4}{5}$ **10**
19. $\frac{5}{9}$ $\frac{1}{2}$ **18**
20. $\frac{1}{4}$ $\frac{3}{5}$ **20**
21. $\frac{3}{10}$ $\frac{1}{4}$ **20**
22. $\frac{1}{6}$ $\frac{4}{5}$ **30**
23. $\frac{7}{6}$ $\frac{5}{9}$ **18**
24. $\frac{3}{8}$ $\frac{1}{9}$ **72**
25. $\frac{5}{6}$ $\frac{1}{15}$ **30**
26. $\frac{1}{7}$ $\frac{1}{3}$ **21**
27. $\frac{2}{5}$ $\frac{1}{8}$ **40**
28. $\frac{1}{12}$ $\frac{2}{3}$ **12**
29. $\frac{1}{4}$ $\frac{1}{7}$ **28**
30. $\frac{1}{9}$ $\frac{2}{3}$ **9**
31. $\frac{1}{3}$ $\frac{2}{9}$ **9**
32. $\frac{3}{8}$ $\frac{1}{7}$ **56**

MENTAL MATH (Extension for Skill 25, page 483)

SOLUTION SEARCH

Look at exercises 1–32.

1. Which exercise has fractions with a least common denominator of 42? **4**
2. Which three exercises have fractions with a least common denominator of 12? **7, 14, 28**
3. Which five exercises have fractions with a least common denominator of 20? **6, 12, 17, 20, 21**

Copymaster S498

SKILL 26 (Use after page 152.)

Here's how

Find the least common denominator.

Change to equivalent fractions.

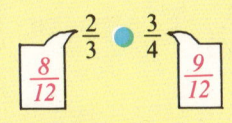

Compare.

$\frac{2}{3} < \frac{3}{4}$

< or >?

1. $\frac{1}{4}$ ● $\frac{3}{4}$ **<**
2. $\frac{3}{5}$ ● $\frac{2}{5}$ **>**
3. $\frac{3}{7}$ ● $\frac{4}{7}$ **<**
4. $\frac{0}{8}$ ● $\frac{5}{8}$ **<**
5. $\frac{5}{4}$ ● $\frac{4}{4}$ **>**
6. $\frac{7}{5}$ ● $\frac{9}{5}$ **<**
7. $\frac{5}{8}$ ● $\frac{7}{8}$ **<**
8. $\frac{7}{3}$ ● $\frac{5}{3}$ **>**
9. $\frac{2}{3}$ ● $\frac{5}{6}$ **<**
10. $\frac{5}{4}$ ● $\frac{3}{2}$ **<**
11. $\frac{2}{7}$ ● $\frac{1}{3}$ **<**
12. $\frac{3}{4}$ ● $\frac{3}{8}$ **>**
13. $\frac{1}{8}$ ● $\frac{1}{6}$ **<**
14. $\frac{3}{8}$ ● $\frac{1}{4}$ **>**
15. $\frac{3}{10}$ ● $\frac{1}{3}$ **<**
16. $\frac{1}{4}$ ● $\frac{2}{5}$ **<**
17. $\frac{1}{4}$ ● $\frac{1}{3}$ **<**
18. $\frac{2}{9}$ ● $\frac{3}{4}$ **<**
19. $\frac{2}{3}$ ● $\frac{3}{4}$ **<**
20. $\frac{5}{8}$ ● $\frac{4}{7}$ **>**
21. $\frac{3}{5}$ ● $\frac{3}{7}$ **>**
22. $\frac{3}{4}$ ● $\frac{5}{6}$ **<**
23. $\frac{8}{9}$ ● $\frac{7}{8}$ **>**
24. $\frac{1}{3}$ ● $\frac{5}{12}$ **<**
25. $\frac{7}{8}$ ● $\frac{5}{6}$ **>**
26. $\frac{2}{3}$ ● $\frac{7}{10}$ **<**
27. $\frac{7}{9}$ ● $\frac{2}{3}$ **>**
28. $\frac{9}{2}$ ● $\frac{9}{4}$ **>**

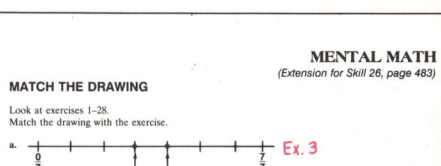

MENTAL MATH (Extension for Skill 26, page 483)

MATCH THE DRAWING

Look at exercises 1–28.
Match the drawing with the exercise.

a. Ex. 3
b. Ex. 7
c. Ex. 9
d. Ex. 17
e. Ex. 27

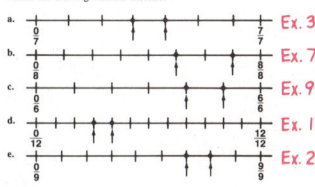

Copymaster S499

Extra Practice **483**

MENTAL MATH
(Extension for Skill 27, page 484)

MATCH THE DRAWING
Look at exercises 1–36.
Match the drawing with the exercise.

a. Ex. 4
b. Ex. 13
c. Ex. 19
d. Ex. 32

Copymaster S499

Here's how

$4 = \dfrac{?}{3}$

Write the whole number over 1 and multiply both numerator and denominator by 3.

$\dfrac{4}{1} = \dfrac{12}{3}$

$2\dfrac{3}{4} = ?$

To change a mixed number to a fraction, multiply the denominator by the whole number and add the numerator.

$2\dfrac{3}{4} = \dfrac{11}{4}$

SKILL 27 *(Use after page 156.)*

Change to thirds.

1. $2\ \dfrac{6}{3}$ 2. $1\ \dfrac{3}{3}$ 3. $4\ \dfrac{12}{3}$ 4. $5\ \dfrac{15}{3}$
5. $3\ \dfrac{9}{3}$ 6. $8\ \dfrac{24}{3}$ 7. $10\ \dfrac{30}{3}$ 8. $6\ \dfrac{18}{3}$

Change to fourths.

9. $3\ \dfrac{12}{4}$ 10. $1\ \dfrac{4}{4}$ 11. $4\ \dfrac{16}{4}$ 12. $2\ \dfrac{8}{4}$
13. $7\ \dfrac{28}{4}$ 14. $9\ \dfrac{36}{4}$ 15. $10\ \dfrac{40}{4}$ 16. $8\ \dfrac{32}{4}$

Change to a fraction.

17. $1\dfrac{1}{3}\ \dfrac{4}{3}$ 18. $2\dfrac{1}{4}\ \dfrac{9}{4}$ 19. $2\dfrac{1}{2}\ \dfrac{5}{2}$ 20. $1\dfrac{1}{5}\ \dfrac{6}{5}$
21. $1\dfrac{2}{3}\ \dfrac{5}{3}$ 22. $2\dfrac{1}{3}\ \dfrac{7}{3}$ 23. $1\dfrac{1}{4}\ \dfrac{5}{4}$ 24. $3\dfrac{1}{4}\ \dfrac{13}{4}$
25. $1\dfrac{1}{2}\ \dfrac{3}{2}$ 26. $2\dfrac{7}{8}\ \dfrac{23}{8}$ 27. $3\dfrac{3}{8}\ \dfrac{27}{8}$ 28. $2\dfrac{5}{6}\ \dfrac{17}{6}$
29. $5\dfrac{1}{4}\ \dfrac{21}{4}$ 30. $2\dfrac{2}{3}\ \dfrac{8}{3}$ 31. $3\dfrac{1}{2}\ \dfrac{7}{2}$ 32. $4\dfrac{5}{8}\ \dfrac{37}{8}$
33. $4\dfrac{4}{5}\ \dfrac{24}{5}$ 34. $3\dfrac{1}{3}\ \dfrac{10}{3}$ 35. $5\dfrac{3}{10}\ \dfrac{53}{10}$ 36. $4\dfrac{1}{5}\ \dfrac{21}{5}$

MENTAL MATH
(Extension for Skill 28, page 484)

MATCH THE DRAWING
Look at exercises 1–32.
Match the drawing with the exercise.

a. Ex. 6
b. Ex. 17
c. Ex. 24
d. Ex. 32

Copymaster S500

Here's how

To change a fraction to a whole number or mixed number, divide the numerator by the denominator.

$\dfrac{18}{3} = 6$

$\dfrac{19}{4} = 4\dfrac{3}{4}$

$\begin{array}{r}4\\4\overline{)19}\\-16\\\hline 3\end{array}$

There are 3 fourths left over.

SKILL 28 *(Use after page 158.)*

Change to a whole number.

1. $\dfrac{4}{2}\ \ 2$ 2. $\dfrac{40}{5}\ \ 8$ 3. $\dfrac{10}{2}\ \ 5$ 4. $\dfrac{9}{3}\ \ 3$
5. $\dfrac{25}{5}\ \ 5$ 6. $\dfrac{8}{2}\ \ 4$ 7. $\dfrac{32}{4}\ \ 8$ 8. $\dfrac{30}{3}\ \ 10$
9. $\dfrac{18}{3}\ \ 6$ 10. $\dfrac{35}{5}\ \ 7$ 11. $\dfrac{16}{4}\ \ 4$ 12. $\dfrac{50}{5}\ \ 10$

Change to a mixed number.

13. $\dfrac{3}{2}\ \ 1\dfrac{1}{2}$ 14. $\dfrac{5}{4}\ \ 1\dfrac{1}{4}$ 15. $\dfrac{5}{3}\ \ 1\dfrac{2}{3}$ 16. $\dfrac{9}{2}\ \ 4\dfrac{1}{2}$
17. $\dfrac{11}{4}\ \ 2\dfrac{3}{4}$ 18. $\dfrac{4}{3}\ \ 1\dfrac{1}{3}$ 19. $\dfrac{13}{5}\ \ 2\dfrac{3}{5}$ 20. $\dfrac{7}{6}\ \ 1\dfrac{1}{6}$
21. $\dfrac{16}{3}\ \ 5\dfrac{1}{3}$ 22. $\dfrac{13}{10}\ \ 1\dfrac{3}{10}$ 23. $\dfrac{5}{2}\ \ 2\dfrac{1}{2}$ 24. $\dfrac{8}{3}\ \ 2\dfrac{2}{3}$
25. $\dfrac{23}{10}\ \ 2\dfrac{3}{10}$ 26. $\dfrac{9}{5}\ \ 1\dfrac{4}{5}$ 27. $\dfrac{13}{12}\ \ 1\dfrac{1}{12}$ 28. $\dfrac{19}{6}\ \ 3\dfrac{1}{6}$
29. $\dfrac{7}{2}\ \ 3\dfrac{1}{2}$ 30. $\dfrac{11}{3}\ \ 3\dfrac{2}{3}$ 31. $\dfrac{19}{4}\ \ 4\dfrac{3}{4}$ 32. $\dfrac{18}{5}\ \ 3\dfrac{3}{5}$

Extra Practice

SKILL 29 (Use after page 160.)

Here's how

simplest form
↓
$\frac{6}{9} = \frac{2}{3}$

simplest form
↓
$3\frac{4}{6} = 3\frac{2}{3}$

simplest form
↓
$\frac{18}{3} = 6$

simplest form
↓
$\frac{14}{4} = 3\frac{2}{4} = 3\frac{1}{2}$

Write in simplest form.

1. $\frac{8}{2}$ 4
2. $\frac{6}{9}$ $\frac{2}{3}$
3. $\frac{10}{15}$ $\frac{2}{3}$
4. $\frac{15}{5}$ 3
5. $\frac{10}{12}$ $\frac{5}{6}$
6. $\frac{6}{15}$ $\frac{2}{5}$
7. $\frac{4}{3}$ $1\frac{1}{3}$
8. $\frac{4}{24}$ $\frac{1}{6}$
9. $\frac{9}{2}$ $4\frac{1}{2}$
10. $1\frac{8}{12}$ $1\frac{2}{3}$
11. $1\frac{3}{6}$ $1\frac{1}{2}$
12. $2\frac{4}{6}$ $2\frac{2}{3}$
13. $1\frac{2}{4}$ $1\frac{1}{2}$
14. $\frac{10}{2}$ 5
15. $\frac{11}{3}$ $3\frac{2}{3}$
16. $3\frac{6}{16}$ $3\frac{3}{8}$
17. $\frac{23}{4}$ $5\frac{3}{4}$
18. $4\frac{4}{10}$ $4\frac{2}{5}$
19. $5\frac{9}{24}$ $5\frac{3}{8}$
20. $\frac{20}{24}$ $\frac{5}{6}$
21. $\frac{36}{6}$ 6
22. $\frac{17}{2}$ $8\frac{1}{2}$
23. $\frac{25}{5}$ 5
24. $\frac{13}{3}$ $4\frac{1}{3}$
25. $\frac{18}{5}$ $3\frac{3}{5}$
26. $\frac{24}{3}$ 8
27. $\frac{12}{15}$ $\frac{4}{5}$
28. $\frac{9}{12}$ $\frac{3}{4}$
29. $6\frac{10}{12}$ $6\frac{5}{6}$
30. $\frac{19}{4}$ $4\frac{3}{4}$
31. $8\frac{12}{16}$ $8\frac{3}{4}$
32. $\frac{17}{5}$ $3\frac{2}{5}$

MENTAL MATH (Extension for Skill 29, page 485)

MATCH THE DRAWING

Look at exercises 1–32.
Match the drawing with the exercise.

a. — Ex. 4
b. — Ex. 9
c. — Ex. 19
d. — Ex. 31

Copymaster S500

SKILL 30 (Use after page 166.)

Here's how

To change a fraction to a decimal, divide the numerator by the denominator.

$\frac{5}{8} = ?$

$$\begin{array}{r} 0.625 \\ 8\overline{)5.000} \\ -4\,8 \\ \hline 20 \\ -16 \\ \hline 40 \\ -40 \\ \hline 0 \end{array}$$

$\frac{5}{8} = 0.625$

$1\frac{5}{8} = ?$

$1\frac{5}{8} = 1.625$

Change to a decimal.

1. $\frac{1}{4}$ 0.25
2. $\frac{3}{4}$ 0.75
3. $\frac{1}{5}$ 0.2
4. $\frac{1}{2}$ 0.5
5. $\frac{9}{10}$ 0.9
6. $\frac{2}{5}$ 0.4
7. $\frac{7}{10}$ 0.7
8. $\frac{4}{5}$ 0.8
9. $\frac{1}{8}$ 0.125
10. $\frac{3}{10}$ 0.3
11. $\frac{7}{4}$ 1.75
12. $\frac{9}{8}$ 1.125
13. $\frac{9}{2}$ 4.5
14. $\frac{3}{8}$ 0.375
15. $\frac{3}{5}$ 0.6
16. $\frac{7}{8}$ 0.875
17. $\frac{9}{4}$ 2.25
18. $\frac{11}{2}$ 5.5
19. $\frac{1}{16}$ 0.0625
20. $\frac{11}{8}$ 1.375
21. $2\frac{1}{2}$ 2.5
22. $3\frac{3}{4}$ 3.75
23. $3\frac{4}{5}$ 3.8
24. $2\frac{7}{8}$ 2.875

Change to a decimal rounded to the nearest hundredth.

25. $\frac{1}{9}$ 0.11
26. $\frac{1}{3}$ 0.33
27. $\frac{4}{3}$ 1.33
28. $\frac{1}{6}$ 0.17
29. $\frac{7}{9}$ 0.78
30. $\frac{17}{6}$ 2.83
31. $\frac{9}{16}$ 0.56
32. $\frac{21}{16}$ 1.31

MENTAL MATH (Extension for Skill 30, page 485)

TIC-TAC-TOE DECIMALS

Fill in the spaces on the tic-tac-toe board with these 9 numbers.

0.4 0.5 0.8 0.75
0.9 0.125 0.7 0.25 0.2

The numbers you filled in are the answers to exercises 1–9. Choose any one of the first nine exercises and solve it. Then mark an X on the answer on the tic-tac-toe board. How many exercises do you have to do to get three X's in a row? In 3 exercises is excellent, in 4 is very good, in 5 is good.

Copymaster S501

Extra Practice **485**

SKILL 31 (Use after page 168.)

Change to a fraction in simplest form.

Here's how
Write as a fraction in simplest form.

$0.75 = \frac{75}{100} = \frac{3}{4}$

Write as a mixed number in simplest form.

$3.4 = 3\frac{4}{10} = 3\frac{2}{5}$

1. 0.4 $\frac{2}{5}$
2. 0.8 $\frac{4}{5}$
3. 0.6 $\frac{3}{5}$
4. 0.1 $\frac{1}{10}$
5. 0.9 $\frac{9}{10}$
6. 0.5 $\frac{1}{2}$
7. 0.3 $\frac{3}{10}$
8. 0.2 $\frac{1}{5}$
9. 0.25 $\frac{1}{4}$
10. 0.75 $\frac{3}{4}$
11. 0.15 $\frac{3}{20}$
12. 0.45 $\frac{9}{20}$
13. 0.375 $\frac{3}{8}$
14. 0.625 $\frac{5}{8}$
15. 0.875 $\frac{7}{8}$
16. 0.125 $\frac{1}{8}$

Change to a mixed number in simplest form.

17. 1.2 $1\frac{1}{5}$
18. 2.3 $2\frac{3}{10}$
19. 1.6 $1\frac{3}{5}$
20. 3.8 $3\frac{4}{5}$
21. 2.4 $2\frac{2}{5}$
22. 7.7 $7\frac{7}{10}$
23. 4.5 $4\frac{1}{2}$
24. 5.9 $5\frac{9}{10}$
25. 3.25 $3\frac{1}{4}$
26. 1.75 $1\frac{3}{4}$
27. 2.12 $2\frac{3}{25}$
28. 4.48 $4\frac{12}{25}$
29. 4.625 $4\frac{5}{8}$
30. 6.875 $6\frac{7}{8}$
31. 2.125 $2\frac{1}{8}$
32. 5.375 $5\frac{3}{8}$

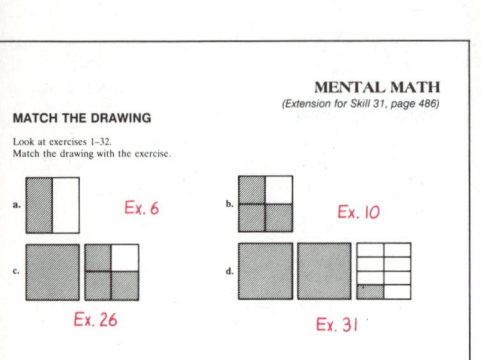

MATCH THE DRAWING
MENTAL MATH (Extension for Skill 31, page 486)
Look at exercises 1–32. Match the drawing with the exercise.

a. Ex. 6
b. Ex. 10
c. Ex. 26
d. Ex. 31

Copymaster S501

SKILL 32 (Use after page 180.)

Give the sum in simplest form.

Here's how
Change to equivalent fractions with common denominators.

$\frac{2}{3} + \frac{3}{4} = \frac{8}{12} + \frac{9}{12}$

Add. Write in simplest form.

$\frac{2}{3} + \frac{3}{4} = \frac{8}{12} + \frac{9}{12}$
$= \frac{17}{12}$
$= 1\frac{5}{12}$

1. $\frac{1}{4} + \frac{1}{4}$ $\frac{1}{2}$
2. $\frac{1}{6} + \frac{1}{6}$ $\frac{1}{3}$
3. $\frac{3}{8} + \frac{1}{8}$ $\frac{1}{2}$
4. $\frac{3}{10} + \frac{3}{10}$ $\frac{3}{5}$
5. $\frac{1}{3} + \frac{1}{6}$ $\frac{1}{2}$
6. $\frac{2}{5} + \frac{3}{10}$ $\frac{7}{10}$
7. $\frac{1}{5} + \frac{3}{10}$ $\frac{1}{2}$
8. $\frac{1}{3} + \frac{1}{4}$ $\frac{7}{12}$
9. $\frac{1}{3} + \frac{3}{7}$ $\frac{16}{21}$
10. $\frac{5}{9} + \frac{1}{6}$ $\frac{13}{18}$
11. $\frac{1}{8} + \frac{3}{4}$ $\frac{7}{8}$
12. $\frac{2}{5} + \frac{1}{4}$ $\frac{13}{20}$
13. $\frac{2}{3} + \frac{1}{9}$ $\frac{7}{9}$
14. $\frac{1}{3} + \frac{1}{2}$ $\frac{5}{6}$
15. $\frac{2}{3} + \frac{1}{8}$ $\frac{19}{24}$
16. $\frac{1}{2} + \frac{1}{6}$ $\frac{2}{3}$
17. $\frac{3}{8} + \frac{1}{6}$ $\frac{13}{24}$
18. $\frac{1}{3} + \frac{2}{5}$ $\frac{11}{15}$
19. $\frac{1}{3} + \frac{5}{8}$ $\frac{23}{24}$
20. $\frac{1}{8} + \frac{1}{6}$ $\frac{7}{24}$
21. $\frac{3}{4} + \frac{1}{2}$ $1\frac{1}{4}$
22. $\frac{1}{6} + \frac{2}{9}$ $\frac{7}{18}$
23. $\frac{5}{16} + \frac{7}{8}$ $1\frac{3}{16}$
24. $\frac{1}{4} + \frac{2}{3}$ $\frac{11}{12}$
25. $\frac{5}{12} + \frac{3}{4}$ $1\frac{1}{6}$
26. $\frac{2}{3} + \frac{11}{12}$ $1\frac{7}{12}$
27. $\frac{3}{7} + \frac{1}{2}$ $\frac{13}{14}$
28. $\frac{7}{8} + \frac{5}{16}$ $1\frac{3}{16}$
29. $\frac{2}{9} + \frac{2}{3}$ $\frac{8}{9}$
30. $\frac{5}{6} + \frac{4}{9}$ $1\frac{5}{18}$
31. $\frac{1}{2} + \frac{3}{5}$ $1\frac{1}{10}$
32. $\frac{9}{16} + \frac{7}{8}$ $1\frac{7}{16}$

SOLUTION SEARCH
Look at exercises 1–32.

1. Which exercise can you solve mentally by thinking
   a. $\frac{2}{6}$ plus $\frac{1}{6}$? 5
   b. $\frac{1}{8}$ plus $\frac{6}{8}$? 11
   c. $\frac{3}{4}$ plus $\frac{2}{4}$? 21
   d. $\frac{2}{9}$ plus $\frac{6}{9}$? 29

2. How would you find the answer to exercise 16 mentally?
   3/6 plus 1/6

Copymaster S502

486 Extra Practice

SKILL 33 (Use after page 182.)

Here's how

Write equivalent fractions with common denominators.

$3\frac{2}{3} = 3\frac{8}{12}$
$+2\frac{3}{4} = +2\frac{9}{12}$

Add the fractions. Add the whole numbers. Write in simplest form.

$3\frac{2}{3} = 3\frac{8}{12}$
$+2\frac{3}{4} = +2\frac{9}{12}$
$\qquad\qquad 5\frac{17}{12} = 6\frac{5}{12}$

Give the sum in simplest form.

1. $2\frac{1}{4} + 3\frac{1}{2}$ $5\frac{3}{4}$
2. $3\frac{1}{8} + 3\frac{3}{4}$ $6\frac{7}{8}$
3. $1\frac{3}{8} + 4\frac{1}{4}$ $5\frac{5}{8}$
4. $5\frac{1}{3} + 2\frac{1}{4}$ $7\frac{7}{12}$
5. $8\frac{5}{9} + 2\frac{1}{6}$ $10\frac{13}{18}$
6. $9\frac{1}{4} + 3\frac{2}{5}$ $12\frac{13}{20}$
7. $6\frac{1}{9} + 8\frac{2}{3}$ $14\frac{7}{9}$
8. $4\frac{1}{6} + 5\frac{1}{2}$ $9\frac{2}{3}$
9. $7\frac{3}{8} + 4\frac{1}{6}$ $11\frac{13}{24}$
10. $8\frac{1}{6} + 5\frac{2}{3}$ $13\frac{5}{6}$
11. $9\frac{1}{2} + 9\frac{3}{8}$ $18\frac{7}{8}$
12. $7\frac{1}{3} + 7\frac{1}{2}$ $14\frac{5}{6}$
13. $6\frac{11}{12} + 8\frac{2}{3}$ $15\frac{7}{12}$
14. $5\frac{2}{3} + 5\frac{2}{5}$ $11\frac{1}{15}$
15. $7\frac{1}{8} + 8\frac{5}{16}$ $15\frac{7}{16}$
16. $6\frac{2}{3} + 9\frac{3}{4}$ $16\frac{5}{12}$
17. $9\frac{2}{5} + 3\frac{3}{10}$ $12\frac{7}{10}$
18. $6\frac{5}{6} + 7\frac{1}{3}$ $14\frac{1}{6}$
19. $8\frac{3}{4} + 6\frac{5}{12}$ $15\frac{1}{6}$
20. $5\frac{1}{6} + 9\frac{2}{9}$ $14\frac{7}{18}$

MENTAL MATH (Extension for Skill 33, page 487)

ESTIMATION HUNT

Look at exercises 1–20.

1. Which exercise has an answer that you can estimate by thinking
   a. 1 plus 4? **3**
   b. 4 plus 6? **8**
   c. 5 plus 6? **14**
   d. 9 plus 7? **13**

2. Estimate. Which exercise has an answer that is the greatest sum? **11**

Copymaster S502

SKILL 34 (Use after page 188.)

Here's how

Write equivalent fractions with common denominators.

$\frac{2}{3} - \frac{1}{6} = \frac{4}{6} - \frac{1}{6}$

Subtract. Write in simplest form.

$\frac{2}{3} - \frac{1}{6} = \frac{4}{6} - \frac{1}{6}$
$\qquad = \frac{3}{6}$
$\qquad = \frac{1}{2}$

Give the difference in simplest form.

1. $\frac{3}{4} - \frac{1}{4}$ $\frac{1}{2}$
2. $\frac{5}{8} - \frac{1}{8}$ $\frac{1}{2}$
3. $\frac{7}{9} - \frac{2}{9}$ $\frac{5}{9}$
4. $\frac{5}{8} - \frac{3}{8}$ $\frac{1}{4}$
5. $\frac{3}{4} - \frac{0}{2}$ $\frac{3}{4}$
6. $\frac{2}{3} - \frac{1}{2}$ $\frac{1}{6}$
7. $\frac{1}{4} - \frac{1}{8}$ $\frac{1}{8}$
8. $\frac{1}{3} - \frac{1}{8}$ $\frac{5}{24}$
9. $\frac{3}{4} - \frac{3}{8}$ $\frac{3}{8}$
10. $\frac{7}{8} - \frac{2}{3}$ $\frac{5}{24}$
11. $\frac{1}{2} - \frac{1}{3}$ $\frac{1}{6}$
12. $\frac{2}{3} - \frac{5}{8}$ $\frac{1}{24}$
13. $\frac{5}{9} - \frac{1}{6}$ $\frac{7}{18}$
14. $\frac{1}{3} - \frac{1}{6}$ $\frac{1}{6}$
15. $\frac{5}{8} - \frac{2}{5}$ $\frac{9}{40}$
16. $\frac{7}{10} - \frac{2}{5}$ $\frac{3}{10}$
17. $\frac{1}{2} - \frac{3}{8}$ $\frac{1}{8}$
18. $\frac{9}{10} - \frac{2}{3}$ $\frac{7}{30}$
19. $\frac{3}{4} - \frac{2}{3}$ $\frac{1}{12}$
20. $\frac{2}{3} - \frac{2}{5}$ $\frac{4}{15}$
21. $\frac{11}{12} - \frac{3}{8}$ $\frac{13}{24}$
22. $\frac{2}{3} - \frac{1}{4}$ $\frac{5}{12}$
23. $\frac{1}{2} - \frac{5}{12}$ $\frac{1}{12}$
24. $\frac{7}{12} - \frac{3}{8}$ $\frac{5}{24}$
25. $\frac{5}{6} - \frac{5}{9}$ $\frac{5}{18}$
26. $\frac{7}{8} - \frac{5}{16}$ $\frac{9}{16}$
27. $\frac{8}{9} - \frac{5}{6}$ $\frac{1}{18}$
28. $\frac{3}{4} - \frac{1}{6}$ $\frac{7}{12}$

MENTAL MATH (Extension for Skill 34, page 487)

SOLUTION SEARCH

Look at exercises 1–28.

1. Which exercise can you solve mentally by thinking
   a. $\frac{2}{8}$ minus $\frac{1}{8}$? **7**
   b. $\frac{7}{10}$ minus $\frac{4}{10}$? **16**
   c. $\frac{6}{12}$ minus $\frac{5}{12}$? **23**
   d. $\frac{8}{24}$ minus $\frac{3}{24}$? **8**

2. How would you find the answer to exercise 26 mentally? $\frac{14}{16}$ minus $\frac{5}{16}$

Copymaster S503

Extra Practice **487**

SKILL 35 *(Use after page 192.)*

Here's how

Change to a common denominator.

$5\frac{1}{3} = 5\frac{2}{6}$
$-2\frac{1}{2} = -2\frac{3}{6}$

Regroup.

$5\frac{1}{3} = 5\frac{2}{6} = 4\frac{8}{6}$
$-2\frac{1}{2} = -2\frac{3}{6} = -2\frac{3}{6}$

Subtract.

$5\frac{1}{3} = 5\frac{2}{6} = 4\frac{8}{6}$
$-2\frac{1}{2} = -2\frac{3}{6} = -2\frac{3}{6}$
$\phantom{5\frac{1}{3} = 5\frac{2}{6} = \ } 2\frac{5}{6}$

Give the difference in simplest form.

1. $4\frac{3}{4} - 3\frac{1}{4}$ $1\frac{1}{2}$
2. $5\frac{5}{8} - 1\frac{3}{8}$ $4\frac{1}{4}$
3. $7\frac{3}{4} - 4\frac{1}{2}$ $3\frac{1}{4}$
4. $8\frac{2}{5} - 2\frac{1}{2}$ $5\frac{9}{10}$
5. $9\frac{1}{3} - 6\frac{3}{4}$ $2\frac{7}{12}$
6. $6\frac{1}{4} - 1\frac{3}{8}$ $4\frac{7}{8}$
7. $8\frac{1}{8} - 3\frac{1}{3}$ $4\frac{19}{24}$
8. $5\frac{2}{3} - 4\frac{7}{8}$ $\frac{19}{24}$
9. $7\frac{3}{8} - 2\frac{2}{3}$ $4\frac{17}{24}$
10. $8\frac{5}{9} - 3\frac{1}{6}$ $5\frac{7}{18}$
11. $6\frac{1}{6} - 4\frac{1}{3}$ $1\frac{5}{6}$
12. $4\frac{3}{8} - 2\frac{1}{2}$ $1\frac{7}{8}$
13. $6\frac{2}{3} - 5\frac{3}{4}$ $\frac{11}{12}$
14. $8\frac{2}{5} - 3\frac{2}{3}$ $4\frac{11}{15}$
15. $7\frac{1}{4} - 6\frac{2}{3}$ $\frac{7}{12}$
16. $5\frac{5}{12} - 2\frac{1}{2}$ $2\frac{11}{12}$
17. $9\frac{5}{9} - 4\frac{5}{6}$ $4\frac{13}{18}$
18. $7\frac{5}{16} - 2\frac{7}{8}$ $4\frac{7}{16}$
19. $6\frac{7}{16} - 3\frac{1}{2}$ $2\frac{15}{16}$
20. $8\frac{3}{10} - 4\frac{4}{5}$ $3\frac{1}{2}$

SKILL 36 *(Use after page 202.)*

Here's how

Multiply numerators and denominators.

$\frac{3}{4} \times \frac{4}{5} = \frac{12}{20}$

Write in simplest form.

$\frac{3}{4} \times \frac{4}{5} = \frac{12}{20} = \frac{3}{5}$

Give the product in simplest form.

1. $\frac{1}{2} \times \frac{1}{3}$ $\frac{1}{6}$
2. $\frac{3}{4} \times \frac{1}{4}$ $\frac{3}{16}$
3. $\frac{2}{3} \times \frac{4}{5}$ $\frac{8}{15}$
4. $\frac{3}{8} \times 2$ $\frac{3}{4}$
5. $3 \times \frac{1}{5}$ $\frac{3}{5}$
6. $\frac{1}{2} \times \frac{1}{4}$ $\frac{1}{8}$
7. $\frac{4}{3} \times \frac{3}{2}$ 2
8. $\frac{3}{4} \times \frac{16}{3}$ 4
9. $\frac{1}{3} \times \frac{3}{8}$ $\frac{1}{8}$
10. $\frac{7}{4} \times \frac{4}{3}$ $2\frac{1}{3}$
11. $\frac{1}{3} \times \frac{4}{5}$ $\frac{4}{15}$
12. $\frac{1}{2} \times \frac{4}{9}$ $\frac{2}{9}$
13. $\frac{5}{8} \times \frac{4}{5}$ $\frac{1}{2}$
14. $\frac{1}{4} \times \frac{8}{5}$ $\frac{2}{5}$
15. $\frac{3}{2} \times \frac{2}{3}$ 1
16. $5 \times \frac{4}{5}$ 4
17. $\frac{5}{6} \times \frac{6}{5}$ 1
18. $\frac{1}{2} \times \frac{2}{3}$ $\frac{1}{3}$
19. $\frac{3}{8} \times \frac{4}{3}$ $\frac{1}{2}$
20. $\frac{2}{3} \times 9$ 6
21. $\frac{1}{3} \times 6$ 2
22. $\frac{5}{8} \times \frac{4}{5}$ $\frac{1}{2}$
23. $\frac{1}{2} \times \frac{4}{5}$ $\frac{2}{5}$
24. $3 \times \frac{8}{3}$ 8

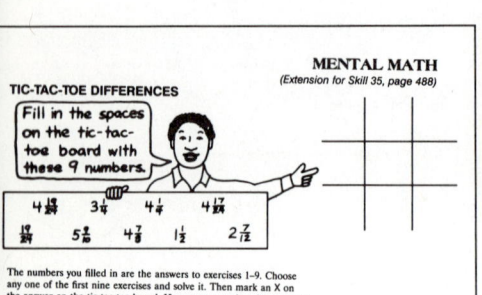

TIC-TAC-TOE DIFFERENCES

MENTAL MATH *(Extension for Skill 35, page 488)*

Fill in the spaces on the tic-tac-toe board with these 9 numbers.

$4\frac{19}{24}$ $3\frac{1}{4}$ $4\frac{1}{4}$ $4\frac{17}{24}$
$1\frac{9}{10}$ $5\frac{7}{18}$ $4\frac{7}{8}$ $1\frac{1}{2}$ $2\frac{7}{12}$

The numbers you filled in are the answers to exercises 1–9. Choose any one of the first nine exercises and solve it. Then mark an X on the answer on the tic-tac-toe board. How many exercises do you have to do to get three X's in a row? In 3 exercises is excellent, in 4 is very good, in 5 is good.

Copymaster S503

SOLUTION SEARCH

MENTAL MATH *(Extension for Skill 36, page 488)*

Look at exercises 1–24.

1. Which two exercises have an answer of 1? 15, 17
2. Which seven exercises have answers that are greater than 1? 7, 8, 10, 16, 20, 21, 24
3. Which exercise has an answer of $\frac{1}{3}$? 18

Copymaster S504

SKILL 37 (Use after page 204.)

Here's how
Divide by the denominator and then multiply the result by the numerator.

$$24 \div 4 \times 3 = 18$$

$\frac{3}{4}$ of 24 = 18

Complete.

1. $\frac{1}{4}$ of 12 = ?   3
2. $\frac{1}{2}$ of 18 = ?   9
3. $\frac{1}{3}$ of 24 = ?   8
4. $\frac{1}{5}$ of 30 = ?   6
5. $\frac{1}{8}$ of 72 = ?   9
6. $\frac{1}{6}$ of 42 = ?   7
7. $\frac{2}{3}$ of 18 = ?   12
8. $\frac{3}{4}$ of 20 = ?   15
9. $\frac{2}{5}$ of 10 = ?   4
10. $\frac{3}{8}$ of 32 = ?   12
11. $\frac{2}{3}$ of 24 = ?   16
12. $\frac{5}{6}$ of 30 = ?   25
13. $\frac{3}{4}$ of 32 = ?   24
14. $\frac{5}{8}$ of 40 = ?   25
15. $\frac{2}{5}$ of 40 = ?   16
16. $\frac{3}{5}$ of 25 = ?   15
17. $\frac{5}{6}$ of 42 = ?   35
18. $\frac{2}{3}$ of 36 = ?   24
19. $\frac{9}{10}$ of 90 = ?   81
20. $\frac{2}{3}$ of 21 = ?   14
21. $\frac{4}{5}$ of 45 = ?   36
22. $\frac{3}{4}$ of 36 = ?   27
23. $\frac{7}{10}$ of 40 = ?   28
24. $\frac{7}{8}$ of 56 = ?   49

MENTAL MATH (Extension for Skill 37, page 489)

SOLUTION SEARCH
Look at exercises 1–24.
1. Which exercise can you solve mentally by
   a. dividing by 5 and multiplying by 3?   16
   b. dividing by 8 and multiplying by 5?   14
   c. dividing by 10 and multiplying by 9?   19
   d. dividing by 8 and multiplying by 7?   24
2. How would you find the answer to exercise 16 mentally?
   Divide by 5 and multiply by 3.

Copymaster S504

SKILL 38 (Use after page 206.)

Here's how
Change to fractions.
$$2\frac{1}{2} \times 1\frac{2}{3} = \frac{5}{2} \times \frac{5}{3}$$

Multiply.
$$2\frac{1}{2} \times 1\frac{2}{3} = \frac{5}{2} \times \frac{5}{3}$$
$$= \frac{25}{6}$$

Write in simplest form.
$$2\frac{1}{2} \times 1\frac{2}{3} = \frac{5}{2} \times \frac{5}{3}$$
$$= \frac{25}{6}$$
$$= 4\frac{1}{6}$$

Give the product in simplest form.

1. $2 \times 1\frac{1}{2}$ 3
2. $1\frac{1}{2} \times 1\frac{1}{3}$ 2
3. $2\frac{2}{3} \times 1\frac{1}{4}$ $3\frac{1}{3}$
4. $1\frac{3}{4} \times 1\frac{3}{4}$ $3\frac{1}{16}$
5. $3 \times 2\frac{1}{3}$ 7
6. $2\frac{1}{3} \times 2$ $4\frac{2}{3}$
7. $2\frac{2}{5} \times 3$ $7\frac{1}{5}$
8. $1\frac{5}{6} \times 2\frac{1}{3}$ $4\frac{5}{18}$
9. $3\frac{1}{4} \times 3\frac{1}{4}$ $10\frac{9}{16}$
10. $4\frac{1}{6} \times 2\frac{1}{3}$ $9\frac{13}{18}$
11. $2\frac{2}{3} \times 2\frac{1}{2}$ $6\frac{2}{3}$
12. $3 \times 4\frac{1}{2}$ $13\frac{1}{2}$
13. $2 \times 1\frac{2}{3}$ $3\frac{1}{3}$
14. $1\frac{1}{2} \times 2\frac{1}{2}$ $3\frac{3}{4}$
15. $3\frac{3}{4} \times 2$ $7\frac{1}{2}$
16. $1\frac{3}{8} \times 2\frac{1}{2}$ $3\frac{7}{16}$
17. $3\frac{3}{4} \times 3\frac{1}{8}$ $11\frac{23}{32}$
18. $1\frac{5}{8} \times 1\frac{5}{8}$ $2\frac{41}{64}$
19. $1\frac{1}{2} \times 1\frac{3}{4}$ $2\frac{5}{8}$
20. $2 \times 2\frac{3}{4}$ $5\frac{1}{2}$
21. $2\frac{2}{3} \times 1\frac{3}{4}$ $4\frac{2}{3}$
22. $4\frac{1}{2} \times 2\frac{3}{8}$ $10\frac{11}{16}$
23. $1\frac{5}{8} \times 6$ $9\frac{3}{4}$
24. $6\frac{3}{4} \times 3\frac{3}{8}$ $22\frac{25}{32}$

MENTAL MATH (Extension for Skill 38, page 489)

ESTIMATION HUNT
Look at exercises 1–24.
1. Which exercise has an answer that you can estimate by thinking
   a. 4 times 3?   17
   b. 3 times 5?   12
   c. 5 times 2?   22
   d. 2 times 6?   23
2. Estimate. Which exercise has an answer that is the greatest product?   24

Copymaster S505

Extra Practice **489**

SKILL 39 (Use after page 209.)

Here's how

$2\frac{3}{4}$ days = __?__ h

Find the hours in 2 days and $\frac{3}{4}$ of a day.

$2\frac{3}{4}$ days = 48 h + 18 h

Add.

$2\frac{3}{4}$ days = 48 h + 18 h
= 66 h

Complete.

1. $1\frac{1}{2}$ days = __?__ h   36
2. $1\frac{1}{4}$ h = __?__ min   75
3. $1\frac{3}{4}$ h = __?__ min   105
4. $2\frac{1}{2}$ min = __?__ sec   150
5. $2\frac{1}{4}$ min = __?__ sec   135
6. $2\frac{1}{3}$ days = __?__ h   56
7. $1\frac{2}{3}$ yd = __?__ ft   5
8. $1\frac{3}{4}$ ft = __?__ in.   21
9. $1\frac{1}{2}$ yd = __?__ in.   54
10. $2\frac{2}{3}$ ft = __?__ in.   32
11. $4\frac{1}{3}$ yd = __?__ ft   13
12. $1\frac{3}{4}$ yd = __?__ in.   63
13. $1\frac{1}{4}$ gal = __?__ qt   5
14. $2\frac{1}{2}$ qt = __?__ pt   5
15. $3\frac{1}{2}$ pt = __?__ c   7
16. $3\frac{1}{2}$ qt = __?__ pt   7

SOLUTION SEARCH

Look at exercises 1–16.

1. Which exercise can you solve mentally by thinking
   a. 120 seconds plus 30 seconds? 4
   b. 12 inches plus 9 inches? 8
   c. 12 feet plus 1 foot? 11
   d. 6 cups plus 1 cup? 15

2. How would you find the answer to exercise 5 mentally?
   120 seconds plus 15 seconds

Copymaster S505

SKILL 40 (Use after page 212.)

Here's how

To divide by a fraction, multiply by its reciprocal.

$\frac{5}{4} \div \frac{3}{2} = \frac{5}{4} \times \frac{2}{3}$
$= \frac{10}{12}$

Write in simplest form.

$\frac{5}{4} \div \frac{3}{2} = \frac{5}{4} \times \frac{2}{3}$
$= \frac{10}{12}$
$= \frac{5}{6}$

Give the quotient in lowest terms.

1. $\frac{3}{4} \div \frac{1}{4}$ 3
2. $\frac{2}{3} \div \frac{1}{3}$ 2
3. $\frac{1}{2} \div \frac{1}{3}$ $1\frac{1}{2}$
4. $\frac{3}{5} \div \frac{1}{5}$ 3
5. $\frac{4}{5} \div 3$ $\frac{4}{15}$
6. $\frac{7}{8} \div \frac{7}{8}$ 1
7. $\frac{5}{6} \div \frac{2}{3}$ $1\frac{1}{4}$
8. $\frac{2}{3} \div \frac{1}{2}$ $1\frac{1}{3}$
9. $\frac{3}{10} \div \frac{4}{5}$ $\frac{3}{8}$
10. $\frac{3}{4} \div \frac{3}{2}$ $\frac{1}{2}$
11. $6 \div \frac{3}{4}$ 8
12. $\frac{5}{8} \div 3$ $\frac{5}{24}$
13. $\frac{5}{6} \div \frac{5}{8}$ $1\frac{1}{3}$
14. $\frac{5}{8} \div \frac{2}{3}$ $\frac{15}{16}$
15. $\frac{2}{3} \div \frac{4}{5}$ $\frac{5}{6}$
16. $\frac{7}{10} \div \frac{1}{5}$ $3\frac{1}{2}$
17. $\frac{2}{3} \div \frac{5}{6}$ $\frac{4}{5}$
18. $\frac{3}{5} \div \frac{6}{1}$ $\frac{1}{10}$
19. $9 \div \frac{4}{3}$ $6\frac{3}{4}$
20. $\frac{1}{2} \div \frac{5}{8}$ $\frac{4}{5}$
21. $\frac{5}{3} \div \frac{3}{2}$ $1\frac{1}{9}$
22. $\frac{1}{2} \div \frac{5}{4}$ $\frac{2}{5}$
23. $\frac{3}{4} \div \frac{3}{8}$ 2
24. $\frac{7}{4} \div \frac{2}{5}$ $4\frac{3}{8}$
25. $6 \div \frac{2}{3}$ 9
26. $\frac{5}{9} \div \frac{2}{3}$ $\frac{5}{6}$
27. $\frac{5}{8} \div \frac{3}{4}$ $\frac{5}{6}$
28. $\frac{3}{5} \div \frac{5}{4}$ $\frac{12}{25}$
29. $\frac{3}{4} \div \frac{5}{8}$ $1\frac{1}{5}$
30. $\frac{9}{4} \div \frac{6}{5}$ $1\frac{7}{8}$
31. $\frac{9}{10} \div \frac{5}{4}$ $\frac{18}{25}$
32. $\frac{8}{3} \div \frac{5}{6}$ $3\frac{1}{5}$

SOLUTION SEARCH

Look at exercises 1–32.

1. Which exercise can you solve mentally by thinking
   a. $\frac{4}{5}$ times $\frac{1}{3}$? 5
   b. $\frac{5}{8}$ times $\frac{3}{2}$? 14
   c. $\frac{5}{3}$ times $\frac{3}{2}$? 21
   d. $\frac{3}{5}$ times $\frac{4}{5}$? 28

2. How would you find the answer to exercise 24 mentally?
   7/4 times 5/2

Copymaster S506

Extra Practice

Here's how

Change to fractions.

$$2\frac{1}{4} \div 4\frac{1}{2} = \frac{9}{4} \div \frac{9}{2}$$

Divide. Write in simplest form.

$$2\frac{1}{4} \div 4\frac{1}{2} = \frac{9}{4} \div \frac{9}{2}$$
$$= \frac{9}{4} \times \frac{2}{9}$$
$$= \frac{18}{36}$$
$$= \frac{1}{2}$$

SKILL 41 (Use after page 214.)

Give the quotient in simplest form.

1. $5 \div 2\frac{1}{2}$ 2
2. $2\frac{1}{2} \div 1\frac{1}{4}$ 2
3. $5 \div 1\frac{1}{4}$ 4
4. $3\frac{1}{2} \div 2$ $1\frac{3}{4}$
5. $10 \div 3\frac{1}{3}$ 3
6. $5\frac{1}{4} \div 3$ $1\frac{3}{4}$
7. $4\frac{1}{6} \div 5$ $\frac{5}{6}$
8. $4\frac{3}{4} \div 2$ $2\frac{3}{8}$
9. $2\frac{1}{3} \div 1\frac{1}{4}$ $1\frac{13}{15}$
10. $2\frac{1}{2} \div 2\frac{1}{2}$ 1
11. $7\frac{1}{2} \div 2\frac{1}{2}$ 3
12. $6\frac{3}{4} \div 3\frac{1}{2}$ $1\frac{13}{14}$
13. $3\frac{1}{2} \div 1\frac{3}{4}$ 2
14. $2\frac{7}{8} \div 3\frac{1}{4}$ $\frac{23}{26}$
15. $4\frac{5}{8} \div 2\frac{2}{3}$ $1\frac{47}{64}$
16. $3\frac{5}{6} \div 2\frac{1}{3}$ $1\frac{9}{14}$
17. $5 \div 1\frac{1}{4}$ 4
18. $5\frac{3}{4} \div 2\frac{2}{3}$ $2\frac{5}{32}$
19. $6\frac{2}{3} \div 5\frac{1}{3}$ $1\frac{1}{4}$
20. $4\frac{7}{8} \div 6\frac{1}{4}$ $\frac{39}{50}$
21. $2\frac{3}{4} \div 5\frac{2}{3}$ $\frac{33}{68}$
22. $10 \div 3\frac{1}{3}$ 3
23. $4\frac{1}{4} \div 3\frac{1}{8}$ $1\frac{9}{25}$
24. $3\frac{1}{2} \div 1\frac{3}{4}$ 2

MENTAL MATH
(Extension for Skill 41, page 491)

ESTIMATION HUNT

Look at exercises 1–24.

1. Estimate. Which four exercises have answers that are less than 1? **7, 14, 20, 21**
2. Estimate. Which two exercises have an answer of 4? **3, 17**
3. Estimate. Which exercise has an answer of about $\frac{1}{2}$? **21**

Copymaster S506

Here's how

Jan is 158 centimeters tall and Alice is 1 meter 47 centimeters tall. How many centimeters taller is Jan? Hint: 100 cm = 1 m

Step 1. Find how many centimeters in 1 meter 47 centimeters.

1 m + 47 cm
= 100 cm + 47 cm
= 147 cm

Step 2. Subtract to find the difference.

158 − 147 = 11

11 centimeters.

SKILL 42 (Use after page 230.)

Solve.

1. David has 8 apples that weigh 2 kilograms. What is the average weight in grams of each apple?
Hint: 1000 g = 1 kg **250 g**

2. Stephanie jogs 2500 meters every day except Sunday. How many kilometers does she jog in a week? Hint: 1000 m = 1 km **15 km**

3. Sara drinks 200 millimeters of juice each day. If juice costs $2.40 a liter, what does she spend each week?
Hint: 1000 mL = 1 L **$3.36**

4. Bill wants to fill 50 paper cups that hold 120 millimeters each for a bike-a-thon. How many liters of water does he need? **6 L**

5. Greg high jumped 1 meter 62 centimeters. Eric high jumped 9 centimeters more than Greg. How many centimeters did Eric jump?
Hint: 100 cm = 1 m **1071 cm**

6. Gayle's cereal contains 2 grams of sodium per box. If there are 10 servings in each box, how many milligrams of sodium are in each serving? Hint: 1000 mg = 1 g **200 mg**

Extra Practice **491**

Here's how

Mary had 27 inches of ribbon. She bought 2 more yards. How many inches of ribbon did she have then? Hint: 1 yd = 36 in.

Step 1. Find how many inches in 2 yards.

$$\begin{array}{r} 36 \\ \times\ 2 \\ \hline 72 \end{array}$$

Step 2. Add to find the total.

$$\begin{array}{r} 72 \\ +27 \\ \hline 99 \end{array}$$

99 inches.

SKILL 43 (Use after page 246.)

Solve.

1. William bought 8 feet 6 inches of rope. Gayle bought 90 inches of rope. How many more inches of rope did William buy? Hint: 12 in. = 1 ft *12 inches*

2. Jean needs 14 yards 2 feet of molding to go around her room. If molding costs $1.20 a foot, what will be the total cost? Hint: 3 ft = 1 yd *$52.80*

3. Allison needs to buy 2 gallons of punch for a party. If punch costs $1.49 a quart, what will be the total cost? Hint: 4 qt = 1 gal *$11.92*

4. Andrew made 3 pints of ice cream. The total cost was $2.52. What was the cost per cup? Hint: 2 c = 1 pt *$.42*

5. Beth bought 1 pound 8 ounces of cheese. How many 4-ounce servings of cheese could she serve? Hint: 16 oz = 1 lb *6 servings*

6. Ted bought 8000 pounds of hay for his horses. If he paid $48 a ton, what was the total cost? Hint: 2000 lb = 1 T *$192*

Here's how

$$\frac{5}{3} = \frac{7}{n}$$

Write the multiplication equation

$$\frac{5}{3} = \frac{7}{n}$$
$$5n = 21$$

Solve the multiplication equation.

$$\frac{5}{3} = \frac{7}{n}$$
$$5n = 21$$
$$n = 4\frac{1}{5}$$

SKILL 44 (Use after page 262.)

Solve.

1. $\frac{n}{6} = \frac{11}{4}$ $16\frac{1}{2}$
2. $\frac{8}{n} = \frac{2}{9}$ 36
3. $\frac{9}{8} = \frac{n}{7}$ $7\frac{7}{8}$
4. $\frac{10}{7} = \frac{4}{n}$ $2\frac{4}{5}$

5. $\frac{5}{n} = \frac{2}{9}$ $22\frac{1}{2}$
6. $\frac{n}{13} = \frac{6}{5}$ $15\frac{3}{5}$
7. $\frac{7}{4} = \frac{n}{6}$ $10\frac{1}{2}$
8. $\frac{5}{8} = \frac{3}{n}$ $4\frac{4}{5}$

9. $\frac{3}{7} = \frac{9}{n}$ 21
10. $\frac{18}{n} = \frac{6}{5}$ 15
11. $\frac{n}{21} = \frac{6}{7}$ 18
12. $\frac{n}{6} = \frac{15}{8}$ $11\frac{1}{4}$

13. $\frac{9}{n} = \frac{3}{13}$ 39
14. $\frac{18}{n} = \frac{9}{16}$ 32
15. $\frac{6}{11} = \frac{24}{n}$ 44
16. $\frac{5}{7} = \frac{9}{n}$ $12\frac{3}{5}$

17. $\frac{n}{7} = \frac{11}{4}$ $19\frac{1}{4}$
18. $\frac{6}{n} = \frac{8}{3}$ $2\frac{1}{4}$
19. $\frac{19}{2} = \frac{n}{5}$ $47\frac{1}{2}$
20. $\frac{10}{13} = \frac{16}{n}$ $20\frac{4}{5}$

21. $\frac{9}{n} = \frac{2}{15}$ $67\frac{1}{2}$
22. $\frac{n}{9} = \frac{4}{3}$ 12
23. $\frac{6}{15} = \frac{30}{n}$ 75
24. $\frac{11}{4} = \frac{n}{8}$ 22

25. $\frac{n}{4} = \frac{6}{8}$ 3
26. $\frac{1}{n} = \frac{7}{21}$ 3
27. $\frac{6}{9} = \frac{n}{3}$ 2
28. $\frac{6}{10} = \frac{18}{n}$ 30

29. $\frac{5}{15} = \frac{3}{n}$ 9
30. $\frac{10}{3} = \frac{n}{12}$ 40
31. $\frac{7}{n} = \frac{10}{9}$ $6\frac{3}{10}$
32. $\frac{n}{8} = \frac{7}{4}$ 14

MENTAL MATH (Extension for Skill 44, page 492)

TIC-TAC-TOE VALUES FOR N

Fill in the spaces on the tic-tac-toe board with these 9 numbers.

$10\frac{1}{2}$ $7\frac{7}{8}$ $22\frac{1}{2}$ 21
$15\frac{3}{5}$ $4\frac{4}{5}$ $2\frac{4}{5}$ $16\frac{1}{2}$ 36

The numbers you filled in are the answers to exercises 1–9. Choose any one of the first nine exercises and solve it. Then mark an X on the answer on the tic-tac-toe board. How many exercises do you have to do to get three X's in a row? In 3 exercises is excellent, in 4 is very good, in 5 is good.

Copymaster S507

Extra Practice

SKILL 45 (Use after page 268.)

Here's how

A trail mix has 2 pounds of mixed nuts to every 3 pounds of dried fruit. How many pounds of mixed nuts would be needed for 12 pounds of dried fruit?

Step 1. Set up a proportion.

nuts → $\frac{2}{3} = \frac{n}{12}$ ← nuts
fruit →   ← fruit

Step 2. Solve.

$3n = 2 \times 12$
$3n = 24$
$n = 8$

8 pounds.

Solve.

1. The ratio of boys to girls in a nature club is 4 to 7. What is the ratio of girls to boys? 7 to 4

2. Each hiker carried 1 pound of supplies for every 8 pounds of body weight. How many pounds of supplies did a 120-pound hiker carry? 15 pounds

3. For a hike, the club decided to take 2 quarts of water for every 3 hikers. How many quarts of water should they take for 21 hikers? 14 quarts

4. During the first 3 hours the club hiked 8 miles. At that rate, how many miles would they hike in 5 hours? $13\frac{1}{3}$ miles

5. Five cups of water was needed for every 2 packages of dried soup. How much water was needed for 16 packages of dried soup? 40 cups

6. After lunch they hiked 7 miles in 3 hours. At that rate, how many hours would it take them to hike 10 miles? $4\frac{2}{7}$ hours

---

SKILL 46 (Use after page 282.)

Here's how

Write the percent as a fraction with a denominator of 100. Write the fraction in simplest form.

$25\% = \frac{25}{100} = \frac{1}{4}$

$33\frac{1}{3}\% = \frac{33\frac{1}{3}}{100}$
$= 33\frac{1}{3} \div 100$
$= \frac{100}{3} \times \frac{1}{100}$
$= \frac{100}{300}$
$= \frac{1}{3}$

Change to a fraction in simplest form.

1. 10% $\frac{1}{10}$
2. 15% $\frac{3}{20}$
3. 40% $\frac{2}{5}$
4. 50% $\frac{1}{2}$
5. 90% $\frac{9}{10}$
6. 60% $\frac{3}{5}$
7. 25% $\frac{1}{4}$
8. 75% $\frac{3}{4}$
9. 20% $\frac{1}{5}$
10. 30% $\frac{3}{10}$
11. 45% $\frac{9}{20}$
12. 85% $\frac{17}{20}$
13. 80% $\frac{4}{5}$
14. 40% $\frac{2}{5}$
15. 48% $\frac{12}{25}$
16. 96% $\frac{24}{25}$
17. 18% $\frac{9}{50}$
18. 24% $\frac{6}{25}$
19. 72% $\frac{18}{25}$
20. 84% $\frac{21}{25}$
21. 150% $1\frac{1}{2}$
22. 125% $1\frac{1}{4}$
23. 175% $1\frac{3}{4}$
24. 120% $1\frac{1}{5}$
25. 180% $1\frac{4}{5}$
26. 250% $2\frac{1}{2}$
27. 275% $2\frac{3}{4}$
28. 225% $2\frac{1}{4}$
29. 160% $1\frac{3}{5}$
30. 200% 2
31. 300% 3
32. 320% $3\frac{1}{5}$
33. $33\frac{1}{3}\%$ $\frac{1}{3}$
34. $66\frac{2}{3}\%$ $\frac{2}{3}$
35. $37\frac{1}{2}\%$ $\frac{3}{8}$
36. $16\frac{2}{3}\%$ $\frac{1}{6}$
37. $87\frac{1}{2}\%$ $\frac{7}{8}$
38. $81\frac{1}{4}\%$ $\frac{13}{16}$
39. $162\frac{1}{2}\%$ $1\frac{5}{8}$
40. $118\frac{3}{4}\%$ $1\frac{3}{16}$

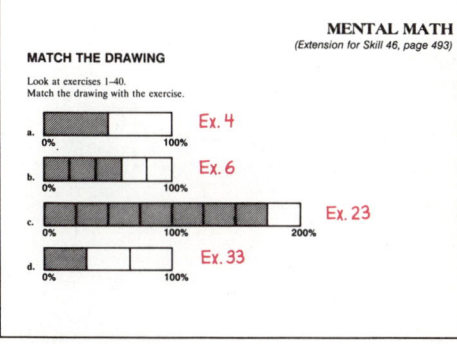

MENTAL MATH (Extension for Skill 46, page 493)

MATCH THE DRAWING

Look at exercises 1–40. Match the drawing with the exercise.

a. Ex. 4
b. Ex. 6
c. Ex. 23
d. Ex. 33

Copymaster S507

Extra Practice 493

SKILL 47 (Use after page 284.)

Here's how

Change to an equivalent fraction with a denominator of 100. Write as a percent.

$$\frac{1}{2} = \frac{50}{100} = 50\%$$

Solve a proportion.

$$\frac{1}{6} = \frac{n}{100}$$
$$6n = 100$$
$$n = 16\frac{2}{3}$$

So $\frac{1}{6} = \frac{16\frac{2}{3}}{100} = 16\frac{2}{3}\%$

Change to a percent.

1. $\frac{1}{5}$ 20% 2. $\frac{1}{4}$ 25% 3. $\frac{1}{2}$ 50% 4. $\frac{4}{5}$ 80%
5. $\frac{3}{4}$ 75% 6. $\frac{1}{3}$ $33\frac{1}{3}$% 7. $\frac{2}{5}$ 40% 8. $\frac{2}{3}$ $66\frac{2}{3}$%
9. $\frac{1}{8}$ $12\frac{1}{2}$% 10. $\frac{3}{5}$ 60% 11. $\frac{3}{8}$ $37\frac{1}{2}$% 12. $\frac{3}{2}$ 150%
13. $\frac{6}{5}$ 120% 14. $\frac{1}{6}$ $16\frac{2}{3}$% 15. $\frac{5}{4}$ 125% 16. $\frac{7}{5}$ 140%
17. $\frac{1}{10}$ 10% 18. $\frac{8}{5}$ 160% 19. $\frac{3}{10}$ 30% 20. $\frac{1}{12}$ $8\frac{1}{3}$%
21. $\frac{5}{2}$ 250% 22. 3 300% 23. $\frac{7}{4}$ 175% 24. $\frac{1}{20}$ 5%
25. $\frac{5}{8}$ $62\frac{1}{2}$% 26. $\frac{7}{10}$ 70% 27. $\frac{11}{8}$ $137\frac{1}{2}$% 28. $\frac{7}{3}$ $233\frac{1}{3}$%
29. $\frac{5}{12}$ $41\frac{2}{3}$% 30. $\frac{5}{3}$ $166\frac{2}{3}$% 31. 2 200% 32. $\frac{7}{12}$ $58\frac{1}{3}$%

SOLUTION SEARCH

MENTAL MATH (Extension for Skill 47, page 494)

Look at exercises 1–32.
1. Which exercise can you solve mentally by thinking of
   a. $\frac{25}{100}$? 2 b. $\frac{75}{100}$? 5
   c. $\frac{10}{100}$? 17 d. $\frac{300}{100}$? 22
2. How would you find the answer to exercise 19 mentally?
   $\frac{30}{100}$

Copymaster S508

SKILL 48 (Use after page 286.)

Here's how

To change a percent to a decimal, move the decimal point two places to the left and remove the percent sign.

$$3\% = 0.03$$
$$17\% = 0.17$$
$$15.2\% = 0.152$$
$$66\frac{2}{3}\% = 0.66\frac{2}{3}$$
$$112\% = 1.12$$

Change to a decimal.

1. 25% 0.25 2. 20% 0.20 3. 45% 0.45 4. 82% 0.82
5. 34% 0.34 6. 65% 0.65 7. 80% 0.80 8. 56% 0.56
9. 75% 0.75 10. 12% 0.12 11. 81% 0.81 12. 62% 0.62
13. 5% 0.05 14. 8% 0.08 15. 2% 0.02 16. 4% 0.04
17. 1% 0.01 18. 3% 0.03 19. 7% 0.07 20. 6% 0.06
21. 150% 1.50 22. 125% 1.25 23. 173% 1.73 24. 225% 2.25
25. 160% 1.60 26. 250% 2.50 27. 300% 3 28. 500% 5
29. 28.5% 0.285 30. 16.2% 0.162 31. 34.6% 0.346 32. 56.8% 0.568
33. 47.2% 0.472 34. 93.9% 0.939 35. 60.5% 0.605 36. 85.2% 0.852
37. $33\frac{1}{3}$% $0.33\frac{1}{3}$ 38. $87\frac{1}{2}$% $0.87\frac{1}{2}$ 39. $16\frac{2}{3}$% $0.16\frac{2}{3}$ 40. $37\frac{1}{2}$% $0.37\frac{1}{2}$

SOLUTION SEARCH

MENTAL MATH (Extension for Skill 48, page 286)

Look at exercises 1–40.
1. Which exercise has an answer of 0.80? 7
2. Which four examples have answers that are less than 0.05?
   15, 16, 17, 18
3. Which six exercises have answers that are greater than 1.50?
   23, 24, 25, 26, 27, 28
4. Which exercise has an answer that is closest to 0.40? 40

Copymaster S508

494 *Extra Practice*

SKILL 49 *(Use after page 287.)*

Here's how

To change a decimal to a percent, move the decimal point two places to the right and write the percent sign.

0.5 = 50%

0.125 = 12.5%

$\mathbf{0.87\frac{1}{2}} = 87\frac{1}{2}\%$

1.475 = 147.5%

Change to a percent.

1. 0.04 4%	**2.** 0.07 7%	**3.** 0.02 2%	**4.** 0.06 6%
5. 0.08 8%	**6.** 0.01 1%	**7.** 0.05 5%	**8.** 0.03 3%
9. 0.09 9%	**10.** 0.56 56%	**11.** 0.49 49%	**12.** 0.66 66%
13. 0.71 71%	**14.** 0.11 11%	**15.** 0.98 98%	**16.** 0.38 38%
17. 0.5 50%	**18.** 0.4 40%	**19.** 0.3 30%	**20.** 0.8 80%
21. 0.2 20%	**22.** 0.6 60%	**23.** 0.1 10%	**24.** 0.9 90%
25. 2.25 225%	**26.** 1.58 158%	**27.** 1.55 155%	**28.** 2.75 275%
29. 3.46 346%	**30.** 8.25 825%	**31.** 7.52 752%	**32.** 5.37 537%
33. $0.12\frac{1}{2}$ $12\frac{1}{2}\%$	**34.** $0.33\frac{1}{3}$ $33\frac{1}{3}\%$	**35.** $0.62\frac{1}{2}$ $62\frac{1}{2}\%$	**36.** $0.16\frac{2}{3}$ $16\frac{2}{3}\%$
37. $1.37\frac{1}{2}$ $137\frac{1}{2}\%$	**38.** $1.66\frac{2}{3}$ $166\frac{2}{3}\%$	**39.** $2.33\frac{1}{3}$ $233\frac{1}{3}\%$	**40.** $2.83\frac{1}{3}$ $283\frac{1}{3}\%$

MENTAL MATH
(Extension for Skill 49, page 495)

SOLUTION SEARCH

Look at exercises 1–40.

1. Which exercise has an answer of 50%? 17
2. Which four exercises have answers that are less than 5%? 1, 3, 6, 8
3. Which six exercises have answers that are greater than 250%? 28, 29, 30, 31, 32, 40
4. Which exercise has an answer that is closest to 100%? 15

Copymaster S509

SKILL 50 *(Use after page 290.)*

Here's how

First change the percent to a fraction or decimal. Then multiply.

75% of 36 $= \frac{3}{4} \times 36$
$= 27$

6.5% of 9 $= 0.065 \times 9$
$= 0.585$

Solve.

1. 50% of 28 = n 14	**2.** 25% of 24 = n 6
3. 40% of 60 = n 24	**4.** 20% of 45 = n 9
5. 80% of 40 = n 32	**6.** 10% of 50 = n 5
7. 75% of 48 = n 36	**8.** 25% of 64 = n 16
9. 100% of 56 = n 56	**10.** 150% of 18 = n 27
11. 125% of 40 = n 50	**12.** 250% of 72 = n 180
13. 14% of 26 = n 3.64	**14.** 23% of 75 = n 17.25
15. 56% of 29 = n 16.24	**16.** 41% of 83 = n 34.03
17. 5.4% of 60 = n 3.24	**18.** 6.5% of 47 = n 3.055
19. 0.25% of 160 = n 0.4	**20.** 0.6% of 314 = n 1.884

MENTAL MATH
(Extension for Skill 50, page 495)

SOLUTION SEARCH

Look at exercises 1–20.

1. Which exercise can you solve mentally by thinking
   a. $\frac{1}{4}$ of 24? 2
   b. $\frac{1}{10}$ of 50? 6
   c. $\frac{5}{4}$ of 40? 11
   d. $\frac{4}{10}$ of 60? 3
2. Which exercise can you solve mentally by adding the number to $\frac{1}{2}$ of the number to get an answer of 27? 10

Copymaster S509

ESTIMATION HUNT

Look at exercises 1–20.

1. Which exercise has an answer that you can estimate by thinking

   a. $\frac{1}{4}$ of 120? 1
   b. $\frac{1}{5}$ of 100? 5
   c. $\frac{1}{10}$ of 50? 7
   d. $\frac{2}{3}$ of 60? 17

2. How would you estimate the answer to exercise 14?

   $\frac{1}{3}$ of 80

MENTAL MATH *(Extension for Skill 51, page 496)*

Copymaster S510

SKILL 51 *(Use after page 294.)*

Here's how

Change to a fraction or decimal and multiply.

20% of $45 = \frac{1}{5} \times 45$
$= 9$

8.2% of $6 = 0.082 \times 6$
$= 0.492$

Solve a proportion.

$16\frac{2}{3}\%$ of $72 = n$

$\frac{16\frac{2}{3}}{100} = \frac{n}{72}$

$100n = 16\frac{2}{3} \times 72$

$= \frac{50}{3} \times 72$

$= 1200$

$n = 12$

Solve.

1. 25% of 116 = n 29
2. 60% of 60 = n 36
3. 100% of 73 = n 73
4. 75% of 12 = n 9
5. 20% of 95 = n 19
6. $87\frac{1}{2}$% of 88 = n 77
7. 8.2% of 50 = n 4.1
8. 6.5% of 35 = n 2.275
9. 2.25% of 304 = n 6.84
10. $37\frac{1}{2}$% of 104 = n 39
11. $62\frac{1}{2}$% of 72 = n 45
12. 3.45% of 156 = n 5.382
13. 7.5% of 56 = n 4.2
14. $33\frac{1}{3}$% of 81 = n 27
15. 0.36% of 100 = n 0.36
16. 0.25% of 200 = n 0.5
17. $66\frac{2}{3}$% of 63 = n 42
18. $8\frac{1}{3}$% of 96 = n 8
19. 0.5% of 38 = n 0.19
20. $16\frac{2}{3}$% of 48 = n 8

TIC-TAC-TOE VALUES FOR N

Fill in the spaces on the tic-tac-toe board with these 9 numbers.

80   160   56   115
2400   44   190   80   38

The numbers you filled in are the answers to exercises 1–9. Choose any one of the first nine exercises and solve it. Then mark an X on the answer on the tic-tac-toe board. How many exercises do you have to do to get three X's in a row? In 3 exercises is excellent, in 4 is very good, in 5 is good.

MENTAL MATH *(Extension for Skill 52, page 496)*

Copymaster S510

SKILL 52 *(Use after page 296.)*

Here's how

60% of $n = 27$

Solve a proportion.

$\frac{part}{whole}\ \frac{60}{100} = \frac{27}{n}\ \frac{part}{whole}$

$60n = 2700$

$n = 45$

Solve.

1. 25% of n = 14   56
2. 50% of n = 19   38
3. 75% of n = 33   44
4. 60% of n = 48   80
5. 80% of n = 64   80
6. 40% of n = 46   115
7. 10% of n = 16   160
8. 5% of n = 120   2400
9. 30% of n = 57   190
10. 20% of n = 65   325

Solve. Round each answer to the nearest tenth.

11. 8.5% of n = 6   70.6
12. 7.5% of n = 11   146.7
13. 20.5% of n = 16.2   79.0
14. 14.2% of n = 37   260.6
15. 34.6% of n = 18.5   53.5
16. 42.8% of n = 10.3   24.1
17. 1.8% of n = 2.4   133.3
18. 4.7% of n = 2.9   61.7
19. 125% of n = 17.6   14.1
20. 250% of n = 31.3   12.5

Extra Practice

SKILL 53 (Use after page 298.)

Here's how

David made 3 baskets out of 9 shots. What percent of the 9 shots did he make?

$$\frac{3}{9} = \frac{n}{100}$$

$$9n = 300$$

$$n = 33\frac{1}{3}$$

He made $33\frac{1}{3}\%$.

Solve.

1. Steve named 43 of the 50 state capitals. What percent of the state capitals did he name? **86%**

2. Karen sold 21 of her 25 tickets to the school talent show. What percent of her tickets did she sell? **84%**

3. Four out of a class of 32 students were absent. What percent of the students were absent? **12.5%**

4. Francy got 36 out of 40 math problems correct. What percent of the problems did she get correct? **90%**

5. In a class election, 20 out of 36 students voted for Andrew. What percent of the class voted for Andrew? **$55\frac{5}{9}\%$**

6. In a school survey, 48 out of 80 students rode a bus to school. What percent of those surveyed rode the bus? **60%**

SKILL 54 (Use after page 300.)

Here's how

How much did Todd spend on a record?

15% of $48 = n

0.15 of $48

= 0.15 × $48

= $7.20

He spent $7.20 for a record.

Solve.

1. What percent of the money was spent on a sweater and book? **$36**

2. How much did Todd spend on a sweater? **$24**

3. How much was spent on a movie? **$4.80**

4. How much more was spent on a book than a movie? **$7.20**

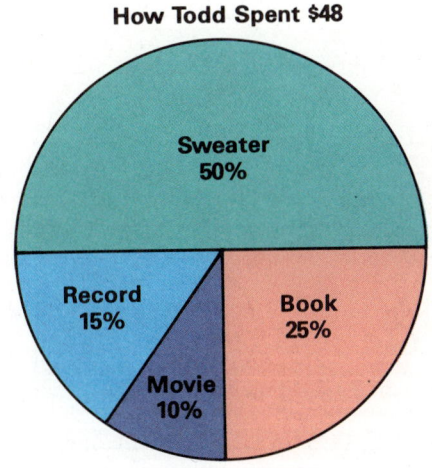

How Todd Spent $48

Sweater 50%
Record 15%
Movie 10%
Book 25%

Extra Practice **497**

SKILL 55 (Use after page 326.)

Here's how

Ned bought a shirt for $12.50 and some socks for $2.50. Sales tax was 5%. What was the total bill?

Step 1. Find the sales tax for $15.

$$\begin{array}{r} \$15 \\ \times\ .05 \\ \hline \$0.75 \end{array}$$

Step 2. Add.
$15 + $0.75 = $15.75

The total is $15.75.

Solve.

1. Lolly bought some shoes for $28 and a sweater for $22. Sales tax was 4%. What was the total bill? $52

2. A radio is on sale at a discount of 25%. The regular price is $36. What is the sale price? $27

3. Which is the better buy, an 18-ounce box of cereal for $1.19 or a 12-ounce box of cereal for $.75? 12-ounce box

4. Paige had $446.75 in her checking account. She wrote a check for $40 and made a deposit of $123.50. What was the balance then? $530.25

5. Jack earns $30 per week. According to his budget he saves 25% of his earnings. How much does he save each week? $7.50

6. Deanna borrowed $800 for 6 months to buy a car. The bank charged her a yearly rate of 14%. How much interest did she owe after 6 months? Hint: $I = p \times r \times t$ $56

SKILL 56 (Use after page 354.)

Give the area. Use 3.14 for π.

Here's how

$A = l \times w$
$A = 8\ m \times 6\ m$
$= 48\ m^2$

$A = \frac{1}{2} \times b \times h$
$A = \frac{1}{2} \times 26\ in. \times 10\ in.$
$= 130\ in.^2$

$A = \pi \times r^2$
$A \approx 3.14 \times 3\ ft \times 3\ ft$
$\approx 28.26\ ft^2$

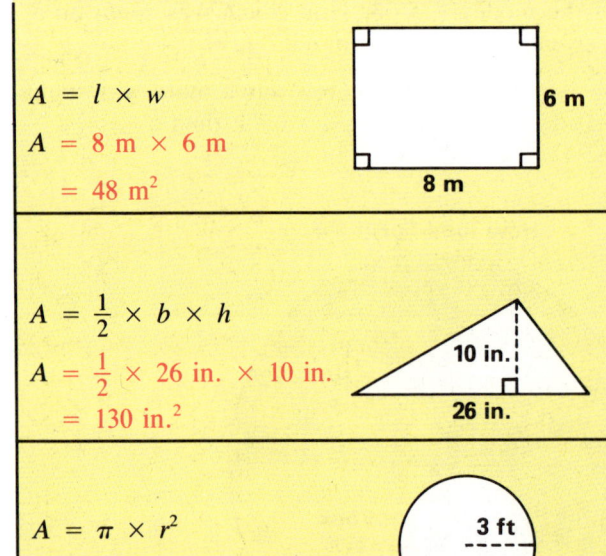

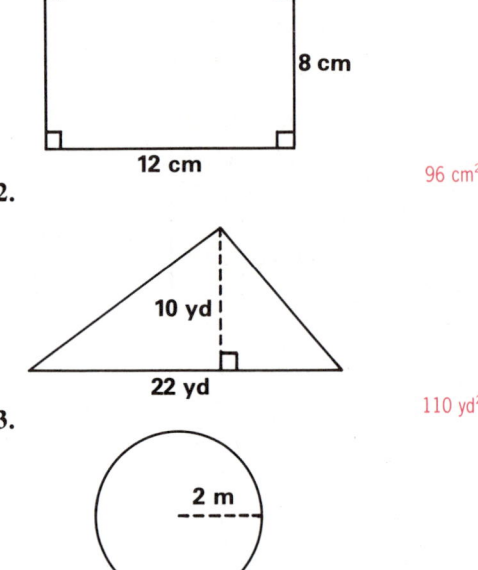

1. 96 cm²
2. 110 yd²
3. 12.56 m²

498 Extra Practice

SKILL 57 (Use after page 376.)

Give the volume. Use 3.14 for π.

Here's how

Rectangular Prism

$V = B \times h$
$V = l \times w \times h$
$\quad = 20 \text{ m} \times 10 \text{ m} \times 5 \text{ m}$
$\quad = 1000 \text{ cm}^3$

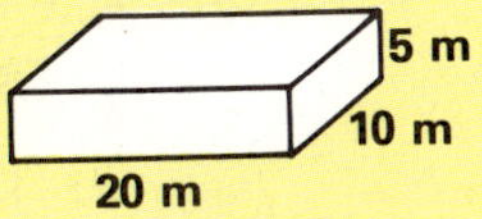

Cylinder

$V = B \times h$
$V = \pi \times r^2 + h$
$\quad = 3.14 \times 2 \text{ cm} \times 2 \text{ cm} \times 4 \text{ cm}$
$\quad = 50.24 \text{ cm}^3$

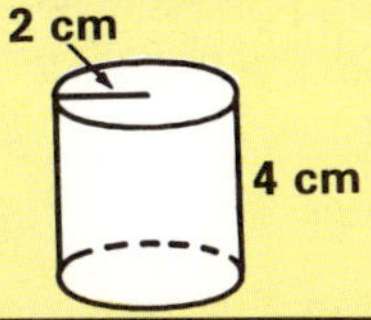

Pyramids and Cones

$V = \frac{1}{3} \times B \times h$
$V = \frac{1}{3} \times l \times w \times h$
$\quad = \frac{1}{3} \times 5 \text{ m} \times 6 \text{ m} \times 4 \text{ m}$
$\quad = 40 \text{ m}^3$

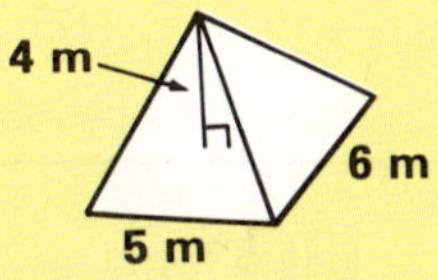

1. (3 m, 2.5 m, 6 m)   45 m³

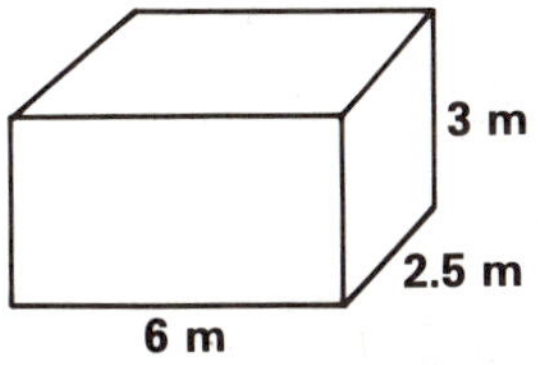

2. (3 cm, 5 cm)   141.3 cm³

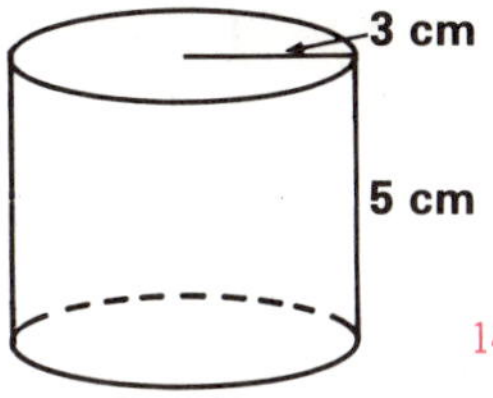

3. (6 in., 3 in.)   169.56 in.³

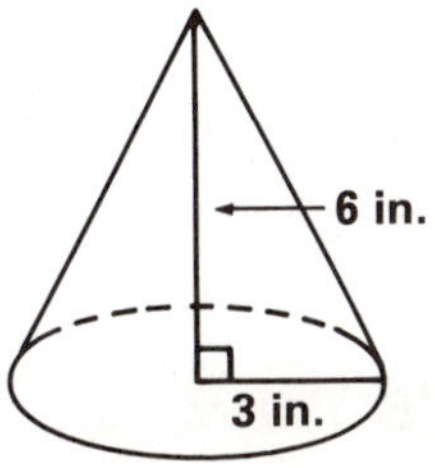

Here's how

A box contains 3 red pencils, 2 blue pencils, and 6 yellow pencils. If you pick a pencil without looking, what is the probability that it will be red?

$P(\text{red}) = \dfrac{\text{number of red pencils}}{\text{number of pencils}}$

$P(\text{red}) = \dfrac{3}{11}$

The probability is $\frac{3}{11}$ or 3 out of 11.

SKILL 58 (Use after page 398.)

Solve.

1. A box contains 5 blue pencils and 4 green pencils. If you pick a pencil without looking, what is the probability that it will be blue? $\frac{5}{9}$

2. A bag contains 2 black marbles, 3 red marbles, and 4 green marbles. If you pick a marble without looking, what is the probability that it will be green? $\frac{4}{9}$

3. A spinner has 8 equally likely outcomes, the whole numbers 1 through 8. You can win the game if you spin a number less than 4. What is the probability that you will win the game on the next spin? $\frac{3}{8}$

4. There are 12 socks in a drawer. Eight of the socks are blue and 4 of the socks are white. If you pick a sock without looking, what is the probability that it will *not* be blue? $\frac{1}{3}$

Extra Practice 499

SKILL 59 (Use after page 410.)

Here's how

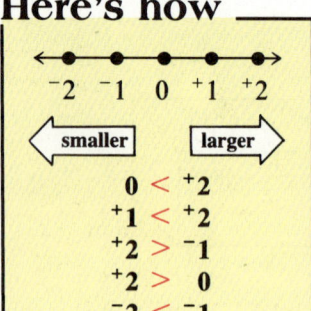

0 < +2
+1 < +2
+2 > -1
+2 > 0
-2 < -1

< or >?

1. +5 ● +8 <
2. +5 ● -8 >
3. -5 ● -8 >
4. -6 ● +1 <
5. +6 ● -1 >
6. +6 ● +1 >
7. 0 ● -5 >
8. 0 ● +5 <
9. -7 ● 0 <
10. +9 ● -4 >
11. -9 ● +4 <
12. +9 ● +4 >
13. -7 ● -2 <
14. -8 ● +3 <
15. 0 ● +6 <
16. +9 ● +4 >
17. -6 ● +6 <
18. -3 ● -5 >
19. -11 ● +10 <
20. +16 ● -17 >
21. +15 ● +17 <
22. 0 ● -12 >
23. +15 ● -15 >
24. -19 ● -13 <
25. -16 ● +11 <
26. -19 ● -18 <
27. +14 ● +17 <
28. -22 ● +22 <
29. -26 ● -21 <
30. +23 ● +24 <
31. +27 ● -20 >
32. -29 ● +23 <
33. -28 ● -26 <
34. 0 ● +32 <
35. -36 ● 0 <
36. +37 ● -31 >

MENTAL MATH
(Extension for Skill 59, page 500)

MATCH THE DRAWING
Look at exercises 1–36.
Match the drawing with the exercise.

a. ———|———→ +10 ———— Ex. 5
b. ←——|——— -10 ———— 0 ———— Ex. 13
c. ———|——— 0 +1 ———— Ex. 17
d. ←—— -30 ———— -20 ———— Ex. 29
e. ———— -1 0 ———————— Ex. 22

Copymaster S511

SKILL 60 (Use after page 412.)

Here's how

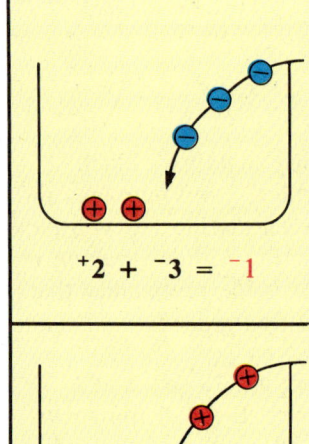

+2 + -3 = -1

-1 + +2 = +1

Give the sum.

1. +3 + +2   +5
2. +3 + -2   +1
3. -3 + -2   -5
4. -1 + +5   +4
5. -1 + -5   -6
6. +1 + -5   -4
7. 0 + -8   -8
8. 0 + +8   +8
9. -6 + 0   -6
10. +4 + -4   0
11. -4 + +4   0
12. +4 + +4   +8
13. +7 + -3   +4
14. -7 + +3   -4
15. -7 + -3   -10
16. -6 + +9   +3
17. +6 + -9   -3
18. -6 + -9   -15
19. +10 + +12 +22
20. +11 + -11 0
21. -17 + +14 -3
22. -19 + -13 -32
23. +18 + -19 -1
24. -16 + +11 -5
25. +15 + +14 +29
26. -18 + -12 -30
27. -11 + +19 +8
28. +17 + -17 0
29. -19 + +14 -5
30. +12 + +15 +27
31. -20 + -20 -40
32. -21 + +25 +4
33. +27 + -22 +5
34. -31 + +31 0
35. 0 + +34 +34
36. -32 + -36 -68

MENTAL MATH
(Extension for Skill 60, page 500)

SOLUTION SEARCH
Look at exercises 1–36.

1. Which four exercises have answers that are greater than +20? **19, 25, 30, 35**
2. Which four exercises have answers that are less than -20? **22, 26, 31, 36**
3. Which exercise has an answer of -1? **23**

Copymaster S511

500   Extra Practice

SKILL 61 (Use after page 415.)

Here's how

To subtract an integer, add the opposite of the integer.

⁺5 − ⁺2 = ⁺5 + ⁻2
         = ⁺3

⁻5 − ⁺2 = ⁻5 + ⁻2
         = ⁻7

⁺5 − ⁻2 = ⁺5 + ⁺2
         = ⁺7

⁻5 − ⁻2 = ⁻5 + ⁺2
         = ⁻3

Give the difference.

1. ⁺4 − ⁻6 ⁺10
2. ⁻4 − ⁺6 ⁻10
3. ⁺4 − ⁺6 ⁻2
4. ⁻7 − ⁻3 ⁻4
5. ⁺7 − ⁻3 ⁺10
6. ⁻7 − ⁻3 ⁻10
7. ⁺8 − ⁺1 ⁺7
8. ⁻8 − ⁺1 ⁻9
9. ⁺8 − ⁻1 ⁺9
10. ⁻7 − 0 ⁻7
11. 0 − ⁺7 ⁻7
12. 0 − ⁻7 ⁺7
13. ⁺5 − ⁺9 ⁻4
14. ⁺5 − ⁻9 ⁺14
15. ⁻5 − ⁺9 ⁻14
16. ⁻5 − ⁻5 0
17. ⁺5 − ⁺5 0
18. ⁻5 − ⁺5 ⁻10
19. ⁺12 − ⁻11 ⁺23
20. ⁻13 − ⁺16 ⁻29
21. ⁻14 − ⁻14 0
22. ⁻16 − ⁺19 ⁻35
23. ⁺18 − ⁺11 ⁺7
24. ⁺19 − ⁺14 ⁺5
25. ⁺17 − ⁻10 ⁺27
26. ⁻10 − ⁺17 ⁻27
27. ⁻16 − ⁻16 0
28. ⁺11 − ⁺15 ⁻4
29. ⁻15 − ⁺18 ⁻33
30. ⁺17 − ⁻14 ⁺31
31. ⁻23 − ⁻27 ⁺4
32. ⁺25 − ⁺25 0
33. ⁻28 − ⁺24 ⁻52
34. 0 − ⁻34 ⁺34
35. ⁻36 − ⁺32 ⁻68
36. ⁺33 − ⁻38 ⁺71

MENTAL MATH (Extension for Skill 61, page 501)

SOLUTION SEARCH

Look at exercises 1–36.

1. Which exercise can you solve mentally by thinking
   a. ⁺7 plus ⁺3? **5**
   b. ⁻5 plus ⁻9? **15**
   c. ⁺11 plus ⁻15? **28**
   d. ⁻23 plus ⁺27? **31**

2. How would you find the answer to exercise 20 mentally?
   ⁻13 plus ⁻16

Copymaster S512

SKILL 62 (Use after page 419.)

Here's how

The product of two integers with the same signs is positive.

⁺2 × ⁺3 = ⁺6
⁻4 × ⁻5 = ⁺20

The product of two integers with different signs is negative.

⁺6 × ⁻3 = ⁻18
⁻5 × ⁺6 = ⁻30

The product of any integer and 0 is 0.

⁺7 × 0 = 0
0 × ⁻4 = 0

Give the product.

1. ⁺3 × ⁺4 ⁺12
2. ⁺3 × ⁻4 ⁻12
3. ⁻3 × ⁻4 ⁺12
4. ⁻6 × ⁺5 ⁻30
5. ⁺6 × ⁺5 ⁺30
6. ⁺6 × ⁻5 ⁻30
7. ⁻5 × ⁺7 ⁻35
8. ⁻5 × ⁻7 ⁺35
9. ⁺5 × ⁺7 ⁺35
10. ⁺2 × ⁻6 ⁻12
11. ⁻6 × ⁺2 ⁻12
12. ⁻6 × ⁻2 ⁺12
13. ⁺7 × 0 0
14. ⁻7 × 0 0
15. 0 × 0 0
16. ⁻9 × ⁻8 ⁺72
17. ⁺8 × ⁻8 ⁻64
18. ⁻6 × ⁺5 ⁻30
19. ⁺4 × ⁺6 ⁺24
20. ⁻5 × ⁺5 ⁻25
21. ⁺6 × ⁻6 ⁻36
22. ⁻5 × ⁻9 ⁺45
23. ⁺7 × ⁻8 ⁻56
24. ⁻8 × ⁺7 ⁻56
25. ⁺9 × ⁺7 ⁺63
26. ⁻7 × ⁻7 ⁺49
27. 0 × ⁺8 0
28. ⁺10 × ⁻6 ⁻60
29. ⁻10 × ⁺6 ⁻60
30. ⁺10 × ⁺6 ⁺60
31. ⁻11 × ⁻11 ⁺121
32. ⁺11 × ⁻11 ⁻121
33. ⁺11 × ⁺11 ⁺121
34. ⁻13 × ⁺14 ⁻182
35. ⁺18 × ⁺15 ⁺270
36. ⁻19 × ⁻13 ⁺247

MENTAL MATH (Extension for Skill 62, page 501)

SOLUTION SEARCH

Look at exercises 1–36.

1. Which two exercises have an answer of ⁺35? **8, 9**
2. Which two exercises have an answer of ⁻56? **23, 24**
3. Which exercise has an answer that is the greatest product? **35**
4. Which exercise has an answer that is the least product? **34**

Copymaster S512

Extra Practice **501**

SOLUTION SEARCH

Look at exercises 1–36.

1. Which three exercises have an answer of $^-7$?
   19, 20, 24
2. Which two exercises have an answer of $^+9$?
   21, 26, 29, 32
3. Which three exercises have an answer of 0?
   16, 17, 18
4. Which exercise has an answer that is the greatest quotient?
   36

MENTAL MATH
(Extension for Skill 63, page 502)

Copymaster S513

Here's how

The quotient of two integers with the same signs is positive.

$$^+12 \div ^+6 = ^+2$$
$$^-21 \div ^-7 = ^+3$$

The quotient of two integers with different signs is negative.

$$^+20 \div ^-4 = ^-5$$
$$^-32 \div ^+8 = ^-4$$

The quotient of 0 divided by any non-zero integer is 0.

$$0 \div ^-6 = 0$$
$$0 \div ^+7 = 0$$

SKILL 63 (Use after page 420.)

Give the quotient.

1. $^+12 \div ^+3$ $^+4$
2. $^+12 \div ^-3$ $^-4$
3. $^-12 \div ^-3$ $^+4$
4. $^-8 \div ^+2$ $^-4$
5. $^+8 \div ^+2$ $^+4$
6. $^+8 \div ^-2$ $^-4$
7. $^-16 \div ^+4$ $^-4$
8. $^-16 \div ^-4$ $^+4$
9. $^+16 \div ^-4$ $^-4$
10. $^-18 \div ^+6$ $^-3$
11. $^+18 \div ^+6$ $^+3$
12. $^+18 \div ^-6$ $^-3$
13. $^-25 \div ^+5$ $^-5$
14. $^-25 \div ^-5$ $^+5$
15. $^+25 \div ^-5$ $^-5$
16. $0 \div ^+4$ 0
17. $0 \div ^-4$ 0
18. $0 \div ^-9$ 0
19. $^+49 \div ^-7$ $^-7$
20. $^-42 \div ^+6$ $^-7$
21. $^-45 \div ^-5$ $^+9$
22. $^+36 \div ^+6$ $^+6$
23. $^-32 \div ^+8$ $^-4$
24. $^+35 \div ^-7$ $^-5$
25. $^-36 \div ^-9$ $^+4$
26. $^+54 \div ^+6$ $^+9$
27. $^-64 \div ^+8$ $^-8$
28. $^+72 \div ^-9$ $^-8$
29. $^-81 \div ^-9$ $^+9$
30. $^+63 \div ^-7$ $^-9$
31. $^-70 \div ^+10$ $^-7$
32. $^+90 \div ^+10$ $^+9$
33. $^-60 \div ^-10$ $^+6$
34. $^-132 \div ^+12$ $^-11$
35. $^+182 \div ^-13$ $^-14$
36. $^+224 \div ^+16$ $^+14$

Here's how

$$2x + 4 = ^-2$$
$$2x + 4 - 4 = ^-2 - 4$$
$$\frac{2x}{2} = \frac{^-6}{2}$$
$$x = ^-3$$

Check.
$$2 \times ^-3 + 4 = ^-2$$

$$\frac{y}{3} - 2 = ^-5$$
$$\frac{y}{3} - 2 + 2 = ^-5 + 2$$
$$3 \times \frac{y}{3} = 3 \times ^-3$$
$$y = ^-9$$

Check:
$$\frac{^-9}{3} - 2 = ^-5$$

SKILL 64 (Use after page 448.)

Solve.

1. $2n + 3 = 17$ 7
2. $3x + 4 = ^-2$ $^-2$
3. $5y - 3 = 12$ 3
4. $4a - 3 = ^-3$ 0
5. $6j + 6 = ^-12$ $^-3$
6. $3w - 10 = ^-1$ 3
7. $5z - 8 = 17$ 5
8. $4b - 4 = ^-40$ $^-9$
9. $\frac{n}{2} + 3 = 2$ $^-2$
10. $\frac{x}{3} + 8 = 7$ $^-3$
11. $\frac{w}{4} - 3 = ^-3$ 0
12. $\frac{c}{5} - 4 = ^-7$ $^-15$
13. $\frac{d}{3} + 2 = ^-1$ $^-9$
14. $\frac{s}{6} - 8 = ^-4$ 24
15. $\frac{b}{4} - 6 = 3$ 36
16. $\frac{f}{5} + 5 = 0$ $^-25$

Extra Practice

GLOSSARY

absolute value (p. 411)   The number of units that a number is from zero.

acute angle (p. 334)   An angle that measures between 0° and 90°.

addends (p. 6)   Numbers that are added.

A.M. (p. 119)   A symbol used for times after 12:00 midnight and before 12:00 noon.

angle (p. 334)   A figure formed by two rays with the same endpoint.

area (p. 348)   The number of unit squares that it takes to cover a region.

associative property of addition (p. 422)   Changing the grouping of the addends does not change the sum.

$$(a + b) + c = a + (b + c)$$

associative property of multiplication (p. 422)   Changing the grouping of the factors does not change the product.

$$(a \times b) \times c = a \times (b \times c)$$

average (p. 82)   The sum of the numbers divided by the number of numbers.

axes (p. 424)   Two perpendicular lines used as a reference for graphing ordered pairs.

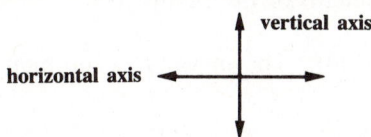

balance (p. 26)   The amount of money remaining in an account.

bar graph (p. 116)   A visual comparison of two or more quantities by representing each by a bar. The length of the bar corresponds to the quantity it represents.

base (p. 138)   When a number is followed by an exponent, the number is called the base of the expression. For instance, 4 is the base of the expression 4^3.

bisect (pp. 338–339)   To divide a line segment or an angle into two congruent parts.

budget (p. 316)   A plan for using your money.

Celsius temperature scale (°C) (p. 236)   The metric temperature scale, in which 0°C is the freezing point of water and 100°C is the boiling point of water.

centimeter (p. 224)   A metric unit of length. 1 centimeter = 0.01 meter.

checking account (p. 318)   An account in which money is deposited and held until the bank is told to pay a certain amount by means of a check.

circle (p. 344)   A curved plane figure with all points a given distance from the center.

circle graph (p. 120)   A circle used to represent all of something being studied, with pie-shaped slices labeled to show how the whole is divided up. The size of each slice corresponds to the percent of the whole that the slice represents.

circumference (p. 344)   The distance around a circle.

commission (p. 308)   The part of the total sales that goes to the salesperson.

Glossary **503**

common denominator (p. 150) A common denominator is a common multiple of the denominators of two fractions.

common factor (p. 144) 2 is a common factor of 4 and 6 because it is a factor of both 4 and 6.

common multiple (p. 150) 30 is a common multiple of 5 and 6 because it is a multiple of both 5 and 6.

commutative property of addition (p. 422) Changing the order of the addends does not change the sum. $a + b = b + a$

commutative property of multiplication (p. 422) Changing the order of the factors does not change the product. $a \times b = b \times a$

composite number (p. 142) A number other than zero that has more than two factors.

compound interest (p. 321) Interest that is computed on the principal plus the interest previously earned at regular intervals.

cone (p. 364) A space figure with one flat face (known as the base) that is a circle and with one other face that is curved.

corresponding sides (p. 270) Sides of similar figures that are proportional.

cross products (p. 260) The cross products for the ratios below are 2×10 and 5×4. Two ratios are equal if their cross products are equal.

$$\frac{2}{5} = \frac{4}{10} \text{ because } 2 \times 10 = 5 \times 4$$

cube (p. 364) A rectangular prism whose six faces are squares.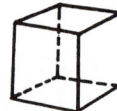

customary system (p. 240) The system of measurement that uses foot, quart, pound, and Fahrenheit temperature.

cylinder (p. 364) A space figure that has two circular bases that are the same size and are in parallel planes. It has one curved face.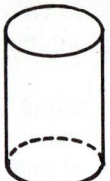

data (p. 114) Pieces of information.

decimal (p. 18) A number such as 3.86 or 0.4 that is written using a decimal point and place value.

decimeter (p. 226) A metric unit of length. 1 decimeter = 0.1 meter.

deductions (p. 309) The amount of money withheld from a person's pay.

degree (p. 334) A unit for measuring angles. This is a 1° (1-degree) angle.

dekameter (p. 226) A metric unit of length. 1 dekameter = 10 meters.

denominator (p.146) In the fraction $\frac{2}{3}$, the denominator is 3.

deposit (p.26) A sum of money put into a checking or savings account.

diameter (p. 344) The distance across a circle through its center. The length of the diameter is twice the length of the radius.

difference (p. 34) The answer to a subtraction problem.

discount (p. 290) An amount subtracted from the regular price of an item.

dividend (p. 102) The number that is divided.

divisor (p. 102) The number that one divides by.

down payment (p. 298) The first amount paid when buying on an installment plan.

equal ratios (p. 258) Ratios that indicate the same rate or comparison. The cross product of equal ratios are equal.

equation (p. 438) A sentence with an equal sign such as $3 \times 9 = 27$ or $8 + x = 10$.

equivalent fractions (p. 146) Fractions that name the same number. $\frac{1}{2}$, $\frac{2}{4}$, and $\frac{3}{6}$ are equivalent fractions.

estimate (p. 12) To use rounded numbers to check whether an answer is correct. To estimate $47 + 32$, you would add $50 + 30$. The sum should be about 80.

evaluate an expression (p. 436) To replace a variable in an expression with one of its values and then complete the indicated arithmetic.

even number (p. 140) Zero and multiples of 2.

expectation (p. 400) The probability of winning times the value of the prize.

exponent (p. 138) A small raised number written after a second number. The exponent tells you how many times the second number is used as a factor. For instance, 4^3 means $4 \times 4 \times 4$.

factorial (p. 388) A shortened notation to show a product.

$$4! = 4 \times 3 \times 2 \times 1$$

read as "4 factorial"

factors (p. 58) Numbers that are multiplied.

Fahrenheit temperature scale (°F) (p. 236) The customary temperature scale, in which 32°F is the freezing point of water and 212°F is the boiling point of water.

finance charge (p. 324) Buying an item on an installment plan costs more than paying cash. The difference is called the finance charge.

flowchart (p. 250) A plan for the solution for a problem with the steps written in boxes, and the boxes connected by arrows, making the order of the steps clear. Flowcharts are often made in planning the strategy for writing a computer program.

formula (p. 342) A general way of expressing a relationship using symbols.

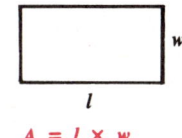

$A = l \times w$

fraction (p. 146) A numeral for part of a group or for part of a region. $\frac{1}{2}$, $\frac{4}{6}$, and $\frac{6}{5}$ are fractions.

frequency table (p. 114) A table showing the number of times different events or responses occur.

gram (p. 232) A metric unit of weight (mass). 1 gram = 0.001 kilogram.

graph (p. 116) A picture used to show numerical information. It can be a bar graph, a picture graph, a circle graph, or a line graph.

greatest common factor (p. 144) The greatest number that is a factor of each of two or more numbers.

8: {1, 2, 4, 8}   12: {1, 2, 3, 4, 6, 12}

4 is the greatest common factor of 8 and 12.

Glossary **505**

gross pay (p. 309)   Total pay before deductions.

hectometer (p. 226)   A metric unit of length.
1 hectometer = 100 meters.

hexagon (p. 340)   A polygon with six sides.

hypotenuse (p. 452)   The side of a right triangle that is opposite the right angle.

indirect measurement (p. 272)   A measurement that is computed from other measurements rather than measured directly.

installment plan (p. 394)   A way of buying expensive items. You pay part of the cost (the down payment) when you get the item and then agree to pay a certain amount each month for a certain number of months.

integers (p. 410)   The numbers . . . , $^-5$, $^-4$, $^-3$, $^-2$, $^-1$, 0, $^+1$, $^+2$, $^+3$, $^+4$, $^+5$, . . .

interest (p. 320)   The amount a borrower pays for using the money.

intersecting lines (p. 336)
Lines that meet at only one point.

invert (p. 212)   To reverse in position. When $\frac{3}{4}$ is inverted, you get $\frac{4}{3}$.

kilogram (p. 232)   A metric unit of weight.
1 kilogram = 1000 grams.

kilometer (p. 226)   A metric unit of length.
1 kilometer = 1000 meters.

least common denominator (p. 150)   The least common multiple of the denominator of two or more given fractions. Each of the fractions can be rewritten using this number as the denominator.

least common multiple (p. 150)   The smallest number that is a multiple of two or more given numbers.

legs of a right triangle (p. 452)   The sides that form the right angle.

line graph (p. 118)   A line graph is commonly used to show how a quantity changes over a period of time. Units of time are marked across the bottom of the graph. Above each time for which data are available a point is located. Its height corresponds to the size of the quantity at that time. When all the points have been located they are connected by line segments.

liter (p. 230)   A metric unit of liquid volume.

loan (p. 322)   Money that is borrowed for a fixed period of time at a set rate of interest.

long word-name (p. 4)   The name for a standard numeral written in words. The long word-name for 6,100,087 is six million, one hundred thousand, eighty-seven.

lower terms (p. 148)   To write a fraction in lower terms, divide the numerator and denominator by a common divisor.

lowest terms (p. 148)   A fraction is in lowest terms if the greatest common factor of the numerator and denominator is 1.

mass (p. 232)   Another name for weight in the metric system.

mean (p. 124)   The average of a set of numbers.

median (p. 126)   After ordering a set of numbers, the median is the middle number. If there are two middle numbers, the median is the average of the two.

meter (p. 226)   A metric unit of length.
1 meter = 100 centimeters.

metric system (p. 226) An international system of measurement that uses meter, liter, gram, and Celsius temperature.

milligram (p. 232) A metric unit of weight (mass). 1 milligram = 0.001 gram.

milliliter (p. 230) A metric unit of liquid volume. 1 milliliter = 0.001 liter.

millimeter (p. 224) A metric unit of length. 1 millimeter = 0.001 meter.

mixed number (p. 156) A number that has a whole-number part and a fraction part. $2\frac{3}{4}$ is a mixed number.

mode (p. 126) The number that occurs most often

multiple (p. 56) A product. 4, 8, 12, 16, 20, and so on, are multiples of 4.

negative number (p. 410) A number that is less than 0.

net pay (p. 309) Pay after deductions; "take-home" pay.

numerator (p. 146) In the fraction $\frac{2}{3}$, the numerator is 2.

obtuse angle (p. 334) An angle that measures between 90° and 180°.

odd number (p. 140) A whole number that is not divisible by 2. The numbers 1, 3, 5, 7, 9, 11, and so on, are odd.

odds (p. 398) The ratio of the number of ways that an outcome can occur to the number of ways that the outcome cannot occur.

opposites (p. 415) Two numbers are opposites if their sum is 0.

ordered pair (p. 424) A pair of numbers that give the location of a point on a grid.

order of operations (p. 94) Rules to assure that an expression has only one value.

origin (p. 424) The point where axes intersect.

outcome (p. 390) A possible result.

parallel lines (p. 336) Lines in a plane that do not intersect.

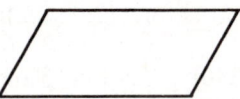

parallelogram (p. 340) A polygon with four sides and two pairs of parallel sides.

passbook (p. 320) A book issued by a bank when a person opens a savings account. It is used by the bank to record deposits, withdrawals, and interest.

pentagon (p. 340) A polygon with five sides.

percent (%) (p. 282) *Percent* means "per hundred." 5% (5 percent) equals $\frac{5}{100}$.

perimeter (p. 343) The distance around a figure; the sum of the lengths of the sides.

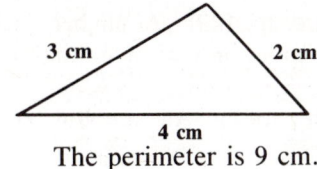
The perimeter is 9 cm.

Glossary **507**

permutation (p. 388) An arrangement of things in a definite order.

perpendicular bisector (p. 338) A line perpendicular to a segment and dividing the segment into two congruent parts.

perpendicular lines (p. 336) Two lines that intersect to form right angles.

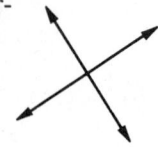

pi (p. 344) The number that is the ratio of the circumference of a circle to its diameter. It is represented by the Greek letter π and is approximately equal to 3.14.

picture graph (p. 120) A bar graph with the bars replaced by representative pictures. Each picture stands for a specified number of units.

P.M. (p. 119) A symbol used for times after 12:00 noon and before 12:00 midnight.

polygon (p. 340) A closed plane figure made up of segments.

 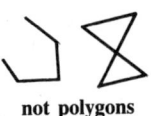
polygons   not polygons

positive number (p. 410) A number greater than 0.

prime factorization (p. 142) The expression of a composite number as a product of prime numbers.

prime number (p. 142) A number greater than 1 that is divisible only by itself and 1.

principal (p. 322) An amount of money borrowed.

prism (p. 364) A space figure that has two bases that are the same size and shape and are in parallel planes. The other faces are all rectangles.

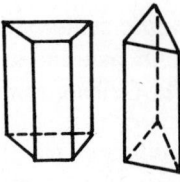

probability (p. 390) The ratio of the number of favorable outcomes to the total number of outcomes.

 probability of picking black = $\frac{3}{5}$

product (p. 58) The answer to a multiplication problem.

proportion (p. 260) An equation stating that two ratios are equal. $\frac{5}{8} = \frac{30}{48}$

protractor (p. 334) An instrument used to measure angles.

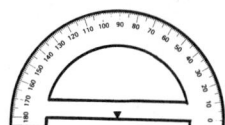

pyramid (p. 364) A solid figure with a face (known as the base) that is any polygon and with all other faces, which are triangles, sharing a common vertex.

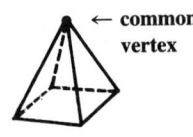
← common vertex

Pythagorean rule (p. 452) In a right triangle, the sum of the squares of the lengths of the legs is equal to the square of the length of the hypotenuse.

quadrant (p. 424) The regions into which the plane is separated by the horizontal and vertical axes.

Quadrant II	Quadrant I
Quadrant III	Quadrant IV

508 *Glossary*

quotient (p. 88) The answer to a division problem.

radius (p. 344) The distance from the center of a circle to the circle. The radius is equal to one half the diameter.

range (p. 126) The difference between the least and greatest numbers.

rate (p. 264) A ratio of two quantities of different kinds. Familiar examples would be $\frac{4 \text{ meters}}{2 \text{ seconds}}$ or "2 meters per second," $\frac{75 \text{ miles}}{3 \text{ gallons}}$ or "25 miles per gallon," and $\frac{16 \text{ grapefruit}}{4 \text{ dollars}}$ or "4 grapefruit per dollar."

ratio (p. 258) A comparison of two quantities by division. Below, the ratio of squares to circles is 4 to 2, 4:2, or $\frac{4}{2}$.

ray (p. 334) A part of a line that starts at a point and extends indefinitely in one direction.

reciprocal (p. 212) Two numbers are reciprocals when their product is 1.

$$\frac{3}{4} \times \frac{4}{3} = 1$$
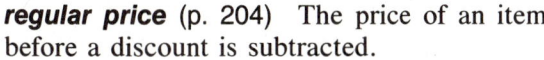
reciprocals

rectangle (p. 340) A polygon with four sides and four right angles.

rectangular prism (p. 364) A prism whose six faces are rectangles.

regular price (p. 204) The price of an item before a discount is subtracted.

remainder (p. 88) In a division problem, the number that is "left over." When the remainder is added to the product of the divisor and quotient, the sum is the dividend.

$$\begin{array}{r} 3 \\ 8\overline{\smash{)}29} \\ -24 \\ \hline 5 \end{array} \leftarrow \text{remainder}$$

right angle (p. 334) An angle whose measure is 90°.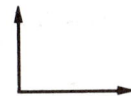

rounding a number (p. 12) To replace a number by another one that is easier to use. You round a number to the nearest ten by choosing the nearest multiple of 10. (5 is rounded up.)

$$13 \to 10 \quad 27 \to 30 \quad 45 \to 50$$

You round a number to the nearest hundred by choosing the nearest multiple of 100.

$$489 \to 500 \quad 1238 \to 1200 \quad 550 \to 600$$

sale price (p. 204) The price of an item after a discount is subtracted.

sample space (p. 392) The set of all possible outcomes of an experiment.

savings account (p. 320) A bank account in which money is deposited or withdrawn. The money earns interest.

scale drawing (p. 268) A drawing of an object such that the ratio of a unit of length on the drawing to a unit of length on the object is fixed.

segment (p. 334) A part of a line that has two endpoints.

short word-name (p. 4) The name for a standard numeral written using both words and numerals. The short word-name for 6,100,087 is 6 million, 100 thousand, 87.

Glossary **509**

side of an angle (p. 334) One of the rays that make up an angle.

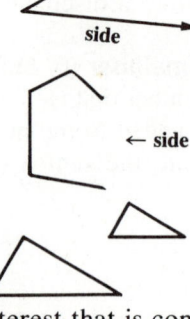

side of a plane figure (p. 340) One of the segments that make up a figure.

similar figures (p. 270) Two figures that have the same shape.

simple interest (p. 322) Interest that is computed by using the formula $I = prt$, where p is principal, r is rate, and t is time.

simplest form (p. 160) A fraction or mixed number is in simplest form if the fraction or fraction-part of the mixed number is less than 1 and in lowest terms.

solution (p. 438) A number that makes an equation true.

solve (p. 438) To find all the numbers that make an equation true.

sphere (p. 364) A round space figure shaped like a basketball. All points on a sphere are the same distance from the center.

square (p. 340) A polygon with 4 sides the same length and 4 right angles.

standard numeral (p. 4) The standard numeral for 3 billion, 24 million, 65 is 3,024,000,065.

statistics (p. 114) A branch of mathematics that studies numerical facts as a basis for drawing general conclusions and making predictions.

substitute (p. 436) To replace a variable with a numeral.

$$7a + 3$$
$$7 \cdot 6 + 3$$

sum (p. 6) The answer to an addition problem.

surface area (p. 368) The sum of the areas of all the surfaces of a solid figure.

terms (p. 148) The numerator and denominator are sometimes called the terms of the fraction.

trapezoid (p. 340) A polygon with 4 sides and exactly 1 pair of parallel sides.

tree diagram (p. 386) A diagram that shows all the possible outcomes of an event.

triangle (p. 340) A polygon with three sides.

triangular prism (p. 364) A prism with two triangular faces in parallel planes.

unit price (p. 312) The cost per unit (weight, volume, etc.) of an item.

variable (p. 434) A symbol—usually a letter—that holds the place for a number.

$$8x + 19 = 23$$
↑
variable

vertex of an angle (p. 334) The common endpoint of the two rays forming the sides.

volume (p. 372) The amount that a space figure holds.

whole number (p. 6) Any of the numbers 0, 1, 2, 3, 4, and so on.

withdrawal (p. 26) A sum of money taken out of a checking or savings account.

withholdings (p. 309) An amount deducted from a person's pay for the payment of taxes.

SYMBOLS

<	is less than	a:b	the ratio of a to b
>	is greater than	%	percent
=	is equal to	°	degree
≠	is not equal to	π	pi
≈	is approximately equal to	P(2)	the probability of the outcome 2
'	foot/feet	⁺5	positive 5
"	inch/inches	⁻5	negative 5

FORMULAS

Perimeter of a square	$P = 4 \times s$
Perimeter of a rectangle	$P = 2 \times (l + w)$
Circumference of a circle	$C = \pi \times d$
Area of a rectangle	$A = l \times w$
Area of a square	$A = s \times s$
Area of a parallelogram	$A = b \times h$
Area of a triangle	$A = \frac{1}{2} \times b \times h$
Area of a circle	$A = \pi \times r \times r$
Volume of a prism	$V = B \times h$
Volume of a rectangular prism	$V = l \times w \times h$
Volume of a cylinder	$V = (\pi \times r \times r) \times h$
Volume of a pyramid	$V = \frac{1}{3} \times B \times h$
Volume of a cone	$V = \frac{1}{3} \times (\pi \times r \times r) \times h$
Interest	$I = p \times r \times t$

Symbols and Formulas

INDEX

Absolute value, 411
Accounts
  charge, 324–325
  checking, 5, 318–319
  savings, 320–321
Acute angle, 334–335
Addition
  of customary units, 248–249
  of decimals, 20–23
  estimation in, 12–13, 22–23
  of fractions, 178–181
  of integers, 412–413
  of mixed numbers, 182–183
  properties, 422
  of whole numbers, 6–7
Addition equations, 438–439
Algebra, 433–460
A.M., 119, 238, 324–325
Analyzing data, 124
Angle
  acute, 334–335
  bisector, 339
  measure of an, 334–335
  obtuse, 334–335
  right, 334–335
Approximately equal to (≈), 344
Area
  of a circle, 354–355
  of a parallelogram, 350–351
  of a rectangle, 348–349
  of a square, 348–349
  surface, 368–369
  of a triangle, 352–353
Associative property
  of addition, 422
  of multiplication, 422
Averages, 82–83, 124–125
Axis/axes, 424–425
Balance of an account, 26, 319–320
Bar graph, 116–117, 119, 128–129, 132
Base of a figure, 350–353, 372–377
Basic counting principle, 386–387
Billion, 4–5
Bills, 324–325
Bisectors
  angle, 339
  constructing, 338–339
  segment, 338
Borrowing money, 322
Budget, 316
Buying a car, 40
Calculators, 21, 22, 26, 38, 39, 41, 46–47, 50, 63, 66, 71, 87, 91, 98, 122, 139, 154, 164, 167, 172, 183, 234, 263, 275, 283, 288, 291, 299, 309, 316, 319, 323, 328, 355, 377, 389, 393–394, 416, 428, 442
Celsius temperature scale (°C), 236–237
Centimeter, 224–229
Change, 65, 76, 179, 291
Chapter Review, 28, 52, 78, 110, 134, 174, 198, 220, 254, 278, 304, 330, 360, 382, 406, 430, 458
Chapter Test, 29, 53, 79, 111, 135, 175, 199, 221, 255, 279, 305, 331, 361, 383, 407, 431, 459
Charge account, 324–325
Checking account, 5, 318–319
Circle, 344–345, 354–355
  area of a, 354
  diameter of a, 344
  circumference of a, 344
  radius of a, 344
Circle graph, 120–121, 128, 132, 152, 282, 300–301
Circumference, 344–345
Collecting, organizing, and analyzing data, 97, 114–121, 127, 227, 239, 423
Commission, 308
Common denominator, 150–151
Common factors, 144
Common multiples, 150
Commutative property
  of addition, 422
  of multiplication, 422
Comparing numbers
  decimals, 42–43
  fractions, 152–153
  integers, 410–411
  whole numbers, 32–33
Comparison buying, 312–313
Compass, 338–339
Composite number, 142–143

Compound interest, 321
Computers
  at the airport, 50
  in architecture, 404
  in automobiles, 196
  in banking, 26
  calculating a raise with, 302
  in cash registers, 76, 172
  computer graphics with, 132, 380
  computer language for, 456
  gardening with, 252
  for home use, 252, 302, 328
  language of LOGO for, 456
  in photography, 428
  in schools, 108
  shopping with, 328
  software package for, 276
  in sports, 358
  in watches, 218
Cone, 364–365, 376–377
Constructing a bisector of an angle, 339
Constructing a bisector of a line segment, 338
Consumer Mathematics
  bills, 324–325
  change, 65, 76, 179, 291
  charge accounts, 324–325
  checking accounts, 5, 318–319
  commission, 308
  comparison buying, 312–313
  compound interest, 321
  coupons, 314–315
  deductions, 309
  discounts, 290–291, 310–311
  down payment, 298–299, 394
  earning money, 308–309
  gross pay, 309
  interest, 320–323
  loans, 322–323
  net pay, 309
  paychecks, 309
  sales, 204–205, 290–291, 310–311, 314–315, 440–441
  savings accounts, 320–321
  take-home pay, 309
  taxes, 309, 402–403
  unit prices, 312–313
Constructing an angle bisector, 339
Constructing a bisector of a line segment, 338
Converting
  customary units, 242–247

  metric units, 228–233
Corresponding sides, 270
Counting principle, 386
Cross products, 260–263
Cube, 364–365, 368–369, 372–373
Cube of a number, 138
Cumulative Skill Practice, 9, 17, 27, 41, 51, 67, 77, 91, 99, 109, 123, 133, 155, 165, 173, 185, 197, 211, 219, 235, 253, 267, 277, 289, 303, 317, 329, 347, 359, 371, 381, 395, 405, 417, 429, 443, 457
Cumulative Test, 30, 54, 80, 112, 136, 176, 200, 222, 256, 280, 306, 332, 362, 384, 408, 432, 460
Cup, 244–245, 248–249
Customary system of measurement, 209, 240–249
Cylinder, 364–365, 374–376
Data, 114
Decimals
  adding, 20–23
  comparing, 42–43
  divided by a decimal, 102–105
  divided by a whole number, 96–97, 100–101
  and fractions or mixed numbers, 166–169
  multiplying, 68–69
  multiplying by powers of 10, 72
  and percents, 286–287
  reading and writing, 14–15
  rounding, 18–19, 69
  subtracting, 44–47
Decimeter, 226
Deductions, 308–309
Degree(s)
  of angle measurements, 334
  of temperature, 236
Dekameter, 226
Denominator, 146
  least common, 150–151
Deposit, 26, 319–321
Diameter, 344–345
Difference, 38–39
Discount, 290–291, 310–311
Distributive property, 422
Dividend, 102
Divisibility rules, 140–141
Division
  by a decimal, 102–105
  of a decimal by a whole number, 96–97

Index **513**

Division (*continued*)
   estimation in, 86–87
   of fractions, 212–213
   of integers, 420–421
   of mixed numbers, 214–215
   with remainder as a decimal, 92–93, 104–105
   with remainder as a fraction, 162–163
   with remainder as a whole number, 88–89
   by 10, 100, 1000, 100–101
   of whole numbers, 82–89, 92–93
   zeros in, 86
Division equations, 446–447
Divisor, 102
Down payment, 298–299, 394
Edge, 364–365
Elapsed time, 209, 238–239
Equally likely outcomes, 390–393
Equal ratios, 258–259
Equations
   addition, 438–439
   division, 446–447
   multiplication, 444–445
   percent, 294, 296, 298
   in problem solving, 300–301
   subtraction, 440–441
   two-step, 448–449
   writing, 442, 447, 449
Equivalent fractions, 146–147
Estimating
   differences, 36–37, 38–39, 46
   in measurement, 226–227
   a percent of a number, 292–293
   products, 60–62, 64, 68
   quantities, 141, 163, 243, 293
   quotients, 84–87
   sums, 12–13, 22–23
Evaluate an expression, 436–437
Even number, 140
Event, 396
Expectation, 400–401
Exponents, 138–139
Expressions
   evaluating, 436–437
   simplifying, 70–71, 94–95
   writing, 419, 434–435
Extra Practice, 471–502
Face, 364–365
Factor, 58
   greatest common, 144–145

Factorial, 388–389
Factorization, 142
Factor tree, 142
Fahrenheit temperature scale (°F), 236–237, 410–411
Finance charge, 324–325
Foot, 242–243, 248–249
Formula
   area, 348, 350, 352, 354
   circumference, 344
   distance, 454
   interest, 322
   perimeter, 342
   volume, 372, 374, 376
Fractions
   adding, 178–181
   comparing, 152–153
   and decimals, 166–169
   dividing, 212–213
   equivalent, 146–147
   multiplying, 202–205
   and percents, 282–285
   and probability, 390–391
   as ratios, 258–259
   reciprocals of, 212–213
   simplifying, 148–149
   subtracting, 186–189
   of whole numbers, 204–205
   and whole numbers or mixed numbers, 156–159
Frequency table, 114–115, 128
Gallon, 244–245, 248–249
Geometry, 334–362
Gram, 232–233
Graphs
   bar, 116–117, 119, 128–129, 132
   circle, 120–121, 128, 132, 152, 282
   line, 118–119, 122, 132
   of ordered pairs, 424–425
   picture, 120–121, 129
Greater than, 32–33, 42–43, 152–153, 410–411
Greatest common factor (GCF), 144–145
Gross pay, 309
Hectometer, 226
Height of a figure, 350–353, 372–377
Hexagon, 340
Horizontal axis, 424–425
Hour, 208
Hypotenuse of a right triangle, 452
Inch, 240–243, 248–249

Income tax, 402–403
Indirect measurement, 272–273
Installment plan, 298–299, 394
Integers
   absolute value of, 411
   adding, 412–413
   comparing, 410–411
   dividing, 420–421
   multiplying, 418–419
   subtracting, 414–415
Interest, 320–323
   compound, 321
   simple, 322–323
Intersecting lines, 336–337
Inverting a fraction, 212
Kilogram, 232–233
Kilometer, 226–229
Least common denominator, 150–152
Least common multiple, 150
Legs of a right triangle, 452
Length
   customary units of, 240–243
   metric units of, 224–229
Less than, 32–33, 42–43, 152–153, 410–411
Line graph, 118–119, 122, 132
Line segment, 334, 338
   bisector of a, 338
Lines
   intersecting, 336–337
   parallel, 336–337
   perpendicular, 336–337
Liquid volume, 230–231, 244–245
Liter, 230–231
Loan, 322–323
Long word name, 4–5
Lower, lowest terms, 148–149
Magic square, 413
Mainframe computer, 302
Maps, 9, 24, 164, 180–181, 193, 227, 268–269, 286–287, 410–411, 455
Mass, 232–233
Mean, 124–125
Measurement, 223–256
   of angles, 334–335
   indirect, 272–273
Median, 126–127
Medical insurance deduction, 309
Mental Math, 7, 21, 35, 45, 56–57, 72–73, 100–101, 178–179, 186–187, 204–205, 208
Meter, 226–229

Metric system of measurement, 224–233
Mile, 242–243
Milligram, 232–233
Milliliter, 230–231
Millimeter, 224–229
Million, 2–5
Minute, 208–209
Mixed numbers
   adding, 182–183
   and decimals, 166–169
   dividing, 214–215
   and fractions, 156–159
   multiplying, 206–209
   and percents, 283
   subtracting, 190–193
Mode, 126–127
Multiple, 56
   least common, 150
Multiplication
   of decimals, 68–69
   estimation in, 60–61, 64, 68
   of fractions, 202–205
   of integers, 418–419
   of mixed numbers, 206–209
   properties, 422
   by 10, 100, 1000, 56, 72
   of whole numbers, 56–65
Multiplication equations, 444–445
Negative integers, 410–415, 418–425
Net pay, 309
Not equal sign ($\neq$), 260–261
Number(s)
   composite, 142
   even, 140
   odd, 140
   prime, 142
Number line, 410
Number theory, 138–145
Numerals
   standard, 4
Numerator, 146
Obtuse angle, 334–341
Odd number, 140
Odds, 398–399
One (1) property of multiplication, 422
Opposites, 414–415
Ordered pairs, 424–425
Order of operations, 94–95
Origin, 424
Ounce, 246–249

Outcome, 390–391, 396, 398–399
Parallel lines, 336–337
Parallelogram, 340–341, 350–351
Paycheck, 309
Payroll deductions, 308–309
Pentagon, 340–341
Percent(s)
  and decimals, 286–287
  estimating with, 12, 36, 60, 84
  and fractions, 282–285
  and mixed numbers, 283
  of a number, 290–295, 298–299
  rounding, 12
  unknown, 298–299
  of an unknown number, 296–299
Perimeter, 342–343, 349, 375
Permutations, 388–389
Perpendicular bisector, 338
Perpendicular lines, 336–337
Pi (π), 344
Picture graph, 120–121, 129
Pint, 244–245, 248–249
Place value, 2–5, 10, 14–15, 42–43, 159, 297, 321
Planning a motor trip, 164
P.M., 119, 238, 324–325
Point, 334
Polygons, 340–341
Positive integers, 410–415, 418–425
Possible outcomes, 390–393
Pound, 246–249
Powers, 138
Prediction, 391, 397
Prime factorization, 142
Prime number, 142–143
Principal, 322–323
Prisms, 364–365, 368–369, 372–373
Probability, 385–408
Problem Solving Practice, 8, 16, 26, 40, 50, 66, 76, 90, 98, 108, 122, 132, 154, 164, 172, 184, 196, 210, 218, 234, 252, 266, 276, 288, 302, 316, 328, 346, 358, 370, 380, 394, 404, 416, 428, 442, 456
Problem Solving Strategies
  analyzing data, 97, 125, 127, 227, 239, 423
  checking answers, 73, 183, 189, 207, 215, 263, 345
  choosing appropriate information to solve a problem, 7, 16, 45, 65, 70, 82, 92–93, 101, 147, 153, 163, 166–167, 173, 179, 187, 190–191, 206–207, 214, 244–246, 258–259, 262–264, 274, 309, 322–325, 366–367, 389–393, 396–397, 400–401, 404, 412, 414, 417–418, 436–437
  choosing the correct operation, 34, 38, 44, 50, 65, 66, 71, 76, 82–83, 87, 90, 93, 96–98, 101, 103, 105, 122, 125, 163, 164, 167, 172, 184, 186, 196, 205, 207, 225, 234, 245, 247, 249, 252, 269, 288, 297, 302, 315, 325, 328, 349, 373, 394, 404, 416, 437, 442
  choosing sensible answers, 130–131
  conducting an experiment and collecting data, 97, 124–125, 127, 227, 239, 391, 423
  estimating measurements, 226, 227, 231
  finding a pattern, 139, 167, 217, 378–379
  following directions, 11, 25, 83, 101, 159, 183, 241, 283, 297, 321, 425
  interpreting data, 114–115, 117, 119, 121, 426–427
  interpreting graphs, 116–123, 128–129, 132, 152–153, 188–189, 237, 282–283, 321, 424–425
  interpreting information, 15
  interpreting remainders, 106–107
  making a drawing, 105, 151, 184, 187, 273, 341, 346, 351, 355–357
  making geometric visualizations, 365–367, 369, 373, 380, 456
  making a list, 143, 179, 216–217, 295
  making a problem, 49, 131, 251
  making a table, 194–195
  number sense, 11, 15, 75, 83
  reading an ad, 3, 12–13, 40, 62–63, 66, 86–87, 95, 154, 158, 204–205, 231, 290–291, 298, 308, 310–315, 346, 370, 394
  reading a graph, 60, 117, 122, 129, 237, 321, 424–425
  reading a map, 8, 24–25, 115, 164, 180–181, 193, 227, 268–269, 286–287, 410–411, 455
  reading scales, 193, 236–237, 265
  reading a sign, 7, 16, 288
  reading tables and charts, 2, 4, 6, 19–20, 23, 33–34, 36, 42–45, 48, 49, 57, 65, 68, 72, 88–90, 104, 114–115, 124–127, 182–183, 192–193, 210, 213, 249, 260–261, 274–275, 284–285, 288, 294–295, 316, 325, 327, 340, 398, 402–403, 416

Problem Solving Strategies (*continued*)
  recognizing when information is missing, 24–25, 66
  recognizing when too much information is given, 24–25
  selecting data from a drawing, 104, 187, 218, 270–273, 336–337, 342–346, 348–357, 364–365, 368–369, 372–377
  solving problems involving more than one step, 16, 69, 71, 83, 98, 154, 164, 203, 210, 234, 247, 269, 288, 327, 370, 394, 404, 416, 442, 448–449
  using computer displayed information, 26, 108, 132, 252, 276, 302, 328, 358, 380, 404, 456
  using equations to solve problems, 438, 441, 442, 445, 447, 449
  using estimation, 61, 85, 141, 293, 300–301
  using a formula, 342–345, 348–355, 372–377, 454–455
  using a guess-and-check strategy, 12–13, 21, 39, 47, 63, 76, 89, 145, 167, 169–171, 291, 328, 355, 377, 401, 419, 427, 437, 451
  using logical reasoning, 3, 33, 37, 45, 48–49, 59, 69, 125, 147, 149, 151, 157, 161, 163, 171, 191, 205, 229, 243, 245, 247, 261, 267, 337, 341, 365, 435
  using a proportion to solve a problem, 264–266, 268–269, 271–272, 276, 298–299
  using scale drawings, 268, 271
  using a tree diagram, 386–387, 392–393, 396–397
  visual thinking, 43, 107, 195, 341, 343, 345, 351, 365, 367, 369, 373
  working backward, 250–251, 313, 403, 413, 415, 421
  writing a simpler problem, 74–75, 87, 285
Product, 58
Projects, 97, 127, 227, 239, 423
Properties
  associative, 422
  commutative, 422
  distributive, 422
  of one (1), 422
  of zero (0), 422
Proportion, 260–265, 268–273
Protractor, 334–335
Pyramids, 364–365, 376–377
Pythagorean rule, 452–453

Quadrant, 424
Quart, 244–245, 248–249
Quotient, 82
Radius, 344–345
Range, 126–127
Rate, 264–265, 322–323
Ratio, 258–259
Ray, 334
Reading decimals, 14
Reading a metric ruler, 224–225
Reading scales, 265
Reading standard numerals, 2–3
Reciprocal, 212–213
Rectangle, 340–341, 342–343, 348–349
Rectangular prism, 364–365, 368–369, 372–373
Regular price, 204–205, 290–291, 310–311, 314–315, 440–441
Remainder, 88–89, 92–93, 104–105, 162–163
Renting an apartment, 154
Right angle, 334–335
Right triangle, 452
Rounding decimals, 18–19, 69
Rounding whole numbers, 10–11, 33, 147
Salary, 308–309
Sale price, 204–205, 290–291, 310–311, 314–315, 440–441
Sales tax, 402–403
Sample space, 392–393, 396
Savings account, 320–321
Scale drawing, 268–269, 271
Seconds, 208
Segment, 240–241, 334
Short word name, 4–5, 14–15
Side of an angle, 334
Side of a plane figure, 340–341
Similar figures, 270–273
Simple interest, 322–323
Simplest form, 160–161
Simplifying expressions, 70–71, 94–95
Simplifying fractions, 160
Skill Test, 462–470
Social Security, 309
Software, 277
Solution, 438
Space figures, 364–369
  volume of, 372–377
Sphere, 364–365
Square, 340–343, 349–350
Square of a number, 138

Index **517**

Square root, 450–451
  table, 453
Standard numerals, 2–5, 14–15
Statistics
  interpreting data, 119, 124–126
  organizing data, 114–115
  presenting data, 116–118, 120
Substitute, 436–437, 444–449
Subtraction
  checking, 38–39
  of customary units, 248–249
  of decimals, 44–47
  estimation, 38–39, 46
  of fractions, 186–189
  of integers, 414–415
  of mixed numbers, 190–193
  of whole numbers, 34–39
Subtraction equations, 440–441
Sum, 6
Surface area, 368–369
Symbols, 511
Take-home pay, 309
Tally, 114
Taxes, 309, 326–327, 402–403
Temperature, 236–237, 410–411
Terms of a fraction, 148
Time, 3, 35, 208–209, 238–239, 322–323, 325
Ton, 246–249
Trapezoid, 340–341
Tree diagram, 386–387, 392–393, 396

Triangle, 340–343, 352–353
  right, 452
Triangular prism, 364–365
Unit price, 312–313
Variable, 434
Vertex of an angle, 334
Vertical axis, 424–425
Volume
  of cones, 376–377
  of cubes, 372–373
  of cylinders, 374–375
  liquid, 230–231, 244–245
  of pyramids, 376–377
  of rectangular prisms, 372–373
Whole numbers
  adding, 6–7
  comparing, 32–33
  dividing, 82–89
  multiplying, 56–65
  reading and writing, 2–5
  rounding, 10–11, 33, 147
  subtracting, 34–39
Withdrawal, 26, 320
Withholdings, 309
Writing equations, 449
Writing expressions, 434
Writing standard numerals, 4–5
Yard, 242–243, 248
Zero property of addition, 422

Photo Credits (*continued*)

Chapter 10: 257: Paul Johnson. 258: Paul Johnson/© D.C. Heath. 259: Paul Johnson/© D.C. Heath. 260: Paul Johnson/© D.C. Heath. 261: Focus on Sports. 262: 1 The Bettmann Archive, Inc. r Wide World Photos. 264: Stuart Cohen/© D.C. Heath. 265: 1 Paul Johnson/© D.C. Heath. r Stuart Cohen. 268: Hilary Wallace/© D.C. Heath. 269: Catherine Ursillo (Photo Researchers). 270: Michael Rusnock.
Chapter 11: 281: Paul Johnson/© D.C. Heath. 282: Paul Johnson/© D.C. Heath. 283: Paul Johnson/© D.C. Heath. 285: © Susan Lapides 1987. 288: Hilary Wallace/© D.C. Heath. 291: Stuart Cohen/© D.C. Heath. 296: © Susan Lapides 1987. 297: Stuart Cohen. 300: Richard Haynes/© D.C. Heath. 302:: Paul Johnson/© D.C. Heath.
Chapter 12: 307: Paul Johnson. 308: Stuart Cohen/© D.C. Heath. 311: tl Ken O'Donoghue/© D.C. Heath. 312: Hilary Wallace/© D.C. Heath. 314: tml Ken O'Donoghue/© D.C. Heath. 315: Stuart Cohen/© D.C. Heath. 316: Stuart Cohen/© D.C. Heath. 320: Stuart Cohen/© D.C. Heath. 322: Hilary Wallace/© D.C. Heath. 323: Stuart Cohen/© D.C. Heath. 326: Richard Haynes/© D.C. Heath. 327: Ken O'Donoghue/© D.C. Heath. 328: Paul Johnson/© D.C. Heath.
Chapter 13: 333: Paul Johnson. 340: Paul Buddle. 344: © Susan Lapides 1987. 346: Stuart Cohen/© D.C. Heath. 349: Stuart Cohen/© D.C. Heath. 354: Georg Gerster (Photo Researchers). 356: Richard Haynes/© D.C. Heath. 358: Stuart Cohen/© D.C. Heath.
Chapter 14: 363: © Susan Lapides 1987. 364: Rick Rizzoto/© D.C. Heath.
367: Paul Johnson/© D.C. Heath. 368: Paul Johnson/© D.C. Heath. 369: Paul Johnson/© D.C. Heath. 370: Stuart Cohen/© D.C. Heath. 374: Paul Johnson/© D.C. Heath. 375: 1 Stuart Cohen/© D.C. Heath; r Hilary Wallace/© D.C. Heath. 380: Paul Johnson/© D.C. Heath.
Chapter 15: 385: © Susan Lapides 1987. 386: Paul Johnson/© D.C. Heath. 387: Paul Johnson/© D.C. Heath. 388: Stuart Cohen/© D.C. Heath. 390: Paul Johnson/© D.C. Heath. 391: Paul Johnson/© D.C. Heath. 394: b Hilary Wallace/© D.C. Heath. 396: Paul Johnson/© D.C. Heath. 397: Paul Johnson/© D.C. Heath. 398: © Susan Lapides 1987. 399: © Susan Lapides 1987. 400: Rick Rizzotto/© D.C. Heath. 404: Paul Johnson/© D.C. Heath. 407: t,m Paul Johnson/© D.C. Heath.
Chapter 16: 409: Richard Hutchings (Photo Researchers). 411: Patrick Morrow (Globe Photos). 416: Stuart Cohen/© D.C. Heath. 420: Stuart Cohen/© D.C. Heath. 426: Focus on Sports. 427: Focus on Sports. 428: Discover Magazine, Time, Inc.
Chapter 17: 433: Paul Johnson. 434: Hilary Wallace/© D.C. Heath. 435: Hilary Wallace/© D.C. Heath. 436: Stuart Cohen/© D.C. Heath. 437: © Susan Lapides 1987. 440: Patrick Morrow (Globe Photos). 441: Stuart Cohen/© D.C. Heath. 442: Stuart Cohen/© D.C. Heath. 446: Paul Johnson. 447: Stuart Cohen/© D.C. Heath. 448: Stuart Cohen/© D.C. Heath. 449: Stuart Cohen/© D.C. Heath. 454: Dr Mueller (Zefa/H. Armstrong Roberts).